Sustainable Polymer Composites and Nanocomposites

Inamuddin · Sabu Thomas
Raghvendra Kumar Mishra
Abdullah M. Asiri
Editors

Sustainable Polymer Composites and Nanocomposites

Volume 1

Editors
Inamuddin
Department of Applied Chemistry
Aligarh Muslim University
Aligarh, India

Sabu Thomas
School of Chemical Sciences
Mahatma Gandhi University
Kottayam, Kerala, India

Raghvendra Kumar Mishra
Mahatma Gandhi University
Kottayam, Kerala, India

Abdullah M. Asiri
Chemistry Department, Faculty of Science
King Abdulaziz University
Jeddah, Saudi Arabia

ISBN 978-3-030-05398-7 ISBN 978-3-030-05399-4 (eBook)
https://doi.org/10.1007/978-3-030-05399-4

Library of Congress Control Number: 2018963283

This Springer imprint is published by the registered company Springer Nature Switzerland AG
The registered company address is: Gewerbestrasse 11, 6330 Cham, Switzerland

Contents

Processing, Characterization and Application of Micro and Nanocellulose Based Environmentally Friendly Polymer Composites

Adriana de Campos, Ana Carolina Corrêa, Pedro Ivo Cunha Claro, Eliangela de Morais Teixeira and José Manoel Marconcini

1 Introduction

The demands for biodegradable products, made from renewable and sustainable resources, and present low environmental impact are increasing by the consumers, industry, and the government [112]. The use of these polymer-based materials is important because of the growing need for using renewable and environmentally friendly resources. Collagen, chitin, starch, poly (hydroxybutyrates) (PHB), poly (hydroxyalkanoates) (PHA), polylact acid (PLA) and polycaprolactones (PCL) are examples of biodegradable polymers of high interest [104].

However, many of these materials have their use limited by high cost, poor physical properties, such as humidity sensibility, structure instability and low mechanical properties [6]. In order to improve these properties, natural fibers such as cellulose in micro and nano scale may be used, maintaining the all bio-based character of the material. Furthermore, since they are renewable, inexpensive and environmentally friendly resource, cellulose fibers and its micro e nanocomponents such as microfibrils and nanocellulose are used to manufacturing biodegradable nanocomposites either as nanofiller in a polymer matrix or as an all cellulose-based component film.

There is a wide variety of cellulose sources that can be applied in composites and nanocomposites field, such as kenaf [28, 61], sisal [20, 21, 29, 30, 31, 110, 114], sugar cane bagasse [54, 85, 96], oil palm [8, 32, 33, 43, 53, 66, 75, 93, 109], cotton [20, 21, 35, 46, 87], curaua [5, 23, 118], etc.

A. de Campos (✉) · A. C. Corrêa · P. I. C. Claro · J. M. Marconcini
Embrapa Instrumentation, São Carlos, SP, Brazil
e-mail: dridecampos@yahoo.com.br

P. I. C. Claro
Federal University of São Carlos, São Carlos, Brazil

E. de Morais Teixeira
Barra do Garças Unit, Federal University of São Carlos, São Carlos, Brazil

Inamuddin et al. (eds.), *Sustainable Polymer Composites and Nanocomposites*,
https://doi.org/10.1007/978-3-030-05399-4_1

According to ISO standards(ISO/TS 20477:2017), cellulosic nanomaterials can be subdivided into two classes:

1. cellulose nano-objects: "discrete piece of material with one, two or three external dimensions in the nanoscale" (1–100 nm). It involves cellulose nanofibers as cellulose nanocrystals (CNCs) and cellulose nanofibrils (CNFs);
2. cellulose nano-structured materials: a material having an internal composition of inter-related constituent parts in which one or more of those parts is in nanoscale. Its includes cellulose microcrystals (CMCs) and microcrystalline cellulose (MCCs), cellulose microfibrils (CMFs) or Microfibrillated Cellulose (MFC), and bacterial cellulose (BCs). They also can be aggregated cellulose nanostructures.

The cellulose nanostructures can be distinguished by their different chemical and physical properties, source, obtaining method and morphology [70]. By adding these nanoscale compounds as filler into polymers even in small quantities, the properties of polymers can be improved, depending on the type of nanocellulose, the dispersion throughout the matrix and interfacial interactions between the nanocellulose and polymeric matrix.

In this chapter, we will first make a brief description of the definition, terminology, and methods of obtaining cellulose nanostructures. Next, we will present procedures used in the functionalization of the cellulose surface to improve the hydrophilic character and the compatibility with polymer matrices. We then present studies of all-cellulosic nanostructured films, types of processing involving the production of bionanocomposites and other important applications of them in non-biocomposite areas.

2 Micro and Nano-cellulose

2.1 *Brief Description*

Cellulose fibers can be extracted from a variety of sources such as wood pulp, residues of some industrial process (sugar cane bagasse,cassava bagasse, coconut, rice, oil palm, soy, etc., that can be used for nanocellulose production (see Fig. 1 and Table 1) and plant fibers (sisal, cotton, curaua, hemp, flax, ramie, jute, etc.). They have been widely applied in several fields such as reinforcement in material sciences, catalysis, biomedical engineering, paints, cosmetics and electronic applications due to their sustainability, biocompatibility and good mechanical properties [103].

In nature, the cellulose chains are packaged in such an orderly manner that compact nanocrystals are formed, which are stabilized by inter and intramolecular hydrogen bonding [4, 67, 88]. These hydrogen bonding make the nanocrystals completely insoluble in water and in most organic solvents and lead to a material with mechanical strength only limited by the forces of adjacent atoms [69]. In the cell wall structures of vegetable plants, those cellulose nanocrystals are joined by segments of amorphous holocellulose to form the micro/nanofibrils that constitute the individual cellulose fibers [41].

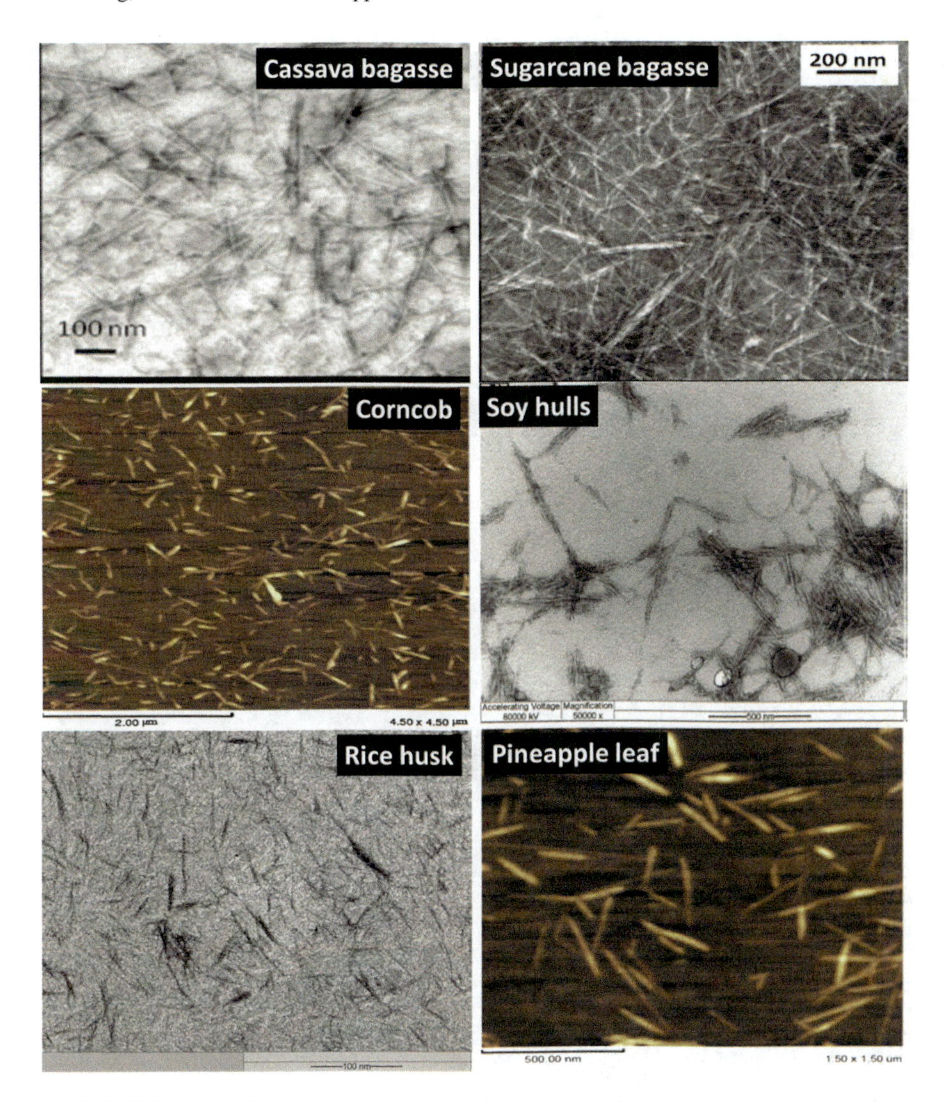

Fig. 1 Cellulose nanofibers extracted from a variety agro- residue

The loss of the hierarchical structure of the cellulose fibers can occur by mechanical, chemical, physical and biological treatments, or by a combination of them, releasing the microfibrils [68, 92]. Depending on these factors, different cellulose nanostructures (or a mix of them) are obtained. These nanostructures present high aspect ratio, with diameters and lengths ranging from units to several microns, excellent mechanical properties, high specific surface area, biodegradability, and biocompatibility [130].

The term *nanocellulose* describes the cellulose fibril or crystallite containing at least one dimension in the nanoscale (1–100 nm). There are many different types of terminologies used for describing nanocelluloses as will be seen next.

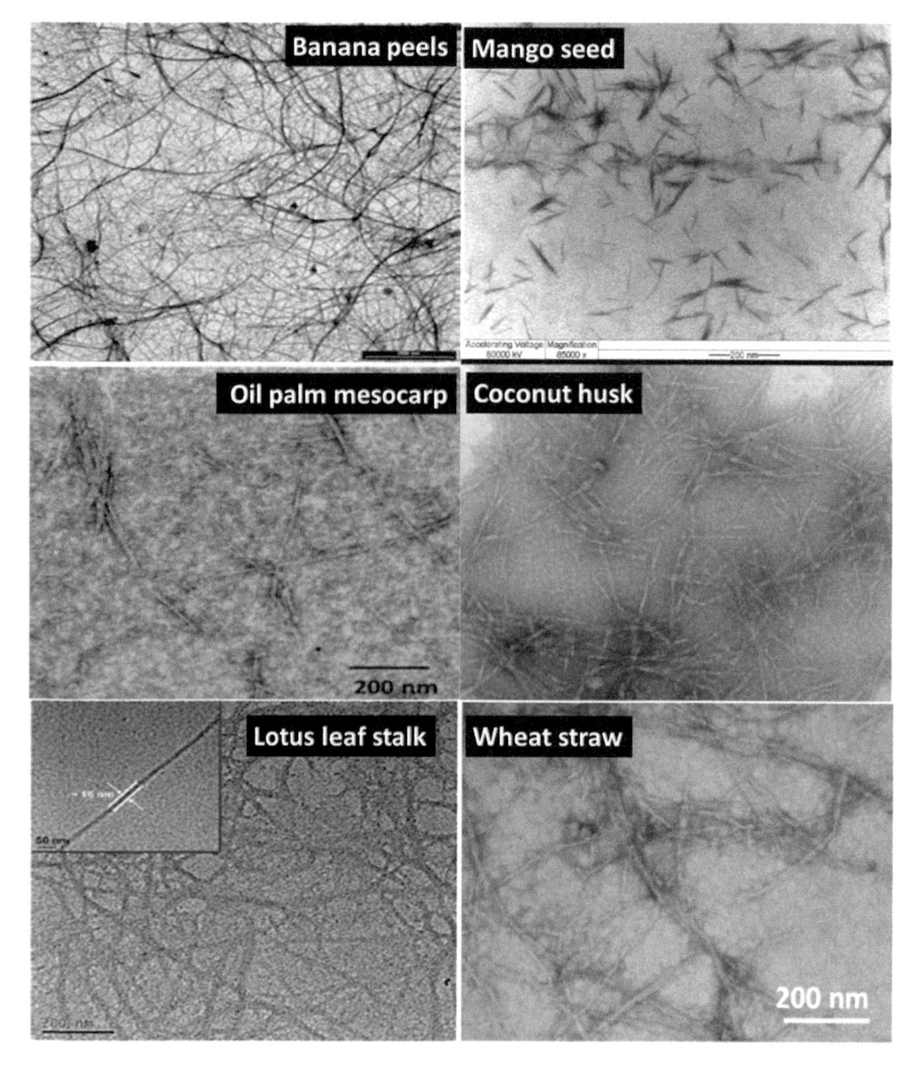

Fig. 1 (continued)

The cellulose fibers disintegrated into micro dimensions are designed as cellulose microfibrils or microfibrillated cellulose (MFC). They are nanofibrils aggregates (bundles) (30–100 cellulose molecules) forming a nanostructured material with diameters of around 3–30 nm and length higher than 1 μm [39, 80]. The MFCs appear as interconnected, nano-fibrillar structure. The MFCs are obtained by mechanical disintegration of fibers which generally involves a combination of processes such as high-pressure homogenization, grinding, ultrasonication and steam explosion, obtaining an aqueous suspension displaying a gel-like behaviour [116].

During the disassembly process, the microfibrils can release more or less their individual constituents, called nanofibrillated cellulose (NFCs), cellulose nanofibers

Table 1 Dimensions characteristics of nanocellulose from agro- residue

Vegetal source	Length (L)/nm	Diameter (D)/nm	L/D	References
Cassava bagasse	1700–360	2–11	76	[34]
Sugarcane bagasse	256 ± 55	4 ± 2	64	[36]
Corncob	211 ± 44	4 ± 1	53 ± 16	[113]
Soy hulls	123 ± 39	2.8 ± 0.7	44	[45]
Rice husk	–	15–20	10–15	[65]
Pineapple leaf	250 ± 52	4.4 ± 1.4	60	[38]
Banana peels	500–330	8–26	50–12	[97]
Mangoo seed	123 ± 22	4.6 ± 2.2	34 ± 19	[56]
Oil palm mesocarp	104 ± 52	9.00 ± 4.00	12	[32, 33]
Coconut husk	194 ± 70	5.5 ± 1.5	39 ± 14	[107]
Lotus leaf stalk	Microns	20 ± 15	40	[16]
Wheat straw	400	5	80	[82]

or nanofibrils (CNFs). These are composed of long and entanglement cellulosic chains and, as well as CMFs, maintain both, their amorphous and crystalline domains. The CNFs present a diameter of the 3–5 nm and length of 500–1000 nm [89]. Khalil et al. [72] disagreed with the "microfibril"term because it does not reflect the real dimensions of the fibril. Nechyporchuk and colleagues [92] reported that both cellulose, microfibrils (bundles) and elementary fibrils are referred to as cellulose nanofibrils. Depending on the production process, both CMFs and CNFs are obtained, and their morphologies also strongly depend on the cellulose source. In both cases, hemicellulose and lignin generally are removed before their productions. These nanostructures maintain the amorphous and crystalline phase of cellulose. They can aggregate to some extent during the drying process. Suspensions of CNFs can also present gel properties even at low cellulose concentrations forming an entangled network structure [92].

Microcrystalline celluloses (MCCs) are also aggregates of multi-sizes cellulose microfibrils. It is found as a fine powder and they are commonly known under the brand name Avicel®. It has a diameter of around 10–50 µm. MCCs are obtained from partially depolymerized pure cellulose, synthesized from the α-cellulose precursor. The MCC can be synthesized by different processes such as reactive extrusion, enzyme-mediated, steam explosion and acid hydrolysis. The MCC is a valuable additive in the pharmaceutical, food, cosmetic and other industries [57, 89]. MCC is characterized by a high degree of crystallinity, typically ranging between 55 and 80% [125]. CCNs can also be prepared from MCCs using NaOH/urea dissolution method and followed by regeneration, neutralization, and ultrasonication [112].

Bacterial cellulose (BCs) or microbial cellulose is a type of cellulose microfibrils produced extracellularly by specific bacteria. The *Acetobacter xylinum* is the most efficient producer of bacterial cellulose and this occurs in a culture medium containing carbon and nitrogen sources [2, 68, 91]. Bacterial cellulose result from direct synthesis, and not from the destruction of the primary structure of cellulose fibers, as in case of CNFs and CNCs. BCs present an average diameter of 20–100 nm and lengths in micrometre, they entangle to form a stable network structured as ribbon-shaped fibrils. BCs do not require any pre-treatment to remove impurities or contaminants such as lignin, pectin, and hemicellulose, i.e. the bacteria produce high-purity cellulose material with a distinct crystallinity of 80–90%. These peculiar properties of BCs make them an attractive material for use in biomedical applications [2, 70, 76, 80, 92].

Cellulose nanocrystals (CNCs) are also called nanocrystalline cellulose (NNCs), cellulose whiskers and rod-like cellulose, the area high crystalline cellulosic material resulting from acid hydrolysis of native cellulose with mineral acids, removing the amorphous phase of cellulose and leaving intact the crystalline phase. In order to obtain CNCs, the native lignocellulosic fibers should be previously submitted to a treatment of delignification process prior to the hydrolysis. CNCs present elongated rod-like aspect and their surface can be negatively charged when sulfuric acid (the most utilized acid) is employed for extraction. The charged surface of CNCs prevents the aggregation in aqueous suspension due to electrostatic repulsion between particles. They are considered a rigid and no defect crystal. Their diameter and length depends on the cellulose source: CNCs present diameters of around 5–30 nm and length of 100–500 nm for plants source, of around 100 nm to several microns for CNCs from tunicate and algae cellulose [2, 80, 92]. Lin and Dufresne [80] have reported several studies that showed the values of CNCs' elastic modulus ranging from 100–206 GPa, values similar to Kevlar and potentially stronger than steel.

Figure 2 shows some examples of nanostructures obtained from bleached sugarcane bagasse. As it can be observed, the type of treatment applied to the same fiber results in different nanostructures, with different crystallinities and thermal behaviour.

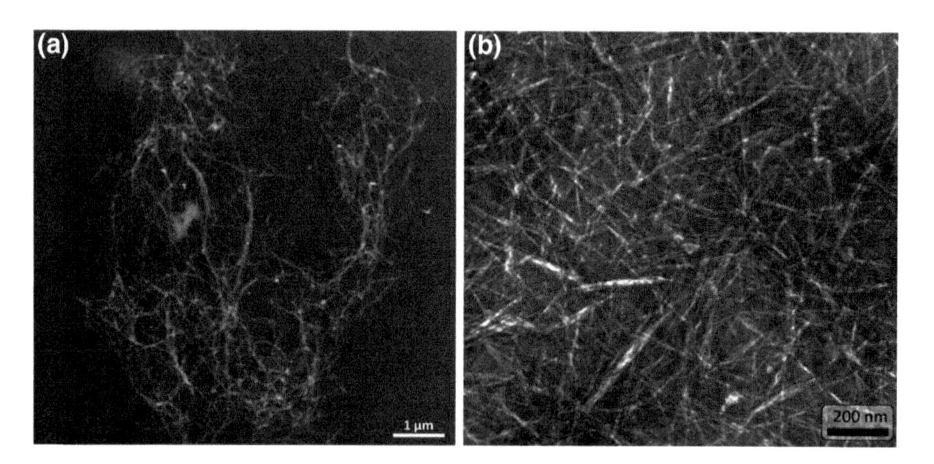

Fig. 2 TEM micrographs showing the morphology of nanostructures from bleached sugar cane bagasse. **a** After enzymatic hydrolysis (10 mg of Enzyme Viscozyme (1000 FBG/g) and 33 mg of Enzyme FiberCare (5000 ECU/g)/g of biomass and reaction time of 3 days at 45 °C) [29]; **b** after acid hydrolysis with sulfuric acid solution (6 M) for 45 min at 45 °C

The thermal properties of these cellulosic materials are important to determine their processing temperature range and use. The thermal degradation of lignocellulosic materials begins with an initial decomposition of the hemicelluloses, followed by lignin pyrolysis, depolymerization, combustion and oxidation of carbides. Thus, the MFCs and NFCs had a thermal degradation temperature higher than the fibers (350 °C) due to the removal of much of the amorphous material. The thermal degradation of CNCs usually starts at temperatures lower than MFCs and NFCs (200–300 °C), due to the presence of sulfate groups in the obtaining of the nanocrystals. The CNCs with lower sulfate content on their surface present higher thermal stability [23, 29, 34, 89, 105]. The combination of sulfuric and hydrochloric acids during the hydrolyses to obtain the CNCs generates nanoparticles with better thermal stability due to the reduced presence of sulfate groups on their surface, also causing a decrease in the stability of CNCs in suspension [105]. Studies have found that CNC obtained by enzymatic hydrolysis exhibited superior thermal stability, compared to CNC obtained by chemical hydrolysis using sulfuric acid [13]. Uschanov et al. [128] studied the esterification of MCCs, CNCs and regenerated cellulose with different kind of long-chain fatty acid as oleic, decanoic, linoleic and tall oil fatty acids (TOFA), a product of a mixture of 10% or less of saturated fatty acids and 90% or more of unsaturated octadecanoic (C18) acids. Thermal stability of CNCs was poorer than that of MCC or regenerated cellulose. They concluded that the modification weakened the thermal stability and the degradation temperature seemed to be dependent on the nature of the fatty acid used. Fatty acid chain length and double bond content affected the reactivity between cellulose and fatty acid; a longer chain length and the increase on double bond content decreased the degree of substitution (DS). Lee and colleagues [77] modified the surface of BC using organic acids (acetic, hexanoic and dodecanoic acids) via esterification reactions. As well as Uschavov et al. [128], they observed that the thermal degradation behaviour of organic acid modified BC sheets decrease with the increase of carbon chain length of the organic acids used. Agustin et al. [3] produced BC esters using different chloride acids and showed that the temperature at maximum weight loss rate (T_{max}) increased after esterification. The thermal stability of CNCs from white and coloured cotton was investigated in dynamic and isothermal (180 °C) conditions under an oxidizing atmosphere [35]. The thermal stability of white cotton CNCs, under dynamic conditions, was slightly higher than of coloured cotton CNCs. However, the colored-CNCs were more thermally stable, in isothermal conditions, than white-CNCs. This behaviour was attributed to lower sulfonation on coloured cotton CNCs surfaces than on white cotton CNCs surface.

Cellulose nanostructures or nanocellulose in general, have gained attention from researchers and industry because of their high Young modulus (130 GPa) [18], which is higher than that of the S-glass (86–90 GPa) and comparable to Kevlar (131 GPa), rendering them good reinforcement in natural and synthetic polymer matrices [103]. The inherent hydrophilic nature of nanocellulose limited its widespread application. Surface modifications of nanocellulose diminish its hydrophilicity which will be briefly discussed further ahead.

2.2 *Obtaining Different Types of Micro and Nano-cellulose by the Mechanical, Chemical and Enzymatic Process*

The nanocellulose materials can be obtained by different processes with the result in specific or a mix of morphologies, physical properties, and consequently different applications.

The mechanical process for extraction involves refining or high shear homogenization, microfluidization and sonication, which result in microfibrils and nanofibrils. Refining and homogenization are performed in the presence of water, producing microfibrils (MFCs)/nanofibrils (CNFs) through a relatively narrow space of a disk apparatus between the rotor and the stator. In the microfluidization process, the suspension is subjected to high pressure to pass through a Y or Z type geometry interaction chamber [132]. Sonication is performed on a fiber suspension to separate the microfibrils or nanofibrils beams from the cell wall of the fibers through cavitation [99]. The cavitation leads to a formation of powerful oscillating high intensive waves. These microscopic gas bubbles expand and implode breaking down cellulose fibers to microfibrils/nanofibrils [105].

The chemical treatment involves strong acid hydrolysis applied to cellulosic fibers allowing dissolution of amorphous domains and therefore longitudinal cutting of the microfibrils which generate cellulose nanocrystals (CNCs) also known as whiskers. During the acid hydrolysis process, the hydronium ions penetrate the cellulose chains in the amorphous regions promoting the hydrolytic cleavage of the glycosidic bonds, under a controlled period of time and temperature, keeping the crystallites intact [35, 40, 124]. Sulfuric acid (H_2SO_4) is generally used as a hydrolyzing agent because its reaction with the surface hydroxyl groups via an esterification process allows the grafting of anionic sulfate ester groups. The presence of these negatively charged groups induces the formation of a negative electrostatic layer covering the nanocrystals and promoting their better dispersion in water [40]. CNCs prepared using hydrochloric (HCl) acid or a mix of HCl/H_2SO_4 for hydrolysis exhibit good thermal but tend to aggregate in water [23]. Their geometrical dimensions depend on the origin of the cellulose source and hydrolysis conditions, but the length is usually in the range of a few hundred nanometers, and the width or diameter is in the range of a few nanometers. An important parameter for cellulose nanocrystals (CNCs) is the aspect ratio, which is defined as the ratio of the length to the diameter (L/d) [40].

Organic acids or a mix of them with mineral acid have been used to extract CNCs and concomitantly to produce carboxylated CNFs and CNCs [17, 64, 119] using mechanical assistance (ultrasound or micro fluidics).

The enzymatic process usually involves bulk of enzymes that act synergistically in the hydrolysis of cellulose since a single enzyme is not able to degrade cellulose [44, 80, 117, 130]. The most used bulks for enzymatic hydrolysis contains predominantly endoglucanase and/or exoglucanase [44, 55, 101, 117]. Celobiohydrolases or exoglucanases are a type of cellulase able to attack cellulose by the end of chains, resulting in cellobiose units. Endoglucanases randomly hydrolyze the amorphous regions, resulting in cellulose nanocrystals, in the single crystal range, since most of the fibers are in the form of crystalline structure entwined in an amorphous cellulose phase.

Both ethanol and nanocelluloses (CNFs and CNCs) were produced using eucalyptus cellulose pulp as raw material for enzymatic hydrolysis route [13]. The solid residues from ethanol production after 24 h of hydrolysis at 50 °C was characterized as CNFs. If the hydrolysis time was increased to 144 h and the temperature reduced to 35–40 °C, CNCs with a crystallinity index of 83%, length of 260 nm and diameter of 15 nm were found in this solid residue. Yarbrough et al. [130] studied the production of nanocellulose of different sizes and aspect ratios using enzymatic treatments (endo- and exoglucanases) with mechanical refinement and acid hydrolysis. The authors related that the majority of commercial cellulase cocktails are optimized for the highest conversion of cellulose to sugars, which is not desired to obtain cellulose nanocrystals. Then, they compared nanocellulose production using *T. reesei*, a classic fungal cellulase system containing predominantly exoglucanases, with that of *C. bescii*, a bacterial enzyme system that contains complex multifunctional enzymes. They showed that CNC produced by *C. bescii*system is more uniform than that produced by the *T. reesei*, after a reaction time of 48 h, due to the difference between the cellulases excreted by *C. bescii* and the cellulolytic agents in fungal excretion of *T. reesei.*

Bacterial celluloses (BCs) are produced by fermentation of low molecular weight sugars using bacteria from *Acetobacter* species. Therefore they are biosynthetic products. *Acetobacter xylinum* produces extracellular cellulose microfibrils to provide a firm floating matrix, allowing the embedded bacteria to stay in close contact with the atmosphere [100]. During the biosynthesis, the glucose chains are produced inside the bacterial body and outgrowth through tiny pores present on the cell envelope. By joining several glucose units, microfibrils are formed and further aggregate as ribbons. BCs are commonly regarded as a material with better biocompatibility than other types of nanocellulose, but their production is a little limited due to high synthesis cost and low yield [80].

In Table 2 is presented morphology and thermal properties of a variety of sources obtained from the different process and its classification.

2.3 Functionalization or Surface Modification of Micro and Nano-cellulose

The chemical modification on the cellulose surface can improve their interaction with apolar matrices, in addition to reducing their hydrophobicity. Esterifications and silanizations are most commonly used in the preparation of cellulose for composite applications [60, 90].

Cellulose can also be modified by the formation of ionic groups on its surface. The oxidation of the cellulose surface, by plasma or corona treatment, can generate carboxylic acids groups that improve their interaction with the matrix in the

Table 2 Morphology and thermal properties of variety of sources of cellulose

Source	Process obtaining	Nanocellulose classification	Morphology	Diameter (nm)	Length (nm)	Crystalinity (%)	Thermal Stability/atmosphere (°C)	References
Eucalyptus pulp	Mineral acid hydrolysis	CNC		10–30	50–100	82	180/air	[123]
		CNC		23 ± 5	200 ± 50	–	163/N_2	[19]
	Enzimatic hydrolysis	CNC		15 ± 6	–	–	323/air	[13]
	Organic acid	CNC		15	275–380	80–82	322/N_2	[17]
	Mechanical grinding	CNF		17 ± 4	1100 ± 50	–	292/N_2	[19]
	Sonification	CNF		30–40	200–250	33	310/air	[123, 124]
	Enzimatic hydrolysis	CNF		21 ± 3	216 ± 86	83	323/air	[13]

(continued)

Table 2 (continued)

Source	Process obtaining	Nanocellulose classification	Morphology	Diameter (nm)	Length (nm)	Crystalinity (%)	Thermal Stability/ atmosphere (°C)	References
Curaua	Mineral acid hydrolysis	CNC	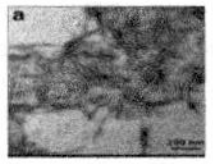	6–10	80–170	81 (H_2SO_4) 83 (H_2SO_4/ HCl) 87 (HCl)	254/air 285/air 295/air	[23]
		CNC		34 ± 5	400 ± 100	–	158/air	[19]
	Enzimatic hydrolysis + sonification	CNF		3.8 ± 1.3	1280 ± 670	78.5	–	[29]
	Mechanical grinding	CNF	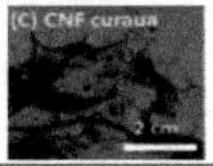	36 ± 8	–	–	303/N_2	[19]
Sugarcane bagasse	Mineral acid hydrolysis	CNC		37.84	220	–	249/N_2	[85]
		CNC	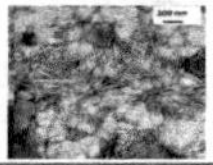	8 ± 3	255 ± 55	70.5	210/air	[36]
		CNC	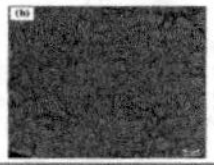	9.8 ± 6.3	280.1 ± 73.3	68.54 ± 1.30	250/N_2	[74]
		CNF-sheet		39 ± 13	–	72.9	179/N_2	[50]
	Enzimatic hydrolysis + sonification	CNF	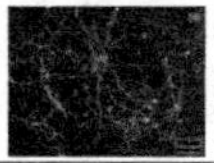	30 ± 7	255 ± 83	72	–	[29]

composites [62]. Modifications by sulfonation, carboxylation or graphitization, in addition to modification by acetylation/alkylation and treatment with silane agents, can also be used. In sulfonation treatments, sulfuric acid solutions in moderate concentration are used, obtaining partial sulfonation on the cellulose surface in aqueous suspension with colloidal appearance, due to repulsive forces of the sulfate groups adhered to the surface of the cellulose. Carboxylation can result in more hydrophilic cellulose surfaces. An effective way of inducing controlled oxidation on the cellulose surface, in order to create carboxylic groups, involves treatment with 2,2,6,6-tetramethylpiperidine-1-oxyl (TEMPO) radical where the hydroxyl groups are selectively converted into carboxylic groups, generating a negative charge on cellulose surface, not aggregating when dispersed in water, forming bi-refringent suspensions [62]. Among methods of modifying polymers, grafting is a versatile method for promoting the polymer in a variety of functional groups. Polymeric materials with good properties can be obtained by grafting, and changing parameters such as polymer type, degree of polymerization and dispersibility in the main and in the side chains, in addition to the density and uniformity of grafts, it can be combined the best properties of two or more polymers in a physical unit, in this case, the cellulose [108].

Lignocellulosic fibers from sugarcane bagasse were chemically modified by Pasquini et al. [95] using dodecanoyl chloride and pyridine, and toluene, octadecanoyl chloride, and pyridine. The modified fibers were incorporated into low-density polyethylene (LDPE) with improved dispersion and surface adhesion to the matrix. However, in spite of the better compatibility of the modified fibers with the matrix, these composites did not present improvements in the mechanical properties than those whose fibers were not treated; this fact can be due to the degradation that the chemical treatment caused to the fibers, reducing its degree of polymerization (DP).

The chemical modification of cellulose nanofibers or nanocrystals follows the same principles as those applied to the fibers, Ljungberg et al. [83, 84] modified the surface of cellulose nanocrystals (CNCs) obtained from tunicates in order to incorporate them in atactic [83] and isotactic polypropylene (PP) [84]. In both cases, the surface treatments in the CNCs were the same; the neutral suspension of CNCs was first dried and redispersed in toluene using ultra Turrax equipment; however, the CNCs did not stand in suspension and decanted. A grafting of PP-g-MA on CNCs surface was also made, but redispersing these grafted CNCs in toluene also precipitated due to the agglomerations. Finally, the aqueous suspension of CNCs was mixed to the surfactant polyoxyethylenenonylphenyl ether phosphate ester (BNA-Ceca ATO Co.) in the ratio of 4:1 surfactant: CNCs, the pH was adjusted to 8 with KOH and the suspension was lyophilized and redispersed in toluene, and this suspension did not precipitate. Subsequently, these suspensions of whiskers in toluene were mixed to the PP solubilized in toluene and films were prepared by casting with the evaporation of the toluene in a vacuum oven. Transparent nanocomposite films were obtained with the introduction of surfactant, resulting in good CNCs dispersion in PP and higher mechanical properties than pure PP films.

Uschanov et al. [128] obtained dispersed CNCs in toluene by modifying their surface with pyridine and toluene sulfonyl chloride (TsCl) solution in an inert atmosphere and adding fatty acids in the same molar concentration of TsCl. The final product was filtered, washed with methanol and ethanol and finally dried in a vacuum oven. However, such modifications have caused a decrease in thermal stability since the degradation temperature depends on the nature of the fatty acid and its degree of substitution on the cellulose surface.

Lif et al. [79] prepared hydrophobic microfibrillated cellulose (MFCs) by adding sodium periodate in the aqueous suspension at room temperature for 1 h. After MFCs were washed with water, they were dispersed in methanol. Octadecylamine and sodium cyanoborohydride was added to the MFCs in methanol, and the solid was washed with methanol, acetone and redispersed in octane. However, in order to disperse these hydrophobic MFCs in an organic solvent, neutral surfactants (without ions) were also used, which gave MFCs dispersed in diesel for up to 30 days.

Stenstad et al. [120] also modified MFCs with cerium-induced grafting; coating with hexamethylenediisocyanate by the introduction of glycidyl methacrylate (GMA) and graphitization of anhydrides. Cerium grafting reactions were carried out in suspensions of MFCs dispersed in HNO_3 solution under an inert atmosphere and adding ceric (IV) ammonium nitrate $(NH_4)_2Ce(NO_3)_6$, followed by the addition of GMA for the polymerization. Cerium (IV) ions are strong oxidizing agents for alcohols with 1,2-glycol groups, forming chelating complexes that decompose forming free radicals in the cellulose, and in the presence of GMA monomers, these radicals enable the formation of grafted polymers on the surface of the fibers, and for each added GMA monomer, an ester group is introduced. The coating with hexamethylenediisocyanate was performed in MFCs dispersed in THF under an inert atmosphere. Hexamethylenediisocyanate and catalyst 1,4-diazabicyclo[2, 2, 2] octane (DABCO) were added and the mixture was stirred for 2 h at 50 °C. Samples were washed with THF and to the isocyanate-coated MFCs suspension, bis-3-amino propylamine and 3-diethylamino propylamine solubilized in THF were added. Grafting diisocyanates promote the formation of a hydrophobic layer on the surface of the microfibril. Isocyanates rapidly react with hydroxyls forming urethane bonds. So reactions must occur in dry solvents and any further reaction should occur immediately after the isocyanate graphitization. The amines were added to the isocyanate-functionalized MFCs to introduce positive charges to its surface because amines readily react with isocyanate forming urea bonds. For grafting anhydrides, diisopropylamine a catalyst was added to the isocyanate-coated MFCs dispersed in THF, under an inert atmosphere. Succinic or maleic anhydrides were dissolved in dry THF (0.8 M concentration) and these solutions were added to the MFCs suspension. With the introduction of anhydrides, vinyl groups were formed on the surface of the fiber and could be a starting point for the polymerization of water-insoluble monomers as an alternative to cerium-induced GMA graphitization.

Siqueira et al. [115, 117] modified the CNCs and MFCs surfaces using a long-chain isocyanate by different methods. After the chemical modification, crystalline structure destruction was not observed. Compared to CNCs, a higher grafting density was necessary to disperse MFCs in a nonpolar liquid medium.

Lin and colleagues [81] extracted CNCs from linter by acid hydrolysis with sulfuric acid (30% v/v) at 60 °C for 6 h, followed by centrifugation and neutralization with ammonia. These whiskers were acetylated with acetic anhydride solution and pyridine. After the reaction, the acetylated cellulose was washed, purified and dried. Subsequently, films were prepared by casting from a mechanical mixture of PLA solubilized in chloroform and acetylated cellulose. The nanocomposite films showed improvement in the mechanical properties of up to 61% at the maximum tensile (with 6% acetylated cellulose) and 40% in the elastic modulus (with 10% acetylated cellulose), when compared to the matrix, due to introduction of filler with high stiffness and good interfacial adhesion with PLA. Improvements in the thermal properties of nanocomposites and increase in crystallinity index were also observed. van der Berg et al. [11] isolated CNCs from tunicates via hydrolysis with sulfuric acid and with hydrochloric acid. The CNCs were dried by lyophilization and redispersed in water and organic solvents such as dimethyl sulfoxide, formic acid, m-cresol, dimethylformamide and dimethyl pyrrolidone. The CNCs were not superficially treated, and even then, they showed good dispersion in these solvents, especially those extracted with sulfuric acid and in the proportion of 1 mg/1 mL of CNCs in the solvent. However, an excessive time was used in the ultrasound to disperse the CNCs in the solvents, being able to cause breakage in the cellulosic chains, reducing their length and, consequently, the aspect ratio (L/D).

Qu and co-workers [102] extracted CNCs from wood pulp with sulfuric acid solution (15%) at 80 °C for 4 h, the mixture was filtered and washed until neutrality, the filtrate was placed in a flask with ethyl alcohol, and acetic acid was added to adjust the pH between 4 and 5. MEMO (3-Methacryloxypropyltrimethoxysilane) was added to modify the cellulose surface with silane agent and to enable its incorporation into PLA by casting. The modified CNCs presented lower thermal stability than the unmodified CNCs because the MEMO modifier presented lower stability than the CNCs, but the morphological integrity of them was maintained. The obtained nanocomposites presented higher tensile strength with 1% by mass of CNCs and 1% of v/v MEMO.

CNCs of ramie fibers were modified by Fischer esterification HCl-catalyzed reaction using di-and tricarboxylic organic acids (malonic, malic and citric acids) [119]. Some properties of modified CNCs were compared to respective CNCs obtained by acid hydrolysis using only HCl. Contrary to what the researchers supposed, a little effect of organic acid pKa was found. The functionality of the free carboxylic acid was introduced to the CNCs surface. The morphology and crystallinity of unmodified and modified CNCs were similar. The results showed that modifying CNCs with bio-based organic acids proved to be an efficient way to introduce carboxylic acid functionality on CNCs surface.

Pommet et al. [100] found a preferential growth of BCs on the surface of the natural fiber than freely in the culture medium composed by fructose, yeast extract, peptone, Na_2HPO_4 and citric acid. They proposed a *green* way to modify natural fibers by attaching bacterial cellulose nanofibers to the surface of these fibers using them as a substrate during the fermentation process of bacterial cellulose. The fermentation process in presence of natural fibers led to the formation of pellicles based on bacterial cellulose, preferably around the natural fibers. An increase in the mechanical strength of the BCs coated fibers was also observed due to the strong hydrogen bonding between the hydroxyl groups present in the BCs and the natural lignocellulosic fibers [48, 68]. So, the coating of bacterial cellulose onto cellulose fibers was considered a new form of controlling the interactions between fibers and polymer matrices because facilitates the good distribution of BCs within the matrix and improve interfacial adhesion between the fibers and the polymer matrix through mechanical interlocking.

The hydrophilic surface of BCs became hydrophobic via esterification reaction with organic acids (acetic, hexanoic and dodecanoic acids) [77]. The authors verified that the degree of surface hydroxyl group substitution decrease with the increase of carbon chain length of the organic acids used.

BCs were acetylated with acetic anhydride in the presence of iodine as a catalyst [58]. The substitution degree (DS) increased when the iodine concentration increased. They verified that the nanostructural morphology preservation is limited by conditions of temperature, time and iodine amount. The better conditions of reaction were 80 °C for 60 min and the amount of catalyst must be less than 0.125 mM. The acetylated BCs showed hydrophobic surface and good mechanical properties which favour the interactions of modified BC and the hydrophobic non-polar polymer matrix. For more drastic reactions conditions, the crystalline structure of BCs was lost.

Thus, there are different types of cellulose/nanocellulose modification reactions as described above. The following three main strategies can be observed: (i) use of a surfactant to functionalize the cellulose/nanocellulose; (ii) chemical reaction and modification of cellulose/nanocellulose in aqueous media and (iii) chemical reaction and modification of cellulose/nanocellulose in an organic solvent.

2.4 All-Based Micro and Nano-cellulose Films

In the literature, there are reported lots of studies of films obtained by casting and still few studies about continuous casting with MFCs, NFCs or CNCs.

Iwamoto et al. [63] studied CNFs films and acrylic composites reinforced with CNFs obtained by casting and extracted from Pinus commercial cellulose pulp. The fibrillation process occurred in the grinder mill with the following numbers of passes: 1; 3; 5; 9; 15 and 30. After 30 passes, the elastic modulus and mechanical tensile strength of CNFs films and acrylic nanocomposites containing CNFs decreased, indicating the decrease of CNFs aspect ratio. CNFs showed a decrease in the crystallinity and in the degree of polymerization with the increase of passes through the grinder, indicating that mechanical shear process causes degradation of cellulose. Therefore, it is necessary to ensure the effective fibrillation of fibers through the grinder, but without cellulose degradation and a decrease of CNFs aspect ratio due to mechanical shearing. In this way, it is necessary to control the fibrilation parameters and consider that each lignocellulosic fiber needs different procedures to obtain the desired final structure.

Siqueira et al. [115] addressed the study of films structured with MFCs by microfluidizer and CNCs by acid hydrolysis obtained from the loofah (*Luffa cylindrical*). The films were obtained by casting, i.e. by water evaporation of aqueous suspension. CNCs had a length of 242 ± 86 nm, the diameter of 5 ± 1 nm and aspect ratio of 47. MFCs had a diameter of 55 ± 15 nm and the length could not be measured. CNCs films achieved a tensile strength of 68 ± 24 MPa and elastic modulus of 2.4 ± 0.2 GPa, while MFCs films showed a tensile strength of 53 ± 19 MPa and elastic modulus of 3 ± 1 GPa. They concluded that CNCs films present greater mechanical resistance than CMFs films.

Sisal CNCs and MFCs were used to prepared cellulosic membranes. The films were obtained by casting at room temperature for five days followed by drying at 60 °C overnight [10]. The water vapour sorption and gas barrier (carbon dioxide, nitrogen, and oxygen) properties of films were evaluated. It was observed that the water diffusion coefficients were higher for CNCs films than for CMFs. This behaviour was associated with the presence of residual lignin, extractive and fatty acids at the surface of MFCs based films. The CNFs films were also much more permeable to gases than MFCs, indicating that gas molecules penetrate slower in CMFs films because of longer diffusion path. Additionally, it was supposed that the entanglements of these long flexible nanoparticles and lower porosity of the films acted as barrier domains, leading to the tortuosity of the diffusion pathway.

Bufalino [14] developed CNFs films from sawdust residues of three Amazonian species (*C. goeldiana, B. parinarioides, and P. gigantocarpa*) and eucalyptus (*E. grandis*). The fibers were pretreated with sodium hydroxide and peroxide to remove lignin and hemicelluloses. Films were produced by conventional casting from CNFs obtained with the following passes in grinder: 10; 20; 30 and 40 passes. There was observed an improvement in tensile strength and elastic modulus on CNFs films obtained from the largest number of passes in the grinder. A decrease in the opacity with the increase in the number of passes through grinder was also observed. The colour of the films varied among the species and was related to the residual lignin, different for each species. Figure 3 shows the colour variation index and transparency of the CNFs films according to the number of passes through the grinder and according to the four species studied. It is concluded that the number of passes through the grinder and the plant species influence the mechanical and optical properties of CNFs films.

Recent works report the use of continuous casting for a scale up production of nanocellulose based films. Claro et al. [19] investigated the morphological structure, thermal and mechanical properties of CNCs and CNFs films from curaua leaf fiber and eucalyptus pulp, obtained by continuous casting. The process of continuous casting produced 6 m of dry nanocellulose film per hour and allows the films to not crack. Figure 4 showed the continuous casting scheme and CNCs/CNFs films obtained by this method.

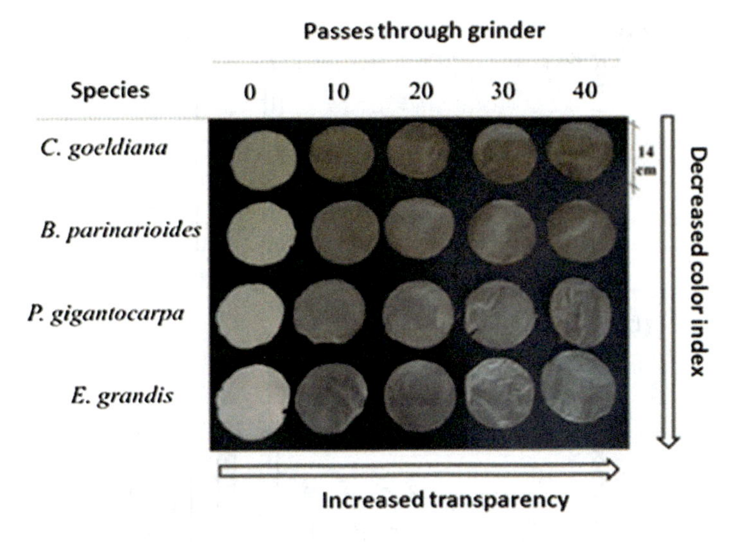

Fig. 3 Variation of color aspect index and transparency of CNFs films in relation to vegetable origin and the number of passes through grinder [111]

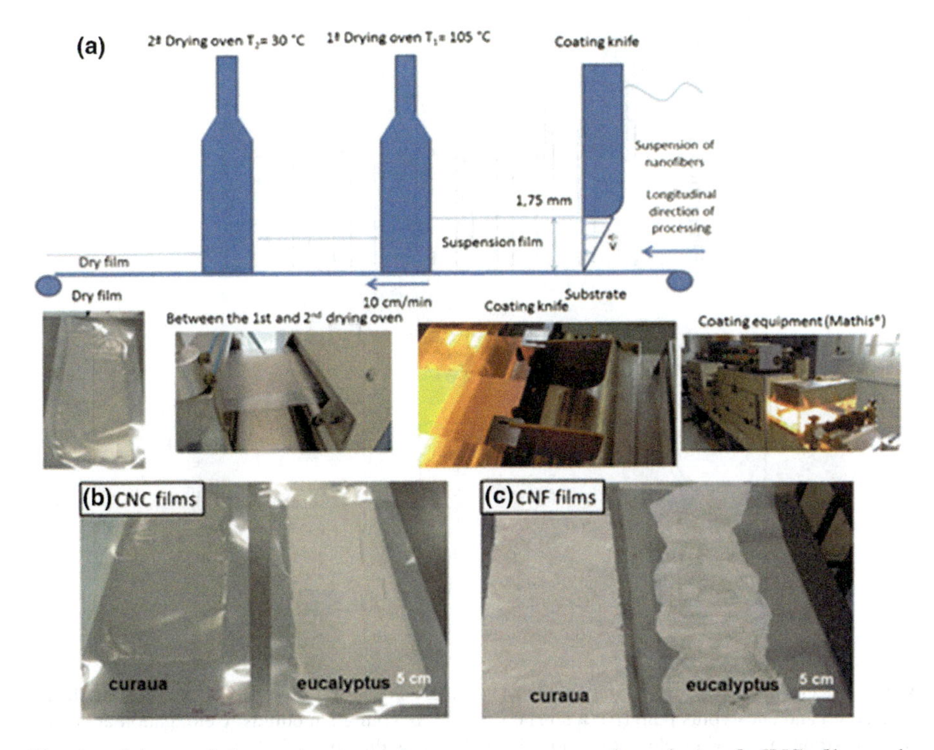

Fig. 4 **a** Scheme of the continuous casting process, curaua and eucalyptus, **b** CNCs films and, **c** CNFs films [19]

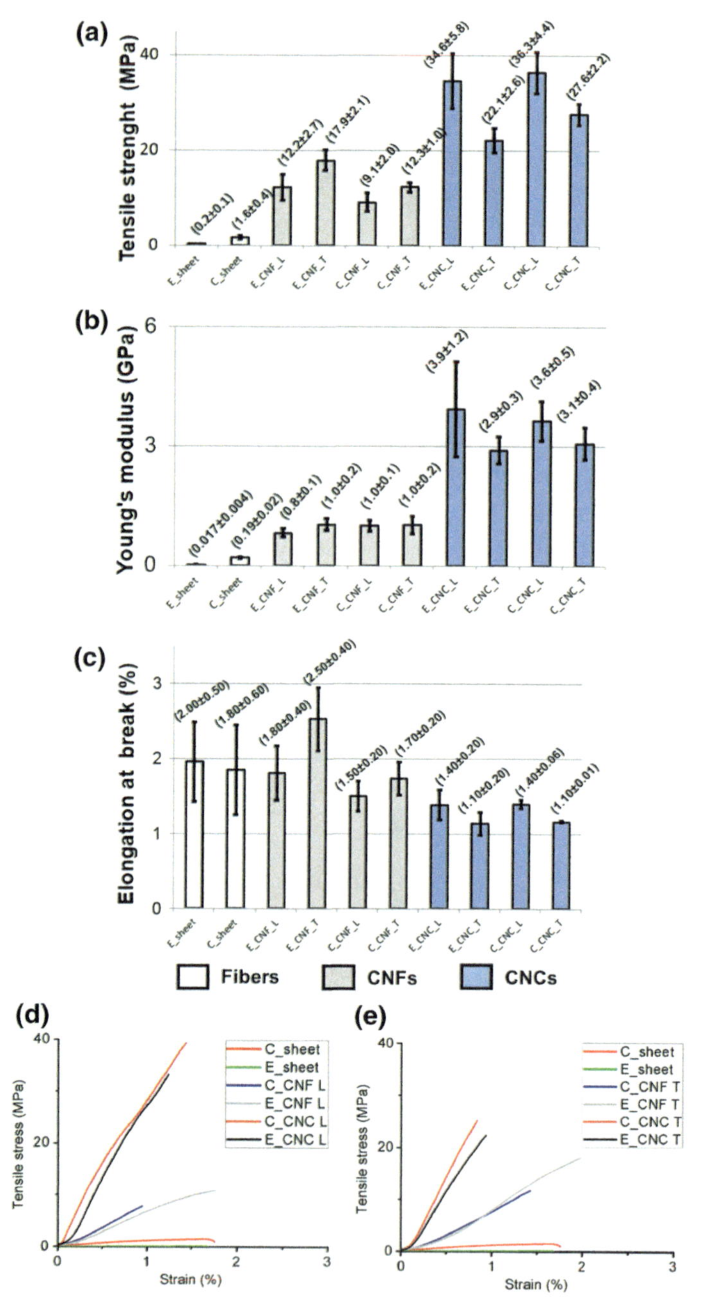

Fig. 5 Mechanical properties under tensile test of fibers sheets and films of CNFs and CNCs from curaua (C) and eucalyptus pulp (E): **a** tensile strength; **b** Young's modulus; **c** elongation at break; **d** stress-strain curve of machine direction (L) and **e** cross direction (T) [19]

CNCs and CNFs suspensions were slowly deposited on a polyester substrate. The coating knife regulates the thickness of the sample and disperses the suspension on a polyester substrate in a pulling speed of 10 cm/min. The equipment presents two drying ovens; the first was set at 105 °C and the second at 30 °C, where the sample undergoes a rapid drying process. CNFs and CNCs films presented mechanical anisotropy due to the orientation of the nanofibers in suspension towards the continuous processing, as shown in Fig. 5.

CNCs films presented higher mechanical resistance (34–36 MPa) in the longitudinal direction of processing and orientation of the nanocrystals. On the other hand, CNFs films presented higher mechanical resistance (12–18 MPa) in the transversal direction of processing and the orientation of the nanofibrils. Therefore, the continuous casting process becomes a viable process for CNCs and CNFs films with different properties and structural morphology than in conventional casting, which is due to the orientation of nanofibers and the rapid drying process.

3 Processing and Applications of Micro and Nano-cellulose Based on Biodegradable Polymers

Micro and nano-cellulose present relative high mechanical properties showing great potential as reinforcement in new biocomposites, innovative bioplastics, and advanced reinforced composite materials [98]. They also present high stiffness, strength, low density and excellent biodegradability and they can be used as a reinforcing material for different polymers, especially biodegradable ones.

Micro and nanocellulose are obtained as dilute suspension and they can be applied to polymer matrix in aqueous solution or freeze-dried aiming better dispersion. CNFs or CNCs are usually dispersed in water, thus, the simplest processing method consists of mixing the cellulose nanostructures in aqueous suspension with water-soluble polymers. This mixture can be cast and the liquid evaporated, resulting in films of nanocomposites, the conventional casting process. This method frequently results in well-dispersed nanostructures, as this avoids the aggregation of the nanomaterial due to the intercalation of the matrix and nanocellulose, preserving its individualized state. This wet casting/evaporation processing method can be extended to other liquids to cover a broader range of polymer matrices [39], but it will be necessary a solvent exchange from water to the polymer solvent, in order to guarantee the dispersion of CNC or CNF through the polymer matrix. However, when it is necessary for high-volume production, the casting or wet processing can be difficult to scale up. The dilute suspension presents stability when it is surface charged by sulfate groups, resulted from acid hydrolysis (CNCs)or from the presence of residual hemicelluloses (CNFs) [39]. But the use of micro/nano cellulose as reinforcement in polymer nanocomposites still presents some challenges and limitations due to their low thermal stability compared to the polymers, their hydrophilicity, the strong hydrogen bonds between adjacent cellulose fibrils, and poor dispersion and compatibility with nonpolar solvents and nonpolar matrices.

In order to overcome some of these problems of dispersion in hydrophobic polymers, the nanocellulose structures can be dispersed in an organic solvent, chemically modified, or grafted with nonpolar molecules. However, the use of them as reinforcements in a wide variety of bio-based polymers, obtained from polysaccharides and proteins, resulted in an increase in the moisture, mechanical and barrier resistance of these materials without compromising their biodegradability [15]. The efficiency of micro/nanocelluloseas reinforcement depends on several factors, such as the good interaction between the polymer and the cellulose, good dispersion, addition of the appropriate amount of filler, among others.

The good interaction between the polymer matrix and the cellulose nanostructures, that is, the good interfacial adhesion or compatibility, together with a good dispersion, are characteristics that provide a more efficient tensile transfer from the polymer matrix to the rigid dispersed phase, resulting in an increase in the mechanical strength of the obtained nanocomposite. The compatibility between the polymer and the micro/nanocellulosecan be more easily achieved when nanocomposites are prepared with polar polymers, i.e. polymers that present polar groups in their chains, such as polyesters, polyether, polyamides, etc., which could be more compatible with the hydroxyls present in the cellulose chains, thus generating higher interfacial adhesion, resulting in much more efficient stress transfer. This good adhesion, allied to randomly dispersed micro/nanocellulose, may also provide a decrease in the permeability of water or other solvents through the polymer as it would hinder the path to be covered by the solvent molecules throughout the nanocomposite.

An important parameter to control is the amount of CNCs that should be introduced to the polymer in order to obtain nanocomposites with improved properties. For fillers with fiber aspects, the percolation threshold is related to the aspect ratio of nanofibres according to the following expression:

$$\Phi_c = \frac{0.7}{L/d}$$

where (L/d) is the aspect ratio of the nanofiber, assuming a cylindrical shape, and (Φ_c) is the percolation threshold in a volumetric fraction. If it is necessary to obtain the massic fraction, Φ_c should be multiplied by the density, which for cellulose is 1.5 g/cm^3 [47]. Below the percolation threshold, few improvements in the properties are observed, but when slightly larger amounts than the percolation threshold are incorporated, a three-dimensional network of nanostructures is formed, where statistically a nanocrystal will touch each other randomly, causing significant improvements in properties, especially in mechanical.

Different techniques have been utilized to produce micro- or nanocellulose-polymer composites: casting-evaporation, melt compounding, electrospinning and solution blow spinning among others.

The processing of materials reinforced with micro/nanocellulose by traditional methods in the molten state is vulnerable and susceptible to agglomeration and poor dispersion of them in the polymeric matrix. In the specified case of CNCs, most studies only found interesting results when they were dispersed in a solvent, and this suspension was mixed with the polymer solubilized in the same solvent, i.e. via casting. Thus, most studies use diluted suspension because lyophilized cellulose nanomaterials aggregate through hydrogen bonding and nanoscale is lost. Normally, lyophilized cellulose nanomaterials incorporated in polymer material request a new dilute suspension by sonication system. The simplest method consists of mixing the dilute suspension of CNC with a polymer material such as starch, for example, making the CNCs well dispersed in the polymer matrix.

Starches are abundant, cheap, biodegradable and a renewable resource material, which makes them attractive, and when together with a plasticizer, under suitable conditions of temperature and shear, the TPS (thermoplastic starch) is formed, which can be molded or mixed with other resins. The major drawback of TPS is its hydrophilicity, in addition to its poor mechanical properties. However, its use, besides reducing costs, can also improve the compatibility of cellulose nanostructures with the polymer matrix, due to the similarity of its chemical structures, which would increase the mechanical properties of the nanocomposite, that is, a compensation of losses due to the use of TPS.

CNCs extracted from cassava bagasse were investigated as reinforcement agent in natural rubber (NR) matrix. The nanoparticles in aqueous suspension were mixed with the NR latex emulsion in fraction varying from 0 to 10 wt% (dry basis). The films were obtained by casting of the mixtures. The favourable interactions between the NR matrix and CNCs filler were confirmed by the relatively high reinforcing effect. An increase from 2.2 MPa for the unfilled matrix to 102 and 154 MPa for the NR film reinforced with the nanofiller was observed [96].

The CNFs were extracted from wheat straw using steam explosion, acidic treatment and high shear mechanical treatment [71]. These nanofibrils were dispersed in regular maize starch (TPS) using glycerol as a plasticizer and high shear mixer. The films were obtained by casting. The results revealed improvement in crystallinity with the addition of CNFs. Mechanical properties increased with the increase of CNFs concentration. Barrier properties also improved with the addition of CNFs up to 10%, but further addition decreased properties due to possible CNFs agglomeration because caused reduction in matrix homogeneity and cohesion. The authors proposed that the increase in CNFs content led to the formation of denser microcrystal network, thereby increasing the mechanical properties. This dense network should decrease the diffusivity through the sample. But this increase in CNFs content may compromise the adhesion level between the nanofiller and the matrix, and the mechanical performance, also causes an increase in the diffusivity of water.

Waxy maize starch nanocrystal (WSNC) and cellulose nanocrystals (CNCs) extracted by acid hydrolysis from microcrystalline cellulose (MCC) were united in order to investigate possible synergistic effects on the normal maize starch matrix plasticized by glycerol [51]. A homogeneous distribution of the nanofillers was demonstrated and the use of CNCs and WSNC upgraded mechanical results, but no significant differences in barrier properties were obtained as compared to the use of only WSNC.

Thomas et al. [122] report another series of studies involving the use of cellulosic nanostructures (MFCs) reinforcing natural rubber (NR) latex matrix. In their specific study, ultra-fine nanocellulose from jute fibers was prepared by steam explosion method and it was used as the reinforcing agent in NR latex along with cross-linking agents, as Zinc-based and sulfur were used during the processing. The nanocomposite films were prepared from pre-vulcanized latex by casting on a glass plate followed by drying at room temperature. The mixture of the aqueous suspension in various proportions of MFCs (0–3 wt%), the latex and the cross-linking agents were done by ball milling, followed by ultra-sonication and drying. The results revealed that the distribution of the filler among the matrix was homogeneous for all the compositions. By adding 2% MFCs, a network by H-bonding interaction was created. The increase of MFCs content in the NR matrix caused a substantial increase in the mechanical properties of the nanocomposite. The vulcanizing agents used for the crosslinking in the NR matrix created a kind of Zn/cellulose complex, forming a network between the layers of NR matrix, improving the dispersion of the MFCs in the NR matrix.

Reports regarding the environmental biodegradability of the starch/CNFs nanocomposites were performed by Babaee et al. [7]. Their study investigated the effect of the addition of unmodified and acetylated CNFs extracted from the kenaf bast fibers (*Hibiscus cannabinus*) in starch glycerol/matrix. The nanofibers were acetylated with acetic anhydride and pyridine under reflux. The nanocomposites were prepared using the solution casting method. The influence of acetylation of CNFs on its biodegradation by white rot fungus (*T. versicolor)* and physicomechanical properties of nanocomposites into the matrix were investigated.

This study showed that both acetylated and non-acetylated CNFs can be used to produce a starch nanocomposite. The mechanical tests showed that the tensile strength and elastic modulus of both nanocomposites increased, in comparison to the matrix, but these improvements were lower for the acetylated ones. The storage modulus and the tan δ peak position of both nanocomposites showed improvement when compared to the matrix. Regardless of the type CNFs, their addition resulted in an increase in the Tg of the nanocomposites. Besides that, the moisture absorption of the nanocomposites reduced by addition of the acetylated nanofibers compared to the non-acetylated one. Furthermore, the fungal biodegradability results showed a longer decomposition period for nanocomposites. But, the acetylated nanocomposite needs a longer time for degradation, and it became more sustainable by replacing the hydroxyl groups with acetyl groups.

Palm oil industry generates a large amount of cellulose-rich residues as oil palm mesocarp fiber (OPMF). Targeting the use of agro-residues as raw materials for cellulose nanocrystals (CNCs) production, de Campos et al. [32, 33] obtained these CNCs from oil palm mesocarp fiber via sulfuric acid hydrolysis and microfluidization, obtaining a stable aqueous suspension and increase in cellulose crystallinity (Fig. 6). The influence of CNCs on properties of cassava starch plasticized with glycerol films was investigated. The reinforcing effect of the CNCswas significant only for loading of up to 6 wt% of CNCs, increasing the elastic modulus. Below percolation threshold, elongation at break was even higher than neat starch

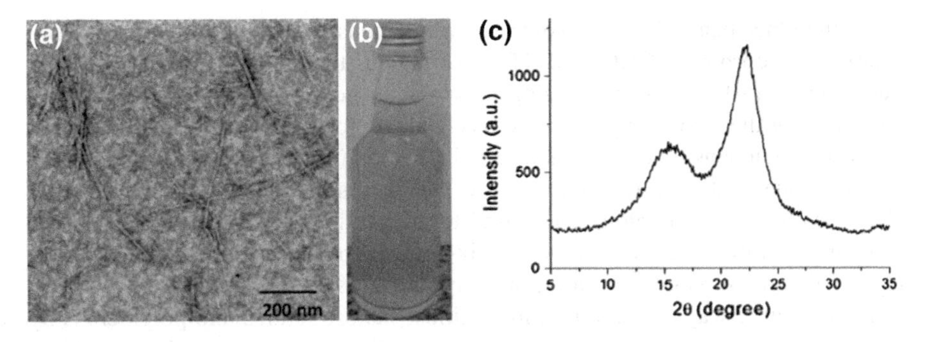

Fig. 6 **a** Scanning transmission electron (STEM) micrograph of cellulose nanocrystals from oil palm mesocarp fibers (OPMF); **b** stable aqueous suspension of CNC-OPMF; **c** XRD patterns of OPMF [32, 33]

films. Above the percolation threshold, there was a formation of a percolating network, leading to CNCs agglomeration and decreasing the mechanical properties of the starch bionanocomposites.

Nanocomposites of chitosan with cellulose are environmentally friendly and the films presented improved physical properties. The CNFs-Chitosan nanocomposites resulted in materials with improved functional properties, in which a wide range of applicability in the field of food packaging, biomedical, water treatment, etc. could be developed. In general, the nanocomposites are obtained by a solvent casting method, electrospinning, and sol-gel transition. Chitosan (cationic) and CNCs (anionic) can be mixed to produce polyelectrolyte complexes using titration. Besides, a two-phase (chitosan and cellulose) based nanocomposites were achieved, and researchers also successfully developed a multi phase material with a high capacity of heavy metal absorption. Furthermore, blends of chitosan and cellulose/nanocellulose resulted in a material with antibacterial activity, metal ions adsorption, odour treatment properties etc. Other commonly used methods for blending are electrospinning, casting and sol-gel transition [1].

Bionanocomposites were developed by casting/evaporation wheat gluten (WG) and (CNCs) from bagasse pulp and TiO_2 nanoparticles [42]. The results demonstrated that CNCs and titanium dioxide nanoparticles improved the mechanical and water vapour barrier properties of gluten films. An optimal content of 7.5% of CNCs and 0.6% of TiO_2 nanoparticles improved the functional properties of WG based materials, according to tensile tests and water resistance of the bionanocomposites. The molecular mobility of amorphous WG chains was not affected by the cellulosic nanofiller. But an increment in Tg of WG/CNC could be verified with TiO_2 nanoparticles incorporation. This behaviour ascribed the strong interfacial interaction between the TiO_2 nanoparticles and the matrix, without disruption in the regularity of the WG chains. Paper sheets coated with the aforementioned nanocomposite exhibited excellent antimicrobial activities i.e. 100, 100 and 98.5% against *S. cervisiae, E. coli, and S. aureus* respectively, for 3 layers coated paper after 2 h of exposure to UVA light illumination.

On the other hand, there are some studies demonstrating that it is possible and feasible to incorporate CNC into TPS matrices in the molten state on a torque rheometer at 140 °C, and also using a twin screw extruder [32, 34, 36]. It was observed that the CNCs improved the mechanical properties of TPS in addition to decreasing their sensitivity to moisture, and any modification of these materials was necessary due to their compatibilities and chemical similarities. Lyophilization of CNCs has been the technique most used to ensure the effective dispersion of the nanoparticles through the polymeric matrix obtained by extrusion, but there are still agglomerations of nanofibers in a polymer matrix [59].

Thus, another way of incorporating cellulose nanostructures into polymers is by the melt processing of polymer nanocomposites. Extrusion and injection-moulding processes are industrially common methods; they are cheap, fast and solvent-free techniques. Due to the hydroxyl groups of cellulose, better results were obtained when cellulose nanostructures were incorporated in polar polymers or starch because strong nanofiller-matrix interactions are expected. The use of polar polymers, such as polyamides, to obtain nanocomposites with CNCs showed promising results in the increase of mechanical properties, however, it was necessary a previous treatment of CNCs, with surface coating with polyamide 6, to increase their thermal resistance in order to support the processing temperatures of polyamides [22]. There are also several studies of poly (lactic acid) (PLA) or poly (Ɛ-caprolactone) to obtain fully biodegradable nanocomposites, because they are polar and biodegradable polymers [29, 37], but they are still more expensive than commodities polymers, such as polyethylene or polypropylene. About the dispersion of hydrophilic cellulose nanostructures in conventional hydrophobic polymers by the extrusion process, it is generally necessary to match the surface properties of the filler and the matrix, i.e. modify the cellulose surface using a surfactant or covalently graft hydrophobic chains with hydroxyls, or also coating the cellulose nanomaterial with chains compatible with the matrix.

In this way, the main issues to overcome for an efficient melt processing of cellulose nanostructures reinforcing polymer nanocomposites are the aggregation of nanocellulose due to the drying process prior to melt processing, the irregular dispersion within the matrix, the low thermal stability, structural integrity after shear pressure of melt processing and orientation towards processing [39].

Film extrusion is a process in which the melt polymer is forced through a planar matrix, in which the film can be formed by blowing or not [59]. When cellulose nanofibers are added to the melt polymer to form films, some problems may arise: film breakdown; thermal degradation of the polymer matrix and/ornanocellulose; alignment of the nanocellulose and increase in viscosity due to the high aspect ratio of the particles [59]. Many studies have shown that the challenge of obtaining extruded polymeric films reinforced with nanocellulose, with improvement on mechanical and barrier properties, is the good dispersion of the nanocellulose throughout the polymer matrix [59]. Martínez-Sanz et al. [86] prepared PLA films reinforced with CNCs by extrusion and observed an elastic modulus of 2.2 GPa and mechanical tensile strength of 61 MPa, much better results than for pure polymer.

Thermoplastic starch (TPS) reinforced with microfibrillated cellulose (MFCs) were prepared via extrusion. The yield strength was improved by ~50% and stiffness by ~250% upon adding 20 wt% MFC compared to neat TPS [78].

de Campos et al. [29] dispersed CNCs in TPS and PCL nanocomposites by the aqueous suspension. The authors first obtained CNCs from sisal by alkali treatment followed by sulfuric acid hydrolysis. The CNCs neutral suspension was dispersed in starch prior to extrusion to obtain TPS and TPS/PCL nanocomposite. They observed greater dispersibility of the CNC with lower concentration. High concentration of CNC in nanocomposite presented agglomeration and compromised mechanical performance, while lower CNC concentration improved the mechanical properties. The displacement and narrowing of the carbonyl band of the blend with 5% CNC showed the interaction between carbonyl groups of PCL with OH groups of CNC, and avoided the interaction between CNCs, preventing their aggregation.

Electrospinning is a technology widely used for fibers formation. This technique uses electrical forces to produce polymer fibers with diameters ranging from 2 nm to several micrometres using bio-based polymer solutions or synthetic polymers is a process that offers capabilities for producing nanofibers and fabrics with controllable pore structure [12]. Electrospun fibers have been applied in various areas, such as, nanocatalysis, tissue engineering scaffolds, protective clothing, filtration, biomedical, pharmaceutical, optical electronics, healthcare, biotechnology, defense and security, and environmental engineering, due to its smaller pores and higher surface area than regular fibers [12]. The technique consists of feeding a polymer solution into a stream of pressurized air using a concentric nozzle. When the aerodynamic forces overcome the solution surface tension, a solution jet jettisons towards a collector and the solvent is evaporated forming polymer fibers that are collected as non-woven mats [27].

Recent studies showed that the use of poly (ethylene oxide) (PEO) as a matrix to obtain nanocomposite fibers containing CNCs and CNFs by electrospinning [73]. The incorporation of CNCs increased the elastic modulus in two times and mechanical tensile strength in 2.5 times in relation to pure PEO [131]. The incorporation of CNFs also increased the elastic modulus and mechanical tensile strength in more than two times in relation to neat PEO, indicating the potential use of cellulose nanofibers as reinforcement in nanocomposites obtained by electrospinning [129].

Solution Blow Spinning (SBS) is a technique for commercial-scale nanofiber production, with lower cost compared to electrospinning [27]. The SBS process is compatible with a wider variety of solvents than electrospinning and eliminates the necessity of using high voltages [26, 94] and it is a great advantage to be more portable, because with the commercial airbrush systems, depositing fibers on a broad range of collectors and surfaces are facilitated[9, 126]. The applications for SBS include their use in sensors and biosensors, wound dressings, tissue sutures, drug delivery materials, filter membranes and adsorbents [27, 94]. da Silva Parize et al. [27] prepared bio-based nanocomposites of PLA and CNCs by SB-Spinning (Fig. 7). CNCs were obtained from Eucalyptus kraft pulp by sulfuric acid hydrolysis (CNC) and esterified with maleic anhydride (CNC_{AM}), they were

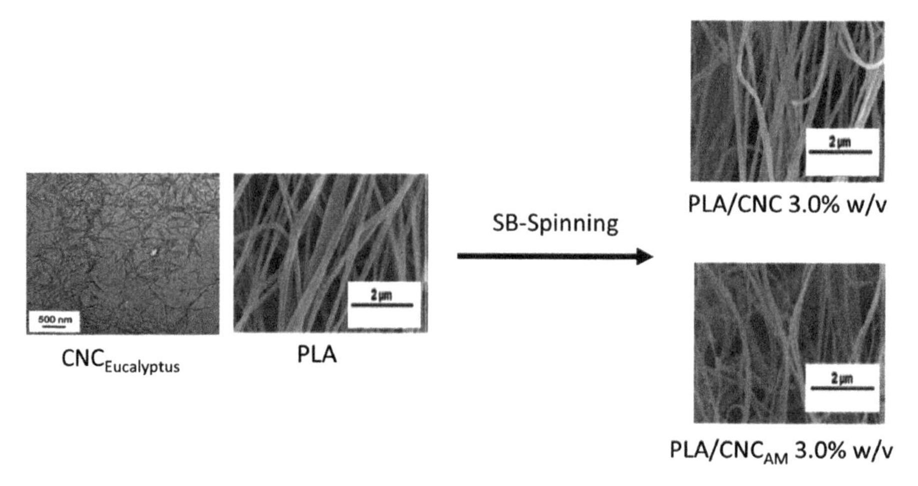

Fig. 7 CNCs from Eucalyptus kraft pulp applied in PLA nanofibers. Adapted from da Parize et al. [27]

applied in PLA solubilized into dimethyl carbonate (DMC) as a solvent. The authors observed that CNCs of both sulfate groups and modified with maleic anhydride acted as nucleating agents and tends to favour the formation of PLA crystals of higher stability since the CNCs of both methods presented crystallinity of around 64%. It is assumed that a fraction of the CNCs are on the surface of the PLA fibers since the hydrophilicity of the composite films increased significantly.

4 General Applications

Nanocellulose and cellulose microfibrils present various potential applications and have the advantage of being derived from natural sources and often a vegetal residue. There are several potential applications, such as barrier [6] for liquid and gaseous materials, reinforcement of plastics and cement, sensors of oil and gas industry, medical devices as special dressings and prostheses [25, 80], in paints, coatings, films and foams, cosmetics, photonics and in electrical as field effect transistors (FET) [49] and electronic industries, solar cells, etc. [24].

The use of nanocellulose as reinforcement is already well understood. Both microfibrils, nanofibrils and cellulose nanocrystals improve especially the mechanical properties as well as thermal properties. The mechanical reinforcement is related to cellulose hydrogen bonding network within the polymer matrix.

MFCs-, CNFs- and CNCs-based polymer nanocomposites provide improvements in barrier properties. These characteristics are dependent on the fibrils morphology of cellulose because their morphology can act as a barrier for the penetration, as well as diffusion of liquid and even gaseous materials into the cellulose-based film [89].

Cellulose-based composite has been applied also as sensor materials. The gas-sensing material can be fabricated from inorganic semiconductor metal oxides, inserting a small amount of metal atom or organic conducting polymer [127]. However, in both systems, they are not entirely satisfactory. The mixture of both and the cellulose can result in a flexible and conductive material. Because cellulose can hold inorganic particles and consequently gain flexibility, being suitable for use as a gas sensor [127].

Lin and Dufresne [80] reported a series of applications of nanocellulose as biomedical materials. They highlighted these nanostructures as a "gif" provides by Nature. Its physical properties, special surface chemistry, biocompatibility, biodegradability and low toxicity make the nanocellulosea potential source of production of diverse biomedical materials as tissue bioscaffolds for cellular culture, drug excipient and drug delivery, immobilization and recognition of enzyme/protein and development in substitutes/medical materials like blood vessel, cartilage and tissue repair

Nanocellulose applied to photonic have the main interest related to have the liquid crystalline behaviour of CNCs which gives rise to iridescent films of defined optical character and because both CNCs and CNFs may form optically transparent stand-alone films. The evaporation of aqueous suspensions from CNCs can form chiral nematic, iridescent and coloured films which depends on polydispersity of the CNC sizes [2].

Another topic of great interest is a nanocellulose-based coating for controlled release of drugs, in the form of membranes. In addition, modified CNC can be used with aromatic groups, which can control the release of amino acids, being a promising candidate in the immobilization of proteins, preserving the structural integrity of the protein and increasing the activity and long-term storage stability [121].

In food industry nanocellulose acts as a food stabilizer. They have a better affinity with water than with oil and in some cases, nanocellulose may be substituted for oil in some cases to produce low-calorie foods acting as a functional food ingredient [52]. The barrier properties increase the protection and preservation of products and can increase the shelf-life of food [76].

5 Conclusions

Microfibrils and cellulose nanofibers can be obtained by several methods and their performance in determined applications depends on the extraction methodology and their dispersion in the matrix. In this way, the improvements that the microfibrils and nanocellulose could give to the polymer nanocomposite depend upon their type and dispersion and also the good interface interactions or compatibility between the microfibrils and nanocellulose and polymer. These are characteristics that provide a more efficient tensile transfer from the polymer matrix to the rigid dispersed phase, resulting in the mechanical improvement of the obtained composites.

References

1. Abdul Khalil HPS, Saurabh CK, Adnan AS et al (2016) A review on chitosan-cellulose blends and nanocellulose reinforced chitosan biocomposites: Properties and their applications. Carbohydr Polym 150:216–226
2. Abitbol T, Rivkin A, Cao Y et al (2016) Nanocellulose, a tiny fiber with huge applications. Curr Opin Biotechnol 39:76–88
3. Agustin MB, Nakatsubo F, Yano H (2016) Products of low-temperature pyrolysis of nanocellulose esters and implications for the mechanism of thermal stabilization. Cellulose 23:2887–2903. https://doi.org/10.1007/s10570-016-1004-0
4. Alemdar A, Sain M (2008) Biocomposites from wheat straw nanofibers: morphology, thermal and mechanical properties. Compos Sci Technol 68:557–565. https://doi.org/10.1016/j.compscitech.2007.05.044
5. Araujo MAM, Sena Neto AR, Hage E et al (2015) Curaua leaf fiber *(Ananas comosus* var. *erectifolius*) reinforcing poly(lactic acid) biocomposites: formulation and performance. Polym Compos 36:1520–1530. https://doi.org/10.1002/pc.23059
6. Azeredo HMC, Rosa MF, Mattoso LHC (2017) Nanocellulose in bio-based food packaging applications. Ind Crops Prod 97:664–671. https://doi.org/10.1016/j.indcrop.2016.03.013
7. Babaee M, Jonoobi M, Hamzeh Y, Ashori A (2015) Biodegradability and mechanical properties of reinforced starch nanocomposites using cellulose nanofibers. Carbohydr Polym 132:1–8. https://doi.org/10.1016/j.carbpol.2015.06.043
8. Bahrin EK, Baharuddin AS, Ibrahim MF et al (2012) Physicochemical property changes and enzymatic hydrolysis enhancement of oil palm empty fruit bunches treated with superheated steam. BioResources 7:1784–1801. https://doi.org/10.15376/biores.7.2.1784-1801
9. Behrens AM, Casey BJ, Sikorski MJ et al (2014) In situ deposition of PLGA nanofibers via solution blow spinning. ACS Macro Lett 3:249–254. https://doi.org/10.1021/mz500049x
10. Belbekhouche S, Bras J, Siqueira G et al (2011) Water sorption behavior and gas barrier properties of cellulose whiskers and microfibrils films. Carbohydr Polym 83:1740–1748. https://doi.org/10.1016/j.carbpol.2010.10.036
11. van der Berg O, Capadona JR, Weder C (2007) Preparation of homogeneous dispersions of tunicate cellulose whiskers in organic solvents. Biomacromol 8:1353–1357. https://doi.org/10.1021/bm061104q
12. Bhardwaj N, Kundu SC (2010) Electrospinning: A fascinating fiber fabrication technique. Biotechnol Adv 28:325–347. https://doi.org/10.1016/j.biotechadv.2010.01.004
13. Bondancia TJ, Mattoso LHC, Marconcini JM, Farinas CS (2017) A new approach to obtain cellulose nanocrystals and ethanol from eucalyptus cellulose pulp via the biochemical pathway. Biotechnol Prog 33:1085–1095. https://doi.org/10.1002/btpr.2486
14. Bufalino L (2014) Filmes de nanocelulose a partir de resíduos madeireiros da Amazônia. UFLA 106
15. Carmona VB, Corrêa AC, Marconcini JM, Mattoso LHC (2015) Properties of a biodegradable ternary blend of thermoplastic starch (TPS), Poly(ε-Caprolactone) (PCL) and Poly(Lactic Acid) (PLA). J Polym Environ 23:83–89. https://doi.org/10.1007/s10924-014-0666-7
16. Chen Y, Wu Q, Huang B et al (2015) Isolation and characteristics of cellulose and nanocellulose from lotus leaf stalk agro-wastes. BioResources 10:684–696
17. Chen L, Zhu JY, Baez C et al (2016) Highly thermal-stable and functional cellulose nanocrystals and nanofibrils produced using fully recyclable organic acids. Green Chem 18:3835–3843. https://doi.org/10.1039/C6GC00687F
18. Chin KM, Sung Ting S, Ong HL, Omar M (2018) Surface functionalized nanocellulose as a veritable inclusionary material in contemporary bioinspired applications: a review. J Appl Polym Sci 135. https://doi.org/10.1002/app.46065

19. Claro PIC, Corrêa AC, de Campos A et al (2018) Curaua and eucalyptus nanofibers films by continuous casting: mechanical and thermal properties. Carbohydr Polym 181:1093–1101. https://doi.org/10.1016/j.carbpol.2017.11.037
20. Corradini E, Teixeira EM, Paladin PD et al (2009) Thermal stability and degradation kinetic study of white and colored cotton fibers by thermogravimetric analysis. J Therm Anal Calorim 97:415–419
21. Corradini E, Imam SH, Agnelli JM, Mattoso LHC (2009a) Effect of coconut, sisal and jute fibers on the properties of starch/gluten/glycerol matrix. J Polym Environ 17:1–9. https://doi.org/10.1007/s10924-009-0115-1
22. Corrêa AC, de Morais Teixeira E, Carmona VB et al (2014) Obtaining nanocomposites of polyamide 6 and cellulose whiskers via extrusion and injection molding. Cellulose 21:311–322. https://doi.org/10.1007/s10570-013-0132-z
23. Corrêa AC, de Teixeira EM, Pessan LA, Mattoso LHC (2010) Cellulose nanofibers from curaua fibers. Cellulose 17:1183–1192. https://doi.org/10.1007/s10570-010-9453-3
24. Costa SV, Pingel P, Janietz S, Nogueira AF (2016) Inverted organic solar cells using nanocellulose as substrate. J Appl Polym Sci 133. https://doi.org/10.1002/app.43679
25. Czaja WK, Young DJ, Kawecki M, Brown RM (2007) The future prospects of microbial cellulose in biomedical applications. Biomacromol 8:1–12
26. da Silva Parize DD, Foschini MM, de Oliveira JE et al (2016) Solution blow spinning: parameters optimization and effects on the properties of nanofibers from poly(lactic acid)/dimethyl carbonate solutions. J Mater Sci 51:4627–4638. https://doi.org/10.1007/s10853-016-9778-x
27. da Silva Parize DD, de Oliveira JE, Williams T et al (2017) Solution blow spun nanocomposites of poly(lactic acid)/cellulose nanocrystals from Eucalyptus kraft pulp. Carbohydr Polym 174:923–932. https://doi.org/10.1016/j.carbpol.2017.07.019
28. Davoodi MM, Sapuan SM, Ahmad D et al (2010) Mechanical properties of hybrid kenaf/glass reinforced epoxy composite for passenger car bumper beam. Mater Des 31:4927–4932. https://doi.org/10.1016/j.matdes.2010.05.021
29. de Campos A, Teodoro KBR, Teixeira EM et al (2013) Properties of thermoplastic starch and TPS/polycaprolactone blend reinforced with sisal whiskers using extrusion processing. Polym Eng Sci 53:800–808. https://doi.org/10.1002/pen.23324
30. de Campos A, Correa AC, Cannella D et al (2013) Obtaining nanofibers from curaua and sugarcane bagasse fibers using enzymatic hydrolysis followed by sonication. Cellulose 20:1491–1500. https://doi.org/10.1007/s10570-013-9909-3
31. de Campos A, Tonoli GHD, Marconcini JM et al (2013) TPS/PCL composite reinforced with treated sisal fibers: property, biodegradation and water-absorption. J Polym Environ 21:1–7. https://doi.org/10.1007/s10924-012-0512-8
32. de Campos A, de Neto ARS, Rodrigues VB et al (2017a) Production of cellulose nanowhiskers from oil palm mesocarp fibers by acid hydrolysis and microfluidization. J Nanosci Nanotechnol 17:4970–4976. https://doi.org/10.1166/jnn.2017.13451
33. de Campos A, Sena Neto AR, Rodrigues VB et al (2017b) Bionanocomposites produced from cassava starch and oil palm mesocarp cellulose nanowhiskers. Carbohydr Polym 175:330–336. https://doi.org/10.1016/j.carbpol.2017.07.080
34. de Morais Teixeira E, Pasquini D, Curvelo AAS et al (2009) Cassava bagasse cellulose nanofibrils reinforced thermoplastic cassava starch. Carbohydr Polym 78:422–431. https://doi.org/10.1016/j.carbpol.2009.04.034
35. de Morais Teixeira E, Corrêa AC, Manzoli A et al (2010) Cellulose nanofibers from white and naturally colored cotton fibers. Cellulose 17:595–606. https://doi.org/10.1007/s10570-010-9403-0
36. de Morais Teixeira E, Bondancia TJ, Teodoro KBR et al (2011) Sugarcane bagasse whiskers: Extraction and characterizations. Ind Crops Prod 33:63–66. https://doi.org/10.1016/j.indcrop.2010.08.009

37. de Morais Teixeira E, de Campos A, Marconcini JM et al (2014) Starch/fiber/poly(lactic acid) foam and compressed foam composites. RSC Adv 4:6616. https://doi.org/10.1039/c3ra47395c
38. dos Santos RM, Flauzino Neto WP, Silvério HA et al (2013) Cellulose nanocrystals from pineapple leaf, a new approach for the reuse of this agro-waste. Ind Crops Prod 50:707–714. https://doi.org/10.1016/j.indcrop.2013.08.049
39. Dufresne A (2018) Cellulose nanomaterials as green nanoreinforcements for polymer nanocomposites. Philos Trans R Soc A Math Phys Eng Sci 376:20170040. https://doi.org/10.1098/rsta.2017.0040
40. Dufresne A, Castaño J (2016) Polysaccharide nanomaterial reinforced starch nanocomposites: a review. Starch/Staerke 1–19. https://doi.org/10.1002/star.201500307
41. Eichhorn SJ, Dufresne A, Aranguren M et al (2010) Review: current international research into cellulose nanofibres and nanocomposites. J Mater Sci 45:1–33. https://doi.org/10.1007/s10853-009-3874-0
42. El-Wakil NA, Hassan EA, Abou-Zeid RE, Dufresne A (2015) Development of wheat gluten/nanocellulose/titanium dioxide nanocomposites for active food packaging. Carbohydr Polym 124:337–346. https://doi.org/10.1016/j.carbpol.2015.01.076
43. Fahma F, Iwamoto S, Hori N et al (2010) Isolation, preparation, and characterization of nanofibers from oil palm empty-fruit-bunch (OPEFB). Cellulose 17:977–985. https://doi.org/10.1007/s10570-010-9436-4
44. Filson PB, Dawson-Andoh BE (2009) Characterization of sugars from model and enzyme-mediated pulp hydrolyzates using high-performance liquid chromatography coupled to evaporative light scattering detection. Bioresour Technol 100:6661–6664. https://doi.org/10.1016/j.biortech.2008.12.067
45. Flauzino Neto WP, Silvério HA, Dantas NO, Pasquini D (2013) Extraction and characterization of cellulose nanocrystals from agro-industrial residue—Soy hulls. Ind Crops Prod 42:480–488. https://doi.org/10.1016/j.indcrop.2012.06.041
46. Forsman N, Lozhechnikova A, Khakalo A et al (2017) Layer-by-layer assembled hydrophobic coatings for cellulose nanofibril films and textiles, made of polylysine and natural wax particles. Carbohydr Polym 173:392–402. https://doi.org/10.1016/j.carbpol.2017.06.007
47. Garcia de Rodriguez NL, Thielemans W, Dufresne A (2006) Sisal cellulose whiskers reinforced polyvinyl acetate nanocomposites. Cellulose 13:261–270. https://doi.org/10.1007/s10570-005-9039-7
48. Gardner DJ, Oporto GS, Mills R, Samir MASA (2008) Adhesion and Surface Issues in Cellulose and Nanocellulose. J Adhesion Sci Technol 22:545–567. https://doi.org/10.1163/156856108X295509
49. Gaspar D, Fernandes SN, De Oliveira AG et al (2014) Nanocrystalline cellulose applied simultaneously as the gate dielectric and the substrate in flexible field effect transistors. Nanotechnology 25. https://doi.org/10.1088/0957-4484/25/9/094008
50. Ghaderi M, Mousavi M, Yousefi H, Labbafi M (2014) All-cellulose nanocomposite film made from bagasse cellulose nanofibers for food packaging application. Carbohydr Polym 104:59–65. https://doi.org/10.1016/j.carbpol.2014.01.013
51. González K, Retegi A, González A et al (2015) Starch and cellulose nanocrystals together into thermoplastic starch bionanocomposites. Carbohydr Polym 117:83–90. https://doi.org/10.1016/j.carbpol.2014.09.055
52. Gómez HC, Serpa A, Velásquez-Cock J et al (2016) Vegetable nanocellulose in food science: a review. Food Hydrocoll. 57:178–186
53. Harmaen AS, Khalina A, Azowa I et al (2015) Thermal and biodegradation properties of poly(lactic acid)/fertilizer/oil palm fibers blends biocomposites. Polym Compos 36:576–583. https://doi.org/10.1002/pc.22974

54. Hassan ML, Bras J, Hassan EA, et al (2012) Polycaprolactone/ Modified Bagasse Whisker Nanocomposites with Improved Moisture-Barrier and Biodegradability Properties. J Appl Polym Sci 1–10. https://doi.org/10.1002/app
55. Henriksson M, Henriksson G, Berglund LA, Lindström T (2007) An environmentally friendly method for enzyme-assisted preparation of microfibrillated cellulose (MFC) nanofibers. Eur Polym J 43:3434–3441. https://doi.org/10.1016/j.eurpolymj.2007.05.038
56. Henrique MA, Silvério HA, Flauzino Neto WP, Pasquini D (2013) Valorization of an agro-industrial waste, mango seed, by the extraction and characterization of its cellulose nanocrystals. J Environ Manage 121:202–209. https://doi.org/10.1016/j.jenvman.2013.02.054
57. Hindi SSZ (2017) Microcrystalline cellulose: the inexhaustible treasure for pharmaceutical industry. Nanosci Nanotechnol Res 4:17–24. https://doi.org/10.12691/nnr-4-1-3
58. Hu W, Chen S, Xu Q, Wang H (2011) Solvent-free acetylation of bacterial cellulose under moderate conditions. Carbohydr Polym 83:1575–1581. https://doi.org/10.1016/j.carbpol.2010.10.016
59. Hubbe MA, Ferrer A, Tyagi P et al (2017) Nanocellulose in thin films, coatings, and plies for packaging applications: a review. BioResources 12:2143–2233
60. Hubbe M, Rojas OJ, Lucia L, Sain M (2008) Cellulosic Nanocomposites: a review. BioResources 3:929–980. https://doi.org/10.15376/biores.3.3.929-980
61. Ibrahim Nor Azowa, Hadithon Kamarul Arifin, Abdan K (2010) Effect of fiber treatment on mechanical properties of kenaf fiber-Ecoflex composites. J Reinf Plast Compos 29:2192–2198. https://doi.org/10.1177/0731684409347592
62. Isogai A, Saito T, Fukuzumi H (2011) TEMPO-oxidized cellulose nanofibers. Nanoscale 3:71–85. https://doi.org/10.1039/C0NR00583E
63. Iwamoto S, Nakagaito AN, Yano H (2007) Nano-fibrillation of pulp fibers for the processing of transparent nanocomposites. Appl Phys A Mater Sci Process 89:461–466. https://doi.org/10.1007/s00339-007-4175-6
64. Jia C, Chen L, Shao Z et al (2017) Using a fully recyclable dicarboxylic acid for producing dispersible and thermally stable cellulose nanomaterials from different cellulosic sources. Cellulose 24:2483–2498. https://doi.org/10.1007/s10570-017-1277-y
65. Johar N, Ahmad I, Dufresne A (2012) Extraction, preparation and characterization of cellulose fibres and nanocrystals from rice husk. Ind Crops Prod 37:93–99. https://doi.org/10.1016/j.indcrop.2011.12.016
66. Kaisangsri N, Kerdchoechuen O, Laohakunjit N (2014) Characterization of cassava starch based foam blended with plant proteins, kraft fiber, and palm oil. Carbohydr Polym 110:70–77. https://doi.org/10.1016/j.carbpol.2014.03.067
67. Kalia S, Boufi S, Celli A, Kango S (2014) Nanofibrillated cellulose: surface modification and potential applications. Colloid Polym Sci 292:5–31. https://doi.org/10.1007/s00396-013-3112-9
68. Kalia S, Dufresne A, Cherian BM t al (2011) Cellulose-based bio- and nanocomposites: a review. Int J Polym Sci 2011
69. Kamel S (2007) Nanotechnology and its applications in lignocellulosic composites, a mini review. Express Polym Lett 1:546–575
70. Kargarzadeh H, Mariano M, Huang J et al (2017) Recent developments on nanocellulose reinforced polymer nanocomposites: a review. Polym (United Kingdom) 132:368–393
71. Kaushik A, Singh M, Verma G (2010) Green nanocomposites based on thermoplastic starch and steam exploded cellulose nanofibrils from wheat straw. Carbohydr Polym 82:337–345. https://doi.org/10.1016/j.carbpol.2010.04.063
72. Khalil HPSA, Davoudpour Y, Aprilia NAS, Mustapha A, Hossain S, Islam N, Dungani R (2014) Nanocellulose-based polymer nanocomposite: isolation, characterization and applications. In: nanocellulose polymer nanocomposites. John Wiley & Sons, Inc., p 273–309. ISBN: 978-1-118-87190-4

73. Kim JH, Shim BS, Kim HS et al (2015) Review of nanocellulose for sustainable future materials. Int J Precis Eng. Manuf Green Technol 2:197–213
74. Lam NT, Chollakup R, Smitthipong W et al (2017) Characterization of cellulose nanocrystals extracted from sugarcane bagasse for potential biomedical materials. Sugar Tech 19:539–552. https://doi.org/10.1007/s12355-016-0507-1
75. Lamaming J, Hashim R, Sulaiman O et al (2015) Cellulose nanocrystals isolated from oil palm trunk. Carbohydr Polym 127:202–208. https://doi.org/10.1016/j.carbpol.2015.03.043
76. Lavoine N, Desloges I, Dufresne A, Bras J (2012) Microfibrillated cellulose—its barrier properties and applications in cellulosic materials: a review. Carbohydr Polym 90:735–764
77. Lee KY, Quero F, Blaker JJ et al (2011) Surface only modification of bacterial cellulose nanofibres with organic acids. Cellulose 18:595–605. https://doi.org/10.1007/s10570-011-9525-z
78. Lendvai L, Karger-Kocsis J, Kmetty Á, Drakopoulos SX (2016) Production and characterization of microfibrillated cellulose-reinforced thermoplastic starch composites. J Appl Polym Sci 133. https://doi.org/10.1002/app.42397
79. Lif A, Stenstad P, Syverud K et al (2010) Fischer-Tropsch diesel emulsions stabilised by microfibrillated cellulose and nonionic surfactants. J Colloid Interface Sci 352:585–592. https://doi.org/10.1016/j.jcis.2010.08.052
80. Lin N, Dufresne A (2014) Nanocellulose in biomedicine: current status and future prospect. Eur Polym J 59:302–325. https://doi.org/10.1016/j.eurpolymj.2014.07.025
81. Lin N, Huang J, Chang PR et al (2011) Surface acetylation of cellulose nanocrystal and its reinforcing function in poly(lactic acid). Carbohydr Polym 83:1834–1842. https://doi.org/10.1016/j.carbpol.2010.10.047
82. Liu Q, Lu Y, Aguedo M et al (2017) Isolation of high-purity cellulose nanofibers from wheat straw through the combined environmentally friendly methods of steam explosion, microwave-assisted hydrolysis, and microfluidization. ACS Sustain Chem Eng 5:6183–6191. https://doi.org/10.1021/acssuschemeng.7b01108
83. Ljungberg N, Bonini C, Bortolussi F et al (2005) New nanocomposite materials reinforced with cellulose whiskers in atactic polypropylene: effect of surface and dispersion characteristics. Biomacromol 6:2732–2739. https://doi.org/10.1021/bm050222v
84. Ljungberg N, Cavaillé JY, Heux L (2006) Nanocomposites of isotactic polypropylene reinforced with rod-like cellulose whiskers. Polymer (Guildf) 47:6285–6292. https://doi.org/10.1016/j.polymer.2006.07.013
85. Mandal A, Chakrabarty D (2011) Isolation of nanocellulose from waste sugarcane bagasse (SCB) and its characterization. Carbohydr Polym 86:1291–1299. https://doi.org/10.1016/j.carbpol.2011.06.030
86. Martínez-Sanz M, Lopez-Rubio A, Lagaron JM (2012) Optimization of the dispersion of unmodified bacterial cellulose nanowhiskers into polylactide via melt compounding to significantly enhance barrier and mechanical properties. Biomacromol 13:3887–3899. https://doi.org/10.1021/bm301430j
87. Miao X, Lin J, Tian F et al (2016) Cellulose nanofibrils extracted from the byproduct of cotton plant. Carbohydr Polym 136:841–850. https://doi.org/10.1016/j.carbpol.2015.09.056
88. El Miri N, Abdelouahdi K, Barakat A et al (2015) Bio-nanocomposite films reinforced with cellulose nanocrystals: rheology of film-forming solutions, transparency, water vapor barrier and tensile properties of films. Carbohydr Polym 129:156–167. https://doi.org/10.1016/j.carbpol.2015.04.051
89. Mishra RK, Sabu A, Tiwari SK (2018) Materials chemistry and the futurist eco-friendly applications of nanocellulose: status and prospect. J Saudi Chem Soc 22:949. https://doi.org/10.1016/j.jscs.2018.02.005
90. Moon RJ, Martini A, Nairn J et al (2011) Cellulose nanomaterials review: structure, properties and nanocomposites
91. Nakagaito AN, Iwamoto S, Yano H (2005) Bacterial cellulose: the ultimate nano-scalar cellulose morphology for the production of high-strength composites. Appl Phys A Mater Sci Process 80:93–97. https://doi.org/10.1007/s00339-004-2932-3

92. Nechyporchuk O, Belgacem MN, Bras J (2016) Production of cellulose nanofibrils: a review of recent advances. Ind Crops Prod 93:2–25
93. Nikmatin S, Syafiuddin A, Irwanto DAY (2017) Properties of oil palm empty fruit bunch-filled recycled acrylonitrile butadiene styrene composites: effect of shapes and filler loadings with random orientation. BioResources 12:1090–1101. https://doi.org/10.15376/biores.12.1.1090-1101
94. Oliveira JE, Moraes EA, Costa RGF et al (2011) Nano and submicrometric fibers of poly(D, L -lactide) obtained by solution blow spinning: process and solution variables. J Appl Polym Sci 122:3396–3405. https://doi.org/10.1002/app.34410
95. Pasquini D, de Teixeira EM, da Curvelo AA et al (2008) Surface esterification of cellulose fibres: processing and characterisation of low-density polyethylene/cellulose fibres composites. Compos Sci Technol 68:193–201. https://doi.org/10.1016/j.compscitech.2007.05.009
96. Pasquini D, de Teixeira EM, da Curvelo AA S et al (2010) Extraction of cellulose whiskers from cassava bagasse and their applications as reinforcing agent in natural rubber. Ind Crops Prod 32:486–490. https://doi.org/10.1016/j.indcrop.2010.06.022
97. Pelissari FM, Sobral PJDA, Menegalli FC (2014) Isolation and characterization of cellulose nanofibers from banana peels. Cellulose 21:417–432. https://doi.org/10.1007/s10570-013-0138-6
98. Peng Y, Gardner DJ, Han Y et al (2013) Influence of drying method on the material properties of nanocellulose I: thermostability and crystallinity. Cellulose 20:2379–2392. https://doi.org/10.1007/s10570-013-0019-z
99. Petersson L, Oksman K (2006) Preparation and properties of biopolymer-based nanocomposite films using microcrystalline cellulose. In: ACS Symposium Series. pp 132–150
100. Pommet M, Juntaro J, Heng JYY et al (2008) Surface modification of natural fibers using bacteria: depositing bacterial cellulose onto natural fibers to create hierarchical fiber reinforced nanocomposites. Biomacromol 9:1643–1651. https://doi.org/10.1021/bm800169g
101. Pääkko M, Ankerfors M, Kosonen H et al (2007) Enzymatic hydrolysis combined with mechanical shearing and high-pressure homogenization for nanoscale cellulose fibrils and strong gels. Biomacromol 8:1934–1941. https://doi.org/10.1021/bm061215p
102. Qu P, Zhou Y, Zhang X et al (2012) Surface modification of cellulose nanofibrils for poly (lactic acid) composite application. J Appl Polym Sci 125:3084–3091. https://doi.org/10.1002/app.36360
103. Ran F, Tan Y (2018) Polyaniline-based composites and nanocomposites. Elsevier, Amsterdam
104. Rodrigues APH, de Souza SD, Gil CSB et al (2017) Biobased nanocomposites based on collagen, cellulose nanocrystals, and plasticizers. J Appl Polym Sci 134:. https://doi.org/10.1002/app.44954
105. Rojas J, Bedoya M, Ciro Y (2015) World' s largest science, technology & medicine open access book publisher current trends in the production of cellulose nanoparticles and nanocomposites for biomedical applications. Cellul Asp Curr Trends 193–228. https://doi.org/10.5772/61334
106. Roman M, Winter WT (2004) Effect of sulfate groups from sulfuric acid hydrolysis on the thermal degradation behavior of bacterial cellulose. Biomacromol 5:1671. https://doi.org/10.1021/bm034519+
107. Rosa MF, Medeiros ES, Malmonge J a., et al (2010) Cellulose nanowhiskers from coconut husk fibers: effect of preparation conditions on their thermal and morphological behavior. Carbohydr Polym 81:83–92. https://doi.org/10.1016/j.carbpol.2010.01.059
108. Roy D, Semsarilar M, Guthrie JT, Perrier S (2009) Cellulose modification by polymer grafting: a review. Chem Soc Rev 38:2046. https://doi.org/10.1039/b808639g
109. Salehudin MH, Salleh E, Muhamad II, Mamat SNH (2014) Starch-based biofilm reinforced with empty fruit bunch cellulose nanofibre. Mater Res Innov 18:S6-322–S6-325. https://doi.org/10.1179/1432891714z.000000000977

110. Santana JS, do Rosário JM, Pola CC et al (2017) Cassava starch-based nanocomposites reinforced with cellulose nanofibers extracted from sisal. J Appl Polym Sci 134:1–9. https://doi.org/10.1002/app.44637
111. Scatolino MV, Bufalino L, Mendes LM et al (2017) Impact of nanofibrillation degree of eucalyptus and Amazonian hardwood sawdust on physical properties of cellulose nanofibril films. Wood Sci Technol 51:1095–1115. https://doi.org/10.1007/s00226-017-0927-4
112. Shankar S, Rhim JW (2016) Preparation of nanocellulose from micro-crystalline cellulose: the effect on the performance and properties of agar-based composite films. Carbohydr Polym 135:18–26. https://doi.org/10.1016/j.carbpol.2015.08.082
113. Silvério HA, Flauzino Neto WP, Dantas NO, Pasquini D (2013) Extraction and characterization of cellulose nanocrystals from corncob for application as reinforcing agent in nanocomposites. Ind Crops Prod 44:427–436. https://doi.org/10.1016/j.indcrop.2012.10.014
114. Siqueira G, Bras J, Dufresne A (2009) Cellulose whiskers versus microfibrils: influence of the nature of the nanoparticle and its surface functionalization on the thermal and mechanical properties of nanocomposites. Biomacromol 10:425–432. https://doi.org/10.1021/bm801193d
115. Siqueira G, Bras J, Dufresne A (2010) Luffa cylindrica as a lignocellulosic source of fiber, microfibrillated cellulose, and cellulose nanocrystals. BioResources 5:727–740. https://doi.org/10.15376/biores.5.2.727-740
116. Siqueira G, Fraschini C, Bras J et al (2011) Impact of the nature and shape of cellulosic nanoparticles on the isothermal crystallization kinetics of poly(ε-caprolactone). Eur Polym J 47:2216–2227. https://doi.org/10.1016/j.eurpolymj.2011.09.014
117. Siqueira G, Tapin-Lingua S, Bras J et al (2010) Morphological investigation of nanoparticles obtained from combined mechanical shearing, and enzymatic and acid hydrolysis of sisal fibers. Cellulose 17:1147–1158. https://doi.org/10.1007/s10570-010-9449-z
118. Souza SF, Lopez A, Cai JHUI, Wu C (2010) Nanocellulose from curava fibers and their nanocomposites. Mol Crust Liq Cryst 522:342–352. https://doi.org/10.1080/15421401003722955
119. Spinella S, Maiorana A, Qian Q et al (2016) Concurrent cellulose hydrolysis and esterification to prepare a surface-modified cellulose nanocrystal decorated with carboxylic acid moieties. ACS Sustain Chem Eng 4:1538–1550. https://doi.org/10.1021/acssuschemeng.5b01489
120. Stenstad P, Andresen M, Tanem BS, Stenius P (2008) Chemical surface modifications of microfibrillated cellulose. Cellulose 15:35–45. https://doi.org/10.1007/s10570-007-9143-y
121. Tan L, Mandley SJ, Peijnenburg W et al (2018) Combining ex-ante LCA and EHS screening to assist green design: a case study of cellulose nanocrystal foam. J Clean Prod 178:494–506. https://doi.org/10.1016/j.jclepro.2017.12.243
122. Thomas MG, Abraham E, Jyotishkumar P et al (2015) Nanocelluloses from jute fibers and their nanocomposites with natural rubber: preparation and characterization. Int J Biol Macromol 81:768–777. https://doi.org/10.1016/j.ijbiomac.2015.08.053
123. Tonoli GHD, de Morais Teixeira E, Corrêa C et al (2012a) Cellulose micro/nanofibres from Eucalyptus kraft pulp: preparation and properties. Carbohydr Polym. https://doi.org/10.1016/j.carbpol.2012.02.052
124. Tonoli GHD, de Morais Teixeira E, Corrêa CC et al (2012b) Cellulose micro/nanofibres from Eucalyptus kraft pulp: Preparation and properties. Carbohydr Polym 89:80–88. https://doi.org/10.1016/j.carbpol.2012.02.052
125. Trache D, Hussin MH, Hui Chuin CT et al (2016) Microcrystalline cellulose: Isolation, characterization and bio-composites application—a review. Int J Biol Macromol 93:789–804
126. Tutak W, Sarkar S, Lin-Gibson S et al (2013) The support of bone marrow stromal cell differentiation by airbrushed nanofiber scaffolds. Biomaterials 34:2389–2398. https://doi.org/10.1016/j.biomaterials.2012.12.020

127. Ummartyotin S, Manuspiya H (2015) A critical review on cellulose: from fundamental to an approach on sensor technology. Renew Sustain Energy Rev 41:402–412. https://doi.org/10.1016/j.rser.2014.08.050
128. Uschanov P, Johansson LS, Maunu SL, Laine J (2011) Heterogeneous modification of various celluloses with fatty acids. Cellulose 18:393–404. https://doi.org/10.1007/s10570-010-9478-7
129. Xu X, Liu F, Jiang L et al (2013) Cellulose nanocrystals vs. cellulose nano fi brils: a comparative study on their microstructures and e ff ects as polymer reinforcing agents. ACS Appl Mater Interfaces 5:2999–3009. https://doi.org/10.1021/am302624t
130. Yarbrough JM, Zhang R, Mittal A et al (2017) Multifunctional cellulolytic enzymes outperform processive fungal cellulases for coproduction of nanocellulose and biofuels. ACS Nano 11:3101–3109. https://doi.org/10.1021/acsnano.7b00086
131. Zhou C, Chu R, Wu R, Wu Q (2011) Electrospun polyethylene oxide/cellulose nanocrystal composite nanofibrous mats with homogeneous and heterogeneous microstructures. Biomacromol 12:2617–2625. https://doi.org/10.1021/bm200401p
132. Zimmermann T, Bordeanu N, Strub E (2010) Properties of nanofibrillated cellulose from different raw materials and its reinforcement potential. Carbohydr Polym 79:1086–1093. https://doi.org/10.1016/j.carbpol.2009.10.045

Extraction of Cellulose Nanofibers and Their Eco/Friendly Polymer Composites

Stephen C. Agwuncha, Chioma G. Anusionwu, Shesan J. Owonubi, E. Rotimi Sadiku, Usman A. Busuguma and I. David Ibrahim

1 Introduction

According to Lee et al. [1], cellulose nanoparticles are rod-like particles that have a diameter in the nanoscale and consist of multiple cellulose chains that are combined with intra- and intermolecular hydrogen bonds. The long discussions that are still on-going about the preservation of our natural resources globally, has led to renewed interest in the natural material [2–10]. The push for renewable materials through researches is not unconnected to the fact that traditional materials like metals, glass, and synthetic polymer increases environmental problems [11–13]. These problems are already viewed as a disaster in waiting and if not checked properly may consume the globe. Problems associated with the usage of these traditional materials include, but not limited to; health problems, waste management issues, land degradation, groundwater contamination etc. [14–16].

To solve these problems, researchers are now focusing on materials from renewable resources. Cellulose, a semi-crystalline polycarbohydrate compound, is

S. C. Agwuncha (✉)
Department of Chemistry, Ibrahim Badamasi Babangida University, Lapai, Nigeria
e-mail: acsjanil22@gmail.com; agwunchas@ibbu.edu.ng

C. G. Anusionwu
Department of Applied Chemistry, University of Johannesburg, Johannesburg, RSA

S. J. Owonubi
Department of Chemistry, University of Zululand, Kwadlangezwa, Kwazulu Natal, RSA

E. R. Sadiku · I. D. Ibrahim
Department of Chemical, Metallurlogical and Material Engineering, Tshwane University of Technology, Pretoria, RSA

U. A. Busuguma
Department of Remedial Science, Ramat Polytechnic, Maiduguri, Nigeria

Inamuddin et al. (eds.), *Sustainable Polymer Composites and Nanocomposites*,
https://doi.org/10.1007/978-3-030-05399-4_2

the most abundant natural polymer on the earth [17–25]. It is derived from ligno-cellulosic biomass [26–29]. It is relatively available in abundance, renewable, biodegradable, eco-friendly, cost-effective and energy efficient [24, 30–32]. It is a biocompatible material obtained from plants, bacteria, algae, and animals too [25, 31–34]. It is composed of anhydroglucouse units linked by chemical β-1,4-glycosidic bonds [35–38].

Basically, cellulose fibers consist of other components like lignin hemicellulose, waxes, oil, and water-soluble inorganic salts [9, 30]. However, the cellulose content remains the main component with its percentage of natural materials ranging from 45 to 90% [40, 41]. In cellulose fibers, the cellulose forms microfibers and are surrounded by lignin and hemicellulose, which are acting as cement in the fibers. A helically cellular cellulose microfiber embedded in cementing material is formed from long-chain cellulose molecules [25, 42]. Through natural biosynthesis, it has been established that about 30–100 individual cellulose molecules come together through the chain extended conformation to form the basic unit in the cellulose fibers [43–45]. These fibers are primarily at Nano-scale [39, 46]. This chapter will be discussing the extraction and isolation of these nanoscale materials and their applications in today's world. The various extractable nanoscale materials will be referred to as cellulose nanoparticles (CNPs).

2 Nano-Scale Structure in Cellulose Fibers

Various types of Nano-scale cellulosic structures have been extracted from cellulose fibers and are classified into different groups based on their shapes, dimensions, functions and preparation methods. These also depend largely on the source of cellulose and the conditioning used in processing the nano-scale structures [3–10, 20–25]. Based on the proposed classification by the technical association of pulp and paper industry (TAPPI), the following nomenclatures are used [35, 47–53]

2.1 Microcrystalline Cellulose

Microcrystalline cellulose (MCC) consists of a large multi-sized aggregate of nanocrystals which are bonded to one another. The commercially available MCC have been found to have spherical or rod-like shapes with sizes ranging from 10 to 200 μm

2.2 Cellulose Microfiberils

Cellulose microfibers [CMF] is said to be made up of multiple aggregates of elementary nanofibers with width ranging between 20 and 100 nm and a length of

0.5–2.0 μm. CMF is produced via the use of intensive mechanical refinement of purified cellulose pulp [54–56]

2.3 *Cellulose Nanofibrils*

Cellulose nanofibers (CNF) are those parts that are made up of stretched bundles of elementary nanofibers made of alternating crystalline and amorphous domains. CNF has dimensions ranged from 20 to 50 nm for the width and 0.5 to 2 μm in length. CNF and CMF have been used interchangeably in many kinds of literature.

2.4 *Cellulose Nanocrystals*

Cellulose nanocrystals (CNC) are rod-like in shape. They are more rigid when compared to CNF. This may be adduced to its high crystalline nature [3, 22, 24]. CNC is also referred to as monocrystalline cellulose, nanowhiskers, nanorods or rod-like cellulose crystals. CNC can vary widely in their geometrical dimension diameter ranging from 5 to 50 nm and 100 to 500 nm in length. This is because CNC is generated by splitting the amorphous domains as well as the breaking of local crystalline contact between the nanofibers done by hydrolysis using highly concentrated acid [57–60].

2.5 *Amorphous Nanocellulose*

Amorphous nanocelluloses (ANC) are obtained by acid hydrolysis of regenerated cellulose followed by ultrasound is integration [61–63]. ANC particles are found to have elliptical shapes with diameters of 50–200 nm. ANC is said to exhibit high accessibility due to their porosity with its irregular pattern, enhanced sorption and thickening ability. However, it has poor mechanical property.

2.6 *Cellulose Nanoyarm*

Cellulose nanoyarn (CNY) is prepared by electrospinning a solution of cellulose or cellulose derivatives [64–66]. Although, knowledge and understanding of CNY are still limited to date because the study on CNY has been very low. However, the majority of the obtained CNY have diameters of 500–800 nm. CNY has low crystalline with very low thermal stability.

3 Source of Cellulose Nanofibers

The different types of nanocellulose particles that were described earlier, in the last section, are all derived from cellulose fibers. It is the mode of preparation that determines the final products. Cellulose fibers are extracted from different sources including plants, animals and bacteria. The source of the cellulose is very important because it influences the type of treatment required and by extension, the size and properties of the extracted cellulose [67–70].

Studies have shown that plant materials vary widely, and so, are the cellulose extracted from them. For example, cellulose from woods has porous antistrophic structure and exhibits a unique combination of high strength, stiffness, toughness and low density [71].

Algae are another source of cellulose. It varies in species. There are green, red, grey and brown types [72]. CMF from an algae cell wall was successfully extracted using a combination of hydrolysis and mechanical refining [4]. Although, other methods have also been used [72–75].

Tunicates are marine invertebrates. Although, research focus has largely been channelled towards Sea squirts (*Ascidiacea*) which is just one of the classes of tunicates. The tunicates produce cellulose in the outer tissue referred to as tunic. It is from the tunic that the tunicin, a fraction of cellulose is extracted [76–81].

Bacterial cellulose is produced from the primary metabolism in certain types of bacteria [7, 14]. *Gluconacetobacter xylinus* is the most widely used species for bacterial cellulose. All these are viable sources of cellulose fibres, emphasizing the bioavailability and renewability of these natural materials [6, 9, 14].

4 Extraction and Isolation of Cellulose Nanoparticle

The extraction method for CNPs can be classified into three main groups, namely mechanical, chemical and biological methods. For the purpose of this chapter, we will concentrate more on the chemical methods and procedures. However, some very important mechanical methods available will be mentioned passively. Such mechanical methods include high-pressure homogenization, microfludization, grinding cryocrushing and ultra-sonication.

High-pressure Homogenization involves the use of force to push a suspension of the cellulose pulp through a very narrow channel using a piston, under a high pressure. It is widely used for large-scale production of CNP. However, this method is associated with problems such as high energy consumption, excessive mechanical damage to the crystalline structure of the CNP and incomplete disintegration of the pulp fibers [35].

Micro fluidization is a mechanical method that uses the application of the constant shear rate to prepare CNPs. Morphological characterization of CNPs from microfludization showed that it produced more homogeneous size distribution than high-pressure homogenization [82]. However, both methods have similar problems.

Grinding involves the passing of the cellulose slurry through two grindstones. One static and the other rotating at approximately 1500 rpm. This helps to apply a shearing stress to the fibers to separating cellulose fiber into nano-sized particles [35].

Cryocrushing involves the freezing and subsequent crushing of water-swollen cellulose fibers using liquid nitrogen. The cryocrushedfibers are then dispersed uniformly in water using disintegrator [35].

5 Chemicals Methods

Chemical method of extraction and isolation of CNPs can be categorized into three steps. These do not involve the mechanical size reduction of the fibers. Step one involves the purification of the raw fiber material to remove the non-cellulosic components, step two involves a controlled chemical treatment to isolate nanocrystals and step three involves the use of chemical or ultrasound treatment to concentrate and dry the CNP. These steps have been summarized in the following order of chemical processes: extractives pre-alkalization, alkalization, acetylation/ bleaching and acid hydrolysis [2]. Whichever theory is adopted, the same processes are involved chemically. We will take the extraction processes step by step.

Step one—removal of no cellulosic components

This step is mostly carried out using alkali solution containing 1–5% sodium hydroxide at elevated temperature, ranging between 30 and 70 °C for 1–4 h, depending on the concentration of the NaOH Solution been used. The alkali treatment removes a certain amount of lignin, hemicellulose, wax and oil. The alkali treatment also helps in depolymerizing the native cellulose structure, defibrillating the external cellulose micro fibrils and by extension exposing the short length crystallites of the cellulose material [45].

Different authors have approached the alkalization of the raw fibers in different ways. This is due to the fact that different plant fibers contain these components in different proportions [17, 24, 26, 29, 35, 83–86]. Therefore, factors such as fiber source, plant history and weather and soil type are very important considerations. In addition, some authors have carried out pre-alkalization using ethanol and water (in the ratio of 1:1, v/v), to remove grease and oil stains on the fibers before the actual alkalization procedure. Table 1 presents different methods used by different authors to carry out the alkalization process [87–95].

Under this step, the bleaching or acetylation of the fibers is done too. The bleaching process is done to further purify the alkaline treated fibers, so as to remove the lignin and hemicellulose left to a large extent, if not completely. These components act as the cementing material in the fibers. Hemicellulose is a water-soluble polysaccharide while lignin is a complex organic compound soluble in alkali medium. Therefore, the bleaching is to improve the quality of the cellulose. Published works have shown that authors have used the different chemical

Table 1 Different methods of alkali treatments of cellulose fibers as published by different authors [24]

Sources	Chemicals/ equipment	Conditions	References
Coconut husk fibers	2 wt% NaOH solution	80 °C, 2 h, mechanical stirring, 2 times	[88]
Kenaf fibers	4 wt% NaOH solution	80 °C, 3 h, mechanical stirring, 3 times	[89]
Kenaf fibers	NaOH-AQ (anthraquinone)	Cook in a digester for 45 min at 160 °C	[90]
Kenaf fibers	12% NaOH, 0.15% AQ	Cook in a digester for 45 min at 120 °C	[91]
Rice husk	Alkali solution (4 wt % NaOH)	Reflux mixture at 120 °C for 1 h in a round bottom flask	[92]
H. sabdariffa fiber	2 wt% NaOH solution	120 °C, 1 h, 15 lbs for	[93]
Soy hull, Wheat straw	17.5% w/w NaOH solution	2 h	[94]
Wood fiber	3 wt% potassium hydroxide	80 °C, 2 h,	[95]
	6 wt% potassium hydroxide	80 °C, 2 h,	

combination to achieve the desired results [3, 6, 8, 25, 26, 31, 83, 84, 90–95]. Some have used sodium chlorite ($NaClO_2$), others have used hydrogen peroxide (H_2O_2), TEMP (2,2,6,6, tetramethylpiperidine-1-oxidant)

Ng et al. [24] listed some various bleaching treatments on fibers after alkalization (Table 2). This table shows that conditions can vary widely, even with the same type of chemical combination. However, bleaching is carried out by heating the alkali-treated fibers in a solution of sodium chlorite to the boiling point under acidic condition [87]. Such conditions (acidic) are created by the use of buffer which can be prepared using NaOH and acetic acid and diluted appropriately. In the presence of the acetic buffer, the $NaClO_2$ is broken down to form chlorine dioxide (ClO_2) leading to the liberation of chlorine dioxide throughout the duration of the bleaching. The use of H_2O_2 had improved the removal of lignin and hemicellulose. This is because H_2O_2 is able to break the linkage formed between the lignin and hemicellulose during the alkalization process which led to the ineffective removal of the lignin. The bleaching can be repeated several times to ensure the complete removal of the cementing components. The resulting white colour fiber after bleaching is an indication that the bleaching process is complete and successful. Literature has shown that the number of times the bleaching steps were repeated has a significant effect on the final properties of the CNP obtained [24]. CNP with more bleaching steps gives CNP that are slightly more thermally stable, better morphologies with higher crystallinity [88]

Table 2 Different methods used for the bleaching process of cellulose fibers [24]

Fiber sources	Chemicals requirement	Temperature/time requirements	References
Coconut	Sodium chlorite, glacial acetic acid	60–70 °C, 1 h, mechanical stirring, 1 and 4 times	[88]
Kenaf	2% $NaClO_2$ and3% acetic acid	70 °C, 180 min	[90]
	1.5% NaOH and 1% H_2O_2	70 °C, 90 min	
	1.25% $NaClO_2$ snd 3%acetic acid	70 °C, 90 min,	
Kenaf	2% $NaClO_2$and 3%acetic acid 1.2% NaOCl	70 °C, 90 min 25 °C, 60 min	[91]
Rice husk	Acetic acidic buffer, aqueous 1.7 wt% $NaClO_2$and distilled water	Reflux in a silicon oil bath at 100–130 °C for 4 h, four times	[92]
H. *sabdariffa*	A solution of NaOH/acetic acid/ sodium hypochlorite mixture	25 °C, one hour	[93]
Wood	Acidified aqueous $NaClO_2$	75 °C, 1 h, 5 times	[95]

Step two—Isolation of CNP

After the alkali and bleaching processes, the new material obtained is referred to as cellulose fibrils (CF). The CF has improved crystalline structures that appeared to be very compact in the nanocrystal form and kept in hydrogen bond interaction among cellulose chains adjacent to each other. The cellulose crystals as a whole are bonded together by the presence of disordered amorphous sections which exhibit imperfect orientation along the axis (as shown in Fig. 1). Therefore, to obtain an isolate with high crystallinity and purity CNP, the disordered amorphous region has to be removed selectively. This is done by the use of chemical and mechanical methods. For the chemical methods, the most common method adopted is the acidic hydrolysis [87, 91, 95, 96]

The acidic hydrolysis method is a very convenient process that has been used in time past and is still being used to obtain single and well-defined cellulose nanocrystals (CNC). The acid attacks the amorphous region preferentially, leaving the crystalline region insoluble [24, 35, 93, 97]. However as reported by some author, the conditions of the acid hydrolysis, may determine to a large extent the properties of the final CNC obtained. These conditions include time, temperature and concentration of the acidic reagents. Table 3 listed some of the commonly used chemicals that have been employed to carry out this hydrolysis process [94, 98–101]. Chemically speaking, by subjecting the bleached cellulose fibrils to the concentrated acid solution, the hydronium ions attacked the oxygen on the glycoside bonds between the two anhydroglucose units through the amorphous region which is seen as structural defects. Hence, a hydrolytic cleavage of the glycosidic bond along the amorphous chain [94, 102]. Also, the glyosidic bond can be cleaved away from the oxygen element. This is followed by two anhydroglucouseunits of separation and splitting into fragments of shorter chains [90, 97].The H_2SO_4 acid solution is most preferred because it forms charged stable crystallites that remain intact and can be easily isolated. If HCl acid solution is used the uncharged CNP

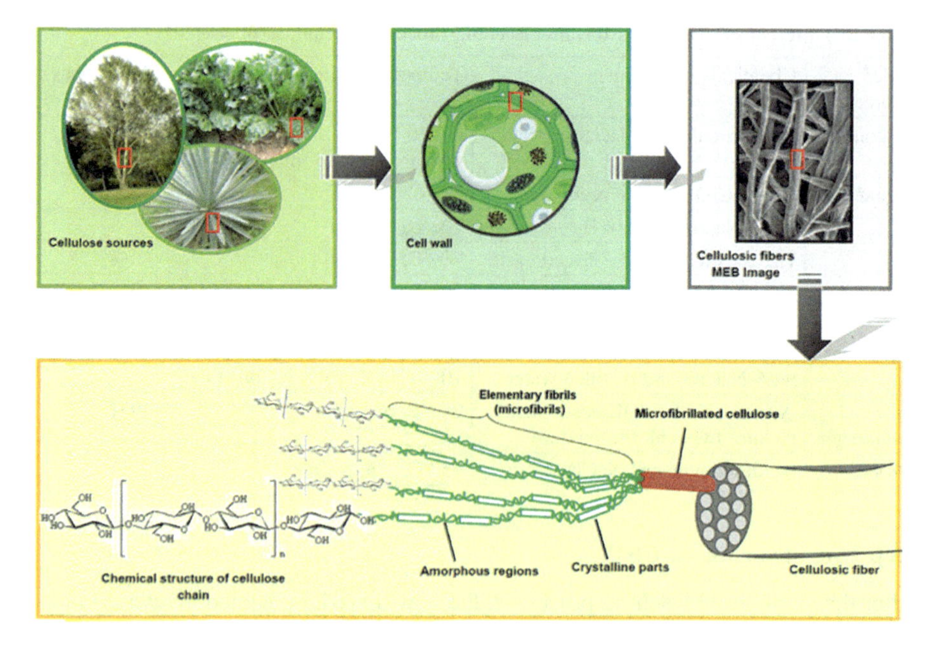

Fig. 1 Schematic diagram emphasizing the sources of cellulose fibers and the nanostructural arrangement of their unit molecules [39]

tend to flocculate in aqueous dispersion making it difficult to isolate from the acidic solution. For H_2SO_4, it reacts with surface hydroxyl groups leading to the formation of the negatively large sulfonic group [103–111]. The condition used for the acidic hydrolysis vary widely and depends largely on the precursors. Ng et al. [24] listed some acidic hydrolysis methods with the condition used by various authors shown in Table 3.

Step three—Mechanical Dialysis and Drying of Isolated CNPs

This stage involves the concentration, isolation and drying of the CNPs obtained from the acid hydrolysis process. Once the hydrolysis process is assumed to be completed, the process needs to be rapidly stopped. Many authors have applied a different method to achieve this objective at this stage. No matter the method employed or condition used, the most important thing here, is to recover most CNP isolated. So authors have argued that hydrolysis alone may not give CNP of uniform sizes. Therefore, additional step mechanical agitation or homogenization may be used to improve uniformity in size [115–117]. To stop the hydrolysis reaction, a large quantity of water or iceberg is added to the reaction to stop any further reaction on the cellulose. Then the solution formed is centrifuge repeated to remove the acidic solution and isolate the CNPs. This is done repeatedly until the supernatant becomes turbid. The supernatant is then homogenized using homogenizer or micro-fluidizer or microfluidics. The aggregate CNPs can also be homogenized using mechanical shearing. However, removing water from the CNP suspensions is

Table 3 Selected procedure used for acid hydrolysis of cellulose nanoparticles isolation [24]

Fiber sources	Chemicals/equipment	Conditions	References
Cotton	47% H_2SO_4 solutions	60 °C, 2 h, strong agitation	[98]
	64–65% (w/w) H_2SO_4 solutions	10 mL/g cellulose at45 °C for 1 h,	[102]
	Conc HCl	20 mL/g cellulose at45 °C for 1 h	[112]
Kenaf	30% H_2SO_4 solutions	80 °C, 4 h, with strong agitation	[101]
	65% H_2SO_4 solutions (preheated)	50 °C, 60 min, constantly mixing using magnetic stirring	[113]
	2.5 M HCl solution	105 °C, 20 min, stirring	[113]
Corncorb	9.17 M H_2SO_4 solutions 15 mL of H_2SO_4 solutions per grams of fibers,	45 °C, 60 min	[114]
*H. sabdariffa*fiber	Oxalic acid	3 h in an autoclave at 20 lbs	[94]
Coconut husk fiber	HNO_3 (0.05 N nitric acid solution)	70 °C for 1 h	[88]
	Concentrated sulfuric acid solution (64 wt%)	10 mL/g cellulose at 45 °C for120, 150, and 180 min	
Rice husk	10.0 mol L1 of sulfuric acid (pre-heated)	strong agitation with a fibre content of 4–6 wt% at 50 °C for40 min	[93]
L. cylindrica	65 wt% sulfuric acid (pre-heated)	mechanical agitation at 50 °C for 40 min	[115]

a delicate process. Therefore, drying the aqueous suspensions and understanding the drying process is highly necessary. The major drying process include

(i) oven drying
(ii) freeze drying
(iii) supercritical drying and
(iv) spray drying [24, 118, 119]

In summary, the process of extracting CNPs from cellulose can be given in the following steps as below and as exemplified in Fig. 2:

i. reduction of the cellulose fiber size into 2–5 mm by suitable mechanical process
ii. alkalization of the chopped fibers using alkaline solution which may be repeated as desired
iii. bleaching of the alkaline treated fibrils, this process may be repeated as desired
iv. acid hydrolysis of the bleached fibrils to isolate CNPs
v. mechanical centrifuge of an acid solution containing the isolated CNPs
vi. drying of the isolated CNPs

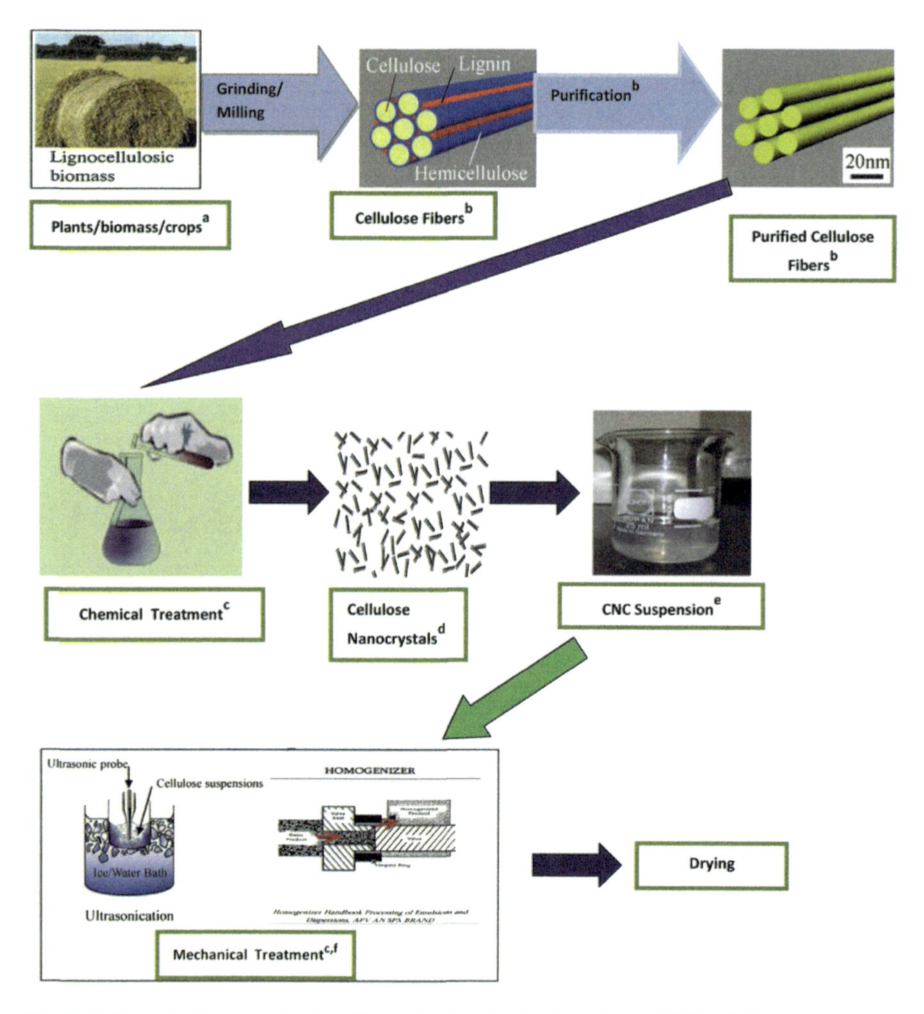

Fig. 2 Schematic diagram showing the main steps in the isolation of CNPs [24]

Feng et al. [83] extracted cellulose nanofibrils from sugarcane bagasse (SCB) and described the method used as environmentally friendly. According to their work, the SCB fibers were treated in a laboratory designed continuous steam explosion extruder [83]. 5 kg of the SCB was wetted with deionized water until the moisture content reached 45% before been transferred to the steam explosion extruder where it was treated four times at 120–130 °C temperature and 1.1–1.5 MPa pressure and then allowed to dry in air. This step was followed by alkali-catalyzed hydrothermal treatment and bleaching with H_2O_2 at 80 °C for 3 h. The bleached fibers were separated, the washed unit it attains pH $\approx$ 7 before it was freeze dried. The freeze-dried bleached fibers were further blended to disintegrate the nanofibrils. This was done by first forming a suspension of the freeze-dried fiber and then blended using a blender at 4800 rpm for 5 min. The suspension is then treated with ultra-sonication in an ice/water bath for 40 min. From the SEM and TEM results obtained, the morphology of the nanofibrils showed uniform diameters, although the CNPs were slightly agglomerate due to the presence of hydrogen bonding on the surface of the CNPs. Also, the thermal analysis showed that the CNP prepared had improved the thermal stability of about 14% increase. They concluded that dilute solution of NaOH at 200 °C elevated temperature was able to remove most, if not all, of the hemicellulose and destroyed the structure of the lignin. The CNPs was obtained with the high-speed dispersion coupled with ultra-sonication. This method completely eliminated the acid hydrolysis step and still gave CNF of uniform diameter.

Neto et al. [120] extracted CNPs from agro-industrial waste using Soya hulls. In their work, the soya hulls were first treated with a diluted solution of NaOH for 4 h at 100 °C under mechanical stirring the treated fiber was washed to remove the alkali solution completely and then dried at 50 °C for 12 h. The alkali treated fibers were then bleached using acetate buffer (prepared with 27 g NaOH and 75 ml glacial acetic acid and diluted to 1L) and sodium chlorite. The bleaching was carried out at 80 °C for 4 h. At the end process, the fibers were washed until it was neutral and then dried at 50 °C. CNPs isolation was done by first blending the bleached fibers and sieved through a 35-mesh before the acid hydrolysis step. The hydrolysis was performed at 40 °C for 30 min under vigorous and constant stirring. On completion, the mixture was diluted with cold water to 10-fold. This is to stop the reaction, then the diluted mixture is centrifuged twice for 10 min at 7000 rpm to remove the excess acid and the precipitated was dialyzed until it is neutral (pH $\approx$ 5–7). Subsequently, the resulting suspension was treated using a disperser for 5 min at 20,000 rpm and sonicated for 5 min before the colloidal suspension was stored in the refrigerator at 4 °C with the addition of some drop of chloroform. From their results, they concluded that the extraction process gave stable CNPs with a high content of cellulose and fairly uniform diameter and aspect ratio.

In other work, Nascimento and Rezende [13] used a combined approach to obtain cellulose nanocrystals (CNC) and nanofibrils (CNF) from elephant grass. Their methods involved repeated alkali and bleaching treatment on the same fibers and using a different concentration of reagents. Furthermore, the isolation temperature, using 60% (w/w) H_2SO_4 solution, was varied from 20 to 60 °C with a 1:30 (g/cm^3) fiber to solution ratio. The results showed that the amount of CNPs obtained decreased as the number of repetition is increased. Also, as the concentration of the chemical reagent was increased, the CNP yield decreases. Furthermore, the extraction time of 20 min was grossly inadequate to isolate the CNP. This is because in all the reagent combination used, none was able to isolate CNP within 20 min. However, with a minimum of 40 min, CNP was isolated. This is dependent on other parameters such as concentration and reagent mixture.

Whether the solution is a concentrated or diluted alkali or acidic solution, the issue of environmental pollution is always there. Therefore, some authors have tagged the use of steam explosion as a green method that is more environmentally friendly [36, 37, 43, 83, 121–125]. In these cases, the fibers were placed in a reactor (e.g. autoclave) at high pressure and temperature of 20 lb and 110 °C respectively for 1 h. Most fibers treated with steam explosion are bleached with a combination of NaOH/acetic acid and sodium hypochlorite solutions [43]. The attempt to avoid the use of chlorite ($NaClO_2$) is due to health and safety concerns.

Many of the focus in cellulose nanocrystal extraction are targeting nonwoody plants such as sugarcane [13, sugar palm [19], rice husk [50], sisal [126–128]. However, the extraction process still remains the same since the extraction process is all about the removal of lignin (i.e. delignification), hemicellulose and all another non-cellulose component.

6 Applications of CNPs

CNPs have unique properties that are very useful to man. Properties like low density, large surface area, high stiffness and crystallinity, high aspect ratio, biocompatibility, biodegradability and renewability [127]. These properties have enabled its application in many industries including medicals, pharmaceuticals, packaging, water treatment, built and construction, and many others [40, 128, 129].

7 Medicals

So many medicals devices are now manufactured using polymer blends, composites and nanocomposites. Some two decades ago, these devices were prepared using only polymer materials. However, the health and environmental risk associated with this practice have necessitated the development of more health and environmentally friendly material. Research has shown that compounding the polymer blends has made these devices, not only environmentally friendly but also biocompatible and biodegradable [22, 26, 40].

Different types of medical devices have been made possible with the use of cellulose nanocrystals. Achaby et al. [26] compounded some selected biopolymers, namely chitosan, alginates and K-carrageenan, with CNC to evaluate their effect on the tensile properties of the polymer film. Although the selected polymers are considered as water-soluble polysaccharides, hence their treatment with water is made easier. Also, CNC hydrolyzed with sulfuric acid has free $OH^-$and anionic sulfate groups on their surfaces. This made the CNC highly dispersible nanomaterial in water. The compatibility between CNC and biopolymers makes the CNC dispersity in the biopolymer uniform [26]. Therefore, films of CNC and biopolymers such as chitosan, alginates and K-carrageenan can be prepared more easily, if all conditions are controlled. In their work, Achaby et al. [26], utilized their properties to prepare a film of CNC and their biopolymer. The CNC-biopolymer composites were found to have smooth faces and good flexibility.

Biopolymer-based composite films are usually known to have low tensile properties. This has limited their possible applications. These drawbacks have been

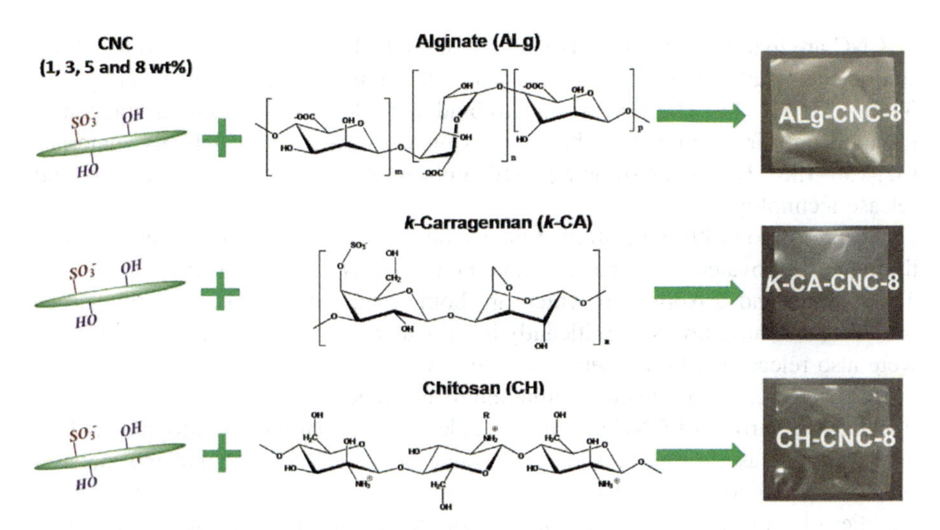

Fig. 3 A schematic representation of nanocomposite films based on CNC as nanofillers and alginate, k-carrageenan and chitosan biopolymer materials [26]

overcome with the compounding of the biopolymers using CNCs as shown by the number of literature on this issue [123–126, 130]. CNC has been reported to have a significantly reinforcing impact on the mechanical properties of the biopolymer. Figure 3 shows samples of films prepared by compounding chitosan, alginate and K-carrageenan with CNC intended for medical applications. This was made possible due to their relatively high aspect ratio and elastic modulus. Base on the application of CNC various medical devices are fabricated using biopolymer-based nanocomposites. For example screen and pin for orthopaedic surgery [131], a scaffold for tissue engineering [35] and many more.

8 Drug Delivery Systems

Drug delivery is a modern technology in which drugs are loaded on special materials and when administrated, are made to gradually release the drug depending on the design condition. Drug delivery system improves the bioavailability of the drug, preventing premature degradation, enhancing uptake, maintain drug concentration and reduce the side effect by treating disease site target cell [132–135]. The uploading of the drugs can be physical if it was just absorbed by the material or chemical if there was any chemical bond formed. Many of the drug delivery materials are pre-design to have these properties for specific target [132, 133]. One of the properties of these materials is hydrophilicity and ability to chemically modify the surface of the selected materials. Cellulose nanocrystals have been found to meet all these requirements.

CNC are hydrophilic materials that are rich in OH groups on their surface. They can be chemically modified with easy and still maintain their biodegradability (50, 34, 36, 22, 41, (1)). The CNP is able to bind and release water-solution molecules through ionic interaction and this enables them to act as drug delivery materials [32, 136–138]. Below are some reported applications of CNPs in drug delivery and release technology.

According to [139], who studied the chemically conjugated molecules to CNCs through the covalent and noncovalent bond, so as to functionalize the CNC. Tetracycline and doxorubicin which are both water soluble and ionizable drugs were found to have bound significantly in large amount to functionalized CNCs and were also released within a period of one day.

Furthermore, cetyl trimethyl ammonium bromide (CTAB) was used to functionalize the surface of CNC in order to make the material more hydrophobic. This led to an increase in the Zeta potential value of the CNC functionalized material (i.e. from −55 to 0 mv). The functionalized CNC were found to bound large qualities of some anticancer drugs namely docetaxel pacitaxel and etoposide. These drugs are hydrophobic and were found to be released in a regular pattern for a period of two days [139].

CTAB was also used to modify the surface of torispherical CNCs. This was to improve the loading capacity for anti-cancer drugs that are water insoluble such as Luteolin and lueoloside [140]. CTAB is commonly used in conjugating nanomaterials in order to improve the hydrophobicity which will eventually lead to improving loading of hydrophobic drugs such as most anticancer drugs [141]. However, this method should be used with caution as it has been shown that CTAB can interact with the phospholipid bilayers of cells [32].

Again, CNCs was modified using cationic porphyrin (por) by [142]. This interesting conjugation which forms suspended particles in aqueous systems was found to be very effective for photodynamic inactivation of staphylococcus aureus and mycobacteriumsmegmatis. However, it showed discrete activity against Escherichia coli. In addition, [143] showed that CNC-Por was also very effective on microorganisms such as Acinetobactorbaumannii.

Curcumin (CUR), an anticancer drug for treating prostate cancer was conjugated to CNC (CNC-CUR) [144]. This particle which was found to be in the nanosize range (5.2 nm) showed a high level of cellular uptake. This led to a maximum ultrastructural change on apoptosis. When the CNC-CUR was compared to the efficacy of CUR alone, The CNC-CUR drug showed increased efficacy in it anticancer activity better than the free CUR. The results obtained showed that conjugating CUR with CNCs improved the drug's availability and also its suitability for the treatment of prostate cancer [144].

Also, [145] described a similar approach for the preparation of drug carrier nanosize CNC for animated biologically active molecules and drugs. A spacer molecule (gamma-aminobutyric acid) was grafted on CNCs through periodate oxidation and Schiff base condensation reaction. However, due to fast releases of the targeting moiety, syringyl alcohol (a releasable linker) was attached to CNC to help control the release to the targeting moiety.

Furthermore, [146] carried out three-step covalent binding procedures that convert CNC into fluorescent labelling nanoparticles. The fluorescent labelling nanoparticles were prepared by conjugating CNC with pyrene (Py-CNC). The prepared nanomaterial was found to be highly selective toward Fe^{3+}. Based on the selectivity, it was suggested that the material might be useful as a chemosensor for Fe^{3+} for applications in chemical, biological and ecological systems. [147] prepared water-soluble photosensitizer–CNC (PS-CNC) by combining CNC with polyaminated chlorine p6. This improved greatly the cancer cell targeting potential of the drug, indicating possible biomedical applications. Subsequently [148], used Chitosan polysaccharide to react with CNC that was modified using TEMPO. This modification helps introduced carboxyl moieties on the CNC surface.

In a review by [138], it was reported that 5-(4,6-dichlorotriazinyl) aminofluorescein (DTAF) was grafted onto cotton–derived CNCs (DTAF-CNCs). This helped to create different charge densities on the surface of the CNC. These Charges influenced the labelling efficacy which was observed to be within nmol/g range. Although, the amount of DTAF bound to the CNCs was found to depend on the surface charge density the amount increases with decreased charge density.

Folic acid (FA) link to CNCs drug delivery system which was observed in an in vitro experiment to have good mediated uptake by rat brain tumour was prepared by [149].

Tosufloxacin osylate (TFLX) was covalently attached to maleate-modified CNC (MA-CNC) surface to prepare a prodrug using L-leucine as a spacer molecule [150]. The investigation was conducted to evaluate the release behaviour in stimulated fluids of the colon, gastric and intestine. Based on the results obtained, the MA-CNC was effective in entrapping the drug with very good release behaviour for colon specificity. Although more research may be required in this area, the MA-CNC has shown good properties for it to be used as a drug delivery system for a colon-treatment.

In all the examples provided above, it can be said that CNPs are finding good application in medicine and medical related fields. The drug delivery system is one of such active area that a lot of researches are ongoing. However, a lot of caution is required at the moment. Many of the drug-related researches are on cancer, HIV/AIDS, tuberculosis and diabetes [151]. Therefore, there is still a lot to be done. The good news here is that it is possible with the right commitment.

9 Industrial Application

Functionalized CNPs have been utilized greatly in the area of filtration for water treatment. Water treatment goal is to improve the aesthetic qualities of domestic and industrial water. Contaminants can be organic or inorganic particulate matters [152, 153]. However, molecular pollutants have defiled the traditional methods of filtration. Hence, the introduction of functionalized materials that can interact chemically with the contaminants at molecular level becomes very important [12, 18].

This method has been used for the removal of organic persistent pollutants. Today domestic water filtration systems are fitted with candles made of functionalized CNPs composites that can remove organic pollutants as well as heavy metals (Fig. 4ai, aii). Also, industrial water treatment system has filtration column too that are packed with nanocellulose materials, functionalized to remove chemicals impurities (Fig. 4b). The presence of OH group on the surface of cellulose nanocrystals has made it possible to modify the CNPs for different functionality [152–154].

Zhang et al. [31] prepared cellulose beads (CBs) with the amino functional group and porous structure for the removal of the active ingredient by adsorption as a new application for nanocellulose particles. The CBs were prepared by the inverse suspension, coupled with a water-in-oil thermal regeneration method. The CBs were then modified using epichlorohydrin, diethylenetriamine and triethylenetetramine. The CBs and modified CBs (MCBs) were then used for adsorption of hyper in (Hy) and 2-o-galloyhperin (Ga), the active ingredient of pyrola (*pyrola incarnata*), a medicinal plant that is traditional to China. The adsorption capacity of the MCBs and ten other commercial resins for Hy and Ga were investigated. From the results obtained, the adsorption capacities of the MCBs were much higher than all the commercial resins. The investigation showed that some of the MCBs had adsorption capacity that were three times higher when compared to the commercial resins. Based on their explanation, the effective interaction between the MCBs and the active ingredient were due to the ability to form chemical bonds such as van de

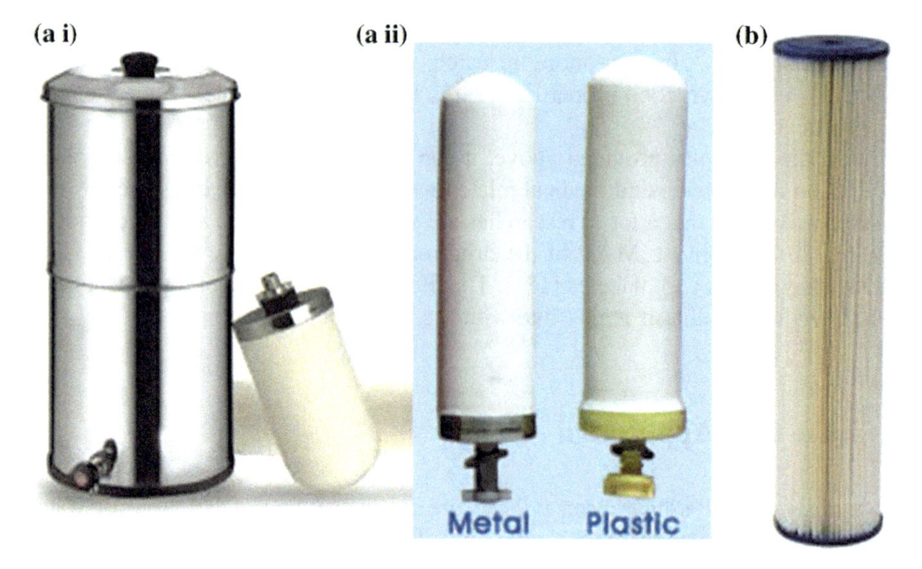

Fig. 4 Typical examples of candles (**a**i and **a**ii) and column (**b**) for domestic and industrial water filtration respectively, researched are on-going to replace these materials with nanocellulose composites [155–157]. Image source: Adapted from google image

Waal forces, hydrogen bonding and electrostatic interactions. Furthermore, the 3D porous structure of the MCBs enhanced the surface area significantly, thereby, providing more sites for adsorption of the active compounds [158–161].

Application of CNPs as adsorption material has aroused interest due to multiple hydroxyl functionalities on the surface of the nanocellulose. These sites offer many active sites for chemical modification. Based on this, CNPs materials have been successfully used in various fields [31]. In recent time, cellulose have been modified with alkenes [162, 163], Oxiranes [164], Amines [165] and carbonyls [166, 167]. Such modification has improved the adsorption efficiencies and physical stability of the CNPs. Reported works have shown that these modified CNPs have been successfully applied to the adsorption of metal ions [168, 169] and organic compounds [170, 171].

10 Preparation of Polymer Composites

The compounding of polymers, either neat or blends of polymers, with CNPs has helped to diversify their possible application. Many polymer materials suffer from water permeability limitation. Some polymers suffer from poor mechanical and thermal properties [172]. However, the addition of CNPs to such polymeric matrix as the filter has helped improved the wettability, or water permeability, mechanical properties (Youngs modulus and impact strength) and thermal stability.

Furthermore, the compounding of polymer with CNPs filters has been shown to improve the degradability of the polymers [30, 36, 55], thereby increasing the possibility of wider applications with little fear of environmental pollution issues. In the building and construction industry concrete waste generated pose a serious disposal problem. However, the substitution of the concrete with biodegradables composites has greatly reduced the problem [173]. This is because many biodegradable composites can either be reused, recycled or compose as manure on farmland.

The aspect ratio for these CNPs has led to improved mechanical properties [88, 129]. The surface area, which is in range of nano, has been reported to be responsible for the improvement in the mechanical properties of the composites. However, the compounding of the polymer matrixes with CNP fillers have conferred on the composites many other properties including oxygen permeability, water absorption and permeability, wettability, e.t.c [7, 14, 17, 20, 24, 25]

10.1 Packaging

The formation of films and preparation of composites using CNPs has revolutionized the packaging industry. The problems of the packaging industry in the past include waste management and environmental pollution [30, 32, 173]. The polymer materials in use then do not undergo degradation easily. Therefore their disposal becomes a serious issue. However, the replacement of these materials with the ones prepared using cellulose fiber extracts, especially CNPs, has solved many of the associated problems [1, 2, 24, 25].

CNP films and composites are applied in different packaging industries. These include (i) film for food and fruit wrapping, which also improve the shelf life of the food items; (ii) multi-layer functionalized packaging items for food extracts like juice extract; (iii) easy-to-serve or take-away pack used in fast food centers and shopping malls; (iv) drug and medical packages.

Nanopaper structure is one of the important applications of MFC suspension which are mainly converted films of MFC [174]. Cellulose gels which are also obtained from MFC can be converted into a film by different methods. Among them are casting method and vacuum filtration. The removal of the water content of the MFC gel leads to the formation of a cellulose nanofiber network. This is made possible because of the inter-fibrillar hydrogen bond formed inside the material. The process involved in the film formation and the quality of MFC used will greatly affect the quality of the film produced. A number of these processes have been developed in order to obtain more uniform films from the MFC. Films prepared by the vacuum filtration method have been observed to have a thickness which varies widely from 60 to 80 μm. However, using the dynamic sheet former, films with homogenous thickness have been obtained [175, 176]. The method is said to be fast and leads to the formation of highly transparent films if the appropriate instrument is used. Solvent exchange process has also been used to prepare a film of good qualities. Such films have displayed porosities of varying thickness ranging between 70 and 90 μm [174]. Other methods used in the preparation of MFC film is the spraying or the classical solvent casting method [177–180]. In all these methods, the concentration of the MFC in the solvents is left below 1 wt%. In whichever method is employed, the objective is to remove the solvent in a controlled manner that combines different conditions of temperature, relative humidity and time. This process often last for 2 h regardless of the method used [181, 182].

10.2 Emulsifiers and Solvent Thickner

CNPs are been used as emulsifiers for so many products these days [179]. CNPs have been used in the preparation of products like paints, liquid detergents and many more [109]. The neutral nature of the CNPs has made their utilization for these purposes very possible. Also, CNPs forms a stable colloid when dissolved in water or suitable solvents. Although, this may depend on the process by which the CNPs were prepared.

11 Conclusion

Cellulose is the most common natural raw material available on earth with a biosynthesis production of 10^{11}–10^{12} tons per year. It has many benefits which include low cost, lightweight, renewable, biodegradable and environmentally friendly. They can be extracted from plants and animals. The extraction and isolation of cellulose nanoparticles (CNPs) from cellulose fibers has led to the discovery of these new renewable, biocompatible, biodegradable, nanosize materials of different kinds and shapes. The processes used in the isolation determine the final properties of these materials. The processes can be categorized as follows: fiber size reduction, alkalization, bleaching, hydrolysis and mechanical isolation. The extraction of nanocellulose particles has led to the isolation of such materials as cellulose nanocrystals, cellulose nanofibrils (CNF) and cellulose microfibrils (CMF). These nanosize materials have different shapes which also affect their chemical properties. In recent time, the CNPs have found very useful applications in our society. They are used as fillers for composites preparation, film formation, biomedicals and industrial materials.

References

1. Lee HR, Kim K, Mun SC, Chang YK, Choi SQ (2018) A new method to produce cellulose nanofibrils from microalgae and the measurement of their mechanical strength. Carbohyd Polym 180:276–285
2. Mondragon G, Fernandes S, Retegi A, Pena C, Algar I, Eceiza A, Arbelaiz A (2014) A common strategy to extracting cellulose nanoentities from different plants. Ind Crops Prod 55:140–148
3. Liu Z, Li X, Xie W, Deng H (2017) Extraction, isolation and characterization of nanocrystalline cellulose from industrial kelp (Laminaria japonica) waste. Carbohyd Polym 173:353–359
4. Wang M, Bi W, Huang X, Chen DDY (2016) Ball mill assisted rapid mechanochemical extraction method for natural products from plants. J Chromatogr A 1449:8–16
5. Ling Z, Zhang X, Yang G, Takabe K, Xu F (2018) Nanocrystals of cellulose allomorphs have different adsorption of cellulase and subsequent degradation. Ind Crops Prod 112:541–549
6. Robles E, Fernández-Rodríguez J, Barbosa AM, Gordobil O, Carreno NLV, Labidi J (2018) Production of cellulose nanoparticles from blue agave waste treated with environmentally friendly processes. Carbohydr Polym. Available online 6 Jan 2018. ISSN 0144-8617
7. Das AM, Hazarika MP, Goswami M, Yadav A, Khound P (2016) Extraction of cellulose from agricultural waste using Montmorillonite K-10/LiOH and its conversion to renewable energy: Biofuel by using Myrothecium gramineum. Carbohyd Polym 141:20–27
8. Guerrero LC, Guzmán SS, Mendoza JS, Flores CA, Camacho OP (2018) Eco-friendly isolation of cellulose nanoplatelets through oxidation under mild conditions. Carbohyd Polym 181:642–649
9. Manzato L, Rabelo LCA, de Souza SM, da Silva CG, Sanches EA, Rabelo D, Mariuba LAM, Simonsen J (2017) New approach for extraction of cellulose from tucumã's endocarp and its structural characterization. J Mol Struct 1143:229–234

10. Liu L, Ju M, Li W, Jiang Y (2014) Cellulose extraction from Zoysia japonica pretreated by alumina-doped MgO in AMIMCl. Carbohyd Polym 113:1–8
11. Harini K, Mohan CC, Ramya K, Karthikeyan K, Sukumar M (2018) Effect of Punica granatum peel extracts on antimicrobial properties in Walnut shell cellulose reinforced Bio-thermoplastic starch films from cashew nut shells. Carbohyd Polym 184:231–242
12. Bali G, Khunsupat R, Akinosho H, Payyavula RS, Samuel R, Tuskan GA, Kalluri UC, Ragauskas AJ (2016) Characterization of cellulose structure of Populus plants modified in candidate cellulose biosynthesis genes. Biomass Bioenerg 94:146–154
13. Nascimento SA, Rezende CA (2018) Combined approaches to obtain cellulose nanocrystals, nanofibrils and fermentable sugars from elephant grass. Carbohyd Polym 180:38–45
14. Faruk O, Bledzki AK, Fink HP (2012) Biocomposites reinforced with natural fibers: 2000–2010. Prog Polym Sci 37:1552–1596
15. Dufresne A (2013) Nanocellulose: a new ageless bionanomaterial. Mater Today 16(6): 220–227
16. Rezende CA, de Lima MA, Maziero P, de Azevedo ER, Garcia W, Polikarpov I (2011) Chemical and morphological characterization of sugarcane bagasse submitted to a delignification process for enhanced enzymatic digestibility. Biotechnol Biofuels 4:54
17. Wang Z, Yao Z, Zhou J, Zhang J (2017) Reuse of waste cotton cloth for the extraction of cellulose nanocrystals. Carbohyd Polym 157:945–952
18. Luo J, Semenikhin N, Chang H, Moon RJ, Kumar S (2018) Post-sulfonation of cellulose nanofibrils with a one-step reaction to improve dispersibility. Carbohyd Polym 181:247–255
19. Ilyas RA, Sapuan SM, Ishak MR (2018) Isolation and characterization of nanocrystalline cellulose from sugar palm fibres (Arenga Pinnata). Carbohyd Polym 181:1038–1051
20. Xu S, Hossain MM, Lau BBY, To TQ, Rawal R, Aldous L (2017) Total quantification and extraction of shikimic acid from star anise (llicium verum) using solid-state NMR and cellulose-dissolving aqueous hydroxide solutions. Sustain Chem Pharm 5:115–121
21. Smyth M, García A, Rader C, Foster EJ, Bras J (2017) Extraction and process analysis of high aspect ratio cellulose nanocrystals from corn (Zea mays) agricultural residue. Ind Crops Prod 108:257–266
22. Ilangovan M, Guna V, Hu C, Nagananda GS, Reddy N (2018) Curcuma longa L. plant residue as a source for natural cellulose fibers with antimicrobial activity. Ind Crops Prod 112:556–560
23. Alila S, Besbes I, Vilar MR, Mutje P, Boufi S (2013) Non-woody plants as raw materials for production of microfibrillated cellulose (MFC): a comparative study. Ind Crops Prod 41:250–259
24. Ng HM, Sin LT, Tee TT, Bee ST, Hui D, Low CY, Rahmat AR (2015) Extraction of cellulose nanocrystals from plant sources for application as reinforcing agent in polymers. Compos B Eng 75:176–200
25. Zhang K, Sun P, Liu H, Shang S, Song J, Wang D (2016) Extraction and comparison of carboxylated cellulose nanocrystals from bleached sugarcane bagasse pulp using two different oxidation methods. Carbohyd Polym 138:237–243
26. El Achaby M, Kassab Z, Barakat A, Aboulkas A (2018) Alfa fibers as viable sustainable source for cellulose nanocrystals extraction: Application for improving the tensile properties of biopolymer nanocomposite films. Ind Crops Prod 112:499–510
27. Chien CH, Zhou C, Wei HC, Sing SY, Theodore A, Wu CY, Hsu YM, Birky B (2018) Feasibility test of cellulose filter for collection of sulfuric acid mists. Sep Purif Technol 195:398–403
28. Vinayaka DL, Guna V, Madhavi D, Arpitha M, Reddy N (2017) Ricinus communis plant residues as a source for natural cellulose fibers potentially exploitable in polymer composites. Ind Crops Prod 100:126–131
29. Sobolciak P, Tanvir A, Popelka A, Moffat J, Mahmoud KA, Krupa I (2017) The preparation, properties and applications of electrospun co-polyamide 6,12 membranes modified by cellulose nanocrystals. Mater Des 132:314–323

30. Khoathane MC, Sadiku ER, Agwuncha SC (2015) Chapter 14—Surface modification of natural fiber composites and their potential applications. In: Thakur VK, Singha AS (eds) Surface modification of biopolymers. Wiley, USA, pp 370–400
31. Zhang DY, Zhang N, Song P, Hao JY, Wan Y, Yao XH, Chen T, Li L (2018) Functionalized cellulose beads with three dimensional porous structure for rapid adsorption of active constituents from Pyrola incarnate. Carbohyd Polym 181:560–569
32. Seabra AB, Bernardes JS, Fávaro WJ, Paula AJ, Durán N (2018) Cellulose nanocrystals as carriers in medicine and their toxicities: a review. Carbohyd Polym 181:514–527
33. Moon RJ, Martini A, Nairn J, Simonsen J, Youngblood J (2011) Cellulose nanomaterials review: structure, properties and nanocomposites. Chem Soc Rev 40(7):3941–3994
34. Tingaut P, Zimmermann T, Sebe G (2012) Cellulose nanocrystals and microfibrillated cellulose as building blocks for the design of hierarchical functional materials. J Mater Chem 22(38):20105–20111
35. Kargarzadeh H, Loelovich M, Ahmad I, Thormas S, Dufresene A (2017) Methods for extraction of nanocellulose from various sources. In: Kargarzadeh H, Ahmad I, Thormas S, Dufresene A (eds) Handbook of nanocellulose and cellulose nano composite, 1st edn. Wiley VCH
36. Nascimento DM, Almeida JS, Dias AF, Figueirêdo MCB, Morais JPS, Feitosa JPA, Rosa MF (2014) A novel green approach for the preparation of cellulose nanowhiskers from white coir. Carbohyd Polym 110(2014):456–463
37. Deepa B, Abraham E, Cherian BM, Bismarck A, Blaker JJ, Pothan LA, Leao AL, de Souza SF, Kottaisamy M (2011) Structure, morphology and thermal characteristics of banana nano fibers obtained by steam explosion. Biores Technol 102(2011):1988–1997
38. Lin J, Miao X, Zhang X, Bian F (2017) Controllable generation of renewable nanofibrils from green materials and their application in nanocomposites. In: Thakur VK, Thakur MK, Kessler MR (eds) Handbook of composites from renewable materials. Wiley-Scrivener publishing, 8, pp 61–102
39. Lavoine N, Desloges I, Dufresne A, Bras J (2012) Microfibrillated cellulose—its barrier properties and applications in cellulosic materials: a review. Carbohyd Polym 90(2):735–764
40. Cherian BM, Leao AL, de Souza SF, Costa LMM, de Olyveira GM, Kottaisamy M, Nagarajan ER, Thomas S (2011) Cellulose nanocomposites with nanofibres isolated from pineapple leaf fibers for medical applications. Carbohyd Polym 86(2011):1790–1798
41. Nascimento DM, Dias AF, Junior CPA, Rosa MF, Morais JPS, Figueiredo MCB (2016) A comprehensive approach for obtaining cellulose nanocrystal from coconut fiber. Part II: environment assessment of technological pathways. Ind Crops and Prod
42. Kalia S, Dufresne A, Cherian BM, Kaith BS, Averous L, Njuguna J et al (2011) Cellulose-based bio- and nanocomposites: a review. Int J Polym Sci 1–35
43. Cherian BM, Leao AL, De Souza SF, Thomas S, Pothan LA, Kottaisamy M (2010) Isolation of nanocellulose from pineapple leaf by steam explosion. Carbohyd Polym 81:720–725
44. Costa LMM, de Olyveira GM, Cherian BM, Leao AL, de Souza SF, Ferreira M (2013) Bionanocomposites from electrospun PVA/pineapple nanofibers/Stryphnodendron adstringens bark extract for medical applications. Ind Crops Prod 41(2013):198–202
45. Abraham E, Deepa B, Pothan LA, Jacob M, Thomas S, Cvelbar U, Anandjiwala R (2014) Extractio of nanocellulose fibrils from lignocellulosic fibres: a novel approach. Carbohyd Polym 86:1468–1475
46. Wang B, Sain M (2007) Isolation of nanofibers from soybean source and their reinforcing capability on synthetic polymers. Compos Sci Technol 67:2521–2527
47. Mariano M, El Kissi N, Dufresne A (2014) Cellulose nanocrystals and related nanocomposites: review of some properties and challenges. J Polym Sci Part B: Polym Phys 52:791–806
48. Li Y, Liu Y, Chen W, Wang Q, Liu Y, Li J, Yu H (2015) Facile extraction of cellulose nanocrystals from wood using ethanol and peroxide solvothermal pretreatment followed by ultrasonic nanofibrillation. Green Chem 00:1–8

49. Luzia F, Fortunati E, Pugliaa D, Lavorgna M, Santulli C, Kenny JM, Torre L (2014) Optimized extraction of cellulose nanocrystals from pristine and carded hemp fibres. Ind Crops Prod 56:175–186
50. Cao Y, Wang WH, Wang QW (2013) Application of mechanical models to flax fiber/wood fiber/plastic composites. BioResources 8(3):3276–3288
51. Jonoobi M, Mathew AP, Oksman K (2012) Producing low-cost cellulose nanofiber from sludge as new source of raw materials. Ind Crop Prod 40:232–238
52. Chen W, Yu H, Liu Y, Hai Y, Zhang M, Chen P (2011) Isolation and characterization of cellulose nanofibers from four plant cellulose fibers using a chemical-ultrasonic process. Cellulose 18:433–442
53. Eichhorn SJ, Dufresne A, Aranguren M, Marcovich NE, Capadona JR, RowanS J, Weder C, Thielemans W, Toman M, Renneckar S et al (2010) Review: current international research into cellulose nanofibres and nanocomposites. J Mater Sci 45:1–33
54. Ahola S, Turon X, Osterberg M, Laine J, Rojas O (2008) Enzymatic hydrolysis of native cellulose nanofibrils and other cellulose model films: effect of surface structure. Langmuir 24 (20):11592–11599
55. Teixeira EM, Bondancia TJ, Teodoro KR, Corrêa AC, Marconcini JM, Mattoso LHC (2011) Sugarcane bagasse whiskers: extraction and characterizations. Ind Crops Prod 33(1):63–66
56. Abe K, Iwamoto S, Yano H (2007) Obtaining cellulose nanofibers with a uniform width of 15 nm from wood. Biomacromol 8(10):3276–3278
57. Li MC, Wu Q, Song K, Lee S, Qing Y, Wu Y (2015) Cellulose nanoparticles: structure–morphology–rheology relationships. ACS Sustain Chem Eng 3(5):821–832
58. Li MC, Wu Q, Song K, Qing Y, Wu Y (2015) Cellulose nanoparticles as modifiers for rheology and fluid loss in bentonite water-based fluids. ACS Appl Mater Interface 7(8): 5006–5016
59. Li W, Yue JQ, Liu SX (2012) Preparation of nanocrystalline cellulose via ultrasound and its reinforcement capability for poly(vinyl alcohol) composites. Ultrason Sonochem 19(3): 479–485
60. Chen WS, Yu HP, Li Q, Liu YX, Li J (2011) Ultralight and highly flexible aerogels with long cellulose I nanofibers. Soft Matter 7(21):10360–10368
61. Ioelovich M (2013) Nanoparticles of amorphous cellulose and their properties. Am J Nanosci Nanotechnol 1(1):41–45
62. Ioelovich M (2014) Cellulose-nanostructured natural polymer. Lambert Academic Publishing, Saarbrücken
63. Ioelovich M (2014) Peculiarities of cellulose nanoparticles. Tappi J 13(5):45–52
64. Kim CW, Kim DS, Kang SY, Marquez M, Joo YL (2006) Structural studies of electrospun cellulose nanofibers. Polymer 47(14):5097–5107
65. Quan SL, Kang SG, Chin IJ (2010) Characterization of cellulose fibers electrospun using ionic liquid. Cellulose 17(2):223–230
66. Stylianopoulos T, Kokonou M, Michael S, Tryfonos A, Rebholz C, Odysseos AD, Doumanidis C (2012) Tensile mechanical properties and hydraulic permeabilities of electrospun cellulose acetate fiber meshes. J Biomed Mater Res 100(8):2222–2230
67. Abdul Khalil HPS, Bhat AH, Yusra AF (2012) Green composites from sustainable cellulose nanofibrils: a review. Carbohyd Polym 87(2):963–979
68. Abdul Khalil HPS, Davoudpour Y, Islam MN, Mustapha A, Sudesh K, Dungani R et al (2014) Production and modification of nanofibrillated cellulose using various mechanical processes: a review. Carbohyd Polym 99:649–665
69. Samir MAS, Alloin F, Dufresne A (2005) Review of recent research into cellulosic whiskers, their properties and their application in nanocomposite field. Biomacromol 6(2):612–626
70. Morais JPS, Rosa MF, de Souza FMM, Nascimento LD, do Nascimento DM, Cassales AR (2013) Extraction and characterization of nanocellulose structures from raw cotton linter. Carbohyd Polym 91(1):229–235

71. Emanuel MF, Ricardo AP, Mano JF, Reisa RL (2013) Bionanocomposites from lignocellulosic resources: properties, applications and future trends for their use in the biomedical field. Prog Polym Sci 38(10–11):1415–1441
72. Imai T, Sugiyama J (1998) Nanodomains of Iα and Iβ cellulose in algal microfibrils. Macromolecules 31(18):6275–6279
73. Kim NH, Herth W, Vuong R, Chanzy H (1996) The cellulose system in the cell wall of micrasterias. J Struct Biol 117(3):195–203
74. Sugiyama J, Harada H, Fujiyoshi Y, Uyeda N (1985) Lattice images from ultrathin sections of cellulose microfibrils in the cell wall of Valonia macrophysa Kütz. Planta 166(2):161–168
75. Hua K, Stromme M, Mihranyam A, Ferraz N (2015) Nanocellulose from green algae modulates the in vitro inflammatory response of monocytes/macrophages. Cellulose 22:3673–3688
76. Elazzouzi-Hafraoui S, Nishiyama Y, Putaux JL, Heux L, Dubreuil F, Rochas C (2008) The shape and size distribution of crystalline nanoparticles prepared by acid hydrolysis of native cellulose. Biomacromol 9(1):57–65
77. Iwamoto S, Isogai A, Iwata T (2011) Structure and mechanical properties of wet-spun fibers made from natural cellulose nanofibers. Biomacromol 12(3):831–836
78. Kimura S, Itoh T (1996) New cellulose synthesizing complexes (terminal complexes) involved in animal cellulose biosynthesis in the tunicate Metandrocarpa uedai. Protoplasma 194(3–4):151–163
79. Peng BL, Dhar N, Liu HL, Tam KC (2011) Chemistry and applications of nanocrystalline cellulose and its derivatives: a nanotechnology perspective. Can J Chem Eng 89(5):1191–1206
80. Sturcová A, Davies GR, Eichhorn SJ (2005) Elastic modulus and stress-transfer properties of tunicate cellulose whiskers. Biomacromol 6(2):1055–1061
81. Zhao Y, Zhang Y, Lindström ME, Li J (2014) Tunicate cellulose nanocrystals: preparation, neat films and nanocomposite films with glucomannans. Carbohydr Polym 117:286–296
82. Ferrer A, Filpponen I, Rodríguez A, Laine J, Rojas OJ (2012) Valorization of residual Empty palm fruit bunch fibers (EPFBF) by microfluidization: production of nanofibrillated cellulose and EPFBF nanopaper. Bioresour Technol 125:249–255
83. Feng YH, Cheng TY, Yang WG, Ma PT, He HZ, Yin XC, Yu XX (2018) Characteristics and environmentally friendly extraction of cellulose nanofibrils from sugarcane bagasse. Ind Crops Prod 111:285–291
84. Abdullah MA, Nazir MS, Raza MR, WahjoediB A, Yussof AW (2016) Autoclave and ultra-sonication treatments of oil palm empty fruit bunch fibers for cellulose extraction and its polypropylene composite properties. J Clean Prod 126:686–697
85. Miao X, Lin J, Tian F, Li X, Bian F, Wang J (2016) Cellulose nanofibrils extracted from the byproduct of cotton plant. Carbohyd Polym 136:841–850
86. Trovatti E, Fernandes SCM, Rubatat L, da-Silva-Perez D, Freire CSR, Silvestre AJD et al (2012) Pullulane nanofibrillated cellulose composite films with improved thermal and mechanical properties. Compos Sci Technol 72:1556–1561
87. Besbesa I, Vilar MR, Boufi S (2011) Nanofibrillated cellulose from alfa, eucalyptus and pine fibres: preparation, characteristics and reinforcing potential. Carbohyd Polym 86:1198–1206
88. Rosa MF, Medeiros ES, Malmonge JA, Gregorski KS, Wood DF, Mattoso LHC et al (2010) Cellulose nanowhiskers from coconut husk fibers: effect of preparation conditions on their thermal and morphological behavior. Carbohyd Polym 81:83–92
89. Kargarzadeh H, Ahmad I, Abdullah I, Dufresne A, Zainudin SY, Sheltami RM (2012) Effects of hydrolysis conditions on the morphology, crystallinity, and thermal stability of cellulose nanocrystals extracted from kenaf bast fibers. Cellulose 19:855–866
90. Jonoobi M, Harun J, Shakeri A, Misra M, Oksman K (2009) Chemical composition, crystallinity, and thermal degradation of bleached and unbleached kenaf bast (Hibiscus cannabinus) pulp and nanofibers. BioResources 4(2):626–639
91. Shin HK, Jeun JP, Kim HB, Kang PH (2012) Isolation of cellulose fibers from kenaf using electron beam. Radiat Phys Chem 81:936–940

92. Johar N, Ahmad I, Dufresne A (2012) Extraction, preparation and characterization of cellulose fibres and nanocrystals from rice husk. Ind Crops Prod 37:93–99
93. Sonia A, Dasan KP, Alex R (2013) Celluloses microfibres (CMF) reinforced poly (ethylene-co-vinyl acetate) (EVA) composites: dynamic mechanical, gamma and thermal ageing studies. Eng Chem 228:1214–1222
94. Alemdar A, Sain M (2008) Biocomposites from wheat straw nanofibers: morphology, thermal and mechanical properties. Compos Sci Technol 68:557–565
95. Chen WS, Yu HP, Liu YX, Chen P, Zhang MX, Hai YF (2011) Individualization of cellulose nanofibers from wood using high-intensity ultrasonication combined with chemical pretreatments. Carbohyd Polym 2011:1804–1811
96. Zimmermann T, Bordeanu N, Strub E (2010) Properties of nanofibrillated cellulose from different raw materials and its reinforcement potential. Carbohyd Polym 79:1086–1093
97. Zainuddin SYZ, Ahmad I, Kargarzadeh H, Abdullah I, Dufresne A (2013) Potential of using multiscale kenaf fibers as reinforcing filler in cassava starch-kenaf biocomposites. Carbohyd Polym 92:2299–2305
98. Maiti S, Jayaramudu J, Dasa K, Reddy SM, Sadiku R, Ray SS et al (2012) Preparation and characterization of nano-cellulose with new shape from different precursor. Carbohyd Polym 98:562–567
99. Aranguren MI, Marcovich NE, Salgueiro W, Somoza A (2013) Effect of the nanocellulose content on the properties of reinforced polyurethanes. A study using mechanical tests and positron annihilation spectroscopy. Polym Test 32:115–122
100. Tee TT, Sin LT, Gobinath R, Bee ST, Hui D, Rahmat AR et al (2013) Investigation of nano-size montmorillonite on enhancing polyvinyl alcohol-starch blends prepared via solution cast approach. Compos Part B 47:238–247
101. Shi J, Shi SQ, Barnes HM, Pittman JCU (2011) A chemical process for preparing cellulosic fibers hierarchically from kenaf bast fibers. BioResources 6(1):879–890
102. Lu P, Hsieh YL (2010) Preparation and properties of cellulose nanocrystals: rods, spheres, and network. Carbohyd Polym 82:329–336
103. Bai W, Holbery J, Li KC (2009) A technique for production of nanocrystalline cellulose with a narrow size distribution. Cellulose 16(3):455–465
104. Filson PB, Dawson-Andoh BE (2009) Sono-chemical preparation of cellulose nanocrystals from lignocellulose derived materials. Bioresour Technol 100(7):2259–2264
105. Araki J, Wada M, Kuga S, Okano T (1998) Low properties of microcrystalline cellulose suspension prepared by acid treatment of native cellulose. Colloids Surf A 142(1):75–82
106. Yu H, Qin Z, Liang B, Liu N, Zhou Z, Chen L (2013) Facile extraction of thermally stable cellulose nanocrystals with a high yield of 93% through hydrochloric acid hydrolysis under hydrothermal conditions. J Mater Chem A 1(12):3938–3944
107. Acharya SK, Mishra P, Mehar SK (2011) Effect of surface treatment on the mechanical properties of bagasse fiber reinforced polymer composite. Bio-Resources 6(3):3155–3165
108. Karimi S, Tahir P, Karimi A, Dufresne A, Abdulkhani A (2014) Kenaf bast cellulosic fibers hierarchy: a comprehensive approach from micro to nano. Carbohyd Polym 101:878–885
109. Habibi Y, Lucia LA, Rojas OJ (2010) Cellulose nanocrystals: chemistry, self-assembly, and applications. Chem Rev 110:3479–3500
110. Ruiz E, Cara C, Manzanares P, Ballesteros M, Castro E (2008) Evaluation of steam explosion pre-treatment for enzymatic hydrolysis of sunflower stalks. Enzyme Microb Technol 42(2):160–166
111. de Souza Lima MM, Borsali R (2002) Static and dynamic light scattering from polyelectrolyte microcrystal cellulose. Langmuir 18(4):992–996
112. Spagnol C, Rodrigues FHA, Pereira AGB, Fajardo AR, Rubira AF, Muniz EC (2012) Superabsorbent hydrogel composite made of cellulose nanofibrils and chitosan-graft-poly (acrylic acid). Carbohyd Polym 87:2038–2045
113. Zaini LH, Jonoobi M, Tahir PMD, Karimi S (2013) Isolation and characterization of cellulose whiskers from kenaf (Hibiscus cannabinus L.) bast fibers. J Biomater Nanobiotechnol 4: 37–44

114. Silverio HA, Neto WPF, Dantas NO, Pasquini D (2013) Extraction and characterization of cellulose nanocrystals from corncob for application as reinforcing agent in nanocomposites. Ind Crops Prod 44:427–436
115. Follain N, Belbekhouche S, Bras J, Siqueira G, Marais S, Dufresne A (2013) Water transport properties of bio-nanocomposites reinforced by Luffa cylindrical cellulose nanocrystals. Membr Sci 427:218–229
116. Liu HY, Liu D, Yao F, Wu QL (2010) Fabrication and properties of transparent polymethylmethacrylate/cellulose nanocrystals composites. Bioresour Technol 101:5685–5692
117. Espino-Perez E, Bras J, Ducruet V, Guinault A, Dufresne A, Domenek S (2013) Influence of chemical surface modification of cellulose nanowhiskers on thermal, mechanical, and barrier properties of poly(lactide) based bionanocomposites. Eur Polym J 49:3144–3154
118. Peng YC, Gardner DJ, Han YS (2012) Drying cellulose nanofibrils: in search of a suitable method. Cellulose 19:91–102
119. Jiang SH, Duan GG, Scheobel J, Agarwal S, Greiner A (2013) Short electrospun polymeric nanofibers reinforced polyimide nanocomposites. Compos Sci Technol 88:57–61
120. Neto WPF, Silvério HA, Dantas NO, Pasquini D (2013) Extraction and characterization of cellulose nanocrystals from agro-industrial residue—Soy hulls. Ind Crops Prod 42:480–488
121. Besbes I, Rei Vilar M, Boufi S (2011) Nanofibrillated cellulose from alfa, eucalyptus and pine fibres: preparation, characteristics and reinforcing potential. Carbohyd Polym 86:1198–1206
122. Besbes I, Alila S, Boufi S (2011) Nanofibrillated cellulose from TEMPO-oxidized eucalyptus fibres: effect of the carboxyl content. Carbohyd Polym 84:975–983
123. El Achaby M, El Miri N, Aboulkas A, Zahouily M, Bilal E, Barakat A, Solhy A (2017) Processing and properties of eco-friendly bio-nanocomposite films filled with cellulose nanocrystals from sugarcane bagasse. Int J Biol Macromol 96:340–352
124. El Miri N, Abdelouahdi K, Barakat A, Zahouily M, Fihri A, Solhy A, El Achaby M (2015) Bio-nanocomposite films reinforced with cellulose nanocrystals: rheology of film-forming solutions, transparency, water vapor barrier and tensile properties of films. Carbohydr Polym 129:156–167
125. El Miri N, El Achaby M, Fihri A, Larzek M, Zahouily M, Abdelouahdi K, Barakat A, Solhy A (2016) Synergistic effect of cellulose nanocrystals/graphene oxide nanosheets as functional hybrid nanofiller for enhancing properties of PVA nanocomposites. Carbohydr Polym 137:239–248
126. Teodoro KBR, Teixeira EM, Corrêa AC, Campos A, Marconcini JM, Mattoso LHC (2011) Whiskers from sisal fibers obtained under different acid hydrolysis conditions: effect of time and temperature of extraction. Polêmeros 21(4):280–285
127. Siqueira G, Abdillahi H, Bras J, Dufresne A (2010) High reinforcing capability cellulose nanocrystals extracted from Syngonanthus nitens (Capim Dourado). Cellulose 17(2): 289–298
128. Silvério HA, Neto WPF, Dantas NO, Pasquini D (2013) Extraction and characterization of cellulose nanocrystals from corncob for application as reinforcing agent in nanocomposites. Ind Crops Prod 44:427–436
129. Islam MT, Alam MM, Zoccola M (2013) Review on modification of nanocellulose for application in composites. Int J Innovative Res Sci Eng Technol 2(10):5451
130. Gandolfi S, Ottolina G, Riva S, Fantoni GP, Patel I (2013) Completechemical analysis of carmagnola hemp hurds and structural features of its components. BioResources 8:2641–2656
131. Klemm D, Schumann D, Udhardt U, Marsch S (2001) Bacterial synthesized cellulose—artificial blood vessels for microsurgery. Prog Polym Sci 26(9):1561–1603
132. Owonubi SJ, Agwuncha SC, Mukwevho E, Aderibigbe BA, Sadiku ER, Biotidara OF, Varaprasad K (2017) Application of hydrogel biocomposites for multiple drug delivery. In: Handbook of composites from renewable materials, vol 6: Nanocomposites: advance applications. Scrivener Publishing, pp 139–166

133. Anirudhan TS, Rejeena SR (2015) Biopolymer-based stimuli-sensitive functionalized graft copolymers as controlled drug delivery systems. In: Thakur VK, Singha AS (eds) Surface modification of biopolymers. Wiley, USA, pp 291–334
134. Zhao Q, Li B (2008) pH-controlled drug loading and release from biodegradable micro capsules. Nanomedicine 4:302–310
135. Tsukagoshi T, Kondo Y, Yoshino N (2007) Preparation of thin polymer film with dontrolled drug release. Colloids Surf B Biointerfaces 57:219–225
136. Charreau H, Foresti ML, Vázquez A (2013) Nanocellulose patents trends: a comprehensive review on patents on cellulose nanocrystals, microfibrillated and bacterial cellulose. Recent Pat Nanotechnol 7:56–80
137. Plackett DV, Letchford K, Jackson JK, Burt HM (2014) A review of nanocellulose as a novel vehicle for drug delivery. Nord Pulp Pap Res J 29:105–118
138. Abitbol T, Palermo A, Moran-Mirabal JM, Cranston ED (2013) Fluorescent labeling and characterization of cellulose nanocrystals with varying charge contents. Biomacromol 14:3278–3284
139. Jackson JK, Letchford K, Wasserman BZ, Ye L, Hamad WY, Burt HM (2011) The use of nanocrystalline cellulose for the binding and controlled release of drugs. Int J Nanomed 6:321–330
140. Qing WX, Wang Y, Wang YY, Zhao DB, Liu XH, Zhu JH (2016) The modified nanocrystalline cellulose for hydrophobic drug delivery. Appl Surf Sci 366:404–409
141. Alkilany AM, Murphy C (2010) Toxicity and cellular uptake of gold nanoparticles: what we have learned so far? J Nanopart Res 12:2313
142. Feese E, Sadeghifar H, Gracz HS, Argyropoulos DS, Ghiladi RA (2011) Photobactericidal porphyrin-cellulose nanocrystals: synthesis, characterization: and antimicrobial properties. Biomacromol 12:3528–3539
143. Carpenter BL, Feese E, Sadeghifar H, Argyropoulos DS, Ghiladi RA (2012) Porphyrin-cellulose nanocrystals: a photobactericidal material that exhibits broad spectrum antimicrobial activity. J Photochem Photobiol 88:527–536
144. Yallapu MM, Dobberpuhl MR, Maher DM, Jaggi M, Chauhan SC (2012) Design of curcumin loaded cellulose nanoparticles for prostate cancer. Curr Drug Metab 13:120–128
145. Dash R, Ragauskas AJ (2012) Synthesis of a novel cellulose nanowhisker-based drug delivery system. RCS Adv 2:3403–3409
146. Zhang L, Li Q, Zhou J, Zhang L (2012) Synthesis and photophysical behavior of pyrene-bearing cellulose nanocrystals for Fe3+ sensing. Macromol Chem Phys 212:1612–1617
147. Drogat N, Granet R, Le Morvan C, Bégaud-Grimaud G, Krausz P, Sol V (2012) Chlorin-PEI-labeled cellulose nanocrystals: synthesis: characterization and potential application in PDT. Bioorg Med Chem Lett 22:3648–3652
148. Akhlaghi SP, Berry RC, Tam KC (2013) Surface modification of cellulose nanocrystal with chitosan oligosaccharide for drug delivery applications. Cellulose 20:1746–1747
149. Dong S, Cho HJ, Lee YW, Roman M (2014) Synthesis and cellular uptake of folic acid-conjugated cellulose nanocrystals for cancer targeting. Biomacromol 15:1560–1567
150. Tang L, Huang B, Li T, Lu Q, Chen X (2014) Functionalized cellulose nanocrystals as a carrier for colon-targeted drug delivery system. Supercond Sci Technol 32:22–28
151. Colacino KR, Arena CB, Dong S, Roman M, Davalos RV, Lee YW (2015) Folate conjugated cellulose nanocrystals potentiate irreversible electroporation-induced cytotoxicity for the selective treatment of cancer cells. Technol Cancer Res Treat 14:757–766
152. Lahiji RR, Boluk Y, McDermott M (2012) Adhesive surface interactions of cellulose nanocrystals from different sources. J Mater Sci 47:3961–3970
153. Cao J, Peng LQ, Du LJ, Zhang QD, Xu JJ (2017) Ultrasound-assisted ionic liquid-based micellar extraction combined with microcrystalline cellulose as sorbent in dispersive microextraction for the determination of phenolic compounds in propolis. Anal Chim Acta 963:24–32

154. Siaueira G, Bras J, Dufresne A (2009) Cellulose whiskers versus microfibrils: Influence of the nature of the nanoparticle and its surface functionalization on the thermal and mechanical properties of nanocomposites. Biomacromol 10(2):425–432
155. Mcallister S (2005) Analysis and comparison of sustainable water filters. United Nations, 22
156. Hassan E, Hassan M, Abou-zeid R, Berglund L, Oksman K (2017) Use of bacterial cellulose and crosslinked cellulose nanofibers membranes for removal of oil from oil-in-water emulsions. Polymers 9(9). https://doi.org/10.3390/polym9090388
157. Voisin H, Bergström L, Liu P, Mathew A (2017) Nanocellulose-based materials for water purification. Nanomaterials 7(3):57. https://doi.org/10.3390/nano7030057
158. El-Nahas AM, Salaheldin TA, Zaki T, El-Maghrabi HH, Marie AM, Morsy SM et al (2017) Functionalized cellulose-magnetite nanocomposite catalysts for efficient biodiesel production. Chem Eng J 322:167–180
159. Lee M, Heo MH, Lee HH, Kim YW, Shin J (2017) Tunable softening and toughening of individualized cellulose nanofibers-polyurethane urea elastomer composites. Carbohyd Polym 159:125–135
160. Xu D, Xiao X, Cai J, Zhou J, Zhang L (2015) Highly rate and cycling stable electrode materials constructed from polyaniline/cellulose nanoporous microspheres. J Mater Chem A 3:16424–16429
161. Zhang J, Li L, Li Y, Yang C (2017) Microwave-assisted synthesis of hierarchical mesoporous nano-TiO_2/cellulose composites for rapid adsorption of Pb2+. Chem Eng J 313:1132–1141
162. Chesney A, Barnwell P, Stonehouse DF, Steel PG (2000) Amino-derivatised beaded cellulose gels: novel accessible and biodegradable scavenger resins for solution phase combinatorial synthesis. Green Chem 2:57–62
163. Chesney A, Steel PG, Stonehouse DF (2000) High loading cellulose based poly (alkenyl) resins for resin capture applications in halogenation reactions. J Comb Chem 2:434–437
164. Weber V, Linsberger I, Ettenauer M, Loth F, Höyhtyä M, Falkenhagen D (2005) Development of specific adsorbents for human tumor necrosis factor-α: influence of antibody immobilization on performance and biocompatibility. Biomacromol 6:1864–1870
165. Heinze T, Liebert T (2001) Unconventional methods in cellulose functionalization. Prog Polym Sci 26:1689–1762
166. Korecká L, Bílková Z, Holèapek M, Královský J, Benes M, Lenfeld J et al (2004) Utilization of newly developed immobilized enzyme reactors for preparation and study of immunoglobulin G fragments. J Chromatogr B 808:15–24
167. Volkert B, Wolf B, Fischer S, Li N, Lou C (2009) Application of modified bead cellulose as a carrier of active ingredients. Macromol Symp 280:130–135
168. He Z, Song H, Cui Y, Zhu W, Du K, Yao S (2014) Porous spherical cellulose carrier modified with polyethyleneimine and its adsorption for Cr(III) and Fe(III) from aqueous solutions. Chin J Chem Eng 22:984–990
169. Monier M, Akl MA, Ali WM (2014) Modification and characterization of cellulose cotton fibers for fast extraction of some precious metal ions. Int J Biol Macromol 66:125–134
170. Wang L, Li J (2013) Adsorption of C. I. reactive red 228 dye from aqueous solution by modified cellulose from flax shive: Kinetics, equilibrium, and thermodynamics. Ind Crops Prod 42:153–158
171. Zhou Y, Min Y, Qiao H, Huang Q, Wang E, Ma T (2015) Improved removal of malachite green from aqueous solution using chemically modified cellulose by anhydride. Int J Biol Macromol 74:271–277
172. Saito T, Kuramae R, Wohlert J, Berglund LA, Isogai A (2013) An ultrastrong nanofibrillar biomaterial: the strength of single cellulose nanofibrils revealed via sonication-induced fragmentation. Biomacromol 14(1):248–253
173. John MJ, Thomas S (2008) Biofibres and biocomposites. Carbohyd Polym 71:343–364
174. Henriksson M, Berglund LA, Isaksson P, Lindström T, Nishino T (2008) Cellulose nanopaper structures of high toughness. Biomacromol 9(6):1579–1585

175. Rodionova G, Lenes M, Eriksen O, Gregersen O (2010) Surface chemical modification of microfibrillated cellulose: Improvement of barrier properties for packaging applications. Cellulose 18(1):127–134
176. Sehaqui H, Liu A, Zhou Q, Berglund LA (2010) Fast preparation procedure for large, flat cellulose and cellulose/inorganic nanopaper structures. Biomacromolecules 11(9):2195–2198
177. Aulin C, Gällstedt M, Lindström T (2010) Oxygen and oil barrier properties of microfibrillated cellulose films and coatings. Cellulose 17(3):559–574
178. Aulin C, Netrval J, Wågberg L, Lindström T (2010) Aerogels from nanofibrillated cellulose with tunable oleophobicity. Soft Matter 6(14):3298–3305
179. Spence KL, Venditti RA, Habibi Y, Rojas OJ, Pawlak JJ (2010) The effect of chemical composition on microfibrillar cellulose films from wood pulps: mechanical processing and physical properties. Biores Technol 101(15):5961–5968
180. Spence KL, Venditti RA, Rojas OJ, Habibi Y, Pawlak JJ (2010) The effect of chemical composition on microfibrillar cellulose films from wood pulps: water interactions and physical properties for packaging applications. Cellulose 17(4):835–848
181. Chinga-Carrasco G, Syverud K (2010) Computer-assisted quantification of the multi-scale structure of films made of nanofibrillated cellulose. J Nanopart Res 12:841–851
182. Blaker JJ, Lee KY, Li X, Menner A, Bismarck A (2009) Renewable nanocomposite polymer foams synthesized from Pickering emulsion templates. Roy Soc Chem 11(9):1321–1326

Synthesis, Characterization and Applications of Polyolefin Based Eco-Friendly Polymer Composites

Akash Deep, Deepanshu Bhatt, Vishal Shrivastav, Sanjeev K. Bhardwaj and Poonma Malik

1 Introduction to Polyolefins

Polyolefins are the term for polymeric materials which ensembles polyethylene plastics of different types including low-density polyethylene (LDPE), linear low-density polyethylene (LLDPE), high-density polyethylene (HDPE), and polypropylene (PP). In another word, polyolefins are the class of synthetic polymers which are formed by polymerization of olefin monomer units. They are commonly produced from the carbon-rich resources such as coal, oil and natural gas. These polymers are prevalently used in a wide array of applications such as food industry, hygiene, medical instrumentation, toys manufacturing, agricultural sector, geotextiles, electronics (as an insulator), apparels and household accessories, sports items, transport systems, ropes and twines, packaging material, etc. Their properties and applicability are highly influenced by molecular properties and branching. In general, Polyolefins can be without any polar group (non-functionalised polymers) or with polar groups (functionalized polymers). Almost all the polyolefin manufacturing processes involve the use of either free radical initiators or coordination catalysts. Currently, polymers bearing polar functionalized side groups are highly desirable materials compared to non-functionalised ones as the former category offers beneficial properties of strong adhesion, toughness, easy print/paintability, good miscibility and favourable rheological properties [1].

A. Deep (✉) · D. Bhatt · V. Shrivastav · S. K. Bhardwaj
Nanotechnology Lab, CSIR-Central Scientific Instrument Organisation (CSIR-CSIO), Chandigarh 160030, India
e-mail: dr.akashdeep@csio.res.in

P. Malik
CSIR-Central Scientific Instrument Organisation (CSIR-CSIO), C-92, CSIO Colony, Chandigarh 160030, India

Inamuddin et al. (eds.), *Sustainable Polymer Composites and Nanocomposites*,
https://doi.org/10.1007/978-3-030-05399-4_3

2 Non-functionalised Polyolefins

Non-functionalised Polyolefins or simply Polyolefins are one of the most essential polymers in modern life. It is impressive to consider that Polyolefins, made from simple monomers containing only carbon and hydrogen, can be used in a wide variety of applications [2]. Polyolefins can be manufactured by the injection molding and extrusion method and they are characterized by excellent rigidity, toughness, and temperature resistance. At present, Polyolefins are the most significant commodity plastics. The large impact of Polyolefins in the market has been created because of their low production costs, relatively low environmental impact, and flexible and tunable physical and mechanical properties [3].

Polyolefins can be broadly classified into two main types, polyethylene, and polypropylene. Polyethylene (PE) are further categorized according to their short and long chain branch (LCB) structure in three major types: low-density polyethylene (LDPE), linear low-density polyethylene (LLDPE), and high-density polyethylene (HDPE) [4].

3 Synthesis of Polyolefins

Low-density polyethylene (LDPE) is produced commonly by free radical polymerization. The manufacturing is carried out in reactors at pressures of 1000–3000 atm and temperatures of 200–275 °C. Under these extreme conditions, the starting material ethylene becomes a supercritical fluid with a density of 0.4–0.5 g/mL. Initiation of reaction can be carried out either by oxygen or peroxides. Here, branching controls the density of high-pressure polyethylenes. In general, as the branching of polymeric chain increases, it leads to reduce the density of the polymer. The rate of propagation steps increases by increasing the pressure and controls the termination and branching steps. Commonly, high pressure favours the production of higher densities, less branching and high molecular weight of polymers.

The synthesis of HDPE and LLDPE is achieved with the help of coordination catalysts [4]. This process is known as coordination polymerization. Most HDPE and LLDPE resins are prepared with either Ziegler-Natta or Phillips catalysts. The use of coordination catalysts is very common for olefin polymerization. Here, the catalytic property depends on the type of transition metal and the geometry and electronic character of the ligands. In most cases, the activation of the catalyst is carried out by some sort of complexes (known as precatalyst or catalyst precursor). The activation is made just prior to the injection of catalyst in the polymerization reactor or inside the reactor itself. The activator further alkylates the pre-catalyst complex to form the active sites and stabilizes the resulting cationic active site. Besides activation, the precursor also works as a Lewis acid (electron acceptor) which helps to scavenge the polar impurities (e.g., oxygen, sulfur, nitrogen compounds) from the reactor.

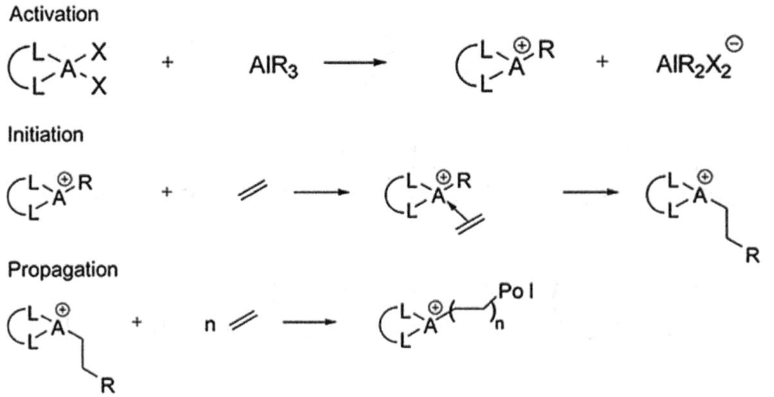

Fig. 1 Activation, initiation and propagation steps for coordination polymerization. Reproduced from Soares et al. [5]

The polymerization process involves two main steps (Fig. 1). The monomer coordination takes place to the active site followed by its insertion into the growing polymer chain.

Prior to insertion, the double bond in the monomer coordinates to the active site of the transition metal. After the insertion into the polymer chain, another olefin monomer can coordinate to the vacant site and the process continues at a fast frequency until a chain transfer reaction takes place. In the case of copolymerization, there is a competition between comonomers for the coordination site and their insertion into the growing polymer chains.

4 Polypropylene

Polypropylene can be produced in a variety of stereochemical configurations (Fig. 2). The isotactic polypropylene has the methyl groups on the same side of the backbone; the syndiotactic candidate carries on alternating sides, and the atactic polypropylene possesses the methyl groups in a random fashion along the chain. Among these, the atactic polypropylene is amorphous and has little commercial value. Both isotactic and syndiotactic polypropylene are semi-crystalline in morphology with high melting temperatures. Isotactic polypropylene is the most dominant material of its class in the market. It is easily produced with the application of heterogeneous Ziegler–Natta, and metallocene catalysts. In comparison to isotactic polypropylene, the syndiotactic polypropylene can be produced only with the help of metallocene catalysts and has much less commercial value.

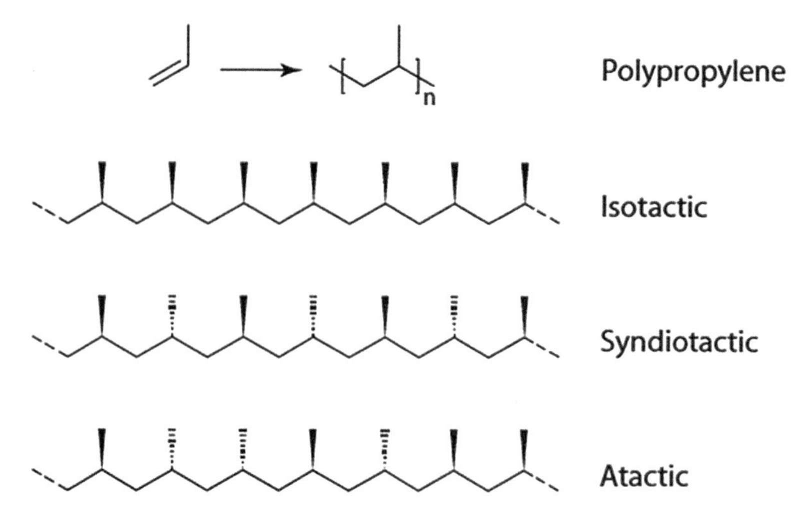

Fig. 2 Main polypropylene types: isotactic, syndiotactic, and atactic. Reproduced from Koltzenburg et al. (2017), [6]

4.1 Controlled Polymerisation of Polyolefins

In comparison to the early generation of multi-site Ziegler-Natta catalysts, modern metallocene catalysts have only one type of catalytically active center, which can readily be fine-tuned to produce some uniform homo- and copolymers with extraordinary properties. These metallocene structures, especially ligand substitution patterns, can then be tailored to control the final polymer microstructure, their molecular weights, end groups, and morphology with utmost precision [7, 8]. The use of metallocene catalysts for the production of Polyolefins became significant since the 1980s. Metallocenes function very actively for olefin polymerization when activated with methyl aluminoxane (MAO), instead of trimethylaluminum (TMA) which was commonly used in a case with Ziegler-Natta catalysts. The presence of MAO in the process enhances the activity of metallocenes by a factor of about 1000 [9–11].

4.1.1 Future of Metallocenes or Single-Site Catalysts

As describes earlier, metallocenes have some unique advantages over conventional Ziegler-Natta catalysts. They allow the realization of tailor-made polymers which were not possible by conventional catalysts. The innovation of single-site catalyst by metallocenes and some other organometallic catalysts from multi-site catalysts was a remarkable event in polyolefin industry. It is estimated that metallocenes will overpass all the other kind of catalytic processes in next few years. However, more varieties of tailored multi-site catalysts should be discovered in near future to catalyze the production of next-generation eco-friendly highly functionalized homopolymers and copolymers.

5 Synthesis of Functionalised Polyolefins with Better Eco-Friendliness

The recent research in the area of Polyolefins has been focussed on making their functionalized forms, e.g., complex, mixtures, block copolymers, and micro-and nanocomposites with inorganic and organic fillers. Such approaches are helpful to realize the development of more efficient and environmentally friendly Polyolefins. Functionalized polyolefins play a fundamental role in improving the properties of morphology and thermal and mechanical behaviours. The functionalized Polyolefins also help in discovering new materials which were otherwise difficult to be obtained by conventional synthesis approaches.

The starting material for the synthesis of polar functionalized polymers is simple polyolefins (e.g., polyethene or polypropylene). An introduction of small amounts of polar functionalities into the alkene chain makes a tremendous effect on the surface properties of resulting polymers [12, 13]. The functionalised Polyolefins, on the basis of their structure, can be classified into four categories (Fig. 3): (a) randomly functionalized copolymers that exist as either branched, (b) linear structures, (c) end-functionalised copolymers, (d) block copolymers, and (e) graft copolymers [14, 15].

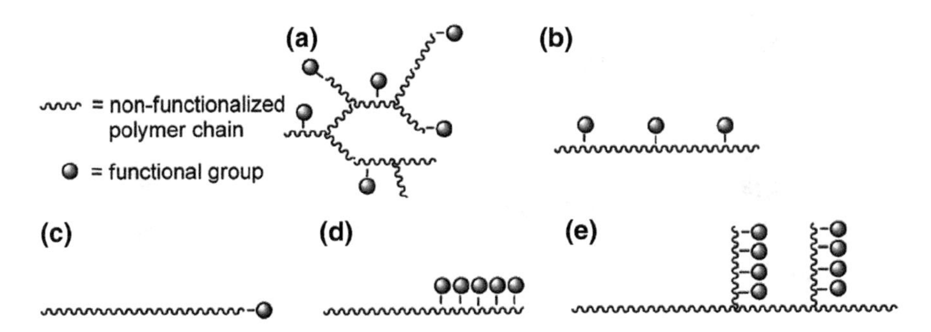

Fig. 3 **a** Randomly functionalized copolymers that exist as either branched, **b** linear structures, **c** end-functionalized copolymers, **d** block copolymers, and **e** graft copolymers. Reproduced from Franssen et al. (2013), [16]

5.1 Randomly Functionalised Copolymers

Randomly functionalized Polyolefins, as the name indicates, have functional groups in an unsystematic way. A slight change in their structure allows the introduction of new properties. Such copolymers can be synthesized in a variety of ways as depicted in Fig. 4.

5.1.1 Polymer Post-functionalization

Polymer post-functionalization is currently one of the most favored approaches. The post-functionalization of polymeric structures can be carried out in the bulk as well as on the surface [12]. For instance, the reaction of free radicals with C–H bonds of a polymer backbone can result in the H abstraction and the subsequent formation of a polymeric radical. In a different approach, insertions of carbenes or nitrenes into the C–H bonds of the polymer backbone is practised. In an industrial setting, surface modification is also processed via plasma, corona and flame treatment [17]. Non-functionalised polyolefins have no reactive groups and therefore their chemical modifications (inert C–H bonds) may generally require harsh conditions [17, 18]. Note that such treatments can also cause many undesirable side reactions to happen; for example, cross-linking or chain scission. This can lead to severely alter the properties of the polymer.

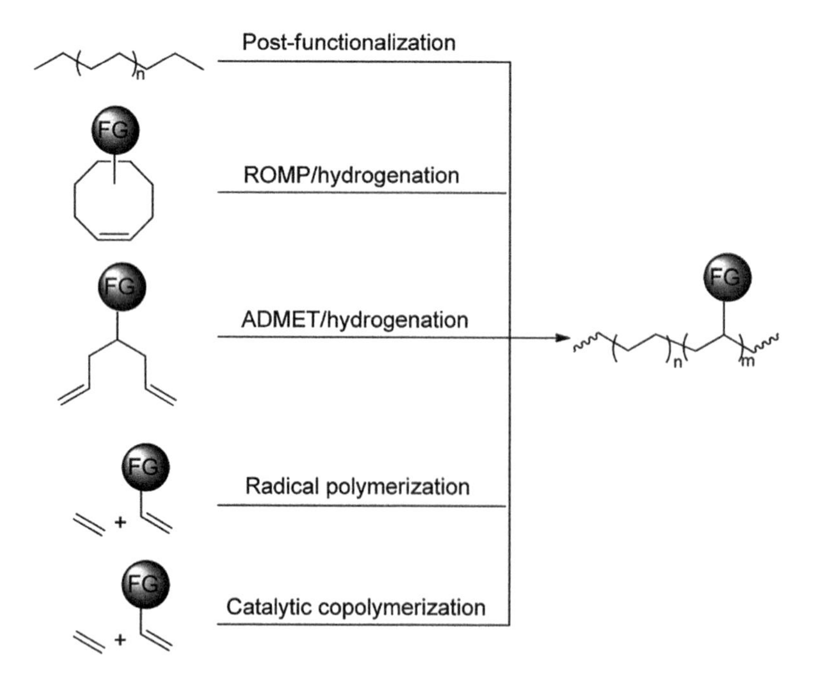

Fig. 4 Different synthesis pathways for the synthesis of randomly functionalized copolymers. Reproduced from Franssen et al. (2013), [16]

5.1.2 Ring-Opening Metathesis Polymerization (ROMP)

ROMP of the functionalized cyclic alkenes, followed by subsequent hydrogenation of the resulting polyalkenamers offer a viable approach to obtain functionalized Polyolefins. The development of more efficient and polar-tolerant catalysts (mainly based on ruthenium) has sparked the growth of ROMP based treatments [19]. The method provides controlled sequence distributions of copolymers which in turn has a major effect on the crystallinity and physical properties of the produced polymers.

5.1.3 Acyclic Diene Metathesis Polycondensation (ADMET)

ADMET process is very much similar to ROMP synthetic route. However, ADMET requires the use of special symmetrically functionalized a,o-diene monomers instead of cyclic monomers. The presence of polar functionalities in monomers leads to catalyst poisoning. As a result, the protection of the functional group is often required prior to polymerization [20, 21]. The recent discovery of certain metal complexes has allowed the production of a wide variety of side group (e.g., alcohol, acetate, ether, ester, amine/amide, halide, etc.) containing functional polymers. ADMET allows gaining precise control over the polymeric microstructure. Hence, it can function as a complementary technique to another process like radical polymerization.

5.1.4 Radical Polymerization

Radical polymerization of olefins and polar vinyl monomers (e.g. acrylates, vinyl acetate and acrylic acid) is a widely applied approach on commercial scales. The resulting copolymers find many applications in our daily life. The radical polymerization of non-polar olefins and polar vinyl monomers is not necessarily feasible due to the poor reactivity of olefins and lack of stability of the generated radicals for smooth polymerization. The product formation is often dependent upon harsh polymerization conditions (such as high temperatures and pressures). Consequently, the formation of highly branched Polyolefins is realized with these processes [22, 23].

5.1.5 Catalytic Routes

Transition-metal (TM) catalyzed routes facilitate the synthesis of functionalized copolymers. These processes allow control over the number of polar functionalities and their distribution along the polymer backbone which is a special advantage of TM catalyzed methods over the radical polymerization reactions [24]. The TM catalyzed routes involve either early transition-metal catalysis or late transition-metal catalysis with both of the methods having their own pros and cons.

5.2 *Chain-End Functionalized Copolymers*

Chain-end functionalization is a key step in the synthesis of Polyolefins. This is, in fact, the starting point for constructing more complex macromolecular architectures including block copolymers and graft copolymers from otherwise unreactive polyolefin chains. Polyolefins bearing a terminal functional group at either one end or both ends (i.e. telechelic polymers) can be prepared by different chain-end functionalization routes. These strategies may contain controlled end-capping of living TM-catalysed polymerization, in situ chain transfer reactions during TM catalyzed coordination polymerization, and modification of preformed unsaturated chain ends [25, 26].

5.2.1 End-Capping of Living Polymerizations

The end-capping of living polymerization is a classical approach for the introduction of terminal functionalities. This approach is rather less suitable for end-functionalization of Polyolefins with polar groups. The process is metal-consuming and requires the use of specific catalysts [27].

5.2.2 Chain-Transfer Reactions

End-functionalisation via chain transfer is a fairly efficient approach because of it being very metal-efficient and attainment of desired end functionality of Polyolefins. The chain-transfer reactions can be carried out in a variety of ways as elaborated in Fig. 5.

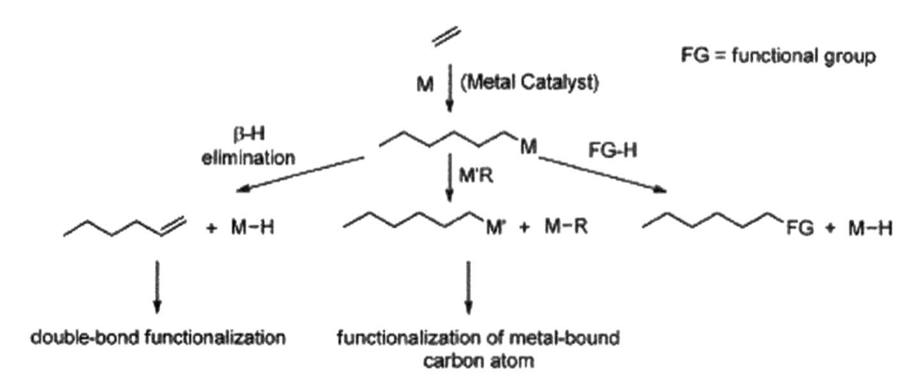

Fig. 5 Overview of chain-transfer reactions used for the end-functionalization of Polyolefins. Reproduced from Franssen et al. (2013), [16]

5.2.3 Functionalisation of Unsaturated Chain Ends

The chemical modification of preformed unsaturated chain-ends of Polyolefins has been used for the introduction of several functionalities. The associated process may involve hydroboration, hydrosilylation and hydroalumination reactions [26]. This approach is not free from certain limitations which ultimately undermine its utility for the synthesis of end-functionalised Polyolefins. As such, the chain-end unsaturation of the preformed Polyolefins should be nearly to quantitative amounts to guarantee complete functionalization.

5.3 Segmented Copolymers: Block and Graft Copolymers

The segmented copolymers can bear a large quantity of functional groups with simultaneous preservation of the original properties of precursors (i.e. non-functionalised polyolefin chains). In this way, the crystallinity, melting point and hydrophobicity of the original Polyolefins are retained even after the functionalization. Both block and graft copolymers are ideally suited to act as compatibilisers in polymer blends. The functionalised segments can ensure desired adhesion to the polar surfaces whereas the non-polar segment can interpenetrate into Polyolefin homopolymer domains [28].

5.3.1 Synthesis of Block Copolymers

The synthesis of polyolefin-based block copolymers is often processed in a multi-step manner that involves the application of two or more mechanistically distinct polymerisation techniques [29]. Most of the popularly used synthesis pathways for block copolymers involve transformation reactions, which are processed via cross-over between distinct polymerization mechanisms. Figure 6 gives an overview of the transformation reactions used for synthesis of block copolymers.

Mechanistic transformations are performed in various living/controlled polymerization for the synthesis of block copolymers. In Fig. 6, the solid lines are indicative of pathways which are suitable for the synthesis of functional Polyolefins. The dashed pathways do not provide desired results.

5.3.2 Synthesis of Graft Copolymers

Graft copolymers can commonly be prepared via transformation reactions. Such reactions involve two mechanistically distinct polymerization mechanisms. The modification of one chain end is more or less straightforward in the of block copolymers. In contrast, the synthesis conditions for graft copolymers are more challenging. These methods require modifications on multiple sites along the

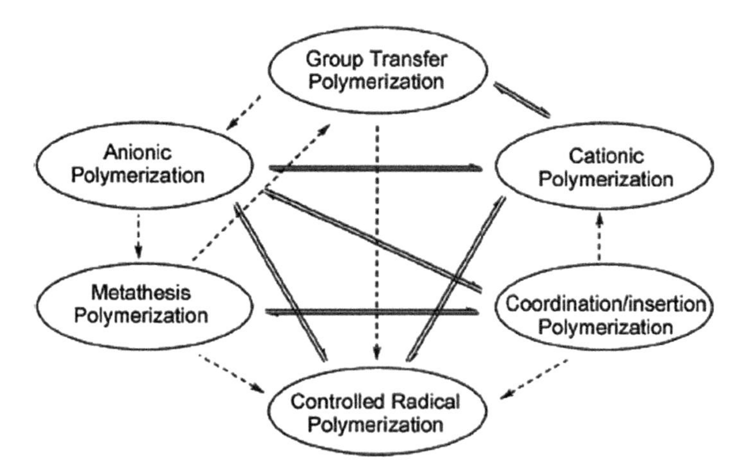

Fig. 6 An overview of different transformation reactions used in the synthesis of block copolymers. Reproduced from Franssen et al. (2013), [16]

backbone of the polymer. In summary, three major routes are used for the synthesis of graft copolymers which include grafting onto, grafting from, and grafting through approaches (as described in Fig. 7) [30, 31].

The "grafting onto" method requires the polymer backbone to be synthesized with randomly placed functionalities. A subsequent step for the coupling of the two functionalities then allows the formation of graft copolymers which have end-functionalized polymers as branches of the main chain. The "grafting from" methods involves the synthesis of a polymer backbone through ETM-catalysed insertion polymerization. The functionalities are thereafter incorporated via direct copolymerization of selected functional monomers in the presence of either ethene or propene (or both) [25, 32]. In the "grafting through" or macro-monomer method, a polymer–oligomer chain (containing a polymerizable end group) is formed which can subsequently be copolymerized to yield graft copolymers [16, 33]. The variations of the degree of polymerization of the side chains and polymer backbone can allow the selection of the length of the grafts and the final grafting density, respectively. At the time, incompatibility of the monomers (e.g., in their viscosities) may pose some difficulties in the grafting though processes. In such cases, the multi-step reaction must be carried out to ensure the desired success.

Overall, a remarkable volume of research work has been carried out over the past decade to prepare functional Polyolefins. As can be expected in the world of chemistry, each method (described above) has its own advantages and limitations. The synthesis of functional Polyolefins in a controlled way with the tuneable amount of functionalities still remains a challenge despite many recent advances in this field. Researchers are working in the direction of improving the quality of new catalysts.

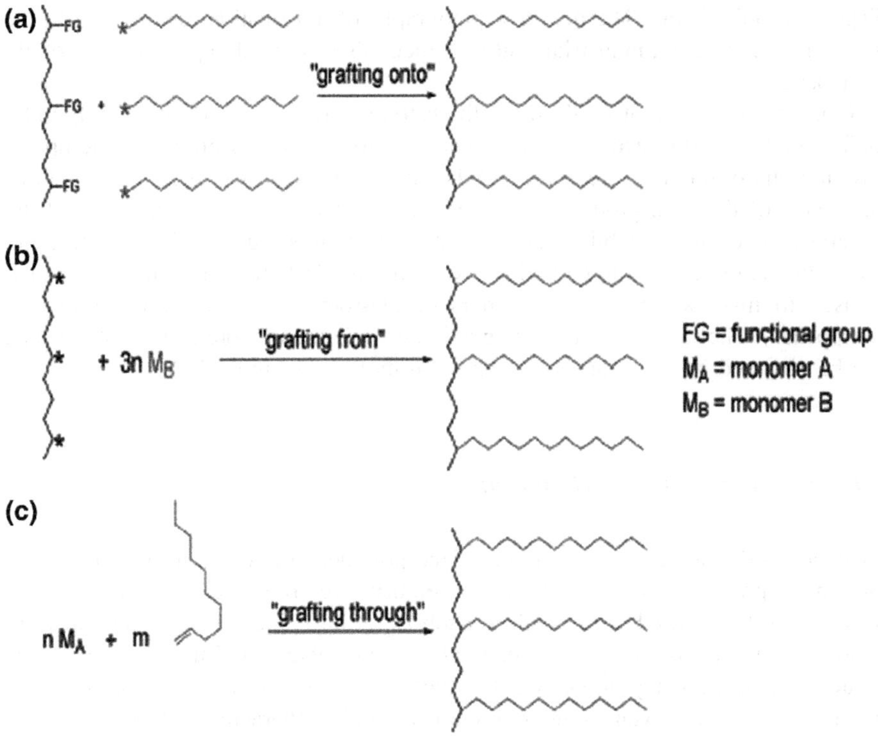

Fig. 7 General methods for the synthesis of graft copolymers. Reproduced from Franssen et al. (2013), [16]

6 Synthesis of Eco-Friendly Polyolefin Composites

With the functionalization of Polyolefins, they become amenable towards the formation of composites with other materials, e.g. natural fibers. The constituent materials define the properties (e.g., strength and stiffness) of the composite. In recent decades, there a great interest has generated toward the development of green composites which shows the advantages of being fully sustainable, biodegradable, and environment-friendly. The addition of biodegradable additives in Polyolefins at the time of their manufacturing may allow the resulting composites to undergo faster degradation after their use. During recycling of such composites, the bacterial action leads to the release of additive and breakage of the long chain of Polyolefin molecules into smaller ones. Thus, it becomes possible to degrade Polyolefins composites with the help of naturally occurring bacteria.

Researchers have proposed the formation of Polyolefin composites with natural materials like wood and natural fibers, wood flour, wood waste, sugar cane, rice hulls, paper waste and industrial fibers. The above mentioned natural sources are easily available and their addition helps in the realization of biodegradation and disposable polyolefins. The next few paragraphs of information are devoted to the discussion on different industrial synthesis methods employed to produce polyolefin composites.

Industry uses different synthesis methods to prepare the composites of polyolefin inefficient and cost-effective manner. The condition, such as processing temperature, the rate of cooling, pressure, and the steps in general influence the properties and shape of the final product to a large extent. In this way, one can control the materials' mechanical stability, crystallinity, elastic modulus, tensile strength, etc. For example, the extrusion method is one of the simplest and oldest methods which is used to mix two or more polymers or elastomers (e.g. wood-thermoplastic composites). The same product having a lighter weight and more complicated shape can be obtained by injection or compression molding methods [34].

6.1 Extrusion and Pultrusion

Extrusion is the process in which plasticized polymers are forced to pass through a die under pre-optimized conditions of temperature, pressure, and dosing. The materials of the desired shape such as such as tubing, pipes, hose, sheet and film, continuous filaments, or coated electrical wire are obtained. Currently, this technique is used in many industries for the production of polyolefin composites because of cost-effectiveness and continuous working characteristics [35–38]. This process allows the formation of a variety of plastic products like bags, pipes, thin sheet, etc. Pultrusion is a technique similar to extrusion with the major difference that the workpiece is pulled from the die. Nowadays it is the most cost-efficient technique for the production of Polyolefins composites. It can also yield continuous production of products with constant and straight cross-sectional profile. The instrument for pultrusion has a heating die, in which the materials get cross-linked with each other and then are released with a desired cross-sectional profile.

6.2 Injection Molding

Though extrusion process is one of the oldest techniques and successful to produce Polyolefin composites, it is limited to form a simple basic structure in forms of pipes, cylinders, sheets, and so on. For complex geometry, alternative techniques have to be used. The injection molding technique overcomes such limitations and is used to produce complex 3D structures. The injection molding is considered as one of the best "net shape" (i.e., the production of desired products using only one process) manufacturing processes. This technique offers the advantages of simplicity, reliability, versatility, and efficiency [39]. One-third of total polymer production is processed by injection molding. The injection molding technology gives advantages like complex structure production and automation. However, some drawbacks are also observed at times. For instance, it is not necessarily suitable for low production purposes due to high-cost of tool, and requirements of auxiliary equipment and mold.

6.3 Calendering

The calendering process to prepared Polyolefin composites involves the passage of two or more materials through a series of heated rollers. Thus, it decreases the thickness of the material and provides the production of thin sheets of about 1 mm thickness. The cost of calendering equipment is high but it can work continuously once the production starts. Calendaring is used to make Polyolefin composites with paper, wool, silk and other natural fibers [40].

6.4 Compression Molding and Thermoforming

Compression molding and thermoforming are also used for the preparation of Polyolefin composites. Compression machine mimics a press oriented vertically with top and bottom halves wholly interlocked with each other. The materials are placed between the above two halves and then compressed under the influence of heat and pressure. Thermoforming is similar to compression molding. The only difference being the presence of additional holes that can be used to alter the air pressure.

7 Improvement of Composite Compatibility Between Polyolefins and Natural Additives

As a common problem, polyolefins are incompatible with other constituents like the natural fiber with which they are going to be blended for the formation of eco-friendly composites. Such a situation can be countered with the application of compatibilizers. For instance, polylactide (PLA)/polyethylene (PE) composites have been prepared with the application of E-GMA and EMA-GMA (glycidyl methacrylate groups) as compatibilizer [41]. PLA's role is to maintain a good mechanical strength and toughness, whereas GMA increases the elastic modulus of the final product [42]. The above technique is useful method when nature fibers (containing –OH group) are used for the Polyolefin composite preparation. The natural fibers are hydrophilic while Polyolefins are hydrophobic, which leads to incompatibility. As another example, bio-polyethene/curaua fiber composite can be processed in presence of hydroxyl-terminated polybutadiene (LHPB) as a compatibility agent. Note that bio-Polyolefins are the biodegradable materials whose olefin monomers are extracted from naturally available sources.

The Polyolefin composites can be modified (stabilized) by various physical, chemical, physical-chemical and mechanical methods. The selection of the appropriate method depends upon the nature of both Polyolefin and the additive as well as the processing conditions deployed during the formation of the composite.

7.1 Chemical Methods

The use of the chemical can stabilize the composite. As already discussed by few examples, surface modification or compatibilization can be achieved by chemical reactions utilizing the hydroxyl groups. Besides this, copolymerization can also be an option. Silanes can couple with a hydroxyl group to form a layer of hydrocarbon which is hydrophobic. Hence the use of silanization chemistry is helpful to reduce wettability of the fiber and improve the stability of composites [43–46]. Similarly, isocyanates also react with the hydroxyl group of natural fibers to form the urethane linkage [47–49]. Since the presence of acetyl group can also decrease the water content or absorption, the acetylation of materials can also stabilize the composites by increasing the hydrophobic behaviour. For example, acetylation of wood increases its interfacial compatibility with polypropylene [50].

Chemical compatibilizers help to stabilize the composites while also simultaneously enhancing certain other properties, such as tensile and mechanical strengths. The processing of natural fibers by benzoylation and maleic acid are some useful examples wherein not only the hydrophobicity of the fiber could be improved but better mechanical strength and thermal stability of the composite (e.g., benzoylated wood/HDPE composites) are realized [51, 52]. The treatment of Jute with succinic acid followed by the composite formation with PP was found to be useful in yielding composites of improved toughness [53].

7.2 Physical Methods

Physical methods have also been demonstrated for improving the compatibility between Polyolefins and natural additives. The treatment with gamma-rays, laser, and plasma cause the polarity of the surface to change. The change in polarity of the surface results in the modification of hydrophobicity nature of constituents. Likewise, the use of enzymes coating and steam explosion process is much simpler and less costly than the chemical methods. However, the composite property can get greatly affected by the treatment condition and duration. The enzyme modification of Abaca and Palm fiber was suggested during its composite preparation with PP [54]. This treatment improved the tensile strength and hydrophobic nature of the composite. Steam explosion process is also helpful to improve the adhesion of fibers with Polyolefin matrices. Furthermore, vibratory ball milling and compression milling are primarily used mechanical means to enhance the compatibility between thermoplastic and lignocellulosic fiber.

8 Characterization of Polyolefins and Composites

8.1 Microstructural Properties

The microstructural properties of Polyolefins and their composites are characterized by their chemical composition, molecular weight, and long or short chain branching. Gel permeation chromatography (GPC) [55], differential scanning calorimetry (DSC) [56], nuclear magnetic resonance spectroscopy (NMR) [57], Viscometry [58], Fourier-transform infrared spectroscopy (FTIR) [4], Osmometry [58], temperature rising elution fractionation (TREF) [59, 60], crystallization analysis fractionation (CRYSTAF) [59, 60], crystallization elution fractionation (CEF) [59, 60], and Raman analysis [61] are some of the techniques used to characterize Polyolefins. These techniques are briefly discussed in the subsequent subsections.

8.1.1 Gel Permeation Chromatography (GPC)

Gel permeation chromatography (GPC), also known as molecular sieving or size exclusion chromatography, is a widely used analytical method for measurement of the molecular weight distribution of olefins and other polymers. It is a fractionation technique where polymer molecules are fractionated according to their molecular size. The instrument for chromatographic procedure usually consists of columns connected in series along with a pump, a sample injection port, and a detector. The columns are packed with a bed of porous particles, also referred to as stationary phase through which the mobile phase containing the mixture of polymers is passed. The column packings have different pore sizes that allow the separation of the molecules by the mechanism of size exclusion. Short chains of molecules penetrate the pores of the packing material, while the larger molecules diffuse through the voids and thus get fractionated faster on the basis of their retention time. Consequently, chains with higher molecular weights will take a shorter time to exit the column set than chains with lower molecular weights. The column elutions are then passed through various detectors intended for monitoring either concentration, viscosity, refractive index or light scattering of the polymer. Finally, the elution times or elution volumes are converted to molecular weight distribution (MWD) using a calibration curve obtained using polystyrene standards. The efficiency of this technique is governed by various factors such as the type of polymer, concentration and molecular weight of the polymer, branching of polymers, type of mobile phase used and the working temperature [4, 62].

High-temperature GPC (HT-GPC) technique based on simplified triple detection can be used for the determination of long chain branching and molar mass distribution in Polyolefins and their composites [55]. An example is given in Fig. 8 to show the overlay of GPC chromatographs of PyO-PP, PyO-POPP40 and distilled pyrolyzed oil PyO-POPP40-A [63]. These samples are of pyrolyzed oil from neat polypropylene, potato-peel powder/polypropylene, and distillate (200 °C fraction) of potato-peel powder/polypropylene components, respectively. A difference in the molecular weight range of PyO-PP and PyO-POPP40 would help confirm the formation of an eco-friendly composite of PP. The average molecular weight (Mw) of PyO-PP, PyO-POPP40 and PyOPOPP40-A oils were estimated as 341, 263 and 248 daltons, respectively. The shift in molecular weight was ascribed to the presence of higher molecular weight oligomers in PyO-PP compared to PyO-POPP40 (presence of 40% potato-peel powder). This addition led to chemical reactions which in turn resulted in the cleavage of high molecular weight oligomers.

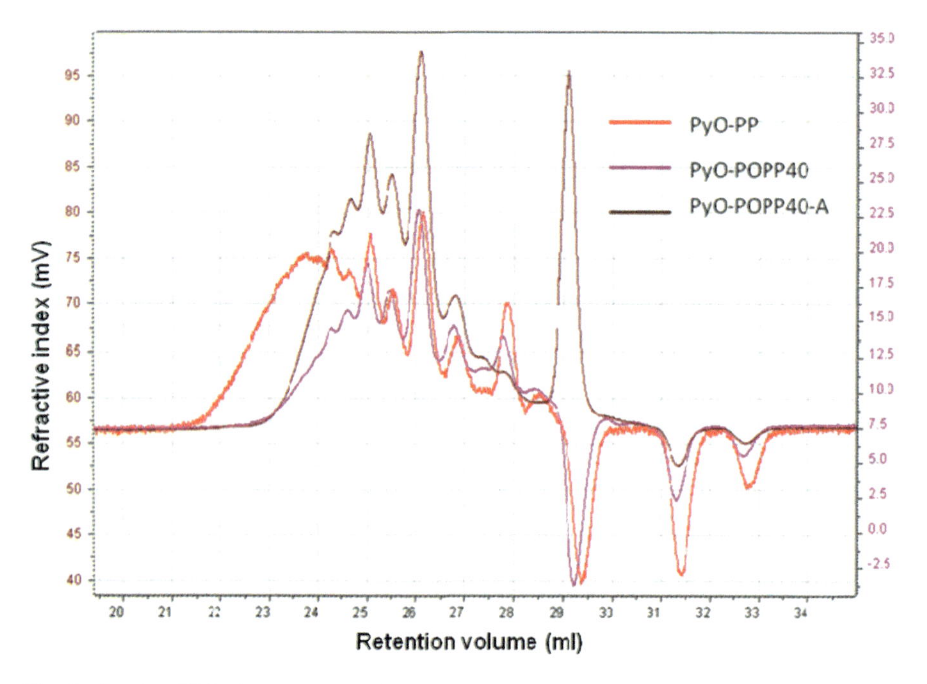

Fig. 8 Overlay GPC chromatogram of pyrolyzed products of polyolefin composites—PyO-PP, PyO-POPP40, and PyO-POPP40-A (Pyrolysed oil from neat PP and POPP40 biocomposite (potato-peel powder/polypropylene) designated as PyO-PP and PyO-POPP40 respectively. A distillate of boiling point up to 200 °C obtained from of PyO-POPP40 designated as PyO-POPP40-A. Reproduced from Sugumaran et al. (2017), [63]

8.1.2 Differential Scanning Calorimetry (DSC)

Differential scanning calorimetry (DSC) is a thermo-analytical technique and was developed by Watson and O'Neill in 1962. DSC thermal fractionation technique is used for the calculation of short-chain branching distribution (SCBD) in olefins. The DSC instrument determines the change in heat flow by measuring the temperature difference between the reference and the sample as a function of temperature. During the DSC analysis, the weighed sample is placed in a sample crucible pan against a reference pan. The sample cell is either heated or cooled at a controlled rate to obtain the melting or the crystallization/glass transition temperature. The heat flow difference between the sample and the reference cells is continuously monitored and displayed as a function of temperature/time. DSC is frequently used to calculate glass transition temperatures, the heat of fusion, crystallization temperatures, and degradation temperatures of various polymeric materials. Further, the integration of the area under the melting peak gives the heat of fusion from which the degree of crystallinity can also be derived [4, 118].

As an example, the DSC properties of a new class of toughened Polyolefin biocomposites (composite of pyrolyzed miscanthus based biocarbon, poly(octene ethylene) (POE) elastomer, and polypropylene (PP)) have been elaborated in Fig. 9 [64]. The DSC curves of composite samples were characterized with the shift in the crystallization peak of PP toward higher temperatures. The effect is found more pronounced with larger biocarbon particles.

8.1.3 Nuclear Magnetic Resonance (NMR)

Nuclear magnetic resonance (NMR) is a commonly used technique for the determination of chemical structures of a variety of polymers. Among different types of NMR, ^{1}H and ^{13}C MNR are frequently utilized for characterization of olefins. These studies determine branching types and help in the identification of

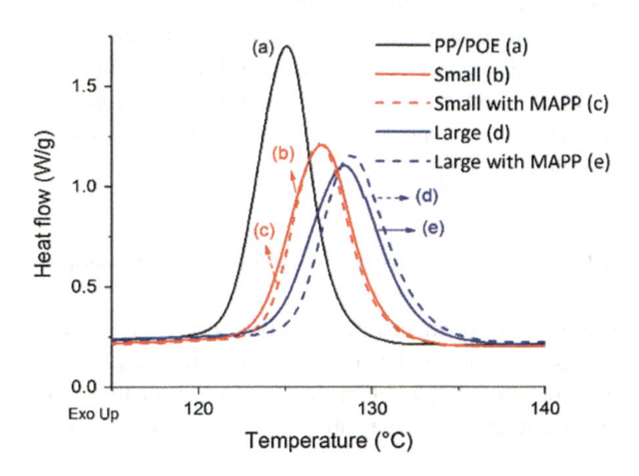

Fig. 9 DSC graph of the first cooling cycle for the Polyolefin matrix and biocomposites. Reproduced from Behazin et al. (2017), [64]

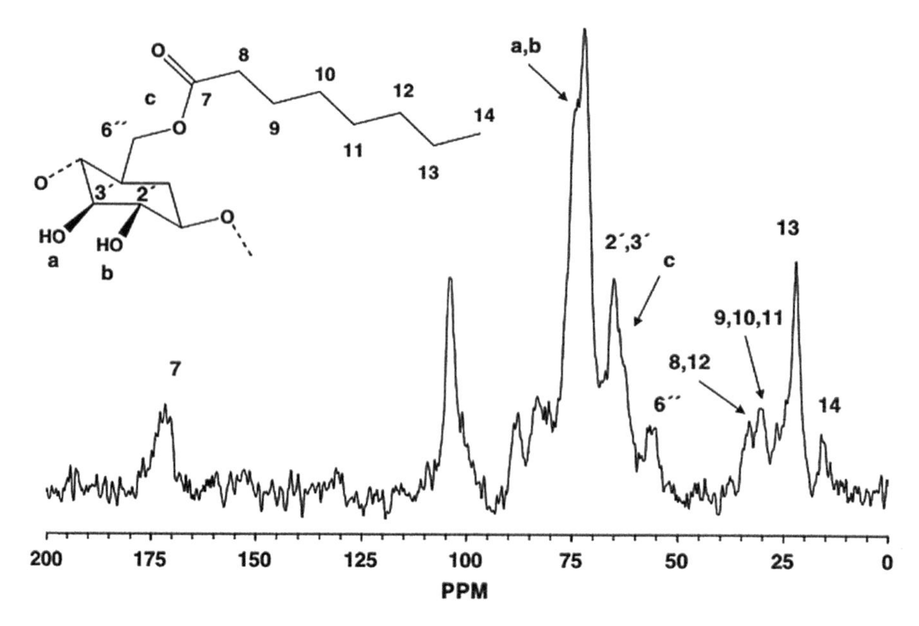

Fig. 10 CP/MAS ^{13}C NMR spectrum for blue agave fiber after esterification. Reproduced from Tronc et al. (2007), [66]

comonomers in the polymeric structures [4]. The change in the number, relative position, number and kind of branching can result in entirely different properties of olefins, and thus the determination of such change becomes very significant. ^{1}H NMR is used for the characterization of copolymers, such as ethylene-1-butene (EBR) and ethylene 1-octene copolymer (EOR). ^{13}C NMR can be used for the quantification of monomer distribution in olefin plastomers/elastomers [65]. Note that ^{13}C NMR may require long analysis durations. Alternatively, high magnetic field coupled improved proton NMR technique provide rapid analysis of complex spectra (1-olefin comonomer) within 1–2 min. The application of NMR in Polyolefin composites can be understood by the example of blue agave fibers which were used to enhance the fiber/HDPE interfacial interaction after their modification by acetylation with acetic anhydride in octanoic acid [66]. The fibers were characterized with a peak at 172 ppm to provide a useful indication of a successful modification (Fig. 10).

8.1.4 Fourier Transform Infrared Spectroscopy (FTIR)

Fourier transform infrared (FTIR) spectrum is produced by a sample when it is exposed to varying IR frequency. During the process of analysis, some frequencies are absorbed while other frequencies pass through and get transmitted. The IR spectrum denotes the absorption and transmission data of sample molecule

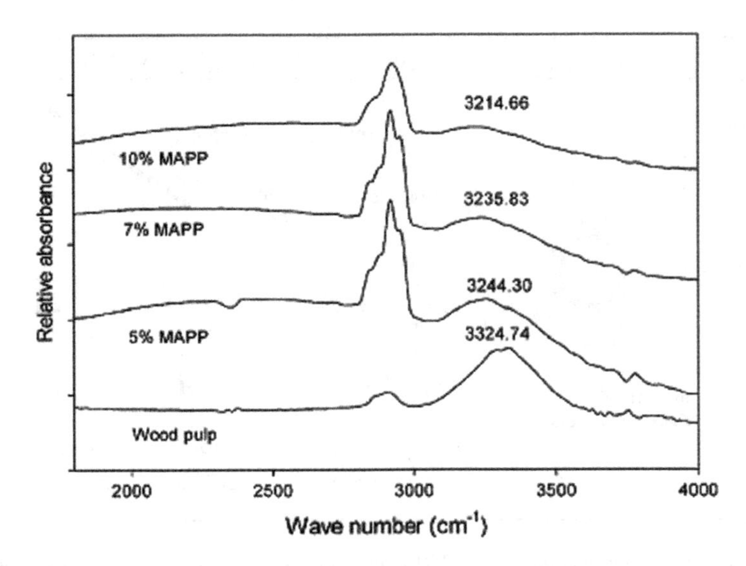

Fig. 11 FTIR spectra of 100% wood pulp and 30% wood pulp-PP bio-composite fiber with varying fractions of MAPP. Reproduced from Awal et al. (2009), [68]

corresponding to frequencies of vibrations between the bonds of the atoms. Since no two compounds reflect exactly similar infrared spectrum, the vibration frequency spectra become fingerprint for a particular compound [4]. FTIR is a very rapid analytical technique for the determination of short chain branching type in olefins. Small dissimilarities in the region of 1300–1400 cm^{-1} have been studied for the identification of different polyethylene types. Researchers have reported a linear relationship between the absorptions ratio at 1378 cm^{-1} (arising from methyl groups deformation) and the branching length in the copolymer [67]. As such, the FTIR ASTM test methods can be used for the quantification of the methyl group content in α-olefin copolymers [4].

Figure 11 shows FTIR spectra of samples of 100% wood pulp and 30% wood pulp-PP bio-composite fiber bearing different fractions of MAPP (maleated polypropylene) [68]. The C–H stretching vibrations (2800–3000 cm^{-1}) were observed in all the samples of pure wood pulp and bio-composite fibers. Particularly, the FTIR spectra of bio-composite fibers were significantly more intensive because of the reason that PP and MAPP would react with pure wood pulp. This reaction influenced the peak intensity of C-H stretching in the bio-composite fibers.

8.1.5 Crystallization Analysis Fractionation (CRYSTAF) and Temperature Rising Elution Fractionation (TREF)

Crystallization analysis fractionation (CRYSTAF) and temperature rising elution fractionation (TREF) are used for the determination of chemical composition

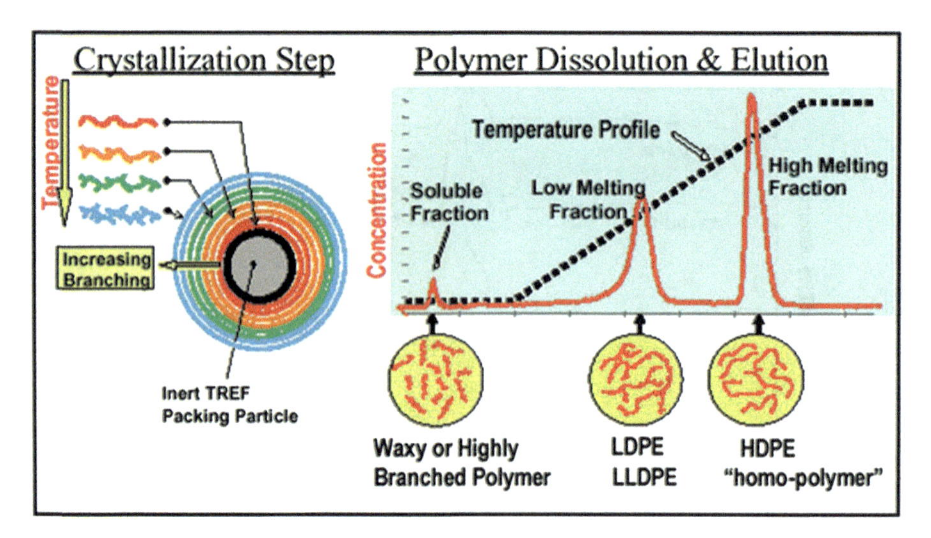

Fig. 12 Schematic presentation of the TREF process which involves crystallization and the dissolution/elution steps. Reproduced from Pasch et al. 2016, [60]

distribution (CCD) and comonomer distribution in olefins. Both the methods study the crystallization temperatures distribution [69, 70]. TREF works on the principle of measuring crystallization and elution temperature cycles followed by computation of the distribution of copolymers in the composition [60] (schematic of the process shown in Fig. 12). The olefins sample is first dissolved in a solvent (maintained at high temperature) and subsequently injected into a column composed of an inert support. The column is then cooled at a very slow rate to achieve crystallization. The chains crystallize in order of their increasing crystallinity. In the next step, elution temperature cycle is introduced to isolate the fractions physically. For this, the temperature of the column is gradually increased while the solvent is passing through. With the increasing temperature, the fractions are dissolved corresponding to their branching. The technique is frequently used to validate the presence of copolymers during the polymerization of olefins [71, 72].

CRYSTAF with simpler hardware design was introduced as a new analytical technique that avoids the additional elution temperature cycle step and physical separation of fractions. It helped to accelerate the CCD analysis of Polyolefins. The process works on crystallization based separation like TREF, but the fractionation takes place during the crystallization step itself. The instrument for CRYSTAF is composed of a filter containing crystallization vessel and an infrared detector [60, 73]. The latter component monitors the concentration of polymer solution after filtration and crystallization of olefins as induced by the decrease in temperature (normally used cooling rate of 0.1 °C/ min) [69, 73]. The presence of alike copolymers in olefins can lead to co-crystallization at a same cooling rate during CRYSTAF analysis. In such cases, TREF is preferred for analyzing copolymers because it is least affected by co-crystallization [4].

8.1.6 Crystallization Elution Fractionation (CEF)

Based on the similar principles of crystallizability, another separation technique was discovered in 2006. This approach comprised the use of a packed column and physical segregation of fractions in the crystallization step (like TREF and CRYSTAF technique). This new analytical approach is known as crystallization elution fractionation (CEF) and is relatively faster while offering higher resolution quantification of the CCD of semicrystalline polymers compared to TREF and CRYSTEF [4, 74]. The instrument of CEF is composed of an injection valve, a packed column, a pump and an IR detector. The sample is injected into the column through a pump flow and then dynamic crystallization process starts at a previously set cooling rate (generally 3 °C/min) and crystallization flow rate. The flow and cooling rates need adjustments such that the column temperature should be equal to room temperature when sample arrives to end of the column. The oven begins the heating program at the end of crystallization process and then the elution flow starts. Finally, the concentration and composition of species are determined through a dual wavelength IR detector attached at the end of the column [60, 75].

8.1.7 Osmometry

Polyolefin molecules are normally comprised of an asymmetric distribution of chain lengths. Some of them may have chains of very high molecular weight. Osmometry technique is used routinely for the assessment of molar mass of Polyolefins. This technique evaluates the number average molecular weight (M_n) [76]. Membrane osmometry (MO) [77] and vapour pressure osmometry (VPO) [58, 76] are the two types of osmometries used for the evaluation of the kinetic data during polymerization and copolymerization reactions. The membrane osmometer analyses the osmotic pressure by employing a pressure transducer. It is commonly used for high molecular weight polymers. However, problems with membrane permeation and polydispersity if polymer samples limit the usage of this technique [59, 77]. Vapour pressure osmometry, on the other hand, can be used for the determination of molecular weight of polymers with M_n less than 20,000 g/mol [58, 76].

8.1.8 Viscometry

Viscometry is a conventional method for the characterization of polymers. The analysis is used to measure the intrinsic viscosity which is related to molecular mass. Viscometry technique uses the universal calibration method to determine the molecular weight distribution, Mark-Houwink coefficients, and branching/structural information of the olefins and copolymers. Viscosity index so obtained, are also used to assign the class type to the olefins [58]. The technique can additionally detect the presence of oligomers and other small molecules. The above points make Viscometry as a preferable option over light scattering methods. Two types of technique, i.e. Ubbelohde viscometer [78] and capillary viscometer [79] are most commonly used. In the former technique, the melted solution is allowed to pass through a vertical capillary under gravity, opposed by capillary forces. The time needed for some defined volume of solution to pass through the capillary is measured [80]. In the capillary viscometer, the solution is passed using inlet pressure at either side of a capillary and the difference in pressure is measured [81].

8.1.9 Raman Spectroscopic Analysis

Raman spectroscopy is a powerful chemical analysis technique owing to unique advantages of being a non-invasive and non-destructive method. Raman spectra of materials are their fingerprint characteristics. As such, it requires no sample preparation and only a small amount of material is sufficient for the analysis. Furthermore, Raman analysis is completed in a short period of time (e.g. few seconds to few minutes). Thus, Raman analysis is highly useful for real timing monitoring of chemical reactions. With reference to plastic materials characterization, Raman spectroscopy can be profitably utilized for the following purposes: in real-time monitoring of polymerization reactions to control the processing time, quantitative compositional analysis of polymer melt streams, and waste plastic characterization in the recycling sector, especially for Polypropylene (PP) and Polyethylene (PE) identification [61].

As such, the initial use of real-time Raman spectroscopy for the estimation of crystallinity of Polyolefins was realized during blown film extrusion process [82–84]. The study established Raman spectroscopy as a robust and non-destructive technique to monitor crystallinity evolution in industrial settings. This analysis offers remote sampling capabilities when coupled via fibre optics. Raman spectroscopic systems (real-time polarized Raman spectroscopy) can also be used to analyze the molecular orientation evolution of trans C–C bond during the blown film extrusion of low-density polyethylene (LDPE). Spectra can be obtained at different locations along the blown film line, starting from the molten state and extending up to the solidified state [82]. In a study, composites of linear low-density poly(ethylene-co-butene) (PE) or maleated linear low-density poly (ethylene-co-butene) (M-PE) and cellulose (CEL), cellulose acetate (CA), cellulose acetate propionate (CAP), or cellulose acetate butyrate (CAB) were synthesized [85]. Raman spectra analysis was used to investigate the extraction of one phase with a selective solvent. Figure 13 shows the collected Raman spectra for M-PE, CAP, and the M-PE-CAP composite before and after extraction with acetone. The samples of PE-CEL and M-PE-CEL composites, for which the polyolefin chains are extracted, show intensity ratio of I_{1373}/I_{1440} in an increasing order increase after extraction. Whereas, the composites prepared with M-PE and CAP or CAB show practically same intensity ratio of I_{1740}/I_{1440} after extraction. The composites prepared with PE were characterized with a decrease in the intensity ratio of I_{1740}/I_{1440} after extraction to reveal that cellulose esters can be more easily removed in the absence of MA.

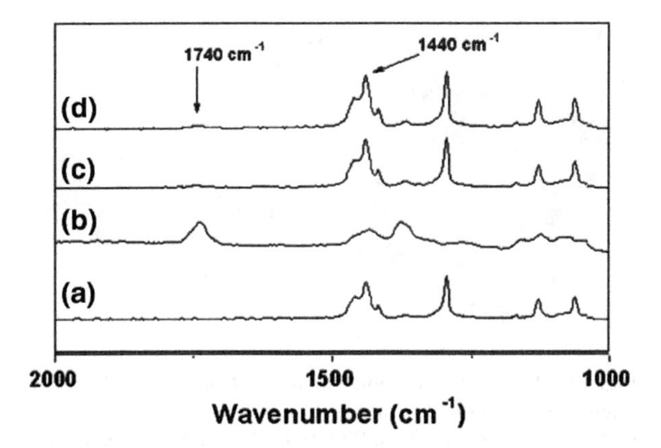

Fig. 13 Raman spectra obtained for **a** M-PE **b** CAP **c** M-PE-CAP composite before extraction with acetone, **d** M-PE–CAP composite after extraction with acetone. Reproduced from Kosaka et al. (2007), [85]

Some researchers have utilized Raman Spectroscopy for the characterization and identification of Polyolefins from the post-consumer plastic waste. The data procured was set up to define quality control logic that could be applied at industrial plant level for Polyolefins recycling [86].

8.2 *Morphological Properties*

8.2.1 Optical Microscopy

Optical microscopy of Polyolefins is frequently used for investigating the details of olefins microstructures. For instance, Fig. 14a, b show the optical microscopic image of a sample of polypropylene particles embedded with catalyst particles. As this example highlight, one can study and assess the completion of the polymerization process [87].

Similarly, an optical photograph of polyolefin biocomposites prepared with waste cardboard (CB) is shown in Fig. 8c [89]. This simple analysis reveals that large shear and compressive forces imparted to the material during the solid-state processing has resulted in major filler size reduction in the polymer composite.

8.2.2 Scanning Electron Microscopy (SEM)

Scanning electron microscope (SEM) technique probes the sample with high-intensity electrons. The collection of subsequently generated secondary electrons allow the study of surface morphology and measurement of particle sizes. Note that secondary electrons are emitted from the surface. SEM images provide useful information on the surface of relatively thicker and bigger samples. At times, SEM is used to produce a good representation of the 3-D structure of the sample

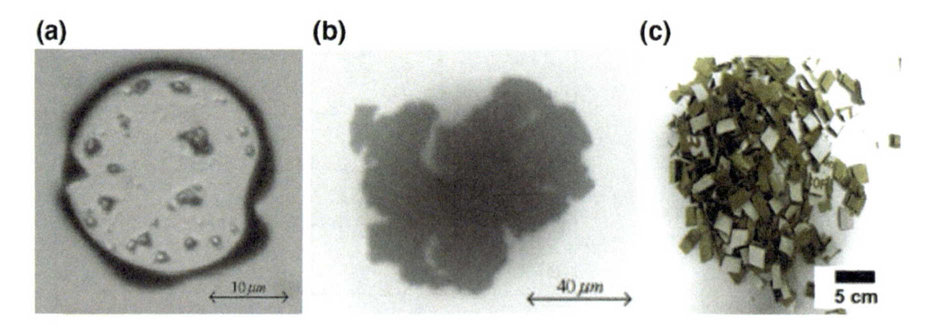

Fig. 14 Optical microscopic images of molten poly(propylene) particles showing catalyst fragments inside the polymer melt: **a** polymer particle produced by reaction of 10 min **b** polymer particle produced after 30 s. Reproduced from Abboud et al. (2005), [88]. **c** Optical photographs of 2–3 cm size waste cardboard (CB) pieces employed in the preparation of a polyolefin/CB composite [87, 89]. Reproduced from Iyer et al. (2015), [89]

and also have some major advantages of higher magnification and greater depth of field over conventional microscopes [90].

SEM is used to analyze the phase morphology of Polyolefins and their composite materials. These analyses can include the study of surface roughness and fracture toughness [91]. SEM can also reveal interface adhesion between the polymer fiber and matrix in which it is used. The measurement of adhesion allows assessing the stress transfer from matrix to fiber. The analysis of phase adherence between fibers and polymeric matrix can also reveal the propagation of the cracks generated during impact tests [92–94]. In some studies, SEM has been used to analyze the effects of addition of organo-montmorillonite and maleated Polyolefins on the phase morphology of the PP/HDPE blend. Such analysis was used to evaluate the change in two-phased morphology of the PP/HDPE blend [95]. As an example of SEM analysis of Polyolefin biocomposite, the investigations on the samples of low-density polypropylene (LDPP) with microcrystalline cellulose (MCC) and PP with cardboard (CB) are shown in Fig. 15 [89]. The fractured surfaces showed no signs of pull-out, highlighting that the filler did not agglomerate during the composite formation process. MCC, as such, can show strong interparticle affinity due to the presence of numerous hydroxyl groups on the cellulose surface but SEM investigations of LDPE/MCC composites showed very good dispersion.

8.2.3 Atomic Force Microscopy (AFM)

Since its invention in 1986, the atomic force microscopy (AFM) has assumed a paramount significance for probing the samples at the nanoscale [96]. Different variations of AFM-based methods have been developed to generate information about electrical, magnetic, morphological, and mechanical parameters from various

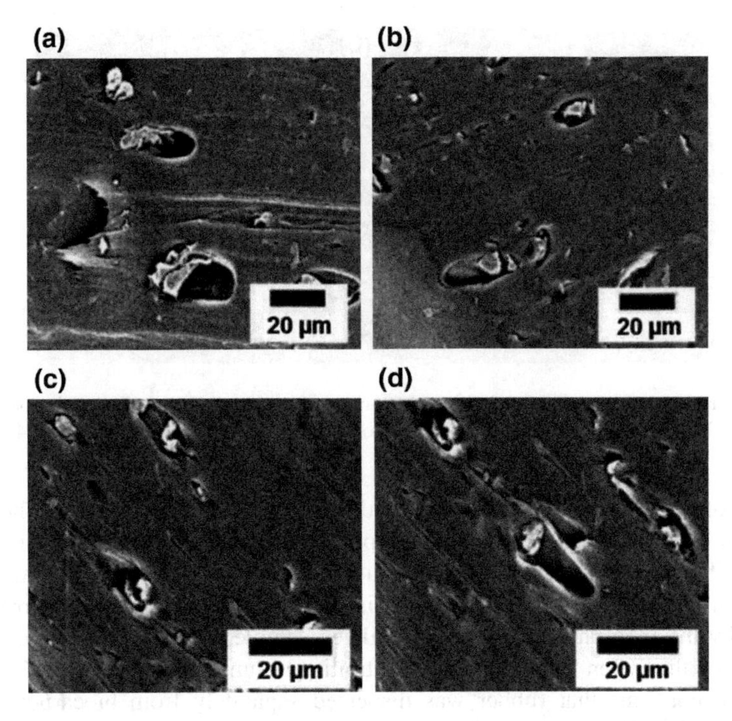

Fig. 15 SEM images of **a** 90/10 wt% LDPE/CB, **b** 85/15 wt% PP/CB, **c** 90/10 wt% LDPE/MCC, **d** 90/10 wt% PP/MCC. Reproduced from Iyer et al. (2015), [89]

materials. AFM has the ability to generate images based on mechanical properties which makes it particularly significant for the analysis of polymers (including Polyolefins). In such analyses, AFM can be operated in tapping mode (or amplitude modulation mode) phase imaging which is one of the most popular and routinely used mode the study of polymers [97]. Some other AFM based techniques used for the quantitative nanomechanical characterizations, include force modulation [97, 98], peak force quantitative nanomechanical mapping (QNM) [99], intermodulation spectroscopy [100], pulsed force mode [97, 101], contact resonance (CR) techniques [97], and force volume imaging [102]. In particular, the contact resonance (CR) techniques (or dynamic contact modes of AFM) were developed to study the elastic properties of stiff materials, such as polymer nanotubes [103, 104]. CR-FM can also be utilized for the study of viscoelastic properties of materials and storage [105, 106] and loss moduli of individual components in blends [99, 107].

The application of AFM studies on polymer composites can be elaborated by citing the example of a novel class of injection-moulded, toughened biocomposites which was engineered from pyrolyzed miscanthus based biocarbon, poly(octene ethylene) (POE) elastomer, and polypropylene (PP) [64]. The addition of maleic anhydride grafted PP (MAPP) controlled the morphology and adhesion between the filler and the matrix. AFM scans of the biocarbon interface in the presence and

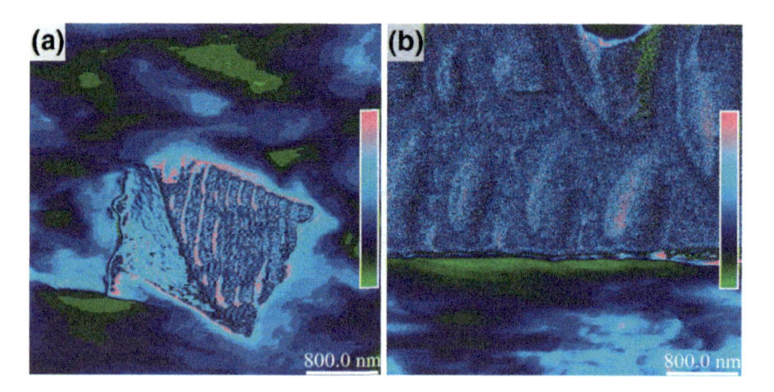

Fig. 16 AFM modulus mapping of a biocarbon–matrix interface in the small particle composites: **a** with 5% MAPP and **b** uncompatibilized. Reproduced from Behazin et al. (2017), [64]

absence of MAPP (Fig. 16) are highlighted by few specific observations. The green colour in both the images represents the phase with the lowest modulus (i.e. POE phase). The blue and pink colours attribute to stiffer phases, i.e, PP, MAPP, and biocarbon particles. Biocarbon particles in the composites are found surrounded mainly by the blue phases in a compatibilized sample, whereas interphase mostly consists of the green phase in uncompatibilized composite. Thus, AFM studies helped to confirms that rubber was dispersed separately from biocarbon in the presence of MAPP.

8.2.4 X-Ray Diffraction (XRD)

XRD studies are used to elaborate crystal forms and dispersibility of different Polyolefins, such as PP and HDPE in the polyolefin blends/composites. XRD analysis can resolve both iso- and syndiotactic crystallinity. The small-angle X-ray scattering (SAXS) has been found useful to determine the lamellar thickness of Polyolefins and crystalline fibril morphology [108]. Whereas, the wide-angle X-ray scattering (SAXS) investigations can reveal the degree of crystallinity and discrimination between crystalline and amorphous phases [108]. SAXS also enables a direct access to the morphology and can quantitatively determine the thicknesses of alternating layers between the crystalline and amorphous regions of the lamellae morphology. The temperature-dependent SAXS measurements are used to verify the location of the order-disorder transition temperatures. SAXS also reveals critical information about the hydrophilic and hydrophobic phase separations in polyolefin composites [39, 109].

The application of XRD analysis for the characterization of ultrahigh molecular weight polyethylene (UHMWPE), which is considered as one of best options for biopolymer materials, is shown in Fig. 17 [110]. The application of UHMWPE as biomaterial implants of human bone can be processed via surface plasma treatment, UV curing, chemical and physical surface treatment, and reinforcements.

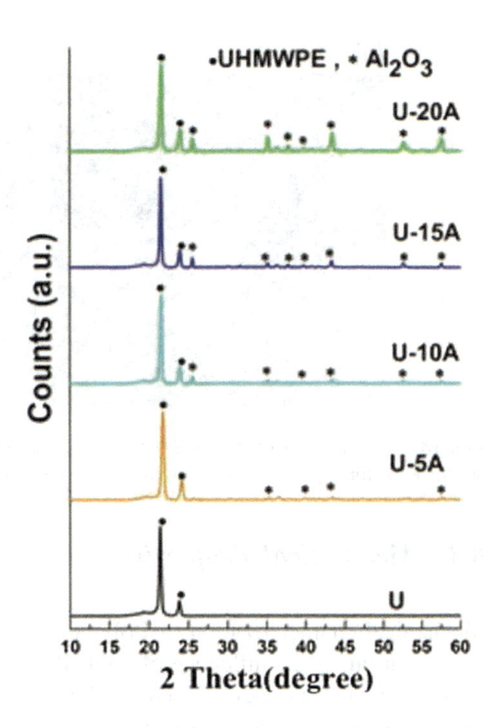

Fig. 17 XRD spectra of UHMWPE-Al_2O_3 composites pellets [110]

Specifically, the reinforcement of metal nanoparticles (Fe, Cu, Ag, Zn,), ceramics (Fe_3O_4, Al_2O_3, ZrO_3, ZnO, AgO, MgO, Si_3N_4) or natural materials serve the purpose well. Figure 17 shows the XRD patterns of one such sample, i.e., Al_2O_3 reinforced UHMWPE compression-moulded composites. The dominant involvement of UHMWPE has suppressed the peaks of Al_2O_3. The presence of UHMWPE is evident with two main characteristic peaks, i.e. at 21.9° and 24.18°. The involvement of Al_2O_3 is suggested with peaks observed at 35, 37.9 and 43.58°. Thus, it can be concluded with XRD analysis that starting phases did not experience any degradation during the compression molding of UHMWPE-Al_2O_3 composites.

8.2.5 Transmission Electron Microscopy (TEM)

TEM also uses electrons like SEM, but here electrons are allowed to pass through a sample. As electrons strongly interact with the sample and need to pass through the sample, the samples are needed to prepare as very thin sections [90]. TEM is used to determine the lamellar thickness of polymers [111]. For instance, the morphology of isotactic polypropylene (iPP) as studied by TEM is shown in Fig. 18. The structure of iPP after non-isothermal melt-crystallization and melt-crystallized at a rate of cooling of 750 K/s have been shown as examples. It is clear that one can study the effect of the application of temperature programs during the annealing process of iP [111]. Similarly, TEM can be applied to study the biodegradation of Polyolefin composites.

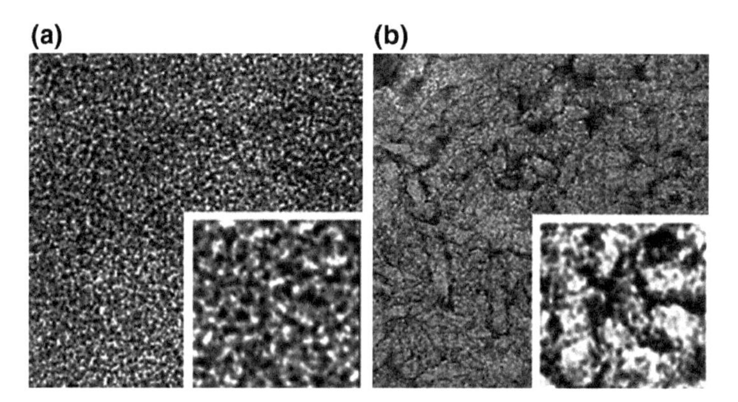

Fig. 18 TEM analysis of iPP, melt-crystallized at 750 K/s, and annealed **a** for a period of 60 min at 393 K; **b** for a period of 60 min at 433 K. Reproduced from Zia et al. [111]

8.3 Mechanical Properties

Mechanical characterization of Polyolefins is important to understand the material behaviour under the influence of applied loading. Mechanical testing machine (e.g. from MTS, Instron) can be used for this purpose, where the material is loaded in its bulk form to evaluate its various mechanical properties, such as yield strength, fracture stress or ultimate strength, elastic modulus or stiffness, Poisson's ratio and elongation [112].

8.3.1 Three-Point Flexural Test

Bending tests or flexural tests determine the mechanical properties of uni-directional composite materials. Flexural modulus and strength are obtained through three-point or four-point test configurations. In the former example, the contact zone between the specimen and cylindrical supports is changed as a result of the rotation of the cross sections in the deformation process. The four-point bending test involves contact between cylindrical loading noses and specimen [113]. For brittle polymers, the study of flexural strength is an essential parameter. The analysis of modulus of elasticity (i.e., the rigidity of the material) is also revealed during the flexural test. The three-point flexural test is a preferred option to evaluate the flexural strength of Polyolefins because of low values of data variation and less procedural complicacy [114, 115].

8.3.2 Tensile Test

Tensile properties of Polyolefins are measured before their applications in actual designs [116]. For instance, the tensile test method is useful to assess the improvement in the in-plane tensile properties of polyolefin matrix composite materials when they are reinforced by the high-modulus fibers. The measurement of tensile properties is important from the point of view of material specifications and quality assurance. The tensile tests can yield information about the ultimate tensile strain, ultimate tensile strength, Poisson's ratio, tensile chord modulus of elasticity and transition strain [117].

8.3.3 Dynamic Mechanical Analysis

Viscoelastic property of Polyolefins can be assessed by dynamic mechanical techniques which are among the most popular one of different available methods. The dynamic mechanical analysis (DMA) can be readily applied to both polymeric solids and liquid samples. In this method, a small cyclic strain is applied to a sample followed by the measurement of the resultant stress response, or sometimes vice-versa [118]. DMA can be used to determine polymer properties like molecular relaxation processes, inherent mechanical or flow properties with respect to time and temperature, viscoelastic transitions or relaxations, glass transition temperature, (T_g), and secondary transitions in amorphous and crystalline polymers [118].

9 Degradation of Polyolefins and Composites

9.1 Ageing and Corrosion

Ageing of plastics is a relatively complex process. Under the influence of harsh and aggressive environments, polymers may experience a change in their material properties due to the involvement of physical and/or chemical processes [119]. Such ageing of polymers can be accounted by the action of sunlight, oxygen, temperature, or other harmful atmospheric emissions [120]. Therefore, the Polyolefins and their composites can also reflect sudden changes in their macroscopic properties after a certain induction period. The ageing of Polyolefins can also result in their catastrophic failure in form of embrittlement and sometimes total disintegration [121, 122].

The corrosion type of degradation of Polyolefins can happen because of their unavoidable interaction with air pollutants (e.g, particulate matter, SO_2, hydrocarbons, N_xO_y, atomic oxygen, ozone, singlet oxygen, etc.). The corrosion can lead to affect the material properties like loss of gloss by abrasion, loss of reinforcement and breakup, crazing and cracking, and or leaching [8]. Sometimes, the electrochemical corrosion can also be observed in the polymer (e.g., in EVA, PE, Butyl Rubber coated metal surfaces) [123–125].

9.2 Chemical Degradation

Stereoregular Polyolefins are often inert toward the attack of chemical reagents. The reagents like alkalis, aqueous acid and basic salts, inorganic and organic acids; or even $KMnO_4$, $K_2Cr_2O_7$, and KNO_3 do not necessarily have much effect on stereoregular Polyolefins [8]. Their chemical degradation is achieved by strong oxidizing agents, such as H_2SO_4 [126] For example, H_2SO_4 can cause rapid deterioration of the mechanical properties of PE through acidolysis [127]. Likewise, it can also be achieved through the action of liquid or gaseous chlorine or fluorine [128, 129]. Specifically, the use of dry chlorine may not cause embrittlement of LDPE but bromine or iodine can get absorbed and diffused through polyethylene (PE) to cause serious harm to the mechanical integrity of Polyolefins [8].

9.3 Biodegradation

The microbial attack does not normally affect the properties of Polyolefins as the associated action limits only to the chain ends. Nonetheless, the paraffin with MW $\leq$ 450 can show some biodegradability, while low MW Polyolefins could be partially degraded [130, 131]. Therefore, strategies have been suggested to first reduce the MW of the Polyolefins with some additional process, which can then follow the action of microbes under suitable conditions [8, 132].

Apart from the other degradation methods, Polyolefins can also be subjected to mechanical [133–135], ultrasonic [136], radiation-induced [137], oxidative [132], photolysis [132], and thermolytic degradation [138] approaches.

10 Applications of Polyolefins and Eco-Friendly Composites

Polyolefins are used in a wide range of applications, ranging from our daily-life materials to high-end industrial products. Among all, Polyethylene is one of the most widely used commodity thermoplastic. The consumption of HDPE is extremely large in many sectors followed by LLDPE and LDPE. They are very commonly employed in the building and construction, electronics, electrical, consumer goods, automotive industries, etc. [139–143]. The application of PE in the manufacturing of various consumer goods includes the production of items like pipes, toys, containers, lids, plastic bags, buckets, covers, plastic wraps, sheets, films, stretch wraps, bottles and extrusion coated paper cartons [144–147]. Isotactic polypropylene (i-PP) is probably the second most important Polyolefin finding applications in many industries just like PE. It is a competitor of HDPE in various areas [148, 149]. When compared with HDPE, i-PP has characteristics of the higher melting point, improved crack resistance, an increased heat deflection temperature, and improved tensile strength. Nonetheless, HDPE is superior to i-PP with regard to thermal stability, exposure to light (UV) and susceptibility toward oxidative degradation. As far as application areas are concerned, i-PP enjoys better acceptability in case of more demanding applications. HDPE is more useful when material specification requires high resistance against thermooxidative degradation.

Another Polyolefin, Polyisobutene (PIB) is not produced on high production scale but is particularly employed in applications like sealants, chewing gum, adhesives, cable filling, synthetic rubber (butyl rubber), roofing membranes, and lubricants. Its low molecular grades variants are viscous oily liquids in nature, while the mid-range is more sticky materials. The high molecular grades PIBs are generated as elastic rubbery materials. Polybutene-1 (PB-1) is a low volume polyolefin which is highly compatible with polypropylene as both of them bear similar molecular structures. Their composites help in to improve the mechanical properties at elevated temperatures. PB-1 is used in applications like the pressure pipes for the supply of hot and cold (drinking) water, and consumer and medical packaging. Its low molecular weight variants are used in the production of paper laminates, adhesives, and sealants to manage tack and peel strength. They are also compatible with a variety of nonpolar resins including PP, PE, SIS, EVA, and SBS. A summary of applications of Polyolefins is presented in Table 1.

The applications of thermoplastic natural fibre-containing eco-friendly composites include their deployment in door and window frames, railings for the parapet wall systems, and decking and furniture material. The applications of eco-friendly composites are expected to increase significantly in near future. The favourable economics of natural fiber composites seems to popularize their use to a great extent as such types of composites can facilitate sustainable development of cost-effective and ecological technologies. The demand for lighter and strong polymer materials is increasing in many sectors, such as environmental friendly

Table 1 Summarized of applications of major categories of Polyolefins

Type	Applications
Low-density polyethylene (LDPE)	Shrink film, carry bags, heavy-duty refuse bags
High-density polyethylene (HDPE)	Crates and boxes, bottles (used in food products, detergents, cosmetics), food containers, industrial wrappings, carry bags, drums for food, beverages, and chemicals
Linear low-density polyethylene (LLDPE)	Stretch film, industrial packaging film, thin-walled containers, heavy-duty medium and small bags
Polypropylene (PP)	Food packaging (e.g. yoghurt and margarine tubs)

automotive components. For instance, the hemp mats in glass-fibre reinforced thermosets have been reported more ecoefficient than the conventional glass-fibre alternative. Likewise, the life cycle of the wood-fibre reinforced PP composites suggests better environmental protection than PP. The addition of biodegradable waste (e.g., rice husks, cotton linter) during the formation of composite materials can also present a viable way of solving the problem of waste disposal.

11 Conclusion

Traditionally used polymers involved synthesis approaches which posed serious health hazards and evolution of toxic gases. Their recycling was also not environmentally sustainable and complete. In this particular context, various classes of polyolefins have enabled an environmentally sustainable use of polymers. Polyolefins synthesized from natural resins generate fewer waste products during production while also helping in safer disposal. The enormous physical properties of polyolefins, such as mechanical strength, durability, elasticity, etc. can be tailored via selection of suitable synthesis conditions and optimized functionalization. Polyolefins support the addition of biodegradable materials (e.g., natural fiber and waste) in them to make composites having features of low-cost, strength, low density, and more importantly, an ecological efficiency. Future development of polyolefins is more focused on achieving lightweight designs, and even better recyclability, reusability.

References

1. Boffa LS, Novak BM (2000) Copolymerization of polar monomers with olefins using transition-metal complexes. Chem Rev 100:1479–1494
2. Soares JB (2007) An overview of important microstructural distributions for polyolefin analysis. In: Macromolecular symposia. Wiley Online Library, pp 1–12
3. Pasch H (2001) Recent developments in polyolefin characterization. In: Macromolecular symposia. Wiley Online Library, pp 91–98
4. Alkhazaal A (2011) Characterization of ethylene/α-olefin copolymers made with a single-site catalyst using crystallization elution fractionation. University of Waterloo
5. Soares JB, McKenna T, Cheng C (2007) Coordination polymerization. Polym React Eng 29–117
6. Koltzenburg S, Maskos M, Nuyken O (2017) Introduction and basic concepts. In: Polymer chemistry. Springer, pp 1–16
7. Storr A, Jones K, Laubengayer A (1968) The partial hydrolysis of ethylalane compounds. J Am Chem Soc 90:3173–3177
8. Vasile C, Seymour RB (2000) Handbook of polyolefins. Marcel Dekker, New York
9. Sinn H (1995) Proposals for structure and effect of methylalumoxane based on mass balances and phase separation experiments. In: Macromolecular symposia. Wiley Online Library, pp 27–52

10. Kaminsky W (1998) Highly active metallocene catalysts for olefin polymerization. J Chem Soc Dalton Trans 1413–1418
11. Bubeck R (2002) Structure–property relationships in metallocene polyethylenes. Mater Sci Eng R Rep 39:1–28
12. Patil AO (2000) Functional polyolefins. Chem Inno 30:19–24
13. Novák I, Borsig E, Hrčková LU, Fiedlerova A, Kleinova A, Pollak V (2007) Study of surface and adhesive properties of polypropylene grafted by maleic anhydride. Polym Eng Sci 47:1207–1212
14. Dong J-Y, Hu Y (2006) Design and synthesis of structurally well-defined functional polyolefins via transition metal-mediated olefin polymerization chemistry. Coord Chem Rev 250:47–65
15. Nakamura A, Ito S, Nozaki K (2009) Coordination-insertion copolymerization of fundamental polar monomers. Chem Rev 109:5215–5244
16. Franssen NM, Reek JN, de Bruin B (2013) Synthesis of functional 'polyolefins': state of the art and remaining challenges. Chem Soc Rev 42:5809–5832
17. Chung TM (2002) Functionalization of polyolefins. Elsevier
18. Boaen NK, Hillmyer MA (2005) Post-polymerization functionalization of polyolefins. Chem Soc Rev 34:267–275
19. Bielawski CW, Grubbs RH (2007) Living ring-opening metathesis polymerization. Prog Polym Sci 32:1–29
20. Opper KL, Fassbender B, Brunklaus G, Spiess HW, Wagener KB (2009) Polyethylene functionalized with precisely spaced phosphonic acid groups. Macromolecules 42:4407–4409
21. Opper KL, Markova D, Klapper M, Müllen K, Wagener KB (2010) Precision phosphonic acid functionalized polyolefin architectures. Macromolecules 43:3690–3698
22. Ouchi M, Terashima T, Sawamoto M (2009) Transition metal-catalyzed living radical polymerization: toward perfection in catalysis and precision polymer synthesis. Chem Rev 109:4963–5050
23. Finch C (1985) Encyclopedia of polymer science and engineering, volume 2, anionic polymerisation to cationic polymerisation editor-in-chief Jacqueline I. Kroschwitz. In: Mark HF, Bikales NM, Overberger CG, Menges G (eds) Wiley-Interscience, New York, pp xxiv+ 814, subscription price£ 175.00 (US 205.00) single volume price £210.00(US 240.00). ISBN 0-471-88786-2 (vol 2), British Polymer Journal, vol 17, pp 377–377
24. Berkefeld A, Mecking S (2008) Coordination copolymerization of polar vinyl monomers H2C=CHX. Angew Chem Int Ed 47:2538–2542
25. Lopez RG, D'Agosto F, Boisson C (2007) Synthesis of well-defined polymer architectures by successive catalytic olefin polymerization and living/controlled polymerization reactions. Prog Polym Sci 32:419–454
26. Yanjarappa M, Sivaram S (2002) Recent developments in the synthesis of functional poly (olefin)s. Prog Polym Sci 27:1347–1398
27. Amin SB, Marks TJ (2008) Versatile pathways for in situ polyolefin functionalization with heteroatoms: catalytic chain transfer. Angew Chem Int Ed 47:2006–2025
28. Matsugi T, Kojoh SI, Kawahara N, Matsuo S, Kaneko H, Kashiwa N (2003) Synthesis and morphology of polyethylene-block-poly (methyl methacrylate) through the combination of metallocene catalysis with living radical polymerization. J Polym Sci Part A: Polym Chem 41:3965–3973
29. Yagci Y, Tasdelen MA (2006) Mechanistic transformations involving living and controlled/living polymerization methods. Prog Polym Sci 31:1133–1170
30. Rezakazemi M, Sadrzadeh M, Matsuura T (2018) Thermally stable polymers for advanced high-performance gas separation membranes. Prog Energy Combust Sci 66:1–41
31. Rezakazemi M, Ebadi Amooghin A, Montazer-Rahmati MM, Ismail AF, Matsuura T (2014) State-of-the-art membrane based CO_2 separation using mixed matrix membranes (MMMs): an overview on current status and future directions. Prog Polym Sci 39:817–861

32. Chung T, Janvikul W (1999) Borane-containing polyolefins: synthesis and applications. J Organomet Chem 581:176–187
33. Kaneyoshi H, Inoue Y, Matyjaszewski K (2005) Synthesis of block and graft copolymers with linear polyethylene segments by combination of degenerative transfer coordination polymerization and atom transfer radical polymerization. Macromolecules 38:5425–5435
34. White JL, Kim EK (1991) Twin screw extrusion: technology and principles. Hanser, Munich
35. Salleh FM, Hassan A, Yahya R, Azzahari AD (2014) Effects of extrusion temperature on the rheological, dynamic mechanical and tensile properties of kenaf fiber/HDPE composites. Compos B Eng 58:259–266
36. Mulinari DR, Voorwald HJ, Cioffi MOH, Da Silva MLC, da Cruz TG, Saron C (2009) Sugarcane bagasse cellulose/HDPE composites obtained by extrusion. Compos Sci Technol 69:214–219
37. Mano B, Araújo J, Spinacé M, De Paoli M-A (2010) Polyolefin composites with curaua fibres: effect of the processing conditions on mechanical properties, morphology and fibres dimensions. Compos Sci Technol 70:29–35
38. AlMaadeed MA, Nogellova Z, Mičušík M, Novak I, Krupa I (2014) Mechanical, sorption and adhesive properties of composites based on low density polyethylene filled with date palm wood powder. Mater Des 53:29–37
39. Zhang M, Shan C, Liu L, Liao J, Chen Q, Zhu M, Wang Y, An L, Li N (2016) Facilitating anion transport in polyolefin-based anion exchange membranes via bulky side chains. ACS Appl Mater Interfaces 8:23321–23330
40. Rao S, Jayaraman K, Bhattacharyya D (2012) Micro and macro analysis of sisal fibre composites hollow core sandwich panels. Compos B Eng 43:2738–2745
41. Brito GF, Agrawal P, Araújo EM, de Mélo TJ (2012) Polylactide/biopolyethylene bioblends. Polímeros 22:427–429
42. Castro D, Ruvolo-Filho A, Frollini E (2012) Materials prepared from biopolyethylene and curaua fibers: Composites from biomass. Polym Testing 31:880–888
43. Jacob M, Francis B, Varughese K, Thomas S (2006) The effect of silane coupling agents on the viscoelastic properties of rubber biocomposites. Macromol Mater Eng 291:1119–1126
44. Donath S, Militz H, Mai C (2004) Wood modification with alkoxysilanes. Wood Sci Technol 38:555–566
45. Fang L, Chang L, Guo W-J, Chen Y, Wang Z (2014) Influence of silane surface modification of veneer on interfacial adhesion of wood–plastic plywood. Appl Surf Sci 288:682–689
46. Xie Y, Krause A, Militz H, Steuernagel L, Mai C (2013) Effects of hydrophobation treatments of wood particles with an amino alkylsiloxane co-oligomer on properties of the ensuing polypropylene composites. Compos A Appl Sci Manuf 44:32–39
47. Spiridon I, Darie RN, Bodîrlău R, Teacă C-A, Doroftei F (2013) Polypropylene-based composites reinforced by toluene diisocyanate modified wood. J Compos Mater 47:3451–3464
48. Bodîrlau R, Teaca C-A, Resmerita A-M, Spiridon I (2012) Investigation of structural and thermal properties of different wood species treated with toluene-2, 4-diisocyanate. Cellul Chem Technol 46:381
49. Kabir M, Wang H, Lau K, Cardona F (2012) Chemical treatments on plant-based natural fibre reinforced polymer composites: an overview. Compos B Eng 43:2883–2892
50. Li Y (2014) Characterization of acetylated eucalyptus wood fibers and its effect on the interface of eucalyptus wood/polypropylene composites. Int J Adhes Adhes 50:96–101
51. Li X, Tabil LG, Panigrahi S (2007) Chemical treatments of natural fiber for use in natural fiber-reinforced composites: a review. J Polym Environ 15:25–33
52. Wei L, McDonald AG, Freitag C, Morrell JJ (2013) Effects of wood fiber esterification on properties, weatherability and biodurability of wood plastic composites. Polym Degrad Stab 98:1348–1361
53. Ahmed AS, Islam MS, Hassan A, Haafiz MM, Islam KN, Arjmandi R (2014) Impact of succinic anhydride on the properties of jute fiber/polypropylene biocomposites. Fibers Polym 15:307

54. Mamun AA, Heim H-P, Beg DH, Kim TS, Ahmad SH (2013) PLA and PP composites with enzyme modified oil palm fibre: a comparative study. Compos A Appl Sci Manuf 53:160–167
55. Ortín A, López E, del Hierro P, Sancho-Tello J, Yau WW (2018) Simplified robust triple detection methods for high temperature GPC analysis of polyolefins. In: Macromolecular symposia. Wiley Online Library, pp 1700044
56. Song S, Fu Z, Xu J, Fan Z (2017) Synthesis of functional polyolefins via ring-opening metathesis polymerization of ester-functionalized cyclopentene and its copolymerization with cyclic comonomers. Polym Chem 8:5924–5933
57. Jung M, Lee Y, Kwak S, Park H, Kim B, Kim S, Lee KH, Cho HS, Hwang KY (2016) Analysis of chain branch of polyolefins by a new proton NMR approach. Anal Chem 88:1516–1520
58. Liu Y, Wang Z, Zhang X (2012) Characterization of supramolecular polymers. Chem Soc Rev 41:5922–5932
59. Monrabal B (2013) Polyolefin characterization: recent advances in separation techniques. In: Polyolefins: 50 years after Ziegler and Natta I. Springer, pp 203–251
60. Pasch H, Malik MI (2016) Advanced separation techniques for polyolefins. Springer
61. Serranti S, Bonifazi G (2010) Post-consumer polyolefins (PP-PE) recognition by combined spectroscopic sensing techniques. Open Waste Manag J 3(1):35–45
62. Soares JB (2004) Polyolefins with long chain branches made with single-site coordination catalysts: a review of mathematical modeling techniques for polymer microstructure. Macromol Mater Eng 289:70–87
63. Sugumaran V, Prakash S, Arora AK, Kapur GS, Narula AK (2017) Thermal cracking of potato-peel powder-polypropylene biocomposite and characterization of products—pyrolysed oils and bio-char. J Anal Appl Pyrol 126:405–414
64. Behazin E, Misra M, Mohanty AK (2017) Sustainable biocomposites from pyrolyzed grass and toughened polypropylene: structure-property relationships. ACS Omega 2:2191–2199
65. Ndiripo A, Albrecht A, Monrabal B, Wang J, Pasch H (2018) Chemical composition fractionation of olefin plastomers/elastomers by solvent and thermal gradient interaction chromatography. Macromol Rapid Commun 39(6):1700703
66. Tronc E, Hernandez-Escobar C, Ibarra-Gomez R, Estrada-Monje A, Navarrete-Bolanos J, Zaragoza-Contreras E (2007) Blue agave fiber esterification for the reinforcement of thermoplastic composites. Carbohyd Polym 67:245–255
67. Zhang Q, Chen P, Xie X, Cao X (2009) An effective method to identify the type and content of α-olefin in polyolefine copolymer by fourier transform infrared-differential scanning calorimetry. J Appl Polym Sci 113:3027–3032
68. Awal A, Ghosh S, Sain M (2009) Thermal properties and spectral characterization of wood pulp reinforced bio-composite fibers. J Therm Anal Calorim 99:695–701
69. Monrabal B (2006) Microstructure characterization of polyolefins. TREF and CRYSTAF. In: Studies in surface science and catalysis. Elsevier, pp 35–42
70. Monrabal B (2015) Separation of ethylene-propylene copolymers by crystallization and adsorption mechanisms. A journey inside the analytical techniques. In: Macromolecular symposia. Wiley Online Library, pp 147–166
71. Takeuchi D, Chiba Y, Takano S, Kurihara H, Kobayashi M, Osakada K (2017) Ethylene polymerization catalyzed by dinickel complexes with a double-decker structure. Polym Chem 8:5112–5119
72. Xue Y-H, Bo S-Q, Ji X-L (2015) Comparison of chain structures between high-speed extrusion coating polyethylene resins by preparative temperature rising elution fractionation and cross-fractionation. Chin J Polym Sci 33:1586–1597
73. Albrecht A, Jayaratne K, Jeremic L, Sumerin V, Pakkanen (2016) Describing and quantifying the chemical composition distribution in unimodal and multimodal ZN-polyethylene using CRYSTAF. J Appl Polym Sci 133(9):43089 (3–8)

74. Monrabal B, Sancho-Tello J, Mayo N, Romero L (2007) Crystallization elution fractionation. A new separation process for polyolefin resins. In: Macromolecular symposia. Wiley Online Library, pp 71–79
75. Monrabal B, Romero L, Mayo N, Sancho-Tello J (2009) Advances in crystallization elution fractionation. In: Macromolecular symposia. Wiley Online Library, pp 14–24
76. Fadeeva V, Tikhova V, Nikulicheva O, Oleynik I, Oleynik I (2010) Composition determination of post-metallocene olefin polymerization catalysts. J Struct Chem 51:186–191
77. Young RJ, Lovell PA (2011) Introduction to polymers. CRC Press
78. Mialon L, Pemba AG, Miller SA (2010) Biorenewable polyethylene terephthalate mimics derived from lignin and acetic acid. Green Chem 12:1704–1706
79. Prut E, Nedorezova P, Klyamkina A, Medintseva T, Zhorina L, Kuznetsova O, Chapurina A, Aladyshev A (2013) Blend polyolefin elastomers based on a stereoblock elastomeric PP. Polym Sci Ser A 55:177–185
80. Yi J, Liu Y, Pan D, Cai X (2013) Synthesis, thermal degradation, and flame retardancy of a novel charring agent aliphatic—aromatic polyamide for intumescent flame retardant polypropylene. J Appl Polym Sci 127:1061–1068
81. Jukić A, Faraguna F, Franjić I, Kuzmić S (2017) Molecular interaction and viscometric behavior of mixtures of polyolefin and poly (styrene-co-dodecyl methacrylate-co-octadecyl methacrylate) rheology modifiers in solution of lubricating base oil. J Ind Eng Chem 56:270–276
82. Gururajan G, Ogale AA (2009) Molecular orientation evolution during low-density polyethylene blown film extrusion using real-time Raman spectroscopy. J Raman Spectrosc 40:212–217
83. Cherukupalli S, Ogale A (2004) Integrated experimental–modelling study of microstructural development and kinematics in a blown film extrusion process: I. Real-time Raman spectroscopy measurements of crystallinity. Plast Rubber Compos 33:367–371
84. Cherukupalli SS, Ogale AA (2004) Online measurements of crystallinity using Raman spectroscopy during blown film extrusion of a linear low-density polyethylene. Polym Eng Sci 44:1484–1490
85. Kosaka P, Kawano Y, Petri H, Fantini M, Petri D (2007) Structure and properties of composites of polyethylene or maleated polyethylene and cellulose or cellulose esters. J Appl Polym Sci 103:402–411
86. Serranti S, Gargiulo A, Bonifazi G (2011) Characterization of post-consumer polyolefin wastes by hyperspectral imaging for quality control in recycling processes. Waste Manag 31:2217–2227
87. McKenna TF, Di Martino A, Weickert G, Soares JB (2010) Particle growth during the polymerisation of olefins on supported catalysts, 1–nascent polymer structures. Macromol React Eng 4:40–64
88. Abboud M, Denifl P, Reichert KH (2005) Fragmentation of Ziegler-Natta catalyst particles during propylene polymerization. Macromol Mater Eng 290:558–564
89. Iyer KA, Flores AM, Torkelson JM (2015) Comparison of polyolefin biocomposites prepared with waste cardboard, microcrystalline cellulose, and cellulose nanocrystals via solid-state shear pulverization. Polymer 75:78–87
90. McMahon G (2008) Analytical instrumentation: a guide to laboratory, portable and miniaturized instruments. Wiley
91. Fang C, Nie L, Liu S, Yu R, An N, Li S (2013) Characterization of polypropylene–polyethylene blends made of waste materials with compatibilizer and nano-filler. Compos B Eng 55:498–505
92. Spiridon I (2014) I. Natural fiber-polyolefin composites. Mini-review. Cellul Chem Technol 48:599–611
93. Fávaro SL, Lopes MS, de Carvalho Neto AGV, de Santana RR, Radovanovic E (2010) Chemical, morphological, and mechanical analysis of rice husk/post-consumer polyethylene composites. Compos A Appl Sci Manuf 41:154–160

94. May-Pat A, Valadez-González A, Herrera-Franco PJ (2013) Effect of fiber surface treatments on the essential work of fracture of HDPE-continuous henequen fiber-reinforced composites. Polym Test 32:1114–1122
95. Chiu F-C, Yen H-Z, Lee C-E (2010) Characterization of PP/HDPE blend-based nanocomposites using different maleated polyolefins as compatibilizers. Polym Test 29:397–406
96. Binnig G, Quate CF, Gerber C (1986) Atomic force microscope. Phys Rev Lett 56:930
97. Yablon DG, Gannepalli A, Proksch R, Killgore J, Hurley DC, Grabowski J, Tsou AH (2012) Quantitative viscoelastic mapping of polyolefin blends with contact resonance atomic force microscopy. Macromolecules 45:4363–4370
98. Radmacher M, Tillmann R, Gaub H (1993) Imaging viscoelasticity by force modulation with the atomic force microscope. Biophys J 64:735–742
99. Young T, Monclus M, Burnett T, Broughton W, Ogin S, Smith P (2011) The use of the PeakForceTM quantitative nanomechanical mapping AFM-based method for high-resolution Young's modulus measurement of polymers. Meas Sci Technol 22:125703
100. Platz D, Tholén EA, Pesen D, Haviland DB (2008) Intermodulation atomic force microscopy. Appl Phys Lett 92:153106
101. Rosa-Zeiser A, Weilandt E, Hild S, Marti O (1997) The simultaneous measurement of elastic, electrostatic and adhesive properties by scanning force microscopy: pulsed-force mode operation. Meas Sci Technol 8:1333
102. Wang D, Fujinami S, Liu H, Nakajima K, Nishi T (2010) Investigation of reactive polymer–polymer interface using nanomechanical mapping. Macromolecules 43:5521–5523
103. Hurley DC (2009) Contact resonance force microscopy techniques for nanomechanical measurements. In: Applied scanning probe methods, vol XI. Springer, pp 97–138
104. Cuenot S, Frétigny C, Demoustier-Champagne S, Nysten B (2003) Measurement of elastic modulus of nanotubes by resonant contact atomic force microscopy. J Appl Phys 93:5650–5655
105. Gannepalli A, Yablon D, Tsou A, Proksch R (2011) Mapping nanoscale elasticity and dissipation using dual frequency contact resonance AFM. Nanotechnology 22:355705
106. Killgore JP, Yablon D, Tsou A, Gannepalli A, Yuya P, Turner J, Proksch R, Hurley D (2011) Viscoelastic property mapping with contact resonance force microscopy. Langmuir 27:13983–13987
107. Proksch R, Yablon DG (2012) Loss tangent imaging: theory and simulations of repulsive-mode tapping atomic force microscopy. Appl Phys Lett 100:073106
108. Deplace F, Wang Z, Lynd NA, Hotta A, Rose JM, Hustad PD, Tian J, Ohtaki H, Coates GW, Shimizu F (2010) Processing-structure-mechanical property relationships of semicrystalline polyolefin-based block copolymers. J Polym Sci Part B: Polym Phys 48:1428–1437
109. Mahanthappa MK, Lim LS, Hillmyer MA, Bates FS (2007) Control of mechanical behavior in polyolefin composites: integration of glassy, rubbery, and semicrystalline components. Macromolecules 40:1585–1593
110. Patel AK, Trivedi P, Balani K (2014) Processing and mechanical characterization of compression-molded ultrahigh molecular weight polyethylene biocomposite reinforced with aluminum oxide. J Nanosci Nanoeng Appl 4:1–11
111. Zia Q, Androsch R, Radusch H-J, Ingoliç E (2008) Crystal morphology of rapidly cooled isotactic polypropylene: a comparative study by TEM and AFM. Polym Bull 60:791
112. S.R. Hartshorn, Structural adhesives: chemistry and technology, Springer Science & Business Media, 2012
113. Mujika F (2006) On the difference between flexural moduli obtained by three-point and four-point bending tests. Polym Test 25:214–220
114. Junior R, Adalberto S, Zanchi CH, Carvalho RVD, Demarco FF (2007) Flexural strength and modulus of elasticity of different types of resin-based composites. Braz Oral Res 21:16–21
115. Chung S, Yap A, Chandra S, Lim C (2004) Flexural strength of dental composite restoratives: comparison of biaxial and three-point bending test. J Biomed Mater Res B Appl Biomater 71:278–283

116. Carlsson LA, Adams DF, Pipes RB (2014) Experimental characterization of advanced composite materials. CRC press
117. Standard AS (2008) Standard test method for tensile properties of polymer matrix composite materials. ASTM D3039/D M 3039:2008
118. Menczel JD, Prime RB (2014) Thermal analysis of polymers: fundamentals and applications. Wiley
119. Rowe RK, Islam M, Hsuan Y (2009) Effects of thickness on the aging of HDPE geomembranes. J Geotech Geoenvironmental Eng 136:299–309
120. Corti A, Muniyasamy S, Vitali M, Imam SH, Chiellini E (2010) Oxidation and biodegradation of polyethylene films containing pro-oxidant additives: synergistic effects of sunlight exposure, thermal aging and fungal biodegradation. Polym Degrad Stab 95:1106–1114
121. Dehbi A, Mourad A, Bouaza A (2011) Ageing effect on the properties of tri-layer polyethylene film used as greenhouse roof. Procedia Eng 10:466–471
122. Baldwin FP, Strate GV (1972) Polyolefin elastomers based on ethylene and propylene. Rubber Chem Technol 45(3):709–881
123. Samimi A (2012) Study an analysis and suggest new mechanism of 3 layer polyethylene coating corrosion cooling water pipeline in oil refinery in Iran. Int J Innov Appl Stud ISSR J 1(2):216–225
124. Castaneda H, Benetton XD (2008) SRB-biofilm influence in active corrosion sites formed at the steel-electrolyte interface when exposed to artificial seawater conditions. Corros Sci 50:1169–1183
125. Kempe M (2011) Overview of scientific issues involved in selection of polymers for PV applications. In: Photovoltaic Specialists Conference (PVSC), 2011 37th IEEE, IEEE, pp 000085–000090
126. Czop M, Biegańska J Impact of selected chemical substances on the degradation of the polyolefin materials 66(4):307–314
127. Pant D (2011) Degradation of various low density polyethylene products on alumina surface with sulphuric acid—DTS technique. J Solid Waste Technol Manag 37:47–54
128. Rubino M, Netramai S, Auras R, Annous BA (2010) Effect of chlorine dioxide gas on physical, thermal, mechanical, and barrier properties of polymeric packaging materials. J Appl Polym Sci 115:1742–1750
129. Eng J, Sassi T, Steele T, Vitarelli G (2011) The effects of chlorinated water on polyethylene pipes. Plast Eng 67:18
130. Shah AA, Hasan F, Hameed A, Ahmed S (2008) Biological degradation of plastics: a comprehensive review. Biotechnol Adv 26:246–265
131. Sheik S, Chandrashekar K, Swaroop K, Somashekarappa H (2015) Biodegradation of gamma irradiated low density polyethylene and polypropylene by endophytic fungi. Int Biodeterior Biodegradation 105:21–29
132. Ammala A, Bateman S, Dean K, Petinakis E, Sangwan P, Wong S, Yuan Q, Yu L, Patrick C, Leong K (2011) An overview of degradable and biodegradable polyolefins. Prog Polym Sci 36:1015–1049
133. Badia J, Strömberg E, Karlsson S, Ribes-Greus A (2012) Material valorisation of amorphous polylactide. Influence of thermo-mechanical degradation on the morphology, segmental dynamics, thermal and mechanical performance. Polym Degrad Stab 97:670–678
134. Badia J, Strömberg E, Ribes-Greus A, Karlsson S (2011) A statistical design of experiments for optimizing the MALDI-TOF-MS sample preparation of polymers. An application in the assessment of the thermo-mechanical degradation mechanisms of poly (ethylene terephthalate). Anal Chim Acta 692:85–95
135. Badia J, Vilaplana F, Karlsson S, Ribes-Greus A (2009) Thermal analysis as a quality tool for assessing the influence of thermo-mechanical degradation on recycled poly (ethylene terephthalate). Polym Test 28:169–175
136. Desai V, Shenoy M, Gogate P (2008) Ultrasonic degradation of low-density polyethylene. Chem Eng Process 47:1451–1455

137. Yoshiga A, Otaguro H, Parra DF, Lima LFC, Lugao AB (2009) Controlled degradation and crosslinking of polypropylene induced by gamma radiation and acetylene. Polym Bull 63:397–409
138. Panda AK, Singh R, Mishra D (2010) Thermolysis of waste plastics to liquid fuel: a suitable method for plastic waste management and manufacture of value added products—a world prospective. Renew Sustain Energy Rev 14:233–248
139. Shen L, Haufe J, Patel MK (2009) Product overview and market projection of emerging bio-based plastics PRO-BIP 2009, Report for European polysaccharide network of excellence (EPNOE) and European bioplastics 243
140. Bilici MK, Yükler Aİ, Kurtulmuş M (2011) The optimization of welding parameters for friction stir spot welding of high density polyethylene sheets. Mater Des 32:4074–4079
141. Sanchis R, Fenollar O, García D, Sanchez L, Balart R (2008) Improved adhesion of LDPE films to polyolefin foams for automotive industry using low-pressure plasma. Int J Adhes Adhes 28:445–451
142. Kanbur Y, Irimia-Vladu M, Głowacki ED, Voss G, Baumgartner M, Schwabegger G, Leonat L, Ullah M, Sarica H, Erten-Ela S (2012) Vacuum-processed polyethylene as a dielectric for low operating voltage organic field effect transistors. Org Electron 13:919–924
143. Siddique R, Khatib J, Kaur I (2008) Use of recycled plastic in concrete: a review. Waste Manag 28:1835–1852
144. Joseph SC, Douglas MF, Butler AF, Bastow DR, Salhus JE, Hartfel MA, inventors; 3M Innovative Properties Co, assignee (2014) Apparatus for spraying liquids, and disposable containers and liners suitable for use therewith. United States patent US 8,628,026.
145. Pace GV, Hartman TG (2010) Migration studies of 3-chloro-1, 2-propanediol (3-MCPD) in polyethylene extrusion-coated paperboard food packaging. Food Addit Contam 27:884–891
146. Kirwan MJ, Plant S, Strawbridge JW (2011) Plastics in food packaging, Food and Beverage Packaging Technology, 2nd edn. pp 157–212
147. Kriegel R, Huang X, Schultheis MW, Bippert DA, Insolia GE, Kolls B, Summerville S (2010) Bio-based polyethylene terephthalate packaging and method of making thereof. Google Patents
148. Yuan X, Matsuyama Y, Chung TM (2010) Synthesis of functionalized isotactic polypropylene dielectrics for electric energy storage applications. Macromolecules 43:4011–4015
149. Arutchelvi J, Sudhakar M, Arkatkar A, Doble M, Bhaduri S, Uppara PV (2008) Biodegradation of polyethylene and polypropylene. Indian J Biotechnol 07(1)

Spectroscopy and Microscopy of Eco-friendly Polymer Composites

Ashish K. Shukla, Chandni Sharma, Syed M. S. Abidi and Amitabha Acharya

List of Abbreviations

AFM	Atomic force microscopy
AT	Nano-Attapulgite
BC	Bacterial cellulose
CASN	Citric acid modified starch nanoparticle
CD	Circular dichroism
CdS QD	Cadium sulfide quantum dot
CHNS	Carbon-hydrogen-nitrogen-sulfur
CNCs	Cellulose nanocrystals
CNS	Cellulose Nanosphere
CPC	Cetylpyridinium chloride
CT/CG	Chitin/cashew gum
DLS	Dynamic light scattering
DMC	Dry-milled corn
DSC	Differential scanning calorimetery
DTPA	Diethylenetriaminepentaacetic acid
EDS	Energy dispersive X-ray spectroscopy
EFPNC	Eco-friendly polymer nanocompositess
FCNTs	Functionalized Carbon nanotube
FESEM	Field emission scanning electron microscope
FTIR	Fourier-transform infrared spectroscopy
GA	Gum arabic
GO	Graphene oxide
HA	Hyaluronan
HRTEM	High resolution transmission electron microscope

A. K. Shukla · C. Sharma · S. M. S. Abidi · A. Acharya (✉)
Biotechnology Division, CSIR-Institute of Himalayan Bioresource Technology, Palampur 176061, Himachal Pradesh, India
e-mail: amitabha@ihbt.res.in

A. K. Shukla · C. Sharma · S. M. S. Abidi · A. Acharya
Academy of Scientific & Innovative Research (AcSIR), CSIR- Institute of Himalayan Bioresource Technology, Palampur 176061, Himachal Pradesh, India

Inamuddin et al. (eds.), *Sustainable Polymer Composites and Nanocomposites*,
https://doi.org/10.1007/978-3-030-05399-4_4

ICP-MS	Inductive coupled plasma mass spectroscopy
kDa	Kilodalton
LOX	Lipoxygenase
MCC	Microcrystalline cellulose
MPa	Megapascal
MPTMS	3-(trimethoxysilyl)-propyl methacrylate
NDs	Nanodiamonds
NF	Nanofibrils
NMR	Nuclear magnetic resonance
NPs	Nanoparticles
PBAT	Poly(butylene adipate-co-terephthalate)
PCL	Poly-caprolactone
PDDF	Pair distance distribution function
PGA	Poly-glycolite
PHA	Polyhydroxyalkanoate
PHBV	(3-hydroxybutyrate-co-3-hydroxyvalerate)
PLA	Polylactic acid
PLGA	Poly(D,L-lactide-co-glycolide)
PVA	Polyvinyl alcohol
RChi	Regenerated chitin
SAXS	Small angle X-ray scattering
SCC	Spherical cellulose container
SEM	Scanning electron microscopy
SLRP	Sequential liquid-lignin recovery and purification
SPI/CNTs	Soy protein isolate-carbon nanotube
SPI–MMT	Soy protein isolate–montmorillonite
SPIONs	Superparamagnetic iron oxide
TEM	Transmission electron microscopy
Tg	Glass-rubber transition
TGA	Thermogravimetric analysis
TG-MS	Thermogravimetric-Mass Spectroscopy
Tm	Melting point
UP	Unsaturated polyester
UV-Vis	Ultraviolet-visible spectroscopy
XRD	X-ray diffraction

1 Introduction

The eco-friendly polymers which are also known as green polymers comprise the materials which are either biodegradable or derived from biobased renewable sources. Therefore, these polymers and their corresponding composites trim down the negative human footprint on the environment. In the current scenario, due to the growing environmental issues; it is indeed necessary to put together the great importance for developing "green materials" in both R&D as well as in industries [1]. Photosynthetic components extracted from plants and wood biomass; such as starch, cellulose, hemicellulose, lignin produced from atmospheric carbon dioxide, can be used as renewable carbon resources [2]. Based on the concept of "carbon neutrality"; when biodegradable polymers such as polylactic acid (PLA), polyhydroxyalkanoate (PHA), polysaccharide derivatives are burnt, they are considered as green materials, because the liberated carbon dioxide is again transformed into biomass [3].

NCs are known as the materials of the 21st century; having a unique design and different properties which are lacking in conventional composites [4]. A number of biodegradable polymers and their layered silicate NCs are widely being used for different applications. NCs, as the term indicates are the composites in which at least one of the dimensions falls under nanometre range (1–100 nm). NCs consist of one or more discontinuous phases which are distributed over single continuous phase. The continuous phase is known as matrix whereas discontinuous phase as reinforcing material. Mechanically, NCs are having high aspect ratio and high surface to volume ratio of the reinforcing material. Green composites are made from both renewable sources based on biopolymers and biofillers (including nano-type fillers), with a positive environmental impact.

NCs are the alternative materials for overcoming the lacunae of microcomposites and monolithic; however, NCs may go through the formulation challenges related to the control of elemental composition in nanocluster. NCs possess multifunctional advantages due to some unique properties like smaller filler size, improved ductility with the same strength, light transmission characteristics, flame retardancy, thermal stability and chemical resistance. Three different techniques are usually followed to prepare eco-friendly polymer NCs (EFPNC) which include solution casting, in situ polymerization, and melt processing. Additionally, as dimension reaches the nanometre level, interactions at phase interfaces improved significantly, and this is important to enhance the properties of the material [1].

Polymers which are used as NCs can be classified based on their (a) native biomass such as plants, animals, microbial and synthetic based and (b) chemical structure (polysaccharide and polypeptide based). Eco-friendly polymers can be extracted from the natural sources mainly by chemical treatment and mechanical method including grinding, milling, sonication, and ultrasonication. The prepared polymer NCs are very useful in making lightweight sensors, producing structural components with a high strength to weight ratio, aerospace, automotive, electronics, biotech, and pharmaceutical industries [3].

2 Isolation of Eco-friendly Polymers

This part of the chapter briefly describes different processes used for the isolation of eco-friendly polymers from their natural sources. The described eco-friendly polymers were classified according to their origin viz., plant, animal, microbial and synthetic based (Scheme 1).

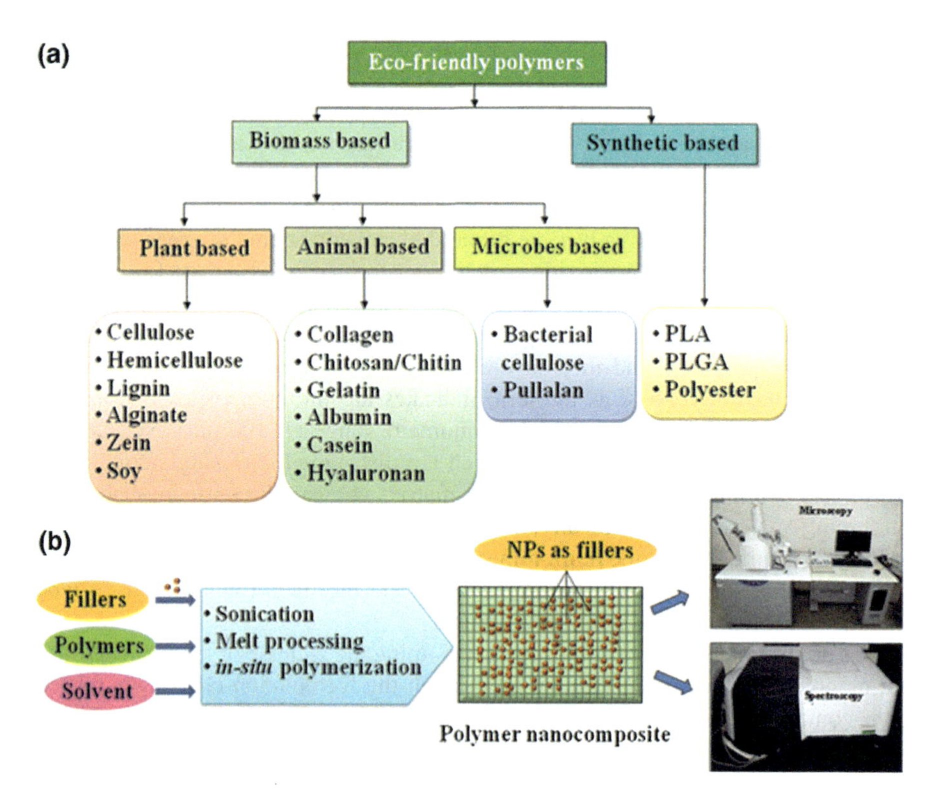

Scheme 1 **a** Pictorial representation of different eco-friendly polymers classified according to their source. **b** Schematic representation of different components of polymer nanocomposites and their characterization techniques

2.1 *Eco-friendly Polymers of Plant Origin*

Plant-based polymers have gained tremendous attention in recent years due to their high sustainability. In the next section, attempts have been made to concisely report different extraction techniques available in the literature for isolation of plant-based biopolymers.

2.1.1 Cellulose Extraction

Cellulose, with $\sim 10^{11}$ tons of annual turnover, is the most abundant natural polymer, containing vital skeletal polysaccharide component, composed of the β(1–4) glycosidic bond [5]. The biosynthesis of cellulose takes place not only in the plant but also in bacteria viz. Acetobacter, Acanthamoeba, Achromobacter and fungi [6, 7]. Cellulose is an extensive linear-chain polymer with a large number of hydroxy groups (three per anhydroglucose unit). Cellulose from its natural precursor sources can be isolated in two stages. The first stage involves pretreatment of raw material for purification and homogenization depending on the source material and desired morphology of cellulose particles. The pretreatments for plant biomass involve the complete or partial removal of matrix materials (hemicellulose, lignin etc.) and the isolation of individual complete fibers. Detailed descriptions of several of these pretreatment methods are already available in literature [8, 9]. The plant cellulose extraction methods include chemical and mechanical treatment as well as a combination of both. The chemical treatment involves acid hydrolysis [8, 10, 11] whereas the mechanical treatment includes high pressure homogenization [12, 13], grinding, cryo-crushing [14] and sonication [15]. Appropriate treatment of cellulose fibers leads to increase in the inner surface, alters degree of crystallinity, breaks hydrogen bonding and increases the reactivity of the cellulose to further facilitate the process of nanocellulose formation [10].

2.1.2 Hemicellulose Extraction

The hemicelluloses usually comprise off 20–30% of the dry weight of wood. It is the monomer of mixed sugar of short polymer length, consist mainly of mannose along with xylose, glucose, galactose and arabinose, where the main chain is connected to β(1–4) glycosidic bonds. The presence of hydroxyl group on the backbone of hemicellulose creates an opportunity for the chemical modification and can lead to the development of new nanobiocomposite. Lignin content was removed from plant biomass using concentrated NaOH and ethanol, and the resulting holocellulose solids were then extracted using NaOH (10%) to provide a filterate rich in hemicellulose. After the addition of ethanol, the precipitates were collected which predominantly contain xylose based hemicellulose [16, 17].

2.1.3 Lignin Extraction

Lignin is one of the abundant biopolymers after cellulose which is derived from plants and their wastes [18] and acts as connecting bridge between cellulose and hemicellulose. Lignin with cellulose combines to form large lignocellulosic biomass which has a high potential for production of biomaterials and chemicals [18–20]. Lignin can be isolated via different techniques viz., physical and chemical pretreatments and oxidation [21], enzymatic cellulose hydrolysis [22], acidification of black liquors (liquid-lignin) through sequential liquid lignin recovery process (SLRP) [23], pre-hydrolysis using kraft pulping [24] and microwave-assisted acidolysis [25]. Lignin has also been isolated from reed straw via thermomechanical process through attritor-type laboratory ball mill [26].

2.1.4 Starch Extraction

Starch is one of the major dietary sources of carbohydrates and it is found abundantly in the form of polysaccharide in plants. This is an important polymer, formed from long chains of α-glucose units joined together by glycosidic linkages containing two types of molecules i.e. amylose and amylopectin [27]. It is commonly found in roots, tubers, cereal grains and also occurs in a variety of foods, fruits and vegetable tissues etc. Starch and its derivatives have been used as capping and reducing agents for the synthesis of many metallic nanoparticles [28]. Starch is mostly isolated from white rice through alkali extraction method [27]. Acid hydrolysis at 37 °C was used to extract starch from peas to form pea starch nanocomposite [29]. Similarly, starch nanocrystals were also prepared from native pea granules through acid hydrolysis using mild stirring [30]. Enzymatic treatment of waxy maize through α-amylase hydrolysis was also one of the approaches used for starch isolation [31]. Few other techniques include ultrasonication [32], combined enzymatic hydrolysis couple with chemical or physical treatment and acidic treatment was also used for starch isolation [33, 34].

2.1.5 Alginate Extraction

Alginate is the derivative of alginic acid widely distributed in the cell wall of brown algae in the form of calcium, magnesium and sodium salt of alginic acid. These are anionic linear polysaccharides containing β-D mannuronic and α-L-guluronic acid residues linked by β 1–4 glycosidic bond forming homo- and heteropolymeric structures. Alginate biopolymer and its NCs are biocompatible and inexpensive which allows them to be used for various biomedical applications [35]. Alginate extraction was done using broken seaweed pieces, which was then stirred with a hot solution of an alkali, usually sodium carbonate. Alginate nanocomposite can be prepared in a two-step procedure based on the ionotropic pre-gelation of polyanion with calcium chloride salt followed by polycationic crosslinking [36].

2.1.6 Zein Extraction

Maize, a major cereal grain throughout the world, is being used for the isolation of zein. Zein is a protein that is exclusively found in corn. However, there are some other proteins with similar prolamin characteristics, which can be isolated from common cereals such as wheat, barley, rye, and sorghum [37]. Zein extractions require a complex balance between yield, quality, and purity. Dry-milled corn (DMC), acts as a good material to extract zein. Wet milling of corns can also be used to create a co-product that is rich in zein [38].

2.1.7 Soy Extraction

The consumption of soy has gained popularity in the world during recent decades due to (i) increased awareness of the consumer, (ii) drive for a healthier lifestyle, (iii) predominance of lactose intolerance cases and (iv) improved processing of soybeans with reduced off-flavors. Common process of soy extraction steps involves soaking of the soybeans, followed by grinding of the materials in cold water. Subsequently, these materials were then filtered and cooked at 100 °C for 30 min. This method can be further modified by grinding the soy extracts in hot water, which has the advantage of lipoxygenase (LOX) inactivation in the final extract [39].

2.2 *Animal-Based Eco-friendly Polymers*

This section includes different biopolymers isolated from animal origin viz., collagen, gelatin, chitin and chitosan, casein and hyaluronan.

2.2.1 Collagen Extraction

Collagen is a fibrous protein which is predominantly present in the connective tissues of animals. It offers a wide range of applications in the food, pharmaceuticals, cosmeceuticals and photographic industries. Before the extraction of collagen, a pretreatment is performed using an acid or alkaline process which depends on the origin of the raw material. Collagen can be extracted by both chemical and enzymatic hydrolysis process. Although, chemical hydrolysis is frequently used in many industries, enzymatic hydrolysis shows more potential in obtaining products having high nutritional significance and improved functionality. Furthermore, the enzymatic processes require less processing time and have the advantage of minimal waste generation [40].

2.2.2 Gelatin Extraction

Gelatin is a fibrous denatured polymer derived from collagen protein. It is an important biodegradable polymer having a broad range of applications in food and pharmaceutical industries. Due to its inherent cross-linking property, it has been widely used for nanoparticle synthesis [41]. There are mainly two types of gelatin viz., Type A and Type B which can be obtained via acid and base treatment [42] respectively in combination with thermomechanical process [43].

2.2.3 Chitin Extraction

Chitin is the second most abundant naturally occurring polysaccharide after cellulose and mostly found in crustaceous shell or in cell walls of fungi. Chitin is consisted of repeated homopolymeric β(1→4) linked *N*-acetyl-D-glucosamine units [44]. It has limited industrial application due to its insolubility in commonly used solvents. Further, the isolation of pure chitin from natural sources is difficult. The general methodologies for isolation of chitin involves three main processes viz., demineralization, decolourisation followed by deacetylation. After initial washing and drying, the raw scales were soaked in HCl (1%) solution for 36 h. These were then washed, dried in oven and kept in NaOH (2N) solution for 36 h to complete demineralization. Finally, these scales were immersed in potassium permanganate solution for 1 h, followed by addition of oxalic acid to achieve decolorization. Chitin isolated from the above process was further subjected to NaOH treatment (50% w/v) to obtain deacetylated chitin [45].

2.2.4 Casein Extraction

Casein is an essential protein which approximately counts for 80% of the total protein content of cow milk. Casein is also termed as globular protein, because it generally forms globules in the milk, and is mainly responsible for white color of the milk. Casein in its native form exists as calcium salt, hence termed as calcium caseinate. Casein can be extracted from reconstituted nonfat powdered milk using acetic acid treatment. This procedure allows the extraction of casein in the form of precipitated mass [46].

2.2.5 Hyaluronan (HA) Extraction

HA is a high-molecular-weight unsulfated polysaccharide. HA is mainly composed of subunits of D-glucuronic acid and *N*-acetyl glucosamine which is coupled by β(1–3) and β(1–4) glycosidic linkages. High water-retention capacity, mucoadhesion property, viscoelasticity, non-immunogenicity and biocompatibility make it an ideal candidate for healthcare applications [47]. HA can be obtained from various animal sources, such as umbilical cords, rooster combs, bovine submaxillary glands and zones of maturing chondrocytes. Muscle and skeleton comprises off ~35% of human HA. Extraction procedure of HA mainly includes dissection of bigeye tuna, followed by repeated thawing and filtration of vitreous at 4 °C to isolate the carbohydrate content which was further precipitated by adding cetylpyridinium chloride (CPC) to the filtrate. The HA-CPC complex was separated by re-suspending the precipitated mass in sodium chloride solution followed by centrifugation. Using a series of continued chemical followed by centrifugation process HA was finally isolated as freeze-dried material [48].

2.3 *Bacterial-Based Eco-friendly Polymers*

2.3.1 Bacterial Cellulose Extraction

Bacterial cellulose shares almost identical chemical structure as that of plant-derived cellulose, though it possesses high degree of polymerization, purity, crystallinity, water holding capacity compared to the later one [49]. This interesting biopolymer has been widely used in food, pharma, paper industry etc. [50]. For the successful production and isolation of BC, several factors such as physiological and nutritional conditions, temperature, incubation time, agitation etc. plays an important role. Standard static condition of bacterial growth is required for the isolation of bacterial cellulose. Glycerol improved the dry mass production of bacterial cellulose approximately 2–3 times as compared to when mannitol and glycerol in combination was used as carbon source [51].

2.3.2 Pullulan Extraction

Pullulan is an extracellular hydrophilic polysaccharide produced by different strains of yeast-like fungus e.g., *Aureobasidium pullulans*. It is synthesized intracellularly at the cell wall and secreted out to the cell surface to form a loose and slimy layer. It comprises of mixed linear linkage of α-D-glucan which consist mainly maltotriose units coupled by α-(1→6) linkages [52]. Pullulanase, is an enzyme belonging to the α-amylase family, identified as glycoside hydrolase and breaks α-(1→6) linkages present in various biopolymers like pullulan, starch and amylopectin [53]. Extraction of pullulan involves the use of fermenter; in which production medium was inoculated with 10% seed culture. Finally, the fermentation was performed at an optimized temperature (27 °C) and air flow condition under optimized revolution (~210 rpm) for one week to isolate pullulan [52].

2.4 *Synthetic Eco-friendly Polymers*

2.4.1 PLA Synthesis

PLA is a synthetic biodegradable polymer of lactic acid with two optically active stereoisomers viz., L(+) and D(−). Isolation of lactic acid in industry is mainly accomplished by fermentation process using bacterial strains of *Lactobacillus* genus viz., *Lactobacillus delbrueckii, L. amylophilus, L. bulgaricus and L. leichmanii* at a pH range between 5.4–6.4 and temperature range of 38–42 °C. PLA can be subsequently produced from lactic acid through polymerization process [54].

2.4.2 PLGA Synthesis

Poly(D,L-lactide-co-glycolide) (PLGA) is amongst the most widely used FDA approved biodegradable polymer for bio-applications since it produces metabolite monomers lactic acid and glycolic acid, after its hydrolysis. This copolymer can be synthesized by treating D, L-lactide and glycolide at 175 °C in the presence of stannous octoate and lauryl alcohol. PLGA-nanoparticles are internalized in cells partly through fluid phase pinocytosis and also through clathrin-mediated endocytosis process [55, 56].

2.4.3 Polyesters Synthesis

Polyesters are a class of polymers which can be synthesized as a result of polycondensation (step-growth polymerization) reaction between dialcohol and diacid/ or diester. Unsaturated polyester (UP) resin is widely used in various industrial applications such as marine, automotive, coatings, storage tanks, piping and construction [57].

3 Spectroscopic and Microscopic Characterization of Biopolymers and Their NCs

This part of the chapter includes different spectroscopic and microscopic techniques which were used for the characterization of the described eco-friendly polymers and their corresponding NCs. For the ease of understanding, the discussion in this section was classified based on the chemical structure of the polymers viz., polysaccharide, polypeptide and synthetic based materials.

3.1 Polysaccharide-Based Biopolymers

Polysaccharide-based biopolymers are composed of long chains of monosaccharide units, which are connected by glycosidic bonds. Polysaccharide NCs have become increasingly important materials over the past decade, since these offer a green alternative to synthetic polymers. These have also been used as composites with hard nanomaterials, such as metal nanoparticles and carbon-based nanomaterials (Fig. 1).

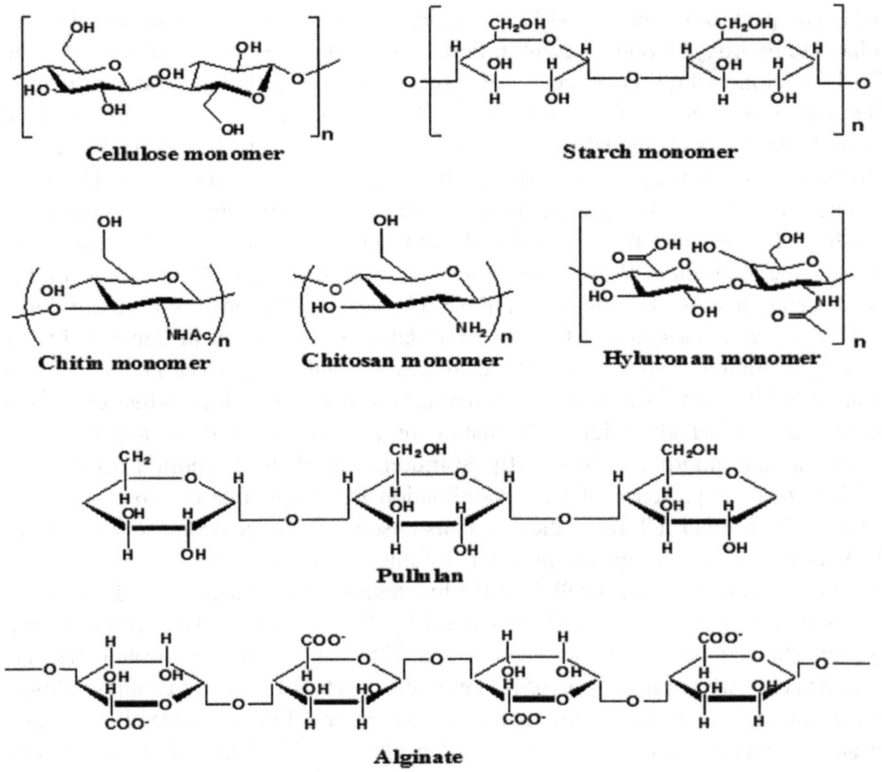

Fig. 1 Chemical structures of different polysaccharide based biopolymers

3.1.1 CHNS Analysis

The elemental analysis of lignin suggested that organosolv and kraft lignin have higher percentage of carbon and lower percentage of nitrogen content compared to other sources of lignin. However, the nitrogen content was higher for aminated lignin of epoxy resin polymer [58]. It was also reported that corn straw and wheat straw contains approximately equal amount carbon, hydrogen, nitrogen, and sulfur [59]. CHNS analysis of modified nanocrystalline cellulose (CNC) revealed higher amount of carbon content and less amount of sulphur and oxygen content as compared to native CNC [60].

3.1.2 FT-IR Spectroscopy

FT-IR has been one of the most common spectroscopic techniques that were used for the identification of different chemical functional groups present in the biopolymers. Numerous studies were already performed on natural fibers by FT-IR

[61]. The peak observed at 3349 cm^{-1} represents the stretching vibration of O–H, related to hydroxyl groups present in lignin, hemicelluloses and cellulose [10]. The C–H stretching vibrations in cellulose, hemicelluloses and lignin were identified at the spectrum range of $\sim$2900 cm^{-1}. The transmittance peak observed at $\sim$1645 cm^{-1} was attributed to the O–H bending vibration which resulted from the interaction between cellulose and the absorbed water molecules [62]. The peak position at $\sim$1462 cm^{-1} present in the spectrum was attributed to the symmetric bending of $-CH_2$ present in sugar backbone whereas the peak at $\sim$1313 cm^{-1} was due to the existence of bending vibration from C–H and C–O bonds in the polysaccharide aromatic rings of cellulose. The peaks observed at $\sim$1216 cm^{-1} and 1157 cm^{-1} were assigned to C–O–C of arylalkyl ether present in lignin and C–H rocking vibration, respectively. The characteristic anhydroglucose chains showed peak at $\sim$1047 cm^{-1} due to C–O stretching. Further, the peak at $\sim$898 cm^{-1} was identified as C–H glycosidic deformation or β-glycosidic linkage and was also known as the amorphous band [10]. Spiridonov et al. [63] reported decrease in –COO– peak in presence of Fe^{3+} coordination for maghemite carboxymethylcellulose NCs. Further, FT-IR studies were also used to evaluate cellulose stability for TiO_2-cellulose nanocomposite prepared by microwave solvothermal process [64]. In another report, a hemicellulose-diethylenetriaminepentaacetic acid (DTPA)-chitosan nanocomposite was characterized by FT-IR studies. The corresponding spectra showed new peaks at $\sim$3030 and 2780 cm^{-1} which indicated the carboxylic acids present in DTPA [65]. The FT-IR spectra of bacterial cellulose almost correspond to the peaks already mentioned for other cellulose species [66]. Foong et al. [67] reported appearance of new IR peaks at $\sim$1720 and 1449 cm^{-1} which corresponds to acetyl and CH_3 groups of PLA and indicated conjugation of PLA on bacterial cellulose surface. A mixed nanocomposite of chitosan and alginate nanoparticles was prepared in presence of glutaraldehyde and characterized using FT-IR. The bands obtained at $\sim$3360–3440 cm^{-1} were assigned as N–H and O–H stretching modes of vibrations for chitosan/alginate NCs. Further, the broadening of IR band confirmed the formation of intermolecular hydrogen bonding for chitosan/alginate NCs [68]. FT-IR spectroscopic analysis of chitin/chitosan revealed that the absorption bands of chitosan were similar to those for standard chitin. IR bands observed in the range of $\sim$3425–2881 cm^{-1} were assigned to N–H stretching for primary amines whereas the bands at $\sim$3425–3422 cm^{-1} were indicative of different vibrations of N–H, O–H and NH_2, present in chitin. When chitin undergoes deacetylation, a higher intensity peak was observed at $\sim$1597 cm^{-1} which suggested effective deacetylation and prevalence of NH_2 group [45]. The chitin and chitosan isolated from *Fusarium solani* present in marine soil revealed the presence of amide I region ($\sim$1657–1642 cm^{-1}), amide II stretching with C–O group ($\sim$1560–1550 cm^{-1}) and amide III region (1381–1375 cm^{-1}) [69]. The herbal nanocomposite of chitosan showed IR peaks of lower intensity at $\sim$3474/3468 cm^{-1} and 1745/1654 cm^{-1}, possibly due to the presence of intermolecular hydrogen bonding between chitosan and corresponding active ingredients [70]. Pullulan nanocomposite prepared using lysozyme nanofibres (LNF) and polysaccharide solutions were characterized by FT-IR and showed characteristic bands for

both the reactants [71]. The existence of both α-(1→4) and α-(1→6) glycosidic linkages in the pullulan structure was confirmed by the presence of IR band at 935 cm^{-1} [72]. Chemometric analysis of lignin isolated from different sources was also studied using infrared spectroscopy [73]. Although there are various sources of lignin or lignocellulosic biomass, the best quality material can be screened by monitoring characteristic FT-IR spectra [74, 75]. IR peaks at ~800, 1350, 1540 and 3300–3500 cm^{-1} were assigned for N–H vibration, C–N vibration, N–H bending vibration, and N–H stretching vibration of aminated lignin respectively. Further, the peak at ~3400 cm^{-1} was assigned for O–H stretching of demethylated lignin [58] while the peak at ~1270 cm^{-1} was designated for aromatic ring vibration of guaiacyl lignin [74, 76].

Similarly, FT-IR spectra were used for the chemical characterization of corn starch, starch nanoparticles (SN) and citric acid modified starch nanoparticles (CASN). The key difference between CASN compared to corn starch and SN was the appearance of a new peak at ~1738 cm^{-1} for an ester group and at ~1017 cm^{-1} for C–O bond stretching of the C–O–C group of the anhydroglucose ring which was exclusively present in CASN. It was being inferred that in corn starch and SN, the oxygen of the C–O–C group could form the hydrogen-bond with the hydroxyl groups and the ester bonds in CASN, shifting C–O bond stretching of the C–O–C group to 1026 cm^{-1} [77].

3.1.3 Powder XRD

Powder XRD technique has been used widely for the identification of crystalline and amorphous nature of the concerned polymers and their composites. The XRD was used to investigate the crystalline structure of the raw fibers, mercerized cellulose, and CNS. It was observed that the XRD patterns of raw fibers and commercial cellulose were clearly different compared to the mercerized cellulose and cellulose nanoparticles. For the raw fibers, a characteristic 2θ band between 13° (101), 17° (101) along with a broad peak at 22.5° (002) were observed, which corresponds to the cellulose-I structure. These results were in agreement with those observed for native cellulose [78]. It was also observed that the peaks for the amorphous region at 14° and 16° were hard to distinguish due to their close proximity [79]. Further, it was found that in case of mercerized cellulose and CNS, the corresponding peaks were shifted to 12°, 20°, and 22° and were related to same crystal planes mentioned earlier. The crystallinity index for raw material, mercerized cellulose and CNS were found to be ~68, 64 and 88%, respectively [80]. In case of TiO_2–cellulose nanocomposite additional peaks at 34.3° (004), 38.2° (004), 48.0° (200) and 70.11° (220) were observed corresponding to the tetragonal structure of TiO_2 and reflected no change in the crystalline structure of the wood cellulose fibers [64]. Observation of a broad peak at 19° in powder XRD indicated the conversion of hemicellulose to xylitol [81]. Xylan hemicellulose (XH)/cellulose fiber NCs showed the presence of crystalline peaks whereas freeze-dried XH powder showed no distinct crystalline peaks [82]. Bacterial cellulose showed three characteristic peaks of cellulose at $2\theta = 14.7°$ [39], 16.4° [39] and 22.6° (200) which were attributed to the elementary cellulose crystalline structure. Carbon sources used for bacterial growth can also affect the degree of crystallinity in bacterial cellulose. Further, the crystallinity of bacterial cellulose produced in agitated condition was found to be lower than the static culture due to the presence of structural disorder [83]. XRD studies of the bacterial cellulose (BC) and drug-loaded BC matrices were carried out by varying the scanning angle in the range of 10° to 60°. The distinct peaks observed in the diffractogram for BC at 14.12°, 16.8°, and 22.72° demonstrated the crystalline structure of BC. The surface modified BC showed peak pattern similar to as-synthesized BC with lower intensity, indicating a reduction in the crystallinity as a result of acetylation. Apart from the peaks at 26.90°, 28.54° and 29.82°, BC-famotidine matrices have strong peaks at 11.58° and 17.76°, confirming the entrapment of drug into matrices. The appearance of distinct peaks in the pattern for BC-tizanidine matrices at 25.06° and 26.46° showed lower crystal growth of the drug nanocomposite [84]. XRD patterns of calcium alginate/graphene oxide NCs revealed an amorphous structure, whereas those of barium alginate/graphene oxide composites indicated the presence of semi-crystalline structure which might have resulted from the preferential binding of barium ions to mannuronic acid blocks [85]. The XRD patterns of chitin and hydrolyzed chitosan confirmed that all chitin samples exhibited strong reflections at a 2θ value between 9 and 10°. XRD spectra further suggested that chitin is amorphous in nature whereas chitosan is crystalline [45]. Chitin/cashew gum (CT/CG) nanocomposite showed sharp peaks at $2\theta = 9.28°$ and 17.54° corresponding to diffractions of CT and CG segments respectively when blended with metal oxide nanoparticles. Furthermore, peaks at 2θ corresponding to 21.65° and 26.13°, indicated the semi-crystalline nature of the blend matrix [86]. The XRD patterns of carbonized lignin showed peaks at 26.7° (002), 43.2° (100/101), 54.7° (004) and 78.1° (110) confirming the structural changes happened in lignin after carbonization whereas untreated lignin did not show above-mentioned peaks, which revealed its non-crystalline nature [87]. Pure pullulan has no crystalline peaks implying that the material is fully amorphous [88]. Unfilled glycerol plasticized and non-plasticized pullulan films were characterized by a broad peak centered at around 19°, typical of fully amorphous materials [89]. Rice starch showed A-type diffraction pattern peaks at a 2θ value of 9.9°, 15.0°, 17.0°, 18.1° and 23.5°. Extensive hydrolysis of rice starch showed less intensity diffraction patterns which indicated that hydrolysis occurred in the amorphous region [31]. A native starch granule was pre-treated with β-amylase and glucoamylase and found that pretreated starch granules have semi-crystalline nature as compared to native starch [33].

3.1.4 NMR

NMR is a one of the strongest technique which is used for characterization of chemical moiety present in most of the compounds. Microcrystalline cellulose (MCC) biopolymer was converted into spherical cellulose container (SCC) by the sonochemical method and investigated by ^{13}C MAS NMR. The spectra depicted that peak of different carbon atoms of the glucose pyranose repeating unit in MCC and SCC were quite similar [90]. ^{1}H-NMR spectra of acetylated and demethylated lignin showed peaks corresponding to 6.0 and 8.0 ppm for two of its important aromatic precursor's viz., syringyl and guaiacyl respectively. Further, the peak at 6.96 ppm was assigned for a proton of C5 position in the 9 carbon units of lignin while 3.82 ppm peak was designated for a proton of methoxy group of the same material [91]. ^{13}C NMR spectrum of the aminated lignin was observed and showed a peak around 130 ppm which corresponds to ortho- and para- positions of the aromatic ring. Further, peaks were observed at δ value 16.5, 17.8, 59.5, 82.2, 136.8 and 137.3 ppm for methyl groups, CH group of the aromatic ring and amine, CH connected to oxygen and amine connected aromatic carbon atoms respectively [58]. 2D NMR analysis of mild wood lignin showed typical lignin substructures such as β-O-4, β-β, β-5, benzaldehyde and cinnamaldehyde units [92]. ^{31}P NMR is a powerful tool for hydroxyl group analysis in lignin biomass. Softwood lignin showed characteristic peaks of hydroxyl groups present in an aromatic moiety of lignin [93].

3.1.5 DLS and Zeta Potential

The stability of CNS is mainly related to the ions present during the acid hydrolysis. For sulfuric acid treatment, the formation of sulfate ester groups allows water dispersion of CNSs and prevents aggregation. The zeta potential of 0.1% CNS in water indicated an average zeta potential of -23.3 ± 3.2 mV [94]. The DLS size of CNSs was found to be in the broad range 30 nm to 1 μm which indicated the anisotropic properties of the CNS suspension which were attributed to the different aggregate states CNSs in solution [95]. Similarly, for hemicellulose broad particle size distribution ranging from 50 to 400 nm was observed. Further, hemicellulose showed negative zeta potential value which can be due to the presence of pectic substances (anionic polysaccharides) and oxidized lignin structure on hemicellulose surface [96]. Silver-chitosan nanocomposite showed particle diameter of $\sim$1553 nm whereas, simple chitosan showed the diameter of 78.8 nm. The zeta potential analysis of silver chitosan nanocomposite revealed that the prepared nanocomposite was negatively charged (−3.4 mV at neutral pH) [97]. Zeta potential and particle size of chitin nanocomposite was found to increase as the amount of chitin (wt%) were enhanced in the corresponding NCs [98]. The hydrodynamic diameter of magnetic nanoparticles coated by a series of carboxymethylated polysaccharides, such as dextran, cellulose, and pullulan was found to be 229, 719, and 330 nm, respectively. The ζ potential in all of these cases had slightly negative values, which increased with increasing magnetite content present in the composite [99].

3.1.6 UV-Vis

The formation of gold-carboxymethylcellulose (CMC) NCs was confirmed by UV-Vis spectroscopy by monitoring the appearance of Au surface plasmon band at 522 nm which was otherwise absent in case of both the precursors [100]. The absorption intensity of Au was found to gradually increase with increasing CMC concentrations in NCs. Moreover, absorption peak width became narrower suggesting uniform size distribution of synthesized Au nanoparticles in presence of carboxymethyl cellulose [101]. The absorption peak of chitosan ZnO nanocomposite was observed at 360 nm, which was having lesser intensity than macrocrystalline ZnO absorption found at 372 nm. This was attributed to the quantum size effect of chitosan-ZnO nanoparticles [102].

3.1.7 SAXS

SAXS was used to determine pair distances distribution function (PDDF) profile related to the shape and the conformational arrangement of macromolecules. The PDDF profile of the CNS before sonication (CNS-BS) and after sonication (CNS-AS) showed different geometries. Interestingly, the non-sonicated CNS exhibited an elongated curve due to agglomerated particles, whereas the CNS-AS exhibited a Gaussian curve corresponding to spherical particles [103].

3.1.8 TGA

TGA was used to characterize the thermal behaviour of raw fibers, mercerized cellulose and CNS and the values were found to be dependent on components present in the plant cell wall. In fact, the cellulose decomposition was reported in the range of 315–400 °C whereas for hemicellulose and lignin it was found to be between 200–315 °C and 160–900 °C respectively [104]. For the raw fiber, mercerized cellulose and CNS, decomposition started at 218, 223 and 209 °C and the maximum degradation was observed at 346, 342 and 326 °C respectively. It was found that compared to the CNS, the thermal stability of raw fibers and mercerized cellulose was higher, possibly because of the presence of sulfate group at the surface of the CNS. The difference in thermal behaviour between mercerized cellulose and raw fibers was explained based on the presence of hemicellulose and lignin material in the raw fibers [105]. Further, it was found that TiO_2–cellulose nanocomposite was more stable than pure cellulose [64]. Xylan-rich hemicellulose (XH) and cellulose nanocomposite showed good thermal stability than freeze-dried XH powder because of the presence of crystal structure [82]. Further, hemicellulose Fe_3O_4 hydrogel NCs were also studied using TGA [106]. Similarly, bacterial cellulose and its NCs were also evaluated for their thermal stability using TGA studies and found that it resembles quite well with the thermal properties of the plant cellulose. Literature reports suggested that the presence of mineral phase has changed the thermal degradation profiles of the bimetallic–alginate nanocomposite samples when compared with the non-mineralized composite. This might have resulted from the presence of mineral phase which improved the thermal stability of the NCs by lowering the rate of alginate decomposition [107]. TGA thermogram of AgCl/chitin nanocomposite suggested a reduction in the area under endothermic peak in DSC and weight loss in TGA; thus confirming better stability for NCs [108]. The aminated lignins were studied in detail to evaluate their thermal stability [58]. TGA curves for freeze-dried raw starch samples suggested that the maximum loss of mass happened around 260–330 °C and for the same composite, it was noticed at ~300 °C [32].

3.1.9 TGMS

Thermogravimetric- mass spectroscopy (TGMS) has been used to understand the pyrolysis mechanism for hemicellulose. Under inert atmosphere and elevated temperature condition carbonaceous material undergo aromatization. The decomposition of the hemicellulose was observed between 200 and 580 °C and the mass spectroscopy (MS) result showed peaks for CO_2 (m/z = 44), CO (m/z = 28), CH_4 (m/z = 16), and H_2O (m/z = 18) [109].

3.1.10 DSC

DSC is one of the best analytical techniques to find the polymer crystallinity. The same technique can also be used to study oxidation reaction as well as other chemical reactions. DSC thermogram of chitosan-alginate (CS-AL) NCs prepared with glutaraldehyde showed one broad endothermic peak at ~112.1 °C for crystallization temperature whereas the glass transition temperature was found to be ~350 °C. The studies suggested that the crosslinker increases the thermal stability of the corresponding NCs [110]. In a different study, glass-rubber transition (T_g), melting point (T_m), and the degree of crystallinity for the synthetic poly (caprolactone) (PCL) was found to be around ~60, 69 °C, and 58%, respectively. For the developed chitin whiskers/PCL composites, it was found that both T_g and T_m was almost independent of the whiskers concentration [111]. Thermal analysis of developed lignin-epoxy resin suggested enthalpy values were directly proportional to the amine content present in the composites [58]. Further, it was also observed that the enthalpy values of lignin may vary depending on their biomass [18].

3.1.11 ICP-MS

Inductive coupled plasma mass spectroscopy is used to determine the concentration of metallic nanoparticles absorbed on the surface of the nanocomposite. In case of silver/cellulose NCs, prepared by different concentration of silver nitrate in presence of α-cellulose, carboxymethyl cellulose and amino-cellulose as stabilizers, showed that the quantity of silver nanocluster present on cellulose sample was directly proportional to the concentration of silver nitrate in the precursor solution [112].

3.1.12 SEM

SEM was used to understand the surface morphology of raw fibers before and after each treatment for cellulose extraction. The morphology of the fibers changed due to the purification process. In general, the raw fibers are constituted by bundles of cellulose microfibrils and are covered by different layers. The structural morphology of microfibrils cannot be visualized, as they are still buried inside lignin and hemicellulose [113]. For mercerized cellulose, the surface was found to be more clean, smooth and the available microfibrils can be clearly seen. Thus, SEM characterization clearly distinct between raw fibers and the mercerized cellulose. Similarly, CNSs obtained after the acid hydrolysis showed a narrow size distribution of particle size in the range of 46 ± 17 nm. SEM image revealed the three-dimensional structure of hemicellulose/chitosan nanocomposite having continuous cell pore structure [106]. Similarly, the SEM studies on BC showed structure composed of a random network of cellulose nanofibers. It was reported that the outer surface consists of dense layers covering an internal microstructure in the shape of honeycombs whereas the inner region is composed of fibers forming large pores [114]. SEM images of the alginate-zinc oxide NCs appeared to be rough which may have resulted from the interaction between the nanoparticles and the alginate matrix. The presence of ZnO-NPs in the alginate matrix was observed as bright spots with a moderate degree of agglomeration. The presence of elemental Zn was also confirmed by EDS analysis. [115]. The diameter of chitin nanowhisker determined by SEM was found to be much higher compared to the actual whiskers diameter. This resulted from a charge concentration effect due to the emergence of chitin whiskers [116]. Drug-loaded pullulan films appeared to have a random distribution of drug particles with no sign of complete phase separation, thus indicating a nanocomposite structure in the films [117]. Carbon microparticles derived from lignin were analyzed using SEM and the data were compared with untreated lignin, thermo-stabilised lignin and carbonized lignin. The average particle size for untreated, thermostabilised and carbonized lignin was found to be 9.1, 7.0 and 5.3 μm respectively [87]. SEM analysis of dilute acid pretreated corn stover revealed the formation of semispherical and spherical structures on its surface which led to the hypothesis that the droplet formations evolved from the lignocellulosic matrix of corn stover during pre-acid treatment [118]. The microstructures of maize, mango, and banana starch granules were observed by SEM imaging. Mango starch granules were found to be spherical and domeshaped with some non-uniform growth and spoting having 5–10 μm particles size, which clearly indicated that the starch particles may have collapsed during drying. In case of maize and starch granules, regular spherical shaped particles were observed. Banana starch granules were elongated and lenticular in shape, and the average longitudinal dimension was found to be 40 μm with a radius around 20 μm [34]. Citric acid modified starch nanoparticles (CASN) and native starch nanoparticles (SN) was found to have a size range between 50–100 nm and 50–300 nm respectively [77]. In one of the reports, the fractured structure of lamella indicated the homogenous dispersity of SN on soy protein matrix [30].

3.1.13 TEM

In TEM, CNS samples showed mainly the presence of aggregates created during the slow drying process [119]. Interestingly, white regions were observed inside the particles and have been related to the entanglement or local twist of the crystalline regions [120]. TEM studies were performed for alginate-ZnO nanocomposite to monitor size, morphology and dispersion of ZnO NPs on the nanocomposite. It was observed that the polymer conjugated ZnO NPs represented an irregular spherical shape with size variations between 20 and 100 nm which was almost comparable to the bare ZnO NPs [115]. Interfacial adhesion between lignin nanoparticles and PVA matrix was observed through TEM. Lignin nanoparticles showed no change in aggregation states even after incorporation of PVA matrix which suggested an interaction of lignin hydroxyl groups with PVA [121]. The TEM images of super-paramagnetic nanoparticles coated by pullalan revealed that the average size for all of the samples varied between 3.4 and 9 nm depending on the amount of magnetite present in the nanocomposite [99]. TEM image of chitosan-ZnO NCs revealed that the particles possess mostly rod-like structure with a size close to 100 nm [102]. HRTEM studies were also used for such characterization [86]. TEM images of pea starch nanoparticles suggested aggregated NPs having a length of ~60–150 nm and width of 15–30 nm [30]. Another report suggested the formation of well dispersed and spherical shaped starch colloids of sizes ~200 nm [32].

3.1.14 AFM

AFM is an imaging technique used for characterizing the surface topology of solid materials. AFM based morphological studies reflected the existence of absorbed nanoparticles on the surface of hemicellulose NCs [96]. Films formed by xylan hemicellulose and chitosan showed a very smooth surface with few nodules. Studies further suggested that cellulose nanofilm have higher roughness factor as compared to cellulose nanofibers [82, 122]. AFM images of GO-alginate biopolymer suggested an average thickness of the sheets was about 1.0 nm, which indicated the formation of single-layered exfoliated GO for the nanocomposite [123]. AFM has been directly applied for the surface analysis of lignin and its polymer composites. The AFM images of hydrothermally pretreated wheat straw and corn stover revealed that the parenchyma cell lining has an aggregate of cellulose fibers and the integrity of microfibers remained intact. The AFM images of pullalan nanocomposite of magnetite nanoparticles revealed that the magnetite nanoparticles had a strong tendency to form agglomerate in the dry state. The magnetite nanoparticles were found uniformly distributed throughout the samples; whereas the aggregation was observed at a higher concentration of magnetite [99]. The AFM images of cross-linked chitosan/chitin NCs showed well isolated nanometer-scale crystals of chitin. The diameter of chitin nanocrystals was found to vary between 13 and 20 nm [124]. In another literature report, the homogeneity of regenerated chitin (RChi) films was investigated by AFM [125].

3.2 Polypeptide-Based Biopolymers

Peptides are emerging as a new class of biomaterials due to their unique chemical, physical, and biological properties. The development of peptide-based biomaterials is driven by the convergence of protein engineering and macromolecular self-assembly. The next section focuses on different spectroscopic and microscopic techniques used for characterization of polypeptide-based polymers (Fig. 2).

3.2.1 CHNS Analysis

Elemental analysis of the collagen waste from goat skins revealed the % of C, N, H and S to be approximately 41.5, 14.7, 7.1 and 0.2%, respectively. Further, XPS and

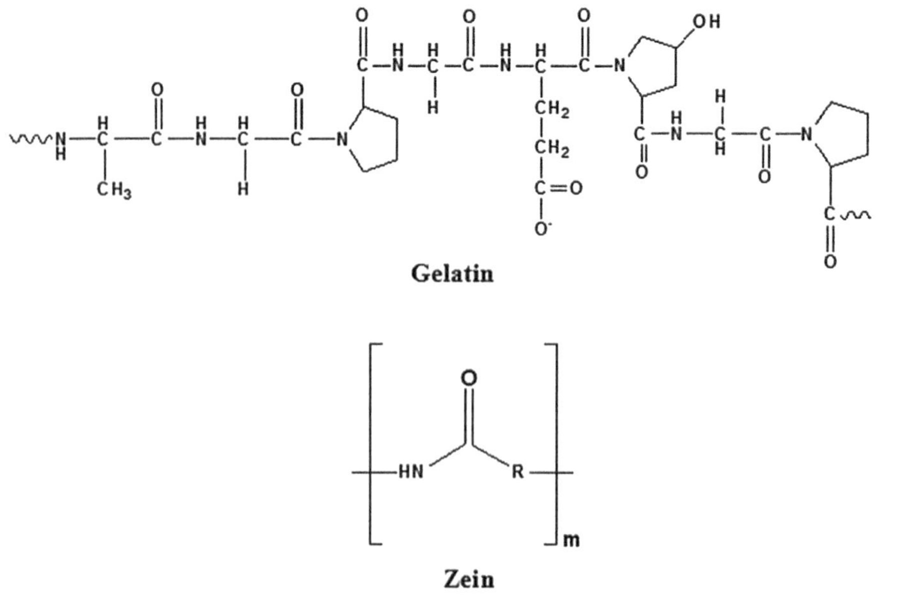

Fig. 2 Chemical structures of different polypeptide based biopolymers

CHNS analysis of carbon materials derived from collagen waste at elevated temperature abundantly showed the presence of carbon, oxygen and nitrogen in the range of 72–82, 6.2–15.4 and 2.9–13.6%, respectively [126]. The amount of zein adsorbed on the surface of montmorillonite (MMT) was determined by CHNS analysis; and it was found that biohybrids based on extracted phase (EXT) showed a lower amount of adsorbed protein than those prepared from precipitate [127].

3.2.2 FT-IR

Collagen is the principal structural constituent present in tissues; found most abundantly in the body. FT-IR spectra of type 1 collagen obtained from bone and skin tissues showed peaks of amide I at 1690, 1660, 1630 cm^{-1} corresponding to stretching vibration of C=O and amide III peak at 1235 cm^{-1} corresponding to stretching vibration of C–N and bending of N–H. Carbohydrate moieties present in collagen exhibited stretching vibration of C–O and C–O–C at 1100–1005 cm^{-1} [128, 129]. In another literature report, it was observed that the collagen-immobilized poly (3-hydroxybutyrate-co-3-hydroxyvalerate) (PHBV) and hydroxyapatite nanocomposite scaffolds showed IR peak at ~1722 cm^{-1} corresponding to vibrational bands of C=O present in PHBV whereas a peak at ~1039 cm^{-1} suggested the presence of PO_4^{3-} group exists in hydroxyapatite. In the same, characteristic peaks were observed at ~1640 and 1572 cm^{-1} for both amide I and amide II bonds of collagen, respectively [130]. The hybrid polymer/inorganic NCs of hyaluronan can be prepared either by in situ or ex situ method using both chemical and physical methods. The FTIR spectrum of HA/ZnO nanocomposite showed the peaks at ~3412, ~1616, ~1411, ~1149, ~1066, ~946, and ~661 cm^{-1}; thus confirming the presence of HA in the nanocomposite. The formation of ZnO nanoparticles was confirmed by distinct peak obtained at 441 cm^{-1}. The asymmetric sulfate stretching vibration band was observed at 1270 cm^{-1} and the peak at 1066 cm^{-1} was attributed to symmetric C–O vibration related to C–O–SO_3 or due to the weak C–OH stretching resulting from ZnO coordination. On the other hand, the peaks at ~1632 and 1433 cm^{-1} corresponded to stretching of (NH) C=O and –COO^- or –OH groups, respectively. These peaks became strong and shifted slightly in the presence of ZnO nanoparticles, indicating that interactions between the (NH)C=O, –COOH, and OH groups of HA and ZnO nanoparticles [47]. Major peaks in FT-IR spectra of gelatin isolated from scales and bones of fish were observed in amide region. Scales gelatin showed amide-I and amide-III peaks along with phosphate stretching peaks which indicated the presence of calcium salts in scales gelatin sample. Interestingly, bone gelatin samples showed N–H stretching vibration of amide A at ~3340 cm^{-1} along with the amide-I, amide-II, and amide-III peaks [131]. FT-IR of glutaraldehyde-crosslinked gelatin nanoparticles showed a peak at 2927 cm^{-1} for asymmetric stretching of –CH_2 groups present in gelatin [132]. The FT-IR spectra of casein-acrylate TiO_2 nanocomposite suggested the successful grafting of acrylate monomers on the casein matrix.

Similar spectroscopic studies were also performed for hollow casein nanospheres [46] or poly(n-butyl acrylate)–casein NCs [133, 134]. Zein-based biodegradable nanopesticide containing geraniol and citronellal as active ingredients were synthesized and characterized using FT-IR. The spectra of zein showed bands between 3100 and 2800 cm^{-1}, which corresponds to –C–H groups present in fatty acids and amino acids. The spectrum of the geraniol loaded zein nanoparticles showed characteristic zein amides I and II peaks along with minimal shifts which may have resulted from the interactions of zein nanoparticles with the corresponding essential oil [135]. Similar FT-IR studies were also performed for soy protein isolate-carbon nanotube (SPI/CNTs) composites [136].

3.2.3 Powder XRD

XRD patterns for silica nanoparticles doped in hydroxyapatite/collagen and hydroxyapatite/gelatin showed crystalline planes for hydroxyapatite which was independent of doping material. Further, the crystallite size of hydroxyapatite gelatin silica nanocomposite and hydroxyapatite collagen silica composite was found to be 10.73 and 4.19 nm respectively [137]. In the X-ray diffraction pattern of HA/ZnO nanocomposite, prominent peaks were observed and assigned to the hexagonal wurtzite structure of nanometer ZnO particles with a degree of crystallinity [47]. The main peak in the XRD diffractogram of zein films with and without oleic acid appeared at the 2θ value of 19° which was susceptible to the number of cellulose nanofibrils (NF) present in the composite [138]. XRD patterns of raw zein and zein nanoparticles suggested amorphous nature for both the material which was further reduced in presence of high flow rate of CO_2 [139]. XRD pattern of gelatin and its corresponding drug loaded counterpart was found to show peaks at $2\theta = 20$ and 22° respectively, suggesting increase in the crystalline nature upon drug entrapment [132]. The XRD pattern for casein calcium phosphate nanocomposite revealed amorphous nature for the material [140].

3.2.4 NMR

Conformational study of a collagen peptide by ^{1}H NMR spectroscopy revealed an interesting temperature dependant 1:1:1 pattern of sharp resonance bands and slightly broader peak at ~6.95 ppm, which was found to become even more broader with decreased temperature conditions. Further, ^{14}N–^{1}H spin-spin couplings were also observed because of quadrupolar relaxation which induces severe resonance broadening [141]. In another study, NMR spectra of collagen were investigated to compare between native and collagen present in biological tissues, such as bone, cartilage and skin. Characteristic signals from all collagen amino acids were obtained with a unique signal at 71.1 ppm, which can be assigned to the C_{γ} carbon of hydroxyproline. The ^{13}C MAS NMR spectra provided supports to resolve the fingerprint region of collagen whereas, ^{31}P cross polarization magic angle spinning

(CPMAS) studies on bone and bone implants have allowed to depict the biomineralization process [142]. ^{1}H-NMR spectra of methacrylic acid modified gelatin composite showed δ values between ~0.86 and 3.57 ppm which indicated characteristic peaks of methyl groups present in the amino acid residue of gelatin [143].

3.2.5 DLS and Zeta Potential

The particle size of zein nanoparticles was found to be ~300 nm which showed a constant decrease with increase in homogenization speed [144]. The zeta potential of zein nanoparticles coupled with gum arabic (GA) was found to be negative because of the presence of carboxylate groups in GA [145]. The DLS measurements revealed that the casein NCs showed the diameter of ~40–65 nm at pH 7.0 and the net charge was found to be negative [46].

3.2.6 UV-Vis and Fluorescence Studies

The presence of characteristic ZnO absorption band at 344 nm confirmed incorporation of ZnO into HA [47]. It was found that with increasing zein concentration, the plasmon resonance band of silver nanoparticles shifted from 458 to 428 nm and confirmed the formation of small size zein-silver nanoparticles [146]. The fluorescence of electrospun zein nanofibers conjugated with CdS QD (Quantum dot) showed emission at 561 nm. The emission intensity was found to be directly proportional to the CdS QD concentration. Further, the uniform fluorescence emission profile confirmed that the nanohybrid structure was stable in nature [147].

3.2.7 Circular Dichroism (CD) Spectroscopy

CD studies were carried out to confirm the presence of triple helix structure for type I collagen obtained from bovine calf skin [148]. Again, collagen obtained from streptococcal proteins showed unfolding of helix structure at 220 nm after denaturation [149]. CD spectroscopy studies were also carried out for gelatin composites [150].

3.2.8 TGA

The thermal stability of collagen was tested using TGA and results indicated that beyond denaturation temperature (T_d) collagen mostly converts into lower molecular weights elements. The specific viscosities started decreasing between 25 and 30 °C for skin collagen and between 30 and 35 °C for bone and muscle collagens [151]. TGA analysis showed that the in situ prepared silver/hyaluronan bionanocomposite increased the thermal and mechanical stability of resultant fibres [152].

Again, the TGA profile for the synthesized soy carbon nanotubes showed a weight loss of about 2% when the temperature was varied from 25 to 450 °C [136].

3.2.9 DSC

Collagen extracted from bovine tendon showed higher denaturation temperature for crosslinked collagen scaffolds due to highly stable triple helix conformation [153]. Collagen helix and their crosslinking nanocomposite showed changes in enthalpy and also in melting temperature [154]. Similar studies were also performed for casein and its developed composites [134].

3.2.10 SEM

SEM characterization of type I collagen extracted from equine tendon has been carried out to observe microscopic changes on the polymer surface [155]. Collagen obtained from bovine tendon suggested the presence of 50–150 μm pores on the nanocomposite surface [153]. FESEM image of HA/ZnO nanocomposite showed that ZnO nanoparticles of 3–8 nm size were present on the HA surface [47]. The majority of zein particles obtained via liquid-liquid dispersion was spherical in shape and possess particles size of less than 200 nm [144]. SEM images of prepared functionalized carbon nanotubes (FCNTs) showed homogeneous dispersion in the modified SPI adhesive, although individual CNTs could be observed as agglomerates [136]. Similar SEM studies were also carried out for SPI–MMT (montmorillonite) bio-nanocomposite [156] and casein NCs [46].

3.2.11 TEM

TEM studies were successfully carried out for HA-ZnO composites which showed the presence of 10–12 nm ZnO particles on the HA surface [47]. Similarly, gelatin nanoparticles of ~100 nm size were also observed [132]. Both zein and its corresponding silver-zein nanocomposite were characterized using TEM and size of the nanoparticles were found to be ~60 nm [146]. Similar studies were also carried out for casein and casein/calcium phosphate NCs [134, 140].

3.2.12 AFM

The AFM studies of collagen isolated from bovine vertebrae showed fibrils having a diameter in the range of 50–200 nm [157]. In another AFM study, the surface morphology of type 1 collagen showed thin film formation on nanocomposite surface [158]. Similar AFM studies were also performed for Zein–GA composites which suggested the formation of uniform spherical particles with average size of

143 nm and height of 43.8 nm [145]. AFM images revealed that casein nanospheres were spherical in shape with an average height of 21.0 ± 1.3 nm [134]. Further data on roughness factor for different casein NCs indicated that higher casein concentration allows homogeneous and smooth film formation [133]. The microstructure of the nanocomposite prepared from CNCs and soyabean oil was also characterized via AFM [159].

3.3 Synthetic Based Polymers

Eco-friendly synthetic based polymers have gained a lot of attention in current scenario due to their versatile nature and various applications in different fields. Different types of green NCs have been derived from modified synthetic polymers such as polylactic acid, poly(D,L-lactide-co-glycolide) (PLA, PLGA) and polyesters. This section draws the attention towards various techniques available for characterization of above mentioned synthetic biodegradable polymers (Fig. 3).

3.3.1 FT-IR

The FT-IR spectrum of PLGA-superparamagnetic iron oxide (SPIONs) nanocomposite with and without BSA showed prominent peak at ~2950–2850 cm^{-1}, assigned to C–H stretching of oleic acid on the SPIONs' surface whereas carboxylic acid present in PLGA showed sharp peaks at 1765–1750 cm^{-1} (C=O stretching), 1300–1090 cm^{-1} (C–C–O stretching), 1190–1085 cm^{-1} (C–O–C stretching) and 3100–2950 cm^{-1} (O–H stretching). However, BSA protein exhibited small signalling peaks at ~1650 cm^{-1} (C–O stretching of amide) and ~1540 cm^{-1} (N–H bending of amide) due to its lower concentration in the corresponding nanocomposite [160]. In another study, FT-IR spectra of PLA NCs showed peaks at ~2992, ~1373, and ~1454 cm^{-1} for stretching vibration, symmetric and asymmetric bending vibration of $-CH_3$ groups respectively. The nanocomposite of PLA poly (butylene adipate-co-terephthalate) nano-attapulgite (PLA/PBAT/AT) exhibited native PLA peak for stretching vibration of C=O group confirming the presence of

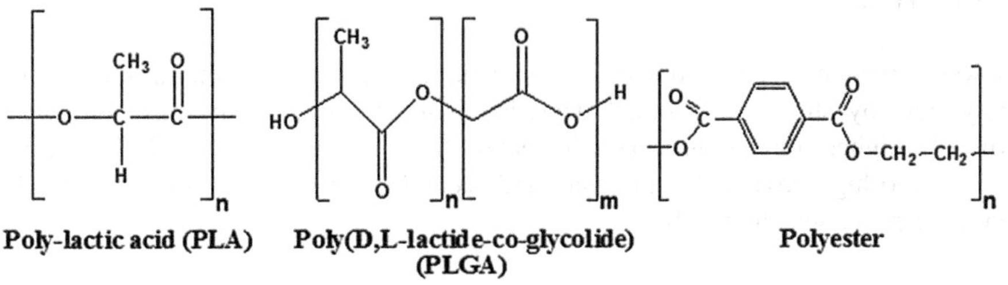

Poly-lactic acid (PLA) Poly(D,L-lactide-co-glycolide) (PLGA) Polyester

Fig. 3 Chemical structures of different synthetic biopolymers

PLA in the composite. Moreover, peaks at ~ 3057 and 740 cm^{-1} were attributed for stretching and bending vibrations of C–H present in benzene [25]. FT-IR spectra were observed for unsaturated polyester/styrene (UP) nanocomposite filled with nanodiamonds (NDs) containing carboxyl and methacrylate functional groups and exhibited a peak at 980 cm^{-1} which was assigned to the C–H out-of-plane bending in polyester molecules. Whereas, peak at 1730 cm^{-1} corresponds to C=O group which remained unchanged in UP/NDs nanocomposite [57].

3.3.2 Powder XRD

XRD pattern of pristine montmorillonite (Mt) and insulin-Mt-PLGA NCs showed a characteristic peak at a 2θ value of 6.4° (001) with a corresponding d spacing of 13.6 Å. However, no XRD pattern was observed for Mt due to its low concentration. Besides this, a hump corresponding to amorphous PLGA matrix also appeared at a 2θ value between 10 and 25°. XRD data depicted that Mt concentration could not influence the encapsulation efficiency of insulin in the composite [91]. XRD pattern observed for PLA and polylactic acid/poly-caprolactone (PLA/PCL) nanocomposite revealed their crystallinity indices to be 31.43, and 17.34% respectively. Further, two peaks of PLA at a 2θ value of ~ 16.4° and ~ 22.6° were observed from the same studies [161].

3.3.3 NMR

^{13}C solid-state NMR spectra of chitosan-PLA modified CNT NCs showed peaks at 65 and between 20 and 22 ppm for chitosan-CNTs and the composite respectively [162]. In another literature report, ^{1}H NMR spectra suggested chemical shift in poly (D,L-lactide-co-glycolide) (PLGA) copolymer with δ value between 1.46–1.68, 4.67–4.90 and 5.13–5.30 ppm assigned to $-CH_3$, $-CH_2$, and –CH functional groups respectively [56]. ^{1}H NMR of unsaturated polyester-styrene cured resin showed the integration of the broad bands at ~ 6.8 and 0.8–4.0 ppm which were assigned to the ring protons of styrene and esterified fumarate residues of aliphatic chain [163].

3.3.4 TGA

It was found that carbon dot (CD) conjugation improved the thermostability of polyesters by shifting the initial degradation temperature of the nanocomposite towards higher temperature. The increased thermostability of the NCs can be ascribed to high cross-linking density and secondary interactions imparted by CD with the polyester chains [164].

3.3.5 DSC

The positive influence of CNTs on the thermo-mechanical properties of unsaturated polyester NCs (UP) was studied using DSC. DSC thermogram provided the information of chain intercalation and thermal transition for the studied NCs. Interestingly, the NCs exhibited a split in the melting endotherm whereas the corresponding polyester showed a single peak, suggesting bond formation between CNT and UP for the former case [57].

3.3.6 SEM

SEM analysis of the nanocomposite prepared using iron oxide and lysine/BSA modified PLGA were studied and was used for protein antigen delivery and immune stimulation in dendritic cells [160]. The surface morphology of the fractured PLA nanocomposite (PLA/PBAT/AT) was observed by SEM studies [25]. In a separate study, the native PLA showed irregular microfibril structure compared to PLA nanocomposite (PLA/PCL) when both were dissolved in the phosphate saline buffer [161].

3.3.7 TEM

The TEM studies suggested that the mean particle size of Fe_3O_4 and Fe_3O_4-3-(trimethoxysilyl)-propyl methacrylate (MPTMS) was 8.8 ± 1.8 and 8.7 ± 1.8 nm respectively. Further, TEM images of Fe_3O_4-MPTMS-PLGA NCs showed 1–2 nm polymer coating thickness [165]. The surface morphology of PLA and PLA nanocomposite (PLA/PBAT/AT) were also studied using TEM [25].

The signature peaks for the described eco-friendly polymers have been summarized in Scheme 2.

4 Future Perspectives

For the sustainable development of the society, next-generation eco-friendly polymer composites are required to produce from renewable sources which possess superior physicochemical and biological properties. Globally BASF, Nature Works, Arkema, Novamont, and Plantic has been found to be the major player which governs the mass production of different polymers like- polylactic acid, starch-based polymer etc. contributing in several industries like—food, healthcare, and agriculture. Moreover, it is expected that this area will witness an increase in the CAGR by a significant amount at the end of 2021. Thus industries are changing their focus toward exploring the possibilities of biodegradable polymers. NCs offer

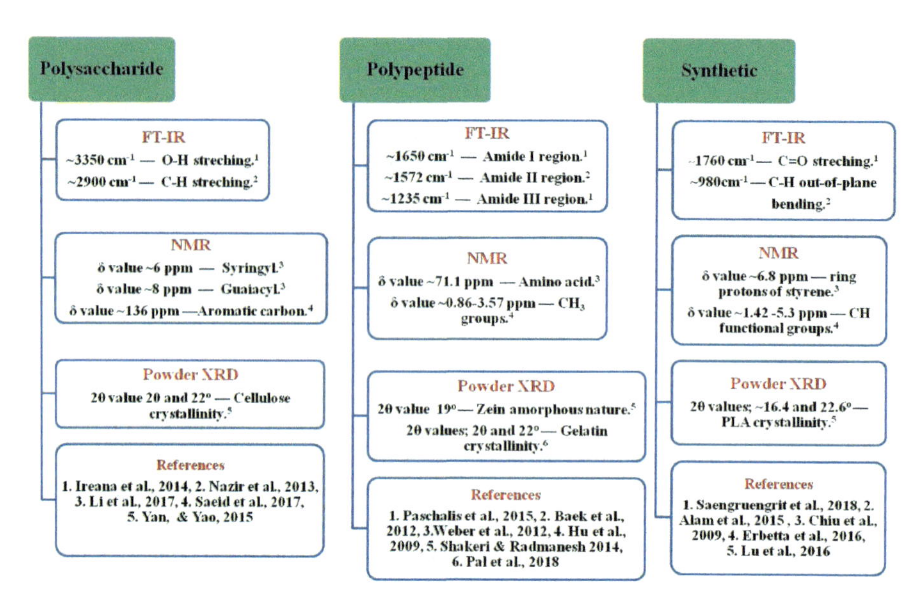

Scheme 2 Illustrative representation of the characteristic FT-IR, NMR and powder XRD peaks observed for different eco-friendly polymers

some great advantages over microcomposites because these possess improved strength and hardness. For the best utilization of the isolated eco-friendly polymers and their corresponding composites, proper characterization needs to be done.

Acknowledgements The authors would like to thank Director, CSIR-IHBT for his constant support and encouragement. AA acknowledges financial assistance in the form of project grant MLP-0201 from CSIR and GAP-0214 (EMR/2016/003027) from DST, Government of India. AKS, CS, SMSA acknowledge Academy of Scientific and Innovative Research (AcSIR) and CSIR-GATE and DBT for their respective JRF fellowship. AKS, CS and SMSA have contributed equally to this book chapter. The CSIR-IHBT communication number of this manuscript is 4254.

References

1. Camargo PHC, Satyanarayana KG, Wypych F (2009) Nanocomposites: synthesis, structure, properties and new application opportunities. Mater Res 12(1):1–39
2. Davidson S (2008) Sustainable bioenergy: genomics and biofuels development. Nat Edu 1(1):175–181
3. Iwata T (2015) Biodegradable and bio-based polymers: future prospects of eco-friendly plastics. Angew Chem Int Ed 54(11):3210–3215
4. Ray SS, Bousmina M (2005) Biodegradable polymers and their layered silicate nanocomposites: in greening the 21st century materials world. Prog Mater Sci 50(8):962–1079
5. Joshi G, Naithani S, Varshney VK, Bisht SS, Rana V, Gupta PK (2015) Synthesis and characterization of carboxymethyl cellulose from office waste paper: a greener approach towards waste management. Waste Manag 38:33–40

6. Singhsa P, Narain R, Manuspiya H (2018) Physical structure variations of bacterial cellulose produced by different Komagataeibacter xylinus strains and carbon sources in static and agitated conditions. Cellulose 25(3):1571–1581
7. Tuli M, Gurumayum S, Kaur S, Nagal S, Attri I (2015) Isolation and screening of cellulolytic fungi by baiting method from soils of Jalandhar. Res J Pharm Biol Chem Sci 6(2):375–380
8. Singla R, Soni S, Kulurkar PM, Kumari A, Mahesh S, Patial V, Yadav SK (2017) In situ functionalized nanobiocomposites dressings of bamboo cellulose nanocrystals and silver nanoparticles for accelerated wound healing. Carbohydr Polym 155:152–162
9. Phanthong P, Reubroycharoen P, Hao X, Xu G, Abudula A, Guan G (2018) Nanocellulose: extraction and application. CRC 1(1):32–43
10. Fatah IYA, Khalil HPS, Hossain MS, Aziz AA, Davoudpour Y, Dungani R, Bhat A (2014) Exploration of a chemo-mechanical technique for the isolation of nanofibrillated cellulosic fiber from oil palm empty fruit bunch as a reinforcing agent in composites materials. Polymers 6(10):2611–2624
11. Shao C, Wang M, Chang H, Xu F, Yang J (2017) A self-healing cellulose nanocrystal-poly (ethylene glycol) nanocomposite hydrogel via diels-alder click reaction. ACS Sustain Chem Eng 5(7):6167–6174
12. Li J, Wei X, Wang Q, Chen J, Chang G, Kong L, Liu Y (2012) Homogeneous isolation of nanocellulose from sugarcane bagasse by high pressure homogenization. Carbohydr Polym 90(4):1609–1613
13. Wang W, Mozuch MD, Sabo RC, Kersten P, Zhu JY, Jin Y (2015) Production of cellulose nanofibrils from bleached eucalyptus fibers by hyperthermostable endoglucanase treatment and subsequent microfluidization. Cellulose 22(1):351–361
14. Alemdar A, Sain M (2008) Isolation and characterization of nanofibers from agricultural residues—wheat straw and soy hulls. Bioresour Technol 99(6):1664–1671
15. Chen L, Wang Q, Hirth K, Baez C, Agarwal UP, Zhu JY (2015) Tailoring the yield and characteristics of wood cellulose nanocrystals (CNC) using concentrated acid hydrolysis. Cellulose 22(3):1753–1762
16. Lindblad Soderqvist M, Albertsson AC, Ranucci E, Laus M, Giani E (2005) Biodegradable polymers from renewable sources: rheological characterization of hemicellulose-based hydrogels. Biomacromolecules 6(2):684–690
17. Muchlisyam JS, Harahap U (2016) Hemicellulose: Isolation and its application in pharmacy. Handbook of sustainable polymers: Processing and applications, p 305–339
18. Dereca W, Nuruddin Md, Mahesh H, Alfred T-N, Shaik J (2015) Extraction and characterization of lignin from different biomass resources. JMRT 4(1):26–32
19. Hu L, Pan H, Zhou Y, Zhang M (2011) Methods to improve lignin's reactivity as a phenol substitute and as a replacement for other phenolic compounds: a brief review. BioResources 6(3):3515–3525
20. Kuhad R, Singh A (2007) Lignocellulose biotechnology: future prospects. I.K. International Publishing House, Delhi
21. Harmsen PFH, Huijgen WJJ, Bermúdez López LM, Bakker RRC (2010) Literature review of physical and chemical pretreatment processes for lignocellulosic biomass. Energy Research Centre of the Netherlands (ECN), ECN-E–10-013
22. Lee SH, Doherty TV, Linhardt RJ, Dordick JS (2009) Ionic liquid-mediated selective extraction of lignin from wood leading to enhanced enzymatic cellulose hydrolysis. Biotechnol Bioeng 102(5):1368–1376
23. Velez J, Thies MC (2016) Liquid lignin from the SLRP process: the effect of process conditions and black liquor properties. J Wood Chem Technol 36:27–41
24. Shi H, Fatehi P, Xiao H, Ni Y (2012) A process for isolating lignin of pre-hydrolysis liquor of Kraft pulping process based on surfactant and calcium oxide treatments. Biochem Eng J68:19–24

25. Zhou L, Santomauro F, Fan J, Macquarrie D, Clark J, Chuck CJ, Budarin V (2017) Fast microwave-assisted acidolysis: a new biorefinery approach for the zero-waste utilisation of lignocellulosic biomass to produce high quality lignin and fermentable saccharides. Faraday Discuss 202:351–370
26. Bychkov AL, Podgorbunskikh EM, Ryabchikova EI, Lomovsky OI (2018) The role of mechanical action in the process of the thermomechanical isolation of lignin. Cellulose 25:1–5
27. Reddy DK, Bhotmange MG (2013) Isolation of starch from rice (*Oryza sativa L.*) and its morphological study using scanning electron microscopy. IJAFST 4(9): 859–866
28. El-Sheikh Manal A (2017) New technique in starch nanoparticles synthesis. Carbohydr Polym 176:214–219
29. Chen Y, Liu C, Chang PR, Anderson DP, Huneault MA (2009) Pea starch-based composite films with pea hull fibers and pea hull fiber-derived nanowhiskers. Polym Eng Sci 49(2): 369–378
30. Zheng H, Ai F, Chang PR, Huang J, Dufresne A (2009) Structure and properties of starch nanocrystal-reinforced soy protein plastics. Polym Composite 30(4):474–480
31. Kim J-Y, Park D-J, Lim S-T (2008) Fragmentation of waxy rice starch granules by enzymatic hydrolysis. Cereal Chem 85(2):182–187
32. Liu D, Wu Q, Chen H, Chang RR (2009) Transitional properties of starch colloid with particle size reduction from micro to nanometer. J Colloid Interface Sci 339(1):117–124
33. LeCorre D, Vahanian E, Dufresne A, Bras J (2012) Enzymatic pretreatment for preparing starch nanocrystals. Biomacromolecules 13(1):132–137
34. Espinosa Solis V, Jane J, Bello Perez LA (2009) Physicochemical characteristics of starches from unripe fruits of mango and banana. Starke 61(5):291–299
35. Haque S, Md S, Sahni JK, Ali J, Baboota S (2014) Development and evaluation of brain targeted intranasal alginate nanoparticles for treatment of depression. J Psychiatr Res 48(1): 1–12
36. Xu X, Qu T, Fan L, Chen X, Gao M, Zhang J, Guo T (2016) Preparation of pH-and magnetism-responsive sodium alginate/Fe_3O_4@ HNTs nanocomposite beads for controlled release of granulysin. RSC Adv 6(113):111747–111753
37. Anderson TJ, Lamsal BP (2011) Zein extraction from corn, corn products, and coproducts and modifications for various applications: a review. Cereal Chem 88(2):159–173
38. Dickey LC, Parris N, Craig JC, Kurantz MJ (2001) Ethanolic extraction of zein from maize. Ind Crops Prod 13(1):67–76
39. Preece KE, Hooshyar N, Zuidam NJ (2017) Whole soybean protein extraction processes: a review. Innovative Food Sci Emerg Technol 43:163–172
40. Schmidt MM, Dornelles RCP, Mello RO, Kubota EH, Mazutti MA, Kempka AP, Demiate IM (2016) Collagen extraction process. Int Food Res J 23(3):913–922
41. Shyni K, Hema GS, Ninan G, Mathew S, Joshy CG, Lakshmanan PT (2014) Isolation and characterization of gelatin from the skins of skipjack tuna (*Katsuwonus pelamis*), dog shark (*Scoliodon sorrakowah*), and rohu (*Labeo rohita*). Food Hydrocoll 39:68–76
42. Anchana D, Kamatchi P, Leela K (2016) Extraction, characterization and application of gelatin from *Carcharhinus amblyrhyncho* and *Sphyraena barracuda*. IOSR-JBB 2(6):40–49
43. Du L, Keplová L, Khiari Z, Betti M (2014) Preparation and characterization of gelatin from collagen biomass obtained through a pH-shifting process of mechanically separated turkey meat. Poult Sci 93(4):989–1000
44. Fernando LAT, Poblete MRS, Ongkiko AGM, Diaz LJL (2016) Chitin extraction and synthesis of chitin-based polymer films from Philippine Blue Swimming Crab (*Portunus pelagicus*) shells. Procedia Chem 19:462–468
45. Kumari S, Rath PK (2014) Extraction and characterization of chitin and chitosan from (*Labeo rohit*) fish scales. Procedia Materials Science 6:482–489
46. Xiao-Zhou S, Hong-Ru W, Mian H (2014) Characterization of the casein/keratin self-assembly nanomicelles. J Nanomater 2014(183815):1–7

47. Namvar F, Azizi S, Rahman HS, Mohamad R, Rasedee A, Soltani M, Rahim RA (2016) Green synthesis, characterization, and anticancer activity of hyaluronan/zinc oxide nanocomposite. OncoTargets Ther 9:4549
48. Amagai I, Tashiro Y, Ogawa H (2009) Improvement of the extraction procedure for hyaluronan from fish eyeball and the molecular characterization. Fish Sci 75(3):805–810
49. Guhados G, Wan W, Hutter JL (2005) Measurement of the elastic modulus of single bacterial cellulose fibers using atomic force microscopy. Langmuir 21(14):6642–6646
50. Huang HC, Chen LC, Lin SB, Hsu CP, Chen HH (2010) In situ modification of bacterial cellulose network structure by adding interfering substances during fermentation. Bioresour Technol 101(15):6084–6091
51. Zeng X, Small DP, Wan W (2011) Statistical optimization of culture conditions for bacterial cellulose production by *Acetobacter xylinum* BPR 2001 from maple syrup. Carbohydr Polym 85(3):506–513
52. Sheoran SK, Dubey KK, Tiwari DP, Singh BP (2012) Directive production of pullulan by altering cheap source of carbons and nitrogen at 5l bioreactor level. ISRN Chemical Engineering 2012:1–5
53. Shehata AN, Darwish DA, Masoud HM (2016) Extraction, purification and characterization of endo-acting pullulanase Type I from white edible mushrooms. J Appl Pharm Sci 6(01): 147–152
54. Jamshidian M, Tehrany EA, Imran M, Jacquot M, Desobry S (2010) Poly-lactic acid: production, applications, nanocomposites, and release studies. Compr Rev Food Sci Food Saf 9(5):552–571
55. Danhier F, Ansorena E, Silva JM, Coco R, Le Breton A, Preat V (2012) PLGA-based nanoparticles: an overview of biomedical applications. J Control Release 161(2):505–522
56. Erbetta CDAC, Alves RJ, Resende JM, de Souza Freitas RF, de Sousa RG (2012) Synthesis and characterization of poly(D, L-lactide-co-glycolide) copolymer. J Biomater Nanobiotechnol 3(02):208
57. Beg MDH, Alam AM, Yunus RM, Mina MF (2015) Improvement of interaction between pre-dispersed multi-walled carbon nanotubes and unsaturated polyester resin. J Nanopart Res 17(1):53
58. Saeid N, Omid Z, Yousef Mojtaba, Saba A, Minoo N (2017) Catalyzed synthesis characterization of a novel lignin-based curing agent for the curing of high-performance epoxy resin. Polymers 9(7):266
59. Le DM, Nielsen AD, Sørensen HR, Meyer AS (2017) Characterisation of authentic lignin biorefinery samples by Fourier transform infrared spectroscopy and determination of the chemical formula for lignin. BioEnerg Res 10(4):1025–1035
60. Abraham E, Kam D, Nevo Y, Slattegard R, Rivkin A, Lapidot S, Shoseyov O (2016) Highly modified cellulose nanocrystals and formation of epoxy-nanocrystalline cellulose (CNC) nanocomposites. ACS Appl Mater Interfaces 8(41):28086–28095
61. Luduena LN, Vecchio A, Stefani PM, Alvarez VA (2013) Extraction of cellulose nanowhiskers from natural fibers and agricultural byproducts. Fibers Polym 14(7):1118–1127
62. Haafiz MM, Eichhorn SJ, Hassan A, Jawaid M (2013) Isolation and characterization of microcrystalline cellulose from oil palm biomass residue. Carbohydr Polym 93(2):628–634
63. Spiridonov VV, Panova IG, Afanasov MI, Zezin SB, Sybachin AV, Yaroslavov AA (2018) Water-Soluble magnetic nanocomposites based on carboxymethyl cellulose and iron (III) oxide. Polym Sci Ser B 60(1):116–121
64. Cardoso GV, Mello LRDS, Zanatta P, Cava S, Raubach CW, Moreira ML (2018) Physico-chemical description of titanium dioxide–cellulose nanocomposite formation by microwave radiation with high thermal stability. Cellulose 25(4):2331–2341
65. Ayoub A, Venditti RA, Pawlak JJ, Salam A, Hubbe MA (2013) Novel hemicellulose–chitosan biosorbent for water desalination and heavy metal removal. ACS Sustain Chem Eng 1(9):1102–1109

66. Badshah M, Ullah H, Khan AR, Khan S, Park JK, Khan T (2018) Surface modification and evaluation of bacterial cellulose for drug delivery. Int J Biol Macromol 113:526–533
67. Foong CY, Hamzah MSA, Razak SIA, Saidin S, Nayan NHM (2018) Influence of poly (lactic acid) layer on the physical and antibacterial properties of dry bacterial cellulose sheet for potential acute wound healing materials. Fibers Polym 19(2):263–271
68. Vijayalakshmi K, Gomathi T, Sudha PN (2014) Preparation and characterization of nanochitosan/sodium alginate/microcrystalline cellulose beads. Der Pharmacia Lettre 6(4): 65–77
69. Krishnaveni B, Ragunathan R (2015) Extraction and Characterization of chitin and chitosan from *F. solani* CBNR BKRR, synthesis of their bionanocomposites and study of their productive application. J Pharm Sci Res 7(4):197–205
70. Shanthi P, Kothai S (2015) Synthesis and characterization of chitosan with incorporated herb —a novel bionano composite. Int J Chemtech Res 8(8):208–214
71. Silva NH, Vilela C, Almeida A, Marrucho IM, Freire CS (2018) Pullulan-based nanocomposite films for functional food packaging: exploiting lysozyme nanofibers as antibacterial and antioxidant reinforcing additives. Food Hydrocoll 77:921–930
72. Mitić Ž, Cakić M, Nikolić GM, Nikolić R, Nikolić GS, Pavlović R, Santaniello E (2011) Synthesis, physicochemical and spectroscopic characterization of copper (II)-polysaccharide pullulan complexes by UV–vis, ATR-FTIR, and EPR. Carbohydr Res 346(3):434–441
73. Xu F, Yu J, Tesso T, Dowell F, Wang D (2013) Qualitative and quantitative analysis of lignocellulosic biomass using infrared techniques: a mini-review. Appl Energy 104:801–809
74. Sills DL, Gossett JM (2012) Using FTIR to predict saccharification from enzymatic hydrolysis of alkali-pretreated biomasses. Biotechnol Bioeng 109(2):353–362
75. Tian X, Rehmann L, Xu CC, Fang Z (2016) Pretreatment of eastern white pine (*Pinus strobes* L.) for enzymatic hydrolysis and ethanol production by organic electrolyte solutions. ACS Sustain Chem Eng 4(5):2822–2829
76. Kubo S, Kadla JF (2005) Hydrogen bonding in lignin: a Fourier transform infrared model compound study. Biomacromolecules 6(5):2815–2821
77. Ma X, Jian R, Chang PR, Yu J (2008) Fabrication and characterization of citric acid-modified starch nanoparticles/plasticized-starch composites. Biomacromolecules 9(11): 3314–3320
78. Poletto M, Pistor V, Zattera AJ (2013) Structural characteristics and thermal properties of native cellulose. In: Cellulose-fundamental aspects. InTech, p 45–68
79. El Oudiani A, Chaabouni Y, Msahli S, Sakli F (2011) Crystal transition from cellulose I to cellulose II in NaOH treated *Agave americana* L. fibre. Carbohydr Polym 86(3):1221–1229
80. Huang HD, Liu CY, Zhou D, Jiang X, Zhong GJ, Yan DX, Li ZM (2015) Cellulose composite aerogel for highly efficient electromagnetic interference shielding. J Mater Chem A 3(9):4983–4991
81. Dietrich K, Hernandez-Mejia C, Verschuren P, Rothenberg G, Shiju NR (2017) One-pot selective conversion of hemicellulose to xylitol. Org Process Res Dev 21(2):165–170
82. Peng XW, Ren JL, Zhong LX, Sun RC (2011) Nanocomposite films based on xylan-rich hemicelluloses and cellulose nanofibers with enhanced mechanical properties. Biomacromolecules 12(9):3321–3329
83. Zhong C, Zhang GC, Liu M, Zheng XT, Han PP, Jia SR (2013) Metabolic flux analysis of *Gluconacetobacter xylinus* for bacterial cellulose production. Appl Microbiol Biotechnol 97 (14):6189–6199
84. Ramírez JAÁ, Hoyos CG, Arroyo S, Cerrutti P, Foresti ML (2016) Acetylation of bacterial cellulose catalyzed by citric acid: use of reaction conditions for tailoring the esterification extent. Carbohydr Polym 153:686–695
85. Chen K, Ling Y, Cao C, Li X, Chen X, Wang X (2016) Chitosan derivatives/reduced graphene oxide/alginate beads for small-molecule drug delivery. Mater Sci Eng C 69:1222–1228

86. Ramesan MT, Siji C, Kalaprasad G, Bahuleyan BK, Al-Maghrabi MA (2018) Effect of silver doped zinc oxide as nanofiller for the development of biopolymer nanocomposites from chitin and cashew gum. J Polym Environ 26(7):2983–2991
87. Köhnke J, Fürst C, Unterweger C, Rennhofer H, Lichtenegger HC, Keckes J, Emsenhuber G, Liebner F, Gindl-Altmutter W (2016) Carbon microparticles from organosolv lignin as filler for conducting poly(lactic acid). Polymers 8(6):205
88. Shivananda CS, Rao BL, Madhukumar R, Sarojini BK, Somashekhar R, Asha, S, Sangappa Y (2016) Silk fibroin/pullulan blend films: preparation and characterization. In: AIP conference proceedings, vol 1731(1). AIP Publishing, p 070013
89. Trovatti E, Fernandes SC, Rubatat L, Freire CS, Silvestre AJ, Neto CP (2012) Sustainable nanocomposite films based on bacterial cellulose and pullulan. Cellulose 19(3):729–737
90. Tzhayik O, Pulidindi IN, Gedanken A (2014) Forming nanospherical cellulose containers. Ind Eng Chem Res 53(36):13871–13880
91. Lal S, Perwez A, Rizvi MA, Datta M (2017) Design and development of a biocompatible montmorillonite PLGA nanocomposites to evaluate in vitro oral delivery of insulin. Appl Clay Sci 147:69–79
92. Goundalkar MJ, Corbett DB, Bujanovic BM (2014) Comparative analysis of milled wood lignins (MWLs) isolated from sugar maple (SM) and hot water extracted sugar maple (ESM). Energies 7(3):1363–1375
93. Pu Y, Cao S, Ragauskas AJ (2011) Application of quantitative 31P NMR in biomass lignin and biofuel precursors characterization. Energy Environ Sci 4(9):3154–3166
94. Mascheroni E, Rampazzo R, Ortenzi MA, Piva G, Bonetti S, Piergiovanni L (2016) Comparison of cellulose nanocrystals obtained by sulfuric acid hydrolysis and ammonium persulfate, to be used as coating on flexible food-packaging materials. Cellulose 23(1): 779–793
95. Astruc J, Nagalakshmaiah M, Laroche G, Grandbois M, Elkoun S, Robert M (2017) Isolation of cellulose-II nanospheres from flax stems and their physical and morphological properties. Carbohydr Polym 178:352–359
96. Kishani S, Vilaplana F, Xu W, Xu C, Wagberg (2018) Solubility of softwood hemicelluloses. Biomacromolecules 19(4):1245–1255
97. Madhusudhan KN, Meghana PB, Vinaya Rani G, Moorthy SM, Mary-Josepha AV (2017) Extraction and characterization of chitin and chitosan from *Aspergillus niger*, synthesis of silver-chitosan nanocomposites and evaluation of their antimicrobial potential. Journal of Advances in Biotechnology 6(3):939–945
98. Mushi NE, Utsel S, Berglund LA (2014) Nanostructured biocomposite films of high toughness based on native chitin nanofibers and chitosan. Front Chem 2(99):1–11
99. Coseri S, Spatareanu A, Sacarescu L, Socoliuc V, Sorin Stratulat I, Harabagiu V (2016) Pullulan: a versatile coating agent for superparamagnetic iron oxide nanoparticles. J Appl Polym Sci 133(5):42926(1–9)
100. Li G, Sun Y, Liu H (2018) Gold-carboxymethyl cellulose nanocomposites greenly synthesized for fluorescent sensitive detection of Hg(II). J Cluster Sci 29(1):177–184
101. Li J, Zhang J, Zha S, Gao Q, Li J, Zhang Q (2017) Fast curing biobased phenolic resins via lignin demethylated under mild reaction condition. Polymers 9(9):428
102. Anandhavelu S, Thambidurai S (2012) Preparation of chitosan-ZnO nanocomposite from chitin polymer. Adv Mat Res 584:234–238
103. Putnam CD, Hamme M, Hura GL, Tainer JA (2007) X-ray solution scattering (SAXS) combined with crystallography and computation: defining accurate macromolecular structures, conformations and assemblies in solution. Q Rev Biophys 40(3):191–285
104. Yang H, Yan R, Chen H, Lee DH, Zheng C (2007) Characteristics of hemicellulose, cellulose and lignin pyrolysis. Fuel 86(12–13):1781–1788
105. Nagalakshmaiah M, Mortha G, Dufresne A (2016) Structural investigation of cellulose nanocrystals extracted from chili leftover and their reinforcement in cariflex-IR rubber latex. Carbohydr Polym 136:945–954

106. Zhao W, Odelius K, Edlund U, Zhao C, Albertsson AC (2015) In situ synthesis of magnetic field-responsive hemicellulose hydrogels for drug delivery. Biomacromolecules 16(8):2522–2528
107. Malagurski I, Levic S, Mitric M, Pavlovic V, Dimitrijevic-Brankovic S (2018) Bimetallic alginate nanocomposites: new antimicrobial biomaterials for biomedical application. Mater Lett 212:32–36
108. Praveen P, Rao V (2014) Synthesis and thermal studies of chitin/AgCl nanocomposite. Procedia Materials Science 5(2014):1155–1159
109. Deng J, Xiong T, Wang H, Zheng A, Wang Y (2016) Effects of cellulose, hemicellulose, and lignin on the structure and morphology of porous carbons. ACS Sustain Chem Eng 4(7): 3750–3756
110. Gokila S, Gomathi T, Sudha PN, Anil S (2017) Removal of the heavy metal ion chromiuim (VI) using chitosan and alginate nanocomposites. Int J Biol Macromol 104:1459–1468
111. Morin A, Dufresne A (2002) Nanocomposites of chitin whiskers from Riftia tubes and poly (caprolactone). Macromolecules 35(6):2190–2199
112. Alahmadi NS, Betts JW, Heinze T, Kelly SM, Koschella A, Wadhawan JD (2018) Synthesis and antimicrobial effects of highly dispersed, cellulose-stabilized silver/cellulose nanocomposites. RSC Adv 8(7):3646–3656
113. Mathew L, Joshy MK, Joseph R (2011) Isora fibre: a natural reinforcement for the development of high performance engineering materials. In: Cellulose fibers: bio-and nano-polymer composites. Springer, Berlin, Heidelberg, p 291–324
114. Uraki Y, Nemoto J, Otsuka H, Tamai Y, Sugiyama J, Kishimoto T, Shimomura M (2007) Honeycomb-like architecture produced by living bacteria, *Gluconacetobacter xylinus*. Carbohydr Polym 69(1):1–6
115. Motshekga SC, Ray SS, Maity A (2018) Synthesis and characterization of alginate beads encapsulated zinc oxide nanoparticles for bacteria disinfection in water. J Colloid Interface Sci 512:686–692
116. PM V, Thomas S (2011) Preparation and characterization of chitin nanowhiskers and their polymer nanocomposites. Int J Polym Technol 3(1):35–44
117. Krull SM, Ma Z, Li M, Davé RN, Bilgili E (2016) Preparation and characterization of fast dissolving pullulan films containing BCS class II drug nanoparticles for bioavailability enhancement. Drug Dev Ind Pharm 42(7):1073–1085
118. Selig MJ, Viamajala S, Decker SR, Tucker MP, Himmel ME, Vinzant TB (2007) Deposition of lignin droplets produced during dilute acid pretreatment of maize stems retards enzymatic hydrolysis of cellulose. Biotechnol Prog 23(6):1333–1339
119. Yan CF, Yu HY, Yao JM (2015) One-step extraction and functionalization of cellulose nanospheres from lyocell fibers with cellulose II crystal structure. Cellulose 22(6):3773–3788
120. Parambath Kanoth B, Claudino M, Johansson M, Berglund LA, Zhou Q (2015) Biocomposites from natural rubber: synergistic effects of functionalized cellulose nanocrystals as both reinforcing and cross-linking agents via free-radical thiol–ene chemistry. ACS Appl Mater Interfaces 7(30):16303–16310
121. Tian D, Hu J, Bao J, Chandra RP, Saddler JN, Canhui Lu (2017) Lignin valorization: lignin nanoparticles as high-value bio-additive for multifunctional nanocomposites. Biotechnol Biofuels 10(1):192
122. Geng L, Peng X, Zhan C, Naderi A, Sharma PR, Mao Y, Hsiao BS (2017) Structure characterization of cellulose nanofiber hydrogel as functions of concentration and ionic strength. Cellulose 24(12):5417–5429
123. Vilcinskas K, Zlopasa J, Jansen K, Mulder FM, Picken SJ, Koper GJ (2016) Water sorption and diffusion in (reduced) graphene oxide-alginate biopolymer nanocomposites. Macromol Mater Eng 301(9):1049–1063
124. Mathew AP, Laborie MPG, Oksman K (2009) Cross-linked chitosan/chitin crystal nanocomposites with improved permeation selectivity and pH stability. Biomacromolecules 10(6): 1627–1632

125. Kittle JD, Wang C, Qian C, Zhang Y, Zhang M, Roman M, Esker AR (2012) Ultrathin chitin films for nanocomposites and biosensors. Biomacromolecules 13(3):714–718
126. Ashokkumar M, Narayanan NT, Reddy ALM, Gupta BK, Chandrasekaran B, Talapatra S, Thanikaivelan P (2012) Transforming collagen wastes into doped nanocarbons for sustainable energy applications. Green Chem 14(6):1689–1695
127. Alcântara AC, Darder M, Aranda P, Ruiz-Hitzky E (2016) Effective intercalation of zein into Na-montmorillonite: role of the protein components and use of the developed biointerfaces. Beilstein J Nanotechnol 7:1772
128. Belbachir K, Noreen R, Gouspillou G, Petibois C (2009) Collagen types analysis and differentiation by FTIR spectroscopy. Anal Bioanal Chem 395(3):829–837
129. Paschalis EP, Gamsjaeger S, Tatakis DN, Hassler N, Robins SP, Klaushofer K (2015) Fourier transform infrared spectroscopic characterization of mineralizing type I collagen enzymatic trivalent cross-links. Calcif Tissue Int 96(1):18–29
130. Baek JY, Xing ZC, Kwak G, Yoon KB, Park SY, Park LS, Kang IK (2012) Fabrication and characterization of collagen-immobilized porous PHBV/HA nanocomposite scaffolds for bone tissue engineering. J Nanomater 2012:1–11
131. Zakaria S, Bakar NHA (2015) Extraction and characterization of gelatin from Black tilapia (Oreochromis niloticus) scales and bones. In: International conference on advances in science, engineering, technology & natural resources (ICASETNR-15), Kota Kinabalu (Malaysia), p 77–80
132. Pal A, Bajpai J, Bajpai AK (2018) Easy fabrication and characterization of gelatin nanocarriers and in vitro investigation of swelling controlled release dynamics of paclitaxel. Polym Bull 75(10):4691–4711
133. Picchio ML, Ronco LI, Passeggi MC, Minari RJ, Gugliotta LM (2017) Poly (n-butyl acrylate)–casein nanocomposites as promising candidates for packaging films. J Polym Environ 26(6):2579–2587
134. Zhang F, Ma J, Xu Q, Zhou J, Simion D, Carmen G, Li Y (2016) Correction to hollow casein-based polymeric nanospheres for opaque coatings. ACS Appl Mater Interfaces 8(24):15856–15856
135. Oliveira JL, Campos EVR, Pereira ADES, Pasquoto T, Lima R, Grillo R, Fraceto LF (2018) Zein nanoparticles as eco-friendly carrier systems for botanical repellents aiming sustainable agriculture. J Agric Food Chem 66(6):1330–1340
136. Sadare OO, Daramola MO, Afolabi AS (2015) Preparation and characterization of nanocomposite soy-carbon nanotubes (SPI/CNTs) adhesive from soy protein isolate. In: Proceedings of the world congress on engineering, vol 2. London, U.K., p 1–3
137. Najafizadeha F, Sadjadia MAS, Fatemib SJ, Mobarakehc MK, Afshard RM (2016) A comparison between biocompatibilities of nanocomposites of silica doped in HA/collagen and those doped in HA/gelatin. OJC 32(3):1551–1557
138. Shakeri A, Radmanesh S (2014) Preparation of cellulose nanofibrils by high-pressure homogenizer and Zein composite films. Adv Mat Res 829:534–538
139. Li S, Zhao Y (2017) Preparation of zein nanoparticles by using solution-enhanced dispersion with supercritical CO_2 and elucidation with computational fluid dynamics. Int J Nanomedicine 12:3485
140. Ding GJ, Zhu YJ, Cheng GF, Ruan YJ, Qi C, Lu BQ, Wu J (2016) Porous microspheres of casein/amorphous calcium phosphate nanocomposite: room temperature synthesis and application in drug delivery. Curr Nanosci 12(1):70–78
141. Consonni R, Santomo L, Tenni R, Longhi R, Zetta L (1998) Conformational study of a collagen peptide by 1H NMR spectroscopy: observation of the 14N–1H spin-spin coupling of the Arg guanidinium moiety in the triple-helix structure. FEBS Lett 436(2):243–246
142. Weber F, Böhme J, Scheidt HA, Gründer W, Rammelt S, Hacker M, Huster D (2012) 31P and 13C solid-state NMR spectroscopy to study collagen synthesis and biomineralization in polymer-based bone implants. NMR Biomed 25(3):464–475

143. Hu X, Ma L, Wang C, Gao C (2009) Gelatin hydrogel prepared by photo-initiated polymerization and loaded with TGF-b1 for cartilage tissue engineering. Macromol Biosci 9(12):1194–1201
144. Zhong Q, Jin M (2009) Zein nanoparticles produced by liquid–liquid dispersion. Food Hydrocoll 23(8):2380–2387
145. Chen H, Zhong Q (2015) A novel method of preparing stable zein nanoparticle dispersions for encapsulation of peppermint oil. Food Hydrocoll 43:593–602
146. Ghazy OA, Nabih S, Abdel-Moneam YK, Senna MM (2015) Synthesis and characterization of silver/zein nanocomposites and their application. Polym Composite 38(S1):E9–E15
147. Dhandayuthapani B, Poulose AC, Nagaoka Y, Hasumura T, Yoshida Y, Maekawa T, Kumar DS (2012) Biomimetic smart nanocomposite: in vitro biological evaluation of zein electrospun fluorescent nanofiber encapsulated CdS quantum dots. Biofabrication 4(2): 025008
148. Consonni R, Zetta L, Longhi R, Toma L, Zanaboni G, Tenni R (2000) Conformational analysis and stability of collagen peptides by CD and by 1H-and 13C-NMR spectroscopies. Biopolymers 53(1):99–111
149. Xu Y, Keene DR, Bujnicki JM, Höök M, Lukomski S (2002) Streptococcal Scl1 and Scl2 proteins form collagen-like triple helices. J Biol Chem 277(30):27312–27318
150. Ahsan SM, Mohan Rao Ch (2016) Structural studies on aqueous gelatin solutions: implications in designing a thermo-responsive nanoparticulate formulation. Int J Biol Macromol 95:1126–1134
151. Muralidharan N, Shakila RJ, Sukumar D, Jeyasekaran G (2013) Skin, bone and muscle collagen extraction from the trash fish, leather jacket (*Odonus niger*) and their characterization. J Food Sci Technol 50(6):1106–1113
152. Abdel-Mohsen AM, Jancar J, Abdel-Rahman RM, Vojtek L, Hyršl P, Dušková M, Nejezchlebová H (2017) A novel in situ silver/hyaluronan bio-nanocomposite fabrics for wound and chronic ulcer dressing: in vitro and in vivo evaluations. Int J Pharm 520(1–2): 241–253
153. Zhang L, Ma D, Wang F, Zhang Q (2002) The modification of scaffold material in building artificial dermis. Artif Cells Blood Substit Biotechnol 30(4):319–332
154. Mizuno K, Hayashi T, Peyton DH, Bächinger HP (2004) Hydroxylation-induced stabilization of the collagen triple helix acetyl-(glycyl-4 (r)-hydroxyprolyl-4 (r)-hydroxyprolyl) 10-nh2 forms a highly stable triple helix. J Biol Chem 279(36):38072–38078
155. Tampieri A, Celotti G, Landi E, Sandri M, Roveri N, Falini G (2003) Biologically inspired synthesis of bone-like composite: self-assembled collagen fibers/hydroxyapatite nanocrystals. J Biomed Mater Res Part A 67(2):618–625
156. Kumar P, Sandeep KP, Alavi S, Truong VD, Gorga RE (2010) Preparation and characterization of bio-nanocomposite films based on soy protein isolate and montmorillonite using melt extrusion. J Food Eng 100(3):480–489
157. Hassenkam T, Fantner GE, Cutroni JA, Weaver JC, Morse DE, Hansma PK (2004) High-resolution AFM imaging of intact and fractured trabecular bone. Bone 35(1), 4–10
158. Zhang J, Senger B, Vautier D, Picart C, Schaaf P, Voegel JC, Lavalle P (2005) Natural polyelectrolyte films based on layer-by layer deposition of collagen and hyaluronic acid. Biomaterials 26(16):3353–3361
159. Song L, Wang Z, Lamm ME, Yuan L, Tang C (2017) Supramolecular polymer nanocomposites derived from plant oils and cellulose nanocrystals. Macromolecules 50 (19):7475–7483
160. Saengruengrit C, Ritprajak P, Wanichwecharungruang S, Sharma A, Salvan G, Zahn DR, Insin N (2018) The combined magnetic field and iron oxide-PLGA composite particles: effective protein antigen delivery and immune stimulation in dendritic cells. J Colloid Interface Sci 520:101–111
161. Lu Y, Chen Y-C, Zhang P-H (2016) Preparation and characterisation of polylactic acid (PLA)/polycaprolactone (PCL) composite microfbre membranes. Fibres Text East Eur 3(117):17–21

162. Carson L, Kelly-Brown C, Stewart M, Oki A, Regisford G, Luo Z, Bakhmutov VI (2009) Synthesis and characterization of chitosan–carbon nanotube composites. Mater Lett 63(6–7): 617–620
163. Chiu HT, Chen SC (2001) Curing reaction of unsaturated polyester resin modified by dicyclopentadiene. J Polym Res 8(3):183–190
164. Hazarika D, Karak N (2016) Biodegradable tough waterborne hyperbranched polyester/ carbon dot nanocomposite: approach towards an eco-friendly material. Green Chem 18(19): 5200–5211
165. Atila Dinçer C, Yildiz N, Karakeçili A, Aydoğan N, Çalimli A (2017) Synthesis and characterization of Fe_3O_4-MPTMS-PLGA nanocomposites for anticancer drug loading and release studies. Artif Cells Nanomed Biotechnol 45(7):1408–1414

Biocompatible and Biodegradable Chitosan Composites in Wound Healing Application: In Situ Novel Photo-Induced Skin Regeneration Approach

Amr A. Essawy, Hassan Hefni and A. M. El-Nggar

List of Abbreviations

Aw	Water absorbency
CCNC	Chitosan-based copper nanocomposite
CF	Chitosan-fibrin
CFU	Colony-forming unit
CMCS	Carboxymethyl chitosan
COX-2	Cyclooxygenase-2
CS-Ag	Chitosan-Ag
CSNPs	Chitosan nanoparticles
CZBs	Chitosan hydrogel/nano-ZnO nanocomposite bandages
DA	Deacetylation
DC	Decoloration
DM	Demineralization
DP	Degree of polymerization
DP	Deproteinization
ECM	Extracellular matrix
FDA	American Food and Drug Administration
FTIR	Fourier transform infrared
IL	Interleukin

A. A. Essawy (✉)
Chemistry Department, Faculty of Science,
Fayoum University, Fayoum 63514, Egypt
e-mail: aae01@fayoum.edu.eg

A. A. Essawy
Chemistry Department, College of Science, Jouf University,
P.O. Box 2014, Sakaka, Aljouf, Kingdom of Saudi Arabia

H. Hefni
Petrochemicals Department, Egyptian Petroleum Research Institute (EPRI),
Cairo, Egypt

A. M. El-Nggar
Chemistry Department, Faculty of Science, Al-Azhar University,
Cairo 11751, Egypt

Inamuddin et al. (eds.), *Sustainable Polymer Composites and Nanocomposites*,
https://doi.org/10.1007/978-3-030-05399-4_5

LPS	Lipopolysaccharide
Mel/CS MS	Melatonin-loaded chitosan-based microspheres
MW	Molecular weight
MyD88	Myeloid differentiation primary-response protein 88
NF-κB	Nuclear factor-kappa
NMR	Nuclear magnetic resonance
nT/COL-CS	Nano-titania/collagen-chitosan
PVA	Polyvinyl alcohol
Q-CF	Quercetin-loaded chitosan–fibrin
ROS	Reactive oxygen species
SSD	Silver sulfadiazine
TLR-4	Toll-like receptor 4
TMC	*N,N,N*-Trimethyl chitosan
TNF-α	Tumor necrosis factor-alpha

1 Introduction

Many types of polymers have diverse biomedical applications since they have unique properties and can be easily tailored. Polymeric materials include natural biopolymers, especially proteins, polysaccharides, polynucleotides, and natural rubber as well as synthetic polymers like polyethers, polyalkenes, polycarbonates, polyesters, and polyamides. The biocompatibility and biodegradability of the polymers is the main reason for their usage in biomedical applications. Biocompatibility term referred to the ability of materials to interact with a living system without revealing the undesirable degree of harm to the living system and biodegradability can be defined as the capability of materials in being degraded by biological activity. Biodegradable polymers are biocompatible; i.e. they do not rise an inflammatory response, and they degrade in vivo by hydrolysis and possible enzyme reactions, afterword they are removed from the body through regular metabolic processes [1]. Biocompatible and biodegradable materials should have important properties to consider when they are used for the medical purpose, in brief; do not trigger inflammatory or toxic response to the body, acceptable shelf life, the degradation period match to the healing or regeneration process, non-toxicity of degraded products, easily released from the body, permeability, appropriate molecular weight, solubility, suitable structure, surface charge and water absorption [2].

Wounds are an abnormal defect in the skin, vary to many types, and each category needs peculiar caring to achieve healing objective [3]. This fact was inspired researchers to develop suitable wound dressings, by considering the use of biocompatible natural and synthetic polymers and their composites as the bases of wound dressings. These polymers overcome the disadvantages of classical dressing materials [4].

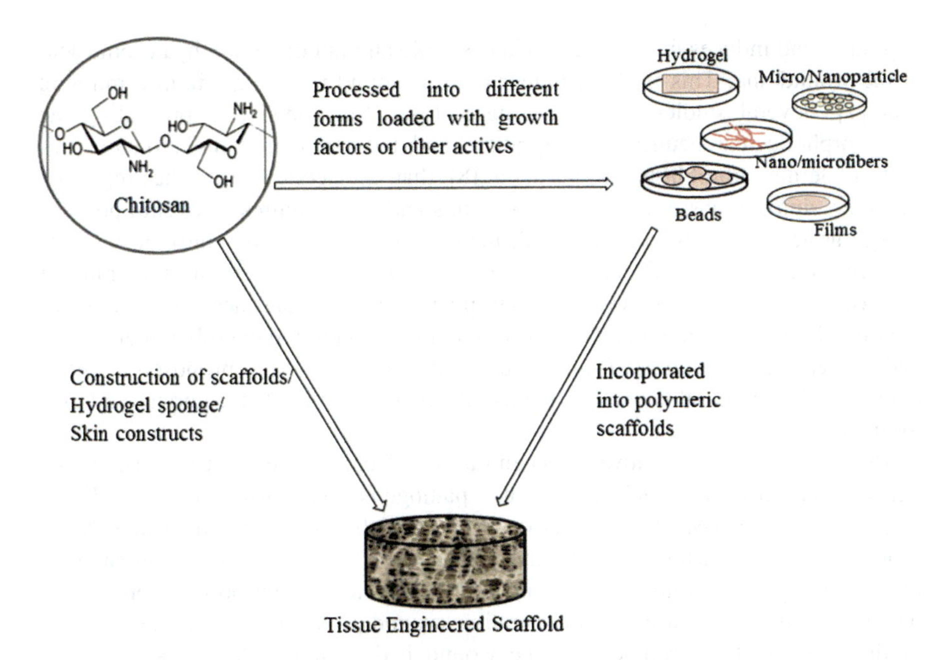

Fig. 1 Schematic representation of the possibilities of processing chitosan into different forms [5]. Copyright 2014. Reproduced with permission from Springer Nature

Recent reports are aiming to develop dressing alternatives to accelerate the wound repairing process. Naturally, derived materials like chitosan and its composites could be used to achieve this target. Chitosan is a renewable basic polysaccharide emerges as a viable structure that can be massively modified to diverse types of bio-based composites have a great potential in biomedical engineering. Chitosan has an interesting capability to be processed into gels, nanofibers, membranes, scaffolds, sponge-like forms, beads, microparticles and nanoparticles (Fig. 1) [5].

The different decorated forms of chitosan take into consideration the adaptation with the requirements of wound healing applications. The dried forms of chitosan have possibly to be hydrated via absorption of wounds exudates forming a hydrogel layer over the wounded area. Chitosan exhibits excellent biocompatibility, biodegradability, nontoxicity, anti-microbial and be used as hydrating agents [6]. Also, chitosan has an adhesive nature, antifungal, bactericidal feature, and oxygen permeability. Because of these immense activities, chitosan and its composites show good positive impacts on wound healing. Earlier reports indicated that chitosan-based dressings can accelerate the repair of diverse tissues and facilitates contraction of wounds. It stimulates cell proliferation. Hemostatic properties of chitosan aids to natural blood thrombosis as well as blocking of nerve endings that reduce pain. Moreover, gradual depolymerization of chitosan releases *N*-acetyl-glucosamine, which initiates fibroblast proliferation, deposits collagen regularly, and induces the synthesis of a desirable amount of natural hyaluronic acid at the wound site. This leads to higher rate of wound repairing within a minimal scar [7]. Several studies enumerate the chitosan functionality in promoting the polymorphonuclear neutrophils migration, and induction of granulation by promoting dermal fibroblasts proliferation [8] that accelerate wound healing. The polymer, chitosan possess basic amino groups and thus acquires an overall positive charge at acidic pH. In common with many cationic polymers, chitosan has pronounced antimicrobial effects due to destabilization of the external membrane of bacteria via interaction between chitosan positive charges and negative charges of the microbial cell membrane. This causes the disruption of the microbial membrane, and subsequent infiltration of intracellular constituents. Studies proposed an alteration in cell permeability via the interaction between the cell membrane and chitosan [9].

Reports conclude that lower concentrations of chitosan exhibit a considerable antibacterial activity against various pathogens like *Escherichia coli* or *Staphylococcus aureus* [7]. Therefore, it finds rich use in the medical field in promoting wound healing. Furthermore, chitosan dressings reveal minimal side effects and provide microbial disinfection. Chitosan is appropriate for wound healing because of its ease of administration, wound protection, water retention that produces a moist environment on the wound beds, without drawbacks of accumulating exudates. In addition, its permeable nature provides appropriate oxygen required for reparative processes.

2 Biodegradable Polymers

Biodegradable materials take a great attention of the researchers aiming to the better suited medical application which will stay inside the body until the healing process is finished and then gradually decomposed. Polymers degradation process is dependent on the polymer properties and its site in the body.

Both synthetic and natural polymers have been widely investigated as biodegradable materials that showed controllable chemical breakdown into non-toxic degradation products. Classification of biodegradable polymers can be explained as illustrated in Fig. 2 [10].

Biodegradation of polymeric biomaterials involves cleavage of hydrolytically or enzymatically sensitive bonds in the polymer leading to polymer decomposition. Polymeric biomaterials could be classified according to degradation mode into enzymatically- and hydrolytically degradable polymers. Enzymatic degradation takes place to almost natural polymers. Natural polymers are the firstly used biodegradable biomaterials in clinical applications. On the other hand, hydrolytically degradable polymers contain in their backbone hydrolytically labile functional groups such as amides, anhydrides, acetal, carbonates, urethanes, esters, orthoesters etc. that have hydrolytic susceptibility [2].

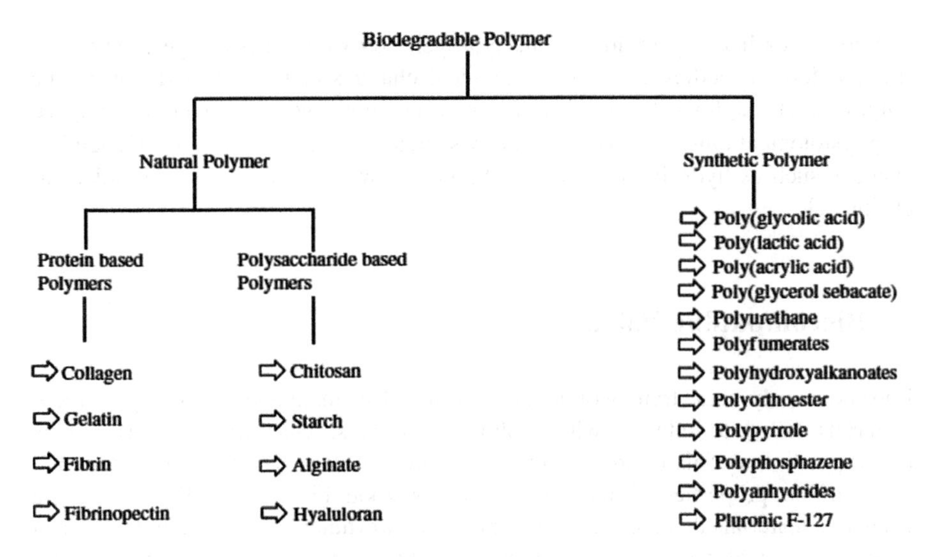

Fig. 2 Classification of biodegradable polymers [10]. Copyright 2016. Reprinted with permission from Springer Nature

The degradation of polymers can go into the different process depending on the polymer structure. The polymer that contains hydrolyzable units directly incorporated in the backbone of the polymer chain slowly degraded to smaller units and can be eliminated from the body. The polymer that has degradable side chains can be a breakdown in the body providing residual polymers with hydroxyl, carboxyl or other hydrating groups making the polymer water–soluble and excreted from the body. Another way depend on a hydrolysable cross-linking unit in the cross-linking polymer where a water-soluble polymer is exerted due to the cleavage of cross-links in polymeric networks (Fig. 3) [1].

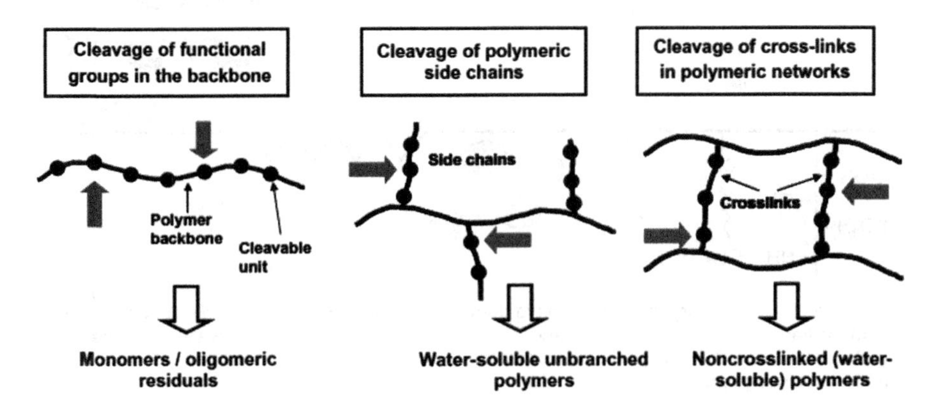

Fig. 3 Degradation routes for biodegradable polymers [1]. Copyright 2017. Reprinted with permission from Springer Nature

Substances like superoxide anion, hydrogen peroxide, macrophages, hypochlorite and foreign bodies can induce chemical changes causing degradation of the polymers. Throughout the process of wound healing, hydrolysis can be catalyzed by physiological ions, such as PO_4^{3-}, or by secreted enzymes. Processes of chemical changes such as hydrolysis can cause the breakdown of polymer macromolecular chains [1].

3 Biocompatible Polymers

Enormous polymeric frameworks can be utilized in medicine, but many of these materials failed to interact with biological systems, wherefore the researchers develop new materials in order to overcome this problem. The most important and precondition properties of any biomaterial is good biocompatibility. There are extensive materials that are, very desirable, but unfortunately rejected by the human tissues, so the biocompatible materials must have an appropriate response for a specific application [11]. Among these biomaterials, polymers are the most widely used for biomedical applications. The features of flexibility and biocompatibility of polymeric biomaterials enable its usage in the wide range of applications. Biocompatible polymers are necessary for repairing, retrieving tissues functionality, delivery of bioactive agents, and tissue engineering [1].

The Physical and chemical properties of the biomaterials; -hydrophilicity, hydrophobicity, ionic groups, the multi-component frame of structurally amorphous and crystalline morphology as well as and the topography, i.e. the surface roughness have a great influence in their compatibility (Fig. 4) [11].

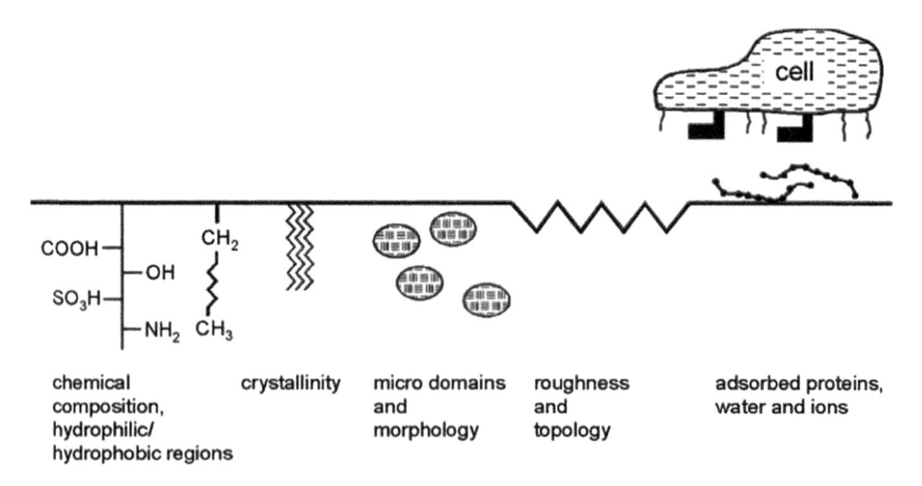

Fig. 4 Interactions of the biomaterial and the bio-system [11]. Copyright 2017. Reprinted with permission from Springer Nature

Biocompatibility examinations of materials perform in vitro and in vivo in order to examine the systematic and local effects of the material on the human body. Evaluation of biocompatibility of biomaterials in vivo includes the assessment of physiologic effects of the body on material and material on the body, that can be achieved by testing the overall biocompatibility of biomaterials; its potential toxicity, biodegradability, the reaction between the tissue and biomaterials and mutagenicity of degraded products, etc. [1]. Biocompatibility of biomaterials can be detected in vitro cell cultures to evaluate their toxicity by measuring the activity of cellular enzymes via colourimetric assays (e.g. MTT assay) or fluorescent measurement [12].

4 Wounds

Wounds are the damage or distraction in the normal anatomical structure and function, and it can be a simple cut in the superficial layer of skin or more deep to extend into subcutaneous tissue. This causes a damage to other structures, for example, tendons, muscles, vessels, nerves, and even bone. Wounds can be resulting from an accidental or surgical operation or they can be a secondary result of the specific disease. Wounds are a principal cause of morbidity that impaired quality of life and acquires a lot of healthcare resources [13]. Different classifications of wounds are based on diverse criteria. Elapsed time is a crucial factor in wound classification. Thus, from a clinical point of view, wounds can be viewed as acute and chronic depending on the consumed time for healing [3]. Acute wounds originate from superficial scratches to deep injuries and heal completely with minimal or no scar formation within a time frame of three weeks [14]. Chronic wounds are those that have failed to proceed through an efficient, convenient and timely reparative procedure to create anatomic and functional integrity of the injured site. Chronic wounds represent a silent epidemic that influences a substantial division of the world population. Patients of chronic wounds oftentimes suffer from "highly branded" diseases for example, certain types of ulcers, diabetes and overweight [15].

Healing wounds early, has a great benefit in reduced or no scar formation, whereas delayed treatment leads to severe hypertrophic scarring directly proportional to the wound closure delay time. Earlier wound healing is also accompanied by lower mortality and high patient compliance [16].

Both chronic and acute skin wounds are susceptible to infection because of sterile loss of the natural hindrance function of the skin, facilitating the development of microbial contamination, inside the wound environment. Microbial contamination is involved in the infection of wounds and hence the failure of those wounds to be healed. Therefore researchers concentrated on the investigation of wound healing materials that can be lead to upgrade in the wound management strategies [17].

The percentage of wound contraction could be measured by tracing the margins of wounds on a transparent paper by a fine tip permanent marker. By taking the traced boundaries, the wound area in square millimetres (mm^2) could be determined plan metrically. At a predetermined time interval, the wounded area of each animal on 0 days starting at 3 h post wounding. Wilson's formula shown below gives an estimate of percent wound contraction for a given wound healed area:

$$\%\,\text{wound contraction} = \frac{0\,\text{day wound area} - \text{unhealed wound}}{0\,\text{day wound area}} \times 100$$

4.1 *Insights to the Wound Healing Process*

Skin is the biggest body organ, working as a hindrance to harmful media, preventing pathogens from penetration into the body. Therefore the early repair of injured skin is a major objective for medical-care experts. Effective wound healing prompts to the rebuilding of tissue integrity and takes place through an organized multistage prepare including a multitude cell types. Skin wound healing is a complicated and dynamic natural process including coordinated interactions between different immunological and biological systems, the typical healing reaction starts quickly after tissue damage and takes several steps.

Some researchers consider that wound healing involves three phases: Inflammation, proliferation and tissue remodelling [18], whereas other researchers believe there are four stages in wound healing: Hemostasis, inflammation, proliferation and tissue remodelling. However, everyone agrees that these phases are connected, suggesting that the wound-healing process is a connected series (Fig. 5) [19]. Normal wound healing timeframe is presented in Fig. 6 [20].

4.1.1 Coagulation and Hemostasis Phase

Firstly hemostasis and a blood coagulation, start immediately after the injury to prevent bleeding when the blood segments spill into the site of injury, the platelets trigger vasoconstriction, to reduce blood loss and come into contact with exposed collagen and different components of the extracellular matrix. This contact triggers the platelets to release clotting factors as well as essential growth factors and cytokines, which provide wound healing cascade via activating the attraction of neutrophils and monocytes into the clot [13].

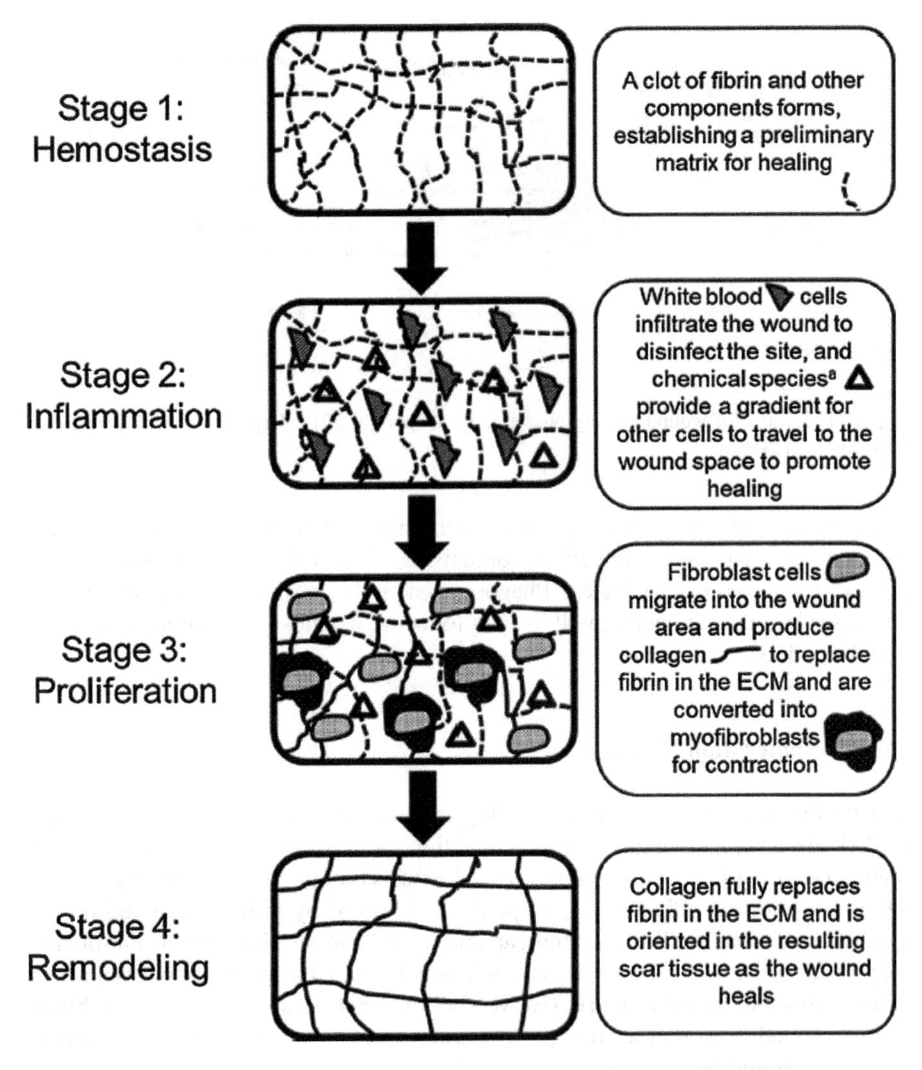

Fig. 5 Wound healing process [19]. Copyright 2016. Reproduced with permission from Springer Nature

4.1.2 Inflammatory Phase

Shortly after hemostasis, the inflammation stage sets in close to a skin injury; the primary inflammatory response is stimulated by chemotactic substances, which attract leukocytes (neutrophils) and monocytes, that its main function is to prevent infection. Neutrophils are white blood cells that migrate across endothelia from local blood vessels and start the debridement via phagocytosis to destroy and expel foreign materials, bacteria and damaged tissue in the wounded area. Monocytes are

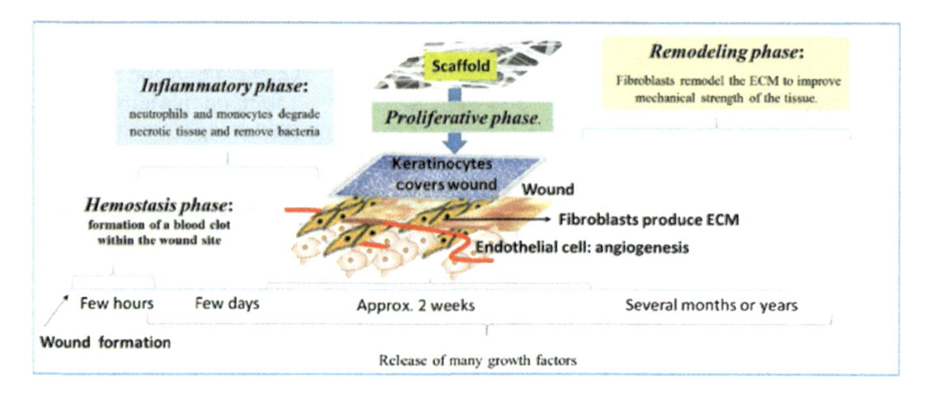

Fig. 6 Timeframe of wound healing phases [20]. Copyright 2017. Reproduced with permission from Springer Nature

white blood cells that develop into macrophages and provide immunological defences against many infectious organisms. The cells of remainders were phagocytosed by macrophages. Phagocytic activity is vital for the subsequent processes because no healing will happen for acute wounds suffering from bacterial imbalance [13].

4.1.3 Proliferation Phase

The proliferative stage in wound healing is distinguished by wound contraction, epithelialization, and angiogenesis, granulation tissue formation. Granulated tissue mainly consists of fibroblast and new blood vessels (Angiogenesis). Once the injury site is scoured, the inflammatory response is followed by proliferation and matrix synthesis, in order to fill the wound gap and reestablish the skin barrier. Fibroblasts migrate into the wound and secrete collagen to start the proliferative stage and deposit new extracellular matrix (ECM). Forming and establishing alternate blood vessels is vital in wound healing [13]. Angiogenesis is one of a critical component of normal wound healing and is stimulated by a vascular endothelial growth factor, resulting in increased vascular permeability, endothelial cell migration, and capillary formation [21]. The proliferation phase includes the initial repair processes for both the epidermal and dermal layers. Fibroblasts, macrophages and vascular tissues coordinately enter the wound to start the formation of a new dermal composite, the granulation tissue. Simultaneously, in a process named re-epithelialization, keratinocytes at the wound edge migrate over the granulation tissue to differentiate the new external layer of the epidermis [22–24].

4.1.4 Remodelling Phase

In remodelling phase, the new collagen matrix becomes more cross-linked and organized due to the wound contraction which brings the wound edge close together and increases the wound tensile strength [13]. During the remodelling phase, fibroblasts organize the collagen matrix and activate the transformation of fibroblasts to myofibroblasts to facilitate wound contraction. The healed wound, at last, enters the maturation phase, and granulation tissue continues to be remodelled and restoration of skin style [25].

Wound healing depends on the right selection of a material for a particular wound to achieve faster healing. These materials are focused to insulate wound from dehydration and promote healing. Different materials can be employed as dressing for healing wounds depending on the cause and kind of wound. Both natural and synthetic materials used in wound healing process can be passive, interactive and bioactive dressings. Passive dressings are permeable materials, such as gauze and tulle dressings, used to cover the wound to renovate its function. Interactive materials are semi-permeable or non-permeable, available in the forms of films, foam, hydrogel, and hydrocolloids. The ideal wound healing materials should achieve a fast cure rate at a reasonable cost with minimal annoyance to the patient [26, 27].

4.2 *Classical Wound Healing*

Classical wound healing agents include topical liquid and semi-solid preparations as well as dry classical dressings. These agents have still had some preferred standpoint in certain clinical settings for wound healing [28]. Classical wound healing dressings provide cover over the wound, including gauze, plasters, bandages (natural or synthetic) and cotton wool, their primary function is to maintain dryness of wounded area by possible evaporation of wound exudates, thus prevent entrance of hazard bacteria into the wound. Gauze dressings produced from cotton fibres, rayon, polyesters give some of the protection against bacterial infection. Different types of sterile gauze dressings are used for absorbing exudates and fluid in an open wound. These dressings need to consecutive change. Gauze dressings are less cost-effective. Because of excrescent wound drainage, dressings become moistened and have a tendency to become adherent to the wound making it painful when removing. Bandages that are produced from natural cotton wool and cellulose differ from synthetic bandages which made out of polyamide in its functions. Tulle dressings, impregnated dressings with paraffin that not adhere to wound surface and suitable for the superficial clean wound. Generally, classical dressings are used for the clean and dry wounds with mild exudate levels or used as secondary dressings. Since classical dressings have some of the limitation such as fail in providing a moist environment to the wound, they have been replaced by modern dressings with more advanced formulations [26]. In topical liquid formulations, for example, the povidone-iodine system is of paramount influence in the initial treatment of wounds as it reduces bacterial infection, thus its incorporation into dressing systems results in controlling or preventing infection. Next treatment of wounded areas by physiological saline solution shows a dual cleansing by removing either the dead tissue or washing the possibly dissolved polymeric dressings adsorbed onto the wounded sites. Saline solution is likewise used to immerse dry wounds during dressing change to help removal with little or no pain. Semi-solid preparations such as silver nitrate used to treat bacterial contamination stay on the surface of the wound for a more extended timeframe compared with solutions [28]. There are a large number of antiseptic agents have cytotoxic properties, but if utilized correctly they can be very effective. Recent reports show that antiseptics can be used selectively as the first line of treatment of infected wounds [29]. The commonly used antiseptics are hydrogen peroxide and iodine-based preparations. Due to the extensive use of the antimicrobial, the incidence of resistance had been increased. In general, topical antimicrobial agents are indicated early in wound management to control contamination and decrease microbial burden [30].

5 Bioactive Materials and Their Progress in Treating Wounds

Bioactive materials are materials which have been designed to induce specific biological activity [31]. Bioactive materials that engaged in wound treatment are developed to facilitate the healing of wounds instead of just to cover it. These materials are focused to insulate the wound from dehydration and promote healing. Bioactive materials play a vital role in the healing process. Biodegradability and biocompatibility are among distinguishable features for bioactive materials processed as dressing agents [4]. These agents are obtained generally from natural tissues like collagen, hyaluronic acid, chitosan, gelatin and alginate. Their usage in healing wounds can be singly or in combination. The collagen basic structure is protein, and because of its unique properties, it has been examined by numerous scientists for its active function in healing wounds. Upon contacting collagen to the wounded tissue, formation of fibroblast and promotion of endothelial migration is achieved. Hyaluronic acid is a glycosaminoglycan segment of extracellular matrix and has unique biological and physicochemical properties. The biocompatibility and biodegradability of collagen and hyaluronic acid are quite similar. Chitosan helps in forming granulated tissue throughout the wound healing proliferative stage. In many ways it was reported that the bioactive materials have superiority to other types of wound healing materials [26]. Perfect wound healing materials ought to keep up a wet domain at the wound interface, allow gaseous exchange, act as a barrier to microorganisms and remove excess exudates. It is necessary for the bioactive wound healing agents to fulfill many requirements such as: safety, non-allergenic, non-adherent to be easily removed, and availability of the raw biomaterials used in their constitution within a minimal processing. In addition, they have antimicrobial properties and promotes wound healing. Recently, many of scientists are dedicate in producing a unique bioactive, wound healing materials, by synthesizing and modifying biocompatible materials helps in promotion of wound healing process. Among these materials, chitosan and its derivatives, of controlled efficacy in repairing wounds are directed to promote healing at the molecular, cellular, and systemic levels. Chitosan is a cost-effective and naturally abundant biological material extracted from invertebrate's skeleton as well as the cell wall of fungi.

Chitosan is non-toxic biodegradable and antimicrobial agents has good biocompatibility, thus it shows positive impacts on wound healing [7]. Naturally originated materials are being widely used in wound healing because of their similarities to the extracellular matrix, typically good bio-characteristics, and cellular interaction. Also, it can be easily formulated to get better cell adhesion and tissue growth. In this way, natural polymers and their applications in wounds for support and healing have attracted much consideration throughout the most recent decade [32]. Using chitosan as an accelerator for wound healing is reported in many studies where chitosan can profitably affect every step of wound healing. Chitosan and its derivatives could accelerate wound healing by improving the functions of inflammatory cells, such as polymorphonuclear leukocytes [33], macrophages, and fibroblasts [34].

6 Chitosan

Chitosan is a linear polysaccharide obtained by deacetylation of chitin. It is a copolymer composed from β(1→4) 2-amino-2-deoxy-D-glucose and β(1→4) *N*-acetyl-2-amino-2-deoxy-D-glucose. Chitin is the abundant component of the skeletal structure of many classes such as the group of invertebrates that include arthropods, molluscs, and annelids. But in animals, chitin is associated with other constituents, such as lipids, calcium carbonate, proteins, and pigments. It has been estimated that the crustacean chitin present in the sea amounts to 1560 million tons [35]. Chitin is also found as a major polymeric component of the cell wall of fungi and algae. Chitin structure is similar to cellulose as shown in Fig. 7 but it has an acetamide group in a position of C-2 instead of a hydroxyl group. Chitosan structure is a deacetylated form of chitin [1].

6.1 Properties of Chitosan

Chitosan has several properties that control its activity, such as solubility, the degree of deacetylation, molecular weight, the degree of crystallinity, and degree of polymerization.

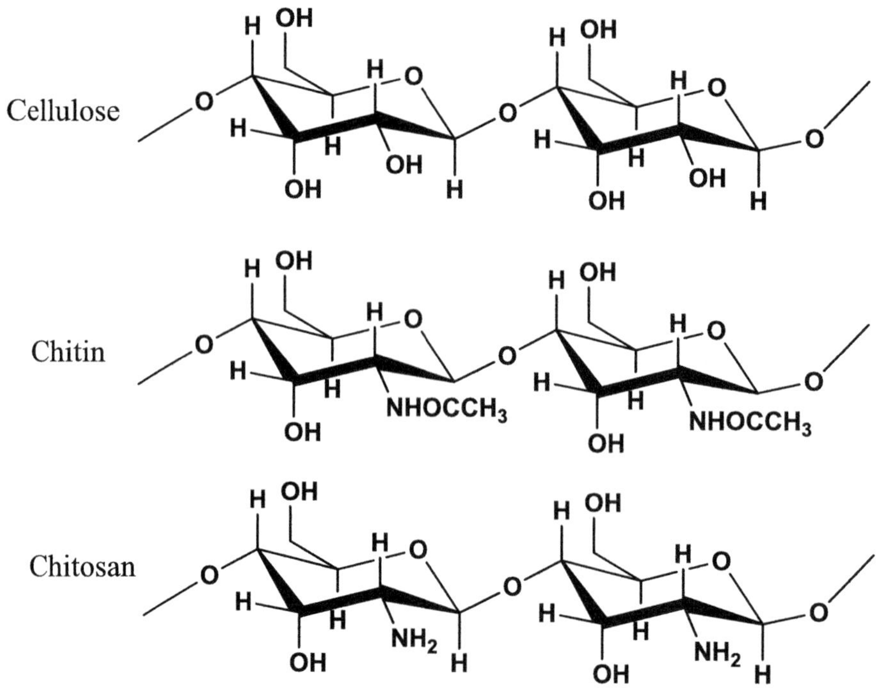

Fig. 7 Chemical structures of cellulose, chitin and chitosan [1]. Copyright 2017. Reproduced with permission from Springer Nature

6.1.1 Solubility of Chitosan

Chitosan is the poly-glucosamine, have a basic nature. It forms salts with acids and produces polyelectrolyte. Chemical modification of chitosan or formulating it as film or fiber is mainly correlated to its solubility in nature. Chitosan easily dissolves in dilute organic or mineral acids via the protonation of free amino groups at pH < 6.5, but it is insoluble in water due to the presence of intra and inter-hydrogen bonds in its structure. Acetic and formic acids are widely used for research and applications of chitosan. On the other hand, chitosan solubility decreases with an increase in molecular weight. Oligomers of chitosan with a degree of polymerization (DP) of 8 or less are water soluble regardless of pH [36].

6.1.2 Degree of N-Deacetylation

Studying the degree of N-deacetylation in chitosan, i.e. the ratio of 2-acetamido-2-deoxy-D-glucopyranose to 2-amino-2-deoxy-D-glucopyranose structural units is important. This ratio has a strong effect on solubility and solution properties of chitosan. Chitin is insoluble in dilute acetic acid. But when chitin is deacetylated to a certain degree (approximately 60% deacetylation) whereupon it becomes soluble in acid, it is referred to as chitosan. Chitosan is the fully or partially N-deacetylated derivative of chitin with a typical degree of acetylation less than 40% [37]. For measuring this ratio, a several methods have been used, such as FTIR spectroscopy [38], H^1 NMR spectroscopy [39], C^{13} solid-state NMR spectroscopy [40], elemental analysis, and thermal analysis [38].

6.1.3 The Molecular Weight (MW)

It is important to mention that the applications of chitosan are correlated with its molecular weight (MW). The chitosan molecular weight depends on its source and deacetylation conditions (time, temperature and NaOH concentration), respectively. Chitosan produced from deacetylation of chitin with a MW over 100,000. Consequently, it is required to reduce the MW by chemical methods to a much lower MW for easy application as a textile finish [38]. The MW of chitosan can be measured by several methods, such as light scattering spectrophotometry [41], gel permeation chromatography [39] and viscometry [41].

Viscometry is the simplest and most rapid methods for determining the molecular weights of polymers [41]. The most commonly used equation relating limiting viscosity values to MW is the Mark-Houwink equation as follows [35]:

$$[\eta] = K \cdot M^{\alpha}$$

where $[\eta]$ is the limiting viscosity number and K, α are constants independent of MW over a wide range of MW. They are dependent on the polymer, temperature, solvent, and, in case of polyelectrolytes, the nature and concentration of the added low-molecular-weight electrolyte.

6.2 *Chitosan Sources and Production*

There are many natural sources to extract chitosan such as mushroom, insects, yeast, the cell wall of fungi and marine shellfish such as lobster, crab, krill, cuttlefish, shrimp, and squid pens. Chitin is the outer protective coating for shellfish, that consists of covalently bound network with some metals, proteins and carotenoids. Crustacean shells composed of 30–50% calcium carbonate, 20–30% chitin, 30–40% proteins and also contain pigments (astaxanthin, canthaxanthin, lutein, and β-carotene), and these percentages change according to species or seasons. Chitin and chitosan are commercially produced from the shrimp, prawn, and crab wastes [42].

Different techniques have been reported for chitosan production. Preceding preparation of chitosan from the crustacean generally consists of four basic steps: demineralization, deproteinization, decolouration, and deacetylation [42].

Before starting, shells were washed by warming water to remove soluble organics, adherent proteins, and other impurities. The shells were then dried in the oven at 70 °C for a period of 24 h until completely dried shells were obtained. Then, the shells are ready to chitosan production [43].

6.2.1 Demineralization (DM)

Demineralization is generally carried out using acids such as hydrochloric acid, nitric acid, acetic acid, or formic acid (up to 10%) at ambient temperature with stirring to dissolve calcium carbonate as calcium chloride [42]. However, hydrochloric acid is the preferred acid and is used at a concentration of 0.2–2 M for 1–48 h at temperatures ranging from 0 to 100 °C with solid-to-solvent ratio of 1:15 (w/v) is usually used and then filtered under vacuum [39].

6.2.2 Deproteinization (DP)

Chitin is combined naturally with protein via covalent bonds, and deproteinization of demineralized shells is usually carried out by alkaline treatment. The shells are treated with sodium or potassium hydroxide (65–100 °C) at the shell to alkali ratio of 1:4 for times ranging from 1 to 12 h. Accordingly, a detachment of protein occurs from the shrimp waste solid component. Relatively high ratios of a solid-to-alkali solution of 1/10 or 1/20 are used with stirring to increase the deproteinization efficiency. After complete deproteinization, simple filtration can easily separate the protein hydrolysate from the solids in the protein slurry [42].

6.2.3 Decoloration (DC)

The deproteinized chitin is a coloured product. For commercial suitability, chitin is decolourized to yield white chitin powder. It is important to assure the inertness of chemicals towards physicochemical or functional properties during the dis-colouration process of chitin. Discolouration firstly starts with the treatment of chitin with acetone and then dried for 2 h at ambient temperature, then a bleaching step using 0.315% (v/v) sodium hypochlorite (NaOCl) solution for 5 min at ambient temperature with a solid to solvent ratio of 1:10 (w/v). Samples were then washed with water and dried in an oven at 60 °C for 2–3 h.

6.2.4 Deacetylation (DA)

The production of chitosan from chitin could be achieved by removing the acetyl groups from chitin in a process known as deacetylation. Various factors that affect the deacetylation process, including temperature, prior treatments applied to chitin isolation, time of deacetylation, alkali concentration, the ratio of chitin to alkali solution, prior treatments applied to chitin isolation, atmosphere (air or nitrogen), the density of chitin, and the particle size. The N-acetyl groups cannot be removed by acidic reagents without hydrolysis of the polysaccharide, thus, alkaline methods must be applied for N-deacetylation. It is generally achieved by treatment with concentrated potassium or sodium hydroxide solution (40–60%) usually at 80–140 °C for 30 min or longer using a solid to solvent ratio of 1:10 (w/v) to remove some or all of the acetyl groups from the polymer. In that process, sodium hydroxide is noted to be the preferred alkali. The resulting chitosan was washed to neutrality with running tap water, rinsed with distilled water, filtered, and dried at 60 °C for 24 h in the oven [42].

6.3 *Modified Chitosan*

The physicochemical properties of chitosan can be amended to more desirable features that meet various applications. Fortunately, the various functional groups in the polymeric domain of chitosan enable its modification (Fig. 8) [42]. Alteration of chitosan properties could be subjected chemically under mild conditions via the reactive hydroxyl and amino groups distributed in chitosan structure. Below, we report on promising examples of amended chitosan that recently take an eminent position in advanced research and have special markets [44].

6.3.1 Thiolated Chitosan

Thiolated chitosans could be formed by the interaction of thiols with the chitosans' primary amino group. There are three types of thiolated chitosans have been formed: chitosan–cysteine conjugates, chitosan–thioglycolic acid conjugates and chitosan–4-thio-butyl-amidine conjugates. The immobilized thiol groups improve

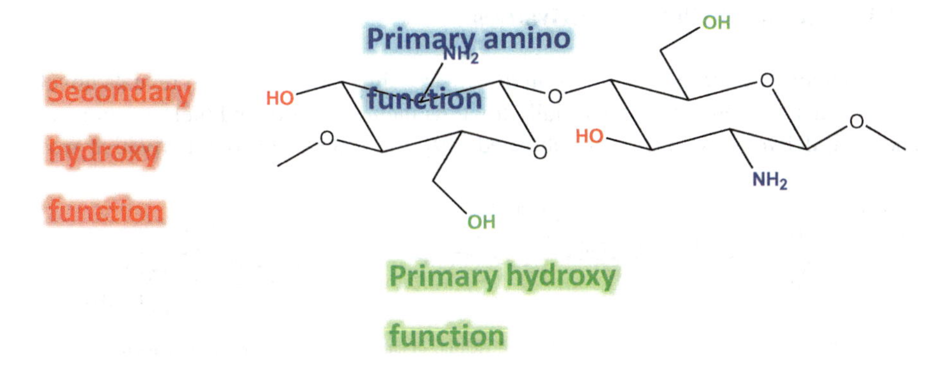

Fig. 8 Functional groups in chitosan that can be modified

many properties of chitosan. Due to the formation of disulfide bonds with mucus glycoproteins, the mucoadhesiveness is aggrandized [6]. For example, the thiolated chitosan rather plain chitosan reveals enhancement by (1.6–3 folds) in the permeation of paracellular markers through intestinal mucosa. Moreover, the thiol groups may react with thiol groups, thus decreases the release rate and maintains the peptide release for a longer time.

6.3.2 *O, N*-Carboxymethyl Chitosan

Another type of modified chitosan is the carboxymethyl chitosan (CMCS) that shows greatly enhanced antimicrobial activity than chitosan. This could be attributed to the inter- and intra-molecular interactions between the carboxyl and amino groups that increase the increased cationic groups [45]. *N*-carboxymethyl chitosan is a glucan type of chitosan carrying pendant glycine groups formed when using glyoxalic acid in the preparation course [46]. The film-forming ability of *N*-carboxymethyl chitosan assists in imparting a pleasant feeling of smoothness to the skin and in protecting it from adverse environmental conditions and consequences of the use of detergents.

6.3.3 Highly Cationic Chitosan

N,N,N-Trimethyl chitosan (TMC) and *N*-[(2-hydroxy-3-trimethylammonium) propyl] chitosan chloride are quaternary soluble derivatives that show effective enhancing properties of transport the peptide, protein, and drugs through mucosal membranes. They have been used as an absorption enhancer, antibacterial agent and gene vector due to its ability to form complexes with anionic gels or macromolecules [47]. TMC was synthesized by reductive methylation of chitosan in an alkaline media at elevated temperature as can be seen in Fig. 9 [48].

6.3.4 PEGylated Chitosan Derivatives

Modification of chitosan and chitosan derivatives through PEGylation reveals novel physicochemical properties, especially towards their solubility and their use in gene and drug delivery. As has been published [49], the PEG-g-chitosan derivative was

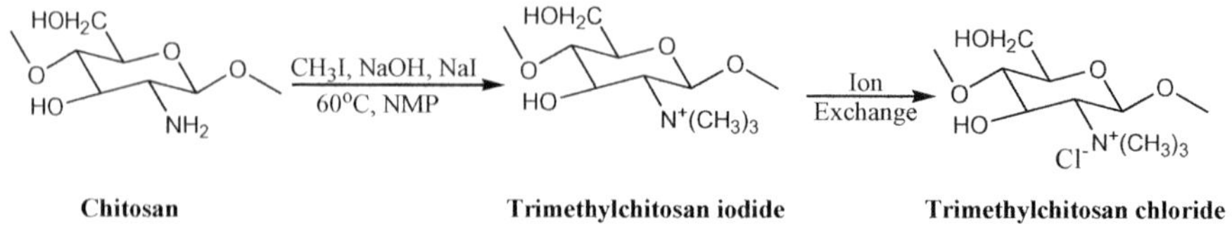

Fig. 9 The preparation methods of TMC

synthesized by functionalizing chitosan using a PEG-aldehyde. Chitosan was first modified with a PEG-aldehyde to yield an imine (Schiff base), which was then reduced to PEG-g-chitosan using sodium borohydride ($NaBH_4$). In recent years many applications, have been suggested for PEGylated chitosans. For example, uses of PEG-g-chitosan for cell adhesion applications and application in wound healing and tissue engineering [50].

7 Chitosan Composites and Their Inherent Biological Properties

Chitosan is a natural cationic copolymer of *N*-acetyl glucosamine and D-glucosamine. There are appropriate properties for chitosan making it suitable for use in several biological fields such as drug delivery and wound healing [51]. The main advantage of chitosan that it can be easily modified or make it in combination via a simple procedure. A new class of biomaterials were obtained from chitosan composites with good mechanical, physicochemical characteristics and useful properties, which can't be accomplished either by chitosan or the combined material alone [52]. Chitosan naturally exists as a composite with proteins and minerals in invertebrates, that keep the integrity of its structural shells [53]. Thus combining of chitosan and protein to form chitosan–protein composites make those composites have biocompatibility. Many such composite scaffolds, for example, chitosan–collagen and chitosan–gelatin [54] have been tested for their ability to promote cells adhesion and proliferation. In the same pattern, chitosan and another glycosaminoglycan such as heparin and hyaluronic acid have been combined to form composites that provide specific cellular adhesion [52]. Chitosan-based wound healing materials, which prepared by combining chitosan with synthetic or natural materials is an effective way to prepare new wound healing materials for enhancing its versatile property. Poly(vinyl alcohol), poly(vinyl pyrrolidone), poly(ethylene oxide), and polyglutamic acid are an example of chitosan composites with synthetic polymers that tested for wound healing applications. Different inorganic materials can be combined with chitosan and give a good benefit in biological applications. Wound healing materials that are prepared from chitosan and nanoparticles, can interact at molecular and cellular level. This gives better antibacterial activity and rapid healing action for the wound [10]. Biocompatibility, biodegradability, hemostatic, bio-adhesion, antimicrobial activity, anti-inflammatory and non-toxicity are inherent biological properties of chitosan and chitosan composites, that give it a great attention in wound healing process [55, 56].

7.1 Non-toxicity

In vivo, administration route affects the extent of chitosan toxicity. Arai et al. reported that oral administration of chitosan within 16 g/kg resulted in lethal dose in mice, whereas it amounts to >15 g/kg in rats. While intra-peritoneal administration shows higher lethal dosage of 30 g/kg in rats and 52 g/kg in mice. Upon subcutaneous administration, the lethal dose amounts to >100 g/kg in mice [57]. However, low toxicity (e.g., loss of appetite) may be the side effect when administrating chitosan orally to rabbits or hens during 34 weeks at doses of 700–800 mg/kg/day As reported by Rao and Sharma [58], there is no pyrogenic effects, no toxicity in mice and no irritation detected for either skin or eyes in rabbits. In addition, almost studies presented chitosan as a safe material that may induce low or minimal toxic effects. Moreover, chitosan shows considerable suitability and safety in food applications and as a pharmaceutical excipient for the parenteral route [59]. Furthermore, concluded data about chitosan toxicity based on human studies are quite limited. A daily dose of 4.5 g from chitosan took by human volunteers did not result in toxic effects [60]. In addition, human trials up to twelve weeks showed no toxic effects, no allergies, and only nausea symptoms and constipation in 2.6–5.4% of subjects [61]. Accordingly, chitosan does not show significant acute toxicity.

7.2 Antimicrobial Activity

The antimicrobial activity of chitosan is influenced by several inherent structural factors including chitosan molecular weight, the degree of deacetylation and consequently the number of available free amine groups. The antimicrobial activity of chitosan increases by decreasing pH as it is correlated to the glucosamine amino group [62]. Nevertheless, the presence of some ions (e.g. Ca^{2+}, Mg^{2+}) [63] could significantly affect the antimicrobial activity of chitosan. Unfortunately, it is difficult to compare the estimated antimicrobial characteristics of chitosans reported in the literature, indeed there is no a reference efficacy [64]. Because of the mutations of microorganisms, a challenge in addressing modified chitosan formulations of increased antimicrobial activity is a continuous demand. Chemically modified chitosan of amended antimicrobial activity could be referred to the primary (C-6) and secondary (C-3) hydroxyl groups on each repeating unit in addition to the amino groups (C-2) on each deacetylated unit. Two mechanisms could be engaged in the antimicrobial action of chitosan [65]. They firstly propose an interaction between the positively charged chitosan and the negatively charged surface groups of the cells. This could affect the permeability and cause a shortage in the fundamental solutes necessary for the cell. The second mechanism refers to inhibition of synthesizing the microbial RNA as a result of interacting with the protonated amino groups of chitosan with the cell DNA and thus inhibit the synthesis of microbial RNA.

7.3 Anti-inflammatory Nature

Anti-inflammatory drugs with high efficiency and fewer side effects are a great target for a large number of researchers. Therefore, it draws attention to the anti-inflammatory drugs from plants that have better effects, fewer side effects and rich resources. Chitosan is one of the most abundant on the earth and naturally occur. Chitosan and chitosan composites used to reduces inflammatory and pain with high efficiency and fewer side effects [66]. Many reports show that the inflammatory response is resulting from lipopolysaccharide, one of the components of the cell wall of gram-negative bacteria. Lipopolysaccharide (LPS) is known to stimulate the production of many local factors, including tumour necrosis factor-alpha (TNF-α), interleukin, and prostaglandin E2, from immune cells (macrophages) and fibroblasts cells in inflamed tissues [67]. It is known that prostaglandin E2 considers a main inflammatory mediator in the inflammatory response. Several researchers show that lipopolysaccharide (LPS) activates macrophages and stimulate the production of proinflammatory cyclooxygenase-2 (COX-2) enzyme, leading to the synthesis of large amounts of Prostaglandin E2 at inflammatory sites [68, 69]. Another pro-inflammatory product that has a critical role in inflammation is the nuclear factor-kappa (NF-κB) that is activated by lipopolysaccharide (LPS) via stimulation of Toll-like receptor 4 (TLR-4) on immune cells. Lipopolysaccharide immune cell activation is facilitated by two proteins known as CD14and MD-2 on the cell surface, then Toll-like receptor 4 (TLR-4) triggers a reaction with the adaptor molecule MyD88 (myeloid differentiation primary-response protein 88). The MyD88-depending signaling cascade leads to the activation of nuclear factor-kappa (NF-κB), that regulates proinflammatory genes that release pro-inflammatory cytokines such as tumor necrosis factor-alpha (TNF-α), interleukin (IL-1, IL-6), which exaggerate the inflammatory response. So, the suppression of nuclear factor-kappa (NF-κB) activation via inactivation of lipopolysaccharide, is a potent strategy for the treatment of inflammation [70, 71].

It is reported in may papers, that chitosan suppresses the production of nuclear factor-kappa (NF-κB) and significantly inhibited Prostaglandin E2 without cytotoxicity [69, 72]. Chitosan as a cationic polymer exhibits an anti-inflammatory effect due to ionic bonding and electrostatic interaction between its positive centers and the highly anionic lipopolysaccharide. However, the therapeutic potential of chitosan, depending on lipopolysaccharide-binding, need to increase the efficiency of interaction between chitosan and lipopolysaccharide. Based on this concept, chemical modification of chitosan was performed to increase the affinity of lipopolysaccharide-chitosan binding [73]. *N*-acetylglucosamine, an anti-inflammatory agent, is the monomer of chitosan and also already exists in the body as hyaluronic acid that plays important role in wound repair. Furthermore, chitosan has a physical analgesic effect on the wound by giving a cooling and soothing effect when applied to an open wound. Therefore, chitosan shows a high anti-inflammatory effect even when comparing it with the synthetic non-steroidal anti-inflammatory drug, such as indomethacin. Due to anti-inflammatory effects of chitosan, these are beneficial for the treatment of acute and chronic inflammation of wounds [66].

7.4 Biocompatibility

This propriety results from the interaction of the positive charge in pure chitosan or chitosan composites with the negatively charged cell membranes due to ionic exchange between the intercellular and extracellular medium [74]. Chitosan and its composites have been investigated in numerous studies in a biological field and in general, they showed high biocompatibility [75]. It was reported in human trials of up to 12 weeks showed no significant clinical symptoms, with no evidence of an adverse response of chitosan indicating to its biocompatibility [61]. Chitosan has been approved by the American Food and Drug Administration (FDA) as a wound dressing material, confirming its biocompatibility [76].

7.5 Biodegradability

An important aspect of the use of polymers in biological systems is their biodegradation process in the body. Many reports illustrated, that chitosan and its composites can be degraded by enzymes such as lysozyme, chitinase, lipases and not produce any harmful byproducts, because these products are oligosaccharides that are either incorporated into glycosaminoglycan and glycoprotein metabolic pathways or easily excreted in urine directly [77, 78]. Chitosan can be degraded by lysozyme via hydrolysis process and the degradation products are nontoxic [59]. The degradation rate of chitosan depends on its physical characterization and the preparation methods. It is important to clear up that rapid degradation of chitosan, may cause an accumulation of amino sugars, resulting in an inflammatory response and, thus, affecting chitosan biocompatibility [74].

7.6 Hemostatic Properties

Chitosan hemostatic dressing is most promising in wound healing process due to an effective response to stop bleeding and its regeneration advantages. Properties of chitosan facilitate attracting red blood cells to chitosan, leading to rapidly blood coagulation, and this was a reason for its approval in the USA for use it in wound healing process [79]. Chitosan as a hemostatic agent would induce clot formation via an interaction between the cell membrane of blood cells and free amino groups chitosan. The positive surfaces of chitosan attract the negative charge of the membranes of erythrocytes and platelets, leading to platelet activation and blood coagulation that reduce bleeding and enhance hemostasis [59]. It is reported recently that chitosan promotes coagulation in vitro, with reducing the time of blood clotting by 40% compared to blood alone, that suggest chitosan can induce blood coagulation because of the interactions between acidic groups of blood cells and free amino groups of chitosan [80]. The hemostatic mechanism of chitosan may be due to three possible actions to stop bleeding, absorption of plasma, coagulation of red blood cells, and platelet aggregation. Absorption features of chitosan is a prominent factor that helps to act as a hemostatic via plasma absorption. In addition to plasma absorption, coagulation of erythrocytes also is one important hemostatic way of chitosan. The clotting of erythrocytes was enhanced in the presence of chitosan due to crosslinking of the erythrocytes with chitosan polymer chains. When chitosan attached to blood, resulting in a change in morphology of erythrocytes. They loss typical shape and seemed to have an unusual affinity towards one another. The main cause of hemostatic effect of chitosan is its ability to the aggregation and activation of platelet, forming an aggregated bloc in irregular shapes [81, 82].

7.7 *Mucoadhesivity*

Mucus surfaces are the moist surfaces, which major components are mucin glycoproteins, lipids, inorganic salts and water, the latter, consists more than 95% of their weight, making them a highly hydrated system. The mucoadhesive molecules have features make them, interact chemically and mechanically with the glycoproteins of the mucus. The favoured of mucoadhesive features are an anionic surface charge, high molecular weight, flexible chains, surface-active properties and affinity to build hydrogen bond, which helps in spreading throughout the mucus layer [83]. The cationic polymers such as chitosan can be absorbed on the mucus surfaces via electrostatic interaction. Several types of polymers have been reported as mucoadhesive due to their ability to interact physically and/or chemically with the mucus surfaces. Moreover, the chitosan muco adhesive properties may be attributable to its cationic character [74]. Chitosan and its composites have a high adhesive force that will promote binding to the negative surface of the mucus surfaces and cover different epithelial surfaces [84]. The mucoadhesion of chitosan can be illustrated by the interaction between negatively charged residues in the mucin glycoprotein and chitosan positively charged amino groups [85]. Mucoadhesion of chitosan and its composites is one of the important characters that promote to improve the wound healing via interaction with the negatively charged cell membranes, thus chitosan can stay close to the wound and prohibits the growth of microorganisms.

8 Modified Chitosan and the Biomedical Engineering of Wound Healings

Over decades, materials applied in wound healing processes are originated from diverse naturally occurring polymers. Moreover, those materials arisen from polysaccharides like chitosan have received the great attention from the worldwide researchers because of its important biological properties. Chitosan helps in every step of wound healing via accelerating the infiltration of inflammatory cells like neutrophils, absorbing exudates, making a barrier against microbes and helps in healing without scar formation. However, the tenuous behaviour of chitosan delimits its usage in a pristine form and pave the way for its modification in many forms such as gels, films, fibers, and scaffolds or blending with other materials to get a more suitable tissue engineering material. Blending of chitosan or its combination either with natural or synthetic polymers reduce cost, improve its handling. Moreover, this enhances the wettability, gas permeability and mechanical properties. The nanoparticles and growth factors blended into chitosan also exhibit better antibacterial activity and minimize the time span for wound healing [10]. Chitosan and modified chitosan activate macrophages, stimulate cell proliferation and tissue organization. Hemostatic properties of chitosan aids to natural blood thrombosis as well as blocking of nerve endings that reduces pain. Gradual depolymerization of chitosan releases *N*-acetyl-β-D-glucosamine, which initiates fibroblast proliferation, deposits of collagen and stimulates synthesis of a higher level from natural hyaluronic acid at the wound site. This plays a pivotal role in wound healing [7]. Chitosan stimulates neutrophil in the inflammatory stage, to clean out the wound site from bacteria. Chitosan activates the inflammatory cells, macrophages, and fibroblasts which helps in wound healing without forming stigma [10].

8.1 Wound Healing Using Chitosan Impregnated Drug

Healing of chronic wounds necessitates a durability in regular drug administration. Prolonged drug release continuously exposes the wounded area to the action of the drug and thus reduce the used dose [86]. Recently, few reports conclude an antibiotic-medicated chitosan dressing to get a controlled drug release system. In this system, optimization of the amount of impregnated drug that provides the controlled and sustained antibiotic action is a demand to be achieved.

8.1.1 Wound Dressing Using Sulfadiazine Loaded Chitosan Nanoparticles

El-Feky et al. [87] reports for silver sulfadiazine (SSD) loaded chitosan nanoparticles (CSNPs) for the controlled-release of SSD into burn wound to control bacterial growth. This work aimed to combine the advantages of the potent antimicrobial and antifungal SSD drug with the advantages of chitosan nanoparticles as drug carrier systems and effective fabric coating material. Generally, the chemical bonding of SSD to textile dressings is not possible. Therefore, no specific concentration or release rate of the sulfadiazine drug from the dressing to the wounded area is precisely identified. Thus, the followed strategy was to encapsulate SSD on a suitable hydrophilic nano-carrier has the possibility to be a coater for the wound. By this way, the drug cytotoxic effect might be prevented and the effective controlled application of SSD on healing wound is enhanced [88].

Water absorbency (Aw) of the wound-dresser is a crucial parameter in the healing process and could be determined by the following equation:

$$Aw = \frac{m_{wet} - m_{dry}}{\mathrm{m_{dry}}} \times 100$$

where $\mathrm{m_{dry}}$ and $\mathrm{m_{wet}}$ are the weights of the dry and wet samples, respectively.

Aw, could be estimated by immersing the investigated dresser in water until reaching an equilibrium where the excess water on the surface of wet samples was drained for 1 h through a calibrated sieve.

The continuous delivery of SSD in its loaded chitosan formulation that could extend over 24 h help to improve patients' compliance by removing the need for multi-daily drug administration and is expected to offer an effective treatment due to the longer undisrupted contact time between the drug and the wounded area.

8.1.2 Wound Dressing Using Simvastatin—Chitosan Microparticles Loaded Polyvinyl Alcohol Hydrogels

Yasasvini et al. [89] prepared a simvastatin—chitosan microparticles loaded with polyvinyl alcohol (PVA) hydrogels via a chemical cross-linking method. The simvastatin loaded chitosan are firstly developed followed by the incorporation of the particles in hydrogels. Depending on the in vitro release profile of the plain drug, the preparative route optimizes the percent composition of the chitosan microparticles and loaded PVA hydrogels. Optimized composition results in high entrapment of simvastatin drug as estimated by the following formula:

$$EE(\%) = \frac{\text{Total amount of drug} - \text{Amount of drug the in supernatent}}{\text{Total amount of drug}} \times 100$$

Adaptation of the composition percent of the constituents of this system provided prolonged release with a minimum amount of drug loaded in the formulation. This reduces the used dose of the drug and maintains its required therapeutic concentration in plasma. Thus, topical wound healing of high efficiency is improved. The simvastatin loaded chitosan revealed a granulated tissue with complete epithelialization was obtained in case of animals treated with the lowest dose hydrogels. On the other side, control animals are in the late fibroblastic state after 21 days whereas the animals treated with the lowest dose of hydrogel acquires a contracted wound profile [89].

8.1.3 Wound Dressing Using Scaffolds of Chitosan-Fibrin (CF) Loaded with Quercetin

Scaffolds composed of quercetin-loaded chitosan–fibrin (Q-CF) due to dry-freeze method were presented by Vedakumari et al. [90]. These scaffolds are non-toxic and showed suitable biocompatibility in tissue engineering applications. Quercetin is found in plants and belongs to flavonoids. It exhibits antiviral, antiulcer, anti-cancer, antibacterial, anti-allergic and anti-inflammatory properties [91]. Moreover, quercetin potentially scavenges oxygen free radicals and shows efficient in vitro and in vivo inhibition of lipids peroxidation. It has been reported that wounds medicated with free radical scavenging drugs showed enhanced healing [92]. The developed Q-CF scaffolds show quick water uptake and consequently quercetin dissolution at the scaffold surface. This leads to an initial rapid release of quercetin for up to 3 h. After that, a slow rate of quercetin release for up to 48 h is further estimated. The slower step could be ascribed to a gradual degradation of chitosan resulting in the formation of interconnected pores on the surface of the scaffolds [90].

Since microbial infection is a major problem during the process of wound healing, hence antimicrobial activity is significant parameter required to be dealt. The Q-CF scaffold demonstrated a higher potential for bactericidal activity against *Escherichia coli* and *Staphylococcus aureus* compared to quercetin free (CF) scaffolds. The noticeable difference in the bactericidal activity of CF and Q-CF scaffolds may be due to a synergistic effect of quercetin and chitosan. The earlier holds strong antibacterial activity against *E. coli* and *S. aureus* [93], whereas the positive charges on the latter interact with the negatively charged microbial cell wall causing discharge of proteinaceous and other intracellular components. It also binds to the nuclear DNA and inhibits the synthesis of mRNA and proteins [94].

Furthermore, a time-dependent monitoring of a group of biochemical parameters due to investigation of the granulation tissues from the Q-CF treated group in comparison to the CF group revealed the following conclusions: Q-CF treated group showed significant increase in the quantity of collagen that causes improved migration of fibroblasts and epithelial cells to the wound site for effective healing. The content of hexosamine was reduced referring to an increase in the extracellular matrix formation. Nevertheless, a prominent difference is noticed among all the three groups. A higher level of total protein content, representing the efficient synthesis and accumulation of proteins in the granulation tissues and a significant increase in the level of uronic acid [95]. These observations highlight the combined role between quercetin, chitosan and fibrin in the induction of fibroblast proliferation and quickness of healing in Q-CF treated wounds.

8.1.4 Wound Dressing Using Melatonin-Loaded Chitosan-Based Microspheres (Mel/CS MS)

Another dry powder formulation of Mel/CS MS was developed by Romić et al. [96] using a spray drying method. Melatonin is a neurotransmitter hormone and biological modifier has pleiotropic bioactivities [97]. Melatonin stimulates antioxidant enzymes, efficient free radical scavenging, and anti-inflammatory properties [98]. Also, melatonin has antibacterial effects that enhance wound healing via its influences on the phases of inflammation, glycosaminoglycan and collagen accumulation in addition to cell proliferation at the wounded site [99]. Moreover, melatonin reduces endothelial dysfunction in burns treatment owing to reactive oxygen species antagonizing and thus, shortens the healing time [100, 101]. In contact with wound exudates, the Mel/CS MS powder renders into the hydrogel. Pluronic-F127 was incorporated in the microspheres matrix to enhance melatonin amorphization and release rate. Furthermore, the entrapped melatonin has potentiated chitosan antimicrobial activity towards Staphylococcus aureus and five clinical isolates S. aureus MRSA strains. The microspheres within relevant concentrations for antimicrobial activity against planktonic bacteria show considerable biocompatibility to skin keratinocytes and fibroblasts.

8.2 Wound Healing Due to Chitosan Composites with Metal or Metal Oxide Nanoparticles

Wound healing could be influenced by copper nanoparticles. The latter could efficiently affect the complex phenomenon of healing involving various cells, cytokines and growth factors. It was found that metallic copper nanoparticles rather than copper ions show better modulation of cells, cytokines and growth factors involved in the process of wound healing. Copper plays role in activating endothelial cells as it stimulates their proliferation and migration. It induces the direct in vitro proliferation of human endothelial cells directly in vitro and it is also considered as a cofactor for many giogenic mediators As mentioned before, chitosan due to its biocompatibility, antifungal/antibacterial properties, and biodegradability is viewed as a beneficial polymeric candidate in healing.

Gopal et al. [102] presented for a chitosan-based copper nanocomposite (CCNC). In their synthetic route, chitosan powder was slowly added to the acetic acid solution (1%) in a normal saline and stirred for 8 h to get (10%) colloidal solution of chitosan. After that, CCNC of 0.3% was prepared by slow addition of an ethanolic dispersion of copper nanoparticles to the chitosan colloidal solution under continuous stirring followed by sonication for 5 min. The healing efficacy of the as-prepared CCNC nanocomposite was tested in the topical treatment of an open excision wound in adult Wistar rats. The relatively fastest wound contraction was detected in the CCNC-treated group during the monitored time frame compared to chitosan and control treated groups.

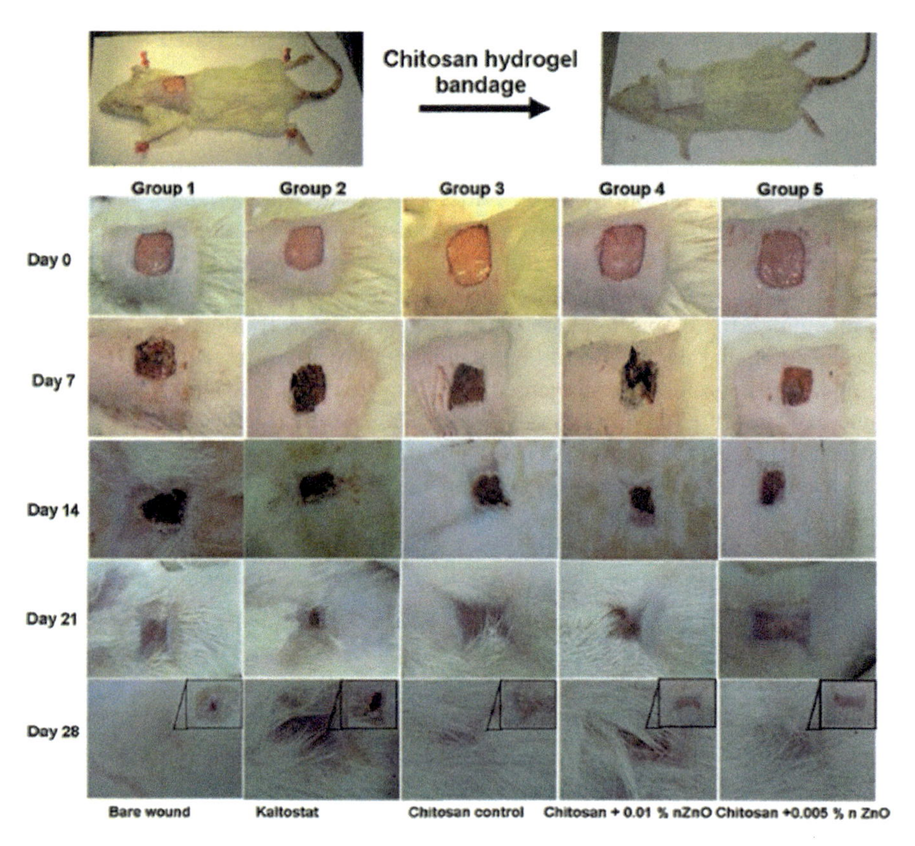

Fig. 10 Photographs of wound healing study [10]. Copyright 2016. Reprinted with permission from Springer Nature

Further composites prepared depending on chitosan and nano-ZnO, that can be used in wound healing. Sudheesh Kumar et al. developed chitosan hydrogel/ nano-ZnO nanocomposite bandages (CZBs) via the mingling of ZnO nanoparticle (nZnO) with chitosan hydrogel, and study the wound healing ability of the as-prepared CZBs on rats. Figure 10 illustrates that chitosan nanocomposites bandages "control group" show excellent healing after 1 and 2 weeks, compared to Kaltostat and bare wound. Wounds treated with chitosan control and CZBs achieved significant closure to ~90% after 2 weeks as compared to Kaltostat treated the wound and bare wounds, which showed 70% closure (Fig. 11a). The chitosan controls and CZBs show enhancing collagen deposition in the wound healing after 4 weeks. CZBs showed controlled degradation, enhanced blood clotting, excellent platelet activation ability, cytocompatibility, and antibacterial activity. All these properties indicate that advanced CZBs can be used for wound healing [10].

Another recent study presented by Fan et al. [103] developed a nano-titania/collagen-chitosan (nT/COL-CS) scaffolds. The stability, safety, and broadens in spectrum antibiosis of TiO_2 making it promising additive in decorating matrices used in biomedical applications. Only the nano-TiO_2 of anatase crystallinity has been reported to have the antimicrobial property [104]. The employed synthetic route implies at first, the preparation of TiO_2 hydrosol as reported by Macwan et al. [105] from tetrabutyl titaniate precursor added to distilled water, ethanol, and hydrochloric acid within a molar ratio 1:200:16:0.3. A second step to finally produce nT/COL-CS was carried out through this technique: 20 mL of COL and CS solution (5 mg/mL) using 0.5 M acetic acid. After that, isovolumetric COL and CS solutions were mixed at 4 °C followed by the addition of appropriate amounts of TiO_2 hydrosol to get (COL-CS/TiO_2) of ratios 1–4. The mixtures were homogenized under continuous stirring at 4 °C for 2 h, then kept for 4 h at 37 °C bath. Then, the mixtures were poured into similar good culture plate and frozen at—30 °C, subjected to vacuum freeze-dried at −50 °C and stored at −20 °C while not in use. The added nano-TiO_2 further binds to collagen and chitosan via hydrogen bonding and serves as a bridge between their molecular chains resulting in an increase in the density of the mesh structure.

The nT-COL/CS reveals a remarkable aggregation of red blood cells on the nT-COL/CS surface that develop into clumps. Therefore, the nT-COL/CS scaffolds

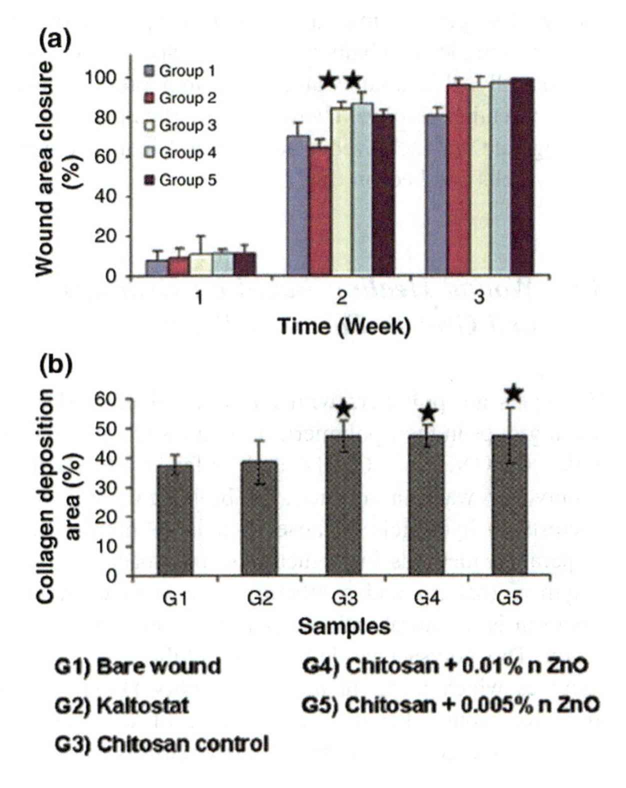

Fig. 11 **a** Evaluation of the wound area closure and **b** study of the collagen deposition area (in both graphs, the star symbols represent the p 0.05 level, indicating that the means are significantly different, compared with the control) [10]

may be conductive to the hemostatic of the wound and may reduce the time of wound repairing. Moreover, the images indicate that the red blood cell aggregation is independent of the content of nano-TiO_2.

Therefore, the biomedical studies on these scaffolds conclude their cleaning and sterilizing effect for the wounded area, as well as, their high swelling property that reduces the wound exudates and the aggregation of red blood cell. These characteristics achieve the requirements of a viable candidate for wound repairing dressing.

Beer et al. [106] and Lara et al. [107] report for silver composites of considerable antimicrobial efficiency and propose a mechanism for this action. However, incorporating Ag to host matrices could enhance the antimicrobial effect. A biocomposite of chitosan/ZnO containing Ag (CS/ZnO/Ag) showed potent antimicrobial, antibacterial, and antifungal properties without any toxic side effects. This may offer a new promising generation from wound healing dressings whereas the blended ZnO nanoparticles have antimicrobial, antibacterial properties in either micro- or nanocomposites that could be beneficial in preventing infectious diseases [2, 108]. However, the higher antibacterial efficiency of ZnO is correlated to the optimum amount of Zn ions precursor during the sol-cast preparation route [109].

Ding et al. [110] developed a bilayer dressing sponge consists of cross-linked chitosan-oxidized Bletillastriata polysaccharide as a base layer with chitosan-Ag (CS-Ag) as the upper layer. The upper layer (CS-Ag) prohibits the bacterial proliferation and avoids direct contact with wound area whereas the lower layer showed less gelling time, more uniform aperture distribution, higher water retention, preferable mechanical strength and improved cell proliferation ability. Additionally, this double layered system provides excellent gas permeation and water retention ability. Therefore, significant acceleration of cutaneous wounds healing rate was achieved where a good mature epidermization with less inflammatory cells reached on day 7.

8.3 Wound Healing Based on Hydrogels and Growth Factor Delivery

Hydrogels are polymers with cross-linked networks, that have hydrophilic functional groups in their polymeric structure such as an amine (NH_2), hydroxyl (–OH), amide (–CONH–, –$CONH_2$) and sulfate (–SO_3H) that make them have a high affinity for water absorption. In biological and medical fields there are great interests to hydrogels because of a large number of convenient properties and preparation methods in medical and pharmaceutical industries. Different types of polymers (natural and synthetic) have been examined in hydrogel preparation. Chitosan is a natural cationic polymer, that has vast benefits at hydrogel modification. This polysaccharide has hydrophilic nature and can be degraded via human enzymes which result in biocompatibility [111]. Chitosan hydrogels are convenient for wound healing due to ease of application, wound protection, water retention, oxygen permeability and flexibility in acquiring the wound geometry.

Chitosan hydrogel was used as dressing to accelerate wound healing [14]. Chitosan hydrogels may be used as a vehicle to deliver proteins such as growth factors into wounds to accelerate healing [112]. Sustained release of growth factors in wounds has some delay, but when using chitosan hydrogels as a carrier, it gives prolonging the period of activity. Growth factors strongly promote cell division, migration, differentiation, and proliferation, and its combination with chitosan hydrogels give the advantages in wound healing process [113].

8.4 Wound Healing via Bioactive Modified Chitosan Based on Photodynamic Therapy

The abuse of antibiotics cause high incidence of bacterial resistance, and therefore, result in a serious problem which threatens the public health. Photodynamic therapy has arisen as a promising and important way in many fields of medical therapy. Recently, many study on photodynamic therapy for wound healing give interesting results that open door for wound healing [114]. Photodynamic therapy technique, depend on photosensitizer, that has been activated by exposure to illumination of suitable wavelength, and in the presence of oxygen produce highly reactive oxygen species that are able to damage the plasma membranes and DNA, leading to bacterial death. Prevention and treatment via photodynamic therapy, has a great advantage comparing to ordinary antimicrobials, because of the dual antimicrobials action of the photosensitizer, that can be attack the infected area via producing reactive oxygen species, and moreover, photosensitizer can be bind directly to cell wall and membranes of bacteria resulting in direct damage to its structures. Thus, bacteria cells have not chance to create resistance [115, 116]. The ideal photosensitizer should be easily synthesized, stable composition, non-toxic, photostable, cost-effective and have eradication power against pathogens [117]. The susceptibility of photosensitizer to eradicate bacteria has been shown to be dependent upon the type of bacteria (Gram-positive or gram-negative). Gram-positive bacteria have a permeable cell membrane which facilitate penetration of neutral, anionic, or cationic photosensitizers into the cell, and otherwise, gram-negative bacteria have an outer membrane that hinder regulate movement of molecules through the cell wall. Therefore, the photosensitizers with cationic charge or using cationic carrier are able to enter through the membrane and penetrate the cell and hence increase the efficacy of photosensitizers against gram-negative and yeast. Thus, negatively charged antimicrobial photosensitizers have narrow spectrum in the treatment of mixed infections, while positively charged photosensitizers such as conjugating polymers or those loaded on a cationic polymer such as chitosan are generally highly effective photosensitizer. Modified chitosan has been chosen, as a suitable delivery system for photosensitizers based on their advantages (e.g. biodegradable, biocompatibility, less toxicity, anti-microbial and hemostatic properties). Also, chitosan has a synergistic effect with photosensitizers, which potentiate their action in Photodynamic therapy [115].

9 Study Case for the Anti-bacterial Activity of Chitosan Grafted Poly(*N*-Methylaniline) Nanoparticles

Antibacterial activity of Chitosan–graft-poly(*N*-Methylaniline) nanoparticle [118] against Escherichia coli (Gram-negative) and Staphylococcus aureus (Gram-positive) as a wound-causing bacteria model was investigated via determination of the growth inhibitory effect of chitosan-graft-poly(*N*-Methylaniline) in broth bacterial suspension with concentration of 10^8 CFU ml^{-1} as a reference for initial colony quantification, that determined by using the plate counting method. To attain this situation, from a fresh bacterial agar plate, transfer the growth bacteria into a sterile capped glass tube containing an appropriate amount of sterile broth solution to make suspension of isolate overnight colonies into broth sterile media. Decimal serial dilutions were made from bacterial suspension, and each of these serial dilutions was cultured in sterile Petri plates. The plates were incubated for 24 h at 37 °C. The plates containing between 30 and 300 colonies were counted, and the colony-forming unit (CFU) was calculated. CFU can be assessed by comparing the optical absorbance of known CFU and unknown CFU via a spectrophotometer at 625 nm. The absorbance will be adjusted to the same range of the McFarland standard 0.5 ($1–2 \times 10^8$ CFU ml^{-1}) with optical absorbance in range (0.08–0.13). Adjust the suspension to be in balance with McFarland Standard 0.5 by adding sterile distilled water or broth. After adjusting the bacterial suspension to a concentration of 10^8 CFU ml^{-1} with sterilized distilled water, the preparing bacterial cell suspension was directly used for the antibacterial tests for Chitosan–graft-poly(*N*-Methylaniline). In each test, 10 ml of the bacterial suspension and 0.1 g of the Chitosan–graft-poly(*N*-Methylaniline) copolymer sample were placed in a sterilized Erlenmeyer flask and carried out in triplicate. The Erlenmeyer flask was shaken (250 rpm), after for 2 h, 1 ml of this suspension was pipetted out from the flask and the change in optical absorbance was recorded [119, 120]. The percentage of inhibition was counted as follows:

$$\text{Inhibition}\,\% = (\text{Control conc.} - \text{Taste conc.})/\text{Control conc.} \times 100$$

E. coli and *S. aureus*, widely used as a pathogenic for wounds, were chosen as the standard bacterium for determining the antibacterial properties of the Chitosan–graft-poly(*N*-Methylaniline) nanoparticle. Growth inhibitory effect (Inhibition %) were used to determine the antibacterial activity of the Chitosan–graft-poly(*N*-Methylaniline) nanoparticle, which is 91.6 and 87.2% for *E. coli* and *S. aureus* respectively. This result indicated to a good antibacterial activity of tested polymer. It was observed that Chitosan–graft-poly(*N*-Methylaniline) nanoparticle showed good antibacterial activity due to the synergistic effect between chitosan and conducting polymer chains grafted onto the chitosan backbone [121, 122].

10 A Case Study in Wound Healing Due to Chitosan Grafted Poly(*N*-Methylaniline) Nanoparticles of Photo-Driven Skin Regeneration

Photodynamic therapy means the combination of a light source with a photosensitizing agent and hence produce reactive oxidizing species able to disinfect microorganisms and promote the process of tissues repairing and regeneration. A promising technique in repairing or treating wounds may ascribe when photodynamic therapy, applied topically to the wounds and the used photoactive dressings irradiated with light of appropriate energy corresponds to the optimum response. The development of nano-polymeric system photosensitizer has a great attention because of the biocompatibility and the accessibility for multi-functionalization of the nanoparticle. Nano-polymeric system based in photodynamic therapy offers several advantages such as the ability to deliver a large amount of photosensitizer to the target area, flexibility toward surface modification for better efficiency, and prevention from microorganisms in the living biological system. Many nano-polymeric systems have been used in photodynamic therapy like chitosan, collagen, albumin and alginate are grouped as natural polymers. Also, synthetic polymers like polyacrylamide and polylactide-polyglycolide co-polymers could be utilized [117]. The copolymer Chitosan–graft-poly(*N*-Methylaniline) is a modified chitosan by grafting with photoactive conducting polymer [118] acting as a potential photosensitizer, that absorbs Ultra violet and visible light, in turn, able to generate reactive oxygen species (ROS). This can reveal an environmentally desirable polymeric photosensitizer. To emphasize the effect of photodynamic therapy based on Chitosan–graft-poly(*N*-Methylaniline) nanoparticles, we applied it in wound creation in animals and investigate the wound repairing output. The effect of Chitosan–graft-poly (*N*-Methylaniline) nanoparticles on wound healing process was tested in dark and in presence of fluorescent lamp illumination with distance 1.5 m, as a visible light source (photodynamic therapy). The wound healing evaluation carried based on rat wound model, a single full-thickness skin excision wounds of 1 cm^2 in diameter was created on the back of each rat. Before wound creation in rats, the skin of the rats was shaved and disinfected using 70% (v/v) ethanol. Then the wounds were treated by Chitosan–graft-poly(*N*-Methylaniline) nanoparticles, and subsequently changed at different time intervals 3, 6, and 9 days after the surgery. The wounds diameters were serially measured using a caliper, and the appearance of the wound was examined visually. The percentage of the remaining wound areas were calculated as area before treatment (A_0) minus area contracted after treatment A_t), divided by area before treatment (A_0), multiplied by 100% as follows:

$$\text{Remaining wound areas}(\%) = [(A_0) - (A_t)]/(A_0) \times 100$$

The test was carried out on three groups, each one consists of four rats, which photodynamic therapy group, dark group, and control (untreated) group. The percentage of the remaining wound areas after the application of Chitosan–graft-poly

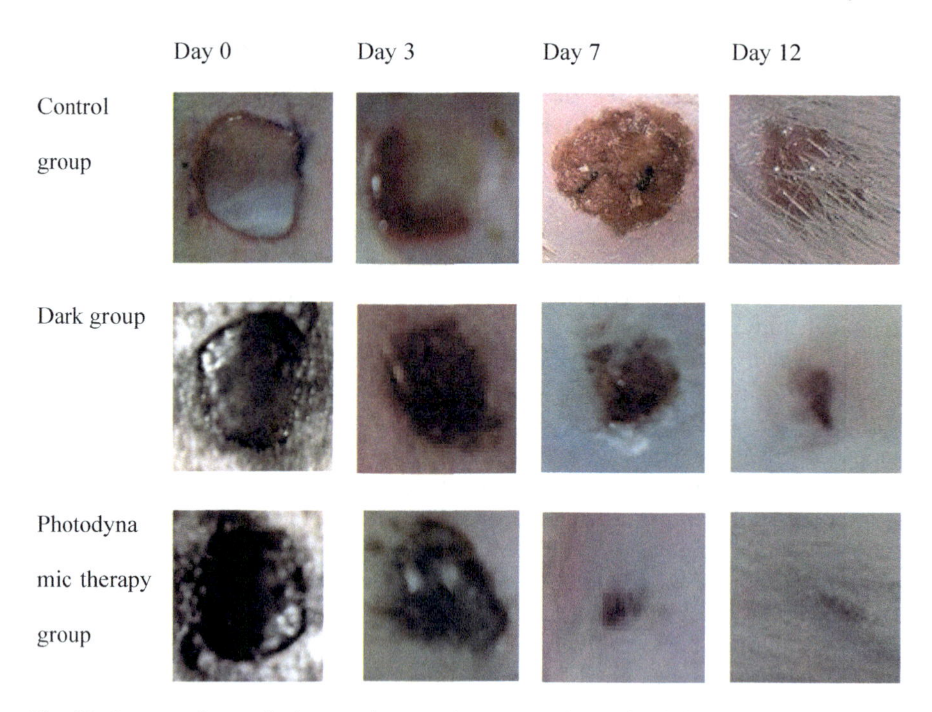

Fig. 12 Images of wounds from authors study case on the back of the rats at different time intervals

(*N*-Methylaniline) nanoparticles was found to decrease in a time-dependent manner in all the groups (Fig. 12). The wounds showed progressive healing up to 12 days for photodynamic therapy group and dark group versus control group. The healing was much better for photodynamic therapy group due to the synergistic effect of, killing bacteria via producing reactive oxygen species after light excitation and the native antibacterial action of chitosan-graft-poly(*N*-methyl aniline) nanoparticles, indicating to its importance in caring of wounds. Photodynamic therapy group and the dark group showed obvious activity in comparison with the control. It was encouraging that results showed significant differences only after 3 days versus control group, and the healing rate of photodynamic therapy was markedly higher than the dark group. After 12 days, the percentage of the remaining wound areas were 43.6, 8.9, and 1.0% for photodynamic therapy group, dark group, and control group, respectively. It suggested that the wound could be healed after treatment by chitosan-graft-poly(*N*-methyl aniline) nanoparticles. And the healing effect of chitosan-graft-poly(*N*-methyl aniline) was significantly better than that of control (untreated) group. It was probably because that photodynamic therapy group and the dark group could availably prevent bacterial invasion and wound moisture to evaporate, which was advantageous for wound healing. Chitosan–graft-poly(*N*-methylaniline) may promote the cell migration and angiogenesis in various stages of wound healing. Moreover, it should be noted that the healing rate of the control group was the slowest by visual monitoring from Fig. 12. After application of chitosan–graft-poly(*N*-methylaniline) nanoparticles in the photodynamic therapy group and dark group, wounds were turned softer due to rapid absorption of wound secretion and adjoined with the wound closely. And at the same time, bacterial infection was prevented by the outer barrier. Moreover, as early as in the 3rd day, new tissue appeared on the wound and granulation tissue was formed, which suggested that wound was in the proliferative phase. By day 12, wound disappeared and hair was seen, which suggested that the generation of hair follicles appeared.

11 Future Prospective

This chapter presents a detailed overview of biocompatible and biodegradable polymeric formulations and their correlation to wounds as well as the insights of wound healing process. Also, highlighting the recent progress in the biomedical treatment of wounds especially the current progress of chitosan-based composites and their applications in the biomedical engineering/management of wound healing. Additionally, the chapter descript the clinical findings that affect the appearance and functionality of the restored tissues taking into consideration the cost and time consumed for wound closure. Despite, a lot of chitosan-based formulations for wound healing purposes was developed, presenting a more decorated chitosan composite with greenly synthesized metallic and metal oxides of immense activity towards wounding areas is still infrequent and represents a novel perspective need to be increasingly dealt. These chitosan-based composites should fulfil many requirements in terms of effectiveness in microorganism disinfection, biocompatibility, and safety. Furthermore, our novel insights in developing chitosan nanografts as an advanced photoactive dressing for wound healing pave the way and opens the door to formulate photostable chitosan-based dressings. In this unconventional strategy, the developed photoactive dressings should safely produce reactive oxygen species able to disinfect microorganisms, promote repair and regeneration of wounded tissues. In this context, future criteria are recommended to investigate in-depth, the in vitro-in vivo correlation, and to conduct trials to human models [2].

References

1. Zivic F et al (2017) Biomaterials in clinical practice: advances in clinical research and medical devices. Springer, Berlin
2. Nair LS, Laurencin CT (2007) Biodegradable polymers as biomaterials. Prog Polym Sci 32(8–9):762–798
3. Mir M et al (2018) Synthetic polymeric biomaterials for wound healing: a review. Progress Biomater 1–21
4. Yudanova TN, Reshetov IV (2006) Modern wound dressings: manufacturing and properties. Pharm Chem J 40(2):85

5. Pandey AR, Singh US, Momin M, Bhavsar C (2017) Chitosan: application in tissue engineering and skin grafting. J Polym Res 24(8):125
6. Liu X et al (2016) In vitro BMP-2 peptide release from thiolated chitosan based hydrogel. Int J Biol Macromol 93:314–321
7. Jayakumar R et al (2011) Biomaterials based on chitin and chitosan in wound dressing applications. Biotechnol Adv 29(3):322–337
8. Howling GI et al (2001) The effect of chitin and chitosan on the proliferation of human skin fibroblasts and keratinocytes in vitro. Biomaterials 22(22):2959–2966
9. Li P et al (2011) A polycationic antimicrobial and biocompatible hydrogel with microbe membrane suctioning ability. Nat Mater 10(2):149
10. Dutta PK (2016) Chitin and chitosan for regenerative medicine. Springer, Berlin
11. Klee D, Höcker H (2000) Polymers for biomedical applications: improvement of the interface compatibility. In: Biomedical applications polymer blends. Springer, pp 1–57
12. Rickert D et al (2006) Biocompatibility testing of novel multifunctional polymeric biomaterials for tissue engineering applications in head and neck surgery: an overview. Eur Arch Oto-Rhino-Laryngol Head Neck 263(3):215–222
13. Velnar T, Bailey T, Smrkolj V (2009) The wound healing process: an overview of the cellular and molecular mechanisms. J Int Med Res 37(5):1528–1542
14. Patrulea V, Ostafe V, Borchard G, Jordan O (2015) Chitosan as a starting material for wound healing applications. Eur J Pharm Biopharm 97:417–426
15. Sen CK et al (2009) Human skin wounds: a major and snowballing threat to public health and the economy. Wound Repair Regeneration 17(6):763–771
16. Shevchenko RV, James SL, James SE (2009) A review of tissue-engineered skin bioconstructs available for skin reconstruction. J Royal Soc Interface rsif20090403
17. Percival SL et al (2012) A review of the scientific evidence for biofilms in wounds. Wound Repair Regeneration 20(5):647–657
18. Li X, Mohan S, Gu W, Baylink DJ (2001) Analysis of gene expression in the wound repair/regeneration process. Mamm Genome 12(1):52–59
19. Jorgensen SN, Sanders JR (2016) Mathematical models of wound healing and closure: a comprehensive review. Med Biol Eng Compu 54(9):1297–1316
20. Kennedy KM, Bhaw-Luximon A, Jhurry D (2017) Skin tissue engineering: biological performance of electrospun polymer scaffolds and translational challenges. Regenerative Eng Transl Med 3(4):201–214
21. Fitzmaurice GJ et al (2014) Do statins have a role in the promotion of postoperative wound healing in cardiac surgical patients? Ann Thorac Surg 98(2):756–764
22. Braiman-Wiksman L, Solomonik I, Spira R, Tennenbaum T (2007) Novel insights into wound healing sequence of events. Toxicol Pathol 35(6):767–779
23. Diegelmann RF, Evans MC et al (2004) Wound healing: an overview of acute, fibrotic and delayed healing. Front Biosci 9(1):283–289
24. Muzzarelli RAA (2009) Chitins and chitosans for the repair of wounded skin, nerve, cartilage and bone. Carbohyd Polym 76(2):167–182
25. Fonder MA et al (2008) Treating the chronic wound: a practical approach to the care of nonhealing wounds and wound care dressings. J Am Acad Dermatol 58(2):185–206
26. Dhivya S, Padma VV, Santhini E (2015) Wound dressings—a review. Biomedicine 5(4)
27. Rivera AE, Spencer JM (2007) Clinical aspects of full-thickness wound healing. Clin Dermatol 25(1):39–48
28. Boateng JS, Matthews KH, Stevens HNE, Eccleston GM (2008) Wound healing dressings and drug delivery systems: a review. J Pharm Sci 97(8):2892–2923
29. Sarabahi S (2012) Recent advances in topical wound care. Indian J Plast Surg: Official Publ Assoc Plast Surg India 45(2):379
30. Lio PA, Kaye ET (2009) Topical antibacterial agents. Infect Dis Clin North Am 23(4): 945–963

31. Vert M et al (2012) Terminology for biorelated polymers and applications (IUPAC recommendations 2012). Pure Appl Chem 84(2):377–410
32. Huang S, Fu X (2010) Naturally derived materials-based cell and drug delivery systems in skin regeneration. J Controlled Release 142(2):149–159
33. Santos TC et al (2007) In vitro evaluation of the behaviour of human polymorphonuclear neutrophils in direct contact with chitosan-based membranes. J Biotechnol 132(2):218–226
34. Ueno H, Mori T, Fujinaga T (2001) Topical formulations and wound healing applications of chitosan. Adv Drug Deliv Rev 52(2):105–115
35. Peniche C, Argüelles-Monal W, Goycoolea FM (2008) Chitin and chitosan: major sources, properties and applications. Monomers, polymers and composites from renewable resources. Elsevier, Amsterdam, pp 517–542
36. Kumar MNVR (2000) A review of chitin and chitosan applications. React Funct Polym 46(1):1–27
37. Pillai CKS, Paul Willi, Sharma CP (2009) Chitin and chitosan polymers: chemistry, solubility and fiber formation. Prog Polym Sci 34(7):641–678
38. Hussein MHM et al (2013) Preparation of some eco-friendly corrosion inhibitors having antibacterial activity from sea food waste. J Surfactants Deterg 16(2):233–242
39. El-Fattah MA et al (2016) Improvement of corrosion resistance, antimicrobial activity, mechanical and chemical properties of epoxy coating by loading chitosan as a natural renewable resource. Prog Org Coat 101:288–296
40. de Velde K, Kiekens P (2004) Structure analysis and degree of substitution of chitin, chitosan and dibutyrylchitin by FT-IR spectroscopy and solid state 13C NMR. Carbohyd Polym 58(4):409–416
41. Wasikiewicz JM, Yeates SG (2013) 'Green' molecular weight degradation of chitosan using microwave irradiation. Polym Degrad Stab 98(4):863–867
42. Ramawat KP, Mérillon J-M (2015) Polysaccharides: bioactivity and biotechnology. Springer, Heidelberg
43. Negm NA et al (2015) Treatment of industrial wastewater containing copper and cobalt ions using modified chitosan. J Ind Eng Chem 21:526–534
44. Muzzarelli RAA, Muzzarelli C (2005) Chitosan chemistry: relevance to the biomedical sciences. In: Polysaccharides I. Springer, Berlin, pp 151–209
45. Vo D-T, Sabrina S, Lee C-K (2017) Silver deposited carboxymethyl chitosan-grafted magnetic nanoparticles as dual action deliverable antimicrobial materials. Mater Sci Eng, C 73:544–551
46. Muzzarelli R, Delben F, Ilari P, Tomasetti M (1994) N-(Carboxymethyl) chitosan, a versatile chitin derivative. Agro-Food-Industry Hi-Tech
47. Wu M, Long Z, Xiao H, Dong C (2016) Recent research progress on preparation and application of *N,N,N*-trimethyl chitosan. Carbohyd Res 434:27–32
48. Mourya VK, Inamdar NN (2009) Trimethyl chitosan and its applications in drug delivery. J Mater Sci Mater Med 20(5):1057
49. Casettari L et al (2012) PEGylated chitosan derivatives: synthesis, characterizations and pharmaceutical applications. Prog Polym Sci 37(5):659–685
50. Hefni HHH et al (2016) Synthesis, characterization and anticorrosion potentials of Chitosan-g-PEG assembled on silver nanoparticles. Int J Biol Macromol 83:297–305
51. Huaixan LN et al (2016) Macroscopic, histochemical, and immunohistochemical comparison of hysterorrhaphy using catgut and chitosan suture wires. J Biomed Mater Res B Appl Biomater 104(1):50–57
52. Hein S, Wang K, Stevens WF, Kjems J (2008) Chitosan composites for biomedical applications: status, challenges and perspectives. Mater Sci Technol 24(9):1053–1061
53. Raabe D et al (2006) Microstructure and crystallographic texture of the chitin-protein network in the biological composite material of the exoskeleton of the Lobster *Homarus americanus*. Mater Sci Eng A 421(1–2):143–153

54. Arpornmaeklong P, Suwatwirote N, Pripatnanont P, Oungbho K (2007) Growth and differentiation of mouse osteoblasts on chitosan-collagen sponges. Int J Oral Maxillofac Surg 36(4):328–337
55. Silva D et al (2013) Chitosan and platelet-derived growth factor synergistically stimulate cell proliferation in gingival fibroblasts. J Periodontal Res 48(6):677–686
56. Zhao R et al (2014) Electrospun chitosan/sericin composite nanofibers with antibacterial property as potential wound dressings. Int J Biol Macromol 68:92–97
57. Costa EM et al (2014) Chitosan mouthwash: toxicity and in vivo validation. Carbohyd Polym 111:385–392
58. Rao SB, Sharma CP (1997) Use of chitosan as a biomaterial: studies on its safety and hemostatic potential. J Biomed Mater Res 34(1):21–28
59. Baldrick P (2010) The safety of chitosan as a pharmaceutical excipient. Regul Toxicol Pharmacol 56(3):290–299
60. Gades MD, Stern JS (2003) Chitosan supplementation and fecal fat excretion in men. Obesity 11(5):683–688
61. Ylitalo R et al (2002) Cholesterol-lowering properties and safety of chitosan. Arzneimittelforschung 52(01):1–7
62. Tamer TM, Valachová K, Mohyeldin MS, Soltes L (2016) Free radical scavenger activity of cinnamyl chitosan schiff base. J Appl Pharm Sci 6:130
63. Fujita M et al (2004) Inhibition of vascular prosthetic graft infection using a photocrosslinkable chitosan hydrogel. J Surg Res 121(1):135–140
64. Eldin MSM, Soliman EA, Hashem AI, Tamer TM (2008) Antibacterial activity of chitosan chemically modified with new technique. J Trends Biomater Artif Organs 22(3):121–133
65. Chung Y-C, Chen C-Y (2008) Antibacterial characteristics and activity of acid-soluble chitosan. Biores Technol 99(8):2806–2814
66. Ahmed S, Ikram S (2016) Chitosan based scaffolds and their applications in wound healing. Achievements Life Sci 10(1):27–37
67. Shoji M et al (2006) Lipopolysaccharide stimulates the production of prostaglandin E2 and the receptor Ep4 in osteoblasts. Life Sci 78(17):2012–2018
68. Grishin AV et al (2006) Lipopolysaccharide induces cyclooxygenase-2 in intestinal epithelium via a noncanonical P38 MAPK pathway. J Immunol 176(1):580–88 (Baltimore, Md. : 1950)
69. Yang E-J et al (2010) Anti-inflammatory effect of chitosan oligosaccharides in RAW 264.7 cells. Open Life Sci 5:95
70. Van Amersfoort ES, Van Berkel TJC, Kuiper J (2003) Receptors, mediators, and mechanisms involved in bacterial sepsis and septic shock. Clin Microbiol Rev 16(3):3 79–414
71. Ma L et al (2016) Anti-inflammatory activity of chitosan nanoparticles carrying NF-KappaB/ P65 antisense oligonucleotide in RAW264.7 macrophage stimulated by lipopolysaccharide. Colloids Surf B 142:297–306
72. Tu J et al (2016) Chitosan nanoparticles reduce LPS-induced inflammatory reaction via inhibition of NF-KappaB pathway in Caco-2 cells. Int J Biol Macromol 86:848–856
73. Naberezhnykh GA et al (2008) Interaction of chitosans and their N-acylated derivatives with lipopolysaccharide of gram-negative bacteria. Biochem Biokhim 73(4):432–441
74. Rodrigues S, Dionisio M, Lopez CR, Grenha Ana (2012) Biocompatibility of chitosan carriers with application in drug delivery. J Funct Biomater 3(3):615–641
75. Xu C et al (2012) Chitosan as a barrier membrane material in periodontal tissue regeneration. J Biomed Mater Res Part B Appl Biomater 100:1435–1443
76. Kean T, Thanou M (2010) Biodegradation, biodistribution and toxicity of chitosan. Adv Drug Deliv Rev 62(1):3–11
77. Leite ÁJ, Caridade SG, Mano JF (2016) Synthesis and characterization of bioactive biodegradable chitosan composite spheres with shape memory capability. J Non-Cryst Solids 432:158–166

78. Mollah MZI et al (2016) Biodegradable colour polymeric film (starch-chitosan) development: characterization for packaging materials. Open J Org Polym Mater 06:11–24
79. Kozen BG et al (2008) An alternative hemostatic dressing: comparison of CELOX, HemCon, and QuikClot. Acad Emerg Med: Official J Soc Acad Emerg Med 15(1):74–81
80. Balan V, Verestiuc L (2014) Strategies to improve chitosan hemocompatibility: a review. Eur Polym J 53
81. Pogorielov MV, Sikora VZ (2015) Chitosan as a hemostatic agent: current state. Eur J Med. Series B 2(1):24–33
82. Whang HS et al (2005) Hemostatic agents derived from chitin and chitosan. J Macromol Sci Part C 45(4):309–323
83. Boddupalli B, Mohammed Z, Nath R, Banji D (2010) Mucoadhesive drug delivery system: an overview. J Adv Pharm Technol Res 1(4):381–387
84. Perchyonok VT et al (2014) Evaluation of nystatin containing chitosan hydrogels as potential dual action bio-active restorative materials: In: Puoci F (ed) Vitro approach. J Funct Biomater 5(4):259–72. http://www.ncbi.nlm.nih.gov/pmc/articles/PMC4285406/
85. Croisier F, Jérôme C (2013) Chitosan-based biomaterials for tissue engineering. Eur Polym J 49(4):780–92. http://www.sciencedirect.com/science/article/pii/S0014305712004181
86. Frykberg RG, Banks J (2015) Challenges in the treatment of chronic wounds. Adv Wound Care 4(9):560–582
87. El-Feky GS, Sharaf SS, El Shafei A, Hegazy AA (2017) Using chitosan nanoparticles as drug carriers for the development of a silver sulfadiazine wound dressing. Carbohyd Polym 158:11–19
88. Agnihotri S, Bajaj G, Mukherji S, Mukherji Soumyo (2015) Arginine-assisted immobilization of silver nanoparticles on ZnO nanorods: an enhanced and reusable antibacterial substrate without human cell cytotoxicity. Nanoscale 7(16):7415–7429
89. Yasasvini S et al (2017) Topical hydrogel matrix loaded with simvastatin microparticles for enhanced wound healing activity. Mater Sci Eng C Mater Biol Appl 72:160–167
90. Vedakumari WS et al (2017) Quercetin impregnated chitosan-fibrin composite scaffolds as potential wound dressing materials—fabrication, characterization and in vivo analysis. Eur J Pharm Sci: Official J Eur Fed Pharm Sci 97:106–112
91. Gomathi K, Gopinath D, Rafiuddin Ahmed M, Jayakumar R (2003) Quercetin incorporated collagen matrices for dermal wound healing processes in rat. Biomaterials 24(16):2767–2772
92. Veerapandian M, Seo Y-T, Yun K, Lee M-H (2014) Graphene oxide functionalized with silver@silica-polyethylene glycol hybrid nanoparticles for direct electrochemical detection of quercetin. Biosens Bioelectron 58:200–204
93. Metwally AM, Omar AA, Harraz FM, El Sohafy SM (2010) Phytochemical investigation and antimicrobial activity of *Psidium guajava* L. leaves. Pharmacognosy Magazine 6(23): 212–218
94. Bhardwaj N, Kundu S (2011) Silk fibroin protein and chitosan polyelectrolyte complex porous scaffolds for tissue engineering applications. Carbohyd Polym 85:325–333
95. Noorjahan SE, Sastry TP (2004) An in vivo study of hydrogels based on physiologically clotted fibrin-gelatin composites as wound-dressing materials. J Biomed Mater Res B Appl Biomater 71(2):305–312
96. Romić MD et al (2016) Melatonin-loaded chitosan/Pluronic® F127 microspheres as in situ forming hydrogel: an innovative antimicrobial wound dressing. Eur J Pharm Biopharm 107:67–79. http://www.sciencedirect.com/science/article/pii/S0939641116302223
97. Reiter RJ, Tan D-X, Fuentes-Broto L (2010) Melatonin: a multitasking molecule. Prog Brain Res 181:127–151
98. Gomez-Florit M, Ramis JM, Monjo M (2013) Anti-fibrotic and anti-inflammatory properties of melatonin on human gingival fibroblasts in vitro. Biochem Pharmacol 86(12):1784–1790

99. Drobnik J (2012) Wound healing and the effect of pineal gland and melatonin. J Exp Integr Med 2(1):3–14
100. Sahib AS, Al-Jawad FH, Al-Kaisy AA (2009) Burns, endothelial dysfunction, and oxidative stress: the role of antioxidants. Annals Burns Fire Disasters 22(1):6–11. http://www.ncbi.nlm.nih.gov/pmc/articles/PMC3188210/
101. Sahib AS, Al-Jawad FH, Alkaisy AA (2010) Effect of antioxidants on the incidence of wound infection in burn patients. Ann Burns Fire Disasters 23(4):199–205
102. Gopal A, Kant V, Gopalakrishnan A, Tandan SK, Kumar D (2014) Chitosan-based copper nanocomposite accelerates healing in excision wound model in rats. Eur J Pharmacol 731:8–19
103. Fan X et al (2016) Nano-TiO_2/collagen-chitosan porous scaffold for wound repairing. Int J Biol Macromol 91:15–22
104. Verdier T, Coutand M, Bertron A, Roques C (2014) Antibacterial activity of TiO_2 photocatalyst alone or in coatings on *E. coli*: the influence of methodological aspects. Coatings 4:670–686
105. Macwan DP, Dave P, Chaturvedi S (2011) A review on nano-TiO_2 sol–gel type syntheses and its applications. J Mater Sci 46:3669–3686
106. Beer C et al (2012) Toxicity of silver nanoparticles—nanoparticle or silver ion? Toxicol Lett 208(3):286–292
107. Lara HH, Ayala-Nunez NV, Ixtepan-Turrent L, Rodriguez-Padilla C (2010) Mode of antiviral action of silver nanoparticles against HIV-1. J Nanobiotechnol 8:1
108. Wang X, Du Y, Liu H (2004) Preparation, characterization and antimicrobial activity of chitosan-Zn complex. Carbohyd Polym 56:21–26
109. Darder M, Aranda P, Ruiz-Hitzky E (2007) Bionanocomposites: a new concept of ecological, bioinspired, and functional hybrid materials. Adv Mater 19:1309–1319
110. Ding L et al (2017) Spongy bilayer dressing composed of chitosan–Ag nanoparticles and Chitosan-*Bletilla striata* polysaccharide for wound healing applications. Carbohyd Polym 157:1538–1547
111. Ahmadi F, Oveisi Z, Samani SM, Amoozgar Z (2015) Chitosan based hydrogels: characteristics and pharmaceutical applications. Res Pharm Sci 10(1):1–16
112. Luca L et al (2011) Injectable RhBMP-2-loaded chitosan hydrogel composite: osteoinduction at ectopic site and in segmental long bone defect. J Biomed Mater Res Part A 96:66–74
113. Dai T, Tanaka M, Huang Y-Y, Hamblin MR (2011) Chitosan preparations for wounds and burns: antimicrobial and wound-healing effects. Expert Rev Anti-Infect Ther 9(7):857–879
114. Chatterjee DK, Fong LS, Zhang Y (2008) Nanoparticles in photodynamic therapy: an emerging paradigm. Adv Drug Deliv Rev 60(15):1627–1637
115. Chen C-P, Chen C-T, Tsai T (2012) Chitosan nanoparticles for antimicrobial photodynamic inactivation: characterization and in vitro investigation. Photochem Photobiol 88(3): 570–576
116. Chien H-F et al (2013) The use of chitosan to enhance photodynamic inactivation against *Candida albicans* and its drug-resistant clinical isolates. Int J Mol Sci 14(4):7445–7456. http://www.ncbi.nlm.nih.gov/pmc/articles/PMC3645695/
117. Gupta A et al (2013) Shining light on nanotechnology to help repair and regeneration. Biotechnol Adv 31(5):607–631
118. Sayyah SM, Essawy AA, El-Nggar AM (2015) Kinetic studies and grafting mechanism for methyl aniline derivatives onto chitosan: highly adsorptive copolymers for dye removal from aqueous solutions. React Funct Polym 96:50–60. http://www.sciencedirect.com/science/article/pii/S1381514815300249
119. Singh G, Joyce E, Beddow J, Mason T (2012) Evaluation of antibacterial activity of ZnO nanoparticles coated sonochemically onto textile fabrics. World J Microbiol Biotechnol 2:106–120
120. Wiegand I, Hilpert K, Hancock REW (2008) Agar and broth dilution methods to determine the minimal inhibitory concentration (MIC) of antimicrobial substances. Nat Protoc 3(2): 163–175

121. Çabuk M, Yusuf A, Yavuz M, Unal H (2014) Synthesis, characterization and antimicrobial activity of biodegradable conducting polypyrrole-graft-chitosan copolymer. Appl Surf Sci 318:168–175
122. Shanmugam A, Kathiresan K, Nayak L (2016) Preparation, characterization and antibacterial activity of chitosan and phosphorylated chitosan from cuttlebone of *Sepia kobiensis* (Hoyle, 1885). Biotechnol Rep 9:25–30

Mechanical, Thermal and Viscoelastic Properties of Polymer Composites Reinforced with Various Nanomaterials

T. H. Mokhothu, A. Mtibe, T. C. Mokhena, M. J. Mochane, O. Ofosu, S. Muniyasamy, C. A. Tshifularo and T. S. Motsoeneng

1 Introduction

Nanotechnology has attracted a considerable attention in science community due to the growing demand to develop high-performance materials for medical, sensors, computing, packaging, textiles, automotive, membrane-based separation, water purification, etc. to make our lives more comfortable. Various nanomaterials such as carbon nanotubes, graphite, metal oxide nanoparticles, clay nanoparticles, and nanocellulose have been extensively investigated due to their good physical, antimicrobial, electrical, thermal, chemical and mechanical properties. In recent years, nanomaterials have been used in applications that require elevated

T. H. Mokhothu
Department of Chemistry, Durban University of Technology,
Durban, South Africa

A. Mtibe (✉) · T. C. Mokhena (✉) · O. Ofosu · S. Muniyasamy
C. A. Tshifularo
CSIR Materials Science and Manufacturing, Polymers and Composites
Competence Area, Nonwovens and Composites Research Group,
Port Elizabeth, South Africa
e-mail: mtibe.asanda@gmail.com

T. C. Mokhena
e-mail: mokhenateboho@gmail.com

T. C. Mokhena · C. A. Tshifularo
Department of Chemistry, Nelson Mandela University, Port Elizabeth, South Africa

M. J. Mochane
Department of Life Sciences, Central University of Technology Free State,
Bloemfontein, South Africa

T. S. Motsoeneng
Department of Polymer Technology, Tshwane University of Technology,
Pretoria, South Africa

Inamuddin et al. (eds.), *Sustainable Polymer Composites and Nanocomposites*,
https://doi.org/10.1007/978-3-030-05399-4_6

mechanical performance such as polymer nanocomposites as a reinforcing element. Polymer nanocomposite is a combination of the polymer matrix and nanomaterials with one, two and/or three dimensions.

Polymers are widely used in various applications due to their low cost, flexibility and easy processing. However, they have inherited some drawbacks such as low tensile properties and poor fracture toughness which limits their applications [20]. Therefore, to address these drawbacks nanomaterials have to be compounded with the polymer matrix. To develop polymer nanocomposites with the required properties depends on the filler properties and dispersion of the filler within the polymer matrix. For instance, to formulate polymer nanocomposites with high conductivity, carbon nanotubes should be used as filler due to its high conductivity [4]. However, it is widely accepted that poor dispersibility causes agglomeration of filler which led to poor interfacial adhesion between a filler and polymer matrix and therefore results in the poor mechanical performance of the resultant material [20]. To improve the dispersion of fillers in a polymer matrix, modification of either filler or polymer should be considered to alter the functional groups of the materials in order to achieve a good interaction between polymer and filler which enhances properties of the resultant materials [95, 96].

Many researchers have investigated the effect of nanomaterials on the properties of polymer nanocomposites [14, 87–89, 93, 94]. However, many studies indicated that the incorporation of nanomaterials enhanced the properties of the resultant polymer nanocomposites [90–92, 101]. For example, [67] reported that the incorporation of nanocellulose enhanced tensile and thermal properties of polyvinyl alcohol (PVA) nanocomposites. It was also widely noticed that the increase in loading showed a positive effect on mechanical properties of nanocomposites.

This book chapter reviews the effect of nanomaterials on the properties of polymer nanocomposites. It is also highlighting the hybridization of fillers and studied their effect on the properties of nanocomposites. Lastly, this chapter highlights the incorporation of nanomaterials in biopolymers and investigated their properties.

2 Mechanical Properties of Nanocomposites

2.1 Mechanical Properties of Polymer Reinforced with Cellulose-Based Nanofillers

Cellulose-based nanofillers are categorized into cellulose nanocrystals (CNCs) and cellulose nanofibres (CNFs). CNFs consist of crystalline and amorphous region whereas CNCs consist of the only crystalline region. In addition, CNFs are web-shaped [58] bundles stabilized by hydrogen bonds while CNCs are rod-like shaped [75]. The diameters of both CNFs and CNCs are in nanoscale and their lengths in microscale [77]. It was also reported that the tensile strength and modulus of CNCs were 14.3–28.6 and 143 GPa, respectively [19]. In addition, CNCs have a low elongation at break, high aspect ratio, and large surface area. CNFs have similar characteristics with CNCs.

Due to the aforementioned extraordinary properties of cellulose-based nanofillers, they have attracted a considerable attention in polymer nanocomposites field as suitable reinforcement using low loading amount. The study reported by Chen et al. [25] revealed that the addition of 10 wt% of CNCs extracted from pea hull fiber in pea starch polymer enhanced tensile strength and elongation at break of the pea starch nanocomposites due to their high aspect ratio. The authors added that the strong adhesion between the two materials led to the improvement of mechanical properties of the resultant nanocomposites. Similar observations were reported in the case of CNFs [13, 45, 48, 49]. In addition, [49] indicated that the increase in CNFs loading in polylactic acid (PLA) led to the enhancement of tensile strength and modulus but, elongation at break was reduced. This was attributed to the good mechanical properties of CNFs and the interaction between CNFs and PLA. In contrast, [47] reported that no significant alterations in tensile properties were observed in melt-spun PLA reinforced with CNCs.

The hydrophilic nature of cellulose-based nanofillers led to poor interaction between nanofillers and the hydrophobic polymer matrix. Hence, the surface modification of these materials is crucial to improving their hydrophobicity, dispersion, and interaction between them and the polymer matrix. The incorporation of acetylated CNCs up to 4.5 wt% resulted in an overall increase in tensile strength and modulus as well as elongation at break. The further increase above that resulted in a decrease in tensile properties. This can be explained by the fact that when loading was less than 4.5 wt%, the distance between CNCs was big and therefore the interaction was weak to form percolation network. However, at higher loadings, a decrease in tensile properties was observed due to agglomeration [114]. On the other hand, [13], reported that the enhancement of tensile properties in thermoplastic starch (TPS) reinforced with unmodified CNFs was due to the formation of hydrogen-bonded nanofibres network, entanglement and strong interfacial adhesion between TPS and CNFs. The authors also reported that tensile properties of TPS reinforced with acetylated CNFs were lower than those of unmodified CNFs reinforced TPS. This was attributed to the lack of fibers-to fibers and fibers-to-polymer matrix interactions due to the surface hydrophobicity in modified CNFs.

2.2 *Mechanical Properties of Polymer Nanocomposites Reinforced with Carbonaceous Nanofillers*

Carbonaceous nanofillers such as carbon nanotubes (CNTs) and graphite have exhibited extraordinary tensile strength and modulus, electrical properties, large surface area, high aspect ratio and low density. Given these extraordinary properties, CNTs can be regarded as ideal candidates for reinforcement in polymer nanocomposites to enhance the mechanical properties, thermal conductivity and electrical properties of the resultant polymer nanocomposites. Considering the properties of CNTs, [20] developed epoxy nanocomposites reinforced with multi-wall carbon nanotubes (MWNTs). The results suggested that the incorporation of pristine MWNTs in epoxy led to a slight increase in tensile modulus but decrease tensile strength. However, incorporation of functionalized MWNTs with polystyrene sulfonate (PSS) and poly(4-amino styrene) (PAS) in epoxy resulted in an increase in both tensile modulus and strength. Similar findings were also reported by Mashhadzadeh et al. [69]. It was also reported that the addition of CNTs in high-density polyethylene (HDPE) resulted in the improvement of its hardness [35]. In contrast, other authors reported that the addition of carbonaceous fillers without modification in polymer matrix results in improvement of tensile properties [4, 74, 109, 111]. Moreover, it is worth noting that the increase in filler loading enhanced mechanical properties of the resultant polymer nanocomposites [4]. In addition, the addition of carbonaceous nanofillers in polymer matrix induce the electrical properties of nanocomposites [50]. Lopez-manchado et al. [63] investigated the effect of thermal reduced graphene oxide on the mechanical properties of plasticized natural rubber with dodecyltrimethylammonium bromide (DTAB). The addition of thermal reduced graphene oxide enhanced the stiffness of plasticised natural rubber.

Numerous researchers have investigated the effect of loading of carbonaceous nanofillers in polymer matrices [2, 50]. For instance, [50] investigated the effect of large aspect ratios and exceptional high mechanical strength MWNTs loading on the mechanical properties of acrylonitrile butadiene styrene. Both tensile strength and modulus increased with increase in loading but, in the case of tensile strength it increased up to 7 wt% and decreased with further increase in loading. The enhancement of mechanical properties was due to the uniform dispersion of MWNTs throughout the polymer matrix. Above 7 wt% loading, the agglomeration of MWNTs was clearly observed which could be the reason for the decline in the tensile strength of the nanocomposites. On the other hand, elongation at break decreased with an increase in loading. In addition, the addition of MWNTs at different loading also improved the electrical conductivity. Liao et al. [60] tested the impact strength of polypropylene (PP) nanocomposites reinforced with MWNTs and hydroxyapatite designed for bone implants. The authors reported in their extensively investigated study that the impact strength of PP reinforced with hydroxyapatite decreases with increasing hydroxyapatite loading. However, the incorporation of MWNTs in PP nanocomposites reinforced with hydroxyapatite enhanced the impact strength due to their flexibility and large strain to failure. Also, it was reported that the inclusion of MWNTs increases the degree of crystallinity of PP nanocomposites which lead to the enhancement of tensile properties and impact strength. Younesi et al. [116] investigated flexural behaviour of low-density polyethylene (LDPE) reinforced with single wall carbon nanotubes (SWNTs) and wood flour. The addition of SWNTs and the modification of LDPE with maleic anhydride enhanced the flexural modulus. However, the flexural modulus increased with increasing loading (1–3 wt%) of SWNTs. The enhancement of flexural properties was due to the high aspect ratio of SWNTs. It was also reported that SWNTs were well dispersed in the polymer and therefore improved the interfacial adhesion between polymer and SWNTs which result in improvement of flexural properties. Furthermore, impact strength also improved when SWNTs were added.

2.2.1 Mechanical Properties of Biopolymers Reinforced with Carbonaceous Fillers

Much research is focusing on the development of biobased and biodegradable products due to their eco-friendliness, sustainability, and biodegradability. Biopolymers are among the materials that have been extensively investigated. Biopolymers which are widely studied include polylactic PLA, TPS, poly(hydroxybutyrate-co-hydroxyvalerate) PHBV, poly(butylene succinate) (PBS), polysaccharides and proteins. Like any other material, biopolymers have inherited some drawbacks such as moisture absorption, difficulty in processability and low properties in comparison to traditional petroleum-based polymers. To address these drawbacks, biopolymers are blended with other polymers or reinforced with stiffer nanofillers. For instance, [97] reported that the addition of 1 wt% CNTs in PHBV reduced water uptake

Table 1 Biopolymers reinforced with carbonaceous nanofillers

Biodegradable polymer	Filler	Publication year	References
Larch lignocellulose	MWNTs	2017	Huang et al. [46]
Poly(butylene succinate-co-adipate) (PBSA)	Halloysite nanotube	2016	Chiu [27]
Polycaprolactone (PCL)	MWNTs	2010	Sanchez-garcia et al. [97]
Hydroxyapatite	MWNTs	2017	Khan et al. [52]
PBSA/maleated polyethylene blend	Halloysite nanotube	2017	Chiu [28]
Poly(lactic-co-glycolic acid)	Carboxylation MWNTs	2011	Lin et al. [61]
Epoxidized natural rubber	MWNTs	2018	Krainoi et al. [55]
Polylactide/poly(ε-caprolactone)	Thermally exfoliated graphene oxide (GO)	2018	Botlhoko et al. [17]
Poly(3-hydroxyalkanoate)	Grafted MWNTs	2015	Mangeon et al. [68]
PLA	Kenaf fibre/ MWNTs	2017	Chen et al. [22]
Poly(l-lactide) (PLLA)/poly (3-hydroxybutyrate-co-4-hydroxybutyrate) (P(3HB-co-4HB)) blend	MWNTs	2017	Gao et al. [38]
TPS	Oxidized MWNTs	2013	Cheng et al. [26]
PLA	CNT	2016	Wang et al. [112]

whereas at increased CNTs loading an increase in water uptake was observed. Similar results were observed in the case of PHBV reinforced with carbon nanofibres. Other studies that investigated the mechanical properties of biopolymers reinforced with carbonaceous nanofillers are summarized in Table 1.

Incorporation of carbonaceous nanofillers into biodegradable polymers improves the mechanical properties of the resultant polymer nanocomposites [55, 97]. For example, the incorporation of MWNTs in epoxidized natural rubber nanocomposites led to enhancement of mechanical properties as shown in Fig. 1.

It was also reported that mechanical properties of epoxidized natural rubber nanocomposites increased with increasing MWNTs loading. It was seen that at 5 wt% MWNTs loading reinforced epoxidized had the highest tensile strength, further increase in loading above 5 wt% MWNTs led to a decrease in tensile strength while modulus was constantly improving. The decrease in tensile strength after 5 wt% MWNTs could be due to CNTs aggregation in the polymer matrix. Conversely, the elongation at break decreased with MWNTs loading [55]. Similar results were reported by Wang et al. [112], in their case they discovered that 3 wt% CNTs loading was the optimum and further increase in CNTs loading led to decrease in tensile properties.

Other researchers [18, 26, 61, 68] incorporated functionalized carbonaceous nanofillers in biopolymers to further enhance mechanical properties of the nanocomposites. Lin et al. [61] incorporated carboxylated MWNTs in poly (lactic-co-glycolic acid) for bone tissue engineering. Morphological properties indicated that the treatment shortened the length of CNTs which tend to avoid agglomeration. This led to enhancement of tensile properties of the resultant nanocomposites by nearly three-fold in comparison to the virgin polymer and by nearly two-fold in comparison to those of polymer reinforced with untreated MWNTs. In addition, functionalized MWNTs based nanocomposites degrade faster than both unfunctionalized MWNTs based nanocomposites and virgin polymer. Similar results were observed in the case of PHBV reinforced with 3 wt% grafted CNTs [68]. In contrast, the addition of functionalized graphene oxide in polylactide/poly(ε-caprolactone) blend led to the decrease in tensile properties. However, tensile properties slightly improved with the increase in loading.

A global research is now moving towards hybridizing two or more fillers to enhance the performance of the material for diversified applications [22, 60, 116]. Chen et al. [22] fabricated PLA nanocomposites reinforced with a combined functionalized kenaf fibres and MWNTs by melt mixing and compression moulding techniques. They reported that tensile properties of the resultant nanocomposites increased due to the interfacial interaction between modified kenaf fibres by 3-glycidoxypropyltrimethoxysilane and PLA which improved the stress transfer and thus, lead to increase in tensile properties. Impact strength results were correlating well with those of tensile properties. The authors reported that the enhancement of impact strength was attributed to the structure of cellulose which tolerates higher deformation under impact.

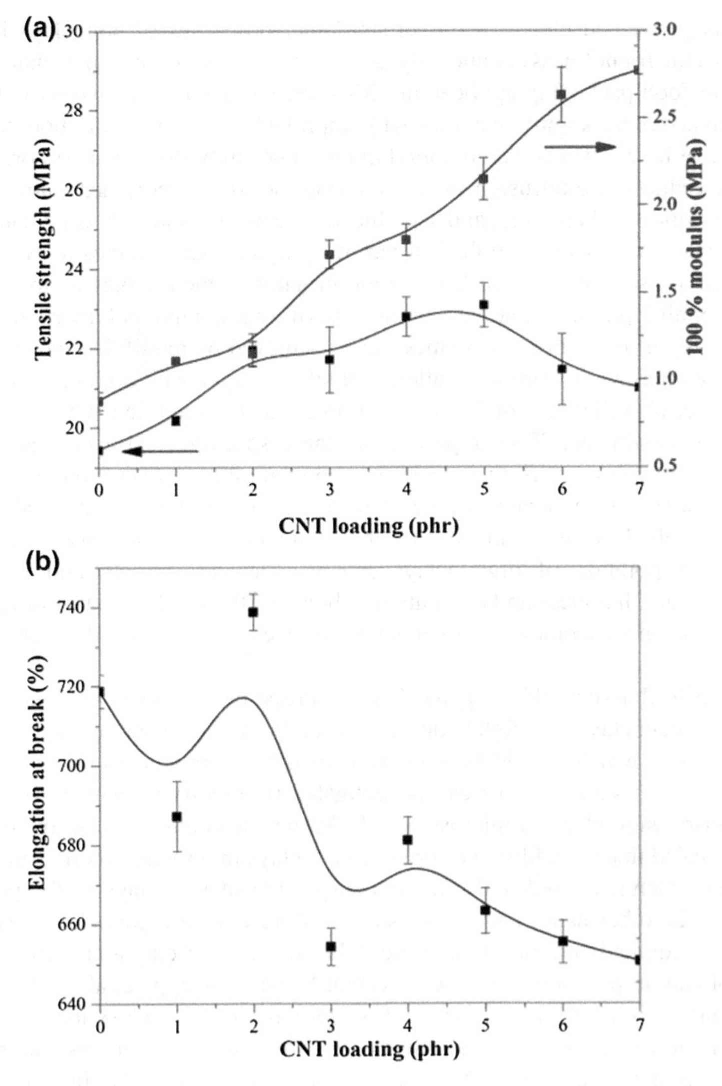

Fig. 1 Tensile strength and modulus of epoxidized natural rubber and their nanocomposites (**a**) and elongation at break of epoxidized natural rubber and their nanocomposites [55], copyright with the permission from Elsevier

2.3 Mechanical Properties of Polymer Reinforced with Nanoclays

In recent years, nanoclays have received considerable attention as reinforcing element in polymer nanocomposites due to their extraordinary properties such as high mechanical properties, large surface area and high aspect ratio, good thermal, optical, magnetic and electrical properties. Other properties of nanoclays include environmental friendliness, abundantly available and non-toxic which make them suitable for food packaging applications. Most recent, nanoclays have been utilized to develop active packaging materials with improved tensile strength, modulus, and elongation at break. It was also reported in the same study that the incorporation of nanoclays reduced the diffusion of water vapour across the polymer matrix [85].

Montmorillonite, bentonite, and sepiolite are the most commonly used nanoclays and they were successfully applied in various polymer nanocomposite systems as nanofillers. These materials are hydrophilic in nature which makes them miscible with hydrophilic polymer matrices. In the case of hydrophobic polymeric matrices, the miscibility of nanoclays and matrix can be achieved by modifying nanoclays by exchanging interlayer of cationic galleries of silicate layer with organic component [86]. Shah et al. [100] reported a study on organoclays modified with quaternary ammonium substituents. They reported that the d-spacing, interlayer spacing and hydrophobicity (parameters that determine compatibility between nanoclays and polymer matrix) of nanoclays increased with increasing chain length and benzyl substituents which result in an increase in exfoliation. This study also discovered that the incorporation of organoclays enhances tensile strength and modulus, flexural strength, hardness and elongation at break of the resultant nanocomposites. However, the incorporation of organoclays showed an inverse effect on impact strength.

Other critical issues affecting mechanical properties of nanocomposites reinforced with nanoclays are higher phase separation and particles aggregation in a polymer matrix which should be prevented to achieve proper reinforce effect of nanoclays. These shortcomings can be mitigated by modifying nanoclays. In one study, transmission electron microscopy (TEM) and scanning electron microscopy (SEM) revealed that the addition of pristine nanoclays in polymer matrix resulted to disordered structures which indicate poor dispersion of nanoclays in the polymer matrix. On the other hand, nanocomposites reinforced with organoclays displayed some fine exfoliated and individual un-exfoliated layers of clay and therefore dispersion of clay in polymer matrix was evident [100]. Malkappa et al. [66] reported that the surface roughness increases with increasing in organoclays loading while; agglomeration was evident and became more visible when organoclays loading was increasing as shown in Fig. 2. Also, tensile strength and modulus increased with increasing organoclays loading whereas elongation at break showed inversely effect. Similar observations were evident in Alcântara et al. [6] study, they also reported that water resistance, biocompatibility, and biodegradation were improved when fibrous nanoclays were incorporated in polysaccharides. In another study, it was reported that tensile strength and modulus (as shown in Fig. 3) of biopolymer blends reinforced with expanded organoclay (EOC) increased linearly with increasing nanoclays loading. However, elongation at break was inversely proportional to nanoclays loading [76].

Recently, [72] fabricated PP/LDPE blends reinforced with organoclays by twin screw extruder and injection moulding and investigated their effect on mechanical properties. They discovered that the impact strength of PP decreased after blending

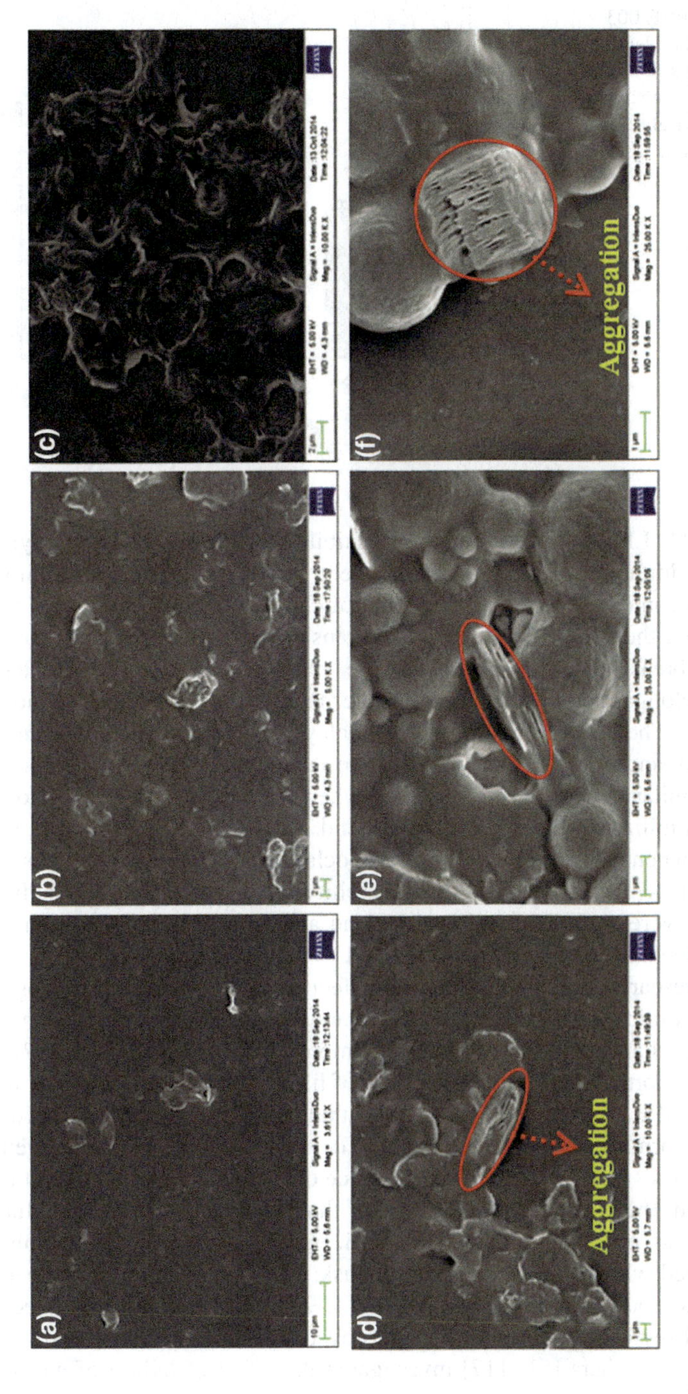

Fig. 2 Field emission electron microscopy (FESEM) images of: **a** water dispersible polyurethane (WDPU)/Cloisite-30B-1 wt%, **b** WDPU/Cloisite-30B-3 wt%, **c** WDPU/Cloisite-30B-5 wt%, **d** WDPU/OKao-1 wt%, **e** WDPU/OKao-3 wt% and **f** WDPU/OKao-5 wt% [66], copyright with the permission from Elsevier

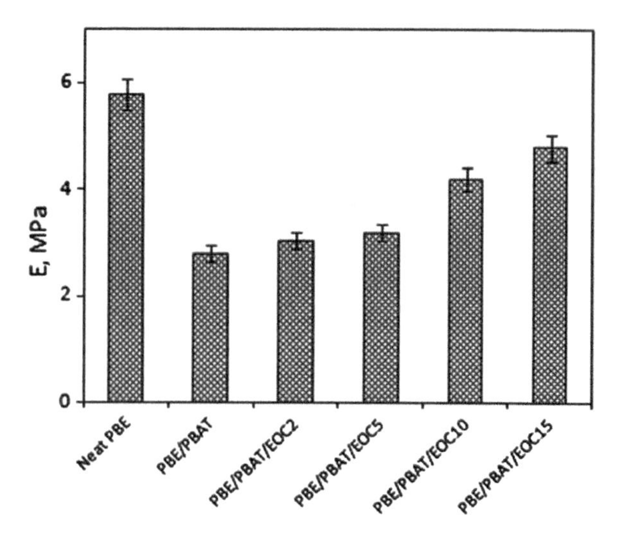

Fig. 3 tensile modulus of neat Natureplast PBE 003 (PBE), PBE/PBAT blend and its nanocomposites reinforced with expanded organoclay (EOC) [54], copyright with the permission from Elsevier

it with 20 wt% LDPE due to poor compatibility between the two polymers. However, the addition of organoclays in the blend led to the deterioration of impact strength which could be due to the restriction of chain mobility. In the same study, they investigated the effect of organoclays on tensile strength, tensile modulus, and elongation at break of polymer blends. The values of tensile strength, tensile modulus, and elongation of the blend were intermediates in comparison to those of neat polymers. The incorporation of organoclays in polymer blends enhanced tensile modulus and elongation at break while tensile strength was deteriorating. Moreover, a similar trend was also observed when the combination of organoclays and the compatibilizers were added to the blend. The highest elongation at break was observed when the combination of organoclays and the compatibilizers were added to the blend this was due to the miscibility of the materials. This indicates that nanoclays were well dispersed in the blend and the interaction between them and both polymers in the blend was improved.

Numerous researchers have reported the effect of hybrid of nanoclays together with other nanofillers such as metal oxide nanoparticles, nanotubes and natural fibres on the mechanical properties of nanocomposites [1, 9, 23, 51, 81, 117]. For instance, [23] reported that the incorporation of fillers (organoclays and rice husk) in recycled HDPE and polyethylene terephthalate (PET) blend enhanced the mechanical performance of nanocomposites. They suggested that the improvement of tensile properties could be due to the presence of organoclays which might carry much load and the fact that clay is stiffer than polymer matrix. The further enhancement was observed when compatibilizers were incorporated in polymer matrix reinforced with organoclays and rice husk. These results were in agreement with the results reported by other researchers [1, 9, 33] and they also reported that the mechanical performance was further improved when natural fibers were alkali treated. Other researchers [51, 117] investigated the effect of hybrid of nanoclays and metal oxide nanoparticles (zinc and silver) on mechanical properties of nanocomposites. They reported that incorporation of nanoparticles enhanced the tensile strength and modulus. In addition, the inclusion of nanoparticles reduced water vapour permeability but, increased water content and density. Moreover, the inclusion of nanoparticles improved the antibacterial activity against Gram-positive S. aureus, Gram-negative *E. Coli*, and foodborne pathogens.

2.3.1 Mechanical Properties of Biopolymers Reinforced with Nanoclays

In recent years, nanocomposites from biopolymers reinforced with nanoclays have been extensively studied due to their low toxicity and biodegradability. Aliphatic polyesters have attracted tremendous attention for diversified applications. Polybutylene succinate (PBS) is among the widely studied aliphatic polyester. Phua et al. [84] investigated the impact of nanoclays on mechanical performance of polybutylene succinate (PBS) nanocomposites. In their study, they investigated mechanical properties to determine the biodegradation behaviour of nanocomposites. Before soil burial, they observed that the mechanical properties improved when organoclays were incorporated. The mechanical properties of nanocomposites deteriorated after soil burial and further reduced with soil burial time. In another study, the incorporation of nanoclays improved the mechanical performance of PLA. It was also reported that the increase in nanoclays loading increased linearly the tensile modulus while inversely effect was observed for elongation at break [43]. In contrast, the inclusion of organoclays in PHBV did not reinforce as anticipated due to the aggregation of clay in a polymer matrix [30]. Thereafter, numerous researchers fabricated nanocomposites from blended aliphatic polyesters with other biodegradable polymers and nanoclays. One of the limitations of biodegradable polymers is low mechanical properties which are not suitable for other applications such as packaging. The poor mechanical properties can be enhanced by blending biopolymers with other polymers or reinforced with stiffer fillers. Ayana et al. [11] and Lendvai et al. [59] blended thermoplastic starch with PLA and PBAT, respectively and subsequently reinforced with nanoclays. Ayana et al. [11] reported that the introduction of nanoclays enhanced tensile properties of the blends with an increase in loading. On the other hand, in the case of TPS/PBAT blend the inclusion of nanoclays did not show any effect on the tensile strength and modulus [59].

Agro-polymers such as starch, cellulose, protein, chitin and chitosan-based nanocomposites have recently attracted more interest due to their renewability, biocompatibility, biodegradability and non-toxicity with outstanding adsorption properties. The major drawbacks of agro-polymers are moisture absorption, difficulty processability, and poor mechanical properties. The inclusion of nanoclays in agro-polymers could result in the reduction of moisture absorption as well as enhance mechanical performance of nanocomposites. Numerous researchers [5, 34] are investigating efforts to overcome the drawbacks of agro-polymers.

The inclusion of nanoclays reduced water absorption and a further increase in nanoclays loading showed a decrease in water absorption [5]. They also reported that tensile strength and elongation at break decreased with increasing loading of nanoclays. Farahnaky et al. [34] prepared gelatin nanocomposites reinforced with nanoclays by a solvent casting method. They showed that tensile modulus increased linearly with increasing nanoclays loading while elongation at break decreased with increasing loading. A similar trend was observed in the case of chitosan nanocomposites reinforced with nanoclays [39, 40].

3 Thermal Properties

3.1 Thermogravimetric Analysis (TGA)

Thermogravimetric analysis is widely used to investigate the thermal degradation of polymer nanocomposites. A typical thermogram of polymer nanocomposite shows a material subjected to heat will suffer mass loss, followed by a sharp drop in mass over a narrow range and subsequently back to the flat slope as reactant is exhausted [12]. TGA and derivative thermogravimetric (DTG) are used to determine the mass loss and degradation of the material at a certain temperature as well as the remaining char content. Typical TGA curves are shown in Fig. 4.

In the published literature, it has been reported that the improvement in thermal stabilities of the nanocomposites is mainly attributed to the reinforcement effect of nanomaterials on polymers [57, 97, 119]. For instance, [3] investigated the effect of CNCs on thermal properties of polyfurfuryl alcohol (PFA) nanocomposites. They reported that the incorporation of CNCs improved the thermal stability of the PFA. Both the neat PFA and the nanocomposites showed two degradation steps at above

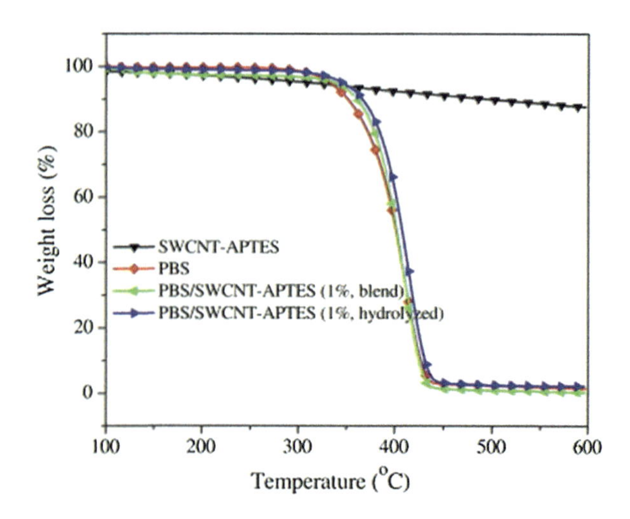

Fig. 4 TGA curves of SWNTs functionalized with acyl aminopropyltriethoxysilane (APTES), neat PBS, PBS/SWNTs-APTES (1%, hydrolyzed) [108], copyright with the permission from Elsevier

200 °C and in the temperature range of 320–400 °C which involves the scission of the weaker chemical bonds. According to Sanchez-garcia et al. [97], high thermal stabilities of PCL was achieved when 1 and 3 wt% of CNTs were incorporated, though increasing CNTs loading above 5 wt% result in filler agglomeration which reduced thermal stabilities of resultant nanocomposites. Similar behaviour was observed in the case of PHBV nanocomposites. Lai et al. [57] reported that the thermal stabilities of PLA nanocomposites increased with increasing in nanoclays loading. Interestingly, char content also increased with increasing in nanoclays loading. Other studies on thermal properties of nanocomposites are listed in Table 2.

Other researchers reported that the functionalization of either filler or polymer matrix enhances thermal stabilities of the resultant materials. For instance, [119] investigated the effect of unmodified and modified CNCs with phthalic anhydride on thermal stabilities of PBSA. They reported that degradation temperature of neat PBSA was above 300 °C due to chain scission and inter and intramolecular transesterification reactions. PBSA reinforced with CNCs modified with phthalic anhydride exhibited higher degradation temperature in comparison to those of PBSA reinforced with unmodified CNCs and neat PBSA. This enhancement was due to the addition of phthalic anhydride. Similar observations were reported in the case epoxy reinforced with GO modified with silane [111]. Majeed et al. [64] studied the incorporation of CNTs in neat LDPE and maleic anhydride grafted polyethylene (MAPE). They reported that MAPE reinforced with CNTs was thermally more stable in comparison to LDPE reinforced with CNTs and neat LDPE. The improved thermal stability with the inclusion of MAPE could result in improved compatibility and better dispersion of CNTs. In another study, they investigated thermal properties of rHDPE/rPET blend mixed with 3 wt% MAPE and 5 wt% ethylene-glycidyl methacrylate and subsequently reinforced with 3 wt% nanoclays and 70 wt% rice husk. Nanocomposites of polymer blends reinforced with nanoclays only exhibited a single step degradation pattern with improved thermal stability. However, the addition of rice husk resulted in three degradation steps which represent moisture evaporation at temperatures ranging from 135 to 145 °C, depolymerisation of hemicellulose and decomposition of cellulose at temperatures ranging from 230 to 370 °C and decomposition of nanocomposites and slightly decomposition of lignin at temperatures ranging from 476 to 482 °C. Incorporation of rice husk and nanoclays in compatibilizing matrix resulted in improvement of thermal stability in comparison to uncompatibilizing matrix.

Nanomaterials have been reported in numerous studies to enhance the thermal stabilities of polymer matrices. In contrast, [112] reported that the inclusion of CNTs did not affect the single stage decomposition pattern of PLA and remain unaltered. However, the incorporation of 1 wt% CNTs in PLA exhibited no alterations in thermal stability in comparison to neat PLA. However, the incorporation of 10 wt% CNTs in PLA resulted in a decrease in thermal stability but, the char content was higher in comparison to that of neat PLA and PLA reinforced with 1 wt% CNTs. The decrease in thermal stabilities after incorporation 10 wt% CNTs was due to the agglomeration of CNTs in PLA. Therefore, they reported that excessive CNTs prevent stress transfer and other superior properties to PLA.

Table 2 TGA degradation temperatures of polymers and nanocomposites

Sample	Degradation temperature (°C)	References
PCL PCL + 10 wt% carbon nanofibres (CNF)	413 412	Sanchez-garcia et al. [97]
PHBV PCL + 10 wt% CNF	286 293	Sanchez-garcia et al. [97]
Poly (acrylic acid) grafted onto amylose (PAA-g-amylose) PAA-g-amylose + 5 wt% graphene oxide (GO)	311 385	Abdollahi et al. [2]
Epoxy Epoxy + 0.5 wt% pristine GO	341 354	Wan et al. [111]
LDPE LDPE + 3 wt% CNTs	479 390	Majeed et al. [64]
Epoxidized natural rubber Epoxidized natu ral rubber + 7 wt% CNTs	430 448	Krainoi et al. [55]
PLA PLA + 10 wt% CNTs	385 370	Wang et al. [112]
rHDPE/rPET (75/25) rHDPE/rPET + 3 wt% clay rHDPE/rPET + 3 wt% clay + 70 wt% rice husk	472 478 481	Chen and Ahmad [23]
PLA PLA + 7 wt% nanoclays	308 324	Fukushima et al. [37]
PP/LDPE (80/20) PP/LDPE + 4 wt% nanoclays	350 380	Mofokeng et al. [72]
Whey protein isolate Whey protein isolate + 3 wt% nanoclays	301 307	Azevedo et al. [10]
Poly(methylmethacrylate) (PMMA) PMMA + 41 wt% CNCs	367 384	Dong et al. [31]
Thermoplastic polyurethane (TPU) TPU + 2.5 wt% CNCs	307 334	Floros et al. [36]
PMMA PMMA + 8 wt% CNCs	183 192	Liu et al. [62]
Epoxy Epoxy + 15 wt,% CNCs	384 388	Xu et al. [113]
PLA PLA + 10 wt% CNCs	351 356	Shi et al. [102]

3.2 *Differential Scanning Calorimetry (DSC)*

Differential scanning calorimetry is a thermal analysis technique that assesses quantitative information on thermal transitions of materials through changes in heat capacity (C_p) by temperature. A sample of known weight (5–10 mg) is subjected to

Table 3 DSC analysis of polymer nanocomposites reinforced with various nanoparticles

Sample	T_m/°C	ΔH_m/J g^{-1}	T_c/°C	ΔH_c/J g^{-1}	χ_c/%	References
PE	138.6	186.06	113.3	–	63.5	Nikkhah et al. [80]
In situ PECN-3%	139.8	174.80	118.8	–	59.6	
In situ PECN-5%	140.2	155.38	122.6	–	53.03	
PP	168.8	95.0	109.7	–	46.8	Baniasadi et al. [15]
In situ PPCN-3%	168.4	90.4	119.7	–	44.5	
In situ PPCN-5%	167.5	89.15	122.5	–	43.9	
PLA	148	27.2	110^a	16.14^b	12	Valapa et al. [110]
PLA-GR-0.3wt%	152.8	27.8	113^a	9.8^b	19.2	
PLA-GR-0.5wt%	152.6	23.3	110^a	6.84^b	17.5	
PP	165.1	–	–	102	48.8	Pedrazzoli et al. [82]
PP-xGnP-3wt%	165.3	–	–	102.1	50.4	
PP-xGnP-5wt%	165.9	–	–	100.4	50.6	
PBS	114.5	79.7	80.1	–	–	Han et al. [44]
PBSSi-3	112.0	77.2	64.3	–	–	
PBSSi-5	110.7	63.5	55.8	–	–	

T_m melting temperature, T_c crystallization temperature, ΔH_m and ΔH_c melting and crystallization enthalpy, χ_c percentage of crystallinity
[a]cold crystallization temperature (T_{cc})
[b]enthalpy of cold crystallization (ΔH_{cc})

heating and cooling through programmed temperature conditions and changes in its heat capacity are tracked as changes in the heat flow. This allows the detection of thermal transitions such as melting temperature (T_m), melt crystallization/cooling temperature (T_{c-}), melting enthalpy (ΔH_m), melt crystallization/cooling enthalpy (ΔH_c), a glass transition (T_g), curing and phase changes. Various studies on polymer matrices reinforced with different nanomaterials/nanoparticles/nanofillers (nanoclays, carbonaceous (carbon nanotubes and graphene), nanocellulose and inorganic oxide) have used DSC analysis to investigate the influence of nanoparticles on their thermal transitions [8, 21, 24, 29, 41, 42, 53, 56, 79, 80, 83, 98, 103–106, 110, 115, 118]. The melting and crystallization behaviours of polymer matrices which affect morphology, mechanical and thermal properties of the resulting nanocomposites are well documented in the literature and some of the undertaken studies investigated the influence of nanoparticles in polymer matrices are summarised in Table 3. Furthermore, the performance of polymer nanocomposites does not only depend on their molecular weight and chemical structure but significantly influenced by their crystallization properties such as crystallization rate, crystallization temperature and crystallinity [118]. In addition, it also reported in the literature that polymer crystallization is usually preceded by heterogeneous or homogeneous nucleation, or self-nucleation and then by crystal growth with respect to crystallization time [106]. These properties can be determined by using various mathematical models reported in literature to determine the crystallization kinetics [41], and the degree of crystallinity (χ_c) which is the most used parameter [8, 16, 21, 24, 98, 107, 118], and is calculated according to Eq. 1.

$$\chi_c = \left(\frac{\Delta H_m / w_p}{\Delta H_m^o} \right) \times 100\% \quad (1)$$

where ΔH_m is the experiment of melting enthalpy of the nanocomposite, w_p is the weight fraction of the polymer in the nanocomposite and ΔH_m^o is the melting enthalpy of 100% crystalline polymer.

Studies on reinforcing polymers such as linear low-density polyethylene (LLDPE), polypropylene (PP), ethylene vinyl acetate (EVA) and poly(lactic acid) with exfoliated or expanded graphite and functionalized graphite nanoplatelets were investigated on their influence on thermal behaviour in polymer nanocomposites [8, 56, 71, 83, 98, 105, 110]. From these studies, incorporation of graphite nanoplatelets in polymer matrices enhanced the melting and crystallization temperature, melting endotherm and the crystallinity of the nanocomposites. For instance, improvement in the crystallinity for graphene (GR)/PLA composite samples were observed up to 0.3 wt% loading in comparison to neat PLA [110], while in binary PP nanocomposites the addition of exfoliated graphite nanoplatelets (xGnP) resulted in a significant increase in the crystallization temperature up to 5 wt% xGnP content [83]. A similar observation was recorded on poly(butylene succinate)/carbon nanotubes nanocomposites (PBS/CNT) [7] and poly(ε-caprolactone) (PCL) blended with a polycarbonate/multi-wall carbon nanotubes masterbatch (PC/MWCNT) [41]. This has demonstrated that the graphene/exfoliated graphite nanoplatelets, MWCNT and CNT dispersed in the composites can facilitate the polymer's crystallization process or act as nucleating agent [8, 98, 105]. On the other hand, slight decreases in the enthalpy and crystallinity were also observed when graphite or carbon nanotubes loadings were increased and this was attributed to agglomeration and poor dispersion of the nanoparticles which restricted polymer chain mobility and reduced the extent of crystallization [41, 70, 83].

In clay reinforced polymer nanocomposites, it has been reported that the presence of small amount of well-dispersed clay nanolayers can act as effective nucleating agents to accelerate crystallization in the polymer, thereby slightly increasing the melting and crystallization temperatures of the polymer composite [15]. On the other hand, the inclusion of clay nanolayers in polymer matrices has been observed to decrease the degree of crystallinity and does not significantly change the thermal transitions of the resulting nanocomposites. This is mainly attributed to the following observations (i) the additions of organoclays into crystalline polymer matrices do not ensure the enhancement of the polymer matrix crystallization rate [29, 103], (ii) that presence of clay nanolayers can form strong polymer-clay network which can limit the mobility of polymer chains and as a result decrease the degree of crystallinity especially at high clay concentration [15, 80, 104]. On the other hand, [73] recently investigated the effects of clay localization and its distribution in an immiscible blend of PP/LDPE on the non-isothermal crystallization and degradation kinetics. The authors observed that the non-isothermal crystallisation analysis for the localization of clay particles in the blend composites had two opposing effects, (i) the poorly dispersed clay particles at the PP/LDPE interface in the non-compatibilized blend composite had no significant effect on the crystallisation temperature of PP but allowed the free movement of PP chains, which resulted in a higher crystallinity of PP than that of PP in the neat blend; (ii) the well-dispersed clay particles in the compatibilized blend composites disrupted the free movement of PP chains, resulting in a lower crystallisation temperature and crystallinity than that of PP in the neat blend.

In the case of inorganic oxides reinforced polymer composites, a heterogeneous nucleation effect was observed to play a significant role in polymer crystallization which can be exploited for the shortening of cycle time during processing [99]. The nanoparticles turn to increasing the crystallization temperature and the rate of the polymer composite, while in other cases the heterogeneous nucleation becomes dominating with increasing filler concentration [106], but was seen to decrease the crystallization activation energy and crystallinity of the polymer with the addition of hydroxyapatite nanorods (HAP) by Zhan et al. [118]. The authors investigated the crystallization and melting properties of PBS composites with titanium dioxide nanotubes (TNTs) or (HAP). This was caused by strong hydrogen bonding interaction that exists between HAP and PBS, which reduced the transport of the PBS macromolecules and as a result lowered the crystallization rate of PBS/HAP composite than that of pure PBS.

On the other hand, PBS/nano-$CaCO_3$ composites showed independence of the crystallization behaviour with increasing nano-$CaCO_3$ content. Furthermore, the nanoparticles had little influence on the crystallization and melting behaviour of PBS. This implied that the nano-$CaCO_3$ might not have played an active role in the heterogeneous nucleation of PBS matrix. For natural fiber reinforced polymer composites, natural fibers we observed to act as nucleating agents promoting crystallization of polymer matrices (16, 32, 107]. For instance, [32] prepared poly (3-hydroxy butyrate) (PHB)/poly lactic acid (PLLA)/Tributyl citrate (TBC) blend reinforced with CNCs to increase the elongation at break of PLLA for food packaging. The well dispersed CNCs and PHB in PLLA matrix acted as bio-nuclei in PLLA matrix to help the crystallization rate and reduce the size of spherulites and thereby improving the elongation at break from 6% for pure PLLA to 40–190% for the composites with CNCs. It is also worth noting that reinforcing with natural fibers could lead to different nucleation activity due to the different surface structure of the fibers [16].

4 Dynamic Mechanical Analysis (DMA)

The dynamic mechanical analysis is a technique that determines the viscous modulus (loss modulus, G''), elastic modulus (storage modulus, G') and damping coefficient ($\tan\delta$) as a function of temperature, time or frequency. The DMA is used to identify transition regions in polymer materials, such as the glass transition temperature (T_g) and to recognize transitions corresponding to other molecular motions which are beyond the resolution of DSC. The viscoelastic properties of the polymer matrices reinforced with various nanofillers have been investigated from the measurements of storage modulus and the loss factor using dynamic mechanical analysis to evaluate the effect polymer/nanofiller interface with changing polymer mobility. These measurements provide indications about the increase in the storage or decrease in modulus of the resulting nanocomposites, shift in the glass transition temperature due to the polymer chain restriction or mobility [44, 65, 82].

Song et al. [105] fabricated exfoliated graphene-based polypropylene nanocomposites with enhanced mechanical and thermal properties prepared by melt blending technique. The exfoliated graphene nanosheets were varied from 0.1 to 5 wt% to evaluate the influence of graphene loading on polypropylene. The storage modulus of the PP/graphene nanocomposites increased with increasing graphene loading up to 1 wt%, while further increases in graphene loading led to a slight decrease in the storage modulus in the entire temperature range (−50–150 °C). The results were in agreement with their tensile modulus and the reduction in the storage modulus was attributed to the plasticization effect of low modulus of PP matrix. The glass transition temperature of PP improved by ~2.5 °C at 0.1 wt% (0.041 vol.%) graphene content, which indicated a restriction in chain mobility of PP. Similar observation was recorded in the study on mechanical and thermal properties of graphite platelet/epoxy composites [115]. At 2.5 and 5 wt%, graphite platelet/epoxy composites showed increased storage modulus (about 8 and 18%) higher than the pure epoxy matrix. As the temperature was increased, both pure epoxy and its composites showed a gradual drop in storage modulus followed by a sudden drop at the glass transition temperature. The drop in modulus is associated with the material transition from a glassy state to a rubbery state as seen in Fig. 5. Furthermore, with increasing graphite contents of 0, 2.5 and 5 wt% the T_{g-} gradually increased to 143, 145 and 146 8 °C, respectively. This was attributed to the good adhesion between the polymer and graphite platelets, which restrict the segmental motion of cross-links under loading.

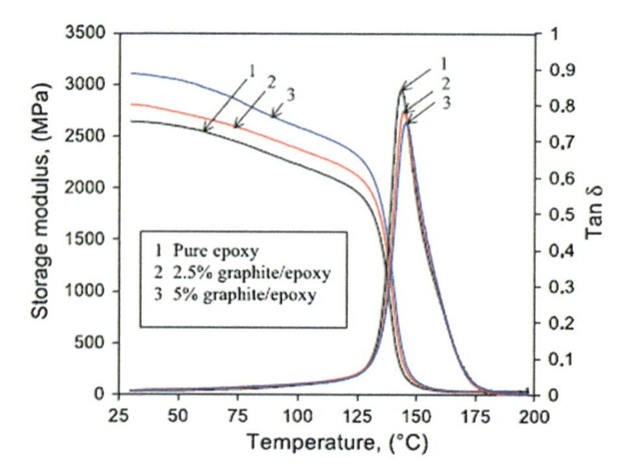

Fig. 5 Dynamic mechanical properties of pure epoxy and its composites [115], copyright with the permission from Elsevier

In a comparison of the effect of expanded graphite (EG) and modified graphite flakes (i-MG) on the physical and thermo-mechanical properties of styrene butadiene rubber/polybutadiene rubber (SBR/BR) blends [65]. A drastic increase in the storage modulus of EG and i-MG loaded SBR/BR composites in the presence of carbon black (CB) was observed in a wide range of temperature compared to the BR based nanocomposites. This was attributed to good dispersion of nanofillers in the rubber blend which increased its stiffness, and as a result, increased the storage modulus of SBR/BR based nanocomposites. In addition, as a result of isocyanate surface modification on graphite sheets (i-MG), higher basal spacing and exfoliated structure of i-MG sheets than EG flakes was achieved. Exfoliated graphite i-MG sheets were uniformly dispersed in different rubber matrices in the presence of CB, and resulted in superior mechanical, dynamic mechanical and thermal properties compared to the EG filled rubber composites. Surface modification of graphite sheets prior to nanocomposite preparation is one of the significant aspects that facilitate the compatibility between the polymer matrix and the nanofiller to form a homogeneous dispersion of nanoparticles and enhance adhesion in the polymer composites. Modification of graphene/graphite sheets by a range of techniques employing various organo modifying agents to improve mechanical properties of polymer nanocomposites are reported in the literature [20, 56, 69]. Among these reported techniques, the nucleophilic addition of organic molecules to the surface of graphene/graphite is an effective way to the bulk production of surface-modified graphene.

Analysis of thermomechanical properties for polymer matrices reinforced with nanomaterials is important in ascertaining the performance of the nanocomposite under stress and temperature. In polypropylene/clay nanocomposites (PPCNs) prepared by in situ polymerization [15], the presence of clay nanolayers dispersed in PP matrix resulted in a significant increase in stiffness (storage modulus) for all nanocomposites with increasing clay content and temperature. The reinforcing effect was at a maximum in the region above the glass transition temperature of the matrix, primarily due to the larger difference in mechanical properties between the filler and the matrix as it changes from the glassy to the rubbery state. Moreover, a marginal increase in T_g with increasing clay concentration (between 1 and 5 wt%) was observed. This was attributed to the interactions between polymer and filler which delay the segmental motion of the chains. Better dispersion of clay particles in a polymer provides greater reinforcement and higher chain immobility, thereby resulting in high storage modulus values [72]. Similar observations were recorded in the extraction of CNCs from flax fibers and their reinforcing effect on poly (furfuryl) alcohol (PFA) [78]. Incorporation of CNCs into PFA matrix resulted in increased storage modulus over the whole temperature range, the loss modulus peak shifted to higher temperatures and the magnitude of the peak decreased due to the presence of CNCs, the glass transition temperature values increased after the inclusion of CNCs into PFA. The overall results implied that the presence of CNCs in PFA improved the stiffness of the composite, restricted polymer chain mobility as a result of good interaction between the polymer and the filler.

5 Melt Rheology Properties

The study on melt rheological properties for polymer materials is very important from the processing point of view. In addition, information on the microstructure of the polymer materials in the melted state can be provided from the melt rheology [7]. Rheological properties are known as mechanical properties of the material that undergoes deformation and flow in the presence of stress. The melt rheology is usually measured from the melt phase of the polymer from its melting temperature at the desired strain that is well within the linear viscoelastic range. Viscoelastic properties of thermoplastic and nanofiller reinforced composites can be measured over a range of frequencies to gauge the rate of the viscosity changes with shear rate. Various studies on the effect of the nanofiller addition, such as carbonaceous [7, 17], natural fibers/cellulose [53], inorganic oxides [24, 82, 106] and nanoclays [72] on the isothermal frequency dependence of the dynamic shear storage modulus (G') and complex viscosity (η^*) were investigated. For instance, [7] prepared PBS/CNTs nanocomposites fabricated by melt mixing and investigated the rheological properties. The viscoelastic properties of PBS/CNTs composites at high frequencies behaved the same while low frequencies the nanocomposites were frequency independent (Fig. 6a, b). The nanocomposites showed gradual changes in the

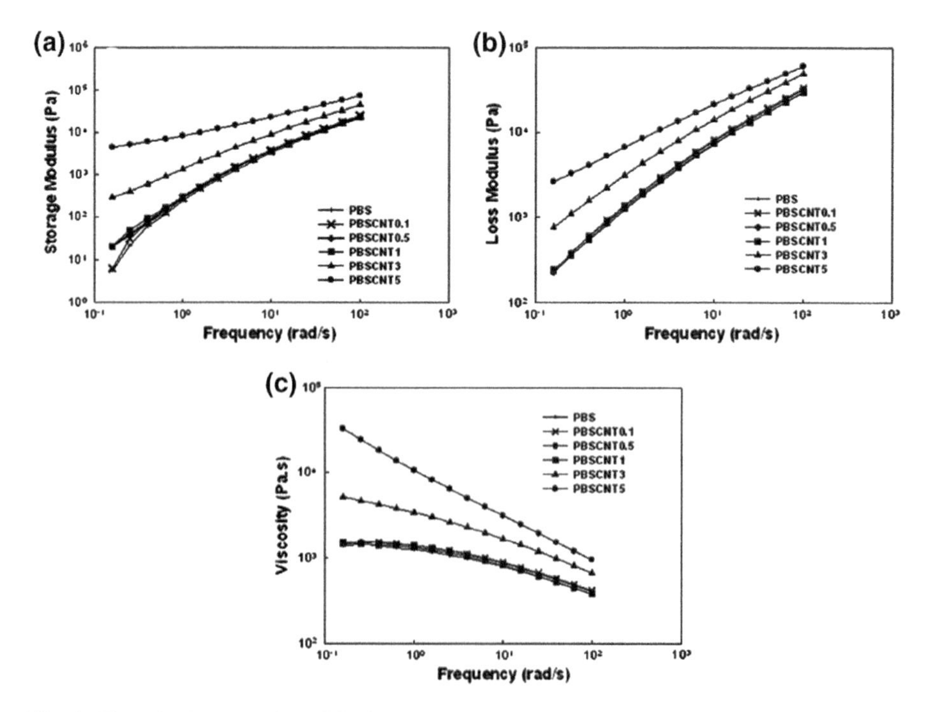

Fig. 6 Viscoelastic properties of PBS and PBS/CNTs composites in the melted state **a** Storage modulus of **b** loss modulus and **c** complex viscosity [7], copyright with the permission from Elsevier

composites from liquid-like to solid-like behaviour, especially for PBS/CNT3 and PBS/CNT5 samples. This was attributed to interactions between nanotubes which formed interconnected structures of CNTs in the PBS/CNTs composites, thereby enhancing the relaxation behaviour and increased the storage modulus. In addition, improvement in the complex viscosity was also observed due to the flow restriction of the polymer by filler in the melted state (Fig. 6c). On contrary, the G' and G'' of pure PBS and PBS/nano-$CaCO_3$ composites with various nano-$CaCO_3$ loadings as a function of frequency exhibited a liquid-like a behaviour for both PBS and $CaCO_3$ filled composites [24]. The results implied that the nano-$CaCO_3$ particles had little influence on the microstructure of PBS; the relaxation mechanism and the microstructure of the PBS/nano-$CaCO_3$ composites mainly depended on the PBS matrix than on nano-$CaCO_3$ content. Furthermore, a similar decrease in G' and η^* was observed for linear-low-density polyethylene reinforced boehmite alumina (LLDPE-BA) [82]. The observation was attributed to the highly branched LLDPE whose chains would tend to get entangled; apparently even poorly bonded plain BA particles fill in the spaces between chain branches and enable an easier flow. Khumalo et al. [54] in a similar study suggested that there was no strong interaction between the BA and LDPE in the nanocomposites.

6 Conclusions

In this chapter, the mechanical, thermal, dynamic mechanical and rheological properties of polymer nanocomposites reinforced with nanomaterials were presented. It can be concluded that nanomaterials have the power to alter the properties of polymer nanocomposites which can be exploited in a variety of applications. By combining nanomaterials with the polymer matrix, novel materials with multi-properties can be achieved. The development of polymer nanocomposite with multi-properties depends on various factors such as filler particle size, surface area, aspect ratio, compositions, purity, crystallinity, shape, properties (mechanical, antibacterial, thermal, electrical etc.), dispersion, filler loading and interaction between filler and polymer matrix. Nanomaterials which were reviewed in this study are nanocellulose, carbonaceous fillers, and nanoclays and they have received significant attention globally particularly due to their extraordinary properties. The major drawback of polymer nanocomposites is a homogeneous dispersion of nanomaterials in the polymer matrix. To address such drawback, functionalization of either filler or polymer is the effective approach to alter the functional groups of the material, thus improving the interfacial adhesion between filler and polymer matrix which enhances the properties of the resultant material. In this chapter, it was noticed that loading nanomaterials in polymer matrix improved their properties and further increase filler loading to threshold amount enhanced the properties polymer. Many researchers indicated that adding filler after threshold amount deteriorates the properties of the material due to agglomeration which led to poor interaction between filler and polymer matrix and therefore, this result in poor properties of the resultant material [55, 112]. The novel polymer nanocomposites with multi-properties can be used in medical, agricultural, food packaging, automotive etc. Much research is required to further develop the commercially available polymer nanocomposites and develop other novel materials.

References

1. Abdellaoui H, Bensalah H, Raji M, Rodrigue D, Bouhfid R, Qaiss AK (2017) Laminated epoxy biocomposites based on clay and jute fibers. J Bionic Eng 14:379–389. https://doi.org/10.1016/S1672-6529(16)60406-7
2. Abdollahi R, Taghizadeh MT, Savani S (2018) Thermal and mechanical properties of graphene oxide nanocomposite hydrogel based on poly (acrylic acid) grafted onto amylose. Polym Degrad Stab 147:151–158. https://doi.org/10.1016/j.polymdegradstab.2017.11.022
3. Ahmad EEM, Luyt AS, Djoković V (2013) Thermal and dynamic mechanical properties of bio-based poly(furfuryl alcohol)/sisal whiskers nanocomposites. Polym Bull 70:1265–1276. https://doi.org/10.1007/s00289-012-0847-2
4. Al-saleh MH (2015) Electrically conductive carbon nanotube/polypropylene nanocomposite with improved mechanical properties. JMADE 85:76–81. https://doi.org/10.1016/j.matdes.2015.06.162
5. Alboofetileh M, Rezaei M, Hosseini H, Abdollahi M (2013) Effect of montmorillonite clay and biopolymer concentration on the physical and mechanical properties of alginate nanocomposite films. J Food Eng 117:26–33. https://doi.org/10.1016/j.jfoodeng.2013.01.042
6. Alcântara ACS, Darder M, Aranda P, Ruiz-hitzky E (2014) Polysaccharide–fibrous clay bionanocomposites. Appl Clay Sci 96:2–8. https://doi.org/10.1016/j.clay.2014.02.018
7. Ali FB, Mohan R (2010) Thermal, mechanical, and rheological properties of biodegradable polybutylene succinate/carbon nanotubes nanocomposites. 1–6. https://doi.org/10.1002/pc.20913
8. An JE, Jeon GW, Jeong YG (2012) Preparation and properties of polypropylene nanocomposites reinforced with exfoliated graphene. Fibers Polym 13:507–514. https://doi.org/10.1007/s12221-012-0507-z
9. Arrakhiz FZ, Benmoussa K, Bouhfid R, Qaiss A (2013) Pine cone fiber/clay hybrid composite: mechanical and thermal properties. Mater Des 50:376–381. https://doi.org/10.1016/j.matdes.2013.03.033
10. Azevedo VM, Silva EK, Pereira CFG, Costa JMG, Borges SV (2015) Whey protein isolate biodegradable films: Influence of the citric acid and montmorillonite clay nanoparticles on the physical properties. Food Hydrocoll 43:252–258. https://doi.org/10.1016/j.foodhyd.2014.05.027
11. Ayana B, Suin S, Khatua BB (2014) Highly exfoliated eco-friendly thermoplastic starch (TPS)/ poly (lactic acid)(PLA)/clay nanocomposites using unmodified nanoclay. Carbohydr Polym 110:430–439. https://doi.org/10.1016/j.carbpol.2014.04.024
12. Azwa ZN, Yousif BF, Manalo AC, Karunasena W (2013) A review on the degradability of polymeric composites based on natural fibres. Mater Des 47:424–442. https://doi.org/10.1016/j.matdes.2012.11.025
13. Babaee M, Jonoobi M, Hamzeh Y, Ashori A (2015) Biodegradability and mechanical properties of reinforced starch nanocomposites using cellulose nanofibers. Carbohydr Polym 132:1–8. https://doi.org/10.1016/j.carbpol.2015.06.043
14. Baheri B, Shahverdi M, Rezakazemi M, Motaee E, Mohammadi T (2015) Performance of PVA/NaA mixed matrix membrane for removal of water from ethylene glycol solutions by pervaporation. Chem Eng Commun 202:316–321. https://doi.org/10.1080/00986445.2013.841149

15. Baniasadi H, Ramazani A, Javan Nikkhah S (2010) Investigation of in situ prepared polypropylene/clay nanocomposites properties and comparing to melt blending method. Mater Des 31:76–84. https://doi.org/10.1016/j.matdes.2009.07.014
16. Běhálek L, Maršálková M, Lenfeld P, et al (2013) Study of crystallization of polylactic acid composites and nanocomposites with natural fibres by DSC method. 1–6
17. Botlhoko OJ, Ramontja J, Ray SS (2017) Thermal, mechanical, and rheological properties of graphite- and graphene oxide-filled biodegradable polylactide/poly (E-caprolactone) blend composites. 45373:1–14. https://doi.org/10.1002/app.45373
18. Botlhoko JO, Ramontja J, Sinha S (2018) Morphological development and enhancement of thermal, mechanical, and electronic properties of thermally exfoliated graphene oxide- fi lled biodegradable polylactide/poly (ε-caprolactone) blend composites. Polymer (Guildf) 139:188–200. https://doi.org/10.1016/j.polymer.2018.02.005
19. Cao X, Xu C, Wang Y, Liu Y, Liu Y, Chen Y (2013) New nanocomposite materials reinforced with cellulose nanocrystals in nitrile rubber. Polym Test 32:819–826. https://doi.org/10.1016/j.polymertesting.2013.04.005
20. Cha J, Jin S, Hun J, Park CS, Ryu HJ, Hong SH (2016) Functionalization of carbon nanotubes for fabrication of CNT/ epoxy nanocomposites. JMADE 95:1–8. https://doi.org/10.1016/j.matdes.2016.01.077
21. Cheewawuttipong W, Fuoka D, Tanoue S, Uematsu H, Lemoto Y (2013) Thermal and mechanical properties of polypropylene/boron nitride composites. Energy Proc 34:808–817. https://doi.org/10.1016/j.egypro.2013.06.817
22. Chen PY, Lian HY, Shih YF, Chen-Wei SM, Jeng RJ (2017) Preparation, characterization and crystallization kinetics of Kenaf fiber/multi-walled carbon nanotube/polylactic acid (PLA) green composites. Mater Chem Phys 196:249–255. https://doi.org/10.1016/j.matchemphys.2017.05.006
23. Chen RS, Ahmad S (2017) Mechanical performance and fl ame retardancy of rice husk/organoclay-reinforced blend of recycled plastics. Mater Chem Phys 198:57–65. https://doi.org/10.1016/j.matchemphys.2017.05.054
24. Chen RY, Zou W, Zhang HC, Zhang GZ, Yang ZT, Jin G, Qu JP (2015) Thermal behavior, dynamic mechanical properties and rheological properties of poly(butylene succinate) composites filled with nanometer calcium carbonate. Polym Test 42:160–167. https://doi.org/10.1016/j.polymertesting.2015.01.015
25. Chen Y, Liu C, Chang PR, Cao X, Anderson DP (2009) Bionanocomposites based on pea starch and cellulose nanowhiskers hydrolyzed from pea hull fibre: effect of hydrolysis time. Carbohydr Polym 76:607–615. https://doi.org/10.1016/j.carbpol.2008.11.030
26. Cheng J, Zheng P, Zhao F, Ma X (2013) The composites based on plasticized starch and carbon nanotubes. Int J Biol Macromol 59:13–19. https://doi.org/10.1016/j.ijbiomac.2013.04.010
27. Chiu F (2016) Fabrication and characterization of biodegradable poly (butylene succinate-co-adipate) nanocomposites with halloysite nanotube and organo-montmorillonite as nano fi llers. Polym Test 54:1–11. https://doi.org/10.1016/j.polymertesting.2016.06.018
28. Chiu F (2017) Halloysite nanotube- and organoclay- fi lled biodegradable poly (butylene succinate-co-adipate)/ maleated polyethylene blend- based nanocomposites with enhanced rigidity. Compos Part B 110:193–203. https://doi.org/10.1016/j.compositesb.2016.10.091
29. Chiu FC, Chu PH (2006) Characterization of solution-mixed polypropylene/clay nanocomposites without compatibilizers. J Polym Res 13:73–78. https://doi.org/10.1007/s10965-005-9009-7
30. Daitx TS, Carli LN, Crespo JS, Mauler RS (2015) Effects of the organic modi fi cation of different clay minerals and their application in biodegradable polymer nanocomposites of PHBV. Appl Clay Sci 115:157–164. https://doi.org/10.1016/j.clay.2015.07.038
31. Dong H, Strawhecker KE, Snyder JF, Orlicki JA, Reiner RS, Rudie AW (2012) Cellulose nanocrystals as a reinforcing material for electrospun poly(methyl methacrylate) fibers: Formation, properties and nanomechanical characterization. Carbohydr Polym 87:2488–2495. https://doi.org/10.1016/j.carbpol.2011.11.015

32. El-hadi AM (2017) Increase the elongation at break of poly (lactic acid) composites for use in food packaging films. Nat Publ Gr 1–14. https://doi.org/10.1038/srep46767
33. Essabir H, Boujmal R, Bensalah MO, Rodrigue D, Bouhfid R, Qaiss AK (2016) Mechanical and thermal properties of hybrid composites: oil-palm fiber/clay reinforced high density polyethylene. Mech Mater 98:36–43. https://doi.org/10.1016/j.mechmat.2016.04.008
34. Farahnaky A, Dadfar SMM, Shahbazi M (2014) Physical and mechanical properties of gelatin–clay nanocomposite. J Food Eng 122:78–83. https://doi.org/10.1016/j.jfoodeng.2013.06.016
35. Ferreira FV, Francisco W, Menezes BRC, Brito FS, Coutinho AS, Cividanes LS, Coutinho AR, Thim GP (2016) Correlation of surface treatment, dispersion and mechanical properties of HDPE/CNT nanocomposites. Appl Surf Sci 389:921–929. https://doi.org/10.1016/j.apsusc.2016.07.164
36. Floros M, Hojabri L, Abraham E, Jose J, Thomas S, Pothan L, Leao AL, Marine S (2012) Enhancement of thermal stability, strength and extensibility of lipid-based polyurethanes with cellulose-based nanofibers. Polym Degrad Stab 97:1970–1978. https://doi.org/10.1016/j.polymdegradstab.2012.02.016
37. Fukushima K, Tabuani D, Camino G (2012) Poly (lactic acid)/clay nanocomposites: effect of nature and content of clay on morphology, thermal and thermo-mechanical properties. Mater Sci Eng, C 32:1790–1795. https://doi.org/10.1016/j.msec.2012.04.047
38. Gao T, Li Y, Bao R, Liu ZY, Xie BH, Yang MB, Yang W (2017) Tailoring co-continuous like morphology in blends with highly asymmetric composition by MWCNTs: towards biodegradable high-performance electrical conductive poly (l-lactide) poly (3-hydroxybutyrate- co-4-hydroxybutyrate) blends. Compos Sci Technol 152:111–119. https://doi.org/10.1016/j.compscitech.2017.09.014
39. Giannakas A, Grigoriadi K, Leontiou A, Barkoula NM, Lavados A (2014) Preparation, characterization, mechanical and barrier properties investigation of chitosan—clay nanocomposites. Carbohydr Polym 108:103–111. https://doi.org/10.1016/j.carbpol.2014.03.019
40. Giannakas A, Vlacha M, Salmas C, Leontiou A, Katapodis P, Stamatis H, Barkoula NM, Ladavos A (2016) Preparation, characterization, mechanical, barrier and antimicrobial properties of chitosan/PVOH/clay nanocomposites. Carbohydr Polym 140:408–415. https://doi.org/10.1016/j.carbpol.2015.12.072
41. Gumede TP, Luyt AS, Hassan MK, Pérez-Camargo RA, Tercjak A, Müller AJ (2017) Morphology, nucleation, and isothermal crystallization kinetics of poly(ε-caprolactone) mixed with a polycarbonate/MWCNTs masterbatch. Polymers https://doi.org/10.3390/polym9120709
42. Gumede TP, Luyt AS, Pèrez-Camargo RA, Müller AJ (2017) The influence of paraffin wax addition on the isothermal crystallization of LLDPE. J App Poly Sci 44398:1–7. https://doi.org/10.1002/app.44398
43. Guo Y, Yang K, Zuo X et al (2016) Effects of clay platelets and natural nanotubes on mechanical properties and gas permeability of Poly (lactic acid) nanocomposites. Polymer (Guildf) 83:246–259. https://doi.org/10.1016/j.polymer.2015.12.012
44. Il HS, Im SS, Kim DK (2003) Dynamic mechanical and melt rheological properties of sulfonated poly(butylene succinate) ionomers. Polymer (Guildf) 44:7165–7173. https://doi.org/10.1016/S0032-3861(03)00673-6
45. Hietala M, Mathew AP, Oksman K (2013) Bionanocomposites of thermoplastic starch and cellulose nanofibers manufactured using twin-screw extrusion. Eur Polym J 49:950–956. https://doi.org/10.1016/j.eurpolymj.2012.10.016
46. Huang J, Zhang S, Zhang F, Guo Z, Jin L, Pan Y, Wang Y, Guo T (2017) Enhancement of lignocellulose-carbon nanotubes composites by lignocellulose grafting. Carbohydr Polym 160:115–122. https://doi.org/10.1016/j.carbpol.2016.12.053
47. John MJ, Anandjiwala R, Oksman K, Mathew AP (2013) Melt-spun polylactic acid fibers: effect of cellulose nanowhiskers on processing and properties. J Appl Polym Sci 127:274–281. https://doi.org/10.1002/app.37884

48. Jonoobi M, Aitomäki Y, Mathew AP, Oksman K (2014) Thermoplastic polymer impregnation of cellulose nanofibre networks: morphology, mechanical and optical properties. Compos Part A Appl Sci Manuf 58:30–35. https://doi.org/10.1016/j.compositesa.2013.11.010
49. Jonoobi M, Harun J, Mathew AP, Oksman K (2010) Mechanical properties of cellulose nanofiber (CNF) reinforced polylactic acid (PLA) prepared by twin screw extrusion. Compos Sci Technol 70:1742–1747. https://doi.org/10.1016/j.compscitech.2010.07.005
50. Jyoti J, Basu S, Pratap B, Dhakate SR (2015) Superior mechanical and electrical properties of multiwall carbon nanotube reinforced acrylonitrile butadiene styrene high performance composites. Compos Part B 83:58–65. https://doi.org/10.1016/j.compositesb.2015.08.055
51. Kanmani P, Rhim J (2014) Physical, mechanical and antimicrobial properties of gelatin based active nanocomposite films containing AgNPs and nanoclay. Food Hydrocoll 35:644–652. https://doi.org/10.1016/j.foodhyd.2013.08.011
52. Khan AS, Hussain AN, Sidra L, Sarfraz Z, Khalid H, Khan M, Manzoor F, Shahzadi L, Yar M, Rehman IU (2017) Fabrication and in vivo evaluation of hydroxyapatite/carbon nanotube electrospun fi bers for biomedical/dental application. Mater Sci Eng, C 80:387–396. https://doi.org/10.1016/j.msec.2017.05.109
53. Khoshkava V, Kamal MR (2014) Effect of cellulose nanocrystals (CNC) particle morphology on dispersion and rheological and mechanical properties of polypropylene/CNC nanocomposites. https://doi.org/10.1021/am500577e
54. Khumalo VM, Karger-Kocsis J, Thomann R (2010) Polyethylene/synthetic boehmite alumina nanocomposites: structure, thermal and rheological properties. eXPPRES Poly Lett 4: 264–274. https://doi.org/10.3144/expresspolymlett.2010.34
55. Krainoi A, Kummerlöwe C, Nakaramontri Y, Vennemann N, Pichaiyut S, Wisunthorn S, Nakason C (2018) Influence of critical carbon nanotube loading on mechanical and electrical properties of epoxidized natural rubber nanocomposites. Polym Test 66:122–136. https://doi.org/10.1016/j.polymertesting.2018.01.003
56. Kuila T, Bose S, Mishra AK, Khanra P, Kim NH, Lee JH (2012) Effect of functionalized graphene on the physical properties of linear low density polyethylene nanocomposites. Polym Test 31:31–38. https://doi.org/10.1016/j.polymertesting.2011.09.007
57. Lai S, Wu S, Lin G, Don T (2014) Unusual mechanical properties of melt-blended poly (lactic acid) (PLA)/clay nanocomposites. Eur Polym J 52:193–206. https://doi.org/10.1016/j.eurpolymj.2013.12.012
58. Lekha P, Mtibe A, Motaung T., Andrew JE, Sithole BB, Gibril M (2016) Effect of mechanical treatment on properties of cellulose nanofibrils produced from bleached hardwood and softwood pulps. Maderas Cienc y Tecnol 18:0–0. https://doi.org/10.4067/s0718-221x2016005000041
59. Lendvai L, Apostolov A, Karger-kocsis J (2017) Characterization of layered silicate-reinforced blends of thermoplastic starch (TPS) and poly (butylene adipate-co-terephthalate). Carbohydr Polym 173:566–572. https://doi.org/10.1016/j.carbpol.2017.05.100
60. Liao CZ, Li K, Wong HM, Tong WY, Yeung KWK, Tjong SC (2013) Novel polypropylene biocomposites reinforced with carbon nanotubes and hydroxyapatite nanorods for bone replacements. Mater Sci Eng, C 33(3):1380–1388
61. Lin C, Wang Y, Lai Y, Yang W, Jiao F, Zhang H, Ye S, Zhang Q (2011) Colloids and surfaces B: biointerfaces Incorporation of carboxylation multiwalled carbon nanotubes into biodegradable poly (lactic-co-glycolic acid) for bone tissue engineering. Colloids Surfs B Biointerfaces 83:367–375. https://doi.org/10.1016/j.colsurfb.2010.12.011
62. Liu H, Liu D, Yao F, Wu Q (2010) Fabrication and properties of transparent polymethylmethacrylate/cellulose nanocrystals composites. Bioresour Technol 101:5685–5692. https://doi.org/10.1016/j.biortech.2010.02.045

63. Lopez-manchado MA, Brasero J, Avil F (2016) Effect of the morphology of thermally reduced graphite oxide on the mechanical and electrical properties of natural rubber nanocomposites. Compos Part B: Eng 87:350–356. https://doi.org/10.1016/j.compositesb.2015.08.079
64. Majeed K, Al M, Almaadeed A, Zagho MM (2018) Comparison of the effect of carbon, halloysite and titania nanotubes on the mechanical and thermal properties of LDPE based nanocomposite films. Chinese J Chem Eng 26:428–435. https://doi.org/10.1016/j.cjche.2017.09.017
65. Malas A, Pal P, Das CK (2014) Effect of expanded graphite and modified graphite flakes on the physical and thermo-mechanical properties of styrene butadiene rubber/polybutadiene rubber (SBR/BR) blends. Mater Des 55:664–673. https://doi.org/10.1016/j.matdes.2013.10.038
66. Malkappa K, Rao BN, Jana T (2016) Functionalized polybutadiene diol based hydrophobic, water dispersible polyurethane nanocomposites: role of organo-clay structure. Polymer (Guildf) 99:404–416. https://doi.org/10.1016/j.polymer.2016.07.039
67. Mandal A, Chakrabarty D (2014) Journal of industrial and engineering chemistry studies on the mechanical, thermal, morphological and barrier properties of nanocomposites based on poly (vinyl alcohol) and nanocellulose from sugarcane bagasse. J Ind Eng Chem 20:462–473. https://doi.org/10.1016/j.jiec.2013.05.003
68. Mangeon C, Mahouche-Chergui S, Versace DL, Guerrouache M, Carbonnier B, Langlois V, Renard E (2015) Reactive & functional polymers poly (3-hydroxyalkanoate)-grafted carbon nanotube nanofillers as reinforcing agent for PHAs-based electrospun mats. React Funct Polym 89:18–23. https://doi.org/10.1016/j.reactfunctpolym.2015.03.001
69. Mashhadzadeh AH, Fereidoon A, Ahangari MG (2017) Surface modification of carbon nanotubes using 3-aminopropyltriethoxysilane to improve mechanical properties of nanocomposite based polymer matrix: experimental and density functional theory study. Appl Surf Sci 420:167–179. https://doi.org/10.1016/j.apsusc.2017.05.148
70. Mochane MJ (2014) Thermal and mechanical properties of polyolefins/Wax Pcm blends prepared with and without expanded graphite
71. Mochane MJ, Luyt AS (2015) The Effect of expanded graphite on the thermal stability, latent heat, and flammability properties of EVA/Wax phase change blends. Polym Eng Sci https://doi.org/10.1002/pen
72. Mofokeng TG, Ray SS, Ojijo V (2018a) Structure—property relationship in PP/LDPE blend composites : The role of nanoclay localization. 46193:1–12. https://doi.org/10.1002/app.46193
73. Mofokeng TG, Ray SS, Ojijo V (2018b) Influence of selectively localised nanoclay particles on non-isothermal crystallisation and degradation behaviour of PP/LDPE blend composites. https://doi.org/10.3390/polym10030245
74. Moradi M, Mohandesi JA, Haghshenas DF (2015) Mechanical properties of the poly (vinyl alcohol) based nanocomposites at low content of surfactant wrapped graphene sheets. Polymer (Guildf) 60:207–214. https://doi.org/10.1016/j.polymer.2015.01.044
75. Motaung TE, Mtibe A (2015) Alkali treatment and cellulose nanowhiskers extracted from maize stalk residues. Mater Sci App 6:1022–1032. https://doi.org/10.4236/msa.2015.611102
76. Moustafa H, Galliard H, Vidal L, Dufresne A (2017) Facile modification of organoclay and its effect on the compatibility and properties of novel biodegradable PBE/PBAT nanocomposites. Eur Polym J 87:188–199. https://doi.org/10.1016/j.eurpolymj.2016.12.009
77. Mtibe A, Linganiso LZ, Mathew AP, Oksman K, John MJ, Anandjiwala RD (2015a) A comparative study on properties of micro and nanopapers produced from cellulose and cellulose nanofibres. Carbohydr Polym 118:1–8. https://doi.org/10.1016/j.carbpol.2014.10.007
78. Mtibe A, Mandlevu Y, Linganiso LZ, Anandjiwala RD (2015b) Extraction of cellulose nanowhiskers from flax fibres and their reinforcing effect on poly (furfuryl) alcohol. 9:1–9. https://doi.org/10.1166/jbmb.2015.1531

79. Murariu M, Dechief AL, Bonnaud L, Paint Y, Gallos A, Fontaine G, Bourbigot S, Dubois P (2010) The production and properties of polylactide composites filled with expanded graphite. Polym Degrad Stab 95:889–900. https://doi.org/10.1016/j.polymdegradstab.2009.12.019
80. Nikkhah SJ, Ramazani A, Baniasadi H, Tavakolzadeh F (2009) Investigation of properties of polyethylene/clay nanocomposites prepared by new in situ Ziegler-Natta catalyst. Mater Des 30:2309–2315. https://doi.org/10.1016/j.matdes.2008.11.019
81. Ortiz AV, Teixeira JG, Gomes MG, Oliveira RR, Díaz FRV, Moura EAB (2014) Preparation and characterization of electron-beam treated HDPE composites reinforced with rice husk ash and Brazilian clay. Appl Surf Sci 310:331–335. https://doi.org/10.1016/j.apsusc.2014.03.075
82. Pedrazzoli D, Ceccato R, Karger-Kocsis J, Pegoretti A (2013) Viscoelastic behaviour and fracture toughness of linear-low-density polyethylene reinforced with synthetic boehmite alumina nanoparticles. Express Polym Lett 7:652–666. https://doi.org/10.3144/expresspolymlett.2013.62
83. Pedrazzoli D, Pegoretti A (2014) Expanded graphite nanoplatelets as coupling agents in glass fiber reinforced polypropylene composites. Compos Part A Appl Sci Manuf 66:25–34. https://doi.org/10.1016/j.compositesa.2014.06.016
84. Phua YJ, Lau NS, Sudesh K, Chow WS, Ishak ZAM (2012) Biodegradability studies of poly (butylene succinate)/organo-montmorillonite nanocomposites under controlled compost soil conditions: effects of clay loading and compatibiliser. Polym Degrad Stab 97:1345–1354. https://doi.org/10.1016/j.polymdegradstab.2012.05.024
85. Ranjan N, Roy I, Sarkar G, Bhattacharyya A, Das R, Rana D, Banerjee R, Paul AK, Mishra R, Chattopadhyay D (2018) Development of active packaging material based on cellulose acetate butyrate/polyethylene glycol/aryl ammonium cation modi fi ed clay. Carbohydr Polym 187:8–18. https://doi.org/10.1016/j.carbpol.2018.01.065
86. Reddy MM, Vivekanandhan S, Misra M, Bhatia SK, Mohanty AK (2013) Biobased plastics and bionanocomposites: current status and future opportunities. Prog Polym Sci 38:1653–1689. https://doi.org/10.1016/j.progpolymsci.2013.05.006
87. Rezakazemi M, Dashti A, Asghari M, Shirazian S (2017) H_2-selective mixed matrix membranes modeling using ANFIS, PSO-ANFIS, GA-ANFIS. Int J Hydrogen Energy 42:15211–15225. https://doi.org/10.1016/j.ijhydene.2017.04.044
88. Rezakazemi M, Ebadi Amooghin A, Montazer-Rahmati MM, Ismail AF, Matsuura T (2014) State-of-the-art membrane based CO_2 separation using mixed matrix membranes (MMMs): an overview on current status and future directions. Prog Polym Sci 39:817–861. https://doi.org/10.1016/j.progpolymsci.2014.01.003
89. Rezakazemi M, Mohammadi T (2013) Gas sorption in H_2-selective mixed matrix membranes: experimental and neural network modeling. Int J Hydrogen Energy 38:14035–14041. https://doi.org/10.1016/j.ijhydene.2013.08.062
90. Rezakazemi M, Razavi S, Mohammadi T, Nazari AG (2011) Simulation and determination of optimum conditions of pervaporative dehydration of isopropanol process using synthesized PVA-APTEOS/TEOS nanocomposite membranes by means of expert systems. J Memb Sci 379:224–232. https://doi.org/10.1016/j.memsci.2011.05.070
91. Rezakazemi M, Sadrzadeh M, Matsuura T (2018) Thermally stable polymers for advanced high-performance gas separation membranes. Prog Energy Combust Sci 66:1–41. https://doi.org/10.1016/j.pecs.2017.11.002
92. Rezakazemi M, Sadrzadeh M, Mohammadi T, Matsuura T (2017b) Methods for the preparation of organic-inorganic nanocomposite polymer electrolyte membranes for fuel cells
93. Rezakazemi M, Shahidi K, Mohammadi T (2012a) Sorption properties of hydrogen-selective PDMS/zeolite 4A mixed matrix membrane. Int J Hydrogen Energy 37:17275–17284. https://doi.org/10.1016/j.ijhydene.2012.08.109

94. Rezakazemi M, Shahidi K, Mohammadi T (2012b) Hydrogen separation and purification using crosslinkable PDMS/zeolite A nanoparticles mixed matrix membranes. Int J Hydrogen Energy 37:14576–14589. https://doi.org/10.1016/j.ijhydene.2012.06.104
95. Rezakazemi M, Vatani A, Mohammadi T (2015) Synergistic interactions between POSS and fumed silica and their effect on the properties of crosslinked PDMS nanocomposite membranes. RSC Adv 5:82460–82470. https://doi.org/10.1039/c5ra13609a
96. Rezakazemi M, Vatani A, Mohammadi T (2016) Synthesis and gas transport properties of crosslinked poly(dimethylsiloxane) nanocomposite membranes using octatrimethylsiloxy POSS nanoparticles. J Nat Gas Sci Eng 30:10–18. https://doi.org/10.1016/j.jngse.2016.01.033
97. Sanchez-garcia MD, Lagaron JM, Hoa SV (2010) Effect of addition of carbon nanofibers and carbon nanotubes on properties of thermoplastic biopolymers. Compos Sci Technol 70:1095–1105. https://doi.org/10.1016/j.compscitech.2010.02.015
98. Sefadi JS, Luyt AS, Pionteck J, Gohs U (2015) Effect of surfactant and radiation treatment on the morphology and properties of PP/EG composites. J Mater Sci 50:6021–6031. https://doi.org/10.1007/s10853-015-9149-z
99. Shafiq M, Yasin T, Saeed S (2012) Synthesis and characterization of linear low-density polyethylene/sepiolite nanocomposites. J Appl Polym Sci 123:1718–1723. https://doi.org/10.1002/app.34633
100. Shah KJ, Shukla AD, Shah DO, Imae T (2016) Effect of organic modi fi ers on dispersion of organoclay in polymer nanocomposites to improve mechanical properties. Polymer (Guildf) 97:525–532. https://doi.org/10.1016/j.polymer.2016.05.066
101. Shahverdi M, Baheri B, Rezakazemi M, Motaee E, Mohammadi T (2013) Pervaporation study of ethylene glycol dehydration through synthesized (PVA-4A)/polypropylene mixed matrix composite membranes. Polym Eng Sci 53:1487–1493. https://doi.org/10.1002/pen.23406
102. Shi Q, Zhou C, Yue Y, Guo W, Wu Y, Wu Q (2012) Mechanical properties and in vitro degradation of electrospun bio-nanocomposite mats from PLA and cellulose nanocrystals. Carbohydr Polym 90:301–308. https://doi.org/10.1016/j.carbpol.2012.05.042
103. Sibeko MA, Luyt AS (2013) Preparation and characterization of vinylsilane crosslinked high-density polyethylene compositesfilled with nanoclays. Polym Compos 34:1720–1727. https://doi.org/10.1002/pc.22575
104. Silva BL, Nack FC, Lepienski CM, Coelho LAF, Becker D (2014) Influence of intercalation methods in properties of clay and carbon nanotube and high density polyethylene nanocomposites. Mater Res 17:1628–1636. https://doi.org/10.1590/1516-1439.303714
105. Song P, Cao Z, Cai Y, Zhao L, Fang Z, Fu S (2011) Fabrication of exfoliated graphene-based polypropylene nanocomposites with enhanced mechanical and thermal properties. Polymer (Guildf) 52:4001–4010. https://doi.org/10.1016/j.polymer.2011.06.045
106. Song X, Zhou S, Wang Y, Kang W, Cheng B (2012) Mechanical properties and crystallization behavior of polypropylene non-woven fabrics reinforced with POSS and SiO_2 nanoparticles. 13:1015–1022. https://doi.org/10.1007/s12221-012-1015-x
107. Sullivan EM, Moon RJ, Kalaitzidou K (2015) Processing and characterization of cellulose nanocrystals/polylactic acid nanocomposite films. Mater 2015:8106–8116. https://doi.org/10.3390/ma8125447
108. Tan L, Chen Y, Zhou W, Ye S, Wei J (2011) Novel approach toward poly(butylene succinate)/single-walled carbon nanotubes nanocomposites with interfacial-induced crystallization behaviors and mechanical strength. Polymer 52:3587–3596. https://doi.org/10.1016/j.polymer.2011.06.006
109. Tarfaoui M, Lafdi K, El MA (2016) Mechanical properties of carbon nanotubes based polymer composites. Compos Part B 103:113–121. https://doi.org/10.1016/j.compositesb.2016.08.016
110. Valapa RB, Pugazhenthi G, Katiyar V (2015) Effect of graphene content on the properties of poly(lactic acid) nanocomposites. RSC Adv 5:28410–28423. https://doi.org/10.1039/C4RA15669B

111. Wan Y, Gong L, Tang L, Wu LB, Jiang JX (2014) Mechanical properties of epoxy composites filled with silane-functionalized graphene oxide. Compos PART A 64:79–89. https://doi.org/10.1016/j.compositesa.2014.04.023
112. Wang L, Qiu J, Sakai E, Wei X (2016) The relationship between microstructure and mechanical properties of carbon nanotubes/polylactic acid nanocomposites prepared by twin-screw extrusion. Compos Part A 89:18–25. https://doi.org/10.1016/j.compositesa.2015.12.016
113. Xu S, Girouard N, Schueneman G, Shofner ML, Meredith JC (2013) Mechanical and thermal properties of waterborne epoxy composites containing cellulose nanocrystals. Polymer (Guildf) 54:6589–6598. https://doi.org/10.1016/j.polymer.2013.10.011
114. Yang ZY, Wang WJ, Shao ZQ, Zhu HD, Li YH, Wang FJ (2013) The transparency and mechanical properties of cellulose acetate nanocomposites using cellulose nanowhiskers as fillers. Cellulose 20:159–168. https://doi.org/10.1007/s10570-012-9796-z
115. Yasmin A, Daniel IM (2004) Mechanical and thermal properties of graphite platelet/epoxy composites. Polymer (Guildf) 45:8211–8219. https://doi.org/10.1016/j.polymer.2004.09.054
116. Younesi H, Farsi M, Rezazadeh Z (2013) Physical, mechanical and morphological properties of polymer composites manufactured from carbon nanotubes and wood flour. Compos Part B 44:750–755. https://doi.org/10.1016/j.compositesb.2012.04.023
117. Zahedi Y, Fathi-achachlouei B, Yousefi AR (2017) Physical and mechanical properties of hybrid montmorillonite/zinc oxide reinforced carboxymethyl cellulose nanocomposites. Int J Biol Macromol. https://doi.org/10.1016/j.ijbiomac.2017.10.185
118. Zhan J, Chen Y, Tang G, Pan H, Zhang Q, Song L, Hu Y (2014) Crystallization and melting properties of poly (butylene succinate) composites with titanium dioxide nanotubes or hydroxyapatite nanorods. J App Poly Sci 40335:1–10. https://doi.org/10.1002/app.40335
119. Zhang X, Zhang Y (2015) Poly(butylene succinate-co-butylene adipate)/cellulose nanocrystal composites modified with phthalic anhydride. Carbohydr Polym 134:52–59. https://doi.org/10.1016/j.carbpol.2015.07.078

Preparation and Characterization of Antibacterial Sustainable Nanocomposites

T. C. Mokhena, M. J. Mochane, T. H. Mokhothu, A. Mtibe, C. A. Tshifularo and T. S. Motsoeneng

1 Introduction

Pathogenic infections have been the major problem in different fields such as agriculture, healthcare, packaging, and wastewater treatment [26, 38]. These infections are known to cause many deaths more than any other cause. Pathogenic infections result from germs which are found at different places such as air, soil, and water. Because germs are found everywhere it is possible to come in contact with them through touching, eating, drinking or breathing something that contains germs. Germs can be categorized into four: bacteria, viruses, fungi, and protozoa. In this chapter, we limit our discussion to bacteria as a form of germs encountered by humans and animals. Currently, different antibiotics are often used to treat an infection caused by various bacteria. It is recognized that bacteria can mutate their genes in such that they become resistant to the commonly used antimicrobial

T. C. Mokhena (✉) · A. Mtibe (✉) · C. A. Tshifularo
CSIR Materials Science and Manufacturing, Polymers and Composites Competence Area, Nonwovens and Composites Research Group, Port Elizabeth, South Africa
e-mail: mokhenateboho@gmail.com

A. Mtibe
e-mail: mtibe.asanda@gmail.com

T. C. Mokhena · C. A. Tshifularo
Department of Chemistry, Nelson Mandela University, Port Elizabeth 6301, South Africa

M. J. Mochane
Department of Life Sciences, Central University of Technology, Free State Private Bag X20539, Bloemfontein, South Africa

T. H. Mokhothu
Department of Chemistry, Durban University of Technology, Durban, South Africa

T. S. Motsoeneng
Chemistry Department, University of South Africa (UNISA), Florida Park, Roodepoort, Johannesburg, South Africa

Inamuddin et al. (eds.), *Sustainable Polymer Composites and Nanocomposites*,
https://doi.org/10.1007/978-3-030-05399-4_7

agents. Thus, these infections can be combated by potent or specific antimicrobial agents that can mitigate, combat and/or eradicate these bacteria.

In recent years, there has been growing interest in employing a wide variety of nanoparticles as new antimicrobial agents [46, 47]. Unlike conventional antibacterial agents, nanoparticles are extremely toxic towards various bacteria and stable in different conditions found in various industries. Different nanoparticles such as silver, copper, and zinc were reported to be exceptionally effective in eradicating various bacteria at fairly low concentrations. This opened doors for their application in different fields such as agriculture, healthcare, packaging, wastewater treatment, textile, and clothing. Although these nanoparticles show high antibacterial efficiency, their use in industrial applications presents several challenges which brought concerns with regard to their impact on the environment. For example metal copper, even though is cheaper than silver, corrode at standard conditions. On the other hand, if these nanoparticles are used alone their release, recovery, and reuse, cost and long-term effect on health and/or environment limit their success in various fields. A suitable solution is to immobilize and control their release to overcome these limitations [48].

Over the past years, research has escalated in employing polymers as a host matrix for nanoparticles [7, 54, 58, 70, 75] in order to control their release and/or to protect these particles. Polymers have attractive properties such as lightweight, low cost, and mold-ability into various shapes which afford their application in different fields foams, structural adhesives and composites, fillers, fibres, films, emulsions, coatings, rubbers, sealing materials, adhesive resins, membranes [53, 55–57, 59–66, 69, 71]. The properties of the resulting nanocomposite material especially antibacterial activity and release rate were found to be directly dependent on the properties of the polymers (e.g. hydrophilicity, crystallinity and molecular weight). Generally, traditional petroleum-based polymers dominate almost all fields; however, their long-term environmental sustainability is now being questioned. This spurred much interest to utilize biodegradable and compostable polymers as alternative replacement owing to their unique properties such as renewability, biodegradability, and their abundant availability. Thus, these polymers mitigate the environmental impact. They are classified into three categories based on their origin: (i) natural polymers such as proteins (e.g. gelatin, soy protein, silk, and collagen), polysaccharides (e.g. hyaluronic acid, chitin and cellulose) and lipids; (ii) chemically synthesized natural raw materials e.g. polylactic acid (PLA), and microbiological produced or genetically modified bacteria (e.g. polyhydroxyalkanoates (PHA) family); and (iii) chemically synthesized from petrochemical products (e.g. aliphatic-co-polyester (PBSA), aromatic-co-polyester (PBSA) polyester amides (PEA), and polycaprolactone (PCL)) [42]. In this chapter, these polymers will all be denoted as biopolymers to avoid any confusion.

The incorporation of the nanoparticles into biopolymers is of significant importance not only with regards to their antibacterial efficiency but also to improve the overall properties of the resulting nanocomposite materials. Different strategies have been employed to prepare anti biocidal biopolymers which can be classified into two, namely wet and dry methods. A wet method involves the addition of the nanoparticles precursor into the biopolymer solution followed by synthesizing of the nanoparticles. Another route is to dissolve a polymer in a suitable solvent and then add nanoparticles into the polymer solution followed by casting and drying. Dry method involves the addition of the nanoparticles into melted polymer followed by mixing under high shear. This method received a considerable interest because it can easily be adopted in the industry since these units are already installed and being used to prepare particulate filled polymer composites.

In this chapter, we review the current development and recent advances in the preparation and characterization of sustainable antimicrobial nanocomposites, the strategies to enhance their antibacterial activity as well as future prospects of these interesting materials. We also highlight the preparation of different antibacterial nanoparticles and recent developments.

2 Synthesis of Different Nanoparticles

Over the past decades, the synthesis of nanoparticles garnered much interest owing to their unique properties such as large surface-to-volume ratio, modified structure and increased activity as compared to macromolecules. Nanoparticles have seen tremendous success in various fields such as optical, electronic, medicine, textile and clothing and drug delivery. Nanoparticles are present in nature, where they are synthesized by natural processes like biodegradation and biomineralization. They are classified according to their origin as either natural-occurring or man-made.

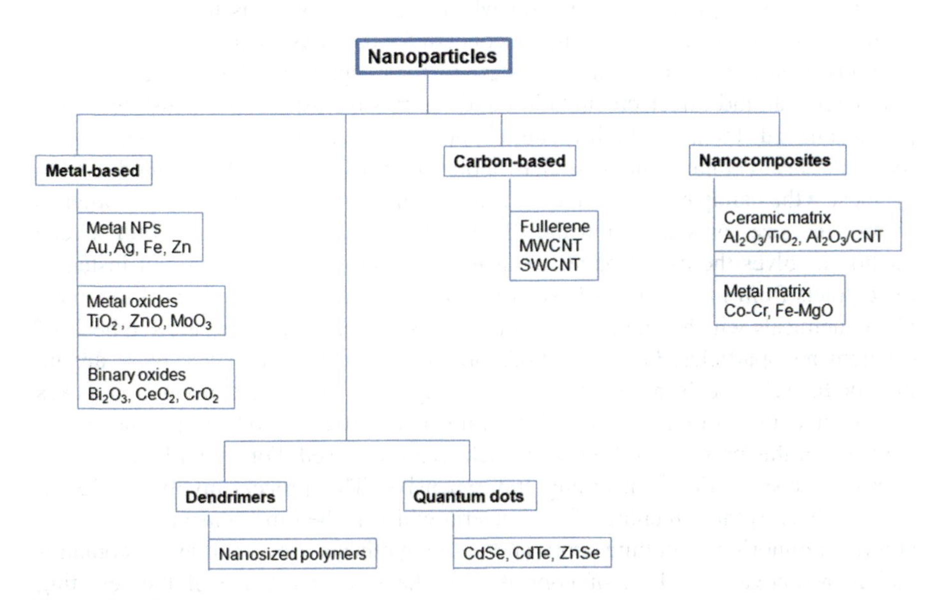

Fig. 1 Classification of nanoparticles by composition

They can further be categorized according to their composition i.e. metal-based, carbon-based, and nanocomposites as well as dendrimers and quantum dots as shown in Fig. 1. In general, there are two approaches used to synthesize various nanoparticles i.e. bottom-up and top-down approach. In a top-down approach, the large macroscopic particles are milled into nano-sized particles. In this case, the synthesized macroscale patterns are reduced into nanoscale particles through plastic deformation. The main disadvantage of this approach is that it cannot be employed for large-scale production since it is an expensive and slow process [2].

2.1 Different Methods Used to Synthesise Nanoparticles

Various nanoparticles synthesis methods have been developed, including physical, chemical and green approaches [9, 39, 40]. The main disadvantages of the physical method are expensive equipment, high temperature and pressure, large area for installation of machines. It involves physical forces which are responsible for the attraction of nano-sized particles to yield large, stable, well-defined nanostructures. The examples of the physical method include colloidal dispersion and basic techniques such as vapour condensation, amorphous crystallization, and physical fragmentation to mention few. Chemical methods involve the use of different toxic chemicals which can be hazardous to the environment and the person handling them. It is recognized that the chemicals employed in chemical approach can bind onto and/or reside in the synthesized nanoparticles (NPs) which can be hazardous especially in medical and packaging applications. In case of food packaging, these chemicals may migrate into the food which can be dangerous to human health. Moreover, the same scenario is often a concern for the physical approach. Chemical methods include chemical emulsion, wet chemical, spray pyrolysis, electrodeposition, chemical and direct precipitation and microwave assisted combustion, while physical include the use of high vacuum in processes such as pulsed laser deposition, molecular beam epitaxy and thermal evaporation. Irrespective of the method used to synthesize the nanoparticles, reducing and stabilizing agents are required in order to control the size (or shape) and the dispersion of the NPs in the medium. Green method involves the use of plants extracts, algae, fungi and bacteria. For instance, the extracts from the plants parts such as roots, leaves, stems, seed and fruits have phytochemicals which can act as reducing and stabilizing agent for the synthesis of different nanoparticles. Green methods are contamination-free since there are no intermediate chemicals involved. The possibility of the large-scale production makes this method the most interesting for future nanoparticles production. The disadvantages of this process are limited to each technique used. For example, the use of micro-organism is time-consuming and expensive. This process involves different steps such as (i) the screening of the bacteria which is the time-consuming process; and (ii) monitoring of culture broth and entire synthesis process to avoid contamination. Moreover, the lack of control over the size and shape of the resulting nanoparticles as well as the cost associated with the media used to grow bacteria add

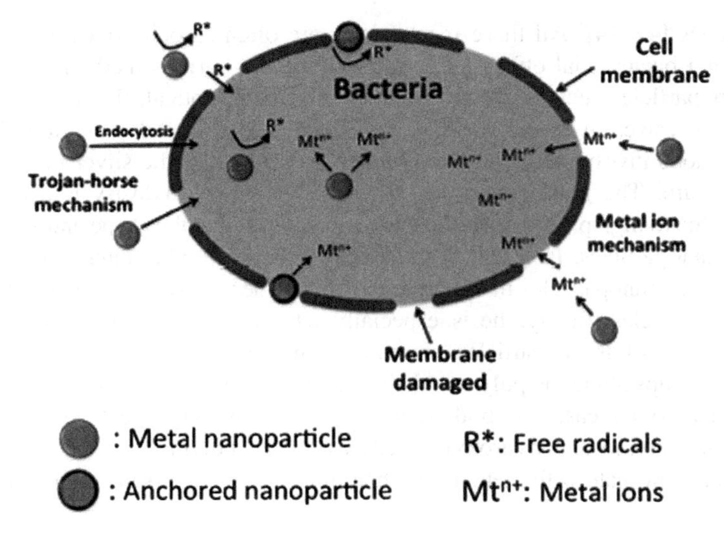

Fig. 2 Schematic presentation based on a summary of the mechanism associated with the antibacterial behaviour of metal nanoparticles (1) "Trojan-horse effect" due to endocytosism processes; (2) attachment to the membrane surface; (3) catalyzed radical formation; and (4) release of metal ions. Redrawn from [46]

to the drawback of this method. However in next sections only synthesize of antibacterial nanoparticles will be discussed. On the other hand, the antibacterial activities of metal nanoparticles are often not well understood. A summary of a possible mechanisms associated with the antibacterial activity of the metal nanoparticles against pathogenic microbes is schematically presented in Fig. 2.

2.2 *Preparation and Antibacterial Mechanisms of the NPs*

2.2.1 Silver Nanoparticles (AgNPs)

The antibacterial properties of silver nanoparticles have been known for more than 2000 years ago. It is highly toxic towards microbes with a wide variety of products that are commercially available because of its broad spectrum of antibacterial activity. It is found in these products in ionic or metallic form. The nanosilver (AgNPs) exhibits high surface area to volume ratio with exceptional antibacterial efficiency towards more than 650 types of bacteria, as well as various viruses and fungi. Moreover, there are 3 proposed mechanisms for the antibacterial activity of AgNPs: (i) the NPs attachment to the cell wall which disturbs the cell membrane permeability or disrupt the cell membrane, (ii) NPs penetration through the cell wall and reacting with thiol and phosphorus-containing compounds which disturbs the respiration and replication processes, and (iii) the release of silver ions by nanoparticles [42–44]. All these mechanisms are often associated with the bacteriostatic and bactericidal effects of AgNPs. There are various methods to prepare silver nanoparticles such as electrochemical, chemical, optical, thermal decomposition, microwave, hydrogel method, sonochemical, UV and gamma radiation. These methods involve reducing an ionic silver salt to metallic silver in an appropriate medium. The most commonly used method in the industry is a chemical reduction due to high production efficiency and control over the shape and size of the resulting nanoparticles. In recent years, there has been growing interest in the synthesis of silver nanoparticles in the presence of polymers to avoid the agglomeration of the nanoparticles after synthesis, especially natural-based polymers. The stability and dispersion of these particles are mostly influenced by the presence of the functional groups along the polymer chain which act as either reducing or stabilizing agent and in other cases as both reducing and stabilizing agent e.g. Chitosan. Chitosan has amino and hydroxyl groups along the polymer chains. The amino groups can coordinate with metal ions when mixed with silver salts while reduction

Table 1 Summary of the synthesize antibacterial nanoparticles

Salt precursor	Synthesize route	Comments	References
AgNPs			
Chemical synthesis			
Silver nitrate	$AgNO_3$ (1 mM, 80 ml) was added drop-wise to an ice-cold $NaBH_4$ 2 mM solution (240 ml) under vigorous stirring	Ag nanoparticles with an average size around 38 nm were obtained and this was confirmed by UV-vis absorption peak at 433 nm for bacterial cellulose/AgNPs and 425 nm for Vegetable Cellulose/AgNPs	[49]
Silver nitrate	AgNPs were prepared by sodium citrate reduction of $AgNO_3$ followed by heating in a furnace at 200 °C for an hour	The resulting AgNPs were a mainly spherical shape with the size of about 50 nm	[83]
Green synthesis			
Silver nitrate	For the first set -the supernatant from soil derived *Pseudomonas putida* MVP2 was mixed with $AgNO_3$ and the second set *P. putida* was directly mixed with $AgNO_3$ to adjust the final concentration of 1 mM and monitored for the reduction of silver solution into AgNPs	For the first step, AgNPs had the size ranging between 5 and 16 nm, while for the second set the $AgNO_3$ treated bacteria showed the number of black spots on the bacterial membrane and cytoplasm which strongly proved that AgNPs were mostly synthesized in the outer membrane of the bacteria	[22]

(continued)

Table 1 (continued)

Salt precursor	Synthesize route	Comments	References
Silver nitrate	A mixture of silver nitrate microcrystalline cellulose was sonicated for 10 min followed by addition of curry leaves (*Murraya koenigii*) extract to reduce silver ions into AgNPs	UV-vis spectra showed Plasmon peak vibration at 430 nm confirming the formation of AgNPs and the resulting spherical AgNPs had a diameter of 10–25 nm with a spherical shape	[87]
Silver nitrate	*Rhizopus stolonifer* aqueous mycelial extract was mixed with $AgNO_3$ solution (1 mM $AgNO_3$ final concentration) and incubated on orbital shaker 180 rpm at 40 °C for two days	AgNPs have a spherical form and diameter about 6.04 nm which was confirmed by UV-vis absorption peak at 420 nm	[1]
ZnNPs			
Zinc nitrate	Mixing crude form with zinc nitrate and heated at 450 ± 10 °C for 30 min followed by calcination at 750 °C	*Euphorbia Jatropha* was used as reducing agent and the resulting hexagonal nanoparticles had a size ranging between 6 and 21 nm	[21]
Zinc nitrate	aloe leaf extracts were mixed with zinc nitrate and heated at 150 °C for 5–6 h. The resulting pale white precipitate was centrifuged for 15 min and dried at 80 °C for 7–8 h Chemical synthesis- Zinc nitrate was dissolved in distilled water and NaOH was added dropwise and then the solution was left overnight. The obtained white precipitate was centrifuge for 10 min and then dried at 80 °C for 6 h. During drying, complete conversion of $Zn(OH)_2$ into ZnO takes place. All the obtained white precipitate were ground	The sizes of the prepared green and chemical ZnO nanoparticles were 40 and 25 nm	[24]
Zinc nitrate hexahydrate	Moringa *Oleifera* extracts as an effective chelating agent. A mixture of the extract and zinc nitrate hexahydrate was left for 18 h then heated at 100 °C and the resulting powder was washed with water to remove any extracts followed by heating at 500 °C for an hour	TEM results showed that highly crystalline zinc oxide nanoparticles with particle size varying from 12.27 and 30.51 nm were obtained.	[37]

(continued)

Table 1 (continued)

Salt precursor	Synthesize route	Comments	References
Chemical synthesis			
Zinc acetate	Chemical synthesis- a mixture of zinc acetate and NaOH (added dropwise) was then calcinated at 300 °C for 6 h	Zinc nanoparticles having an average size of 60 nm was obtained	[85]
CuNPs			
Chemical synthesis			
Cu $(NO_3)_2.3H_2O$	Synthesized ligand benzildiethylenetriamine was used to synthesize the nanoparticles	By reducing the concentration of the reducing agent (ligand) led to an increase in nanoparticles size. The size of the particles ranged between 15 and 63 nm depending on the concentration of the ligand	[12]
$CuSO_4.5H_2O$	Ascorbic acid was used as a reducing agent while starch was used as a stabilizing agent	The obtained Cu and Cu_2O nanoparticles were cubic in shape with a mean size of 28.73 and 25.19 nm, respectively	[30]
Green synthesis			
$CuSO_4$	A Gram-negative bacterium belonging to the genus *Serratia* was isolated from the midgut of *Stibara* sp., an insect of the Cerambycidae family of beetles and used to synthesis CuNPs from 1–48 h	The resulting nanoparticles were polydispersed and vary from 10–30 nm in diameter	[25]
$CuSO_4.5H_2O$	Native cyclodextrins as a stabilizing agent and ascorbic acid reducing agent	The resulting pattern from selected area electron diffraction (SAED) and lattice Fringes confirmed that crystalline structure of Cu-NPs with face-centred cubic (FCC) with a (111), (200) and (220) lattice planes of Cu The Cu-NPs depends on the type of native cyclodextrin (NCD) and the obtained nanoparticles were spherical with a size between 2 and 33 nm. The smaller Cu-NPs were obtained with α-NCD (viz. 4 nm), while the nanoparticles obtained with β-NCD showed narrow size distribution having a size of 6.5 nm	[20]

(continued)

Table 1 (continued)

Salt precursor	Synthesize route	Comments	References
$Cu(NO_3)_2$	*Garcinia mangostana* leaf extract as a reducing agent was mixed with copper nitrate and heated at 70 °C for an hour	Cu nanoparticles with the mean particle size of 28.9 nm were obtained	[50]
$Cu(NO_3)_2$	The extracts of *E. prostrate* was mixed with copper acetate and stirred for 24 h at room temperature without external energy	Monodisperse and spherical particles with sizes ranging from 28 to 45 nm and (mean, 36 ± 1.2 nm) were obtained	[15]

of silver ion is coupled with oxidation of hydroxyl group. Table 1 highlights recent studies based on the production of the silver nanoparticles.

2.2.2 Zinc Nanoparticles

Zinc oxide received a lot of attention owing to its special characteristics such as anti-corrosion, antibacterial, and has low electrons conductivity and excellent heat resistance. It has been prepared via physical and chemical methods [2, 28]. The chemical methods include the reaction of zinc with alcohol, vapour transport, hydrothermal synthesis and precipitation method. In order to avoid the complications that are bound to physical and chemical methods, the green synthesis of zinc oxide has been employed for the past years. There is still a lot of controversy about the mechanism behind the antibacterial activity displayed by zinc nanoparticles [2, 28]. It was postulated that the toxicity mechanism result from intercellular reactive oxygen species (ROS) generation and Zn^{2+} release. ROS-include superoxide anion, hydrogen peroxide and hydroxide-are harmful to bacterial cells, while the released Zn^{2+} disrupts important metabolic pathways. Since the morphology of the zinc nanoparticles-which depends on the synthesis technique-plays significant role in their antibacterial activity. There has been a lot of interest in recent years towards the development of novel synthesis methods and applications. Green synthesis methods received more interest because of their unique attractive properties-including environmentally friendlier, cost-effective, biocompatible and safe. Plants, algae, fungi and bacteria have been employed as a green method to produce zinc nanoparticles. Figure 3 illustrate the synthesis of zinc oxide nanoparticles by using different sources. Recent studies based on the use of different bacteria, fungi, and plants parts that have been used to prepare zinc nanoparticles are summarized in Table 1.

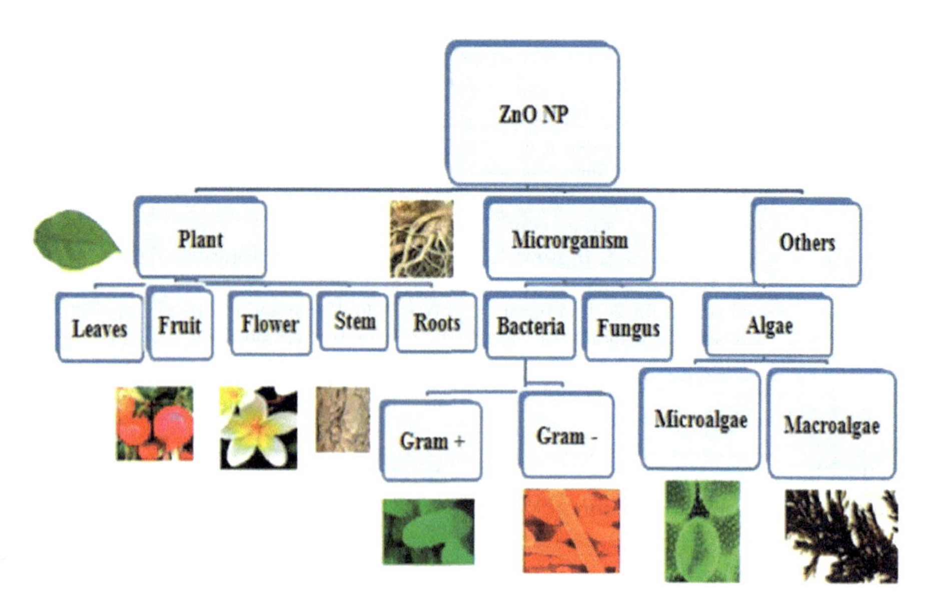

Fig. 3 Zinc oxide nanoparticles synthesis by using different sources [2]

2.2.3 Gold Nanoparticles (AuNPs)

Gold nanoparticles garnered much interest in different fields such as wastewater treatment, catalysis, biosensing, optics and therapeutic applications owing to its unique features which include good antibacterial activity, optical and photothermal properties. There are two proposed mechanisms with regard to antibacterial activity of AuNPs (i) a change in the membrane potential and prevention of ATPase activities resulting in a decline in cellular metabolism, and (ii) inhibiting the binding of subunits of the ribosomes of tRNA resulting in a collapsing in biological processes [86]. There is a paradigm shift towards green, cost-effective and controllable methods to synthesize gold nanoparticles. Similarly to other nanoparticles, the use of plants extracts, bacteria and fungi have been employed to synthesize gold nanoparticles. For an example, typical fresh leaves of *Piper guineense* (Fig. 4) were used for the synthesise of gold nanoparticles [79]. On the other hand, natural polymers were also employed as stabilizing and reducing agents to synthesize gold nanoparticles. Gold nanoparticles having diameters ranging between 50 and 200 nm were synthesized by Shih et al. [78] using chitosan as stabilizing and reducing agent, while carboxymethyl chitosan was used by Laudenslager et al. [72]. The resulting gold nanoparticles had an average size of 22.9 nm.

Fig. 4 Leaf and fruit of *Piper guineense* [79]

2.2.4 Copper Nanoparticles

Copper is the most important element due to its role in living organisms. It plays an important role in the transportation of oxygen during electron transport chain and iron homeostasis. Its antibacterial activity is associated with cellular damage after contact between Cu^{2+} and bacteria membrane [18]. This led to the development of novel synthesis techniques for producing nano-sized copper particles. The electrochemical is the widely used method to produce copper nanoparticles (CuNPs) [27]. This is as a result of the efficiency of this method to produce large quantities of the nanoparticles at a short period of time. CuNPs can also be synthesized through several techniques such as chemical reduction, laser ablation, sol-gel processing, and thermal reduction [18]. The nanoparticles synthesized from these techniques have different antibacterial activity. Furthermore, the reaction medium composition and stabilizers had an effect on the size of the resultant nanoparticles. Satyvaldiev et al. [74] investigated the synthesis and biological activity of copper nanoparticles under different medium (alkaline and ammonia) and in the presence of a stabilizer (gelatin). The FESEM images (Figs. 5 and 6) in both media with and without gelatin were found to be spherical in shape. However, the insets in Fig. 5a, b showed that Cu nanoparticles in an alkaline medium had smaller particle size than in ammonia medium. The addition of gelatin reduced the particle size more, which was associated with an increase of nanoparticle stability in the presence of gelatin. The recent studies based on the synthesis of the CuNPs and their properties are summarized in Table 1.

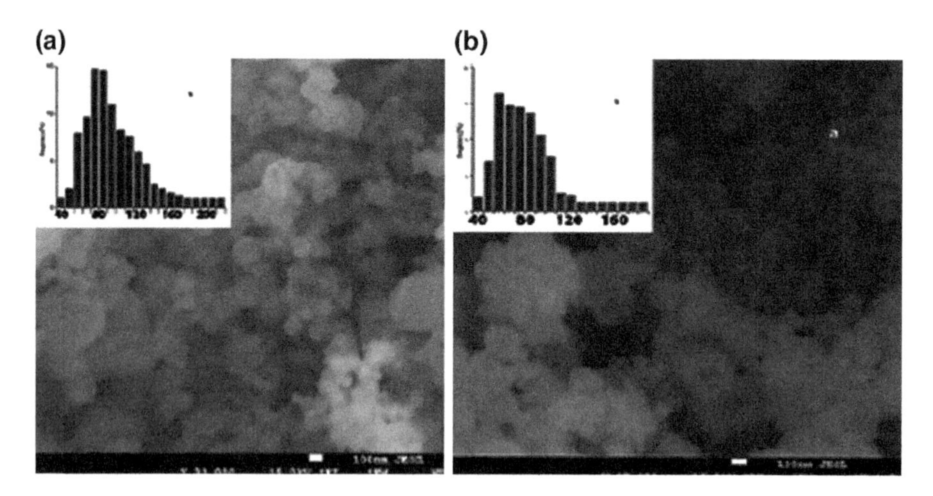

Fig. 5 SEM images of Cu nanopowders produced in an alkaline medium (**a**) and Cu nanopowders obtained in an alkaline medium in the presence of gelatin (**b**) with their particle size histograms in their insets [74]

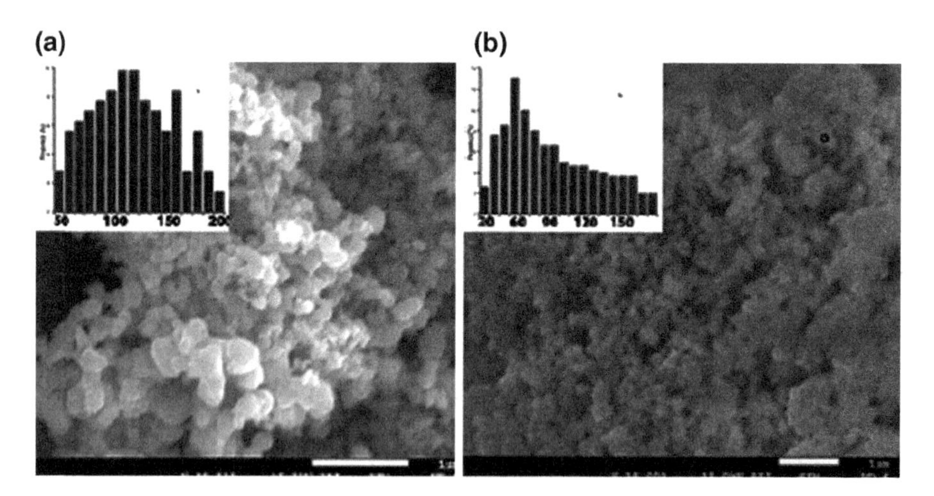

Fig. 6 SEM images of Cu nanoparticles obtained in an ammonia medium (**a**) and Cu nanoparticles obtained in the presence of gelatin in an ammonia medium (**b**) with their particle size histograms in their insets [74]

2.2.5 Carbon-Based Nanoparticles

Carbon nanotubes have attractive properties such as high mechanical strength, large surface area, high thermal stability (>700 °C), high aspect ratio (length to diameter) and electrical conductivity [16, 52]. They have been used in wide variety of industrial applications such as energy storage, construction, and entertainment.

They also have excellent antimicrobial activity [13, 29, 35]. The antibacterial activity of carbon nanotubes is still unknown. It is proposed that the antibacterial activity is induced by physical membrane perturbation and oxidative stress. It is reported that the antibacterial activity is influenced by CNT distribution, diameter, length, and electronic structure [13, 29]. CNT are classified into single and multiwall carbon nanotubes depending on the number of graphite layers. Single-walled CNT is composed of graphene sheet wrapped into a cylinder, while multi-walled CNT consists of two or more concentrically mixed cylinders. Large-scale production of CNT is achieved through chemical vapour deposition. Other techniques employed to produce CNTs include electric arc discharge and laser ablation. Three most important aspects of the synthesis of CNT are carbon source, energy, and a metal catalyst.

Graphite nanosheets-known as graphene is one of the promising carbonaceous materials with exceptional properties such as electrical; thermal; mechanical; barrier and flame retardant properties [8, 33, 73, 90]. It is obtained from expandable natural graphite by different exfoliation process such as chemical or electrochemical, mechanical, and thermal and/or chemical. The mechanical technique includes the use of scotch/adhesive tape to repeatedly peel off graphene layer from graphite crystal until individual atomic planes are obtained. Beside scotch tape, epitaxial growth of graphene on silicon carbide was also reported. In this case, silicon carbides heated to high temperatures above 1100 °C to reduce it into graphene. Yet another method to produce graphene includes CVD of hydrocarbons on a metal or metal-coated substrates such as Ni films and Cu foil. Large-scale graphene production involves exfoliation of graphite via acid oxidation (e.g. nitric acid/sulphuric acid) into expandable graphite oxide (GO), followed by reduction of GO (e.g. by hydrazine, $NaBH_4$) and annealing in argon/hydrogen to yield graphene sheets. The disadvantage of the exfoliation using oxidation/reduction from GO is the quality of the resulting graphene as compared to scotch tape, epitaxially grown on SiC and CVD. The solution methods, however, have been employed for the production of graphene ribbons through cutting open nanotubes in the presence of potassium permanganate and sulphuric acid. The solution methods such as filtration, solution casting, electrophoretic deposition and Langmuir-Blodgett deposition afford the formation of graphene-based functionalized films.

2.2.6 Clay Minerals

Clay minerals are also known as layered silicates because of their stacked structure of 1 nm silicate sheets which are in nanoscale with a variable basal distance. They often used in human and health formulations like excipients or active substances because of the unique features which include chemical inertness and low or zero toxicity [35]. Clay is classified into various classes: smectites, vermiculite, montmorillonite, halloysite, palygorskite and many more. Smectites are planar dioctahedral and trioctahedral 2:1 clay minerals having a layer charge between −0.2 and −0.6 per formula unit which contains hydrate exchangeable cations. Smectite 2:1

layer unit is made up of one alumina octahedral sheets sandwiched between two silica tetrahedral sheets. Montmorillonite is generally defined as dioctahedral smectites. The term vermiculite is used to define planar dioctahedral and trioctahedral with a layer charge between −0.6 and −0.9 per formula unit which contains hydrated exchangeable cations. Halloysite is commonly defined as a 1:1 aluminosilicate structure with its size depending on the deposit, while palygorskite is 2:1 layered structure consists of two-dimensional tetrahedral sheets. The antibacterial activity of clay is proposed to take place through clay permeation into the cell membrane, damage the cell wall, and disturbs the natural processes of the cell which results in eradication of microorganisms.

3 Preparation of Antibacterial Nanocomposites

Depending on the intended application several processing techniques such as melt compounding, solution casting, in situ polymerization and electrospinning were employed to prepare antibacterial nanocomposites. Solution casting involves dissolving the polymer and antibacterial agent in a suitable solvent, casting and drying or evaporation of the solvent [85, 88]. This process is often limited to laboratory-scale and/or polymers which are soluble in available solvents. The heat-sensitive polymers and/or antibacterial agents are usually prepared via solution casting. Melt compounding involves mixing the antibacterial agent and polymer at the temperature above melting temperature of the polymer under high shear. These techniques include single/twin screw extruder, injection moulding, melt mixer and hot melt presser. They can be used either alone or in combination in order to prepare the resulting nanocomposite product [10]. The melt compounding is limited to polymers that can melt with antibacterial agents which are not sensitive to high temperatures. In most cases, the antibacterial agent ends up in the amorphous part of the polymer, thus the crystallinity of the polymer plays a major role on the dispersion a well as the release rate of the antibacterial agent from the system. Nevertheless, these techniques are most significant importance from the commercial viewpoint. The other interesting preparation method of the antibacterial nanocomposites is electrospinning. This technique is cheaper and has a potential for large-scale production. Electrospinning is capable of producing multifunctional nanofibre from natural and synthetic polymers, polymer blends and composites [10]. The resulting fibres have unique properties such as high surface area, inter- and intrafibrous pores as well as strong adhesive force, good air filtration, high adhesion barrier activity and heat resistance which rare of significant importance towards biomedical applications. The resulting electrospun scaffolds have a similar shape to human skin tissues; hence they are often employed in tissue engineering. However, the above-mentioned properties render these fibers opportunity to be applied in other fields such as water and air filtration, energy storage and packaging. Coating of the antibacterial agents onto the polymeric matrix has also been adopted by several authors [23, 41]. All these preparation techniques result in different morphologies and thus, influence the antibacterial efficiency of the resulting nanocomposite material. In this case, the contact between the nanoparticles with the microbes is of significance in order to promote the antibacterial activity of the resulting composite material. It is recognized that in case of metal nanoparticles the ions release is the main contributor to the eradication of the microbes [46]. However, for nanoparticles such as carbon nanotubes, the main contribution comes from the contact with microbes by piercing the membrane of the microbe. Thus the preparation method plays a major role in the antibacterial behaviour of the resulting composites as well as the type of polymer. In summary the antibacterial behavior of the antibacterial nanocomposites is directly dependent on these factors: (i) adsorption of microbes on the polymer surface; (ii) polymer-type to allow water penetration to reach the embedded nanoparticles to allow their dissolution and realization; (iii) contact between nanoparticles with available microbes; and (iv) physical and structural properties of the nanoparticles.

4 Antibacterial Nanocomposites

4.1 Silver/Biopolymer Nanocomposites

Silver nanoparticles nanocomposites are often prepared by either ex situ or in situ [17]. In situ involves the inclusion of silver ionic solution, reducing agent and polymeric material into one system followed by reduction of silver ionic solution into silver nanoparticles, while ex situ involves the synthesis of particles beforehand followed by addition of AgNPs into the polymeric material. In situ preparation of silver nanoparticles was carried out by mixing $AgNO_3$ (1 mM, 80 ml) and an ice-cold $NaBH_4$ 2 mM solution (240 ml) drop-wise to under vigorous stirring in the presence of bacterial cellulose (BC) as well as vegetable cellulose (VC) [49]. A significant antibacterial activity for nanocomposite samples against both gram-negative (*K. pneumonia*) and gram-positive (*S. aureus* and spore-forming *B. subtilis*) tested bacteria, by the action of both BC- and VC-based Ag nanocomposite samples. [31] prepared cellulose/AgNPs nanocomposite by in situ method. In this case, ethylene glycol, cellulose and silver nitrate were mixed together and exposed to microwave radiation. The resulting nanocomposite material displayed high antimicrobial efficiency (Fig. 7) against tested bacteria viz. *Escherichia coli* and *Staphylococcus aureus*.

Ex situ was used to prepare PBAT/AgNPs nanocomposites for packaging applications via solution casting [84]. AgNPs were prepared by chemical synthesis using sodium citrate as reducing agent. It was reported that increase in nanofiller wt% improves mechanical, oxygen permeability, and antimicrobial properties, but at a higher loading the films become stiff. [76] reported on the synthesis of AgNPs using tocopherol as reducing and capping agents in order to prepare PBAT/AgNPs composite films for food packaging. The films were prepared through solution casting

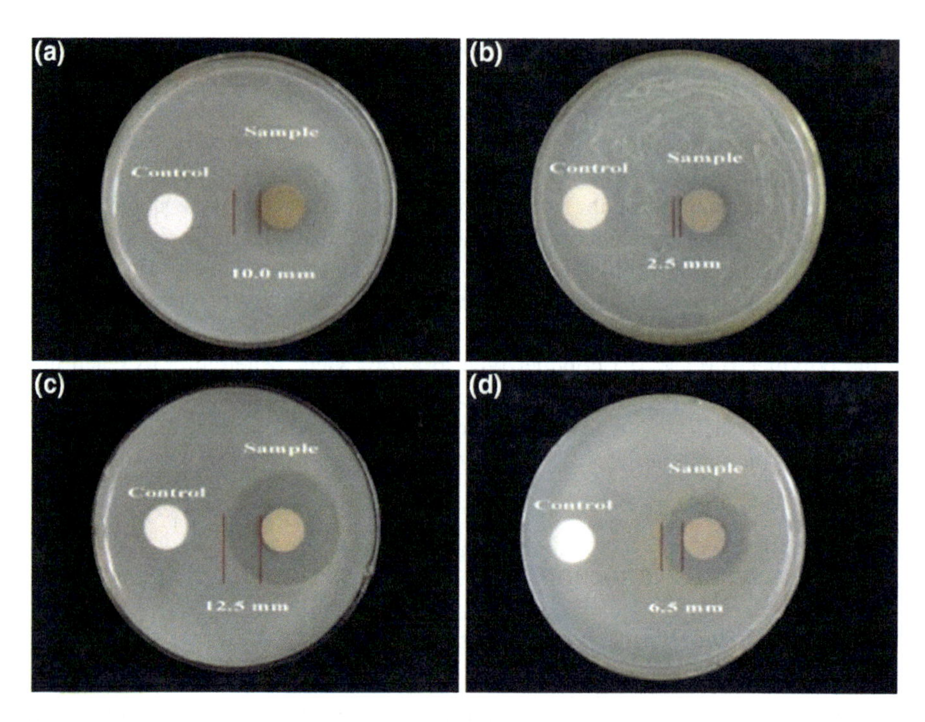

Fig. 7 Antimicrobial activities of the cellulose–Ag nanocomposites: **a**, **c** *Escherichia coli* and **b**, **d** *Staphylococcus aureus* [31]

method and the addition of AgNPs leading to rougher surfaces as compared to smooth PBAT. However, the synthesized AgNPs were spherical with a diameter of 10–50 nm. It was found that the PBAT/AgNPs exhibited good antibacterial activity against both food-borne pathogenic bacteria (i.e. *E. coli* and *L. monocytogenes*), but showed stronger antibacterial activity towards *E. coli* than *L. monocytogenes* since the Gram-negative bacteria are more vulnerable to AgNPs than Gram-positive ones. On the other hand, mechanical strength, water vapour barrier property, and water contact angle of the PBAT films improved significantly after addition of AgNPs making this material a good candidate for the potential for food packaging especially looking at the concentration of silver used in this case i.e. 10 mg silver in 4 g PBAT. [77] studied the effect of silver nanoparticles obtained from different synthesis routes viz. metallic silver (AgM) or silver zeolite (AgZ) prepared under vigorous stirring for 12 h, trisodium reduced ($AgNP^{C}$) and laser ablated ($AgNP^{LA}$) on the properties of alginate as host matrix. The resulting silver zeolite (AgZ) nanoparticles were cubical in shape with the size of 2–5 μm, whereas the shapes of $AgNP^{C}$, $AgNP^{LA}$, and AgM were almost spherical, and the size was in the range of 50–200 nm. It was found that the prepared composites via solution casting were flexible, smooth, and free-standing, but the composites were rougher than pristine alginate films. The minimum inhibitory concentration (MIC) or minimum bactericidal concentration (MBC) of $AgNP^{C}$, AgZ, $AgNO_3$, $AgNP^{LA}$, and AgM against *E. coli* were

3.125/6.25, 3.125/12.5, 1.562/6.25, >50/>50, >50/>50 μg/mL, respectively, and those against *L. monocytogenes* were 6.25/12.5, 6.25/12.5, 3.25/6.25, >50/>50, >50/ >50 μg/mL, respectively. The degree of antimicrobial activity was greatly influenced by the type of silver particles. It was found that among the silver particles included groups, alginate/$AgNP^{C}$, alginate/AgZ, and alginate/$AgNO_3$ composite films exhibited potent antibacterial activity against both foodborne pathogenic bacteria, however, alginate/$AgNP^{LA}$ and alginate/AgM composite films did not show distinctive antibacterial activity. It was postulated that the low or negligible antibacterial activity of alginate/$AgNP^{LA}$ and alginate/AgM composite films may be regarded to the $AgNP^{LA}$ strongly capped with poly vinyl pyrrolidone (PVP) and interacting tightly with the alginate resulting in the prevention of oxidation of AgNPs to form Ag ions, and that the stable metallic silver of AgM was not oxidized easily to release silver ions.

4.2 Zinc/Biopolymer Nanocomposites

Zinc nanoparticles have been used to reinforce different biopolymers, especially for packaging applications. Poly(butylenes adipate-co-terephthalate) PBAT is one of the biopolymers that has been reinforced with zinc nanoparticles for different applications. PBAT is an aliphatic/aromatic copolyester synthesis from 1.4-butanediol, adipic acid, and terephthalic acid. It is a biodegradable polymer having excellent compatibility with other biodegradable polymers such as aliphatic polyesters and starch-based polymers. Due to its unique properties which include biodegradability and biocompatibility, PBAT has been used in packaging films, agricultural films and compost bags. In the case of packaging, different fillers are often added to improve their properties such as barrier properties, and mechanical properties. One of the most important aspects of food packaging is the antibacterial efficiency of the polymer being used. Different antibacterial nanoparticles are often added not only to improve the antibacterial activity but also the barrier properties [85]. Venkatesan and Rajeswari [85] prepared PBAT/ZnO nanocomposites through solution casting using chloroform as a solvent. ZnO was prepared using a chemical method whereby zinc acetate was mixed with NaOH (by adding NaOH drop by drop) followed by calcination at 300 °C for 6 h in a furnace. The antibacterial activity was tested using *E. coli* and *S. aureus*. It was reported that the inhibition zone increased with increase in zinc nanoparticles content regardless of bacteria type. The diameter of the inhibition zones for nanocomposites containing 1, 3, 5 and 10 wt of ZnO nanoparticles were 11.0, 11.8, 12.7, 13.3, and 14.1 mm for *E. coli* and 11.0, 12.0, 13.5, 13.9, and 15.1 mm against *S. aureus*, respectively.

Zinc nanoparticles were also incorporated into PHBV for packaging application by Ana and Angel [4]. Nanocomposites were prepared through solution casting. In this case, the nanoparticles were firstly dispersed in chloroform and then mixed with polymer solution also dissolved in chloroform followed by sonication, casting and drying. Using TEM and SEM, it was found that the ZnNP were well dispersed in the polymer matrix. This was ascribed to the interactions between the –OH groups of the ZnO surface and the polar moieties of the biopolymer preventing nanoparticle aggregation while improving the compatibility between the filler and matrix phases. However, at the higher content of the nanoparticles well-dispersed clusters were obtained. This was also confirmed by an increase in mechanical properties up to 4 wt% of the filler with Young modulus reaching ~ 1.7 GPa and tensile strength of ~ 30 MPa, whereas elongation at break decreased from 4% (for neat polymer) to ~2.9%. The nanocomposites exhibited antibacterial activity against human pathogenic bacteria, and the effect on *E. coli* was stronger than on *S. aureus* due to their difference in thickness of their membranes. Gram-positive bacteria usually have one cytoplasmic membrane and a thick wall composed of multilayers of peptidoglycan, whereas the Gram-negative have a more complex cell wall structure, with a layer of peptidoglycan between the outer membrane and the cytoplasmic membrane. Hence *E. coli* is more susceptible to antibacterial agents than *S. aureus*.

Augustine et al. [6] prepared electrospun polycaprolactone/ZnNPs nanocomposite membrane as biomaterials with antibacterial and cell adhesion properties. The incorporation of ZnNPs reduced the size of resulting electrospun nanofibers and these nanofibers became ore rougher with an increase in ZnNPs content due to their agglomeration. At a low content of ZnNPs viz. up to 1 wt%, tensile strength and modulus increased linearly, while further increase in ZnNPs content led to a decrease in tensile strength and modulus of the nanocomposite membrane. This was attributed to the agglomeration of the nanoparticles because of their high surface energy resulting in their agglomerate and thus led to poor dispersion. It was also reported that the nanocomposite membrane had good antibacterial activity against *E. coli* and *S. aureus*, but at low NPs content i.e. below 5 wt% the antibacterial activity was statistically insignificant. Strong antibacterial activity was recognized at 5 and 6 wt% of the nanoparticles. This was ascribed to the fact that at low NPs content, the NPs are trapped within the polymeric matrix and thus, those in direct contact with bacterial cells are very few in number. Interestingly, the nanocomposites membranes showed 100% cell viability. Agustin and Padmawijaya [5] investigated the effect of glycerol and zinc oxide on the antibacterial activity of biodegradable bioplastics from chitosan-kepok banana peel starch. It was reported that the biodegradability rates decreased (Table 2) as the zinc oxide concentration in banana peel starch-chitosan bioplastic, thus composite bioplastic material will inhibit bacterial growth. The resultant biocomposites have a significant potential to be used for food packaging by having the biodegradable properties and also inhibit bacterial growth.

Table 2 Biodegradable test of chitosan-starch bioplastics [5]

Zinc oxide (%)	Degradation time (minutes)
1	35
3	42
5	72

4.3 Gold/Biopolymer Nanocomposites

Mendoza et al. [38] exploited the reductive and stabilizing action of chitosan to prepare gold/chitosan nanocomposites. In this case, the precursor was introduced dropwise into chitosan solution (with chitosan as a stabilizing and reducing agent) and stirred for 4 h at 60 °C. Spherical gold nanoparticles embedded in chitosan were obtained as confirmed by maximum surface Plasmon resonance at ~525 nm. The authors reported that the size of the nanoparticles increased with an increase in precursor' concentration i.e. 1 mM yielded spherical NPs with a diameter of 14 ± 5 nm while for 2 mM was 14 ± 3 nm. They found that the composite exhibited superior bactericidal ability against both bacteria (*S. aureus* and *E. coli*) models without showing cytotoxicity on human cells at the concentrations tested.

4.4 Copper/Biopolymer Nanocomposites

[11] investigated antimicrobial nanocomposites and electrospun coatings based on poly(3-hydroxybutyrate-co-3-hydroxyvalerate) and copper oxide nanoparticles for active packaging and coating applications. A blend of poly (hydroxybutyrate-co-hydroxyvalerate (PHBV) composed of PHBV3 (3 mol valerate) and PHBV18 (18 mol valerate) was prepared by melt mixing and used as a control. On the other hand, the nanocomposites were prepared via two different coating technology: (i) two different content of CuNPs (i.e. 0.1 and 0.05%) were melt mixed with the blend of PHBV3 pellet (86 wt%) and unpurified PHBV18 powder (14 wt%) followed by melt-pressing at 180 °C for 3 min, (ii) PHBV3 film was coated with PHBV18/CuNPs electrospun ultrathin fibers mats containing 0.05% followed by hot-pressing at 150 °C for 2 min (without pressing) (denoted ES-0.05%). Mechanical properties for all prepared samples were not significantly influenced by the addition of NPs. Water Vapour Permeability (WVP), however, increased by the addition of the NPs, more especially where fiber mats were put as an antibacterial coating onto PHBV3 matrix. This was ascribed to the more hydrophilic character of the electrospun mats with PHBV/CuNPs as compared to neat PHBV3 film used as a substrate in the coated system. In the case of oxygen permeability (OP), the addition of 0.05% NPs reduced OP by 34.2% which was related to the additional tortuous path created by the well-distributed and dispersed NPs. The OP of the ES-005% was higher than its counterpart prepared by melt compounding. This was ascribed to the lower crystallinity of ES-005%. For antibacterial activity, a reduction of about 5 log colony-forming unit/mL of *S. enterica* was recorded for those films prepared with 0.05% CuO by melt-mixing and no viable count of bacteria were recorded either for nanocomposites films containing 0.1% CuO or the 0.05% ES coating structure after 24 h of exposure. No viable counts of *L. monocytogenes* were recorded in any of the samples after 24 h of exposure. The effectiveness for inactivation of *L. monocytogenes* was attributed to structural and chemical

compositional differences between cell surfaces of Gram-positive and Gram-negative bacteria. Bio-disintegration tests showed that the coated structures were fully biodegraded in a period of 35 days at the composting condition.

Mary et al. [36] coated cellulose with copper (II) ions and CuNPs to evaluate their release from the system. Cellulose was chemically modified by periodate-induced oxidation followed by covalent attachment of biopolymer chitosan (CAC). The Cu(II) was immobilized to CAC by its immersion in Cu(II) solution (CBCAC), while CuNPs were produced by borohydride-induced reduction of Cu(II). In this case, CAC was immersed in Cu(II) solution followed by immersion in sodium borohydride solution for 24 h and then dried in a vacuum chamber at 50 °C (NCLCAC). It was reported that the release of Cu(II) from all prepared samples was depended on the concentration of Cu(II) in the system i.e. high release was recorded for higher Cu(II) and/or CuNPs concentration. Similarly, the antibacterial activity against *E. coli* was found to be dependent on the concentration of Cu(II) and CuNPs. The radius of inhibition zone increases with an increase in the copper and CuNPs contents of loading solutions. This was attributed to the biocidal action of Cu(II) ions, which are released from the fibers which can be related to the copper content in the loading solutions increasing, with the amount bound to fibers also increases and thus, the antibacterial action of resulting fibers becomes more effective, thus resulting in the formation of “zone of inhibition” with greater area (or radius).

4.5 Carbon-Based/Biopolymer Nanocomposites

[19] prepared antibacterial carboxymethyl chitosan (CMC)/carbon nanotubes (CNT) nanocomposites via solution casting. CMC-CNTs 20 recorded highest antibacterial efficacy as compared to other composites with inhibition zone diameter of 22.3 ± 0.21 mm against *S. aureus* and of 21.3 ± 0.72 mm against *E. coli* corresponding to 23.2 ± 0.12 mm and 22.5 ± 0.63 mm for ampicillin and gentamicin, respectively. The MIC value for CMC-CNT20 was 1.95 μg/mL against both *S. aureus* and *E. coli* corresponding to 0.98 and 1.95 μg/mL for antibiotics ampicillin and gentamicin, respectively. It was postulated that the highly hydrophilic composite potent result from penetration inside the microbe and cause osmotic imbalances, which enhanced the mode of growth inhibition. Moreover, the presence of CNTs has a synergistic effect in destroying the microbial cell membrane and suppressing the microbial growth. Elsewhere in the literature, it was reported that functionalization of CNT and polymer led to the formation of ester linkages from the condensation of the carboxylic acid of acrylic acid-grafted poly(butylene adipate-co-terephthalate) (PBAT-g-AA) with the hydroxyl groups of multi-hydroxyl functionalized multi-walled carbon nanotubes (MWCNT-OH) [89]. This resulted in strong antibacterial activity against *E. coli* (BCRC 10239) when compared to poly(butylenes adipate-co-terephthalate) (PBAT)/multiwalled carbon nanotubes (MWCNTs) nanocomposites. This was attributed to electrostatic

interactions between the composite and bacterial strains since *E. coli* with an extracellular capsule carry a less negative charge and are less prone to adsorption on the positively charged surface of PBAT-g-AA/MWCNT-OH. An et al. [3] prepared composite blend of polylactic acid/polyurethane (PLA/PU) reinforced with graphene oxide (GO) through solution mixing. In this case, the authors prepared composites based on two different concentration of GO i.e. 3wt% (PLA/PU3) and 5wt% (PLA/PU5) to evaluate their antibacterial efficiency against gram-negative and gram-positive bacteria. It was reported the incorporation of 5 wt% of GO into PLA/PU reduced *E. coli* and *S. aureus* growth up to 100%, while 3 wt% reduced *E. coli* growth up to 100% and *S. aureus* growth up to 99% after 24 h of incubation. Moreover, after 4 h of incubation with PLA/PU3 and PLA/PU5 composites at 37 °C, the antibacterial activity for *E. coli* was 54 and 91%, whereas 54 and 89% were recorded for *S. aureus*, respectively. This confirms the antibacterial efficacy of GO which was dependent on its content as shown in Fig. 8. It can be seen that after 4 h of incubation of both bacteria (viz. *E. coli* and *S. aureus*) lost their original appearance. This was attributed to the effect of either oxidative stress or physical disruption related to the carbon nanomaterials. Despite the antibacterial efficiency of graphene nanosheets, it can also serve as substrate or matrix for the antimicrobial agent [91].

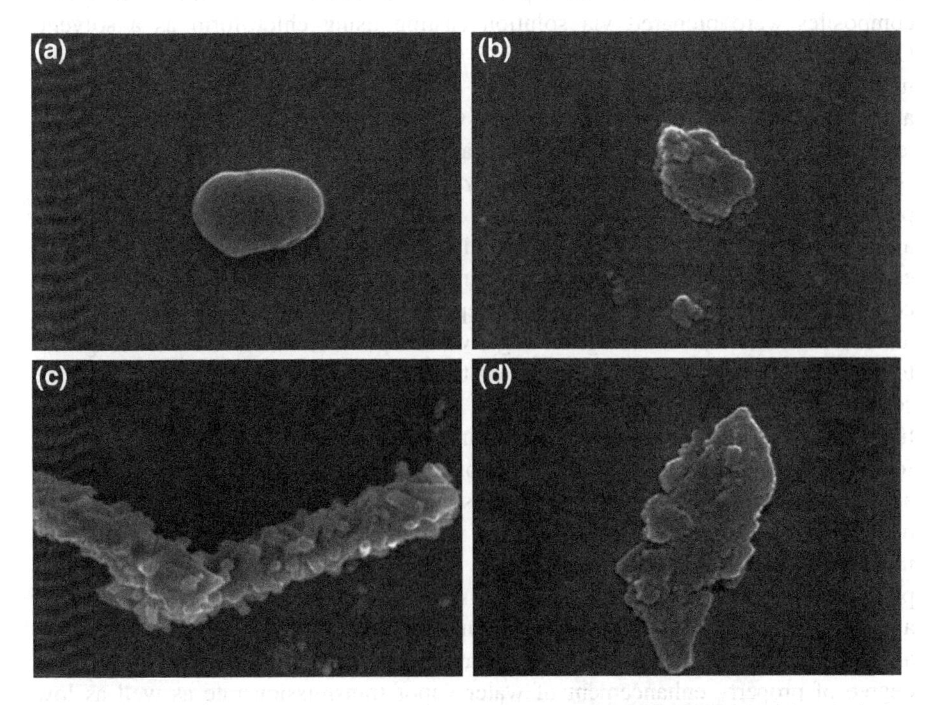

Fig. 8 SEM images of *S. aureus* attached to **a** glass plate and **b** PLA/PU/G5 for 4 h incubation at 37 °C SEM images of *E. coli* attached on **c** glass plate and **d** PLA/PU/GO (5%) for 4 h incubation at 37 °C. Reprinted from [3]

4.6 Clay Minerals/Biopolymer Nanocomposites

In most cases, clay is usually added to different polymeric materials to improve the mechanical and thermal properties of the resulting nanocomposite product. In case of packaging, clay minerals were found to improve the barrier property which is one of the important aspects [17]. The most used method of preparation is solution intercalation. It is recognized that the modification of clay minerals plays an important role in their antibacterial activity especially the presence of quaternary ammonium groups [67, 68]. On contrary, [51] reported that unmodified montmorillonite displayed good antibacterial activity against *S. aureus* and *E. faecalis* antibacterial. The authors mentioned that this scenario could be related to the preparation method of the composites and composites. The nanocomposites were prepared through solution casting. Despite unmodified MMT not showing any antibacterial activity, the nanocomposites showed good antibacterial activity towards *S. aureus* and *E. faecalis* as well as lactose-positive bacteria from *Enterobacteriacea* family. This was attributed to a very good homogeneity obtained from the preparation method. A comparison between three nanoclays i.e. Cloisite 20A (dimethyl di(hydrogenated tallowalkyl) quaternary ammonium), Cloisite 30B (bis-(2-hydroxyethyl) methyl(hydrogenated tallowalkyl) quaternary ammonium) and unmodified Cloisite Na^+ was reported by Sothornvit et al. [80]. PLA/nanoclay composites were prepared via solution casting using chloroform as a solvent. On contrary to the previous study, it was reported in this study that the unmodified-based composites showed no antibacterial activity towards gram negative and gram-positive bacteria. Despite Cloisite 30B nanoclay showing antibacterial activity towards both gram negative and gram positive bacteria, its composites showed bacteriostatic against *L. monocytogenes*, while Cloisite 20 A based composites displayed no antibacterial activity. However, it is worth mentioning that the antibacterial activity of the nanoclays also depends on the polymer properties especially hydrophilicity which plays a major role for the bacteria to enter the composite material. Rhim et al. [67] reported that Cloisite 30B showed good antibacterial activity towards gram negative and positive bacteria when using chitosan as host matrix. This was attributed to the hydrophilicity of chitosan and the solvent used to prepare the composites i.e. acetic acid. [45] modified clay with cetyl trimethylammonium bromide (CTAB) to improve its antibacterial efficiency. It was found that the CTAB modified clay composites exhibited good antibacterial activity against *B. subtilis* and *S. aureus*. This was attributed to the presence of long chain hydrophobic alkyl and cationic charge of a quaternary ammonium group in modified clay. Zones of inhibition diameter of PBAT/modified nanoclay nanocomposites were 11.2, 13.7, 12.0 mm against *B. subtilis* and 11.1, 13.5, 11.5 mm against *S. aureus* with loading 2, 4, and 8 wt%, respectively. Besides good antibacterial efficacy, PBAT/modified nanoclay composites showed the greater degree of property enhancement of water vapor transmission rate as well as low degradation rate when compared to unmodified clay composites. The latter was attributed to the smooth surfaces obtained from modified clay composites.

The antibacterial activity for nanoclay (rectorite modified with CTAB) under different conditions i.e. weak acid, under weak acid, water, and weak basic condition was reported by Wang et al. [88]. It was reported that the nanocomposites displayed good antibacterial activity against both gram negative and positive bacteria in all media. This was ascribed to high affinity and the strong interaction between hot matrix and modified nanoclay resulting in adsorption and immobilization capacity of modified nanoclay and the antimicrobial activity of host matrix i.e. quaternized chitosan.

5 Hybrid Biopolymer Nanocomposites

A combination of two or more nanoparticles is often introduced into polymeric materials with the aim of overcoming the limitations of one filler by other filler [32]. In some cases, the introduction of the second nanofiller may also improve the antibacterial efficiency of the resulting nanocomposite material. Li et al. [32] developed green nanocomposites material based on PLA/silver/titanium nanoparticles hybrid composite. It reported that the hybrid composites showed a significant improvement in the thermal stability and mechanical properties, while the antibacterial efficiency increased with increase in the concentration of silver nanoparticles as second filler. Vasile et al. [82] recently reported on the preparation of PLA reinforced with Cu-doped ZnO powder functionalized with AgNPs nanocomposites with melt compounding method. It was stated that the optimum composition was PLA/ZnO: Cu/Ag 0.5 to give suitable mechanical and thermal properties, good barrier properties to ultraviolet light, water vapour, oxygen and carbon dioxide, antibacterial activity and low migration amount of nanoparticles into food simulants. These results suggested that the prepared nanocomposites have potential to be used in food packaging.

In situ preparation of silver nanoparticles was carried out by mixing silver ionic solution with cellulose nanocrystals and then introduced into a blend of PLA/PBAT [34]. The antibacterial activity of the PLA/PBAT/NCC-Ag nanocomposite films was studied by using a testing protocol similar to Kirby-Bauer disc diffusion test and it was reported that thickness of the inhibition rings in the growth of *Gram-negative Escherichia coli* and *Gram-positive Staphylococcus aureus* is around 2.6 and 1.8 mm, respectively. On the other hand, the killing efficiency of the composites against *Escherichia coli* was determined to be 99.7% obtained from plate count method. [81] investigated the antibacterial property of a poly (lactic acid) nanosilver-doped multiwall carbon nanotube nanocomposite. It was reported that the antibacterial activity (Fig. 9) declined with increasing in MWCNT-Ag content in the nanocomposites. Furthermore, the authors claimed that the MWCNT-Ag synergy was able to transfer the antimicrobial properties of the PLA.

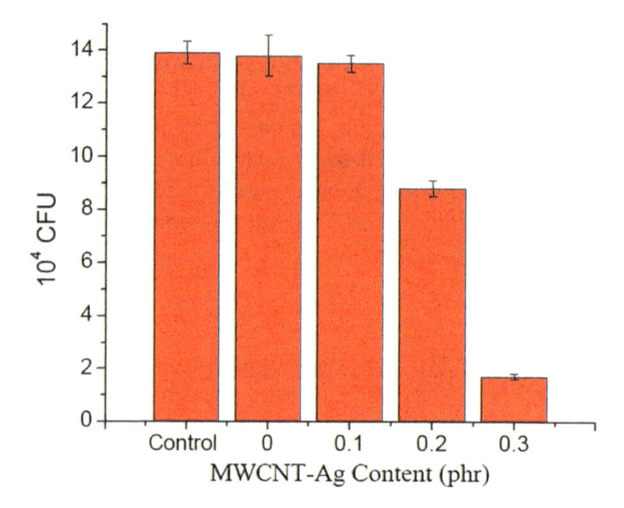

Fig. 9 Antibacterial activity of the PLA and PLA/MWCNT-Ag nanocomposites particles in the PLA matrix [81]

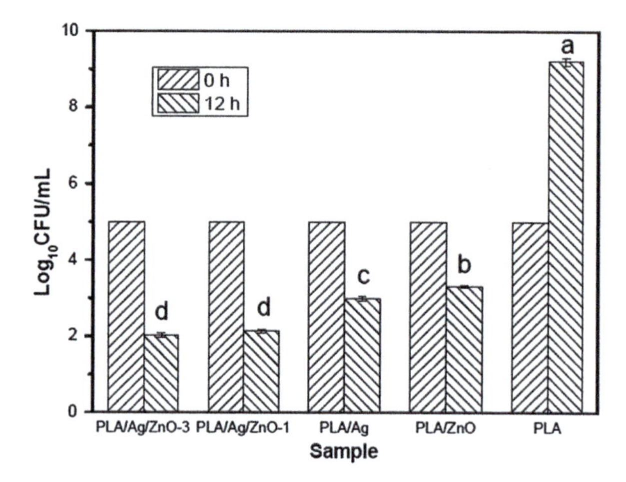

Fig. 10 Antimicrobial activity of pure PLA and PLA nano-composite films. Values followed by different superscript letters (a–e) in the same column were significantly different ($p < 0.05$), where a is the highest value[14]

Solvent volatizing method was used for the preparation of antimicrobial active based on PLA with nanosilver (Ag) and nano-zinc oxide (ZnO) [14]. The antimicrobial properties of both nanoparticles and their synergy were investigated for inhibition of *E. coli* growth. It was noted by the authors that the control film, neat PLA film, had no antimicrobial activity (Fig. 10) to inhibit the growth of E *coli*. The addition of two nanoparticles separately reduced the $\log_{10}$ CFU/mL values (Fig. 10) of the PLA nano-composite. This was attributed to the nanoparticle being able to release the surface of the films through the micro-voids, formed in the PLA nanocomposites film by nanoparticles and restrained the growth of *E. coli*. The synergy between ZnO and Ag decreased the $\log_{10}$ CFU/Ml values further in comparison to single nano.

6 Conclusion and Future Recommendations

It can be concluded that the use of nanoparticles as antibacterial agents not only improve the antibacterial efficacy but contribute to the overall properties of the resulting nanocomposites. Amongst all synthesized routes for antibacterial nanoparticles, green methods using abundantly available natural products is of significant importance from biomedical and packaging point of view. These techniques avoid the use of toxic chemicals which can reside within nanoparticles that can be hazardous to human and animals. The possibility of large-scale production using green methods makes this technique one of the most interesting for future research. Even though nanoparticles possess high antibacterial properties with low toxicity towards mammalian cells, general concerns about the potential hazard to the environment and human beings need to be addressed. At the current moment, very little information is available to assess the migration of nanoparticles and possible hazard to human beings and the environment despite the claims that nanoparticles synthesized from green routes have a lower impact on the environment. Among all metallic nanoparticles, silver nanoparticles are the most studied which calls for more research on other cheaper metals having similar properties as silver nanoparticles. Hybridization or combination of two or more nanoparticles incorporated into the common polymer and/or polymer blends serve as one of the promising subjects for the future especially in biomedical in order to eradicate multidrug resistant bacteria.

References

1. Abdel Rahim K, Mahmoud SY, Ali AM et al (2017) Extracellular biosynthesis of silver nanoparticles using *Rhizopus stolonifer*. Saudi J Biol Sci. https://doi.org/10.1016/j.sjbs.2016.02.025
2. Agarwal H, Venkat Kumar S, Rajeshkumar S (2017) A review on green synthesis of zinc oxide nanoparticles—An eco-friendly approach. Resour Technol. https://doi.org/10.1016/j.reffit.2017.03.002
3. An X, Ma H, Liu B, Wang J (2013) Graphene oxide reinforced polylactic acid/polyurethane antibacterial composites. J Nanomater. https://doi.org/10.1155/2013/373414
4. Ana MD, Angel LD (2014) ZnO-reinforced poly (3-hydroxybutyrate- co -3-hydroxyvalerate) bionanocomposites with antimicrobial function for food packaging. J Mol Sci, Int. https://doi.org/10.3390/ijms150610950
5. Agustin YE, Padmawijara (2017) Effect of glycerol and zinc oxide addition on antibacterial activity of biodegradable bioplastics from chitosan-kepok banana peel starch. https://doi.org/10.1088/1757-899x/223/1/012046
6. Augustine R, Malik HN, Singhal DK, et al (2014) Electrospun polycaprolactone/ZnO nanocomposite membranes as biomaterials with antibacterial and cell adhesion properties. J Polym Res. https://doi.org/10.1007/s10965-013-0347-6
7. Baheri B, Shahverdi M, Rezakazemi M et al (2015) Performance of PVA/NaA Mixed matrix membrane for removal of water from ethylene glycol solutions by pervaporation. Chem Eng Commun 202:316–321. https://doi.org/10.1080/00986445.2013.841149

8. Bayer IS (2017) Thermomechanical properties of polylactic review for biomedical applications. https://doi.org/10.3390/ma10070748
9. Botlhoko OJ, Ramontja J, Ray SS (2017) Thermally shocked graphene oxide-containing biocomposite for thermal management applications. RSC Adv 7:33751–33756. https://doi.org/10.1039/c7ra05421a
10. Castro-Mayorga J, Fabra M, Cabedo L, Lagaron J (2016) On the use of the electrospinning coating technique to produce antimicrobial polyhydroxyalkanoate materials containing in situ-stabilized silver nanoparticles. Nanomaterials. https://doi.org/10.3390/nano7010004
11. Castro Mayorga JL, Fabra Rovira MJ, Cabedo Mas L et al (2018) Antimicrobial nanocomposites and electrospun coatings based on poly(3-hydroxybutyrate-co-3-hydroxyvalerate) and copper oxide nanoparticles for active packaging and coating applications. J Appl Polym Sci. https://doi.org/10.1002/app.45673
12. Chandra S, Kumar A, Tomar PK (2014) Synthesis and characterization of copper nanoparticles by reducing agent. J Saudi Chem Soc 18:149–153. https://doi.org/10.1016/j.jscs.2011.06.009
13. Chen H, Wang B, Gao D et al (2013) Broad-spectrum antibacterial activity of carbon nanotubes to human gut bacteria. 2735–2746. https://doi.org/10.1002/smll.201202792
14. Chu Z, Zhao T, Li L et al (2017) Characterization of antimicrobial poly (lactic acid)/ nano-composite films with silver and zinc oxide nanoparticles. Materials (Basel). https://doi.org/10.3390/ma10060659
15. Chung IM, Rahuman AA, Marimuthu S et al (2017) Green synthesis of copper nanoparticles using *Eclipta prostrata* leaves extract and their antioxidant and cytotoxic activities. 18–24. https://doi.org/10.3892/etm.2017.4466
16. De Azeredo HMC (2009) Nanocomposites for food packaging applications. Food Res. Int. https://doi.org/10.1016/j.foodres.2009.03.019
17. De Azeredo HMC (2013) Antimicrobial nanostructures in food packaging. Trends Food Sci Technol https://doi.org/10.1016/j.tifs.2012.11.006
18. Din MI, Arshad F, Hussain Z, Mukhtar M (2017) Green adeptness in the synthesis and stabilization of copper nanoparticles : catalytic, antibacterial, cytotoxicity, and antioxidant activities. https://doi.org/10.1186/s11671-017-2399-8
19. El-ghany NAA (2017) Antimicrobial activity of new carboxymethyl chitosan–carbon nanotube biocomposites and their swell ability in different pH media. J Carbohydr Chem 0:1–14. https://doi.org/10.1080/07328303.2017.1353610
20. Espinoza-go H, Alonso-nu G, Sua J (2017) A green synthesis of copper nanoparticles using native cyclodextrins as stabilizing agents. J Saudi Chem Soc 341–348. https://doi.org/10.1016/j.jscs.2016.10.005
21. Geetha MS, Nagabhushana H, Shivananjaiah HN (2016) Green mediated synthesis and characterization of ZnO nanoparticles using Euphorbia Jatropa latex as reducing agent. J Sci Adv Mater Devices. https://doi.org/10.1016/j.jsamd.2016.06.015
22. Gopinath V, Priyadarshini S, Loke MF et al (2017) Biogenic synthesis, characterization of antibacterial silver nanoparticles and its cell cytotoxicity. Arab J Chem. https://doi.org/10.1016/j.arabjc.2015.11.011
23. Gopiraman M, Jatoi AW, Hiromichi S et al (2016) Silver coated anionic cellulose nanofiber composites for an efficient antimicrobial activity. Carbohydr Polym. https://doi.org/10.1016/j.carbpol.2016.04.084
24. Gunalan S, Sivaraj R, Rajendran V (2012) Green synthesized ZnO nanoparticles against bacterial and fungal pathogens. Prog Nat Sci Mater Int. https://doi.org/10.1016/j.pnsc.2012.11.015
25. Hasan SS, Singh S, Parikh RY, Dharne MS (2008) Bacterial synthesis of copper/copper oxide nanoparticles bacterial synthesis of copper/copper oxide nanoparticles. https://doi.org/10.1166/jnn.2008.095
26. Huang KS, Yang CH, Huang SL et al (2016) Recent advances in antimicrobial polymers: a mini-review. Int J Mol Sci 17(9):1578. https://doi.org/10.3390/ijms17091578

27. José De Andrade C, Maria De Andrade L, Mendes MA, Oller Do Nascimento CA (2017) An overview on the production of microbial copper nanoparticles by bacteria, fungi and algae. Glob J Res Eng
28. Judith P, Espitia P (2012) Zinc oxide nanoparticles : synthesis, antimicrobial activity and food packaging applications. Food Bioprocess Technol 1447–1464. https://doi.org/10.1007/s11947-012-0797-6
29. Kang S, Pinault M, Pfefferle LD et al (2007) Single-walled carbon nanotubes exhibit strong antimicrobial activity. Langmuir 23(17):8670–8673. https://doi.org/10.1021/la701067r
30. Khan A, Rashid A, Younas R, Chong R (2015) A chemical reduction approach to the synthesis of copper nanoparticles. Int Nano Lett. https://doi.org/10.1007/s40089-015-0163-6
31. Li SM, Jia N, Ma MG, et al (2011) Cellulose-silver nanocomposites: Microwave-assisted synthesis, characterization, their thermal stability, and antimicrobial property. Carbohydr Polym. https://doi.org/10.1016/j.carbpol.2011.04.060
32. Li W, Zhang C, Chi H et al (2017) Development of antimicrobial packaging film made from poly(lactic acid) incorporating titanium dioxide and silver nanoparticles. Molecules. https://doi.org/10.3390/molecules22071170
33. Li X, Xiao Y, Bergeret A, et al (2014) Preparation of polylactide/graphene composites from liquid-phase exfoliated graphite sheets. https://doi.org/10.1002/pc.22673
34. Ma P, Jiang L, Yu M et al (2016) Green antibacterial nanocomposites from Poly (lactide)/ Poly (butylene adipate -co-terephthalate)/nanocrystal cellulose-silver nanohybrids
35. Martynková GS, Valášková M (2014) Antimicrobial nanocomposites based on natural modified materials: a review of carbons and clays. J Nanosci Nanotechnol. https://doi.org/10.1166/jnn.2014.8903
36. Mary G, Bajpai SK, Chand N (2009) Copper (II) Ions and copper nanoparticles-loaded chemically modified cotton cellulose fibers with fair antibacterial properties. https://doi.org/10.1002/app
37. Matinise N, Fuku XG, Kaviyarasu K et al (2017) Applied Surface Science ZnO nanoparticles via Moringa oleifera green synthesis: Physical properties & mechanism of formation. Appl Surf Sci 406:339–347. https://doi.org/10.1016/j.apsusc.2017.01.219
38. Mendoza G, Regiel-Futyra A, Andreu V et al (2017) Bactericidal effect of gold-chitosan nanocomposites in coculture models of pathogenic bacteria and human macrophages. ACS Appl Mater Interfaces 9:17693–17701. https://doi.org/10.1021/acsami.6b15123
39. Mochane MJ, Luyt AS (2015) Synergistic effect of expanded graphite, diammonium phosphate and Cloisite 15A on flame retardant properties of EVA and EVA/wax phase-change blends. J Mater Sci 50:3485–3494. https://doi.org/10.1007/s10853-015-8909-0
40. Mochane MJ, Luyt AS (2015) The effect of expanded graphite on the thermal stability, latent heat, and flammability properties of EVA/wax phase change blends. Polym Eng Sci 55:1255–1262. https://doi.org/10.1002/pen.24063
41. Mokhena TC, Jacobs NV, Luyt AS (2018) Nanofibrous alginate membrane coated with cellulose nanowhiskers for water purification. Cellulose 25. https://doi.org/10.1007/s10570-017-1541-1
42. Mokhena TC, Jacobs V, Luyt AS (2015) A review on electrospun bio-based polymers for water treatment. Express Polym Lett 9. https://doi.org/10.3144/expresspolymlett.2015.79
43. Mokhena TC, Luyt AS (2017) Development of multifunctional nano/ultrafiltration membrane based on a chitosan thin film on alginate electrospun nanofibres. J Clean Prod 156:. https://doi.org/10.1016/j.jclepro.2017.04.073
44. Mokhena TC, Luyt AS (2017) Electrospun alginate nanofibres impregnated with silver nanoparticles: preparation, morphology and antibacterial properties. Carbohydr Polym 165. https://doi.org/10.1016/j.carbpol.2017.02.068
45. Mondal D, Bhowmick B, Mollick MMR et al (2014) Antimicrobial activity and biodegradation behavior of poly(butylene adipate-co-terephthalate)/clay nanocomposites. J Appl Polym Sci. https://doi.org/10.1002/app.40079
46. Palza H (2015) Antimicrobial polymers with metal nanoparticles. Int J Mol, Sci

47. Palza H, Quijada R, Delgado K (2015) Antimicrobial polymer composites with copper micro- and nanoparticles: effect of particle size and polymer matrix. J Bioact Compat Polym. https://doi.org/10.1177/0883911515578870
48. Phogat N, Khan SA, Shankar S, et al (2016) Fate of inorganic nanoparticles in agriculture. Adv. Mater. Lett
49. Pinto RJB, Marques PAAP, Neto CP, et al (2009) Antibacterial activity of nanocomposites of silver and bacterial or vegetable cellulosic fibers. Acta Biomater. https://doi.org/10.1016/j.actbio.2009.02.003
50. Prabhu YT, Rao KV, Sai VS, Pavani T (2017) ORIGINAL ARTICLE A facile biosynthesis of copper nanoparticles: a micro-structural and antibacterial activity investigation. J Saudi Chem Soc 21:180–185. https://doi.org/10.1016/j.jscs.2015.04.002
51. Rapacz-Kmita A, Pierchała MK, Tomas-Trybuś A et al (2017) The wettability, mechanical and antimicrobial properties of polylactide/montmorillonite nanocomposite films. Acta Bioeng Biomech. https://doi.org/10.5277//abb-00820-2017-02
52. Review CNA, Gonçalves C (2017) Poly (lactic acid) composites containing. 1–37. https://doi.org/10.3390/polym9070269
53. Rezakazemi M, Dashti A, Riasat Harami H et al (2018) Fouling-resistant membranes for water reuse. Environ Chem Lett 1–49. https://doi.org/10.1007/s10311-018-0717-8
54. Rezakazemi M, Ebadi Amooghin A, Montazer-Rahmati MM et al (2014) State-of-the-art membrane based CO < inf > 2</inf > separation using mixed matrix membranes (MMMs): an overview on current status and future directions. Prog Polym Sci 39:817–861. https://doi.org/10.1016/j.progpolymsci.2014.01.003
55. Rezakazemi M, Khajeh A, Mesbah M (2017) Membrane filtration of wastewater from gas and oil production. Environ Chem Lett 1–22. https://doi.org/10.1007/s10311-017-0693-4
56. Rezakazemi M, Mohammadi T (2013) Gas sorption in H < inf > 2</inf > -selective mixed matrix membranes: experimental and neural network modeling. Int J Hydrogen Energy 38:14035–14041. https://doi.org/10.1016/j.ijhydene.2013.08.062
57. Rezakazemi M, Razavi S, Mohammadi T, Nazari AG (2011) Simulation and determination of optimum conditions of pervaporative dehydration of isopropanol process using synthesized PVA-APTEOS/TEOS nanocomposite membranes by means of expert systems. J Memb Sci 379:224–232. https://doi.org/10.1016/j.memsci.2011.05.070
58. Rezakazemi M, Sadrzadeh M, Matsuura T (2018) Thermally stable polymers for advanced high-performance gas separation membranes. Prog Energy Combust Sci 66:1–41. https://doi.org/10.1016/j.pecs.2017.11.002
59. Rezakazemi M, Sadrzadeh M, Mohammadi T (2017b) Separation via pervaporation techniques through polymeric membranes
60. Rezakazemi M, Sadrzadeh M, Mohammadi T, Matsuura T (2017) Methods for the preparation of organic-inorganic nanocomposite polymer electrolyte membranes for fuel cells
61. Rezakazemi M, Shahidi K, Mohammadi T (2012) Sorption properties of hydrogen-selective PDMS/zeolite 4A mixed matrix membrane. Int J Hydrogen Energy 37:17275–17284. https://doi.org/10.1016/j.ijhydene.2012.08.109
62. Rezakazemi M, Shahidi K, Mohammadi T (2012) Hydrogen separation and purification using crosslinkable PDMS/zeolite A nanoparticles mixed matrix membranes. Int J Hydrogen Energy 37:14576–14589. https://doi.org/10.1016/j.ijhydene.2012.06.104
63. Rezakazemi M, Shahidi K, Mohammadi T (2015) Synthetic PDMS composite membranes for pervaporation dehydration of ethanol. Desalin Water Treat 54:1542–1549. https://doi.org/10.1080/19443994.2014.887036
64. Rezakazemi M, Shahverdi M, Shirazian S et al (2011) CFD simulation of water removal from water/ethylene glycol mixtures by pervaporation. Chem Eng J 168:60–67. https://doi.org/10.1016/j.cej.2010.12.034
65. Rezakazemi M, Vatani A, Mohammadi T (2015) Synergistic interactions between POSS and fumed silica and their effect on the properties of crosslinked PDMS nanocomposite membranes. RSC Adv 5:82460–82470. https://doi.org/10.1039/c5ra13609a

66. Rezakazemi M, Vatani A, Mohammadi T (2016) Synthesis and gas transport properties of crosslinked poly(dimethylsiloxane) nanocomposite membranes using octatrimethylsiloxy POSS nanoparticles. J Nat Gas Sci Eng 30:10–18. https://doi.org/10.1016/j.jngse.2016.01.033
67. Rhim J-W, Hong S-K, Park H-M, N.g PKW (2006) Preparation and Characterization of Chitosan-Based Nanocomposite Films with Antimicrobial Activity. J. Agric. Food Chem. 54, 16, 5814-5822. https://doi.org/10.1021/jf060658h
68. Rhim JW, Park HM, Ha CS (2013) Bio-nanocomposites for food packaging applications. Prog Polym Sci. https://doi.org/10.1016/j.progpolymsci.2013.05.008
69. Rostamizadeh M, Rezakazemi M, Shahidi K, Mohammadi T (2013) Gas permeation through H2-selective mixed matrix membranes: experimental and neural network modeling. Int J Hydrogen Energy 38:1128–1135. https://doi.org/10.1016/j.ijhydene.2012.10.069
70. Sadeghi A, Nazem H, Rezakazemi M, Shirazian S (2018) Predictive construction of phase diagram of ternary solutions containing polymer/solvent/nonsolvent using modified Flory-Huggins model. J Mol Liq 263:282–287. https://doi.org/10.1016/j.molliq.2018.05.015
71. Sadrzadeh M, Rezakazemi M, Mohammadi T (2017) Fundamentals and measurement techniques for gas transport in polymers
72. Laudenslager MJ, Schiffman JD, Schauer CL (2008) Carboxymethyl chitosan as a matrix material for platinum, gold, and silver nanoparticles. 2682–2685. https://doi.org/10.1021/bm800835e
73. Sengupta R, Bhattacharya M, Bandyopadhyay S, Bhowmick AK (2011) A review on the mechanical and electrical properties of graphite and modified graphite reinforced polymer composites. Prog Polym Sci 36:638–670. https://doi.org/10.1016/j.progpolymsci.2010.11.003
74. Satyvaldiev AS, Zhasnakunov ZK, Omurzak E, Doolotkeldieva TD, Bobusheva ST, Orozmatova GT, Kelgenbaeva Z (2018) Copper nanoparticles : synthesis and biological activity. https://doi.org/10.1088/1757-899x/302/1/012075
75. Shahverdi M, Baheri B, Rezakazemi M et al (2013) Pervaporation study of ethylene glycol dehydration through synthesized (PVA-4A)/polypropylene mixed matrix composite membranes. Polym Eng Sci 53:1487–1493. https://doi.org/10.1002/pen.23406
76. Shankar S, Rhim J (2016) LWT—food science and technology tocopherol-mediated synthesis of silver nanoparticles and preparation of antimicrobial pbat/silver nanoparticles composite films. LWT - Food Sci Technol 72:149–156. https://doi.org/10.1016/j.lwt.2016.04.054
77. Shankar S, Wang LF, Rhim JW (2016) Preparations and characterization of alginate/silver composite films: effect of types of silver particles. Carbohydr Polym. https://doi.org/10.1016/j.carbpol.2016.03.026
78. Shih CM, Shieh YT, Twu YK (2009) Preparation of gold nanopowders and nanoparticles using chitosan suspensions. Carbohydr Polym 78:309–315. https://doi.org/10.1016/j.carbpol.2009.04.008
79. Shittu KO, Bankole MT, Abdulkareem AS et al (2017) Application of gold nanoparticles for improved drug efficiency
80. Sothornvit R, Rhim JW, Hong SI (2009) Effect of nano-clay type on the physical and antimicrobial properties of whey protein isolate/clay composite films. J Food Eng. https://doi.org/10.1016/j.jfoodeng.2008.09.026
81. Tsou CH, Yao WH, Lu YC, et al (2017) Antibacterial property and cytotoxicity of a poly (lactic acid)/nanosilver-doped multiwall carbon nanotube nanocomposite. Polymers (Basel) 9. https://doi.org/10.3390/polym9030100
82. Vasile C, Râpă M, Ştefan M et al (2017) New PLA/ ZnO: Cu/ Ag bionanocomposites for food packaging. 11:531–544
83. Venkatesan R, Rajeswari N (2017) TiO_2 nanoparticles/poly(butylene adipate-co-terephthalate) bionanocomposite films for packaging applications. https://doi.org/10.1002/pat.4042
84. Venkatesan R, Rajeswari N, Tamilselvi A (2018) Antimicrobial, mechanical, barrier, and thermal properties of bio-based poly (butylene adipate-co-terephthalate) (PBAT)/Ag_2O nanocomposite films for packaging application. https://doi.org/10.1002/pat.4089
85. Venkatesan R, Rajeswari N (2016) ZnO/PBAT nanocomposite films : investigation on the mechanical and biological activity for food packaging. https://doi.org/10.1002/pat.3847

86. Vimbela GV, Ngo SM, Fraze C, Yang L, David A Stout DA (2017) Antibacterial properties and toxicity from metallic nanomaterials. Int J Nanomedicine 12:3941–3965. https://doi.org/10.2147/ijn.s134526
87. Vivekanandhan S, Christensen L, Misra M, Kumar Mohanty A (2012) Green process for impregnation of silver nanoparticles into microcrystalline cellulose and their antimicrobial bionanocomposite films. J Biomater Nanobiotechnol. https://doi.org/10.4236/jbnb.2012.33035
88. Wang X, Du Y, Luo J, et al (2009) A novel biopolymer/rectorite nanocomposite with antimicrobial activity. Carbohydr Polym. https://doi.org/10.1016/j.carbpol.2009.01.015
89. Wu CS (2009) Antibacterial and static dissipating composites of poly(butylene adipate-co-terephthalate) and multi-walled carbon nanotubes. Carbon N Y 47:3091–3098. https://doi.org/10.1016/j.carbon.2009.07.023
90. Wu D, Cheng Y, Feng S, et al (2013) Crystallization behavior of polylactide/graphene composites crystallization behavior of polylactide/graphene composites. https://doi.org/10.1021/ie4004199
91. Yan X, Li F, Di Hu K et al (2017) Nacre-mimic reinforced Ag@reduced graphene oxide-sodium alginate composite film for wound healing. Sci Rep. https://doi.org/10.1038/s41598-017-14191-5

Extraction of Nano Cellulose Fibres and Their Eco-friendly Polymer Composite

Bashiru Kayode Sodipo and Folahan Abdul Wahab Taiwo Owolabi

1 Introduction

Global green revolution aimed at mitigating the negative ecological effect of polymeric materials has led to research and development into various sustainable and renewable eco-friendly materials to replace petroleum-based materials [1–3].

Nano fibre cellulose (NFC) is one of the most attractive renewable materials for advanced applications. This is due to its mechanical and physical properties. NFC consists of flexible cellulosic nano-material with lateral dimensions of about 10–100 nm diameter and several micrometres long [4]. They are described as long flexible nanofilaments composed of a crystalline and amorphous portion [5]. Nanocellulose can be divided into three types of materials: (I) nano-fiber cellulose (NFC), (II) nanocrystals cellulose (NCC) or cellulose nanowhiskers (CNWs), and (III) bacterial cellulose (BC). However, a report in this book chapter covers only that of NFC and typical Tem micrograph of NFC is shown in Fig. 1.

Unlike Petroleum based and synthetic polymer nanocomposite, NFC polymer nanocomposite has many advantages due to low weight, reduced tool wearing, recyclable and biodegradable properties [6]. Moreover, in order to develop a fully eco-friendly polymer nanocomposite, the use of a reinforcement derived from renewable biomass is needed. Petroleum-based polymers have gained attention due to a variety

The original version of this chapter was inadvertently published with the incorrect author sequence and corresponding author. The correction to this chapter is available at https://doi.org/10.1007/978-3-030-05399-4_48

B. K. Sodipo (✉)
Department of Physics, Kaduna State University, Kaduna, Nigeria
e-mail: bashirsodipo@gmail.com

F. A. W. T. Owolabi
Pulp and Paper division, Federal Institute of Industrial Research Oshodi, Oshodi, Nigeria
e-mail: fathok2375@gmail.com

Inamuddin et al. (eds.), *Sustainable Polymer Composites and Nanocomposites*,
https://doi.org/10.1007/978-3-030-05399-4_8

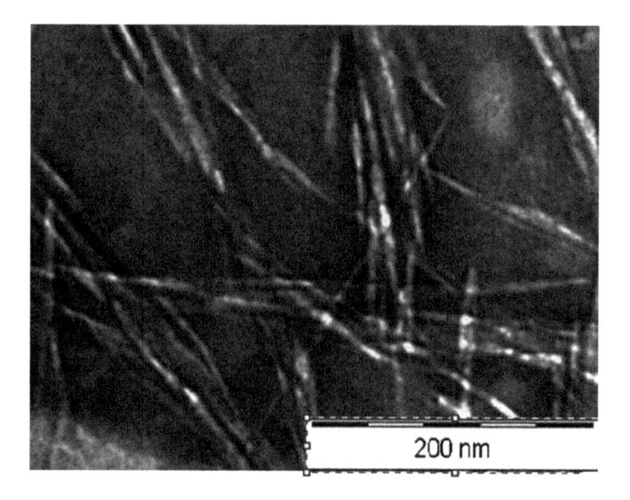

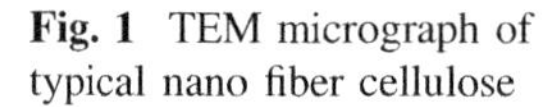
Fig. 1 TEM micrograph of typical nano fiber cellulose

of products, e.g. plastic materials having several superior properties (e.g. water repellence and formability). Despite the wide acceptability of the petroleum-based polymers, there are major disadvantages associated with petroleum-based polymers. These include; crude oil as a non-renewable resource, the products are not bio-degradable, increasing oil prices, dwindling oil resources and a high focus on sustainability.

Similar to the petroleum-based polymer, the global concern towards the potential hazard of the use of the synthetic fibres such as glass (or carbon, aramid, etc.) fibres result in release of CO_2 into the atmosphere (global warming), along with some other health hazardous gases like NO_x and SO_x and dust [7, 8]. In addition, dust and fragments are generated when recycling conventional plastic composites by grinding them down and constitute disposal problem either to landfill or by incineration [9].

All these factors have motivated a renewed interest in bio-based polymers. For the past two decades, research and development on the utilization of the most abundant biopolymer on earth, such as cellulose have resulted in a variety of products, e.g. cellophane, rayon, nitrocellulose (used in gunpowder), adhesives and lacquer [10]. Substantial breakthrough is recorded through the production of the biodegradable composite. The exceptional breakthrough recorded from the properties of the biopolymers has attracted more research findings due to the growing environmental awareness among the consumers. In addition, the natural fibres have high specific strength and modulus, high sound attenuation of lignocellulosic-based composites, non-food agricultural based economy, relatively reactive surface comparatively, easy processability and economical [4].

Previous use of a cellulose-based polymer as ropes, paper, timber for housing structures and its recent use in the field of biopolymers, has proven the ubiquitous and abundant nature of cellulose fibers. The global annual biomass production of cellulose has been reported as about 1.5×10^{12} tons [11]. Natural fibers otherwise referred to as cellulosic fibers are everywhere throughout the world in plants such as grasses, seeds, stalks, and woody vegetation. Apart from the massive availability of cellulose source of NFC, its application in polymer composite has been gaining universal attention for its environmentally friendly nature and its mechanical reinforcement property [12]. Apart from wood which has been widely exploited commercially as cellulose-based natural bio-resource, other cellulosic materials gaining wide attention include, plants, bacteria, non-wood such as hemp, flax, jute, ramie, cotton and agro-industrial wastes because they contain a natural polyphenolic polymer, lignin, in their structure. The fiber of the cellulose bio-polymers composed of bundles of microfibrils stabilized laterally by inter and intra-molecular hydrogen bonding. In contrast, the use of natural fibers can minimize harmful pollutants, and their eventual breakdown is environmentally benign.

Despite the fact that natural fibers, come from renewable animal or plant sources, they usually lack the high-performance characteristics of many synthetic fibers [13] (Khalil et al. 2016). However, natural fibers fillers or reinforcement in polymeric matrix composites provides positive environmental benefits with respect to ultimate disposability and raw material use [14]. Bio-composites which could be micro or nano-based composite are a family of materials consisting of a polymeric matrix reinforced derived from renewable sources or biodegradables.

2 Production of Cellulose Nanofibrils

NFC can be manufactured basically from Pulping processes which include; Mechanical pulping, Homogenization, Chemical pulping, Steam explosion, High-intensity ultrasonication etc. Furthermore, they can be prepared from a number of different cellulosic sources through an energy-dependent process entailing three major pathways as shown in Fig. 2.

NFC is obtained after a strong mechanical shearing applied on cellulose slurry which is pumped through a homogenizer or grinder device. NFC displayed higher specific area which leads to a higher amount of hydrogen interactions compared with other cellulose fiber-based suspension, and they give gel-like structure at solid content as low as between 2 and 5% [15]. So far, several devices based on a high-pressure homogenizing system such as homogenizer system, micro fluidizer, grinder and recently refiners' devices have been developed to increase the production yield and the quality of the NFC [16].

Several pre-treatments schemes needed in order to obtain fibers with decreased fibrillation energy have been proposed. This include, enzymatic pre-treatment [15], TEMPO-mediated oxidation pre-treatment [17], Carboxymethylation and acetylation [18] and alkaline peroxide pre-treatment [14]. The pre-treatment methods are aimed

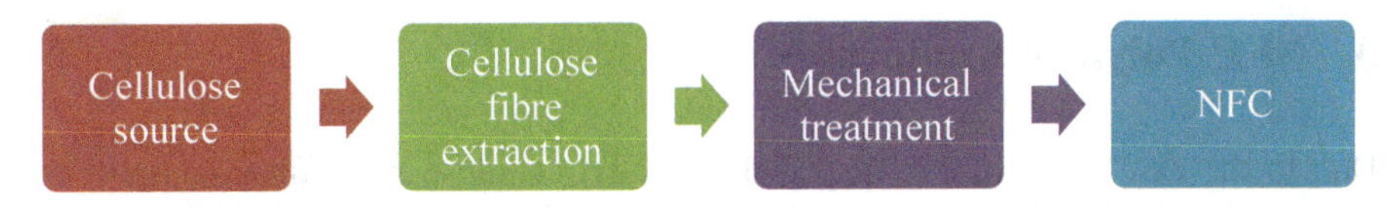

Fig. 2 Nanofibrillated cellulose production flow chat

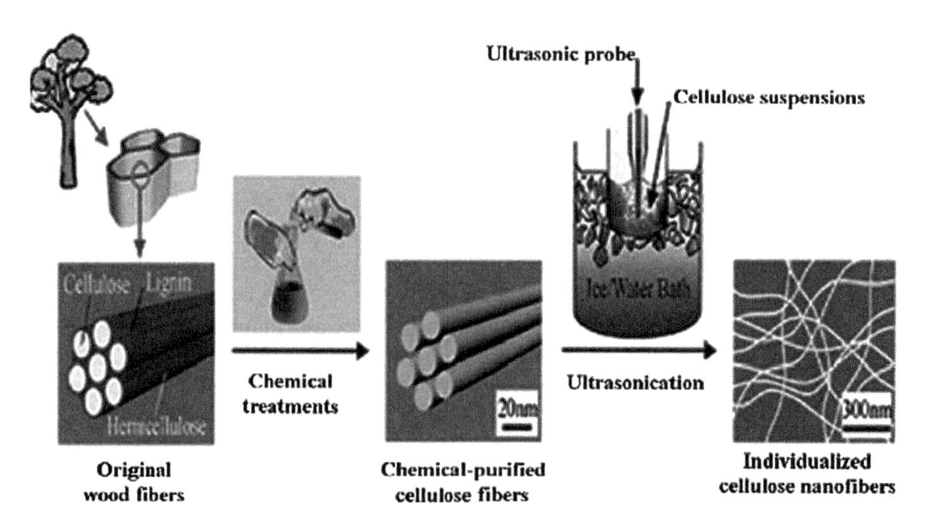

Fig. 3 Schematic process for cellulose nanofibres production [68] copyright permission

at: (a) limiting the hydrogen bonding, (b) increasing the repulsive charge, and (c) decrease the degree of polymerization DP or the amorphous link between individual MFCs. The first and the most common mechanical treatments for NFC production include Gaulin homogenizer and micro fluidizer, grinding process, cryocrushing, electrospinning, energy consumption and new processes. The common mechanical devices used in NFC production are shown in Fig. 3.

According to Syyerud and coworkers (2011) the less energy that is utilized, the less the fibrillation of cellulose fibres and the amount of nanofibrilsproduced. Since the first successful isolation of cellulose microfibrils was reported by Turbak et al. [19] with the aid of Gaulin laboratory homogenizer, several pathways have been successfully proposed for producing cellulose nanofibrils [15, 17].

All of these methods comprise, Mechanical treatments, e.g. cryocrushing, grinding, high pressure homogenizing, chemical treatments and enzyme-assisted hydrolysis [20]. Other processes include TEMPO-mediated oxidation on the surface of microfibrils with mild mechanical treatment, electrospinning methods and ultrasonic technique. In addition, cellulose fibersare subjected to homogenization, steam explosion, and high-intensity ultra-sonication to disperse the agglomeration of the microfibers [12]. These methods lead to different types of nanofibrillar materials, depending on the cellulose raw material and its pre-treatment and more importantly, depending on the disintegration process itself.

3 NFC Polymer Composite

Due to the ease of nanofibre dispersal in water, cellulose nano fibre composites have explored primarily solvent casting as means of processing nanocomposites. In another development, Oksman et al. [21] reported the processing of cellulose

Table 1 Summary of recent nano fibrilated cellulose-thermoplastic composites and their applications [69]

Polymer component	Manufacturing technique	Applications
Polyethylene glycol	PEG-g-CNF ribbons by stretching hydrogel	Ultra-high tensile strength and modulus for optoelectronic and medical devices
Amorphous dialcohol cellulose	Oxidation + reduction of CNF surface	Barrier films
Polyethylene	Extrusion	High-performance cellulosics environmentally friendly HDPE, Evaluation of cotton filler in LDPE
Thermoplastic starch	Solution casting	Decreased water sensitivity, thermally stable starch
Maize amylopectin	Solution casting	Continuous papermaking
Polyvinyl amine	Layer by layer	Self-healing polymer films
Polyacrylamide	Solution casting	Films with good mechanical, optical thermal and oxygen barrier properties
PVA	Solution casting	Flexible displays, optical devices, packaging and automobile windows, food packaging
Carboxymethyl cellulose	Solution casting	Edible coatings and packaging materials
Poly(butylene adipate-co-terephthalate)	Injection molding	Light-weight and high-performance materials for defense, infrastructure, and energy

nanocomposites using twin-screw extrusion to produce different thermoplastic polymers composite with cellulose nanofibres. Researchers have successfully exploited the use of NFC as biodegradable polymers with improved brittleness, low thermal stability and poor barrier properties [22]. NFC has been successfully used in the field of engineering, biomaterials and medical care. The potential of NFC in research and development of novel research in engineering reinforcement, medical devices and applications in healthcare and veterinary medicine have also been reported. Significant improvements in properties, disposal and recycling problems, combined with environmental and societal concerns have further revealed the potential of NFC [4].

The common nano fibrilated cellulose-thermoplastic composites and their applications are tabulated in Table 1.

4 Challenges of NFC Polymer Composite

In spite of the global trends in the use of NFC in polymer composite, their application as fillers and strength reinforcement in polymer composite has been saddled with a lot of challenges which borders on the intrinsic physical properties.

These physical properties including a high number of hydroxyl groups, leading to strong hydrogen interactions between two nanofibrils and to the gel-like structure once produced [23]. In addition, the high hydrophilicity of this material makes it vulnerable to form agglomerates in petrochemical polymers.

As the dispersal of cellulose NFC in organic solvents is essential, consequently, surface modification of NFC is of utmost importance in order to improve compatibility with a wider variety of matrices. The reactivity of NFC in polymer composite is achievable by its surface modification chemically, to reduce the number of hydroxyl interactions and also to increase the compatibility with several matrices. So, many methods have been proposed for cellulose surface modification [24], which consist of corona or plasma discharges [25], surface derivatization [26], graft copolymerization [27] and use of surfactant [28]. Some approaches aiming at surface modification in order to hydrophobizenano cellulosic include: Acetylation [29], Silylation [30], Grafting [31], grafting in situ catalyzed ring-opening polymerization [32] and the use of coupling agents [33]. Due to strong hydrogen bonding interactions between cellulose hydroxyl groups, it is challenging to obtain well separated NFC in organic solvents, especially for non-polar solvents.

5 Poly-lactic Acid (PLA) Based Nanocellulosic Composites

PLA is biodegradable, thermoplastic and aliphatic polyester derived from renewable resources such as starch. Also, it appears as one of the best sustainable alternatives to petrochemical-derived products [34]. PLA has been found to have good stiffness and strength. It can be processed with conventional plastic processing machinery and is being used in several applications, such as food packaging, water, and milk bottles, and degradable plastic bags as well as in automotive applications [35]. Products made from PLA are biodegradable, eco-friendly and potentially compostable [36, 37]. The performance of PLA can be greatly enhanced by the addition of nano-reinforcements.

Recent studies on PLA had shown that the biopolymer has good mechanical properties, thermal plasticity, and biocompatibility, and is readily fabricated, thus being a promising polymer for various end-use applications [38–40]. However, PLA, similar to polystyrene, is a comparatively brittle and stiff polymer with low deformation at the break and low impact strength. Dufresne et al. [41], reported that the overall mechanical performance of nanocomposites, depends on six factors: (a) Crystallinity of the matrix (b) aspect ratio of additive, (c) volume fraction of additive (d) adhesion and compatibility between the polymer matrix and additives, (e) the orientation of additives (f) stress transfer efficiency of additives. The potential reinforcement efficacy of nanofibre was carried out with PLA using mechanically fibrillated nanofibres [34]. The result shows that Young's modulus and tensile strength of the PLA was increased by 40 and 25%, respectively with the addition of micro-fibrillated cellulose (MFC), without a reduction of yield strain at a fibre content of 10 wt% [13].

Ali Abdulkhani et al. [42] investigated the effect of morphological, thermal, mechanical and barrier properties of PLA based biocomposites prepared with embedded CNF-Ac using a solvent casting procedure. The report has it that, the tensile strength (TS), elastic modulus (EM) and elongation percentage (E) were significantly increased for the prepared cellulose nanocomposites with 3 and 5 wt% CNF-Ac. The reinforcement of PLA with CNF-Ac caused a slight increase in glass transition and melting temperatures. The mechanical tests of PLA and its nanocomposite films results in an improvement in the mechanical properties of PLA composites by the addition of acetylated cellulose nanofibers [34]. This observation is born out of the fact that there was an increase incompatibility between the moieties.

Meanwhile, NFC has shown to be a promising reinforcement of PLA composites. Their contribution to the biodegradability with improved barrier properties has been of immense advantage. The application of PLA as a potential biopolymer to substitute the conventional petroleum-based plastics is gradually gaining the interest of researchers in the area of polymer biocomposite. PLA products are used in packaging films (for textiles and non-wovens), packaging with good barrier properties and low heat-seals. Other areas of interest are a paper coating, fibres, and a host of moulded articles [4].

6 Polyhydroxyalkanoate (PHA) Based Nanocellulosic Composites

PHA are polyesters produced naturally by numerous microorganisms. Different monomers of PHA can be combined to give materials various properties. The availability of PHAs has necessitated much research in the area of biosynthesis, microstructure, thermal and mechanical properties. Research focuses on the application of PHAs in recent years has been driven by its renewable resources and the similarity of PHA physical properties to those of conventional plastics [43, 44]. In the bid to reducing their hydrophilicity, cellulose nanocrystals from microfibrils cellulose were successfully topochemically trimethylsilylated. PHAs are used in packaging films like bags, containers and paper coatings [45]. Analogous applications in conventional commodity plastics include disposable items such as razors, utensils, nappies, feminine hygiene products, cosmetic containers, shampoo bottles and cups [46].

7 Starch-Based Nanocellulosic Composites

Starch-based nanocellulosic as biodegradable thermoplastic materials has offered great potential application in food packaging or biomedical industry. Among the advantages of starch, films are its application as excellent intermediates for transporting antimicrobials and antioxidants. Different starch sources for starch-based nanocellulose composite have been reported. Polysaccharide sources for starch nanofillers and nanocomposites include flax, ramie, cassava bagasse, wheat straws, regular maize, and chitin, chitosan, among others. Starch comprises of a linear polymer amylose, and a branched polymer amylopectin, with α-(1–6)-linked branch points [49]. Starch sourced from variety of crops such as corn, wheat, rice and potato can be blended with biodegradable polymers such as PHB [50], PLA [51], PCL [52] and chitosan [53]. Apart from its wide availability, starch is a source of biodegradable biopolymer which is readily available at low cost when compared with most synthetic plastics [54]. It has been reported that MFC and biodegradable cellulose have also been reported as promising candidates for starch reinforcement [55]. Some different methods for processing both starch matrix and nanocomposites include solution casting method [56] and extrusion technique [57].

8 NFC Polymer Composite in Thermoplastics Materials

Thermoplastic polymers were compounded with biomaterials to reduce production costs while maintaining original properties. The development of environmentally friendly plastics for production of composites and nanocomposites is ultimately promising. Aliphatic biodegradable polyesters, such as polylactic acid (PLA), polycaprolactone (PCL), poly(3-hydroxybutyrate) (PHB), and polyglycolic acid (PGA), have been widely compounded with different materials to produce green composites. PLA is one of the most promising alternatives to typical plastics and has gained much attention mainly due to its biodegradability. The advantages of cellulose nanocomposite materials in polymer composite compared with conventional composites is that at low reinforcement levels, there is superior thermal, mechanical and barrier properties as well as their improved recyclability, transparency and low weight [21, 22]. Researchers have explored the concept of bio-derived nanocomposites as a route to the development of bioplastics or bioresins with better properties [21, 58]. There is some common eco-friendly nanocomposite which includes: PLA, PHB, and Starch based nanocellulosic composites. This recent shift to "green" composites have necessitated the coupling of various kinds of natural fibers to biodegradable resins such as PLA and modified starch to reinforce plant-derived, polymeric matrix materials and improve their mechanical properties [59].

The fundamental compatibility challenges in the nanocomposite preparations are polymer matrix is hydrophobic while natural fibers are generally hydrophilic. Hence this barrier usually causes non-uniform dispersion of the fibers within the matrix and poor mechanical properties have to be broken to allow for proper coupling of the two or more composite fragments. To overcome this challenges of compatibility and grafting between fibers and thermoplastic matrices in composites production, surface treatment by the use of additives is adopted [9]. The common additives used include chemical coupling agents or compatibilizers Maleated polyethylene (MAPE) [60], carboxylated polyethylene (CAPE) [61], titanium derived mixture (TDM), Maleic anhydride polypropylene (MAPP), [62], calendaring, [63] thermal treatment [64], reaction with methanol melamine, isocyanates, triazine, silane [65] and mercerization of the matrix [66]. Despite all these possibilities, a better understanding of the molecular structure and interfacial interaction between the matrix and the fibres and the relationship between the structure and property is very important in this area of research [48].

9 NFC Polymer Composite in Automotive

Khalfallah et al. [67] reported that automotive parts industry is highly selective in terms of the matrix characteristics. This means that matrices with good visco-elastic properties, high thermal stability is required for meeting automotive specifications. Nanocomposites have been used in several applications in automotive industries such as various vehicle types of door handles, door panels, instrument panels, parcel shelves, headrests, roofs, upholstery and engine covers and intake manifolds and timing belt covers [68]. The use of green nanocomposite in impellers and blades for vacuum cleaners, power tool housings, mower hoods and covers for portable electronic equipment such as mobile phones are receiving interesting attention [60, 69]. In addition, the recycling by combustion of lignocellulosics filled composites is easier in comparison with inorganic fillers systems. Therefore, the possibility of using lignocellulosic fillers in the plastic industry has received considerable interest. Automotive applications display strong promise for natural fibre reinforcements. 2–5 Potential applications of agrofibre based composites in railways, aircraft, irrigation systems, furniture industries, and sports and leisure items are currently being investigated [35, 70].

10 Conclusions

Cellulose nanofibres have been seen as stimulating potential reinforcements in nanocomposites. Their potential application in medicine, automobile, and construction has been attributed to their size and the ability to undergo surface chemical modification. Several methods channelled towards cellulose nanofibres extraction from cellulose sources have been categorized as chemical and mechanical treatment. In order to reduce the mechanical energy, enzymatic or chemical pre-treatment methods are inevitable. The strength properties of cellulose nanofibres composite compete favourably with other engineering materials, hence, could be useful in high-end technological applications. This study revealed that dispersion of NFC is a very critical step to promote remarkable percolation of NFC by interacting with each other, and with the surrounding matrix, in a way that greatly enhances the mechanical properties of the resultant material. Due to compatibility problems of nanocellulosic materials and hydrophobic matrices, it can be anticipated that nanocomposites based on hydrophilic matrix polymers will be easier to commercialize. In order to achieve improved mechanical properties in polymer nanocomposites, good filler-matrix interaction is essential. Moreover, in order to develop a fully eco-friendly polymer nanocomposite, the use of a reinforcement derived from renewable biomass is needed. However, by combining the mechanical treatment with certain pre-treatments, various works have shown that it can decrease energy consumption significantly.

Acknowledgements The authors are thankful to the Federal Institute of Industrial Research Oshodi Nigeria and the Kaduna State University, Nigeria for their role in the successful completion of this book chapter.

References

1. Brown TD, Dalton PD, Hutmacher DW (2016) Melt electrospinning today: an opportune time for an emerging polymer process. Prog Polym Sci 56:116–166
2. He K, Huo H, Zhang Q, He D, An F, Wang M, Walsh MP (2005a) Oil consumption and CO_2 emissions in China's road transport: current status, future trends, and policy implications. Energy Policy 33(12):1499–1507
3. He MC, Xie HP, Peng SP, Jiang YD (2005b) Study on rock mechanics in deep mining engineering. Chin J Rock Mechan Eng 24(16):2803–2813
4. Trache D, Hussin MH, Chuin CTH, Sabar S, Fazita MN, Taiwo OF, Hassan TM, Haafiz MM (2016) Microcrystalline cellulose: isolation, characterization and bio-composites application —a review. Int J Biol Macromole 93:789–804
5. Abe K, Iwamoto S, Yano H (2007) Obtaining cellulose nanofibers with a uniform width of 15 nm from wood. Biomacromolecules 8(10):3276–3278
6. Cheung HY, Ho MP, Lau KT, Cardona F, Hui D (2009) Natural fibre-reinforced composites for bioengineering and environmental engineering applications. Compos B Eng 40(7): 655–663
7. Mansor MR, Sapuan SM, Zainudin ES, Nuraini AA, Hambali A (2013) Hybrid natural and glass fibres reinforced polymer composites material selection using analytical hierarchy process for automotive brake lever design. Mater Des 51:484–492
8. Marsh G (2003) Next step for automotive materials. Mater Today 6(4):36–43 (Elsevier)
9. Balakrishnan H, Hassan A, Imran M, Wahit MU (2012) Toughening of polylactic acid nanocomposites: a short review. Polym-Plast Technol Eng 51(2):175–192
10. Abe K, Nakatsubo F, Yano H (2009) High-strength nanocomposite based on fibrillated chemi-thermomechanical pulp. Compos Sci Technol 69(14):2434–2437
11. Klemm D, Kramer F, Moritz S et al (2011) Nanocelluloses: a new family of nature-based materials. AngewandteChemie Int Ed 50:5438–5466
12. Abdul Khalil HPS, Bhat AH, IreanaYusra AF (2012) Green composites from sustainable cellulose nanofibrils: a review. Carbohydr Polym 87:963–979
13. Haafiz MM, Hassan A, HPS AK, Owolabi AF, Marliana MM, Arjmandi R, Inuwa IM, Fazita MR, Nurul MR (2017) Cellulose nanowhiskers from oil palm empty fruit bunch biomass as green fillers. Cellulose-Reinforced Nanofibre Compos 241
14. Owolabi AWT, Ghazali A, Wanrosli WD, Abbas FMA (2016) Effect of alkaline peroxide pre-treatment on microfibrillated cellulose from oil palm fronds rachis amenable for pulp and paper and bio-composite production. BioResources 11(2):3013–3026

15. Paakko M, Ankerfors M, Kosonen H, Nykanen A, Ahola S, Osterberg M (2007) Enzymatic hydrolysis combined with mechanical shearing and high-pressure homogenization for nanoscale cellulose fibrils and strong gels. Biomacromolecules 8(6):1934–1941
16. Pandey JK, Kumar AP, Misra M, Mohanty AK, Drzal LT, Singh RP (2005) Recent advances in biodegradable nanocomposites. J Nanosci Nanotechnol 5:497–526
17. Saito T, Nishiyama Y, Putaux JL, Vignon M, Isogai A (2006) Homogeneous suspensions of individualized microfibrils from TEMPO-catalyzed oxidation of native cellulose. Biomacromolecules 7(6):1687–1691
18. Aulin C, Ahola S, Josefsson P, Nishino T, Hirose Y, Österberg M et al (2009) Nanoscale cellulose films with different crystallinities and mesostructures—their surface properties and interaction with water. Langmuir 25(13):7675–7685
19. Turbak AF, Snyder FW, Sandberg KR (1983) Microfibrillated cellulose, a new cellulose product: properties, uses, and commercial potential. J Appl Polym Sci 28:815–827
20. Wang YX, Tian HF, Zhang LN (2010) Role of starch nanocrystals and cellulose whiskers in synergistic reinforcement of waterborne polyurethane. Carbohydr Polym 80(3):665–671
21. Oksman K, Mathew AP, Bondeson D, Kvien I (2006) Manufacturing process of cellulose whiskers/polylactic acid nanocomposites. Compos Sci Technol 66:2776–2784
22. Sorrentino A, Vittoria GGV (2007) Potential perspectives of bionanocomposites for food packaging applications. Trends Food Sci Technol 18:84–95
23. Lamaming J, Hashim R, Sulaiman O, Leh CP, Sugimoto T, Nordin NA (2015) Cellulose nanocrystals isolated from oil palm trunk. Carbohydr Polym 127:202–208
24. John MJ, Thomas S (2008) Biofibres and biocomposites. Carbohydr Polym 71(3):343–364
25. Bataille P, Ricard L, Sapieha S (1989) Effects of cellulose fibers in polypropylene composites. Polym Compos 10:103–108
26. Hafren J, Zou WB, Cordova A (2006) Heterogeneous 'organoclick' derivatization of polysaccharides. Macromol Rapid Commun 27:1362–1366
27. Gruber E, Granzow C (1996) Preparing cationic pulp by graft copolymerisation. 1. Synthesis and characterization. Papier 50:293
28. Bonini C, Heux L, Cavaille JY, Lindner P, Dewhurst C, Terech P (2002) Rodlike cellulose whiskers coated with surfactant: a small-angle neutron scattering characterization. Langmuir 18:3311–3314
29. Kim DY, Nishiyama Y, Kuga S (2002) Surface acetylation of bacterial cellulose. Cellulose 9:361–367
30. Gousse C, Chanzy H, Cerrada ML, Fleury E (2004) Surface silylation of cellulose microfibrils: preparation and rheological properties. Polymer 45:1569–1575
31. Stenstad P, Andresen M, Tanem BS, Stenius P (2008) Chemical surface modifications of microfibrillated cellulose. Cellulose 15:35–45
32. Habibi Y, Heux L, Mahrouz M, Vignon MR (2008) Morphological and structural study of seed pericarp of *Opuntia ficus-indica* prickly pear fruits. Carbohydr Polym 72(1):102–112
33. Lu J, Askeland P, Drzal LT (2008) Surface modification of microfibrillated cellulose for epoxy composite applications. Polymer 49:1285–1296
34. Iwatake A, Nogi M, Yano H (2008) Cellulose nanofibre-reinforced polylactic acid. Compos Sci Technol 68(9):2103–2106
35. Behrens BA, Doege E, Reinsch S, Telkamp K, Daehndel H, Specker A (2007) Precision forging processes for high-duty automotive components. J Mater Process Technol 185 (1):139–146
36. Kosior E, Braganca RM, Fowler P (2006) Lightweight compostable packaging: literature review. Waste Resour Action Program 26:1–48
37. Kyrikou I, Briassoulis D (2007) Biodegradation of agricultural plastic films: a critical review. J Polym Environ 15(2):125–150
38. Haafiz MM, Eichhorn SJ, Hassan A, Jawaid M (2013) Isolation and characterization of microcrystalline cellulose from oil palm biomass residue. Carbohyd Polym 93(2):628–634

39. Haafiz MM, Hassan A, Khalil HA, Fazita MN, Islam MS, Inuwa IM, Marliana MM, Hussin MH (2016) Exploring the effect of cellulose nanowhiskers isolated from oil palm biomass on polylactic acid properties. Int J Biol Macromol 85:370–378
40. Ray SS, Yamada K, Okamoto M, Fujimoto Y, Ogami A, Ueda K (2003) New polylactide/layered silicate nanocomposites. 5. Designing of materials with desired properties. Polymer 44(21):6633–6646
41. Dufresne A, Kellerhals MB, Witholt B (1999) Transcrystallization in mcl-PHAs/cellulose whiskers composites. Macromolecules 32(22):7396–7401
42. Abdulkhani A, Hosseinzadeh J, Dadashi S, Mousavi M (2015) A study of morphological, thermal, mechanical and barrier. properties of PLA based biocomposites prepared with micro and nano sized cellulosic fibers. Cell Chem Technol 49(7–8):597–605
43. Evans JD, Sikdar SK (1990) Biodegradable plastics: an idea whose time has come? Chem Technol 20:38–42
44. Plackett D, Vázquez A (2004) Natural polymer sources. In: Baillie Caroline (ed) Green composites polymer composites and the environment. Woodhead Publishing Ltd/CRC Press LLC, Cambridge, pp 123–153
45. Kunioka M, Tamaki A, Doi Y (1989) Crystalline and thermal properties of bacterial copolyesters:poly(3-hydroxybutyrate-co-3-hydroxyvalerate) and poly(3-hydroxybutyrate-co-4-hydroxybutyrate). Macromolecules 22:694
46. Ma X, Yu J, Ma Y (2005) Urea and formamide as a mixed plasticizer for thermoplastic wheat flour. Carbohydr Polym 60:111. Yang J-H, Yu J-G, Ma X (2006) Preparation and properties of etylenebisformamide. Carbohydr Polym 63(2006):218
47. Abdul Khalil HPS, Hanida S, Kang SCW, NikFuaad NA (2007) Agro-hybridcomposite: the effects on mechanical and physical properties of oil palm fiber(EFB)/glass hybrid reinforced polyester composites. J Reinf Plast Compos 26:203–218
48. Adeosun SO, Lawal GI, Balogun SA, Akpan EI (2012) Review of green polymer nanocomposites. J Miner Mater Charact Eng 11(04):385
49. Dufresne A (2003) Interfacial phenomena in nanocomposites based on polysaccharide nanocrystals. Compos Interfaces 10(4–5):369–388
50. Lai SM, Don TM, Huang YC (2006) Preparation and properties of biodegradable thermoplastic starch/poly(hydroxyl butyrate) blends. J Appl Polym Sci 100:2371–2379
51. Jang WY, Shin BY, Lee TX, Narayan R (2007) Thermal properties and morphology of biodegradable PLA/starch compatibilized blends. J Ind Eng Chem 13:457–464
52. Sarazin P, Li G, Orts WJ, Favis BD (2008) Binary and ternary blends of polylactide, polycaprolactone and thermoplastic starch. Polymer 49:599–609
53. Durango AM, Soares NFF, Benevides S, Teixeira J, Carvalho M, Wobeto C et al (2006) Development and evaluation of an edible antimicrobial film based on yam starch and chitosan. Packaging Technol Sci 19:55–59
54. Ma PC, Siddiqui NA, Marom G, Kim JK (2010) Dispersion and functionalization of carbon nanotubes for polymer-based nanocomposites: a review. Compos A Appl Sci Manuf 41(10): 1345–1367
55. Mondragón M, Arroyo K, Romero-García J (2008) Biocomposites of thermoplastic starch with surfactant. Carbohyd Polym 74:201–208
56. Piyada K, Waranyou S, Thawien W (2013) Mechanical, thermal and structural properties of rice starch films reinforced with rice starch nanocrystals. Int Food Res J 20:439–449
57. Hietala M, Mathew AP, Oksman K (2013) Bionanocomposites of thermoplastic starch and cellulose nanofibers manufactured using twin-screw extrusion. EurPolym J 49:950–956
58. Plackett D, Andersen TL, Pedersen WB, Nielsen L (2003) Biodegradable composites based on L-polylactide and jute fibres. Compos Sci Technol 63:1287–1296
59. Ray SS, Okamoto M (2003) Polymer/layered silicate nanocomposites: a review from preparation to processing. Prog Polym Sci 28(11):1539–1641
60. Pandey JK, Ahn SH, Lee CS, Mohanty AK, Misra M (2010) Recent advances in the application of natural fibre based composites. Macromol Mater Eng 295(11):975–989

61. Lee SY, Kang IA, Doh GH, Yoon HG, Park BD, Wu Q (2008) Thermal and mechanical properties of wood flour/talc-filled polylactic acid composites: effect of filler content and coupling treatment. J Thermoplast Compos Mater 21(3):209–223
62. Qu P, Gao Y, Wu G, Zhang L (2010) Nanocomposites of poly (lactic acid) reinforced with cellulose nanofibrils. BioResources 5(3):1811–1823
63. Okubo K, Fujii T, Thostenson ET (2009) Multi-scale hybrid biocomposite: processing and mechanical characterization of bamboo fibre reinforced PLA with microfibrillated cellulose. Compos A Appl Sci Manuf 40(4):469–475
64. Kim JP, Yoon T-H, Mun SP, Rhee JM, Lee JS (2006) Wood-polyethylene composites using ethylene-vinyl alcohol copolymer as adhesion promoter. Bioresource Biotechnol 97:494–499
65. Rong MZ, Zhang MQ, Liu Y, Yang GC, Zeng HM (2001) The effect of fibre treatment on the mechanical properties of unidirectional sisal-reinforced epoxy composites. Compos Sci Technol 61:1437–1447
66. Qin C, Soykeabkaew N, Xiuyuan N, Peijs T (2008) The effect of fibre volume fraction and mercerization on the properties of all cellulose composites. Carbohydr Polym 71:458–467
67. Khalfallah M, Abbès B, Abbès F, Guo Y, Marcel V, Duval A, Vanfleteren F, Rousseau F (2014) Innovative flax tapes reinforced acrodur biocomposites: a new alternative for automotive applications. Mater Des 64:116–126
68. Chen W, Yu H, Liu Y, Chen P, Zhang M, Yunfei H (2011) Individualization of cellulose nanofibers from wood using high-intensity ultrasonication combined with chemical pre-treatments. Carbohyd Polym 83:1804–1811
69. Tiffany A, Rivkin A, Cao Y, Nevo Y, Abraham E, Ben-Shalom T, Lapidot S, Shoseyov O (2016) Nanocellulose, a tiny fiber with huge applications. Curr Opin Biotechnol 39:76–88
70. Herrera N, Salaberria AM, Mathew AP, Oksman K (2016) Plasticized polylactic acid nanocomposite films with cellulose and chitin nanocrystals prepared using extrusion and compression molding with two cooling rates: effects on mechanical, thermal and optical properties. Compos A Appl Sci Manuf 83:89–97

Static and Dynamic Mechanical Properties of Eco-friendly Polymer Composites

Bernardo Zuccarello

1 Introduction

The increasing sensitivity to environmental protection and the recent laws against environmental pollution that were implemented because of the production of high amounts of synthetic materials based on petroleum chemistry, have led to a widespread attention toward biocomposites, i.e. to eco-friendly polymer composites produced by an eco-sustainable or renewable matrix reinforced by natural fibres. If properly combined with 'green' matrixes or biopolymers, natural fibres could enable for partially or completely renewable biocomposites to be produced. These biocomposites can be easily biodegraded at their end of life by composting and are therefore used as improver/fertilizer in agriculture terrains. Biodegradable polymers and natural fibres are extremely attractive because they can substitute the synthetic matrix obtained by the petroleum industry, and can produce composite materials with interesting mechanical properties such as good tensile strength, sufficient stiffness, and high toughness. Moreover, many natural fibres exhibit other properties that are highly regarded in the industrial field, e.g., low damageability, good thermal and acoustic insulation, low skin irritability, high availability in the current market, low embodied energy, and extremely low cost. Natural fibres have been used to reinforce thermoplastic matrixes (polypropylene, polyethylene, polyurethane, polystyrene, PVC, etc.) that are characterized by higher toughness and easier recyclability, and thermosetting matrixes (polyester, phenolic, and epoxy resins, etc.) that exhibit better mechanical characteristics, but lower recyclability and environmental compatibility.

Despite such interesting properties, biocomposites are hitherto used only for non-structural applications (filling material, soundproofing, thermal insulation, packaging etc.) in various fields of industrial production, packaging and automotive

B. Zuccarello (✉)
Viale delle Scienze, 90128 Palermo, Italy
e-mail: bernardo.zuccarello@unipa.it

Inamuddin et al. (eds.), *Sustainable Polymer Composites and Nanocomposites*,
https://doi.org/10.1007/978-3-030-05399-4_9

(large manufacturers such as BMW, Volvo, Mercedes-Benz, Ford, GM, Toyota, etc., increasingly use of biocomposites for the manufacturing of dashboards, insulating elements, doors, backs, etc.), naval industry, and civil constructions (panels, sandwich etc.), where their lightness and low cost, both inferior to that of any composite material reinforced by synthetic fibres, are particularly appreciated. In typical non-structural applications, such biocomposites are constituted by green thermosetting (partially bio-based) or thermoplastic (recyclable and/or renewable) matrixes, reinforced by short or discontinuous randomly oriented natural fibres; they are typically manufactured by moulding or extrusion processes, and are characterized by relatively low mechanical strength combined with sufficient stiffness.

Although various recent works have been devoted to the implementation of high-performance biocomposites that can be used for structural applications (self-bearing or load-bearing panels, etc.), and also by the preliminary improvement of the fibre properties [1–18], the development of eco-friendly or renewable high-performance biocomposites reinforced by natural fibres is an objective expected by the scientific community; however, it has not yet been fully achieved.

The interesting synthesis of biocomposites hitherto are reported in various review articles published in high-quality journals devoted to materials and composites [18–34]. The primary objective of this chapter is to present both the static and dynamic mechanical properties of eco-friendly polymer composites described in literature, as well as to present the reader some scientific background to correctly evaluate the matrix/fibre adhesion and interpret the mechanical behaviour of such materials. This can also be achieved using recent micromechanical models developed for particular biocomposites; further, it can be extended to the entire family of eco-friendly polymer composites reinforced by natural fibres. Therefore, the reader will have sufficient knowledge on biocomposites and their capacity to substitute traditional materials as metals and fiberglass.

2 Constituent Materials: Polymer Matrixes and Natural Fibres

As mentioned above, eco-friendly polymer composites can be obtained using several types of matrixes and various reinforcing natural fibres. Obviously the static and the dynamic properties of such biocomposites are strictly related to the mechanical characteristics of the constituent materials as well as to the primary functional parameters such as the matrix-fibre adhesion, manufacturing process, and fibre volume fraction. Considering the materials used as the matrix, the analysis of the numerous research works reported in literature shows that they have considered both thermosetting and thermoplastic polymers characterized by variable environmental impact, from green matrixes obtained using an eco-friendly manufacturing process [1, 2, 35–39], to partially bio-based matrixes [40–44], to completely renewable matrixes obtained using proper biopolymers, such as PLA and the like [45–47].

Consequently, the environmental impact of a composite included in the wide family of the polymer matrix biocomposites reinforced by natural fibre can vary in a wide range; in practice, the renewability can vary from approximately 20–30% (biocomposite produced by 20–30% of the weight of natural fibres mixed to traditional synthetic matrixes) to approximately 100% (biocomposites produced by renewable matrix reinforced by natural fibres obtained by renewable extraction processes).

2.1 Static Mechanical Properties

The static mechanical properties of the material that constitute biocomposites are generally performed using tensile strength, although various researchers have also used the proper flexural tests on the matrixes. In general, the matrixes and fibres exhibit an elasto-plastic behaviour with variable plastic phase; in terms of experimental scattering, the matrixes show values aligned with those of typical plastics with standard deviations less than 3–5%, whereas a higher scattering characterizes the single fibre tests used for the fibre characterization. As an example, Fig. 1 shows the results of champagne tensile tests performed on one lot containing 10 fibres of agave sisalana [36, 37].

This case shows an almost ideal linear elastic behaviour with the ultimate stresses that vary from approximately 600 MPa to approximately 1150 MPa; the tensile modulus varies instead from approximately 20 GPa to approximately 60 GPa. Considering that the experimental evidence regarding the tensile test results show a good accordance with the Gaussian distribution, we can conclude that the tensile strength (having in this case a mean value of approximately 690 MPa) is typically characterized by the standard deviation of approximately 10%, i.e. 2–3 times that of polymer matrix materials. Such relatively higher experimental scattering of the fibre characteristics forewarned researchers that it can lead to components with variable mechanical characteristics, although this

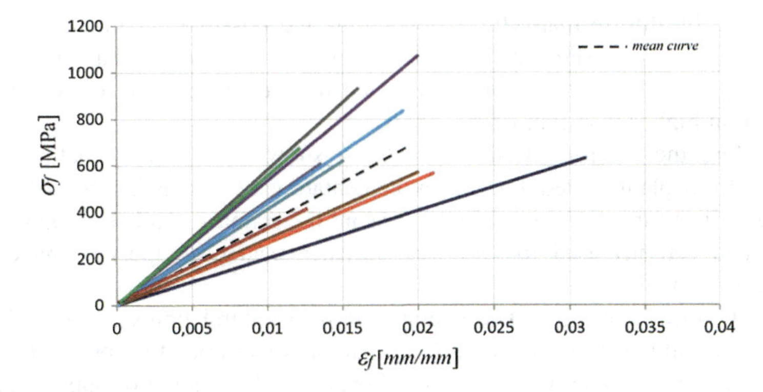

Fig. 1 Tensile tests results on single fibre of agave sisalana, and relative mean curve

conjecture is not true because its presence in a generic component of thousands of fibres leads to an obvious statistical mean such that, as widely confirmed by the experimental evidence, the mechanical properties of different biocomposite specimens have standard deviations comparable with those of other composite materials.

Synthetically, Table 1 shows the mechanical characteristics of the primary polymer matrixes used in literature [25, 35, 36, 45–48] for the manufacturing of biocomposites reinforced by various natural fibres.

Table 1 shows that the tensile strength and relative tensile modulus of various thermosetting matrixes (epoxy, polyester, vinylester, phenolic, etc.) are within a relatively small range: the tensile strength from 35 to 100 MPa; the Young modulus from 2.0 to 4.8 GPa. The failure strain, however, is within a wider range, with variations of approximately one order of magnitude: from 1 to 8%, approximately.

Regarding the synthetic thermoplastic matrixes (PP, LDPE, HDPE, PS, and Nylon), Table 2 shows that the tensile strength varies from approximately 10–95 MPa, whereas more significant variations affect the tensile modulus that varies from approximately 0.1 GPa (LDPE) to approximately 3.9 GPa (Nylon 6.6). As shown, the failure strain for thermoplastic resins can become extremely high; in practice, it can vary from 1% (PS) to extremely high values of approximately 600–800% (PP, LDPE).

Regarding the primary biodegradable matrixes in Table 3, except for PCL, PEA, SPI and starch, the tensile mechanical properties of such polymers are comparable with those of the synthetic thermosetting resins: tensile strength from approximately 25 MPa to approximately 60 MPa; tensile modulus from 0.35 GPa to approximately 6 GPa; failure strain from 1.4 to 9%. Unlike these, PCL, PEA, SPI, and starch exhibit relative low tensile strength (5–20 MPa), low tensile modulus (0.1–0.85 GPa), and high failure strain (approximately 30–235%). It is noteworthy that although the failure strain is often neglected by various authors in the prediction of the mechanical performance of biocomposites, as it has been clearly observed in [1, 2, 35–39], it influences significantly the damage mechanisms of the biocomposites and, consequently, the actual mechanical strength. As an example, in a unidirectional biocomposite laminate, a matrix failure strain less than that of the reinforcing fibres leads to a premature damage of the biocomposite that occurs prior the tensile fibre failure, by diffuse matrix/fibre debonding and possible delamination; consequently, the biocomposite tensile strength can be much lower than that can be obtained by the full utilization of the fibre tensile strength using a matrix having a failure strain higher than that of the fibres.

Regarding the reinforcing fibres, Table 4 shows the primary natural fibres extracted from plants, used for the implementation of interesting biocomposites reported in literature. For a useful comparison, the primary synthetic fibres (glass, aramid, and carbon) used to reinforce polymer matrix composites, have been reported as well.

Table 4 shows that the tensile strength of such natural fibres can vary in two orders of magnitude, from the values comparable or inferior to those of the thermosetting matrixes, as shown by the alfa fibres (tensile strength of approximately) to values comparable with that of the synthetic fibres. This is also shown by the

Table 1 Mechanical properties of thermosetting polymers used as matrix in biocomposites

Polymer	Density (g/cm^3)	Tensile strenght (MPa)	Tensile modulus (GPa)	Elongation (%)	Impact strength (J/m^2)	References
Polyester resin	1.2–1.5	40–90	2.0–4.5	2.0	0.2–3.2	[25]
Polyester resin	1.0–1.4	41–90	2.1–4.4	2.0–2.6	0.9	[47]
Vinyl ester resin	1.2–1.4	69–83	3.1–3.8	4.0–7.0	2.5	[25]
Epoxy resin	1.1–1.4	35–100	3.0–6.0	1.0–6.0	0.3	[25]
Epoxy resin	1.1–1.4	45–90	2.3–3.1	2.0–8.0	0.4	[47]
Phenolyc resin	1.3	35–62	2.8–4.8	1.5–2.0	0.35	[25]
Green epoxy	1.1	50	2.5	2.5	–	[35, 36]

Table 2 Mechanical properties of thermoplastic polymers used as matrix in biocomposites

Polymer	Density (g/cm^3)	Tensile strenght (MPa)	Tensile modulus (GPa)	Elongation (%)	Impact strength (J/m^2)	References
PP	0.9	26–41	0.9–1.8	15–700	21.4–267.0	[25, 47, 48]
LDPE	0.9	8–78	0.4	90–800	>854.0	[25, 48]
HDPE	0.9	14–38	0.4–1.5	2–130	26.7–1068.0	[25]
PS	1.0–1.1	25–69	2.8–5.0	1.0–2.5	1.1	[25, 48]
Nylon 6	1.1	43–79	2.9	20–150	42.7–160.0	[25]
Nylon 6,6	1.1	12–94	2.5–3.9	35–100	16.0–654.0	[25]

Table 3 Mechanical properties of biodegradable polymers used as matrix in biocomposites

Polymer	Density (g/cm^3)	Tensile strenght (MPa)	Tensile modulus (GPa)	Elongation (%)	References
PLA	0.9–1.3	21–60	0.3–3,8	2.5–8.0	[45–48]
PHB	1.1–1.3	21–40	0.9–4.0	5.0–8.4	[45–48]
PHBV	1.2	26	1.0–2.4	1.4–25	[46, 48]
PGA	1.5	60	6.0	1.5	[45]
PCL	1.1	21	0.2	300.0	[45]
PEA	1.2	16	0.4	85.0–119.0	[46]
SPI	1.2–1.5	6	0.1	170.0–236.0	[46]
Starch	1.0–1.4	5–6	0.1–0.8	31.0–44.0	[45–47]

curaua fibres (tensile strength of approximately 3000 MPa). Similarly, the tensile modulus varies in the two orders of magnitude range, from approximately 1.44 GPa for pineapple to approximately 128 GPa for ramie. Regarding the failure strain, it varies from approximately 1% for 'rigid' fibres such as flax, hemp, jute, abaca, and bagasse; approximately 8–50% for "deformable" fibres such as oil palm, piassava, and coir. Consequently, we can conclude that this wide family of natural fibres includes (a) "high failure strain" fibres that cannot be used to reinforce thermosetting matrixes (having strain failures of less than 10%, see Table 1), the rigid polystyrene (PS, see Table 2), the most biodegradable resins such as PLA, PHB, PHBV, and PGA, as well as "low modulus" fibres that cannot be used to reinforce polymer matrixes having relatively high modulus. Therefore, a fibre classification more useful than the classical one based on the vegetable component from which they are extracted (seed, fruit, bast, stem, leaf, etc.), is one that is based on the tensile modulus that governs the reinforcing effects and, consequently, the failure strain and subsequently the tensile strength.

To obtain high-performance biocomposites, two basic requirements must be satisfied: (1) a good fibre reinforcing effect where the fibres should have a tensile modulus at least 10 times higher than that of the matrix, (2) the failure strain of the fibre should be inferior to that of the matrix to fully exploit the fibre strength without premature matrix failure. Considering these requirements, similar to synthetic fibres, natural fibres can be divided into three classes: low modulus (LM) fibres that include fibres with a tensile modulus less than 10 GPa (coir, low modulus sisal, cotton, low modulus pineapple, oil palm, low modulus palf, piassava), intermediate modulus (IM) fibres that include fibres with a tensile modulus in the range of 10–40 GPa (jute, bamboo, low modulus flax, bagasse, low modulus kenaf, low modulus ramie, sisal, and low-modulus curaua), high modulus (HM) fibres that include fibres with a modulus higher than 40 GPa (high-modulus pineapple, high-modulus flax, high-modulus hemp, high-modulus jute, high-modulus kenaf, high-modulus ramie, high-modulus palf, high-modulus curaua). The LM natural fibres can be used to reinforce low-modulus polymer matrixes such as PE, PP, low-modulus PLA, low-modulus PHB, PCL, PEA, SPI, and starch, whereas the IM natural fibres can reinforce most polymer matrixes except those

Table 4 Mechanical properties of the various natural fibres used to reinforce biocomposites

Origin	Fiber	Density (g/cm^3)	Tensile strenght (MPa)	Tensile modulus (GPa)	Elongation (%)	References
Seed or fruit	Cotton	1.5–1.6	287–800	5.5–12.6	3.0–12.0	[25, 45–47]
	Coir	1.1–1.5	95–593	2.8–6.0	15.0–51.4	[25, 45–47, 49]
	Pineapple	0.8–0.6	180–627	1.4–82.0	3.2–14.5	[2, 6]
	Oil palm	0.7–1.5	248	3.2	25.0	[5]
	Flax	1.4–1.5	343–2000	27.6–103.0	1.0–3.5	[25, 45–47, 49]
Bast or stem	Hemp	1.4–1.5	270–920	23.5–90.0	1.0–4.0	[25, 46, 47, 49]
	Jute	1.3–1.5	320–860	10.0–60.0	1.0–1.8	[25, 45–47, 49]
	Kenaf	1.4–1.5	195–930	14.5–66.0	1.3–5.5	[25, 46, 47, 49]
	Ramie	1.0–1.6	400–1000	24.5–128.0	1.2–4.0	[25, 45–47, 49]
	Sisal	1.3–1.5	363–790	9.0–39.5	2.0–7.0	[25, 35, 45–47, 49]
Leaf	Abaca	1.5	400–980	6.2–20.0	1.0–10.0	[46, 49]
	Henequen	1.2	430–570	10.1–16.3	3.7–5.9	[46]
	Banana	1.3	500–800	12.0–32.0	1.5–9.0	[45, 46]
	Palf	0.8–1.6	180–1627	1.4–82.5	1.6–14.5	[46]
	Curauà	1.4	87–3000	10.5–96.0	1.3–6.0	[45, 46, 49]
	Bamboo	0.6–1.1	140–800	11.0–32.0	2.5–3.7	[46, 47, 49]
Cane, grass and reed	Bagasse	1.3	222–290	17.0–27.1	1.1–10.0	[45, 46, 49]
	Softwood	1.5	1000	40.0	4.4	[25]
Wood	Wood	2.5–2.6	1000–3500	70.0–85.0	0.5–4.8	[25, 46, 47]
Glass	E	2.5	4570	86.0	2.8	[25, 46]
	S	1.4	3000–3150	63.0–67.0	3.3–3.7	[25, 46]
Aramid	Kevlar	1.4	4000	230.0–240.0	1.4–1.8	[25, 46]
Carbon	LS	1.8	4400–4800	225.0–260.0	1.8–1.9	[47]
Carbon	HM	1.9	3600–3900	390–410	0.9–1.0	[47]

having a modulus higher than 4–5 GPa such as some PSs, thermosetting matrixes, and biodegradable PGA. Obviously, the HM natural fibres can reinforce any polymer matrix except those having a lower failure strain. As an example, the coir fibres (LM), having a modulus in the range of 3–6 GPa and a strain failure in the range of 15–51% (see Table 4), can be used advantageously to reinforce the LDPE that has a tensile modulus less than 0.2 GPa (less than 1/10 of the fibre modulus, see Table 1) and a failure strain that is always higher than 90% (approximately 2–6 times the failure strain of the fibres); however, it cannot be used to reinforce a green epoxy having comparable stiffness and a lower failure strain (approximately 2%, see Table 1). Unfortunately, such elementary rules are not always adhered to in literature, as can be observed from Ref. [1] that reported coir fibre being used to reinforce PP having a tensile modulus of approximately 1–2 GPa, i.e. comparable with that of the fibres, and a failure strain of less than 15%; consequently, it is not surprising if the relative biocomposite has mechanical properties (tensile strength in the range of 25–30 MPa, and tensile modulus in the range of 1.1–1.25 GPa) that are comparable with those of the simple matrix (tensile strength in the range of 26–34 MPa, and tensile modulus of approximately 1–2 GPa). Further, as mentioned above, a rigid fibre cannot be always advantageously used to reinforce a relatively deformable matrix; for example, an epoxy resin having a tensile modulus of 3 GPa and a strain failure of approximately 2% can be reinforced advantageously by flax, jute, hemp, kenaf, ramie, banana, and curaua with a mean tensile modulus higher than 30 GPa and a failure strain of less than 2%, but cannot be reinforced by sisal, bamboo, isora, alfa, piassava, and softwood because they always show a strain failure higher than 2%. An exemplary demonstration of such important feature is reported in Ref. [1], where the authors considered agave Marginata fibres extracted by rolling (without any fibre treatment), and clearly showed how the substitution to the green epoxy (having strain failure of 2%) with a PLA matrix (having strain failure of approximately 5%) allowed a biocomposite having a tensile strength of 103.6 MPa to be transformed into a biocomposite having a tensile strength of 188.3 MPa, i.e. to obtain a strength increment of approximately 83%.

Finally, the density of the polymer matrixes used for the manufacturing of biocomposites vary in an extremely limited range, from approximately 0.9–1.5 g/cm^3 whereas the density of the natural fibres falls within the range of 0.6 (low density pineapple)–1.6 g/cm^3 (cotton and palf), and is always significantly inferior to the density of synthetic glass fibres (approximately 2.6 g/cm^3); consequently, polymer biocomposites reinforced by natural fibres are always lighter than the fibre glass composites that they intend to substitute in several fields of industrial production: automotive, construction, nautical, packaging, etc.

2.2 *Dynamic Mechanical Properties*

Unlike the static properties that can be defined for both matrixes and fibres, the dynamic properties, i.e. impact and fatigue strength, can be analysed only for the polymer matrixes that can be subjected to both impact and fatigue tests; such particular service loadings, in fact, cannot be applied to a single fibre. Tables 1, 2 and 3 show how the impact strength, determined typically by Charpy or Izod test, is relatively low for the thermosetting matrixes (from 0.15 to approximately 3 J/m^2) that, as it is well known, exhibit brittle behaviour. The impact strength is instead significantly higher, from one to three orders of magnitude, for thermoplastic matrixes (from 20 to approximately 1000 J/m^2) that exhibit a ductile behaviour with higher ultimate strains. As expected by considering the experimental evidence for synthetic composites, the reinforcing by natural fibres allow a user to increase significantly the impact strength of the thermosetting matrixes, being more limited to the reinforcing fibre contribution for thermoplastic resins (unlike the thermosetting matrices, the stress concentration effect prevails on the reinforcing effect).

Similarly, the fatigue strength that is relatively low for a generic polymer matrix (fatigue ratio of less than 0.25), can be increased significantly by the introduction of natural fibres, especially in the cases of high module fibres. Obviously, the actual improvement in the dynamical properties is strictly related to various parameters, e.g., fibre length (short/long fibre) and its orientation (random or aligned), fibre volume fraction, laminate setup, etc.

3 Fibre/Matrix Adhesion

As mentioned above and widely shown in literature, the mechanical properties of generic biocomposites are strictly related to the fibre/matrix adhesion; in fact, it influences significantly the load transmission, from matrix to fibres, as well as the peculiar damage mechanisms that in many cases can include the premature phenomena of fibre/matrix debonding, pull-out, and delamination, with important consequence on the mechanical strength of the biocomposites in the static and dynamic loading conditions. As it is well known, for short fibre composites, a low matrix-adhesion can lead to the easy debonding of the transversal fibres along with the possible pull-out of the longitudinal fibres. In the long fibre composite, instead, a poor fibre-matrix adhesion can lead to the debonding phenomena, but only when significant and diffuse matrix defects appear [49]. Unfortunately, little research has been devoted to the accurate analysis of the actual fibre/matrix adhesion of biocomposites. Typically, the low adhesion has been summarily deducted from the observed low mechanical performance of the analysed biocomposites, ad also when they were actually related to the low quality of the manufacturing process (presence of diffuse voids and/or direct contact between adjacent fibres, etc.), or to a low reinforcing effect owing to a low fibre/matrix elastic modulus mismatch. However, many studies have been devoted to the improving of the fibre/matrix adhesion by proposing various surface fibre treatments such as NaOH treatment (mercerization) [18], the addition of nanomaterials [50], and the addition of a silane coupling agent [14]. As it is well known, the fibre/matrix adhesion is typically analysed through the pull-out test that comprises the measurement of the load that lead to the pull-out of a single fibre partially embedded on the matrix. In general, such a test is performed by embedding partially a single fibre into a cylinder of matrix having the proper diameter, and by computing the mean shear stress at the fibre/matrix interface by dividing the relieved pull-out load to the interface area. The ratio between the computing mean shear stress and the ultimate shear stress of the matrix is considered, as well as the goodness index of the matrix/fibre adhesion. Unfortunately, as widely demonstrated in [1], such an approach yields an underestimated evaluation of the fibre/matrix adhesion because the interface shear stress distribution is not uniform and, most importantly, such nonuniformity tends to increase significantly in the pull-out tests with the stiffness mismatch between the fibre and the surrounding matrix cylinder. Consequently, considering this in the pull-out test, the ratio of the transversal sections of the fibre and matrix can be extremely different from that of the analysed biocomposite. subsequently, the interface shear stress distribution can be extremely different, and the matrix/fibre adhesion evaluation performed by considering the mean shear stress will not be reliable. As explained in [1], a reliable evaluation can be instead obtained by considering the ratio r_a between the maximum shear stress τ_{max} that occurs at the fibre/matrix interface free point in the pull-out incipient condition, and the matrix ultimate shear stress $\tau_{m,u}$. Such a ratio is an index that varies from 0 (null adhesion) to 1 (perfect adhesion), and it is related to the pull-out load $P_{pull\text{-}out}$ experimentally detected, by the simple formula [1]:

$$r_a = \frac{\tau_{\max}}{\tau_{m,R}} = \frac{P_{pull-out}\lambda}{2\pi d\tau_{m,R}}\left[\frac{1-S}{1+S}\cdot \mathrm{Tanh}\left(\lambda\frac{l}{2}\right) + \frac{(d/D)}{\mathrm{Tanh}(\lambda l/2)}\right] \tag{1}$$

where d is the fibre diameter, D is the diameter of the matrix cylinder as shown in Fig. 2, S is the well-known stiffness unbalancing defined by the ratio between the fibre stiffness and the surrounding matrix cylinder stiffness, whereas λ is the base parameter of the bi-material couple analysed, related to the geometry and the elastic moduli of the coupled materials [1].

Equation 1 shows that the maximum shear stresses are proportional with the fibre stiffness, and inversely proportional to the matrix stiffness. For a given matrix, the fibre stiffness has a noticeable influence on the pull-out strength of the biocomposite, as shown in Fig. 3. The latter shows the influence of the tensile modulus of various agave fibres (it increases from the MDN type to MPN type) on the interface shear stress distribution of a green epoxy (Fig. 3a) and a PLA (Fig. 3b) matrix biocomposite.

Figure 3a shows that for the green epoxy matrix, the maximum specific shear stresses vary from 1.85 to 2.87 N/mm, an increase of approximately +55%, corresponding to the transition from the stiffest MPN fibres to the less stiff MDN fibres. Figure 3b shows that by substituting green epoxy with PLA, the maximum shear stresses decrease (by approximately 40%), being included between 1.09 for the stiffer MPN fibres to 1.74 N/mm for the less stiff MDN fibres.

The values of the index r_a reported in Table 5 for the abovementioned agave fibres reinforcing epoxy and PLA, corroborate clearly how, in accordance with the

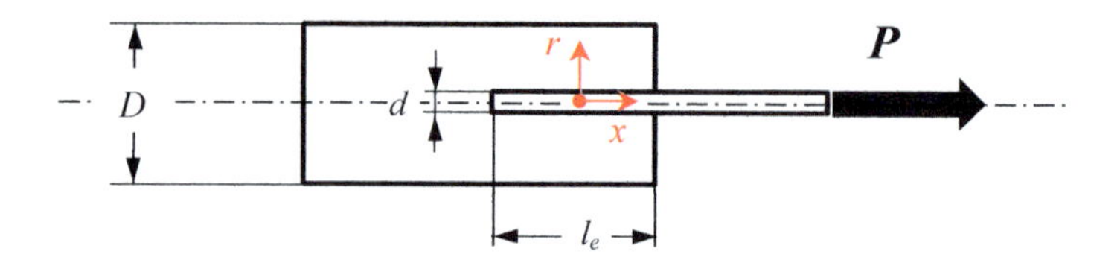

Fig. 2 Experimental layout of single-fiber pull-out test

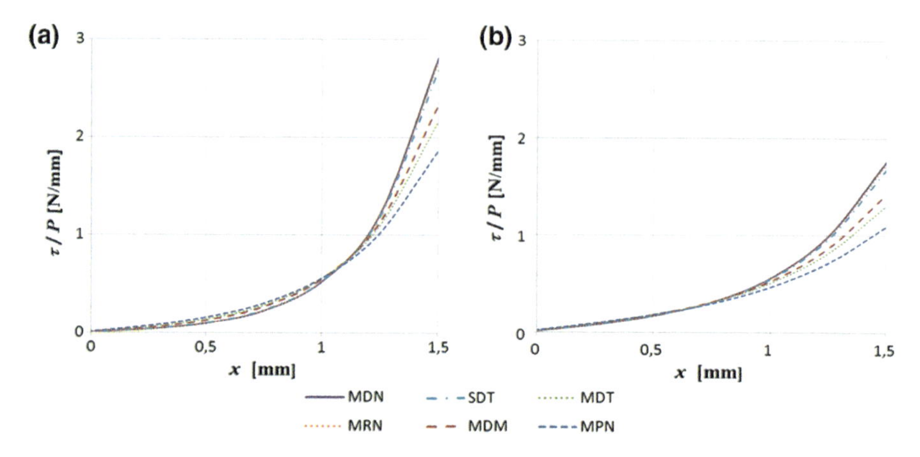

Fig. 3 Fibre-matrix interface shear stress distribution for **a** green epoxy and **b** PLA matrix reinforced by various agave fibres

findings above, the increase in the pull-out load does not correspond necessarily to an increase in the fibre-matrix adhesion, but a significant increase can be also be associated with the increase in the fibre stiffness (and subsequently the unbalancing of S that compares with Eq. 1).

Table 5 shows that for a given fibre, i.e., the PLA, the pull-out loads are always higher than the epoxy ones. Such a result has caused several authors to erroneously state that the PLA/agave adhesion is better than that of the epoxy/agave combination [35], whereas the higher pull-out load observed is solely due to the lower stiffness of the PLA compared to the epoxy. Further, Table 5 shows that particular fibre treatments such as mercerization (MDT and SDT fibres) lead to improvements in the fibre-matrix adhesion (increasingly higher for the PLA), although significant pull-out strength increments are obtained by increasing the fibre stiffness, as shown in the stiffer MPN fibres that are not subjected to any fibre treatment.

As an example, Fig. 4 shows the typical trend of the pull-out test curve, observed for natural fibres: a first elastic-linear step, corresponding to the loading phase of the embedded fibre, followed by a second decreasing step characterized by the fibre sliding that occurs in a discontinuous manner until the complete extraction.

Table 5 Results of the pull-out tests on green epoxy and PLA reinforced by various agave fibres

Fiber	Matrix	E_f (GPa)	$P_{pull\text{-}out}$ (N)	$\tau_{\max}$ (MPa)	r_a	$\bar{\tau}_m$ (MPa)	$r_a^{(*)}$
MDN	Epoxy	9.2	8.1	23.12	0.64	6.44	0.17
MRN	Epoxy	9.3	9.2	28.09	0.72	7.88	0.21
MPN	Epoxy	18.7	13.7	26.26	0.73	10.90	0.30
MDM	Epoxy	12.7	13.5	32.28	0.89	10.74	0.29
MDT	Epoxy	14.5	14.2	31.52	0.87	11.30	0.31
SDT	Epoxy	9.8	10.1	27.85	0.77	8.04	0.22
MDN	PLA	9.2	10.3	17.61	0.46	8.20	0.21
MRN	PLA	9.3	10.5	17.83	0.46	8.36	0.22
MPN	PLA	18.7	16.1	16.75	0.44	12.81	0.33
MDM	PLA	12.7	21.9	30.26	0.79	17.43	0.45
MDT	PLA	14.5	23.2	29.20	0.76	18.47	0.48
SDT	PLA	9.8	18.3	30.06	0.79	14.57	0.38

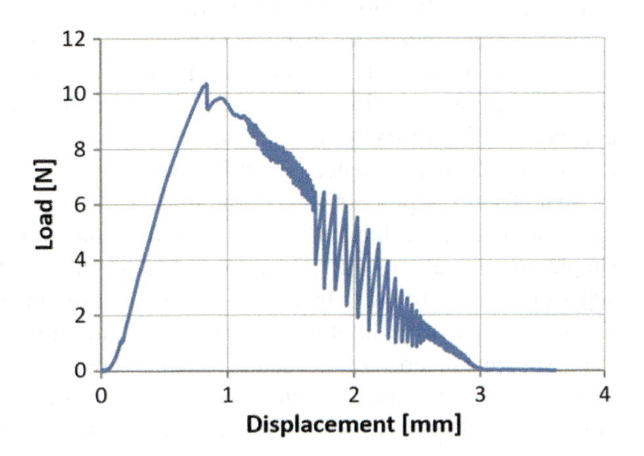

Fig. 4 Typical single fibre pull-out curve observed from test on natural fibres

4 Static Mechanical Properties

One of the primary objectives of the abundant research works reported in literature regarding polymer matrix composites is the analysis of the mechanical properties, primarily the tensile properties, although when the biocomposite is considered for sandwich and/or structural panels, the flexural properties have also been analysed. As mentioned above, the mechanical properties of a polymer matrix biocomposite are strictly related to the particular type and distribution of the fibres; hence, in the following, the static properties are reported and discussed by separating the biocomposites into three different classes: short fibres of randomly oriented biocomposites, MAT discontinuous fibres biocomposites, and long fibre biocomposites. The so-called class of woven biocomposites is neglected because the use of coarse woven fabrics typically manufactured for the production of craft items is not adequate for the manufacturing of good quality biocomposites to be used in the industrial field.

4.1 Random Short Fibres Biocomposites

Natural short fibres of polymer matrix composites are the class of biocomposites that are used more in the industrial field, owing to its particular properties of low cost, easy manufacturing, and sufficient stiffness; in general, they are manufactured by low-cost extrusion or injection/compression moulding processes and are used for non-structural applications in various industrial fields, in which they are particularly appreciated for their light weight, low cost, and high renewability that allow the environmental problems related to their final disposal to be overcome. As demonstrated in [37], a three-dimensional (3D) random distribution of the reinforcing fibres leads to better mechanical properties with respect to a two-dimensional (2D) random distribution that, therefore, should be avoided. However, the manufacturing process is an important parameter that can influence significantly the quality of such biocomposites, because the presence of internal defects as voids and/or the direct contact between the fibres (i.e. fibre wetting defects) can lead to a significant decrease in the mechanical strength, affected by the easy propagation of local damage mechanisms. Hence, several research works devoted to the optimization of the manufacturing process have been published [1, 2, 35–38]. In general, such short fibre biomaterials exhibit a tensile strength comparable or inferior to that of the simple matrix, along with a tensile modulus that, instead, can be significantly superior to the matrix one. In other words, the reinforcing by randomly oriented short fibres overcomes the limitation of plastics related to their low stiffness, although improvements in terms of mechanical strength are again expected from the industrial users to extend the use of such materials to semi-structural and structural applications. As explained in [37], the mechanical strength of such materials can be potentially improved by four distinct actions:

(1) increase the fibre-matrix adhesion to contrast both the debonding phenomena affecting the transverse fibres (occurring at the initial phase of the local damage), and the partial pull-out of longitudinal fibres occurring in the final phase of the damage process;
(2) increase the fibres length (l), assuring that it is greater than the critical length (l_c): in this case, the composite failure should be caused by fibre failure and not by premature pull-out;
(3) increase the stiffness of the fibres to increment the load fraction supported by the fibres themselves, and decrease the mean stress supported by the matrix, with a consequent reduction of both the transverse debonding and the subsequent matrix failure phenomena;
(4) increase the fibre volume fraction, to decrease, as in the previous action (3), the mean stress on the matrix, with the same advantages.

Action (1) has been considered in literature by various authors, as it is corroborated by the large number of research works [3, 4, 6, 7, 13–19] devoted to the fibre surface treatments aimed to improve the fibre-matrix adhesion. Unfortunately, this approach leads to limited increases in the biocomposite strength (always less than approximately 10–20%), because the strength to the transverse debonding, that is the local damage mechanism with which the damage of short fibre biocomposite starts (see Fig. 5), is always limited by the relatively low matrix strength that is also significantly influenced by the overall quality of the composite, i.e. by voids and the direct contacts between the fibres, and are equivalent to microcracks in the matrix and fibre-matrix interface, respectively.

In detail, the SEM micrograph of Fig. 5b corroborates that after the initial debonding of transverse fibres, the typical damage mechanism involves both the pull-out and tensile failure of the fibres aligned with the applied load.

Unfortunately, action (2) is not effective for natural fibre biocomposites because, owing to the viscosity of the matrix, the manufacturing processes used for the thermosetting (simple matrix-fibres mixing) and thermoplastic matrix (extrusion or similar processes), yields a fibre curvature that increases with fibre length and leads to a significant reduction in the biocomposite strength. Hence, the experimental evidence shows that the maximum strength of the biocomposites is achieved using a fibre length in the range of 3–6 mm [1, 2, 17, 21, 35–38].

Regarding action (3), it is noteworthy that increasing the fibre stiffness by fibre treatment yields limited results with non-negligible increase in the environmental impact; therefore, such an action is not advisable.

For the abovementioned reasons, the only potentially effective action that can increase the strength of the natural short fibre biocomposites is the increment of the fibre volume fraction (action 4), by the contemporary use of a proper manufacturing process that permits the defects concentration to be limited, which tends to increase with V_f. Consequently, as demonstrated in [37], similar results can be obtained by implementing the proper compression moulding process characterized by a curing cycle under a suitable combined pressure. In practice, such an optimized process

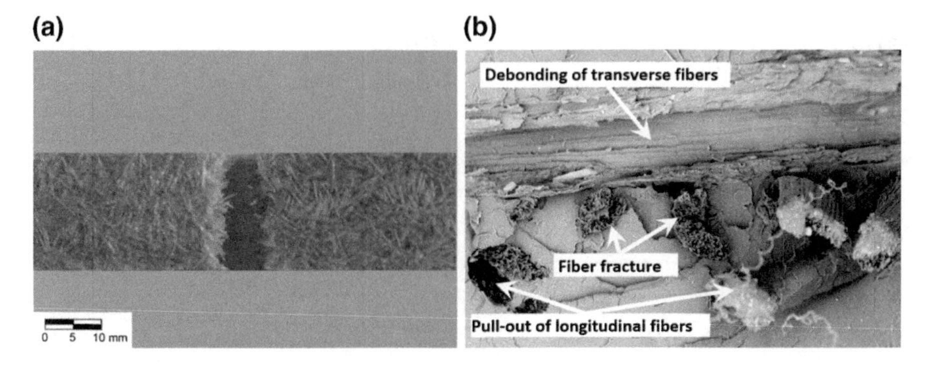

Fig. 5 **a** Typical failure of short fibre biocomposites subject to tensile load and **b** SEM micrograph of the relative fracture surface

should allow the increase in V_f above the typical values (10–20%) of literature, without significant increases in the defects concentration.

As an example, Fig. 6 shows the optimal curing processes (p_c and t_g are the curing pressure and gelling time, respectively) versus V_f, for a green epoxy matrix biocomposite reinforced by sisal fibres [37].

Table 6 shows the static mechanical characteristic of the most random short fibre biocomposites reported in literature.

By varying the fibre type (agave, hemp, jute, sisal, coir, abaca, flax), the maximum tensile strength corresponds to the critical fibre volume fraction $V_{f,crit}$ that falls within the range of 20–35% In practice, the tensile strength varies with V_f as indicated in Fig. 7. Although it refers to agave fibres [37], it can be qualitatively applied to any short natural fibre biocomposites.

Figure 7 shows that the ultimate tensile stress $\sigma_{b,u}$ decreases in the range $0 < V_f < V_{f,min}$, obeying to the simple relationship:

$$\sigma_{b,u} = \sigma_{m,R} \cdot (1 - V_f) \quad (2)$$

whereas for $V_f \geq V_{f,min}$, the tensile strength follows the relationships obtained by considering a [0/±60] quasi-isotropic laminate model [37]:

$$\sigma_{b,u} = \left\{ \left[C_\sigma \cdot \sigma_{pull-out} \cdot V_f + \sigma_{m,u} \cdot (1 - V_f)\right]^{0^\circ} + 2\left[(\sigma_{m,u} \cdot r_a / K)\right]^{\pm 60^\circ} \right\}/3 \quad (3)$$

in which C is a calibration coefficient introduced to consider the biocomposite defects (primarily owing to the incomplete wetting of the free fibres), whereas the subscripts f and m refer to the fibre and matrix, respectively.

The optimization of the manufacturing process can lead to an appreciable improvement in the tensile strength, up to approximately 50% higher than that of the simple matrix, as shown in Fig. 8 for the particular case of epoxy-sisal biocomposite manufactured by an optimized compression moulding [37].

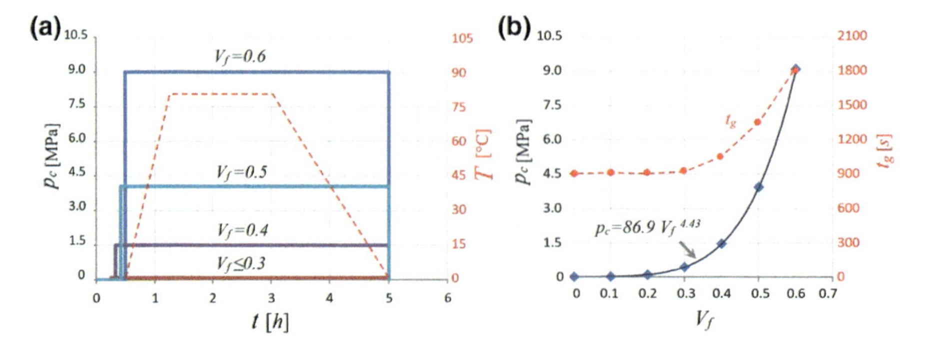

Fig. 6 **a** optimal thermo-mechanical curing processes for short agave fibre biocomposites and **b** relative optimal relationship between p_c and t_g parameters versus V_f

Table 6 Mechanical properties of the short fibre biocomposites reported in literature (matrix property in brackets)

Matrix	Fiber	V_f (%)	Tensile strenght (MPa)	Tensile modulus (GPa)	Elongation (%)	Manufacturing process	Impact strength (kJ/m^2)	References
PP	Agave C	10–30	23–27	1.7–2.2	4–8	Extrusion	2.8–4.1	[55]
PP	Abaca	30	41(30)	5(1.8)	–	Extrusion/injection	–	[49]
PP + PPgMA5%	Agave C	10–30	30–32	1.9–2.5	4–9	Extrusion	2,8	[55]
PP	Agave R	10–30	23–28	1.8–2.2	5–8	Extrusion	2.7–4.2	
PP + PPgMA5%	Agave R	10–30	25–28	1.8–2.3	4–8	Extrusion	3,2	
MAPP 5%	Coir	30	25–31	1.1–1.25	–	Compression moulding		[25]
PLA	Hemp	40	45–55(32)	57–74	–	Compression moulding	9–25	
PLA	Abaca	30	79(65)	3.8	–	Extrusion coating and injection moulding	–	[49]
PBS	Jute	20	27(17)	1.9	4.3	Compression moulding	–	[25]
LLDPE	Flax	10	13–14(13)	–	–	Hand layup	–	
HDPE	Flax	10	20(19)	–	–	Hand layup	–	
HDPE	Hemp	20–40	37–30(22)	3.2–7	–	Twin-screw	–	
Starch	Jute	30	25	2.5	–	Extrusion	–	[35]
PLA	Ramie	30	72	–	–	Injection moulding	–	
	Jute	30	78	9	–	Injection moulding	–	
	Flax	30	50	8	–	Injection moulding	–	
	Hemp treated	52	–	–	–	Injection moulding	–	
PTB	Hemp	30	60	7	–	Extrusion	–	
PTP	Hemp	25	62	9.5	–	Extrusion	–	
PHBV	Jute	30	35	7	–	Extrusion	–	
PLLA	Flax	30	99	8	–	Extrusion	–	

(continued)

Table 6 (continued)

Matrix	Fiber	V_f (%)	Tensile strenght (MPa)	Tensile modulus (GPa)	Elongation (%)	Manufacturing process	Impact strength (kJ/m^2)	References
PHB	Flax	30	38	–	–	Extrusion	–	
Epoxy	Optmized agave	35	75	6	1.2	Hand lay-up	–	
PP	Jute	30	50	6	–	Compression moulding	–	
	Glass	30	80	4.8	–	Compression moulding	–	
	Hemp	40	52	6	–	Compression moulding	–	
	Kenaf	40	28	–	–	Compression moulding	–	
	Flax	30	38	5	–	Compression moulding	88	
Epoxy	Agave opt	50	62	8	–	Hand lay-up	–	
Epoxy	Agave americana	35	37	2.6	1.2	Extrusion	–	[37]
Polyester	Sisal	30	60	2	–	Compression moulding	–	
HDPE	Agave	20	27	0.9	–	Compression moulding	–	
Epoxy	Sisal	10	67	2.5	2.2	Extrusion	–	
Epoxy	Sisal	35	77	10.5	1.2	Compression moulding	–	
HDPE	Flax	20–40	39–32(23)	3.5–6.5	5.2–5.0	Hand lay-up	–	[25]
Epoxy	Agave MPN	30	36	1.9	1.9	Extrusion	–	[2]
PLA	Agave MPN	30	46	2.9	2.1	Extrusion	–	
PP	Hemp + glass	15/5	52–60(30)	3.7–4.4(1.1)	–	Injection moulding	–	[49]
Polyester	Glass/Sisal	/	176	–	–	Compression moulding	–	[9]
Polyester	Glass/Jute	/	229	–	–	Compression moulding	–	[9]
Polyester	Glass/Sisal/Jute	/	200	–	–	Compression moulding	–	[9]

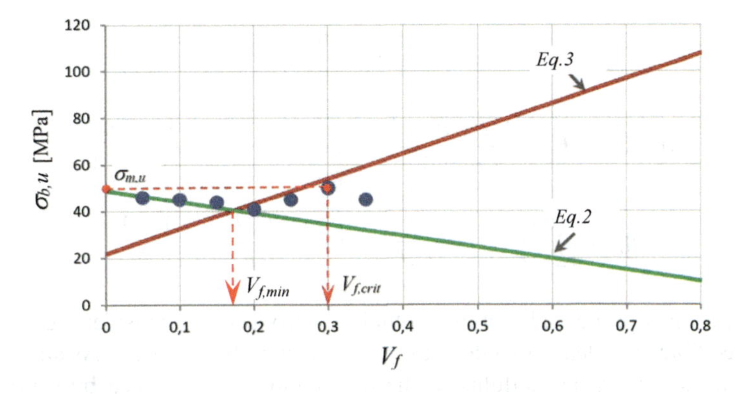

Fig. 7 Strength of short agave fibre biocomposites versus V_f and comparison with theoretical models

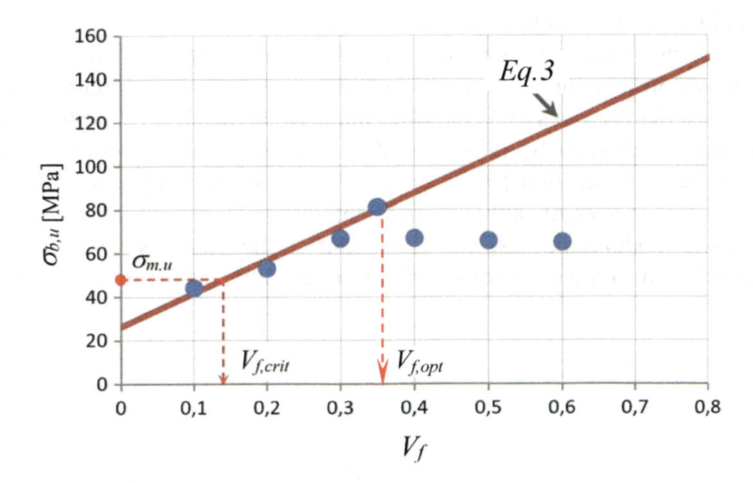

Fig. 8 Strength of short agave fibre biocomposites obtained by an optimized manufacturing process, versus V_f and comparison with the theoretical model of Eq. 3

As shown, using an optimized compression moulding manufacturing process allows the tensile strength to increment significantly; further, in the range $V_{f,crit} < V_f < V_{f,opt}$, the tensile strength obeys again the abovementioned quasi-isotropic model (Eq. 3).

In terms of the tensile modulus, Table 6 shows that the maximum stiffness corresponds to the higher volume fractions that fall within the range of 30–60%. In other words, the maximum tensile strength and the maximum tensile modulus corresponds to different fibre volume fractions that have to be properly chosen if the primary reinforcing objective is to increase the matrix strength or, in turn, increase the matrix stiffness. In general, the tensile modulus varies with V_f in good accordance with the following modified Halpin–Tsai model [37]:

$$E_b = \frac{3}{8}E_L + \frac{5}{8}E_T \quad (4)$$

$$\text{with } E_L = E_m \frac{1 + 2(l/d)\eta_L V_f}{1 - \eta_L V_f} \quad \text{and} \quad E_T = E_m \frac{1 + 2\eta_T V_f}{1 - \eta_T V_f} \quad (5\text{–}6)$$

$$\eta_L = \frac{(C_E E_f/E_m) - 1}{(E_f/E_m) + 2(l/d)}; \quad \eta_T = \frac{(C_E E_f/E_m) - 1}{(E_f/E_m) + 2} \quad (7\text{–}8)$$

in which C_E is a proper calibration coefficient introduced to consider the curvature of the free fibres that leads to a decrease in the actual fibre stiffness. As an example, Fig. 9 shows the tensile modulus of the biocomposites reinforced by short agave fibres versus V_f, along with Eq. 4 and the non-modified Halpin–Tsai model.

As shown, the introduction of the calibration coefficient C_E allows a reliable theoretical relationship to be obtained that can be used for the prediction of the tensile stiffness at the design stage.

Finally, hybrid biocomposites reinforced by natural and synthetic fibres have been proposed in literature by various authors, as sisal/glass, jute/glass, and sisal/ jute/glass (see the last rows of Table 6) to improve the mechanical performance of biocomposites reinforced by only natural fibres. The examination of the results shows that the hybridization allows appreciable increments (up to approximately 25%) of the tensile strength, although more interesting results are obtained in terms of the tensile modulus because the significant elastic modulus mismatch of the fibres do not lead to a contemporary failure of the different reinforcing fibres.

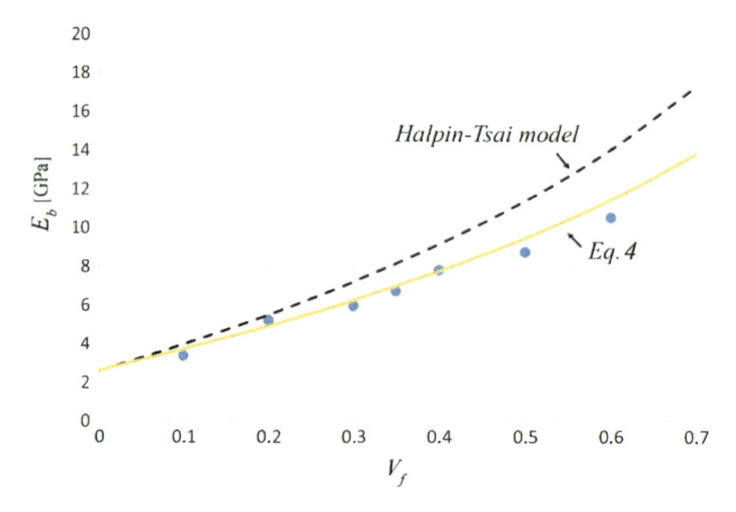

Fig. 9 Tensile modulus of short agave biocomposites and comparison with the theoretical model of Eq. 4

4.2 MAT Biocomposites

With the term 'MAT biocomposites', most authors refer to biocomposites reinforced by 2D MAT fabrics composed of discontinuous and randomly oriented natural fibres . The use of such fabrics enables the user to manufacture biocomposites laminate using hand lay-up or similar processes (hot pressed, etc.). Interesting results have been reported in [25] that consider polypropylene-hemp biocomposites, and in [37] that consider green epoxy/sisal biocomposites (see Table 7).

From such works, it is shown that although using MAT fabrics allows for an easy manufacturing process and a laminate of relatively high dimensions, it produces biocomposites with mechanical strength inferior to that obtained by 3D randomly oriented short fibres (in general, slightly inferior tensile strength compared to that of the simple matrix). This is attributable to the higher defect concentration, especially to the partial fibre wetting owing to the preliminary mutual fibre contact that occurs during the MAT fabric manufacturing. Further studies are required to implement new manufacturing processes that would overcome such drawbacks.

4.3 Long Fibre Biocomposites

As it is well known from micromechanics [51, 52], reinforcement with aligned long fibres is a unique method that can potentially yield high-performance biocomposites. Using such method, biocomposites having tensile strength 3–10 times that of the simple matrix, and tensile modulus that obeys the modified rule of mixture can be obtained. However, a good fibre alignment is an important feature, such that high-quality natural long fibre biocomposites can be obtained only using unidirectional fabrics with well-aligned fibres [35]. Further, limited misalignment can yield significant mechanical properties reduction. Hence, the mechanical properties of similar biocomposites reported in literature vary often in a wide range; for example, as shown in Table 8, the tensile strength of unidirectional epoxy/sisal biocomposites with a high volume fraction can vary from 150 to approximately 470 MPa according to the accuracy of the manufacturing process and fibre alignment.

The influence of the matrix/fibre adhesion on the mechanical properties of long fibre biocomposites is relatively negligible, except for a high percentage of matrix defects or significant delamination [5]. As mentioned above, the ratio between the failure strain of the matrix and fibres, along with the fibre alignment are basic parameters that govern the tensile strength. Unfortunately, accurate unidirectional fabrics of natural fibres are not available in the current market; therefore, the practical exploitation of high-performance biocomposites depends on the future industrial production of such products; for example, Fig. 10 shows the accurate 'stitched' unidirectional fabrics properly produced in laboratory [35].

Table 7 Mechanical properties of MAT biocomposites reported in literature

Matrix	Fiber	V_f (%)	Tensile strenght (MPa)	Tensile modulus (GPa)	Elongation (%)	Manufacturing process	Impact strength (kJ/m^2)	References
PP	Hemp	40–50	51–57(60)	6–7(2.7)	–	Hot pressed	–	[25]
Green epoxy	Optimized agave	30	50(50)	6(3)	0.8	Optimized compression molding	–	[37]
Green epoxy	Optimized agave	35	45(50)	7(3)	0.7	Optimized compression molding	–	[37]

Table 8 Mechanical properties of long fibre biocomposites reported in literature

Matrix	Fiber	V_f (%)	Tensile strenght (MPa)	Tensile modulus (GPa)	Elongation (%)	Manufacturing process	Impact strength (kJ/m^2)	References
PP	Glass	20	64	4.1	–	Hand lay-up	14	[55]
Epoxy	Sisal	47–52	157–180	13.0–15.0	–	Vertical/axial injection	40–47	[56]
Cornstarch	Curaua	69.3	216	13.0	1.5	Compr. moulding DM	–	[57]
		67.0	275	29.0	1.2	Compr. moulding PF	–	
		69.9	327	36.0	1.2	Compr. moulding PS	–	
Epoxy	Jute	40	179			Compression moulding	–	[58]
Epoxy	Sisal	40	177			Compression moulding	–	
Epoxy	Banana	40	102			Compression moulding	–	
PP	Jute	40	38(33)	1.3(0.9)	–	Carding/hot pressing	–	[25]
PP	Kenaf	30–40	32(32)	1.2(0.9)	–	Carding/hot pressing	–	
Epoxy	ArcticFlax	54	275	39.0	–	Hand lay-up	–	[35]
	Sisal treated	73	412	6.0	–	Hand lay-up	–	
	Sisal untreated	77	327	10.0	–	Hand lay-up	–	
Soy protein	Sisal	50	180	3.5	–	Hand lay-up	–	
Polyester	Flax	60	310	28.0	–	Hand lay-up	–	
PCL	Curaua	70	220	13.5	–	Hand lay-up	–	
Green epoxy	Agave	30–70	210–470	13–30	1.8–2.1	Opt. compr. molding	–	
Green epoxy	Agave MPN	30	230	11.1	1.9	Hand lay-up	–	[2]
PLA	Agave MPN	30	241	10.0	2.1	Compression molding	–	

Fig. 10 **a** unidirectional 'stitched' fabric of agave fibre obtained in laboratory

Figure 11 shows the primary parameter of the optimized compression moulding process used to manufacture high-performance unidirectional epoxy/ sisal biocomposites.

The experimental analysis performed in [35] has shown that the use of accurate manufacturing processes leads to high-quality biocomposites whose tensile characteristics obey the modified rule of mixture (MRoM). The longitudinal tensile strength is expressed by the well-known formulas:

$$\sigma_{L,R} = C_d \sigma_{f,u} V_f + \sigma_m(\varepsilon_{f,u})(1 - V_f) \quad (9)$$

in which C_d is the correction coefficient that primarily considers fibre damage and residual misalignment. Figure 12 shows the good accordance between Eq. 9 and the experimental evidence, with deviation of less than 10%.

Figure 13 shows the good accordance between the longitudinal Young modulus and the value predicted by a ROM properly corrected by introducing a corrective calibration coefficient for the fibre modulus [35].

Finally, the example of fracture surface and the SEM micrograph reported in Fig. 14 corroborates how the damage mechanism of high-quality unidirectional biocomposites involves the fracture orthogonal to the longitudinal axis and the expected fibre tensile failure without premature damage mechanisms as pull-out and/or debonding.

Finally, it is remarkable to observe how high-performance biocomposites exhibit tensile strengths of up to approximately 500 MPa (see Fig. 10 as an example or Table 8), i.e. comparable with that of typical construction materials such as steels (400–500 MPa) and good GFRP ($\approx$500 MPa). Further, such a maximum strength is higher than that of many aluminium alloys widely used in the aeronautics and automotive industry (260–410 MPa), and of typical epoxy-glass short fibre composites (50–150 MPa). In terms of the specific tensile strength, it is noteworthy that

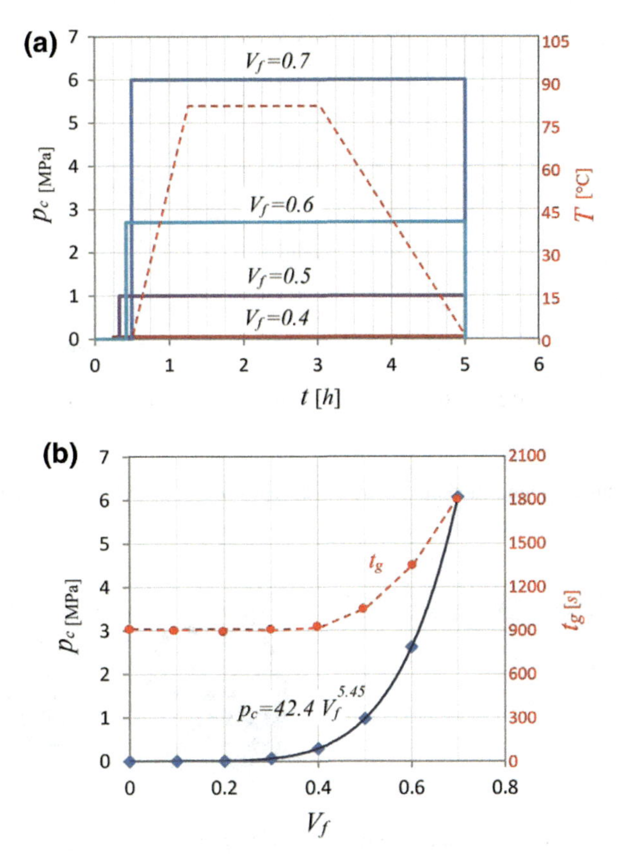

Fig. 11 **a** Diagrams of an optimal curing process for long fibre biocomposites and **b** optimal curing parameters versus V_f

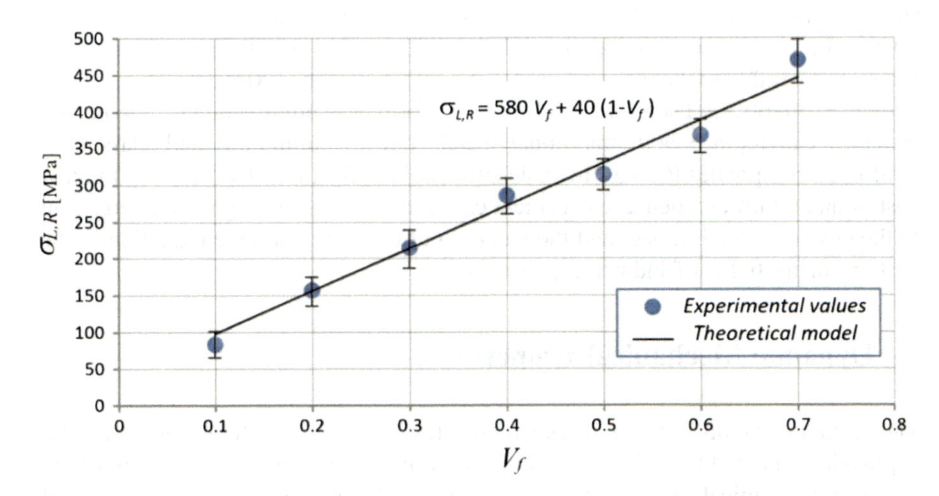

Fig. 12 Tensile strength of long agave fibre biocomposites versus V_f and theoretical model (MROM)

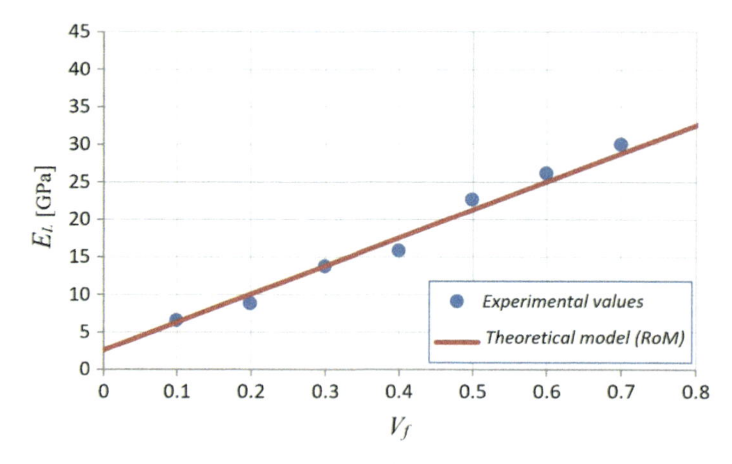

Fig. 13 Longitudinal young modulus of the green epoxy/sisal fibre biocomposites versus fibre volume fraction

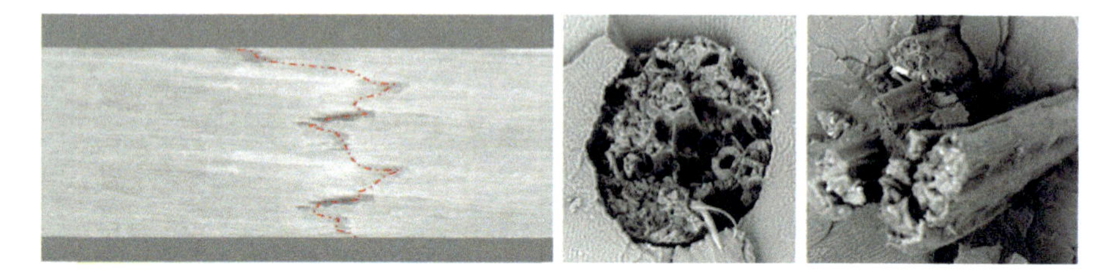

Fig. 14 Typical fracture surface and relative SEM micrographs of the long fibre biocomposites under tensile load

the strength index of approximately $\sigma_{L,R}/\rho = 35 \times 10^3$ m are reached, i.e. values superior not only to that of typical steels (between 5.8×10^3 m and 10.6×10^3 m), but also to that of good aluminium (15×10^3 m approximately). It is also comparable to that of good unidirectional GFRP (approximately 35×10^3 m with $V_f = 50$–60%). Therefore, such high performance unidirectional biocomposites [35] can replace both traditional metals (steel, aluminium) and glass fibre reinforced composites (GFRP) in static structural applications. Considering the low cost of natural fibres, such a replacement can lead not only to a significant reduction in the environmental impact and the weight, but also a remarkable reduction in the costs in many fields of industrial production.

5 Dynamic Mechanical Properties

The need to extend the use of biocomposites for semi-structural and structural applications has led to an increased interest in the analysis and optimization of the dynamic mechanical performance of such materials, especially in the presence of fatigue and impact loading. Considering the impact strength, the research works reported in literature (see also Tables 6, 7 and 8) show that the natural fibres have a relatively high strength to impact, especially the fibres (agave, etc.) constituted by subfibers that, like aramidic fibres, lead to high fracture energy. In more detail, the reinforcing by natural fibres yields appreciable increments of the impact strength of the thermosetting matrixes (see Tables 6, 7 and 8), up to one order of magnitude, whereas negligible improvements are obtained for the thermoplastics matrixes. Further studies are however necessary for a more reliable analysis of the impact strength of most biocomposites, by varying the primary parameters as fibre volume fraction, manufacturing process, etc.

Considering the fatigue strength, a few research works have been reported hitherto, and further studies are necessary to deeply understand the fatigue behaviour of biocomposites. However, the research activities reported in literature, has shown firstly that, as expected, the fatigue behaviour of the polymer matrix biocomposites is strictly related to various parameters such as matrix fatigue strength, fibre orientation, laminate lay-up, and manufacturing process. As observed from Table 9, the fatigue properties of good quality short fibre biocomposites is relatively good with respect to metal and other technical materials, which exhibits a fatigue ratio of approximately 0.5; the fatigue ratio of the analysed biocomposites, in fact, varies in the range of 0.7–0.8. Owing to the high fatigue ratio, we can conclude that the reinforcing of polymeric matrixes with natural fibres allows the user to increment significantly the fatigue strength of the matrixes; considering that the fatigue ratio of the matrixes varies in the range of 0.15–0.30 (typical plastics), whereas the fatigue ratio of the biocomposites varies in the range of 0.70–0.80. It follows that for a typical biocomposite having static strength equal to that of the matrix, the fibre reinforcing leads to a significant improvement in the fatigue strength, from 3 to 5 times the value of the simple matrix. As an example, the HDPE matrix shown in Table 6 exhibits a fatigue limit of approximately 5 MPa whereas the corresponding biocomposite exhibits a fatigue limit of approximately 20–24 MPa, i.e. approximately 4–5 times the value of the simple matrix. Low quality biocomposites obtained by compression moulding, lead instead to low fatigue strength with fatigue ratio less than 0.55 (see last rows of Table 9). Therefore, it is possible to state that although the reinforcing polymer matrixes by the natural fibres do not lead in general to appreciable improvements in the static strength, good quality biocomposites leads to interesting strength increments in terms of fatigue strenght.

Regarding the long fibre biocomposites, only a few authors have performed fatigue analysis and further studies are necessary to gain a complete knowledge of the fatigue behaviour of such materials. As an example, Table 10 shows some data relative to polyester and epoxy reinforced by sisal, with and without alkali treatment [53], and by hemp, jute, kenaf and flax with different lay-ups.

As shown in Table 10, the unidirectional (UD) long natural sisal fibre biocomposites exhibit fatigue ratio in the range of 0.40–0.58, i.e. lower than that of good-quality short fibre biocomposites; similar results are shown for jute, flax and kenaf, whereas better results are obtained for good quality biocomposites reinforced

Table 9 Fatigue properties of short fibre biocomposites reported in literature

Matrix	Fiber	V_f (%)	Tensile strenght (MPa)	Fatigue limit 10^6 cycles (MPa)	Manufacturing process	References
HDPE	Hemp	20–40	29–30(28)	22–24(0.80)	Injection moulding	[59]
HDPE (moisture)	Hemp	20	27.5	19,5(0.70)	Injection moulding	
Polyester (Al_2O_3)	Hemp/Jute	0	24–41(23)	16.5–31(0.75)	Injection moulding	[60]
PBT	Glass 0°	30	80(60)	40(0.5)	Injection moulding	[61]
PBT	Glass 18°	30	78(60)	35(0.45)	Injection moulding	
PBT	Glass 45°	30	50(60)	25(0.50)	Injection moulding	
PBT	Glass 90°	30	48(60)	24(0.50)	Injection moulding	
PA6 + rubber 10%	Glass0°	35	55(150)	27(0.50)	Injection moulding	
PA6 + rubber 10%	Glass 18°	35	50(150)	25(0.50)	Injection moulding	
PA6 + rubber 10%	Glass 45°	35	33(150)	19(0.55)	Injection moulding	
PA6 + rubber 10%	Glass 90°	35	30(150)	16(0.50)	Injection moulding	
PA6 + rubber 10%	Short birch	40	55(35)	22(0.45)	Injection moulding	[62]
HDPE	Short birch	50	37(30)	15(0.40)	Compres. moulding	[63]
Polyester	Hemp	55	50(32)	19(0.38)	Compres. moulding	[64]
Polyester	Glass	55	200(32)	50(0.25)	Compres. moulding	[64]
Polyester	Jute	40	47(32)	23.5(0.50)	Compres. moulding	[70]

Table 10 Fatigue properties of long fibre biocomposites reported in literature

Matrix	Fiber	V_f (%)	Tensile strenght (MPa)	Fatigue limit 10^6 cycles (MPa)	Manufacturing process	References
Polyester	Sisal (UD)	68.2	255(89)	100(0.40)	Moulding hot press	[53]
Polyester	Sisal (UD)	64.4	225(89)	120(0.54)	Moulding hot press	
Epoxy	Sisal (UD)	71.5	350(95)	139(0.40)	Moulding hot press	
Epoxy	Sisal (UD)	68.5	340(95)	159(0.47)	Moulding hot press	
Epoxy	Flax $[\pm45]_{16}$	55	75(60)	44(0.58)	Hand lay-up	[66]
Polyester	Sisal $[0/90]_S$	50	102(32)	56(0.55)	Low press. molding	[67]
Polyester	Hemp	40	170(32)	130(0.76)	Compres. moulding	[68]
Polyester	Jute	40	225(32)	110(0.48)	Compres. moulding	[68]
Epoxy	Flax	40	150(60)	65(0.43)	Compres. moulding	[68]
Epoxy	Kenaf/Kevlar	35	216(73)	130(0.60)	Compres. moulding	[71]

by hemp. However, owing to the high static strength increment from the long fibre, it is possible to state that the reinforcement with long natural fibres leads to significant fatigue strength increment of the polymer matrix that is always higher than that obtained by short fibre biocomposites. Considering that the static strength of good-quality long fibre biocomposites is in general 3–5 times that of the matrix one, it is possible to state that the reinforcing by long natural fibre leads to an increment in the fatigue life of the matrix by approximately one order of magnitude. As an example, the reinforcing by sisal of a polyester matrix [53] having a fatigue strength of approximately 12 MPa, leads to a biocomposite with a fatigue strength of 120 MPa (see also Table 10). Further, in the fatigue load condition, the scattering of the experimental data of biocomposites are comparable with those of other materials such as metals or synthetic composites, and a statistical approach involving simple models such as the normal or Weibull distribution, can be used for a better prediction of the fatigue behaviour [54]. Finally, it is noteworthy that such results are extremely attractive especially by considering the low cost of the natural fibres; unlike synthetic composites, the cost of the biocomposites is lower than that of the simple matrix, i.e. unlike synthetic fibres, the fatigue properties increment obtained by natural fibres is not associated with a significant cost increment.

6 Conclusions

The critical analysis of the static and dynamic mechanical properties of polymer matrix composites reported in literature has shown that such properties are strictly affected by the corresponding properties of the matrix and fibre, as well as by the actual fibre/matrix adhesion whose correct estimation requires the consideration of the nonuniform shear stress distribution that occurs at the fibre/matrix interface. Such a distribution is significantly influenced by the fibre/matrix stiffness ratio such that the pull-out load increases with the fibre stiffness, and such a parameter is not a reliable parameter to the correct estimation of the fibre/matrix adhesion.

In general, the mechanical strength of the short fibre biocomposites is comparable with that of the simple matrix and an appreciable increment can be obtained only by using the proper manufacturing processes that allow the user to increase the fibre volume fraction (until approximately 35% at least) without increasing the typical defects such as voids and incomplete wetting of the free fibres. Hence, recent research activities have shown that a strength increment of approximately 60% can be obtained, as it has been demonstrated for green epoxy/sisal biocomposites. Further, a 3D random distribution of the fibres is always better than a 2D random distribution; therefore, the latter should be avoided during the biocomposite manufacturing. Moreover, although the use of MAT fabrics (2D discontinuous random fibres) allows the user to manufacture high-dimension laminates with hand lay-up or similar manufacturing processes, the diffuse wetting defects lead to mechanical strength that is always less than that of the randomly oriented short fibre biocomposites. Consequently, this configuration is not advisable if the maximum mechanical strength is the primary objective of the reinforcement. However, the best static and dynamic mechanical performances are shown by the long fibre biocomposites. Such configuration is less affected by the limited matrix/fibre adhesion and, if an optimized manufacturing process is used along with the appropriate matrix/fibre couple (having a sufficient stiffness mismatch and adequate failure strains), it can lead to unidirectional high-performance biocomposites with a tensile strength up to approximately 500 MPa, with a tensile modulus of approximately 30 GPa or higher. In this case, an important influence parameter is the fibre alignment and the use of free fibre must be avoided; instead, it should be substituted by high-quality unidirectional fabrics. The low specific weight of the natural fibres (always less than that of the synthetic fibres) leads to long fibre biocomposites with a strength index superior to that of typical steels, aluminium alloys, and GFRP; consequently, these biocomposites can replace such materials (metals an composites) in various industrial applications.

References

1. Zuccarello B, Zingales M (2017) Toward high performance renewable agave reinforced biocomposite: optimization of fiber performance and fiber-matrix adhesion analysis. Compos B 122:109–120
2. Zuccarello B, Scaffaro R (2017) Experimental analysis and micromechanical models of high performance renewable agave reinforced biocomposites. Compos Part B 119:141–152
3. Herrera-Franco PJ, Valadez-Gonzalez A (2005) A study of the mechanical properties of short natural-fiber reinforced composites. Compos B 36:597–608
4. Sreekumar PA, Joseph K, Unnikrishnan G, Thomas S (2007) A comparative study on mechanical properties of sisal-leaf fiber reinforced polyester composites prepared by resin transfer and compression moulding techniques. Compos Sci Technol 67:453461
5. Cirello A, Zuccarello B (2006) On the effects of a crack propagating toward the interface of a bimaterial system. Eng Fract Mech 73:1264–1277
6. Murherjee PS, Satyananrayana KG (1987) Structure and properties of some vegetable fibres, part 1. Sisal fibres. J Mater Sci 19:3925–3934
7. Belaadi A, Bezazi A, Bourchak M, Scarpa F, Boba K (2014) Novel extraction techniques, chemical and mechanical characterization of agave Americana L. natural fibres. Compos Part B 66:194–203
8. Chand N, Hashimi SAR (1993) Mechanical properties of sisal fibres at elevated temperatures. J Mater Sci 28:6724–6728
9. Chan N, Verma S, Khazanchi AC (1989) SEM and strength characteristic of acetylated sisal fiber. J Mater Sci Lett 8:1307–1309
10. Silva FA, Chawla N, Filho RDT (2008) Tensile behavior of high performance natural (sisal) fibers. Compos Sci Technol 68:3438–3443
11. Thomason JL, Carruthers J, Kelly J, Johnson G (2008) Fibre cross-section determination and variability in sisal and lax and its effects on fibre performance characterization. Compos Sci Technol 71:1008–1015
12. Belaadi A, Bezazi A, Bourchak M, Scarpa F, Zhu C (2008) Thermochemical and statistical mechanical properties of natural sisal fibres. Compos Part B 67:481–489
13. Kaewkuk S, Sutapun W, Jarukmjorn K (2013) Effects of interfacial modification and fiber content on physical properties of sisal fiber/polypropylene composites. Compos Part B 45:544–549

14. Bisanda ETN, Ansell MO (1999) The effect of silane treatment on the mechanical and physical properties of sisal-epoxy composites. Compos Sci Technol 41:165–168
15. Joseph K, Thomas S, Pavithran C (1996) Effect of chemical treatment on the tensile properties of short sisal fiber-reinforced poly-ethylene composites. Polymer 37:5139–5149
16. Singh B, Gupta M, Verma A (1996) Influence of fibre surface treatment on the properties of sisal-polyester composites. Polym Compos 17:910–918
17. Mylsamy K, Rajendran I (2011) Influence of alkali treatment and fibre length on mechanical properties of short agave fibre reinforced epoxy composites. Mater Des 32:4629–4640
18. Kim JT, Netravali AN (2010) Mercerization of sisal fibers: effect of tension on mechanical properties of sisal fiber and fiber-reinforced composites. Compos Part A 41:1245–1252
19. Sood M, Dwivedi G (2017) Effect of fiber treatment on flexural properties of natural fiber reinforced composites: a review. Egyp J Pet https://doi.org/10.1016/j.ejpe.2017.11.005
20. Furqan A, Heung SC, Myung KP (2015) A review: natural fiber composites selection in view of mechanical, light weight, and economic properties. Macromol Mater Eng 300:10–24
21. Li Y, Mai Y-W, Ye L (2000) Sisal fibre and its composites: a review of recent developments. Compos Sci Technol 60:2037–2055
22. Omrani E, Menezes PL, Rohatgi PK (2015) State of the art on tribological behavior of polymer matrix composites reinforced with natural fibers in the green materials world. Eng Sci Technol 21:165–175
23. Koronis G, Silva A, Fontul M (2013) Green composites: a review of adequate materials for automotive applications. Compos Part B 44:120–127
24. Nabi D, Jog JP (2017) Natural fiber polymer composites: a review. Adv Polym Technol 18 (4):351–363
25. Ku H, Wang H, Pattarachaiyakoop N (2011) A review on the ensile properties of natural fiber reinforced polymer composites. Compos Part B 42:856–873
26. Jagadeesh D, Kanny K, Prashantha, K (2017) A review on research and development of green composites from plant protein-based polymers. Polym Compos 38:1505–1518
27. Bharath KK, Basavarajappa S (2015) Applications of biocomposite materials based on natural fibers from renewable resources: a review. Sci Eng Compos Mater 23:1–10
28. Ramesha M, Palanikumarb K, Hemachandra Reddy KK (2017) Plant fibre based bio-composites: sustainable and renewable green materials. Renew Sustain Energy Rev 79:558–584
29. Sharath Shekar HS, Ramachandra M (2018) Green composites: a review. Mater Today Proc 5:2518–2526
30. Benzait Z, Trabzon L (2018) A review of recent research on materials used in polymer-matrix composites for body armor application. J Compos Mater, pp 1–23 https://doi.org/10.1177/0021998318764002
31. Sanjay MR, Madhu P, Jawaid M, Sentamaraikannan P, Senthil S, Pradeep S (2018) Characterization and properties of natural fiber polymer composites: a comprensive review. J Clean Prod 172:566–581
32. Khan T, Sultan MTBH, Ariffin AH (2018) The challenges of natural fiber in manufacturing, material selection, and technology application: a review. J Reinf Plast Compos 37:770–779
33. Layth M, Ansari NMN, Pua G, Jawaid M, Saiful Islam M (2015) A review on natural fiber reinforced polymer composite and its applications. Int J Polymer Sci, 15 https://doi.org/10.1155/2015/243947
34. Nunna S, Chandra PR, Shrivastava S, Jalan A (2012) A review on mechanical behaviour of natural fiber based hybrid composites. Reinf Palstics Compos 31(11):759–769
35. Zuccarello B, Marannano G, Mancino A (2018) Optimal manufacturing of high performance biocomposites reinforced by sisal fibers. Compos Struct 194:575–583
36. Mancino A, Marannano G, Zuccarello B (2017) Analisi del comportamento meccanico di diverse varietà di fibre di agave e dei relativi biocompositi ecosostenibili. In: Proceedings of the 46° Aias national congress 2017
37. Zuccarello B, Marannano G (2018) Random short fiber biocomposites: optimal manufacturing process and reliable theoretical models. Mater Des 149:87–100

38. Zuccarello B, Mancino A, Marannano G (2017) Implementation of eco-sustainable biocomposite materials reinforced by optimized agave fibers. Struct Int Proc 8:526–538
39. Pantano A, Zuccarello B (2017) Numerical model for the characterization of biocomposites reinforced by sisal fibers. Struct Int Proc 8:517–525
40. Di Landro L, Janszen G (2014) Composites with hemp reinforcement and bio-based epoxy matrix. Compos B Eng 67:220–226
41. Chang KH (2015) Development of a bio-based composite material from soybean oil and keratin fibers. J Polym Sci 95:1524–1538
42. Haq M, Burgueño R, Mohanty AK, Misra M (2008) Hybrid bio-based composites from blends of unsaturated polyester and soybean oil reinforced with nanoclay and natural fibers. Compos Sci 68(15):3344–3351
43. Feldman M, Bledzki K (2014) Bio-based polyamides reinforced with cellulosic fibres—processing and properties. Compos Sci Technol 100:113–120
44. Patel M, Marini L (2015) Life-cycle assessment of bio-based polymers and natural fiber composites. Wiley Online Publications. https://doi.org/10.1002/3527600035
45. Satyanarayana KG, Arizaga GGC, Wypych F (2009) Biodegradable composites based on lignocellulosic fibers—an overview. Prog Polym Sci 34:982–1201
46. Yan Libo, Kasal Bohumil, Huang Liang (2016) A review of recent research on the use of cellulosic fibers, their fibre fabric reinforced cementitious, geo-polymer and polymer composites in civil engineering. Compos B 92:94–132
47. Dicker MPM, Duckworth PF, Baker AB, Francois G, Hazard MK, Weaver PM (2014) Green composites: a review of material attributes and complementary applications. Compos Part A 56:280–289
48. Mohanty AK, Misra M, Drzal LT (2005) Natural fibers, biopolymers and biocomposites. CRC Press
49. Faruk O, Bledzki AK, Fink HP, Sain M (2012) Biocomposites reinforced with natural fibers: 2000-2010. Progr Polim Sci 37:1552–1596
50. Ramzy A, Beermann D, Steuernagel L, Meiners D, Ziegmann G (2014) Developing a new generation of sisal composite fibers for use in industrial application. Compos Part B 66:287–298
51. Agarwal BD, Broutman LJ (1998) Analysis and performance of fiber composites. Wiley, New York
52. Barbero EJ (1999) Introduction to composite materials design. Taylor & Francis, Ann Arbor, MI
53. Towo AN, Ansell MP (2008) Fatigue and evaluation and dynamic mechanical thermal analysis of sisal fibre-thermosetting resin composites. Compos Sci Technol 68:925–932
54. Torres JP, Vandi LJ, Deidt M, Heitzmann MT (2017) The mechanical properties of natural fibre composite laminates: a statistical study. Compos Part A 98:99–104
55. Langhorst AE, Burkholder J, Long J, Thomas R, Kiziltas A, Mielewski D (2017) Blue-agave fiber-reinforced polypropylene composites for automotive applications. Bioresources 13:820–835
56. Yan L, Hao M, Yiou S, Qian L, Zhuoyuan Z (2015) Effect of resin inside fiber lumen on the mechanical properties of sisal fiber reinforced composites. Compos Sci Technol 108:32–40
57. Gomes A, Matsuo T, Goda K, Ohgi J (2007) Development and effect of alkali treatment on tensile properties of curaua fiber green composites. Compos Part A 38:1811–1820
58. Singh JIP, Dhawan V, Singh S, Jangid K (2017) Study of effect of surface treatment on mechanical properties of natural fiber reinforced composites. Mater Today 4:2793–2799
59. Fotouh A, Wolodko JD, Lipsett MG (2014) Fatigue of natural fiber thermoplastic composites. Compos B 62:175–182
60. Bendigeri C, Jwalesh HN (2016) Review on fatigue behavior of polymeric biomaterials with natural fibers. Int J Adv Eng Res Sci 3(2):2349–6495
61. Mortazavian S, Fatemi A (2017) Fatigue of short fiber thermoplastic composites: a review of recent experimental results analysis. Int J Fatigue 102:171–183

62. Mejri M, Toubal L, Cuillière JC, Francois V (2017) Fatigue life and residual strength of a short-natural-fiber-reinforced plastic vs. nylon. Compos B 110:429–441
63. Bravo A, Toubal L, Koffi D, Erchiqui F (2018) Gear fatigue life and thermomechanical behavior of novel green and bio-composite materials VS high-performance thermoplastics. Polym Testing 66:403–414
64. Shahzad A, Isaac DH (2014) Fatigue properties of hemp and fiber composites. Polym Compos 35:1926–1934
65. Belkacemi C, Bezzazi B (2014) Quasi-static mechanical characterization and fatigue of a composites laminates. Adv Appl Sci Res 5(3):328–335
66. Boughera B, Sawi IE, Fawaz Z, Maraghni F (2015) Investigation and modeling of the fatigue damage in natural fiber composites. In: TMS middle east—mediterranen materials congress on energy and infrastructure systems 2015, Doha, Qatar
67. Belaadi A, Bezazi A, Maache A, Scarpa F (2014) Fatigue in sisal fiber reinforced polyester composites: hysteresis and energy dissipation. Proc Eng Part B 74:325–328; 67:481-489
68. Mahboob Z, Bougherara H (2018) Fatigue of flax-epoxy and other plant fibre composites: critical review and analysis. Compos Part A 109:440–462
69. Saman S, Sahrba MJ, Leman Z, Sultan MTH, Ishak MR, Cardona F (2016) Tension-compression fatigue behaviour of plain woven kena/Kevlar hybrid composites. BioResource 11(2):3575–3586
70. Milanese AC, Cioffi MOH, Woorwald HJC (2012) Thermal and mechanical behavior of sisal/ phenolic composites. Compos Part B 43:2843–2850
71. Etaati A, Pather S, Fang Z, Wang H (2014) The study of fiber/matrix bond strength in short hemp polypropilene composites from dynamic mechanical analysis. Compos B 62:19–28

Synthesis, Characterization, and Applications of Hemicellulose Based Eco-friendly Polymer Composites

Busra Balli, Mehmet Harbi Calimli, Esra Kuyuldar and Fatih Sen

1 Introduction

Generally, composites offer very promising solutions to the ongoing problems of the world[1–90]. Fuel cells [30–32, 90] materials for catalysis [79, 33–35, 80–86, 90–96], capacitors [20], solar cells [25, 26], thermopower applications [2], sensors [9, 10, 13, 42, 90, 94] are some examples of these solutions. Materials are also used corporately in combining their superior properties. Metal-metal combinations, polymer-metal combinations and their some hybrids with carbon-based materials are used for variety of nanomaterial applications [6, 7, 14–17, 23, 55, 78–84].

Composite materials are also widely used in developing appropriate materials for various applications as conductive polymer and other electronically active material composites. Graphene and graphene oxide [4, 27, 45–47], carbon nanotubes (CNTs) [14–17, 45–47, 79–85, 93, 95], activated carbon (AC) [14–17, 33], vulcanized carbon (VC) [30–32], carbon black [9, 10] and graphene derived materials which have different structure and morphologies as reduced graphene oxide [3, 30–32, 37, 45–47], graphene nanosheets, graphene nanoribbons, graphene nanoplatelets may thought as active materials for different applications. Also, a wide variety of nanoparticles are utilized in composite systems [34, 35, 73, 74, 85–89, 91–93, 95, 96]. Besides, biodegradable polymers or polymer composites (PCs) produced using renewable materials are referred to as eco-friendly polymer composites (EFPCs). Since they have a carbon-neutral lifecycle, their use as an alternative to petroleum-based materials can result in the reduction of carbon dioxide emissions. Having a net zero carbon footprint, EFPCs are also good materials for the envi-

B. Balli · E. Kuyuldar · F. Sen (✉)
Sen Research Group, Biochemistry Department, Faculty of Arts and Science, Dumlupınar University, Evliya Çelebi Campus, 43100 Kütahya, Turkey
e-mail: fatih.sen@dpu.edu.tr

M. H. Calimli
Tuzluca Vocational School, Igdir University, Igdir, Turkey

Inamuddin et al. (eds.), *Sustainable Polymer Composites and Nanocomposites*,
https://doi.org/10.1007/978-3-030-05399-4_10

ronment and surrounding eco-systems [97–109]. It is an unavoidable fact that the world is currently facing serious challenges with regard to the environmental, social and political crisis. Another pressing issue that needs to be tackled by people is that of energy. Today, a large part of the world depends on fossil fuels to meet the demand for energy [110–115]. However, fossil-based energy sources are becoming exhausted at a great speed. Therefore, it is imperative that not only scientific organizations but also public and private sectors focus on the innovation of eco-friendly materials, also referred to as "green materials." This research area has recently witnessed important accomplishments thanks to scientists and engineers who have realized that making environmentally responsible decisions is the ultimate way to protect the environment and to prevent energy crises [108]. Figure 1 shows the world energy consumption according to energy sources.

Production of environmentally friendly materials requires different fields to participate in the process. These fields engage in sustainable chemistry and work on biodegradable and bio-based materials structurally, chemically, and physically [116–120]. They also pay close attention to make sure that environmentally hazardous substances are not included in any stage of the design and production process. The fields with which we are concerned are EFPCs, sustainable chemistry, natural resources and eco-friendly engineering processes, etc. Animal and plant products can be used to manufacture many natural polymers. These materials have various appealing features. They are ubiquitous, do not contain synthetic substances, can be recycled and consumed by microorganisms. In the last ten years, numerous studies have been carried out on the production of polymeric materials (blends, composites, and nanocomposites). These studies mostly conducted experiments polymers and fillers were mixed. When resin systems are integrated with reinforcing materials, EFPCs achieve excellent features. The combination of resin systems and reinforcing fiber/filler particles results in the composite material integrating the features of resin features and of fibers/fillers. In a composite material, the resin matrix distributes the load on the composite between fiber/filler particles.

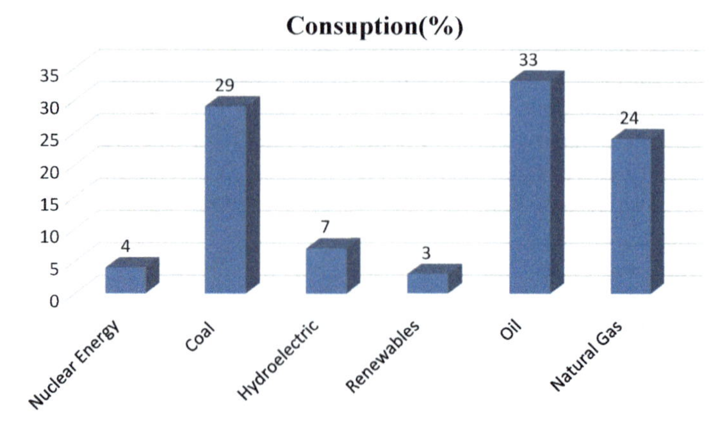

Fig. 1 The primary energy consumption in the world

In addition, it prevents fibers from being damaged due to abrasion and impact and enhances the load-bearing capacity of the composite. The resulting composite has high hardness and resistance and low density and is easy to use. Therefore, it is a better material than metals to be used in the manufacture of various products.

EFPCs have the high specific strength and specific stiffness, high fracture, abrasion, impact, corrosion and fatigue resistance, and low cost.

The greatest appeal of eco-friendly composites is that they are environmentally friendly, completely degradable and sustainable. Baillie addressed the construction and biorhythm evaluation of green composites in detail [8]. EFPCs might be employed effectively for various purposes such as short life-cycle goods manufactured in large quantities and disposable goods. Cellulose, NR, polysaccharides, starch, chitin, proteins, collagens/gelatin, and lignin are natural and biodegradable chemical compounds while polyamides, polyvinyl alcohol, polyvinyl acetate, polyglycolic acid, and polylactic acid are synthetic compounds. New fibre-reinforced materials are made from renewable resources and used in almost all areas [107].

Being superior to synthetic raw materials, EFPCs can play a significant role in the elimination of environmental problems, especially caused by plastic waste, and in the sustainable development of eco-friendly and cost-effective technologies [69]. Numerous studies are conducted on EFPCs to find innovative ways to manufacture eco-friendly materials from biodegradable polymers and renewable sources.

Eco-friendly Polymer Nanocomposites (EPNs) have also drawn enormous attention owing to their potential application in the field of agriculture such as product waste management [118]. EPNs can also be employed to improve the environmental compatibility of recycling processes.

There is the classification of six types of the most common EPN, including (i) EPN with Green Fillers and (ii) EPN with Green-Base-Composite such as—EPN from Cellulose,—EPN from Thermoplastic Starch,—EPN from Polylactic Acid,—EPN from Polymer Mixture,—EPN from Others (Gelatin, PHB, Chitosan).

2 Definitions About Hemicellulose and Derivatives

Of natural polymers, lignocellulosic is likely to be a very important material as it can be converted into biofuels and bioproducts. Mainly found in lignocellulose, hemicelluloses are the second most abundant renewable material in nature, easy to use, has film-forming features, good biocompatibility, and biodegradability [118, 119]. Hemicelluloses contain a very large quantity of free hydroxyl groups, which are perfect for chemical modification. Materials with exclusive features can be manufactured using chemically modified hemicelluloses in order to improve the biopolymers [39, 75, 101]. In recent years, hemicelluloses-based biocompatible films have been popular owing to reduced cost, oxygen barrier features and ease of access [18, 51].

Natural biopolymers are superior to synthetic ones in the sense that the former is more cost-effective, biocompatible, harmless and biodegradable. Being the non-cellulose cell-wall polysaccharides of agricultural and forest plants, hemicelluloses are regarded as unlimited and sustainable materials for the manufacturing of biopolymers and biomaterials [29, 70].

However, hemicelluloses contain a very large amount of hydroxyl groups distributed along the central and terminal region. In addition, these hydroxyl groups generate intermolecular and intramolecular hydrogen bonds [70, 77]. Hemicellulosic materials are non-hydrophobic, semi-crystalline and hygroscopic and have low mechanical features [36]. Due to these disadvantages, hemicelluloses films cannot be used widely. Plasticizers are generally used to enhance the mechanical features of films consist of hemicellulose [43, 44, 120].

As the main hemicelluloses in hardwood and annuals, xylan-type hemicelluloses are regarded as appropriate sources to achieve films. Emulsification and coating processes were employed to achieve arabinoxylan-based films [71, 72]. Lignin or glycerol was used to produce pure xylan, yet, self-supporting films [43, 44]. Table 1 shows the content of some lignocellulosic materials.

Plasticized films from aspen glucuronoxylan exhibit low oxygen permeability and therefore might be ideal for food packaging materials [49]. Up to 40% xylan in wheat gluten was used to produce composite films [57]. In addition, chemical modifications were performed in order to reduce the moisture sensitivity of

Table 1 Some lignocellulosic materials and their contents

Natural fibre	Cellulose (wt %)	Hemicellulose (wt %)	Lignin (wt %)	Total (wt %)
Cotton	89.7	1.0	2.7	93.4
Flax	80	13	2	95
Hemp	74.1	7.6	2.2	83.9
Sugar cane	51.8	27.6	10.7	90.1
Bamboo	54.6	11.4	21.7	87.7
Coconut	51.3	11.7	30.7	93.7
Wheat straw	38	36	22	96
	38.0	29.0	15.0	82
	33.2	24.0	15.1	72.3
	28.8	39.1	18.6	86.5
Rape straw	36	37	24	97
	37.6	31.4	21.3	90.3
Spruce + bark	42	27	26	95
	41.0	24.3	30.0	95.3
	50.8	21.2	27.5	99.5
Poplar wood	50	30	≤ 20	≤ 100

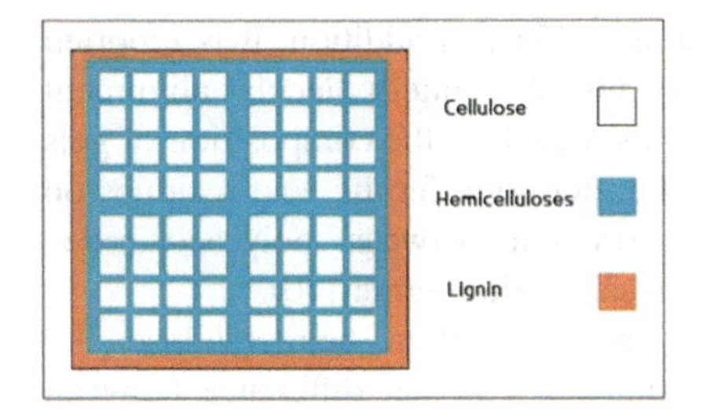

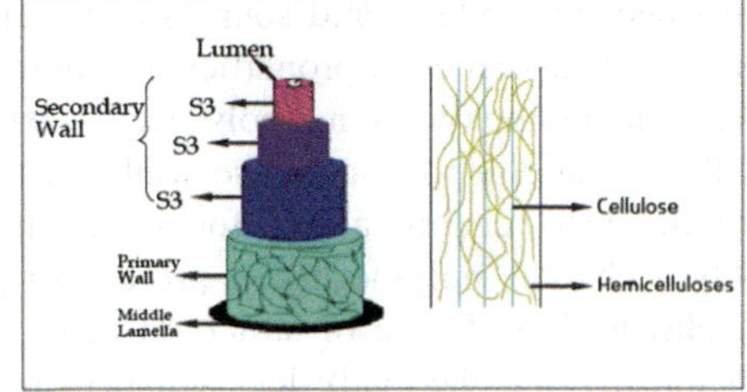

Fig. 2 The schematic representation of hierarchical structure of wood cell

hemicellulose-based films [50, 52]. The molecular structure of hemicelluloses had a large effect on film formation, structure, mechanical features, and moisture content. Side groups separated the xylan or mannan chains, and therefore, hemicellulose where the backbone was least substituted with arabinose or galactose was the most interactive one [29, 53, 66, 102].

Hemicelluloses can also be described in more detail and seen schematically in Fig. 2. They are the second most abundant polysaccharides in biomass. They are biosynthesized by trees and terrestrial plants. Despite the fact that they are in large quantity, hemicelluloses are little used by industries. Hemicelluloses are not completely degraded in most of the lignocellulosic refining processes, but they are removed or further used as feed raw material. Nevertheless, some methods have been developed to separate and isolate high molar mass hemicelluloses. Alkalines are used to extract hemicelluloses from plants. Water is also used to extract some parts of hemicelluloses. Forests and agricultural areas provide a large number of raw materials and studies have been and are being conducted to find potential application areas for it. Potential areas of use of hemicelluloses are paper, packaging, coatings, hydrogels, absorbents, and emulsifiers. As a group of polysaccharides, hemicelluloses have a positive effect on cell wall flexibility by cross-linking cellulose and lignin. They compose the hydrophilic component of the cell wall. They have categorized into different groups according to their structure: xylans, mannans, galactans, arabinans, and b-glucans. Xylans and mannans, the two most common ones, are used as components of polymer blends, PCs, and nanocomposites. As plant mannan gums have similar chemical features to that of hemicelluloses, they have also been categorized as hemicelluloses [73].

2.1 Properties

Most hemicelluloses support cell walls as do celluloses [100]. Thanks to the ability of hemicelluloses to absorb a large body of water, cell walls become less brittle and, more elastic and resistant to bending. Hemicelluloses also form absorbent molecular networks around cellulose crystallites, which enable the cell walls to transfer intermediate products and sources of nourishment [54]. In addition, it is reported that some fundamental properties of arabinoxylans allow intermolecular alignment between arabinoxylans and polysaccharides, which enables the composition of gels with several components in the wall matrix. Furthermore, ferulic acid residues on arabinoxylan chains allow for covalent interaction between polysaccharide–polysaccharide, polysaccharide-protein, or polysaccharide-lignin [40].

Having less degree of structural order and polymerization, hemicelluloses are chemically and thermally less stable than celluloses. Another difference between hemicelluloses and celluloses is that the former is soluble in alkalines, the feature of which is widely used to section various polysaccharides in samples which do not contain lignin. Nevertheless, this process leads to the removal of ester groups and decrease in solubility. Some hemicelluloses are able to be dissolved in water partially or completely [5]. Hemicelluloses gain hydrophilic property thanks to various free hydroxyl groups in them [38]. The more stable hemicelluloses are, the more they are dissolved in water and less tightly bind to cellulose. On the other hand, the more sporadic the side chains of molecules are, the less they are dissolved in water and more tightly bind to cellulose [60]. A reduction in polymerization generally leads to an increase in solubility [105]. Acetylated hemicelluloses can also be dissolved in dimethyl sulfoxide, formamide, and *N,N*-dimethylformamide. Almost all plants contain hemicelluloses, which are, therefore, found in diets. Cell-wall polysaccharides are important as dietary fibers and contribute to lipid metabolism. Therefore, their consumption helps feces pass through the digestive tract [5, 99, 108]. There is a classification of hemicelluloses such as (i) Softwood hemicelluloses;—Galactoglucomannan (Mannans),—Arabinoglucuronoxylan (Xylans),—Arabinogalactan,—Pectins (ii) Hardwood hemicelluloses such as—Glucuronxylan, -Glucomannan, (iii) Grasses like -Arabinoxylan-main.

3 Hemicellulose Based Composites and Their Applications

Environmentally friendly polymer composites (EFPCs) are used in the manufacture of packaging materials, films, sanitation products, bottles, fishing nets, cutlery, trays and various other products; sensing, adsorbing and many other applications [1, 19, 24, 28, 41, 61, 64, 99, 106, 109].

Expanding their areas of application, most eco-friendly polymers possess perfect features. Hemicelluloses contain a very large quantity of free hydroxyl groups distributed along the backbone and side chains, which are perfect for chemical modification. Materials with exclusive features can be manufactured using chemically modified hemicelluloses in order to improve the biopolymers [39, 101, 119]. In recent years, hemicellulose-based nanocomposite films (HBNCFs) have been popular owing to reduced cost, oxygen barrier features and ease of access [75]. In addition, these hydroxyl groups generate intermolecular and intramolecular hydrogen bonds [70, 77]. Hemicelluloses are hydrophilic and HBNCFs are semicrystalline and hygroscopic and have low mechanical features [36]. Due to these disadvantages, hemicelluloses films cannot be used widely. Plasticizers are generally used to improve the mechanical features of hemicellulose films [43, 44, 51, 120].

Owing to the fact that hemicellulose-based biomaterials are biocompatible, harmless and biodegradable, they have become of great interest for the manufacture of pharmaceutical, tissue engineering, and food packaging products. HBNCFs have attracted the greatest attention due to their oxygen barrier properties. However, they are not very good at forming films and have low performance. Plasticizers or hydrophobization is generally needed to turn hemicelluloses into appropriate materials for packaging products. Rather than concentrating on materials themselves, most studies on hemicelluloses address converting polymers into sugars [109]. Peng et al. suggested an efficient and feasible technique to manufacture superior HBNCFs. He combined cellulose nanofibers (CNFs) and xylan (XH) films as plasticizers were also included in the process. Sugar composition was xylose (89.38%), arabinose (5.75%), glucose (1.87%), galactose (0.66%) glucuronic acid (1.78%) and galacturonic acid (0.55%). Morphological tests indicated that XH film surface and CNF-reinforced NCF consisted mainly of nodules (10–70 nm) and that CNFs were embedded in the XH matrix. Aggregates started to develop in an interval of arabinose/xylose ranging from 0.23 to 0.31 in an aqueous solution and when the arabinose substituents were exhausted, unsubstituted xylan chains formed a steady connection, which shows that the intermolecular interaction could be adjusted by manipulating the substituted groups or by including a second component that establishes an interaction with the molecular chains of hemicelluloses. Film formation was better, and the tensile strength improved greatly due to high aspect ratio and strong interactions between CNF and the XH matrix. Fang et al. [39] employed the freeze-thaw method in order to use hemicelluloses from bamboo holocellulose, PVA, and chitin nanowhiskers to manufacture a hybrid hydrogel. FTIR and NMR results showed that the gelation process witnessed physical crosslinking instead of a chemical reaction. PVA functioned as a hydrogel scaffold while hemicelluloses were firmly hydrophilic and showed hydrogen-bonding features when chitin nanowhiskers functioned as a cross-linker. According to atomic force microscopy (AFM) images of chitin, whisker size was about 200 nm in length and 40 nm in width. The mechanical features of hydrogels were greatly enhanced by an increase in the rate of chitin nanowhiskers. The hydrogels are potential materials to be utilized in the manufacture of tissue engineering products. Laminated films, one layer of which is polyester and the other of which is made of carbon nanotubes and hemicellulose, have been patented in Japan [58]. Polymers were supplied by removing hemicellulose from *Abelmoschus manihot* or *Hydrangea paniculata.* The homogenous distribution of carbon nanotubes in hemicellulose films was leading to obtaining mechanically strong films [109].

3.1 Composite Formation and Characterization with Layered Silicates

For nanocomposite formation, polymer chains should be dispersed in the galleries between silicate layers in order to obtain two distinct classes of lamellar nanocomposites; intercalated and exfoliated as represented schematically in Fig. 3. Intercalation takes place with a little polymer permeating into the galleries, which leads to silicate layers expanding in a limited manner. This way, a systematic structure with multiple layers separated by several nanometers is obtained. Large-scale polymer diffusion causes silicate layers to exfoliate or delaminate. An exfoliated nanocomposite contains nanometer thick platelets that are uniformly dispersed all over the polymer matrix. On the contrary, when the polymer and silicate do not form a homogeneous mixture when mixed, the layers do not apart, and they function as agglomerates or tactoids.

Total diffusion of clay platelets in a polymer increases the number of usable reinforcing materials to carry an applied load and to avert the formation of cracks. The pairing of the enormous surface area of the clay and the polymer matrix enables stress to be shifted to the reinforcement stage, which leads to mechanical improvement. Furthermore, the impenetrable clay layers require a convoluted track for a permeant to travel across the nanocomposites as seen in Fig. 4. Thanks to obstructed diffusion tracks along nanocomposites, the barrier property and flame retardancy of polymer-clay nanocomposites are improved, and they are more resistant to chemicals and uptake less solvent [104].

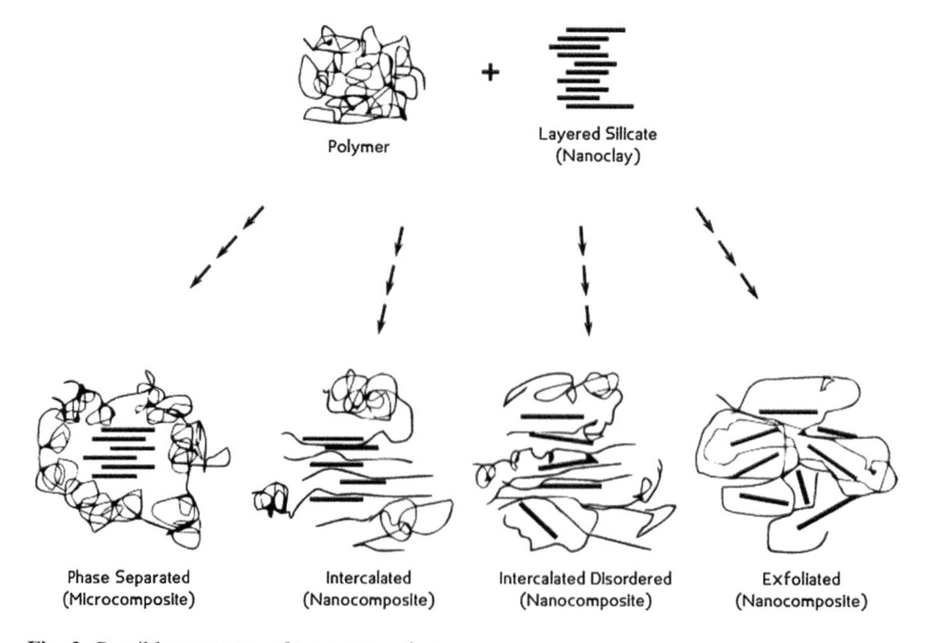

Fig. 3 Possible structures of nanocomposites

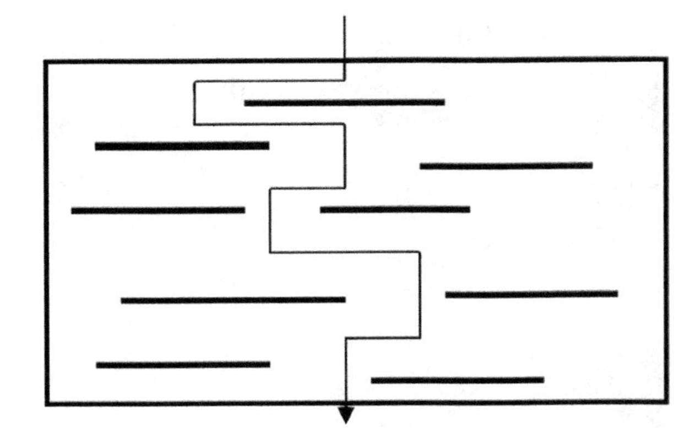

Fig. 4 Proposed model for the tortuous zigzag diffusion path in a polymer-clay nanocomposite when used as a gas barrier

3.2 Packaging Materials

The food producers have a wide range of natural recipes that are used in packaging applications and have a variety of needs that ensure the quality and safety of food. The first goal of this industry is assuring the protection of food products from the many factors related to the environment, gases, water etc. The most important factor in food packages is oxygen since oxygen has a very important material in the various chemical industries because oxygen mostly affects the expiration date of foods, fruits, vegetables, microbial growth etc. Nowadays, in order to cancel this effect of oxygen, Al plates and/or foil and some of the polymers such as polyethylene vinyl alcohol (EVOH), polyvinylidene chloride (PVDC) etc. are mainly used as packaging materials. Water, then again, is one of most important material, the catalyst that affects the chemical reactions and/or enzymatic changes and known as the principal food ingredient.

Biomass has become of greater interest for food packaging applications. The aim of most studies is to alter hemicelluloses and enhance the mechanical features of HBNCFs converted into a quaternary form. To this end, those studies focus on assuming an environmentally friendly and feasible approach. Chen et al. Integrated quaternized hemicelluloses, chitosan, and MMT to produce films. They analyzed the morphological structure and optical, physical, mechanical and thermal features of the films. Another aim of the study was to determine the ideal ratio in order to produce HBNCFs. The principal endeavour will present chitosan-based materials.

Peng et al. [70] reported that the molecular makeup of hemicelluloses had a great effect on film formation, mechanical features, and moisture content. The side groups made xylan or mannan chains separate, indicating that the hemicellulose which had the biggest interaction was the one in which arabinose or galactose substituted the backbone the least [53, 100, 102]. This was a pioneer study in the sense that it aimed to incorporate cellulose nanofibers into XH in order to manufacture mechanically improved nanocomposite films. SEM and AFM images pointed out that surface of XH and CNF-reinforced nanocomposite films consisted

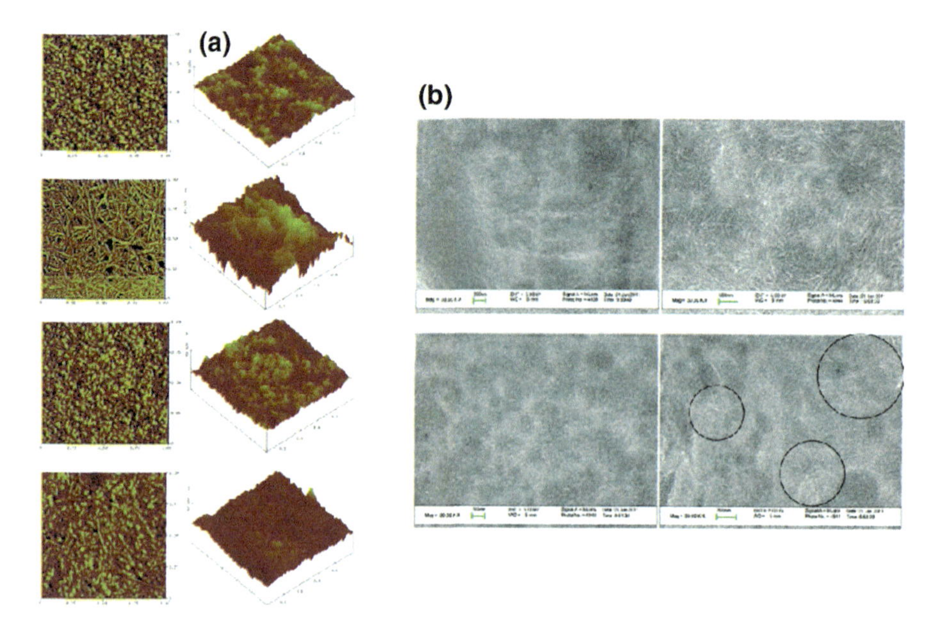

Fig. 5 Properties of CNF-reinforced nanocomposite films

mainly of nodules with a diameter of 10–70 nm and that CNFs were embedded in the XH matrix as shown in Fig. 5. The XH film had a peak at $2\theta = \sim 18°$, indicating that it was semicrystalline. The thermal stability of the nanocomposite film was better than that of the XH film. Film formation was better, and the tensile strength improved greatly due to high aspect ratio and strong interactions between CNF and the XH matrix, which points to an efficient and feasible technique for the manufacturing of superior HBNCFs [77].

3.3 *Other Applications*

Xylan is used in a wide variety of medical applications. Glyceronoxylene, xylitol or sorbitol-based films are widely used in the field of food packaging because their oxygen permeability is very low and there is a potential for further spread in the future [49]. Xylooligosaccharides (XOS) would hydrolysis results from claiming xylan-type hemicelluloses for guaranteeing possibility to a significant number of fields including chemical, sustenance and pharmaceutical commercial enterprises. For case, in the nourishment industry, XOS has an assortment of incredible physiological properties such as bringing down cholesterol levels, moving forward gastrointestinal function and the organic accessibility of calcium, and decreasing the risk of colon cancer [62, 76].

In order to make further improvements on the properties of the all-cellulose composites, it is recommended to use cellulose nanofibers that extracted from Napier grass fibers as reinforcement in cellulose matrix in further studies. According to the properties of composite films, it can be concluded that all these cellulose composites are suitable for using in warping, mulching and sorbent applications [97]. Another example of using natural fibers as reinforcement was seen in the study of Maniraman et al. [65]. Furcraea foetida (FF) plant was extracted and investigated. The structure of natural fibers generally longer, they have smaller diameter values and low spiral angle of the cellulose arrangement [56]. That quality of a bio-composite relies on the properties of the fiber and the resin and additionally on the quality from claiming their interfacial holding. So, to in front of fabricating bio-composites, additionally, it is a necessity to learn about those properties of fibers. There are many kinds of literature about natural fibers and their properties. They enlighten that celluloses, hemicelluloses and non-cellulosic components (lignin, wax, and pectin) are constructed in the natural fibers. But, the bio-fiber yielding plants grow only in particular regions based on environmental circumstances [11]. New assortments of fibers have appropriate properties for use as reinforcement are needed to be identified and developed. A lot of leaf fibers, such as those mined from Artisditahystrix, Sansevieria cylindrica, Tamarindus Indica L and sansevieria ehrenbergii have been used as an unsurpassed reinforcement for polymeric matrices [65].

Also, Kozlowski and Przybylak carried out a research and review to collect fire performance data for several types of eco-friendly polymer composites reinforced by cellulosic fibers [59].

Glucomannan type's materials can be seen as hemicellulosic materials which have film-forming properties. It is generally used in wooden as a "green" absorption material (gels, etc.) and it can be isolated from various method turbines or production fibreboard. Modification of new materials including non-hydrophilic and thermal character also can bring about the xylan which has high molecular weight displaying a stepped forward capability to form films. Film-forming is essential so as for xylan if you want to form self-supporting barrier films to be used in food packaging.

3.4 Basic Components of Composite Materials

The composites consist essentially of the matrix and reinforcing materials that give it physicochemical properties. The matrix part is the basic component of the composite and supports the physicochemical and biological properties of the reinforcing material. Matrixes are the main component to protect the reinforcements of the composites from abrasion and environmental influences. At the same time, it provides composites stability by allowing the matrix fibers and filler materials to coexist. A reaction with synergistic effect occurs by the matrix loading of the reinforcing materials. Thus, thanks to the reinforcement material and matrix, composite materials with the desired properties are prepared. Due to the physicomechanical properties of the polymer composites, they are of great interest because they have a much higher application area than the individual polymeric structures. The main materials reinforced with composite matrices are metallic oxides, clay, fiber constructions and carbonates. These reinforcing materials give physicomechanical and biological properties to the composites [12, 98].

The structure of lignocellulosic stocks forms polysaccharide and lignin. When lignocellulosic raw materials are introduced into biorefinery and subjected to a series of treatments, products such as cellulose, hemicellulose and lignin are formed. Cellulose, hemicellulose and lignin products are raw materials of lignocellulosic composites. These substances are the main components of animal feed and lumber resources. Cellulose is naturally present as a composition of lignin and hemicellulose fibers. Lignin protects carbohydrates from biological attack and is infectious by very few microorganisms [22, 68].

3.5 The Effect of Fibers on Composite Materials

Fibers play a very important role in the development of multifunctional composite materials. The shrinking the size of any composite material causes a reduction in material defects. The fibrous materials have a higher surface area, a larger surface area, compared to bulk materials. Many studies have shown that multifunctional composite properties depend on properties such as fiber length, volume, fiber type, fiber orientation, etc. These features can be briefly summarized as follows:

- Composite materials composed of long fiber structures have higher mechanical resistance than short fibers. Whereas short fiber composites are less costly and easier to process than long fibers [117].
- Increasing the amount of fiber loading to a certain amount of composite material, resulting in an increase in composite material properties. However, after this particular value, there is a decrease in the composite properties due to the interlocking between the matrix and fibers [63].
- Fiber loading method make a significant impact on composite material properties. The lowest value of Young's modulus was found to be 45° and 60° in the fibers and there have been seen discrepancies between theoretical and experimental values [48].
- Generally, the fibers are made of beeswax, gelatin starch and oily materials to improve adhesion and retention properties [67].

4 Conclusions

In this chapter; the synthesis and characterization of Eco-friendly polymer composites (EFPCs) and their usage in various applications such as food packaging, medical applications, gas barriers etc. have been examined in detail and as it can be seen these polymer composites are getting more and more attention to science and technology due to their superior properties such as biodegradability, renewability, high strength and stiffness, low cost and ecologically friendly formation. Eco-friendly polymer composites (EFPCs) from natural components provide a potential alternative to other commercial composite materials without compromising strength, stiffness and abrasion resistance properties to a variety of industrial and other applications.

References

1. Abdelwahab NA, Al-Ashkar EA, Abdel-Ghaffara MA (2015) Preparation and characterization of eco-friendly poly(p-phenylenediamine) and its composite with chitosan for removal of copper ions from aqueous solutions. Trans Nonferrous Met Soc China 25:3808–3819
2. Abrahamson JT, Sen F, Sempere B et al (2013) Excess thermopower and the theory of thermopower waves. ACS Nano 7(8):6533–6544
3. Aday B, Yıldız Y, Ulus R et al (2016) One-pot, efficient and green synthesis of acridinedione derivatives using highly monodisperse platinum nanoparticles supported with reduced graphene oxide. New J Chem 40:748–754
4. Akocak S, Sen B, Lolak N et al (2017) One-pot three-component synthesis of 2-amino-4H-chromene derivatives by using monodisperse Pd nanomaterials anchored graphene oxide as highly efficient and recyclable catalyst. Nano-Struct Nano-Objects 11:25–31
5. Alén R (2000) Structure and chemical composition of wood. In: Stenius P (ed) Forest products chemistry. Fapet Oy, Helsinki, pp 12–57
6. Ayranci R, Baskaya G, Guzel M et al (2017) Carbon-based nanomaterials for high-performance optoelectrochemical systems. Chem Select 2(4):1548–1555
7. Ayranci R, Baskaya G, Guzel M et al (2017) Enhanced optical and electrical properties of PEDOT via nanostructured carbon materials: a comparative investigation. Nano-Struct Nano-Objects 11:13–19
8. Baillie C (2004) Polymer composites and the environment. Green Compos 1–8
9. Baskaya G, Esirden I, Erken E et al (2017) Synthesis of 5-substituted-1H-tetrazole derivatives using monodisperse carbon black decorated Pt nanoparticles as heterogeneous nanocatalysts. J Nanosci Nanotechnol 17:1992–1999
10. Baskaya G, Yıldız Y, Savk A et al (2017) Rapid, sensitive, and reusable detection of glucose by highly monodisperse nickel nanoparticles decorated functionalized multi-walled carbon nanotubes. Biosens Bioelectron 91:728–733
11. Batra SK (1985) Handbook of fiber science and technology. In: Lewin M, Pearce EM (eds) Fiber chemistry, vol IV. Marcel Dekker, New York, pp 727–808
12. Bloor D, Donnelly K, Hands PJ, Laughlin P, Lussey D (2005) A metal–polymer composite with unusual properties. J Phys D Appl Phys 38:2851–2860
13. Bozkurt S, Tosun B, Sen B et al (2017) A hydrogen peroxide sensor based on TNM functionalized reduced graphene oxide grafted with highly monodisperse Pd nanoparticles. Anal Chim Acta 989C:88–94
14. Celik B, Baskaya G, Karatepe O et al (2016) Monodisperse Pt(0)/DPA@GO nanoparticles as highly active catalysts for alcohol oxidation and dehydrogenation of DMAB. Int J Hydrogen Energy 41:5661–5669

15. Celik B, Erken E, Eris S et al (2016) Highly monodisperse Pt(0)@AC NPs as highly efficient and reusable catalysts: the effect of the surfactant on their catalytic activities in room temperature dehydrocoupling of DMAB. Catal Sci Technol 6:1685–1692
16. Celik B, Kuzu S, Erken E et al (2016) Nearly monodisperse carbon nanotube furnished nanocatalysts as highly efficient and reusable catalyst for dehydrocoupling of DMAB and C1 to C3 alcohol oxidation. Int J Hydrogen Energy 41:3093–3101
17. Celik B, Yildiz Y, Erken E et al (2016) Monodisperse palladium-cobalt alloy nanoparticles assembled on poly (*N*-vinyl-pyrrolidone) (PVP) as highly effective catalyst for the dimethylammine borane (DMAB) dehydrocoupling. RSC Adv 6:24097–24102
18. Chen GG, Qi XM, Guan Y et al (2016) High strength hemicellulose-based nanocomposite film for food packaging applications. ACS Sustain Chem Eng 4:1985–1993
19. Chen L, Wu P, Chen M et al (2018) Preparation and characterization of the eco-friendly chitosan/vermiculite biocomposite with excellent removal capacity for cadmium and lead. Appl Clay Sci https://doi.org/10.1016/j.clay.2017.12.050
20. Chen T, Dai L (2013) Carbon nanomaterials for high-performance supercapacitors. Mater Today 16:272–280
21. Ciolacu D, Popa VI (2011) Cellulose allomorphs: structure. Accessibility and reactivity. Nova Science Publisher, New York, pp 1–3 (Chapter 1)
22. Clark J, Deswarte F (2008) Introduction to chemicals from biomass. Wiley, UK
23. Dasdelen Z, Yıldız Y, Eris S et al (2017) Enhanced electrocatalytic activity and durability of Pt nanoparticles decorated on GO-PVP hybride material for methanol oxidation reaction. Appl Catal B 219C:511–516
24. Davachi SM, Shekarabi AS (2018) Preparation and characterization of antibacterial, eco-friendly edible nanocomposite films containing Salvia macrosiphon and nanoclay. Int J Biol Macromol 113:66–72
25. Demir E, Savk A, Sen B et al (2017) A novel monodisperse metal nanoparticles anchored graphene oxide as counter electrode for dye-sensitized solar cells. Nano-Struct Nano-Objects 12:41–45
26. Demir E, Sen B, Sen F (2017) Highly efficient nanoparticles and f-MWCNT nanocomposites-based counter electrodes for dye-sensitized solar cells. Nano-Struct Nano-Objects 11:39–45
27. Demirci T, Celik B, Yıldız Y et al (2016) One-pot synthesis of hantzsch dihydropyridines using highly efficient and stable PdRuNi@GO catalyst. RSC Adv 6:76948–76956
28. Divsalar E, Tajik H, Moradi M et al (2018) Characterization of cellulosic paper coated with chitosan-zinc oxidenanocomposite containing nisin and its application in packaging of UF cheese. Int J Biol Macromol 109:1311–1318
29. Ebringerova A, Hromadkova Z, Heinze T (2005) Hemicellulose. Adv Polym Sci 186:1–67
30. Eris S, Daşdelen Z, Sen F (2018) Enhanced electrocatalytic activity and stability of monodisperse Pt nanocomposites for direct methanol fuel cells. J Colloid Interface Sci 513:767–773
31. Eris S, Daşdelen Z, Sen F (2018) Investigation of electrocatalytic activity and stability of Pt@f-VC catalyst prepared by in-situ synthesis for methanol electrooxidation. Int J Hydrogen Energy 43(1):385–390
32. Eris S, Daşdelen Z, Yıldız Y et al (2018) Nanostructured polyaniline-rGO decorated platinum catalyst with enhanced activity and durability for methanol oxidation. Int J Hydrogen Energy 43(3):1337–1343
33. Erken E, Esirden İ, Kaya M et al (2015) A rapid and novel method for the synthesis of 5-substituted 1H-tetrazole catalyzed by exceptional reusable monodisperse Pt NPs@AC under the microwave irradiation. RSC Adv 5:68558–68564
34. Erken E, Pamuk H, Karatepe O et al (2016) New Pt(0) nanoparticles as highly active and reusable catalysts in the C1–C3 alcohol oxidation and the room temperature dehydrocoupling of dimethylamine-borane (DMAB). J Cluster Sci 27:9

35. Erken E, Yildiz Y, Kilbas B et al (2016) Synthesis and characterization of nearly monodisperse Pt nanoparticles for C1 to C3 alcohol oxidation and dehydrogenation of dimethylamine-borane (DMAB). J Nanosci Nanotechnol 16:5944–5950
36. Escalante A, Gonçalves A, Bodi A et al (2012) Flexible oxygen barrier films from spruce xylan. Carbohydr Polym 87:2381–2387
37. Esirden İ, Erken E, Kaya M et al (2015) Monodisperse Pt NPs@rGO as highly efficient and reusable heterogeneous catalysts for the synthesis of 5-substituted 1H-tetrazole derivatives. Catal Sci Technol 5:4452–4457
38. Fang JM, Sun RC, Fowler P et al (1999) Esterification of wheat straw hemicelluloses in the *N,N*-dimethylformamide/lithium cloride homogeneous system. J Appl Polym Sci 74:2301–2311
39. Fang JM, Sun RC, Tomkinson J et al (2000) Acetylation of wheat straw hemicellulose B in a new non-aqueous swelling system. Carbohydr Polym 41:379–387
40. Fincher GB, Stone BA (1986) Cell walls and their components in cereal grain technology. In: Pomeranz Y (ed) Advances in cereal science and technology. American Association of Cereal Chemists Inc., St Paul, pp 207–295
41. Galindo-Rosales FJ, Martínez-Aranda S, Campo-Deaño L (2015) CorkSTFlfluidics—a novel concept for the development of eco-friendly light-weight energy absorbing composites. Mater Des 82:326–334
42. Giraldo JP, Landry MP, Faltermeier SM et al (2014) A nanobionic approach to augment plant photosynthesis and biochemical sensing using targeted nanoparticles. Nat Mater 13:400–408
43. Goksu EI, Karamanlioglu M, Bakir U et al (2007) Production and characterization of films from cotton stalk xylan. J Agric Food Chem 55:10685–10691
44. Goksu EI, Karamanlioglu M, Bakir U et al (2007) Production and characterization of films from cotton stalk xylan. J Agric Food Chem 55(26):10685–10691
45. Goksu H, Celik B, Yıldız Y et al (2016) Superior monodisperse CNT-supported CoPd (CoPd@CNT) nanoparticles for selective reduction of nitro compounds to primary amines with $NaBH_4$ in aqueous medium. Chem Select 1(10):2366–2372
46. Goksu H, Yıldız Y, Celik B et al (2016) Eco-friendly hydrogenation of aromatic aldehyde compounds by tandem dehydrogenation of dimethylamine-borane in the presence of reduced graphene oxide furnished platinum nanocatalyst. Catal Sci Technol 6:2318–2324
47. Goksu H, Yıldız Y, Celik B et al (2016) Highly efficient and monodisperse graphene oxide furnished Ru/Pd nanoparticles for the dehalogenation of aryl halides via ammonia borane. Chem Select 1(5):953–958
48. Graupner N, Ziegmann G, Wilde F et al (2016) Procedural influences on compression and injection moulded cellulose fibre-reinforced polylactide (PLA) composites: influence of fibre loading, fibre length, fibre orientation and voids. Compos Part an Appl Sci Manuf 81:158–171
49. Grondahl M, Eriksson L, Gatenholm P (2004) Material properties of plasticized hardwood xylans for potential application as oxygen barrier films. Biomacromol 5(4):1528–1535
50. Grondahl M, Gustafsson A, Gatenholm P (2006) Surface- and bulk-modified galactoglucomannan hemicellulose films and film laminates for versatile oxygen barriers. Macromolecules 39:2718–2721
51. Hansen NM, Plackett D (2008) Sustainable films and coatings from hemicelluloses: a review. Biomacromol 9:1493–1505
52. Hartman J, Albertsson AC, Sjoberg J (2006) Surface- and bulk-modified galactoglucomannan hemicellulose films and film laminates for versatile oxygen barriers. Biomacromol 7:1983–1989
53. Hoije A, Sternemalm E, Heikkinen S et al (2008) Material properties of films from enzymatically tailored arabinoxylans. Biomacromol 9:2042–2047
54. Izydorczyk MS, Biliaderis CG (1995) Cereal arabinoxylanase: advances in structure and physicochemical properties. Carbohydr Polym 28:33–48

55. Karatepe O, Yildiz Y, Pamuk H et al (2016) Enhanced electro catalytic activity and durability of highly mono disperse Pt@PPy-PANI nanocomposites as a novel catalyst for electro-oxidation of methanol. RSC Adv 6:50851–50857
56. Kathiresan M, Pandiarajan P, Senthamaraikannan P et al (2016) Physicochemical properties of new cellulosic artisditahystrix leaf fiber. Int J Polym Anal Charact 21(6):663–668
57. Kayseriliolu BS, Bakir U, Yilmaz L et al (2003) Use of xylan, an agricultural by-product, in wheat gluten based biodegradable films: mechanical, solubility and water vapor transfer rate properties. Bioresour Technol 87:239–246
58. Khalil HPSA, Davoudpour Y, Islam MN et al (2014) Production and modification of nanofibrillated cellulose using various mechanical processes: a review. Carbohydr Polym 99:649–665
59. Kozłowski R, Władyka-Przybylak M (2008) Flammability and fire resistance of composites reinforced by natural fibers. Polym Adv Technol 19:446–453
60. Lawther JM, Sun RC, Banks WB (1995) Extraction, fractionation, and characterization of structural polysaccharides from wheat straw. J Agric Food Chem 43:667–675
61. Li Y, Liu H, Dai K et al (2015) Tuning of vapor sensing behaviors of eco-friendly conductive polymercomposites utilizing ramie fiber. Sens Actuators B 221:1279–1289
62. Lin Q, Li H, Ren J (2017) Production of xylooligosaccharides by microwave-induced, organic acid-catalyzed hydrolysis of different xylan-type hemicelluloses: optimization by response surface methodology. Carbohyd Polym 157:214–225
63. Liu W, Luo L, Xu S, Zhao H (2014) Effect of fiber volume fraction on crack propagation rate of ultra-high toughness cementitious composites. Eng Fract Mech 124:52–63
64. Luo Z, Zhang J, Zhuang C et al (2016) An eco-friendly composite adsorbent for efficient removal of Cu^{2+} from aqueous solution. J Taiwan Inst Chem Eng 60:479–487
65. Manimaran P, Senthamaraikannan P, Sanjay MR et al (2018) Study on characterization of Furcraea foetida new natural fiber as composite reinforcement for lightweight applications. Carbohyd Polym 181:650–658
66. Mikkonen KS, Rita H, Helen H et al (2007) Effect of polysaccharide structure on mechanical and thermal properties of galactomannan-based films. Biomacromol 8(10):3198–3205
67. Mittal G, Rhee K, Mišković-Stanković V, Hui D (2018) Reinforcements in multi-scale polymer composites: processing, properties, and applications. Compos B Eng 138:122–139
68. Mukrimin SG (2013) Utilization of hazelnut husk as biomass. Sustain Energy Technol Assess 4:72–77
69. Okamoto M (2004) Biodegradable polymer/layered silicate nanocomposites: a review. J Ind Eng Chem 10(7):1156–1181
70. Peng XW, Ren JL, Zhong LX (2011) Nanocomposite films based on xylan-rich hemicelluloses and cellulose nanofibers with enhanced mechanical properties. Biomacromol 12:3321–3329
71. Peroval C, Debeaufort F, Despre D et al (2002) Edible arabinoxylan-based films. 1. Effects of lipid type on water vapor permeability, film structure, and other physical characteristics. J Agric Food Chem 50(14):3977–3983
72. Phan TD, Debeaufort F, Peroval C et al (2002) Arabinoxylan–lipid-based edible films and coatings. 3. Influence of drying temperature on film structure and functional properties. J Agric Food Chem 50(8):2423–2428
73. Sahin B, Aygün A, Gündüz H et al (2018) Cytotoxic effects of platinum nanoparticles obtained from pomegranate extract by the green synthesis method on the MCF-7 cell line. Colloids Surf B 163:119–124
74. Sahin B, Demir E, Aygün A et al (2017) Investigation of the effect of pomegranate extract and monodisperse silver nanoparticle combination on MCF-7 cell line. J Biotechnol 260C:79–83
75. Salam A, Pawlak JJ, Venditti RA et al (2011) Incorporation of carboxyl groups into xylan for improved absorbency. Cellulose 18:1033–1041

76. Samanta AK, Jayapal N, Jayaram C et al (2015) Xylooligosaccharides as prebiotics from agricultural by-products: production and applications. Bioact Carbohydr Dietary Fibre 5 (1):62–71
77. Schroeter J, Felix F (2005) Melting cellulose. Cellulose 12:159–165
78. Sen B, Akdere EH, Savk A et al (2018) A novel thiocarbamide functionalized graphene oxide supported bimetallic monodisperse Rh-Pt nanoparticles (RhPt/TC@GO NPs) for Knoevenagel condensation of aryl aldehydes together with malononitrile. Appl Catal B 225 (5):148–153
79. Sen B, Kuzu S, Demir E et al (2017) Highly efficient catalytic dehydrogenation of dimethly ammonia borane via monodisperse palladium-nickel alloy nanoparticles assembled on PEDOT. Int J Hydrogen Energy 42(36):23307–23314
80. Sen B, Kuzu S, Demir E et al (2017) Highly monodisperse RuCo nanoparticles decorated on functionalized multiwalled carbon nanotube with the highest observed catalytic activity in the dehydrogenation of dimethylamine borane. Int J Hydrogen Energy 42(36):23292–23298
81. Sen B, Kuzu S, Demir E et al (2017) Hydrogen liberation from the dehydrocoupling of dimethylamine-borane at room temperature by using novel and highly monodispersed RuPtNi nanocatalysts decorated with graphene oxide. Int J Hydrogen Energy 42(36):23299–23306
82. Sen B, Kuzu S, Demir E et al (2017) Monodisperse palladium-nickel alloy nanoparticles assembled on graphene oxide with the high catalytic activity and reusability in the dehydrogenation of dimethylamine-borane. Int J Hydrogen Energy 42(36):23276–23283
83. Sen B, Kuzu S, Demir E et al (2017) Polymer-Graphene hybride decorated Pt nanoparticles as highly efficient and reusable catalyst for the dehydrogenation of dimethylamine-borane at room temperature. Int J Hydrogen Energy 42(36):23284–23291
84. Sen B, Lolak N, Paralı Ö et al (2017) Bimetallic PdRu/graphene oxide-based catalysts for one-pot three-component synthesis of 2-amino-4H-chromene derivatives. Nano-Struct Nano-Objects 12:33–40
85. Sen F, Boghossian AA, Sen S et al (2013) Application of nanoparticle antioxidants to enable hyperstable chloroplasts for solar energy harvesting. Adv Energy Mater 3(7):881–893
86. Sen F, Ertan S, Sen S et al (2012) Platinum nanocatalysts prepared with different surfactants for C1 to C3 alcohol oxidations and their surface morphologies by AFM. J Nanopart Res 14:922–926
87. Sen F, Gokagac G (2007) Different sized platinum nanoparticles supported on carbon: an XPS study on these methanol oxidation catalysts. J Phys Chem C 111:5715–5720
88. Sen F, Gokagac G (2007) The activity of carbon supported platinum nanoparticles towards methanol oxidation reaction—role of metal precursor and a new surfactant, *tert*-octanethiol. J Phys Chem C 111:1467–1473
89. Sen F, Gokagac G (2014) Pt nanoparticles synthesized with new surfactans: improvement in C1–C3 alcohol oxidation catalytic activity. J Appl Electrochem 44(1):199–207
90. Sen F, Karataş Y, Gülcan M et al (2014) Amylamine stabilized platinum (0) nanoparticles: active and reusable nanocatalyst in the room temperature dehydrogenation of dimethylamine-borane. RSC Adv 4(4):1526–1531
91. Sen F, Ozturk Z, Sen S et al (2012) The preparation and characterization of nano-sized Pt–Pd alloy catalysts and comparison of their superior catalytic activities for methanol and ethanol oxidation. J Mater Sci 47:8134–8144
92. Sen F, Sen S, Gokagac G (2011) Efficiency enhancement in the methanol/ethanol oxidation reactions on Pt nanoparticles prepared by a new surfactant, 1,1-dimethyl heptanethiol, and surface morphology by AFM. Phys Chem Chem Phys 13:1676–1684
93. Sen F, Sen S, Gokagac G (2013) High performance Pt nanoparticles prepared by new surfactants for C1 to C3 alcohol oxidation reactions. J Nanopart Res 15:1979
94. Sen F, Ulissi ZW, Gong X et al (2014) Spatiotemporal intracellular nitric oxide signaling captured using internalized, near-infrared fluorescent carbon nanotube nanosensors. Nano Lett 14(8):4887–4894

95. Sen S, Sen F, Boghossian AA et al (2013) The effect of reductive dithiothreitol and trolox on nitric oxide quenching of single walled carbon nanotubes. J Phys Chem C 117(1):593–602
96. Sen S, Sen F, Gokagac G (2011) Preparation and characterization of nano-sized Pt–Ru/C catalysts and their superior catalytic activities for methanol and ethanol oxidation. Phys Chem Chem Phys 13:6784–6792
97. Senthil Muthu Kumar T, Rajini N, Obi Reddy K et al (2018) All-cellulose composite films with cellulose matrix and Napier grass cellulose fibril fillers. Int J Biol Macromol 112:1310–1315
98. Shah N, Ul-Islam M, Khattak WA, Park WJ (2013) Overview of bacterial cellulose composites: a multipurpose advanced material. Carbohyd Polym 98(2):1585–1598
99. Sinha Ray S (2013) Environmentally friendly polymer nanocomposites. Woodhead Publishing Limited, p 2856
100. Sjöström E (1993) Wood chemistry fundamentals and applications. Academic Press Inc., San Diego, p 293
101. Söderqvist Lindblad M, Ranucci E, Albertsson AC (2001) Biodegradable polymers from renewable sources. New hemicellulose-based hydrogels. Macromol Rapid Commun 22:962–967
102. Sternemalm E, Höije A, Gatenholm P (2008) Effect of arabinose substitution on the material properties of arabinoxylan films. Carbohydr Res 343:753–757
103. Sun RC, Sun XF, Tomkinson J (2004) Hemicelluloses and their derivatives. In: Gatenholm P, Tenkanen M (eds) Hemicelluloses: science and technology. ACS symposium series, vol 864. American Chemical Society, Washington, pp 2–22
104. Tang XZ, Kumar P, Alavi S (2012) Recent advances in biopolymers and biopolymer-based nanocomposites for food packaging materials. Crit Rev Food Sci Nutr 52:426–442
105. Tenkanen M (2004) Enzymatic tailoring of hemicelluloses. In: Gatenholm P, Tenkanen M (eds) Hemicelluloses: science and technology. ACS symposium series, vol 864. American Chemical Society, Washington, pp 292–311
106. Tersur Orasugh J, Ranjan Saha N, Rana D et al (2018) Jute cellulose nano-fibrils/hydroxypropylmethylcellulose nanocomposite: a novel material with potential for application in packaging and transdermal drug delivery system. Ind Crops Prod 112:633–643
107. Thakur VK, Thakur MK (2015) Eco-friendly polymer nanocomposites. Process Prop 75:579
108. Thomas S, Visakh Aji PM, Mathew P (ed) (2012) Advances in natural polymers: composites and nanocomposites, vol 18. Springer Science & Business Media, p 426
109. Won JP, Kang HB, Lee SJ et al (2012) Eco-friendly fireproof high-strength polymer cementitious composites. Constr Build Mater 30:406–412
110. Yildiz Y, Erken E, Pamuk H et al (2016) Monodisperse Pt nanoparticles assembled on reduced graphene oxide: highly efficient and reusable catalyst for methanol oxidation and dehydrocoupling of dimethylamine-borane (DMAB). J Nanosci Nanotechnol 16:5951–5958
111. Yıldız Y, Esirden İ, Erken E et al (2016) Microwave (Mw)-assisted synthesis of 5-substituted 1H-tetrazoles via [3 + 2] cycloaddition catalyzed by Mw–Pd/Co nanoparticles decorated on multi-walled carbon nanotubes. Chem Select 1(8):1695–1701
112. Yıldız Y, Kuzu S, Sen B et al (2017) Different ligand based monodispersed metal nanoparticles decorated with rGO as highly active and reusable catalysts for the methanol oxidation. Int J Hydrogen Energy 42(18):13061–13069
113. Yıldız Y, Onal Okyay T, Gezer B et al (2016) Monodisperse Mw-Pt NPs@VC as highly efficient and reusable adsorbents for methylene blue removal. J Cluster Sci 27:1953–1962
114. Yildiz Y, Onal Okyay T, Sen B et al (2017) Highly monodisperse Pt/Rh nanoparticles confined in the graphene oxide for highly efficient and reusable sorbents for methylene blue removal from aqueous solutions. Chem Select 2(2):697–701
115. Yıldız Y, Pamuk H, Karatepe O et al (2016) Carbon black hybride material furnished monodisperse platinum nanoparticles as highly efficient and reusable electrocatalysts for formic acid electro-oxidation. RSC Adv 6:32858–32862

116. Yıldız Y, Ulus R, Eris S et al (2016) Functionalized multi-walled carbon nanotubes (f-MWCNT) as highly efficient and reusable heterogeneous catalysts for the synthesis of acridinedione derivatives. Chem Select 1(13):3861–3865
117. Zhang D, He M, Qin S, Yu J (2017) Effect of fiber length and dispersion on properties of long glass fiber reinforced thermoplastic composites based on poly(butylene terephthalate). RSC Adv 7:15439–15454
118. Zhang Y, Pitkänen L, Douglade J et al (2011) Wheat bran arabinoxylans: chemical structure and film properties of three isolated fractions. Carbohydr Polym 86:852–859
119. Zhong LX, Peng XW, Yang D et al (2013) Longchain anhydride modification: a new strategy for preparing xylan films. J Agric Food Chem 61:655–661
120. Zhu Ryberg Y, Edlund U, Albertsson AC (2011) Conceptual approach to renewable barrier film design based on wood hydrolysate. Biomacromol 12:1355–1362

Impact of Nanoparticle Shape, Size, and Properties of the Sustainable Nanocomposites

Thandapani Gomathi, K. Rajeshwari, V. Kanchana, P. N. Sudha and K. Parthasarathy

1 Introduction

Nanotechnology has imposed a hopeful approach to resolve many technological impasses incurred in various the branches of science and technology. The field of nanotechnology is an interdisciplinary area and it is one of the most popular areas of current research and development basically in all disciplines. It obviously includes polymer science and technology [78]. It can be defined as "the design, characterization, production and application of structures, devices and systems by controlling shape and size at the nanoscale" (<100 nm) [108]. According to Drexler, "Nanotechnology is the principle of manipulation of the structure of matter at the molecular level. It entails the ability to build molecular systems with atom-by-atom precision, yielding a variety of nanomachines [131, 72]."

In the past decades, major research on the production of nanocomposites includes the use of synthetic polymeric materials, which have become a present dispute [85, 37, 135]. The dispute is because of the shortage of the natural resources and also the environmental concerns for their non-biocompatibility, cross contaminations and toxicity risks [4, 22, 63, 101, 129]. Renewable and sustainable

T. Gomathi (✉) · P. N. Sudha
Department of Chemistry, D.K.M. College for Women, Vellore, Tamil Nadu, India
e-mail: drgoms1@gmail.com; chemist.goms@gmail.com

K. Rajeshwari
Department of Chemistry, Adhi College of Engineering, Walajabad, Tamilnadu, India

V. Kanchana
Department of Chemistry, Sree Sastha Institute of Engineering and Technology, Chennai, Tamil Nadu, India

K. Parthasarathy
Siddha Central Research Institute, Central Council for Research in Siddha, Ministry of AYUSH, Arumbakkam, Chennai, Tamil Nadu, India

Inamuddin et al. (eds.), *Sustainable Polymer Composites and Nanocomposites*,
https://doi.org/10.1007/978-3-030-05399-4_11

materials are of great concern in scientific research and industry which has overcome the disputes over the use of synthetic materials. These materials include biomass as industrial and agricultural residues and energy crops. The preparation of polymer composites using natural polymers has advantages over synthetic sources, particularly as a solution to the environmental problems [109]. Also, the green chemistry explores chemistry techniques and methodologies so as to reduce or eliminate the use of hazardous materials to human health or to the environment [116].

Therefore, sustainable green nanocomposites are these days broadly researched due to the want for transformation to nanocomposites from biodegradable polymers, complete eco-friendly conservation and reduction of carbon dioxide release [2]. The nanocomposite is the term, which is extensively used to describe a broad range of materials with one of the dimension is in nano range. Blends of polymers and nanoparticles, commonly called "polymer nanocomposites" (PNC), have garnered much attention due to the possibility of dramatic improvement of polymeric properties with the addition of a relatively small fraction of nanoparticles [49, 157, 122, 32, 145, 146, 132]. The use of agricultural wastes has attracted many researchers for the production of nanocomposites [128]. Polymer nanocomposites (PNCs) are made by dispersing nanoparticles (NPs) into polymer matrices [112]. The use of PNCs is anticipated to enhance producing speed and utilization with increased biocompatibility [109]. This chapter will provide the physicochemical and biological importance of sustainable nanocomposites in various fields of applications with the impact of its properties.

2 Composites

Composites, the wonder materials composed of matrix phase and dispersed phase and have become an essential part of today's material. They are widely used in diverse areas of applications such as transportation, construction, electronics and consumer products. Composites are the materials that consist of two or more chemically and physically different phases separated by a distinct interface. Matrix phase is the primary phase having a continuous character, usually more ductile and less hard phase. Dispersed (reinforcing) phase is the second phase which is embedded in the matrix and it is usually stronger than the matrix. When this different phases combined, it will lead to the material with more useful structural or functional properties non-attainable by any of the constituent alone [45]. They also offer unusual combinations of properties such as stiffness, strength, and weight which was difficult to attain separately by the individual components [48, 49]. The progress in the field of materials science has taken a new led since the advent of nanocomposites.

2.1 Nanocomposites

Nanocomposites are a relatively new class of materials with ultrafine phase dimensions, typically of the order of a few nanometers. Nanocomposites are the composites which are size dependent with at least one of the phases shows dimensions in the nanometer range [123] and are different from those of the atomic and bulk counterparts [45]. The size plays a vital role in the nanocomposites with unique properties typically not shared by their microcomposite and, therefore, offer new technology and business opportunities [48, 49]. According to the type of matrix materials that are used for the preparation of nanocomposites, the nanocomposites are classified into following three classes [34]: Polymer Matrix Nanocomposites, Ceramic Matrix Nanocomposites and Metal Matrix Nanocomposites.

In the current economic and environmental climate, it is more critical to develop sustainable composites utilizing nanotechnology. But the polymers which have excellent properties such as low cost, lightweight, easy processing, corrosion resistance, ductility and high durability the productions of sustainable nanocomposites are achievable. These properties can arrive from the polymer more over than ceramics and metals. Also on comparing with metals and ceramics, polymers have poor electrical, thermal and mechanical properties, low coordination number and lightweight atoms, poor gas barrier, heat resistance and fire performance properties. Therefore, polymers can find many more applications as engineered materials [141].

2.2 Sustainable Polymeric Nanocomposites

Polymer nanocomposites (PNCs) are two-phase materials with high industrial importance, in which the polymers are reinforced by nanoscale fillers. All thermoplastics, thermosets, and elastomers have been used to make sustainable polymer nanocomposites. As a result of increasing focus concerning on the environment, the development of new environmentally friendly or biodegradable polymers has come on the role and these materials are called environmentally friendly PNCs (EFPNCs). The field of applications EFPNCs is immense [118, 13], because of their lightweight, high chemical resistance, and low dielectric constant, high interfacial area, unique shape- and size-dependent, tunable properties [110]. Also, PNCs have to provide the scope for the improvement of various mechanical, thermal, optical, rheological, magnetic, and electrical [95] even at a very low loading of nanoparticles. Due to the advantage of nanotechnology nowadays, nanoscale materials have been used as fillers instead of conventional microscale materials and also provide better properties [110].

Nanoparticles, such as nanochitosan, nanocellulose, carbon nanotubes, clays or layered silicates, nano silica, graphene, and nano calcium carbonate are widely used in the polymer nanocomposites in the development of sustainable polymer nanocomposites to alter the chemical, electrical, mechanical, optical, and thermal properties of biodegradable polymer matrices [94]. However, nanocomposites prepared using biodegradable polymers and clays materials are much attractive and impressive because of its enhancements in the physicochemical properties.

3 Preparation Methods

There is a great challenge over the past decade in preparing the uniform nanocomposite using hydrophobic (water repelling) polymer matrix and hydrophilic (water absorbing) fibers. Because this combination resulted in the non-uniform dispersion of fibers within the matrix with poor mechanical properties [11, 66]. Traditionally, there are three methods available for the preparation of nanocomposites. The methods are solution casting, melt blending, and in-situ polymerization.

Nowadays, nanocomposites are engineered with a variety of sizes, shapes, and chemistries. Also, for preparing nanocomposites, initially the nanoparticles are synthesized and modified for its surface area and exposed to the physical and biological application in our current research paradigm. In the literature reports, many researchers adopted various techniques for the fabrication and modification of nanomaterials. In the case of biomedical application, the nanostructures material to be capable of navigating the body, infecting and transforming cells, or detecting and repairing diseased cells. Therefore to engineer such kind of biocompatible material, the better understanding of the size, shape, surface area and functional properties of the nanomaterials and its interaction with the system are needed. Hence the fabrication methods include solvent casting, phase separation, drawing-processing, template-assisted synthesis, self-assembly, and electrospinning techniques.

3.1 Electrospinning

Recently, electrospinning methods have achieved widespread attention as an easy, leading and adaptable technique for the fabrication of fibers matrix in nanometer scale. Electrospinning technique is a preferred technique because of the delivering of nanofibers with the high surface area to volume ratio and a large number of inter-/intra fibrous voids capable for various applications over solvent casting and phase-separation [120]. From the literature, it was reported that the electrospinning technique has helped for the advancements in many applicable areas like bioengineering, environmental protection, electronics, optics, sensors and catalysis [24, 28, 70, 74].

Especially in biomedical applications, nanofibers fabricated via electrospinning methods leads to the materials with high surface area which is essential for surface interchanges and highly interconnected porous structure for the diffusion and transport of fluid, essential nutrients and drugs for tissue engineering applications to achieve a great success in that field [130]. This type of fibrous mats can be produced using electrospinning method by electrically charging a suspended droplet of polymer melt or solution and control over physical nature of the fibrous mats with suitable thickness and composition along with the porosity of the (nano) fibrous mesh [125] which is available in the electrospinning instrument. The physico-chemical properties such as large surface-to-volume ratio, flexibility, surface functionality, mechanical properties, and the freedom on materials' design [51, 59] are possessed by the resulting ideal membranes for the above specified important applications.

3.2 One Step in-Situ Polymerization Method

Zou et al. [165], reported one step in-situ polymerization method for the preparation of polymeric nanocomposites. This technique involves the dispersion followed by bulk polymerization. During dispersion of nanofillers in the monomer(s), the nanofillers was always modified initially to increase the interaction or good dispersion between the polymers. Easy handling and better performance are the important advantages of this method. Using this method, [54], synthesized graphene, GO, and functionalized GO—Epoxy nanocomposites, Dash et al. [33] synthesized poly(anthranilic acid) (PAnA)/MWCNT composites and Wu and Liu [158] prepared PS/MWCNTs.

3.3 Thermal Spray Synthesis

Thermal spray processing is a well-proven technique for the development of nanostructured coating [64]. This thermal spray technique is considered as an effective method for the preparation of nanocomposites because agglomerated nanocrystalline powders formed during synthesis of nanomaterials are melted, accelerated and quenched very rapidly in a single step. Due to rapid melting and solidification process, there will be the retention of a nanocrystalline phase and even amorphous structure. This retention of the nanocrystalline structure leads to enhanced wear behaviour, greater hardness, and reduced coefficient of friction compared to other conventional coatings methods.

3.4 Sol-Gel Method

It is a bottom-up approach using for the preparation of polymeric nanocomposites. The term sol-gel is associated with two relations steps, sol (a colloidal suspension) and gel (3D interconnecting network formed between phases) [119]. During engineering the nanostructured material using this technique, the solid nanoparticles is dispersed initially in the monomer solution. The resulting colloidal suspension (sol) from polymerization reactions forms an interconnecting 3D network which extends throughout the liquid between phases (gel followed by the hydrolysis procedure [137]. Here, the polymer serves as a nucleating agent and promotes the growth of layered crystals. As the crystals grow, the polymer seeps between layers and hence nanocomposite is formed.

4 Biophysical Properties

Polymer nanocomposites always exhibit enhanced physical properties than individual materials [80]. These properties do not depend solely on the chemical structure of the matrix but also on other factors like processing steps which determine the orientation of the molecules in the final morphology and the effect of fillers. The enhanced of properties can be arrived even at small amounts of nanoparticles loadings in the matrix and were browbeaten at the economic degree for decades already [13].

Many parameters of the nanoparticles can also play a position inside the reinforcement of the polymer matrix. The parameters are nanoparticle shape and size, loading and dispersion, interaction, mobility, temperature, entanglement of the polymers, diploma of polymerization, elasticity, surface chemistry, and biopersistence [75]. The effect of particle morphology is the main parameter during synthesis and for the synthesis of nanoparticles, the solutions of polymeric materials used should contain size-confined, nanosized pools of inter- and intramolecular origin. Also for optimizing the ultimate properties of the nanocomposites for various application the understanding of the relationship between processing, morphology and functional properties of nanocomposites will be very helpful for predicting properties of nanocomposite systems.

4.1 Structure

As structural members, composite materials are used as crashworthiness, protective armours in air and space vehicles. Nanocomposites materials can be produced both by complex and simple processes. For preparing the nanocomposites with the preferred structure the first essential condition is the formation of homogeneous phase by the nanoparticles in the polymer matrix [71]. The complexity of structure and the factors figuring out it trade from one nanocomposite to the other, consequently the structure is mentioned according to the kind of reinforcement. The nanostructure phases of nanocomposite structure can be defined as:

zero-dimensional (embedded cluster), 1D (one-dimension; e.g. nanotubes), 2D (nanoscale coatings) and 3D (embedded network).

Particulate reinforcements have dimensions which might be about same in all guidelines. Large particle and dispersion-strengthened composites are the 2 sub-classes of particle-reinforced composites [136]. Depending upon the reinforcements, the PNCs have structures like molecular layers, films, membranes, plates, particles, network, nanotubes and nanofibers, ribbon, tape, rosette, cage-like and tubular morphologies have been synthesized [71, 50].

4.2 *Shape*

Numerous attempts had been made to distribute spherical nanoparticles in a polymer matrix with the maximum one of a kind strategy, due to the fact that aggregation is the main issue in those composites. The most regularly used method is the use of conventional technologies to homogenize the previously organized particles into the matrix polymer. According to the shape of the crystallites or grains, the nanomaterials are broadly classified into four categories such as (1) Three dimensional (3D) nanomaterials like zeolites or molecular sieves with cage-like nanopores, (2) Two dimensional (2D) nanomaterials such as thin films, quantum well, (3) One dimensional (1D) nanomaterials such as quantum wire/quantum rod etc., (4) Zero-dimensional (0D) nanomaterials such as quantum dots [134]. In 3D, nanomaterials are composed of equiaxed nanometer-sized grains. In this case, the electrons are delocalized and move freely in all directions (x, y and z). In 2D, the electrons are delocalized in two directions (x and y) and confined in one direction (z). In 1D, the electrons are delocalized in one direction (z) and confined into two directions (x and y). Zero dimension exhibits confinement in all three spatial directions and there is no delocalization. Usually, particles don't have any preferred directions and are specially used to improve properties or lower the cost of isotropic materials [17]. The shape of the reinforcing particles can be spherical, cubic, platelet, or regular or irregular geometry.

4.3 *Size*

Lately, there has been an extremely good deal of attempt to use both the shape and the size of nanoparticles to goal precise cell uptake mechanisms, biodistribution styles, and pharmacokinetics [75]. Even though the physicochemical properties of nanosized complexes and structures have an effect on their function [40], it will likely be critical to apprehend how the physicochemical properties relate to biological interactions and functions in order to appropriately materialize the capability of nanotechnology. For instance, the shape at once impacts uptake into cells: rods display the best uptake, followed with the aid of spheres, cylinders, and cubes [53].

NP size may additionally have an effect on the uptake efficiency and kinetics, the internalization mechanism and additionally the subcellular distribution. A size-dependent uptake in different cell lines has been reported for Au [10, 30, 156], mesoporous silica [91], polystyrene [147] and iron oxide NPs [61], with the maximum cellular uptake at a NP core size in the range of 30–50 nm, which suggests that there is an optimal size for active uptake. Gratton et al. [53] determined this ordering using synthesized nanoparticles larger than 100 nm. In studies with sub-100-nm nanoparticles, spheres show an appreciable advantage over rods [30, 114].

4.4 Surface Area/Morphology

Surface area is the important parameter to enhance the applications of nanocomposites. The surface area of the nanoparticles is more essential for the improvements of the properties, although, the high surface area and surface attraction can also result in agglomeration. The nanofillers may not give expected results if they exist as micrometre-sized agglomerates. The two main aspects of mixing viz, dispersion and distribution, facilitate the breakdown of the filler agglomerates into aggregates and particles, as well as distributing the particles uniformly throughout the polymeric matrix, without affecting the particle size and final properties. The improved properties can be obtained if there is a relatively better dispersion of nanoparticles in the polymer matrix. A good dispersion enhances the number of possible interaction between the polymer and the filler. Although, the introduction of nanoparticles into polymer or polymer blends have benefited in improving various properties, the polymer that interacts with the nanofiller plays the determining role in the final property. The high specific area associated with nanoparticles enhance the chances of adsorption of the macromolecular chains onto their surfaces and influences the structural evolution and phase separation. The effective dispersion of the fillers in the polymer matrix and the improvement of polymer-filler interactions are two key challenges in the field of polymer nanocomposites.

4.5 Applications

Size, shape, and surface morphology play pivotal roles in controlling the physical, chemical, biological, optical, and electronic properties of nanoscopic materials. The high surface energy of these particles makes them extremely reactive, and most systems undergo aggregation without protection or passivation of their surfaces [44, 155, 32, 144]. The use of nanoparticles (NPs) in the biomedical field is one of the most important branches of nanobiotechnology.

4.6 Biological Applications

The biological applications of nanocomposites mainly depend on the specific structure−activity relationships between NP shape and certain common biological endpoints [154, 152]. The variables such as size, shape, composition, ligand chemistry and coverage can influence the toxicity of nanoparticles to organisms. This can be avoided by analyzing the unique effects shown by each variable. It is a highly challenging job to study the impact of each parameter, but it can be studied by changing one variable and keeping others as constant [19].

From the biological viewpoint, the majority of microorganisms such as bacteria and virus appears in different shapes. The majority of bacterias and viruses appears to be spherical in shape. The other shapes such as filamentous, crescent, twisted, bullet- and rod-shaped forms have also been observed in nature [75]. Khatamiana et al. [73] has worked on binary hybrids of chitosan-zeolite nanocomposites various medical applications such as tissue engineering and industrial applications such as antimicrobial food packaging, which exhibit dose-dependent toxicity to cells and very low toxicity was observed up to 0.2 5 and 0.1 mg/mL concentrations.

Lifeng et al. [113] and Ma and Lim [93] reported that the nanochitosan exhibited higher antibacterial activity and antifungal activity. The antibacterial activity is due to the fact that the negatively charged surface of the bacterial cell wall interacts effectively to a greater degree with the polycationic nanochitosan (NH_3^+) and hinder the growth of the microorganism [164]. This suggested also suggested that chitosan nanoparticles might also be able to diffuse into a fungal cell and hence disrupt the synthesis of DNA as well as RNA. This interaction caused disruption of the microbial cells, which then changed their metabolism and led to cell death [83].

4.7 Drug Delivery/Gene Delivery

The use of nanocomposite in drug delivery application depends on many factors such as size, shape, surface morphology and its composition [87]. The particle size is the prime factor which determines the half-life of drug clearance in tissues [65]. To arrive at a tumour-specific nanoparticle design, it is first essential to understand the relationships between nanoparticle size and shape with its transport. The remarkable factor in the competence of most of the drug delivery system is particle size and surface area. These factors will enhance the solubility and bioavailability, an additional capability to cross the blood-brain barrier (BBB), go through the pulmonary system and be absorbed through the endothelial cells of the skin [76]. The effect of particle size on nanoparticle pharmacokinetics, margination, extravasation, and binding has been comprehensively studied. Early studies evaluating the effect of liposome size on pharmacokinetics have identified the size range of 60–100 nm to be ideal for maximizing blood circulation half-life [88, 100, 104].

When the particles size of the nanoparticles gets smaller, the surface area of the nanoparticles gets larger, which results in the adherence of the drug molecules to the surface of the particles. The above literature assessment and discussion leads to the importance of size optimization of nanoparticles composites for the effective delivery of the drug molecules. It was predicted that approximately 100 nm size of drug carrier will be effective. Due to the high diffusive component of transport for smaller nanoparticles, there may be significant particle washout from a tumour over time [142]. On the other hand, larger nanoparticles have higher long-term tumour accumulation. Therefore caution must be taken with size selection, however, as pharmacokinetic and margination performance does not necessarily correlate to faster extravasation kinetics. For instance, the apparent permeability of a nanoparticle to the vascular endothelial wall decreases as the radius of the particle increases [162]. The particle size and distribution of nanoparticles can be controlled by different synthesis method and modified various synthesis parameters for example by adjusting the pH and ionic strength in coprecipitation, a commonly used method [68].

The shape also plays a major role in not only nanoparticle pharmacokinetics, but also intravascular transport, binding, and accumulation at the site of a tumour. Hence, the basic ideas about nano-bio interactions will be helpful in engineering the nanostructured material for the specific design of carrier to ensure the highest possible delivery efficiency. In many cases, nanoparticles enter the cell after binding to the receptor target. Once bound, several factors can dictate the behaviour of nanomaterials at the nano-bio interface.

Li et al. [87] reported that effect of nanoparticles shapes using a coarse grain model with self-consistent field theory and Flory theory. They developed PEGylated in different shapes such as nanospheres, nanorods, nanocubes, and nanodisks with the same surface area of 201 nm^2 to ensure an equal number of PEG molecules on the surface regardless of shape [87], polyethylene glycol (PEG) coating improves the biocompatibility and circulation time [150].

Veiseh et al. [148] studied the effect of surface charge on internalization on different cell lines by MNPs. The report tells that the material can bind with non-targeted cells, resulting in a specific internalization which was due to the surface phenomenon of the MNPs. [98], utilized Chitosan functionalized magnetite doped luminescent rare earth nanoparticles (Fe_3O_4@LaF_3: Ce^{3+}, Tb^{3+}/chitosan nanoparticles) in targeted delivery of paclitaxel for lung cancer.

An et al. [6], used glycopolymer modified magnetic mesoporous silica nanoparticles for Magnetic resonance imaging (MRI) scan and controlled drug delivery of anticancer drug 5-fluorouracil. The hybrid Fe_3O^{4-} carboxymethyl chitosan nanoparticles [87] finds the medical application of tumor-targeted delivery of rapamycin.

The other, nanoparticles coated with polymers or surfactants like polyethylene glycol polyethylene oxide, polyoxamer, poloxamine, and polysorbate 80 [8], single-walled carbon nanotubes (SWCNTs) and multiwalled carbon nanotubes

(MWCNTs) [115, 124] have been confirmed valuable in drug delivery applications. [25, 103] applied Buckminsterfullerene C60 (spherical molecule) and its derivatives are utilized for the treatment of cancer [69].

4.8 Wound Healing

Wound healing is a multifaceted and major biological response that helps in the restoration of tissue consistency and body functions [3]. The main point in wound healing is a physiological situation favourable to tissue repair and rejuvenation in the shortest time with less pain, restlessness, and scarring to the patient. Wound healing follows a complex sequence of events at the cellular level, which replaces the devitalized tissues and restores the natural characteristics of the skin that are conciliation by an injury (Matthews and Rawlings [99]. A recent study of biomaterials on wound healing has gained special interest because of their inimitable chemical, physical, and biological properties.

Nanofibers are key materials for wound healing applications because of their surface area. There are several methods for producing nanofibers such as drawing, template synthesis, phase separation, self-assembly, and electrospinning. A range of polymeric, metallic, and ceramic nanomaterials are investigated for wound healing applications.

Silver has been used as an antimicrobial since the 1800s but in the last two decades, interest in silver for wound treatment resurged [42]. Silver nanoparticles impregnated in the biocompatible polymeric matrix were therefore utilized for antimicrobial wound healing material [97]. TiO_2 and ZnO nanoparticles are extensively used in the cosmetic and pharmaceutical industry as UV shield and also as a wound healing material (Murphy and Evans [102]. Antimicrobial peptides formulated with gold Nanodots inhibit the multiplication of multidrug-resistant bacteria pathogen and also promotes wound healing in an animal model [86]. The drug delivery of curcumin using various nanomaterial-based vehicles has been studied in wound healing acting as anti-biotic, anti-viral and antioxidant properties [21].

The size and concentration of nanoparticles will influence the wound healing characteristics. Graphene oxide is used as antibacterial in the field of biomedicine. Graphene related materials have shown potential results in the area of tissue engineering and wound healing [6].

4.9 Tissue Engineering

Tissue engineering (TE) is a multidisciplinary field focused on the development and application of knowledge in chemistry, physics, engineering, life and clinical sciences to the solution of critical medical problems, as tissue loss and organ failure [82]. In tissue engineering, the scaffold prepares a three-dimensional (3D) construct for cell attachment and migration, proliferation, and formation of an extracellular matrix (ECM), deliver and retain cells and biochemical factors enable diffusion of vital cell nutrients, as well as a carrier of growth factors or other biomolecular signals [9]. Many biomedical applications require nanoparticles with low toxicity, biocompatibility, and stability with high preferential accumulation in the target tissue or organ [79].

Scaffolds for tissue engineering should have good mechanical properties, suitable biodegradability, and most importantly good biocompatibility [9, 117]. Particularly, the surface properties of the material define the interactions between the cells and the material and, consequently, affect cell adhesion [36].

Nanomaterials and nanotechnology are being progressively integrated into tissue engineering to ease the development of various tissue treatment and substitutes to restore missing, damaged and injured tissues and organs (Fig. 1). The triumph of nanomaterials in tissue engineering is contingent upon the inflammatory responses they elicit *in vivo* [96] and also be utilized for a number of biomedical and biotechnological applications such as drug delivery, enzyme immobilization and DNA transfection etc. [84]. Nanotechnology offers hybrid materials with potentials at the molecular level [35].

Yadav et al [161] suggested that Graphene-based chitosan nanocomposites have been used in tissue engineering application for wound healing purpose. A nano-HAp/collagen composite was fabricated into a three-dimensional scaffold from nano-HA powders and natural collagen by simply blending the two

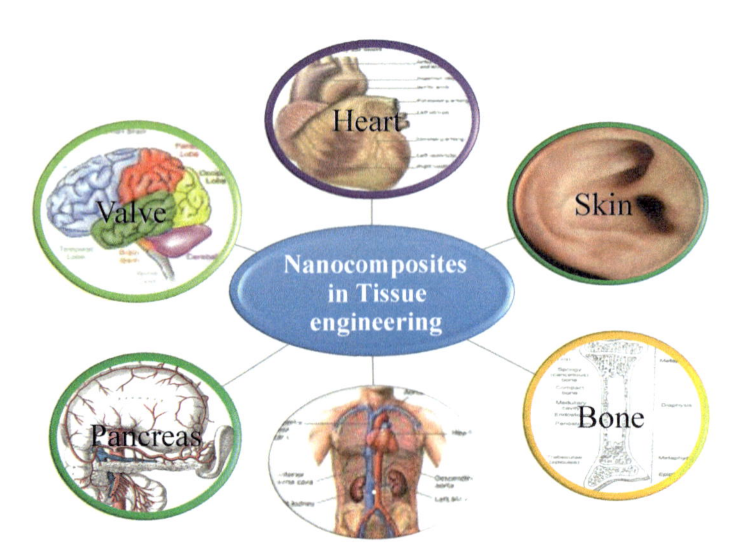

Fig. 1 Polymer nanocomposites in tissue engineering applications

components together. The composite of CNPs has also been generated for bone tissue engineering.

The naturally derived silk fibroin nanoparticles find numerous applications in various application mainly in bone regeneration and dentistry nanospheres and nanofibrous scaffolds [111, 153, 57]. Fonseca-Santos and Chorilli [43] reported the faster cell attachment on graphene/CS nanocomposite films displaying good biocompatibility and non-cytotoxicity. Similarly, [29, 151], prepared CS/GO nanocomposites containing nano-hydroxyapatite composites with improved mechanical properties. These materials can show high cell proliferation rate [133] and [149]. Chitosan-gold nanoparticles have managed to supplement osseointegration of dental implants. Nanomaterials enhance the wound healing and burn curing. Gold and silver metal nanoparticles prove to be a marvellous property such as low toxicity in vivo, bacteriostatic and bactericidal activities [107].

Silver nanoparticles based bandage supports the recovery of severe partial thickness burns. The benefit of silver nanoparticles-based dressings, even for a prolonged period, does not have a negative effect on the proliferation of fibroblasts and keratinocytes, which leads the healing of the wound [58]. The implant of Collagen with silver nanoparticles proves to have the antibacterial activity and makes it an appropriate component for wound dressing material.

According to [52], Gold nanoparticles are biocompatible, cell viable, hence used in drug delivery and wound healing process. The cross-linking of collagen with gold nanoparticles which allowed the ease assimilation of biomolecules like growth factors, peptides, cell adhesion molecules by their immobilization at the Gold surface without altering of collagen structure. The combination of gelatin chitosan with gold nanoparticles shows safe and excellent wound healing property. [143], suggested that bio-composite scaffolds containing chitosan/ nanohydroxyapatite/ nano-copper-zinc have appropriate morphological and physical quality in Bone Tissue Engineering (BTE).

4.10 Water Treatment

Wastewater usually contains several poisonous and unsafe contaminants, such as heavy metals, organic impurities, dyes, pesticides, and inorganic salts etc., that being discharged from various industrial effluents and domestic resources [5]. The water reservoirs containing these substances causes many chronic or acute diseases, therefore these contaminants must be removed and discarded from the wastewater before being discharged into the nearest water bodies [1]. The methods for removing pollutants from the wastewater was given in Fig. 2. Many researchers suggested that the nanocomposites have excellent adsorption capability, selectivity, and steadiness than nanoparticles. So, they find a range of applications in multiple areas.

The adsorption of different pollutants such as heavy metal ions, dyes, and pesticides from the polluted water with the help of nanocomposites has alarmed

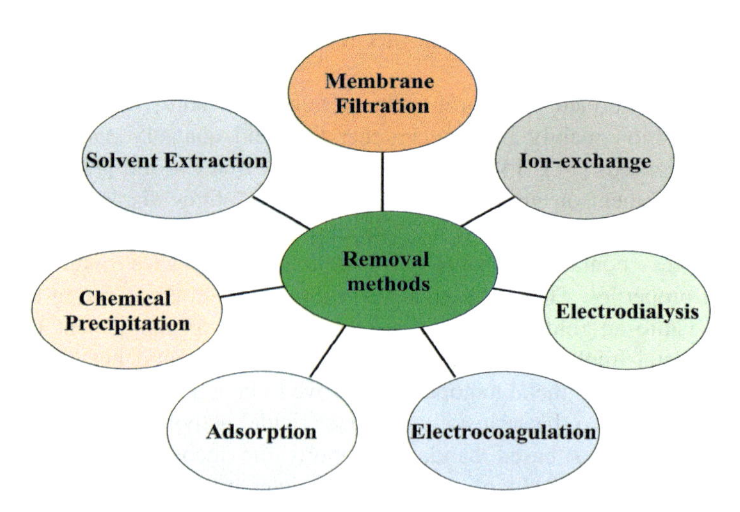

Fig. 2 Removal methods for wastewater treatment

important interest. In this chapter, various nanocomposites and its blends utilized for treatment of wastewater have been highlighted such as graphene, CNT, nanoclays [41, 126], nanochitosan [89], nano alginate [20, 38]. Bai et al. [12] suggested that the Graphene with certain polysaccharides (such as alginates, starch, cellulose, chitosan) forms nanocomposites with improved mechanical properties. Starch/graphene nanocomposite films with 0.2–3.0 wt% graphene platelets were prepared via aqueous solutions casting which was used as an adsorbent. Apart from this, it was also utilized to use in the selective determination of iodide in seafood samples. The heavy metals like Au(III) and Pd(II) were adsorbed on cross-linked CS/GO nanocomposites, with maximum adsorption capacities were observed [127].

Long and Yang [90], has found CNTs to be exceptional adsorbent for the removal of lead, cadmium and organic 1, 2- DCB from water, being a potential medium in wastewater treatment. Nanoclays also find applications in remediating heavy metals from the effluents which are readily available, eco-friendly, economical chemical substances and recent studies reveal the above findings [77]. The nanoclay combination with cationic starch was studied to relate the relationship between polymer and nanoclay in atrazine reduction.

The nano adsorption technique, have been effectively used to investigate the removal of pollutants. Remediation of wastewater using several nano-materials and nano-adsorbent is being concentrated recently [163]. Nano-adsorbent has high adsorption capacity on the surface of the nano-material [81].

Recently development on carbon-based nano-materials (CNMs) in the form of carbon nanotubes, carbon nanoparticles and carbon nanosheets has gained more attention for treatment of wastewater [26]. Other than this silicon, nanoclays, polymer-based nano-materials, nanofibers and aerogels nano-material is also used as nano-adsorbents which are discussed in the various literature. [47] reported that oxide based nano-particles have high surface area for the effective wastewater treatment applications [46], cerium oxides, Titanium oxide/dendrimers composites [15, 14] zinc oxides, magnesium oxide [55], manganese oxides, copper oxides, Zinc oxides [39] and ferric oxides [160, 159]. Gupta et al. [56] done Surface modification of Fe_2O_3 nanoparticles with 3-aminopropyltrimethoxysilane. According to Gupta et al. [56], the modification of these nano-adsorbents shows high affinity for the removal of different pollutants such as Cr^{3+}, Co^{2+}, Ni^{2+}, Cu^{2+}, Cd^{2+}, Pb^{2+}, and As^{3+} simultaneously from wastewater.

The nanoparticles of manganese oxide show high adsorption ability due to their morphological and polymorphic structure [92]. Generally, they were used for the removal of various heavy metals such as arsenic from wastewater [154, 152].

Ihsanullah Al-Khaldi et al. [62] used widely Carbon-based nanomaterials organic materials in the field of removing heavy metals in recent decades, due to its nontoxic nature and high sorption abilities. Earlier activated carbon was used as sorbents, but its drawback to remove heavy metals at micro levels other nanoparticles such as carbon nanotubes, fullerene, and graphene are synthesized and used as nanosorbents [6].

By using CNT such as MWCNT [60] and SWCNT were used to several studies reported the removal of heavy metals such as Pb(II) and Mn(II) [139], alumina supported on CNT adsorbed Cu(II) Pb(II) from its aqueous solution [55, 138].

Chen et al [24, 28] and [27], synthesized Inorganic nano-adsorbent the highly ordered $Mg(OH)_2$ nanotube arrays inside the pores anodic alumina membranes to form $Mg(OH)_2/Al_2O_3$ composite membranes. And these membranes are used to adsorb Nickel ions from wastewater with high efficiency. Chandra et al. [23] accounted magnetite-graphene adsorbents with a particle size of $\sim$10 nm which gives a high binding capacity of different oxidation states As^{3+} and As^{5+}, and the rate of binding capacity increases with increase in adsorption sites in the graphene composite.

Therefore nanocomposites have enormous characteristics predominantly for the effective removal of pollutant from wastewater even at micro level concentration, with high selectivity, sensitivity and adsorption capability.

4.11 Agriculture

Nanotechnology applications in the agriculture could play a fundamental role. The main application of nanocomposites in agricultural is to reduce the application of plant protection products, minimize nutrient losses in fertilization and increase yields through optimized nutrient management. Nanoparticles technology is a fast-growing strategy to tackle [106] specific treatment in the agricultural field. The goal of this imaging nanoparticles is to reduce the number of unnecessary troubles in agriculture sectors [140].

The new nanosensors such as active carbon nanotubes, nanofibers, and fullerenes also relevant implications for application in agriculture in particular for soil analysis, easy biochemical sensing and control, water management and delivery, pesticide and nutrient delivery. Research peoples are concentrating in the field of biopolymer composites in order to improve the properties of individual polymers. Generally, silicate, clay and titanium dioxide (TiO_2) added to biopolymers to improve mechanical, barrier properties, other functions, and applications, thereby acts as an antimicrobial agent, biosensors and oxygen scavenger [121]. Antifungal activities of polymers based copper nanocomposites against pathogenic fungi have been reported [31]. A functional hybrid nanocomposite based on the intercalation of two herbicides anions (2,4-dicholorophenoxy acetate and 4-chlorophenoxy acetate) with zinc – aluminium layered double hydroxide [16].

A number of studies reported that the nanoparticles mediated plant transformation has the potential for genetic modification in plants by observed plant reactions after contact. Specifically in specific agricultural problems in plant-pathogen interactions and provide new ways for the crop protection [105]. Nanosilver is widely in agriculture field because of its specific properties [67]. Nanotechnology develops effectiveness, safety, patient adherence, as well as reducing health care costs [7]. Nanoparticles not only influence the health agricultural plant growth, it also associated with null use of hazardous insecticides and pesticides by acting as the alternative product [18].

The application of nanotechnology would be highly promising in agricultural research to put forward not only the detection of plant diseases and also in the analysis and assessment of the use of nanoparticles in plant tissues.

5 Conclusion

Nanostructured materials have a greater impact in various fields. The ability to engineer nanostructured materials has already demonstrated great value. In the next decade, it will be important to elucidate how the physicochemical properties of nanomaterials and their by-products interact with biological systems. This chapter will substantially have an effect on our capacity to engineer new generations nanomaterials with nontoxic and specified properties for various applications. This chapter also provides the essential ideas on basic concepts of structural properties of the material successful nanomaterials to design.

References

1. Abdullahassan MA, Souabi S, Yaacoubi A, Baudu M (2006) Removal of surfactant from industrial wastewaters by coagulation flocculation process. Int J Environ Sci Technol 3(4): 327–332
2. Adeosun SO, Lawal GI, Balogun Sambo A, Akpan Emmanuel I (2012) Review of green polymer nanocomposites. J Miner Mater Charact Eng 11(4):483–514
3. Ajayan PM, Schadler LS, Braun PV (2006) Nanocomposite science and technology, Wiley, New York, NY, USA

4. Amass W, Amass A, Tighe BA (1998) Review of biodegradable polymers: Uses, current developments in the synthesis and characterization of biodegradable polyesters, blends of biodegradable polymers and recent advances in biodegradation studies. Polym Int 47: 89–144
5. Amuda OS, Amoo IA, Ajayi OO (2006) Performance optimization of coagulation/ flocculation process in the treatment of beverage industrial wastewater. J Hazard Mater 129:69–72
6. An J, Zhang X, Guo Q, Zhao Y, Wu Z, Li C (2015) Glycopolymer modified magnetic mesoporous silica nanoparticles for MR imaging and targeted drug delivery. Colloids and Surfaces A. Physicochemical Eng Aspects 482:98–108
7. Anwunobi AP, Emeje MO (2011) Recent application of natural polymers in nanodrug delievery. J Nanomedic Nanotechnol. S 4:002
8. Araujo L, Lobenberg R et al (1999) Influence of the surfactant concentration on the body distribution of nanoparticles. J Drug Target 6:373–385
9. Armentano I, Dottori M, Fortunati E, Mattioli S, Kenny JM (2010) Biodegradable polymer matrix nanocomposites for tissue engineering : a review. Polym Degrad Stab 95:2126–2146. https://doi.org/10.1016/j.polymdegradstab.2010.06.007
10. Arnida, Malugin A, Ghandehari H (2010) Cellular uptake and toxicity of gold nanoparticles in prostate cancer cells: a comparative study of rods and spheres. J Appl Toxicol 30(3): 212–217
11. Ashori A (2008) Wood-plastic composites as promising green-composites for automotive industries. Bioresour Biotechnol 99:4661–4667
12. Bai J, Zhong X, Jiang S, Huang Y, Duan X, (2010) Graphene nanomesh nature nanotechnology (5)3:190–194
13. Balazs AC, Emrick T, Russell TP (2006) Nanoparticle polymer composites: where two small worlds meet. Science, 314(5802), 1107–1110 https://doi.org/10.1126/science.1130557
14. Barakat MA, Al-Hutailah RI, Hashim MH, Qayyum E, Kuhn JN (2013) Titania supported silver-based bimetallic nanoparticles as photocatalysts. Environ Sci Pollut Res 20(6):3751–3759
15. Barakat MA, Ramadan MH, Alghamdi MA, Al-Garny SS, Woodcock HL, Kuhn JN (2013) Remediation of Cu (II), Ni (II), and Cr (III) ions from simulated wastewater by dendrimer/ titania composites. J Environ Manage 117:50–57
16. Bashi AM, Haddawi SM, Dawood AH (2011) Synthesis and characterizations of two herbicides with Zn/Al layered double hydroxide nanohybrides. J Kerbala Univ 9(1):9–16
17. Bednarcyk BA (2003) Compos B 34:175–197
18. Begum N, Sharma B, Pandey RS (2010) Evaluation of insecticidal efficacy of calotropis procera and annona squamosa ethanol extracts against musca domestica. J Biofertil Biopestici 1:101
19. Buchman JT, Miranda J, Gallagher, Chi-Ta Yang, Xi Zhang, Miriam O P, Krausee Rigoberto, Hernandez, Galya Orr, (2016) Research highlights: examining the effect of shape on nanoparticle interactions with organisms. Environ Sci Nano https://doi.org/10.1039/ c6en90015a
20. Cao K, Jiang Z, Zhao J, Zhao C, Gao C, Pan F, Wang B, Cao X, Yang J (2014) Enhanced water permeation through sodium alginate membranes by incorporating graphene oxides. J Membr Sci 469:272–283
21. Castangia I, Nácher A, Caddeo C, Valenti D, Fadda AM, Díez-Sales O, Ruiz-Saurí A, Manconi M (2014) Fabrication of quercetin and curcumin bionanovesicles for the prevention and rapid regeneration of full-thickness skin defects on mice. Acta Biomater 10(3):1292–1300
22. Chandra R, Rustgi R (1998) Biodegradable polymers. Prog Polym Sci 23:1273–1335
23. Chandra V, Park J, Chun Y, Lee JW, Hwang IC, Kim KS (2010) Water-dispersible magnetite-reduced graphene oxide composites for arsenic removal. ACS Nano 4:3979–3986
24. Chen L, Bromberg L, Hatton TA, Rutledge GC (2007) Catalytic hydrolysis of p-nitrophenyl acetate by electrospun polyacrylamidoxime nanofibers. Polymer 48(16):4675–4682

25. Chen Z, Mao R et al (2012) Fullerenes for cancer diagnosis and therapy: preparation, biological and clinical perspectives. Curr Drug Metab 13(8):1035–1045
26. Chen GC, Shan XQ, Wang YS, Wen B, Pei ZG, Xie YN, Liu T, Pignatello JJ (2009) Adsorption of 2,4,6-trichlorophenol by multiwalled carbon nanotubes as affected by Cu(II). Water Res 43(9):2409–2418
27. Chen CZ, Zhou ZW (2004) The preparation of nano-ZnO and its middle infrared-ultraviolet-visible light absorption properties. J Funct Mater 35:97–98
28. Chen X, Mao SS (2007) Titanium dioxide nanomaterials: synthesis, properties, modifications and applications. Chem. Rev 107:2891–2959
29. Cheung RC, Ng TB, Wong JH, Chan WY (2015) Chitosan: an update on potential biomedical and pharmaceutical applications. Mar Drugs 13:5156–5186
30. Chithrani BD, Ghazani AA, Chan WC (2006) Determining the size and shape dependence of gold nanoparticle uptake into mammalian cells. Nano Lett 6(4):662–8
31. Cioffi N, Torsi L, Ditaranto N et al (2004) Antifungal activity of polymer- based copper nanocomposite coatings. Appl Phys Lett 85(12):2417–2419
32. Crandall BC (ed) (1996) Nanotechnology MIT Press, Cambridge
33. Dash MP, Tripathy M, Sasmal A, Mohanty GC, Nayak P (2010) Poly(anthranilic acid)/multi-walled carbon nanotube composites: spectral, morphological, and electrical properties. J Mater Sci 45(14) 3858–3865
34. Deborah DL, Chung (2002) Composite materials, functional materials for modern technologies, Springer-Verlag London Ltd, UK
35. Deepachitra R, Nigam R, Prohit SD, et al (2014) In vitro study of hydroxyapatite coatings on fibrin functionalized/pristine graphene oxide for bone grafting. Mater Manuf Process 30(6): 804–811
36. Dobkowski J, Kołos R, Kamiński J, Kowalczyńska HM (1999) Cell adhesion to polymeric surfaces: experimental study and simple theoretical approach. J Biomed Mater Res 47: 234–242
37. Drzal LT (2010) Sustainable biodegradable green nanocomposites from bacterial bioplastic for automotive applications. http//www.egr.msu.edu/cmsc/biomaterials/index.html (accessed on 20 August 2010)
38. Fan J, Shi Z, Lian M, Li H, Yin J (2013) Mechanically strong graphene oxide/sodium alginate/polyacrylamide nanocomposite hydrogel with improved dye adsorption capacity. J Mater Chem A 1. 51:7433–7443
39. Fang J, Fan H, Ma Y, Wang Z, Chang Q (2015) Surface defects control for ZnO nanorods synthesized by quenching and their anti-recombination in photocatalysis. Appl Surface Sci 332:47–54
40. Feldherr CM, Lanford RE, Akin D (1992) Signal-mediated nuclear transport in simian virus 40–transformed cells is regulated by large tumor antigen. Proc Natl Acad Sci USA 15:11002–11005
41. Floody MC, Theng B, Reyes P, Mora M (2009) Natural nanoclays: applications and future trends–a Chilean perspective. Clay Miner 44:161–176
42. Fong J, Wood F (2006) Nanocrystalline silver dressings in wound management: a review. Int J Nanomed 1(4):441–449
43. Fonseca-Santos B, Chorilli M (2017) An overview of carboxymethyl derivatives of chitosan: Their use as biomaterials and drug delivery systems. Mater Sci Eng C Mater Biol Appl 77:1349–1362
44. Freeman RG, Grabar KC, Allison KJ, Bright RM, Davis JA, Guthrie AP, Hommer MB, Jackson MA, Smith PC, Walter DG, Natan MJ (1995). SERS Substrates. Science 267
45. Gangopadhyay R, De A (2000) Conducting polymer nanocomposites: a brief overview 608–622
46. Gao C, Zhang W, Li H, Lang L, Xu Z (2008) Controllable fabrication of mesoporous MgO with various morphologies and their absorption performance for toxic pollutants in water. Cryst Growth Des 8:3785–3790

47. Geng B, Jin Z, Li T, Qi X (2009) Kinetics of hexavalent chromium removal from water by chitosan-Fe0 nanoparticles. Chemosphere 75(6):825–830
48. Giannees BEP (1996). Polymer layered silicate nanocomposites. 29–35
49. Giannelis EP (1996) Adv Mater 8:29
50. Gleiter H (2000) Nanostructured materials: basic concepts and microstructure. Acta Mater 48:1
51. Graham K, Schreuder-gibson H, Gogins M (2003) Incorporation of electrospun nanofibers into functional structures. Tech Assos Pulp Pap Ind 1–16
52. Grant SA, Spradling CS, Grant DN (2014) Assessment of the biocompatibility and stability of a gold nanoparticle collagen bioscaffold. J Biomed Mater Res, Part A 102:332–339
53. Gratton SE, Ropp PA, Pohlhaus PD, Luft JC, Madden VJ et al (2008) The effect of particle design on cellular internalization pathways. Proc Natl Acad Sci USA 105:11613–11618
54. Guo Y, Bao C, Song L, Yuan B, Hu Y (2011) In situ polymerization of graphene, graphite oxide, and functionalized graphite oxide into epoxy resin and comparison study of on-theflame behavior. Ind Eng Chem Res 50:7772–7783
55. Gupta VK, Agarwal S, Saleh TA (2011) Synthesis and characterization of alumina-coated carbon nanotubes and their application for lead removal. J Hazard Mater 185:17–23
56. Gupta VK, Tyagi I, Sadegh H, Shahryari-Ghoshekand R, Makhlouf ASH, Maazinejad B (2015) Nanoparticles as adsorbent; a positive approach for removal of noxious metal ions: a review. Sci Technol Dev 34:195
57. He P, Sahoo S, Ng KS, Chen K, Toh SL, Goh JC (2013) Enhanced osteoinductivity and osteoconductivity through hydroxyapatite coating of silk-based tissue-engineered ligament scaffold. J Biomed Mater Res A 101:555–566
58. Heydarnejad MS, Rahnama S, Mobini-Dehkordi M, Yarmohammadi P, Aslnai H (2014) Sliver nanoparticles accelerate skin wound healing in mice (Musmusculus) through suppression of innate immune system. Nanomed J 1(2):79–87
59. Huang Z, Zhang Y, Kotaki M, Ramakrishna S (2003) A review on polymer nanofibers by electrospinning and their applications in nanocomposites. Compos Sci Technol 63:2223–2253. https://doi.org/10.1016/S0266-3538(03)00178-7
60. Huang ZN, Wang XL, Yang DS (2015) Adsorption of Cr(VI) in wastewater using magnetic multi-wall carbon nanotubes. Water Sci Eng 8(3):226–232
61. Huang J, Bu L, Xie J, Chen K, Cheng Z, Li X, Chen X (2010) Effects of nanoparticle size on cellular uptake and liver MRI with polyvinylpyrrolidone-coated iron oxide nanoparticles. ACS Nano 4(12):7151–60
62. Ihsanullah Al-Khaldi FA, Abusharkh B, Khaled M, Atieh MA, Nasser MS, Saleh TA, Agarwal S, Tyagi I, Gupta VK (2015) Adsorptive removal of cadmium (II) ions from liquid phase using acid modified carbon-based adsorbents. J Molecul Liq 204:255–263
63. Jamshidian M, Tehrany EA, Imran M et al (2010) Poly-lactic acid: Production, applications, nanocomposites, and release studies. Compr Rev Food Sci Food Saf 9:552–571
64. Jiang H, Lau M, Tellkamp VL, Lavernia EJ (2000) Synthesis of nanostructured coatings by high velocity oxygen-fuel thermal spraying. In: Nalwa HS (ed) Handbook of nanostructured materials and nanotechnology, Academic Press, San Diego, CA, USA
65. Jin R, Lin B, Li D, Ai H (2014) Super paramagnetic iron oxide nanoparticles for MR imaging and therapy: design considerations and clinical applications. Curr Opin Pharmacol 18:18–27
66. John MJ, Thomas S (2008) Biofibres and biocomposites. Carbohyd Polym 71:343–364
67. Jolanta P, Marcin B, Zygmunt K (2011) Nanosilver—making difficult decisions. Ecol Chem Eng 18(2)
68. Jolivet JP, Henry M, Livage J (2000) Metal oxide chemistry and synthesis: from solution to solid state. Wiley, New York
69. Kaminskas LM, Boyd BJ et al (2011) Dendrimer pharmacokinetics: the effect of size, structure and surface characteristics on ADME properties. Nanomedicine 6(6):1063–1084

70. Katepalli H, Bikshapathi M, Sharma CS, Verma N, Sharma A (2011) Synthesis of hierarchical fabrics by electrospinning of PAN nanofibers on activated carbon microfibers for environmental remediation applications. Chem Eng J 171(3):1194–1200
71. Keledi G, Hari J, Pukanszky B (2012) Polymer nanocomposites: structure, interaction, and functionality. Nanoscale 4:1919
72. Khan WS, Ceylan M, Asmatulu R (2012) Effects of nanotechnology on global warming. In: ASEE midwest section conference, Rollo, MO, 19–21, Sep 2012, p 13
73. Khatamiana M, Divband B, Daryana M (2016) Preparation, characterization and antimicrobial property of Ag+- nano Chitosan/ZSM-5: novel Hybrid Biocomposites. Nanomed J 3(4):268–279
74. Kijeńska E, Prabhakaran MP, Swieszkowski W, Kurzydlowski KJ, Ramakrishna S (2012) Electrospun bio-composite P (LLACL)/ collagen I/collagen III scaffolds for nerve tissue engineering. J Biomed Mater Res B Appl Biomater 100(4):1093–1102
75. Kinnear C, Moore Thomas L, Rodriguez-Lorenzo L, Rothen-Rutishauser B, Petri-Fink A (2017) Form follows function. Nanoparticle shape and its implications for nanomedicine. Chem Rev 117:11476–11521
76. Kohane DS (2007) Microparticles and nanoparticles for drug delivery. Biotechnol Bioeng 96(2):203–209
77. Kokabi M, Sirousazar M, Hassan ZM (2007) PVA–clay nanocomposite hydrogels for wound dressing. Eur Polym J 43:773–781
78. Kudumula KK (2016) Scope of polymer nano-composite in bio-medical applications. IOSR-JMCE 13(5):18–21
79. Kumar A, Gupta M (2005) Synthesis and surface engineering of iron oxide nanoparticles for biomedical applications. Biomaterials 26:3995–4021. https://doi.org/10.1016/j.biomaterials.2004.10.012
80. Kumar SK, Krishnamoorti R (2010) Annu Rev Chem Biomol Eng 1:37
81. Kyzas GZ, Bikiaris DN, Seredych M, Bandosz TJ, Deliyanni EA (2014) Removal of dorzolamide from biomedical wastewaters with adsorption onto graphite oxide/ poly (acrylic acid) grafted chitosan nanocomposite. Bioresour Technol 152:399–406
82. Langer R, Vacanti J (1993) Tissue engineering science (80) 260:920–6
83. Leceta I, Guerrero P, Ibarburu I, Duenas MT, de la Caba K (2013) Characterization and antimicrobial analysis of chitosan based films. J Food Engineering 116(4):889– 899
84. Lee KY, Mooney DJ (2012) Alginate: properties and biomedical applications. Prog Polym Sci 37:106–126
85. Leja K, Lewandowicz G (2010) Polymer biodegradation and biodegradable polymers—a review. Polish J Environ Stud 19:255–266
86. Leu JG, Chen SA, Chen HM, Wu WM, Hung CF, Yao YD, Tu CS, Liang YJ (2012) The effects of gold nanoparticles in wound healing with antioxidant epigallocatechin gallate and α-lipoic acid. Nanomedicine. 8(5):767–775
87. Li et al (2015) Nanoscale 7:16631–16646. https://doi.org/10.1039/C5NR02970H
88. Litzinger DC et al (1994) Effect of liposome size on the circulation time and intraorgan distribution of amphipathic poly (ethylene glycol)-containing liposomes. Biochim Biophys Acta 1190(1):99–107
89. Liu D, Zhu Y, Li Z, Tian D, Chen L, Chen P (2013) Chitin nanofibrils for rapid and efficient removal of metal ions from water system. Carbohydr Polym 98:483–489
90. Long RQ, Yang RT (2001) Carbon nanotubes as superior sorbent for dioxin removal. J Am Chem Soc 123(9):2058–2059
91. Lu F, Wu SH, Hung Y, Mou CY (2009) Size effect on cell uptake in well-suspended, uniform mesoporous silica nanoparticles. Small Jun 5(12):1408–1413
92. Luo T, Cui J, Hu S, Huang Y, Jing C (2010) Arsenic removal and recovery from copper smelting wastewater using TiO_2. Environ Sci Technol 44(23):9094–9098
93. Ma Z, Lim LY (2003) Uptake of chitosan and associated insulin in Caco-2 cell monolayers: a comparison between chitosan molecules and chitosan nanoparticles. Pharm Res 20(11): 1812–1819

94. Mago G, Jana SC, Ray SS, Mcnally T, Mcnally T (2012) Polymer nanocomposite processing, characterization, and applications. https://doi.org/10.1155/2012/924849
95. Mago G, Ray SS, Shofner ML, Wang S, Zhang J (2013) Polymer nanocomposite processing, characterization, and applications. J Nanomater https://doi.org/10.1155/2014/403492
96. Malafaya PB, Silva GA, Reis RL (2007) Natural-origin polymers as carriers and scaffolds for biomolecules and cell delivery in tissue engineering applications. Adv Drug Deliv Rev 59:207–233
97. Maneerung T, Tokura S, Rujiravanit R (2008) Impregnation of silver nanoparticles into bacterial cellulose for antimicrobial wound dressing. Carbohydr Polym 72:43–51
98. Mangaiyarkarasi R, Chinnathambi S, Karthikeyan S, Aruna P, Gane- san S (2016) Paclitaxel conjugated $Fe_3O_4.LaF_3:Ce^{3+},Tb^{3+}$ nanoparticles as bifunctional targeting carriers for Cancer theranostics application. J Magn Magn Mater 399:207–215
99. Matthews FL, Rawlings RD (1999) Composite materials: engineering and science: Elsevier, Amsterdam, The Netherland
100. Mayer LD et al (1989) Influence of vesicle size, lipid composition, and drug-to-lipid ratio on the biological activity of liposomal doxorubicin in mice. Cancer Res 49(21):5922–5930
101. Mohanty AK, Misra M, Hinrichsen G (2000) Biofibres, biodegradable polymers and biocomposites: an overview. Macrmol Mater Eng 276(277):1–24
102. Murphy PS, Evans GRD (2012) Advances in wound healing: a review of current wound healing products. Plast Sur Int 190436:1–8
103. Murugesan S, Mousa SA et al (2007) Carbon inhibits vascular endothelial growth factor- and fibroblast growth factor-promoted angiogenesis. FEBS Lett 581:1157–1160
104. Nagayasu A, Uchiyama K, Kiwada H (1999) The size of liposomes: a factor which affects their targeting efficiency to tumors and therapeutic activity of liposomal antitumor drugs. Adv Drug Deliv Rev 40(1–2):75–87
105. Nair R, Varghese SH, Nair BG, Maekawa T, Yoshida Y et al (2010) Nanoparticulate material delivery to plants. Plant Sci 179:154–163
106. Nanjwade BK, Derkar GK, Bechra HM, Nanjwade VK, Manvi FV (2011) Design and characterization of nanocrystals of lovastatin for solubility and dissolution enhancement. J Nanomedic Nanotechnol 2:107
107. Okamoto M, John B (2013) Synthetic biopolymer nanocomposites for tissue engineering scaffolds. Prog Polym Sci 38:1487–1503
108. Ovissipour M, Roopesh SM, Rasco BA, Sablani SS (2014) Engineered nanoparticles (ENPs): applications, risk assessment, and risk management in the agriculture and food sectors, In: Wang S (ed) Food chemical hazard detection: development and application of new technologies Wiley, Chichester, UK. https://doi.org/10.1002/9781118488553.ch7
109. Pandey JK, Chu WS, Lee CS et al (2007) Preparation characterization and performance evaluation of nanocomposites from natural fiber reinforced biodegradable polymer matrix for automotive applications. Presented at the international symposium on polymers and the environment: emerging technology and science, bioenvironmental polymer society (BEPS), Vancouver, WA, USA, 17–20 October 2007
110. Panupakorn P, Chaichana E, Praserthdam P, Jongsomjit B (2013) Polyethylene/clay nanocomposites produced by in situ polymerization with zirconocene/MAO catalyst. J Nanomater https://doi.org/10.1155/2013/154874
111. Park KE, Jung SY, Lee SJ, Min BM, Park WH (2006) Biomimetic nanofibrous scaffolds: preparation and characterization of chitin/silk fibroin blend nanofibers. Int J Biol Macromol 38:165–173
112. Paul DR, Robeson LM (2008) Polymer 49:3187
113. Qi L, Xu Z, Jiang X, Hu C, Zou X (2004) Preparation and antibacterial activity of chitosan nanoparticles. Carbohydr Res 339(15):2693–2700
114. Qiu Y, Liu Y, Wang LM, Xu LG, Bai R et al (2010) Surface chemistry and aspect ratio mediated cellular uptake of Au nanorods. Biomaterials 31:7606–7619
115. Rastogi V, Yadav P et al (2014) Carbon nanotubes: an emerging drug carrier for targeting cancer cells. J Drug Deliv 670815

116. Raveendran P, Fu J, Wallen SL (2003) Completely green synthesis and stabilization of metal nanoparticles. J Am Chem Soc 125:13940–13941. https://doi.org/10.1021/ja029267j
117. Ravichandran R, Sundarrajan S, Venugopal JR, Mukherjee S, Ramakrishna S (2012) Advances in polymeric systems for tissue engineering and biomedical applications. Macromol Biosci 12:286–311. https://doi.org/10.1002/mabi.201100325
118. Ray SS (2013) Environmentally friendly polymer nanocomposites, types, processing and properties. Woodhead Publishing Series in Composites Science and Engineering. 44, Woodhead Publishing Ltd
119. Reddy RJ (2010) Preparation, characterization and properties of injection molded graphene nanocomposites, Master's thesis, Mechanical Engineering, Wichita State University, Wichita, Kansas, USA
120. Reneker DH, Fong H (2006) Polymeric nanofiber. American Chemical Society Publishers, Washington. 1–6
121. Rhim J, Park HM, Ha CS (2013) Bionanocomposites for food packaging application. Prog Polym Sci 38:1629–1652
122. Roco MC, Williams S, Alivisatos P (eds) (2000) nanotechnology research directions: IWGN workshop report vision for nanotechnology in the next decade, Kluwer Academic Publishers, Dordrecht
123. Roy R, Roy R, Roy D (1986) Alternative perspectives on "quasi-crystallinity", non-uniformity and nanocomposites. Mater Lett 4:323–328
124. Sanginario A, Miccoli B et al (2017) Carbon nanotubes as an effective opportunity for cancer diagnosis and treatment. Biosensors 7(1):9
125. Savva I, Krekos G, Taculescu A, Marinica O, Vekas L, Krasia-christoforou T (2012) Fabrication and characterization of magnetoresponsive electrospun nanocomposite membranes based on methacrylic random copolymers and magnetite nanoparticles. J Nanomater 9. https://doi.org/10.1155/2012/578026
126. Shahidi S, Ghoranneviss M (2014) Effect of plasma pretreatment followed by nanoclay loading on flame retardant properties of cotton fabric. J. Fusion Energ 33:88
127. Singh V, Joung D, Zhai L, Das S, Khondaker SI, Seal S (2011) Graphene based materials: past, present and future. Prog Mater Sci 56:1178–1271
128. Sinha SR, Bousmina M, MaiY YuZ (2006) Eds Biodegradable polymer/layered silicate nanocomposites. Polymer nanocomposites. Woodhead Publishing and Maney Publishing, Cambridge, England, pp 57–129
129. Siracusa V, Rocculi P, Romani S et al (2008) Biodegradable polymers for food packaging: a review. Trends Food Sci Technol 19:634–643
130. Stamatialis DF, Papenburg BJ, Giron M, Bettahalli SM, Schmitmeier S, Wessling M (2008) Medical applications of membranes. Drug Delivery Artif Organs Tissue Eng 308:1–34. https://doi.org/10.1016/j.memsci.2007.09.059
131. Stander L, Theodore L (2011) Environmental implications of nanotechnology—an update. Int J Environ Res Public Health 8:470–479
132. Starr FW, Glotzer SC, Dutcher JR, Marangoni AG (eds) (2004) Soft materials, structure and dynamics, Marcel Dekker, New York
133. Sultana N, Mokhtar M, Hassan MI, Jin RM, Roozbahani F, Khan TH (2015) Chitosan-based nanocomposite scaffolds for tissue engineering applications. Mater Manuf Process 30: 273–278
134. Suryanarayana C (1994) Structure and properties of nanocrystalline materials. Bull Mater Sci 17:307
135. TPA Plast global engineering nanocomposite polymers. http://www.tpacomponents.com/uploads/pdf/en/0305_EN.pdf (accessed on 20 August 2010)
136. Tabiei A, Aminjikarai SB (2009) Compos Struct 88:65–82
137. Tanaka (2004) Polymer nanocomposites as dielectrics and electrical insulation-perspectives for processing technologies material characterization and future applications, IEEE Trans Dielectr Electr Insul 11:5

138. Tang X, Zhang Q, Liu Z, Pan K, Dong Y, Li Y (2014) Removal of Cu (II) by loofah fibers as a natural and low-cost adsorbent from aqueous solutions. J Mol Liq 199:401–407
139. Tarigh GD, Shemirani F (2013) Magnetic multi-wall carbon nanotube nanocomposite as an adsorbent for preconcentration and determination of lead (II) and manganese (II) in various matrices. Talanta 115:744–750
140. Thomas S, Waterman P, Chen S, Marinelli B, Seaman M et al (2011) Development of secreted protein and acidic and rich in cysteine (SPARC) targeted nanoparticles for the prognostic molecular imaging of metastatic prostate cancer. J Nanomedic Nanotechnol 2:112
141. Thostenson ET, Li C, Chou TW (2005) Nanocomposites in context. Compos Sci Technol 65:491–516
142. Toy R et al (2013) Multimodal in vivo imaging exposes the voyage of nanoparticles in tumor microcirculation. ACS Nano 7(4):3118–3129
143. Tripathi A, Saravanan S, Pattnaik S, Moorthi A, Partridge NC, Selvamurugan N (2012) Bio-composite scaffolds containing chitosan/nano-hydroxyapatite/nano-copper-zinc for bone tissue engineering. Int J Biol Macromol 50:294–299
144. Ulman A (1996) Formation and structure of self-assembled monolayers. Chem Rev 96 1533–1554. https://doi.org/10.1021/cr9502357
145. Vaia RA, Giannelis EP (2001) MRS Bull 26:394
146. Vaia RA, Giannelis EP (eds) (2001) Polymer nanocomposites. American Chemical Society, Washington
147. Varela JA, Bexiga MG, Åberg C, Simpson JC, Dawson KA (2012) Quantifying size-dependent interactions between fluorescently labeled polystyrene nanoparticles and mammalian cells. J Nanobiotechnol 10(1):39
148. Veiseh O, Gunn JW, Zhang M (2010) Design and fabrication of magnetic nanoparticles for targeted drug delivery and imaging. Adv Drug Deliv Rev 62(3):284–304
149. Venkatesan J, Kim SK (2014) Nano-hydroxyapatite composite biomaterials for bone tissue engineering—a review. J Biomed Nanotechnol 10:3124–3140
150. Vllasaliu et al (2014) Expert Opin. Drug Delivery 11:139–154. https://doi.org/10.1517/17425247.2014.866651
151. Wan Y, Chen X, Xiong G, Guo R, Luo H (2014) Synthesis and characterization of three-dimensional porous graphene oxide/sodium alginate scaffolds with enhanced mechanical properties. Mater Express 4:429–434
152. Wang J, Byrne JD, Napier ME, De Simone JM (2011) More effective nanomedicines through particle design. Small 7:1919–1931
153. Wang X, Wenk E, Matsumoto A, Meinel L, Li C, Kaplan DL (2007) Silk microspheres for encapsulation and controlled release. J Control Release 117:360–370
154. Wang HQ, Yang GF, Li QY, Zhong XX, Wang FP, LiZ S, Li YH (2011) Porous nano-MnO_2 Large scale synthesis via a facile quick-redox procedure and application in a supercapacitor. New J Chem 35:469–475
155. Wang R, Yang J, Zheng Z, Carducci MD, Jiao J, Seraphin S, (2001) Dendron-Controlled Nucleation and Growth of Gold Nanoparticles. Angew Chem Int Edi 40(3) 549–552
156. Wang SH, Lee CW, Chiou A, Wei PK (2010) Size-dependent endocytosis of gold nanoparticles studied by three-dimensional mapping of plasmonic scattering images. J Nanobiotechnol 8(1):33
157. Wypych G (1999) Handbook of fillers, 4th edn. ChemTec Publishing, Toronto
158. X Wu, P Liu (2010) Polymer grafted multiwalled carbon nanotubes via facile in-situ solution radical polymerisation. J Exp Nanosci 5(5):383–389. https://doi.org/10.1080/17458080903583956
159. Xu Z, Gu Q, Hu H, Li F (2008) A novel electrospun polysulfone fiber membrane: application to advanced treatment of secondary bio-treatment sewage. Environ Technol 29:13–21
160. Xu D, Tan X, Chen C, Wang X (2008) Removal of Pb (II) from aqueous solution by oxidized multiwalled carbon nanotubes. J Hazard Mater 154:407–416

161. Yadav M, Rhee KY, Park SJ (2014) Synthesis and characterization of grapheneoxide/carboxymethylcellulose/alginate composite blend films. Carbohydr Polym 110:18–25
162. Yaehne K et al (2013) Nanoparticle accumulation in angiogenic tissues: towards predictable pharmacokinetics. Small 9(18):3118–3127
163. Zhang Y, Shen Z, Dai C, Zhou X (2014) Removal of selected pharmaceuticals from aqueous solution using magnetic chitosan: Sorption behaviour and mechanism. Environ Sci Pollut Res 21:12780–12789
164. Zheng L, Abhyankar W, Ouwerling N, Dekker HL, van Veen H, van der Wel NN, Roseboom W, de Koning LJ, Brul S, de Koster CG (2016) Bacillus subtilis spore inner membrane proteome. J Proteome Res 15:585–594
165. Zou H, Wu SS, Shen J (2008) Polymer/silica nanocomposites: preparation, characterization, properties, and applications. Chem Rev 108:3893–3957

Polymeric Composites as Catalysts for Fine Chemistry

P. SundarRajan, K. GracePavithra, D. Balaji and K. P. Gopinath

1 Introduction

For the past three decades, conjugated polymers (CPs) have been studied intensively ever since the discovery of the conducting polyacetylene (PAc) in 1976 [41]. Among CPs, Polypyrrole (PPy) and polyaniline (PANI) are often studied. Later, a variety of CPs has been developed and extensively used due to their high conductivity, as listed in Table 1. These CPs are normally stable in air, widespread availability, cheap, chemically inert, and can be effectively prepared as a host for incorporating various catalysts which having intriguing properties. New generations of these materials find its application in the various sectors such as chemical sensors, biosensors, energy storage (super capacitors, dielectric capacitors, batteries, solar cells, fuel cells), biomedical devices, optical devices, electromagnetic interference shielding, anticorrosion and antistatic coatings, electro active devices and catalysis.

In recent times, these CPs was recognized as suitable candidate for supporting heterogeneous catalysts (e.g., noble metals) as well as for basic research and practical applications owing to their intrinsic properties, for example, chemical stability, superior electrical conductivity, special optical properties, better carrier mobility, enhanced electrochemical activity, reusability, high accessible surface area, chemical functionalities (like solvation, templating effect, wettability and so on) and bio-compatibility [14, 28]. Most of the catalytic processes which are

P. SundarRajan · K. GracePavithra · D. Balaji · K. P. Gopinath (✉)
Department of Chemical Engineering, Sri Sivasubramaniya Nadar College of Engineering, Rajiv Gandhi Salai (OMR), Kalavakkam 603110, Tamil Nadu, India
e-mail: gopinathkp@ssn.edu.in

P. SundarRajan
e-mail: ksundarp@gmail.com

Inamuddin et al. (eds.), *Sustainable Polymer Composites and Nanocomposites*,
https://doi.org/10.1007/978-3-030-05399-4_12

Table 1 Some reported conducting polymers [1, 12]

Polymer	Label	Band gap (eV)	Conductivity (S cm^{-1})	Year[a]
Polyacetylene	Pac	1.5	10^3 to 1.7×10^5	1977
Poly(p-phenylene)	PPP	–	500	1979
Poly(p-phenylenevinylene)	PPV	2.5	3 to 5×10^3	1979
Polypyrrole	PPy	3.1	10^2 to 7.5×10^3	1979
Polyaniline	PAni	3.2	0–200	1980
Polythiophene	PT	2.0	10 to 10^3	1981
Polyfurane	PF	–	–	1981

[a]Reported for the first time (in the year) as conducting polymer

followed in industrial applications are found to be so complex because individual catalysts cannot meet the demand in terms of catalytic activity, selectivity and resistance towards deactivation. Hence, the introduction of composite catalyst, which has at least two components clubbed together attracted many researchers [60]. Among composite catalyst, CPs based composites show outstanding performance as an effect of synergistic performance which is derived from individual component anchored on them. Therefore, in this chapter, the applications of polymer composite as a catalyst in wider fields such as fuel cells, cross-coupling reaction, photocatalysis and 4-nitrophenol reduction are discussed briefly.

2 Electrocatalytic Activity of Polymer Composite

Low-temperature fuel cells are receiving substantial interest due to an eco-friendly approach of direct electrochemical oxidation of hydrogen/alcohols (mostly, low molecular weight) into either H_2O or CO_2 or both, which produces electricity in the fuel cell. For oxidative and reduction reactions in a fuel cell, platinum or platinum catalysts supported on a conductive material are frequently utilized as electrode materials [4, 7]. In those catalysts, the surface to volume ratio of metal particles is very high which promotes the accessible area for the reactions. The cost of fuel cell operation usually relies upon the morphology and dispersal behaviour of these metal particles because they are assumed as a key part in decreasing the loading rate of the catalyst. The main prerequisites of a proper catalyst support to be used in a fuel cell are

(a) appropriate void fractions for supporting gas flow,
(b) high stability during fuel cell working environments,
(c) high surface to volume ratio, for achieving high metal dispersion, and
(d) high specific conductance.

Currently, in low-temperature fuel cells, carbon, specifically Vulcan XC-72 carbon blacks, is usually preferred for supporting electrocatalyst nanoparticles because of its substantial specific conductance and surface area [3]. A primary issue associated with the utilization of carbon blacks as a cathode catalyst support in a fuel cell is their less resistivity towards corrosion due to the electrochemical oxidation of carbon surface. The carbon support instability leads to platinum particle coalesce and platinum detachment which results in reducing platinum surface area. On the other hand, the similar corrosion was to be found higher in the anode catalyst, during the reversal of the cell voltage caused by fuel starvation. Likewise, platinum catalysts appear to stimulate the carbon corrosion rate [4, 42, 49].

In addition, the existence of more numbers of micropores will leads to the low accessible surface area for the dispersion of particles and uneven transport of the fuel to the surface. Thus, carbon blacks having a huge surface area comprising mainly micropores (<1 nm) becomes a burden to act as a catalyst support. Eventually, carbon does not conduct protons that restricts the performance achievements. A proton-conducting polymer (e.g., Nafion®) is typically blended with a catalyst for the maximum usage of the catalyst because it mainly assists the transport of protons within the catalyst layer. In response to that, several substitutes of electrocatalyst supports are being investigated. Among them, mesoporous carbon and carbon gels received a greater attention as a fuel catalyst supports due to the presence of high quantity of mesopores and high surface area, which permits for high flow of reactant and metal dispersion [4].

Catalysts deposited on aforementioned carbons exhibited catalytic activity much higher than similar catalyst deposited on carbon black support. Their stability during fuel cell operational condition is nearly the same as that of carbon blacks. Lately, carbon-based nanostructures, for example, carbon nanofibers (CNFs) and carbon nanotubes (CNTs) were explored as a supporting material for a catalyst in fuel cells. Platinum/bimetallic Pt-based catalysts deposited on CNT and CNF exhibited catalytic activity much higher than similar catalysts deposited on carbon blacks support, because of the exceptional morphology and features like high chemical stability, high accessible surface area and high electrical conductivity [3]. Investigations performed using carbon-based nanostructures in polymer electrolyte membrane fuel cell (PEMFCs) conditions demonstrated that these nanostructures can be more robust and can replace the traditional carbon black [54]. Though, these supporting materials do not inhibit corrosion due to the oxidation of carbon surface instead merely reduce the rate.

In such a case, non-carbon materials have been explored as a support for the catalyst. Conducting oxides are developing as a suitable candidate for catalyst support due to their oxidative resistance. Additionally, during fuel cell operation, these materials exhibited electrochemical and thermal stability and delivered remarkable resistance towards corrosion in different electrolytic media. Unlike carbon, conducting oxides cease to improve electro-catalysis, but rather acts just as a mechanical support, in some cases several metal oxide supports will be able to serve as co-catalysts. Certainly, it is notable that several metal oxides, for example, SnO_2, WO_3, and RuO_2 can able to enrich the catalysis of platinum for oxidation of alcohol [20, 37, 40]. However, a major issue associated with the ceramic oxides is their low surface area, which influences the metal dispersion and, thereby resulting in a reduced catalytic activity. Additionally, the electrical conductivity of some metal oxides, for example, TiO_2 and SnO_2 are low at temperatures beneath 200 °C.

Depending on the morphology features, conducting polymers (CPs) have been employed in a fuel cell as a supporting material for catalysts. Normally, CPs satisfies the main prerequisites of a proper catalyst support to be used in a fuel cell that is mentioned earlier. Some of the CPs are proton- as well as electron-conducting materials, so they can be alternative to catalyst incorporated with Nafion® and deliver improved performance. The most commonly used CPs are heterocyclic polymers such as polypyrrole (PPy), polythiophene (PTh), polyaniline (PAni) and their derivatives. Table 1 shows the conductivities of some common conjugated polymers. The catalyst activity seems to be higher in PPy and PAni that is mainly due to the synergistic effect of the host matrix and the metal particles [3, 14]. Under fuel cell environment, polymer-supported catalyst usually exhibits a satisfactory stability (e.g., catalytic activity and film integrity). But, in both PAni and PPy, chemical degradation was found during metal doping and the catalysis. In addition, under both oxidizing and reducing conditions, the deposition of the catalyst particles resulted in the degradation of electrical conductivity of the polypyrrole.

And the intermediate products (aldehydes) formed during alcohol oxidation degraded the polyaniline. Furthermore, the use of PAni and PPy as electrocatalyst supports are limited due to loss of electrical conductivity [4].

In summary, the utilization of either carbon or ceramic or polymer materials as a catalyst support in a fuel cell is not totally satisfactory. Accordingly, composite of polymer–carbon, ceramic–carbon and polymer-ceramic materials have been proposed in the most recent years as catalyst supports in fuel cells. These composite may have more reasonable properties than the individual materials when used as a supporting material for the catalyst. Gomez-Romero [12] divided them into two major categories, as per the inherent features of the host and guest phases. Thus, organic-inorganic (OI) materials signify composites with organic hosts and inorganic guests, whereas in inorganic-organic (IO) materials, the organic phase is guest to an inorganic host. In OI hybrids, generally, the inorganic molecules will contribute their chemical activity and requires the structural support from CPs to form a useful solid material. On the other hand, in IO hybrids, the inorganic phase will be providing the structural task even though embedded CPs can also imprint their polymeric nature onto the materials acquired.

2.1 Composite Polymer-Carbon Black Supports

Usually, the electrochemical activity of catalysts supported on composite polymer–carbon blacks (normally Vulcan XC-72), is higher than that of the similar catalysts supported on either carbon or polymer. This enhancement is mainly due to the higher accessible surface area and electrical conductivity of the support and the polymer/electrolyte interface with simpler charge-transfer which permitting the use of almost all the deposited metal nanoparticles. Xu et al. [61] basically examined the influence of PAni on carbon characteristics (IO material) and found the optimum mass ratio of PAni to C (PAni: C = 0.25:1). An excess amount of PAni will reduce the conductivity of the composite, whereas a limited amount of PAni will reduce the anti-poisoning ability of the catalyst. Subsequently, the produced CO poisoning intermediate compound contributes 61.5% of the reduction in methanol oxidation current on the Pt/C catalyst, after 200 potential cycles, however, just 20% reduction is observed while using the Pt/PAni–C composite catalyst. This implies that the anti-poisoning ability of Pt/PAni–C was three folds greater than the Pt/C. In addition, the presence of PAni encourages ability of the catalyst to absorb more water and generates an active oxy-compound (Pt–OH), which elevates oxidation of carbon monoxide into carbon dioxide.

The work of Wu et al. [55] addressed the impact of using conventional carbon particles (Vulcan XC-72) along with CP (PAni) (OI material). They demonstrated that the integration of carbon particles into PAni film not just increases the electron conductivity, yet additionally reduces charge-transfer resistance across PAni/electrolyte interfaces. In both cases, for methanol oxidation, the activity of platinum supported on PAni–C (polymer-carbon) composite was considerably higher than that of the single host component.

Mokrane et al. [35] synthesized conducting polypyrrole (PPy)/C (Vulcan XC-72) composite material with different PPy/C ratios via chemical polymerization method. They noticed that the composite firmly affected the electrochemical activity of supported platinum toward the oxygen reduction reaction (ORR) in acid medium. The variation of the PPy/C ratio decides the so-called substrate effect for electrocatalysis. Thus, ORR is indirectly proportional to PPy content in the composite.

2.2 *Composite Polymer-CNT Supports*

Among many varieties of carbons, carbon nanotubes were commonly used as the carbon material in polymer-carbon composite due to their attracting features. To understand the different characteristics of polymer–CNT composites, comprehensive knowledge of the morphology, properties, and chemistry of CNT is significant. CNTs are prepared by single sheets of hexagonally arranged carbon atoms, called graphene and represented in the form of 3D cylindrical nanostructures. There are two basic classifications of CNT, that is, single-walled carbon nanotubes (SWCNTs) and multi-walled carbon nanotubes (MWCNTs) [4]. SWCNT can be pictured as a rolled-up graphene sheet (tubular), which comprises of benzene molecules, including hexagonal rings of carbon atoms, whereas MWCNT contains a stack of graphene sheets that are rolled up into concentric cylinders. The unique properties related with CNT is usually influenced by various parameters, such as nanotube synthetic method, quality of nanotubes, chirality, size, shape, alignment of nanotubes, defect density, and degree of crystallinity [36]. In most cases, polymer–CNT composites are OI materials, where CNT are dispersed in the polymer matrix to enhance the electrical and mechanical properties of the polymer. The probability of attaining high conductivity at low CNT content makes them a suitable candidate for many potential applications. Moreover, in order to achieve the desired CNT properties, the system is designed in such a manner to create a uniform dispersion of CNT in the composites with high stability.

The major problems associated with the synthesis of polymer-CNT composites exist in the effective integration of CNT into a polymer matrix, the control over the alignment of CNT in the matrix and the evaluation of the dispersion. Accordingly, several techniques have been proposed for the incorporation of CNT in the polymer matrix includes melt mixing, in situ polymerization, electrospinning, chemical functionalization of the carbon nanotubes and solution mixing [43]. Besides, at existing phase of technology growth, CNT production methods are too expensive and not reasonable for manufacturing at pilot scale. Additionally, because of the complex synthesis approach, the quality of CNT differs from one supplier then onto the next and even with the same vendors at various times, making them unreliable for pilot-scale use [31].

Baikeri and Maimaitiyiming [5] revealed that poly(9,9-dioctyl fluorine-alt-2-amino-4,6-pyrimidine) (oligomer) acts as an effective dispersant for SWCNTs. Pt particle was doped on the Py-SWCNT films by H_2PtCl_6 via coordination reaction. Furthermore, the incorporation of SWCNT also lead to higher catalytic activity due to the higher accessible surface area and the electrical conductivity of Py-SWCNT composite was considerably higher. Zhu et al. [68] prepared homogeneous PAni-MWNT nanocomposites by functionalizing the MWCNT by the means of diazotization reaction. The 4-carboxylicbenzene group was altered on MWCNT surface via a C–C covalent bond, which helps the dispersion of carbon nanotubes in aniline. Later, electrochemical polymerization was performed by cyclic voltammetry in sulphuric acid containing aniline and 0.8 wt% MWNT. The functionalization of the MWNT can prevent separation of microscopic phase in the nanocomposite and thereby ensure the adaptability of CNTs in the PAni matrix. Compared to pure PAni film, the Pt-modified PAni–MWNT composite exhibited long-term stability with a higher activity for formic acid oxidation. On this basis, the Polymer–CNT composites can serve as excellent host matrices for fuel cell catalysts.

2.3 *Composite Polymer-Ceramic Supports*

A recently developing field of materials, so-called "nanohybrid" or "nanocomposite" materials, produced by coordinating interactions of different organic and inorganic materials at the molecular level to obtain new materials with enhanced features and unique functions. This strategy has been effectively utilized currently for the production of new nanocomposite materials by a redox intercalation approach to get hybrid lamellar transition-metal oxides possessing improved synergistic activity. It is renowned that intercalation prompts alteration in the spacing between the layered structures. For instance, a few conducting polymers, for example, PAni, PPy, and PTh are known to oxidatively polymerize in presence of strong oxidizing transition-metal oxides (like V_2O_5).

In low-temperature fuel cells, ceramic materials (includes carbides and metal oxides which acts as a carbon-alternative) have been studied as oxidation resistant supports for catalysts [4]. During fuel cell operational condition, these materials are highly stable and exhibited exceptional resistance towards corrosion (under different electrolytic medium). In some cases, various metal oxide supports also acted as co-catalysts. Mesoporous ceramic materials are effectively utilized as a support of fuel cell catalyst because they possess high accessible surface area, large pore volumes with controllable sizes. On these bases, PANI-doped mesoporous metal oxides have been studied as anode materials in fuel cell applications.

Maiyalagan and Viswanathan [30] prepared a stable conducting PEDOT-V_2O_5 nanocomposite material via the intercalation of poly(3,4ethylenedioxythiophene) (PEDOT) in V_2O_5 matrix. PEDOT-V_2O_5 was used as the support for Pt in formaldehyde reduction method. For methanol oxidation, the electrochemical activity and stability of the Pt/PEDOT-V_2O_5 electrode were greater than that of Pt/C electrode, under the electrochemical operating conditions. Pang et al. [39] synthesized PAni-SnO_2 composites via chemical polymerization of aniline in presence of SnO_2. The Pt has supported the PAni-SnO_2 matrix and on SnO_2 for comparative study. The characterization study using XRD (X-ray diffraction) indicated that the nanoparticles Pt were evenly deposited on PAni-SnO_2 than the other one. In comparison with the Pt/SnO_2 electrode, the Pt supported-PAni-SnO_2 electrode exhibited better electrochemical characteristics (larger electrochemical surface area (ESA), better anti-poisoning ability and higher electrocatalytic activity for methanol oxidation) under the same operating parameters.

3 Catalysing Cross-Coupling Reactions

In the coupling reactions like Mizoroki-Heck, Suzuki-Miyaura and Sonogashira-Hahihara reactions, homogeneous palladium catalysis has garnered vast significance. This kind of catalysis offers high turnover numbers (TON), high reactivity and frequently provides high yields and selectivity. With the help of ligands (such as phosphines, amines, carbenes, dibenzylideneacetone (dba), etc.), the Pd catalysts properties can be enhanced.

Recent development in ligand-free Pd catalysts has replaced ligand assisted techniques. In another aspect, homogeneous catalysis has various disadvantages, such as difficulty in reuse or recycling the catalyst. This resulted in the loss of precious metal and ligands. Additionally, the impurities and the residual metals in the end-products have to be expelled [11]. These issues are to be resolved in order to catalyse the coupling reactions in the industry using homogeneous Pd-catalyst which are still an as challenging task. Heterogeneous Pd catalysis was found to be a feasible alternative. In this technique, Pd is attached to a solid support, such as zeolites and molecular sieves, activated carbon, metal oxides (mainly alumina or silica and also ZnO, MgO, ZrO_2, TiO_2), alkali, clays and alkaline earth salts ($BaSO_4$, $CaCO_3$, $SrCO_3$, $BaCO_3$), organic polymers, porous glass or polymers implanted in porous glass. In another aspect, Pd can also be converted to composite material by attaching it directly to a solid support; or by covalently bounding Pd to the supports with the help of ligands. These two methods of solid support permits to reuse or recover the heterogeneous catalyst after the process until the catalyst get deactivated. However, heterogeneous catalysis requires more extreme reaction conditions than homogeneous catalysis without causing a problem to the stability of the catalysts, because Pd catalyst is often thermally stable.

3.1 Suzuki-Miyaura Reactions

For the modern synthetic organic chemistry especially in the synthesis of biaryl compounds, Suzuki-Miyaura reaction was turned into the backbone. In recent times, the biaryl compounds are synthesized by catalysing arylation through Suzuki-Miyaura reaction using Pd/C catalyst and they also exist in heterocyclic. Furthermore, Suzuki-Miaura reactions can utilize for coupling variety of organic compounds which are different from aryl compounds (such as alkynes, alenes or alkenes). Marck et al [32] reported about the Pd/C catalysed Suzuki reaction for the first time in 1994.

Generally, in catalytic applications, some of the expected outcomes are evenly dispersion of nanoparticles, easy recovery and reusability, and control in particle size. However, nanoparticles often get aggregated and influenced the selectivity and catalytic activity of the catalyst. Hence, these nanoparticles should be anchored on a host complex such as macromolecular organic ligands or polymer. Esmaeilpour et al. [10] investigated the Suzuki–Miyaura reactions using $Fe_3O_4@SiO_2$-polymer-imid-Pd magnetic nanocatalyst without added phosphine ligands. So-produced Pd nanoparticles provided better accessibility for reactants without getting aggregated. In addition, they offered the easy separation of the catalyst using a magnetic field, short reaction times, easy purification, higher product yields and reduced Pd leaching. Sun et al. [46] prepared the magnetic polymer-supported catalyst by dispersing of Pd on the surface of the orange-like Fe_3O_4/polypyrrole (PPy) composite and investigated their application in Suzuki cross-coupling reaction in water. The PPy provided two functions: (i) protection of the magnetic particles against corruption and oxidation by acids and oxygen; (ii) presence of numbers of functional groups (–NH–) on the surface facilitated the immobilization of catalytic active species. Furthermore, the easy separation of magnetic Fe_3O_4 (seeds in composite) was achieved and reused for 6 times during a reaction turn without any loss.

3.2 *Heck Cross-Coupling Reactions*

The first Mizoroki-Heck cross-coupling reactions were discovered separately by Mizoroki et al. [33] and Heck and Nolley [15] where the palladium-catalysed arylation of olefins. In presence of palladium catalyst, an aryl (pseudo) halide reacts with an alkene along with a base to form an arylated alkene. This reaction has been employed in many fields, including fine chemicals syntheses, bioactive compounds, drug intermediates, natural products, antioxidants, UV absorbers, and other industrial applications Esmaeilpour et al. [10].

For Heck cross-coupling reactions, soluble palladium compounds such as phosphine palladium complexes are found to be the effective catalyst [21, 45]. The advancement of the catalyst with non-phosphine ligands [18, 29] gathered the attention of many researchers as phosphine ligands are found to be unfit to our environment due to its cost, toxicity, sensitivity to air and moisture. In recent studies, several efficient eco-friendly matrices were reported for supporting heterogeneous catalysts (Palladium) such as carbon nanotubes/nanofibers, clay, ionic liquids, silica, zeolites, metal oxides, graphene, d-glucosamine, magnetic materials and polymers. For the efficient separation of catalyst for recycling and organic end-products, immobilization of metal catalyst on solid matrix considered to be an efficient tool. In most cases, polymers are used a solid matrix because it provides different combinations of metal bonding to the matrix of the polymer by a non-covalent or covalent bonding, through hydrogen bridges, hydrophobic or specific fluorous interactions as well as ionic bonding.

Many types of research recently reported that under phosphine-free conditions polymer-anchored palladium composites were active for the Heck reaction. Under the phosphine-free condition, Islam et al. [18] examined the catalytic activity of Pd (II) supported on poly (N-vinyl carbazole) in the heck reaction (cross-coupling reaction) of terminal alkenes with aryl halides. The polymer-anchored metal complex found to be air-stable, non-polluting and active under different reaction parameters while optimization of the reaction conditions. The simple reaction conditions and in operation, higher yield, easy regeneration of the catalyst and rapid conversion makes them a suitable candidate for industrial application. Sarkar et al. [44] prepared a heterogeneous poly(hydroxamic acid) Pd(II) complex by utilizing corn-cob cellulose waste as a solid matrix with a reusable ability. Cellulose (biopolymer) is considered as a supporting material due to various attracting features such as low-density, inexpensive, insolubility, better stability while using organic solvents and wide availability. Under ambient reaction conditions, the cellulose-poly(hydroxamic acid)-Pd(II) catalysed the reaction of aryl/heteroaryl halides and arenediazonium tetrafluoroborate with a different olefins and without a loss during recycling of the catalyst.

3.3 *Sonogashira-Hagihara Reaction*

The sonogashira cross-coupling reaction is well-known for the construction of c-c bonds, especially for the formation of alkynes. Generally, the catalyst used for this type of transformation includes Pd/C, CELL–Pd(0), Pd(dmba)Cl(PTA), $PdCl_2(PCy_3)_2$, $PdCl_2/PPh_3$ and $PdCl_2(PPh_3)_2$ together with CuI as co-catalyst [25, 34]. In recent decades, modification in the conventional Sonogashira protocol has been done. Notable among them are phase transfer and copper-free condition; and catalyst utilization, which includes N-heterocyclic carbene (NHC) ligands mostly used for reaction with less reactive bromo- and chloroarenes. The consequence of various solvent such as ionic liquids, aqueous-organic solvent mixtures in presence of water-soluble phosphine ligands is explored. (Bhattacharya and Sengupta [6].

Tamami et al. [48] developed a catalytic system by anchoring palladium nanoparticle on poly (N-vinyl imidazole) (PVI) grafted silica through an eco-friendly approach. This catalytic system showed exceptional activity in copper-free Sonogashira-Hagihara reaction of phenylacetylene with aryl halides, under short reaction times with high yields. Furthermore, seven consequence cycles demonstrated that the polymer-supported catalyst retained its activity and recyclability without any loss. Heravi et al. [17] combined copper- and solvent-free Sonogashira coupling for the different reaction of alkynes with several aryl halides in presence of recyclable and reusable $PdCl_2$ catalyst which is supported on modified poly(styrene-co-maleic anhydride). In addition to higher activity, the catalyst also produced a wide variety of coupling products with remarkable yields. Moreover, without any pre-activation steps, the catalyst was reused for a minimum of five consecutive cycles.

4 Photocatalytic Degradation of the Pollutant

Due to the properties like higher catalytic activity, structure-based optical and electronic property, rapid reaction rate, and higher surface area, nanoparticles have high in potential as catalyst as well as redox active media, which attracted many researchers in designing of photo/chem-catalytic materials with greater efficiency especially for the purification of water and gases which are contaminated. Nanosized semiconductor materials such as nano-TiO_2 [24, 50], ZnO [38], CdS [67], and CdO [47], zero-valence metals such as Cu^0 [27, 57] and Fe^0 [51, 52] and bimetallic nanoparticles such as Fe/Pd [53], Fe/Ni [56], and Pd/Sn [26] are commonly used catalytic nanoparticles. For a variety of contaminants which includes polychlorinated biphenyls (PCBs), halogenated aliphatics, organochlorine pesticides, azo dyes [66], halogenated herbicides and nitroaromatics are degraded using a catalyst or redox reagents. But the limitations are seen in the segregation of fine particles from the aqueous suspensions and in catalyst recovering. This limit has been rectified by immobilizing the nanoparticles onto polymer support (such as porous resins [26], polymeric membranes [9, 47, 50] and ion exchangers [27]) with minimum particle loss and coalesce. Table 2 summarizes some polymer-supported nanocomposites for photocatalytic degradation of pollutants from the various aqueous environment.

In degradation of organic pollutants, nano-TiO_2 usually plays as a catalyst. Ameen et al. [2] done a research on degradation of methylene blue (MB) dye by poly 1-naphthylamine (PNA)/TiO_2 nanocomposite prepared by in situ polymerization, where enhanced photocatalytic activity was observed. The photodegradation efficiency of MB colour might be due to the effective charge separation of the electrons (e^-) and hole (h^+) pairs at the interfaces of PAN and $TiO_{2.}$ Some bimetals (such as Cu^0, Fe^0, Fe/Pd, Pd/Sn, Ni/Fe, etc.) and nanoscale metals are found to be very efficient in degradation of different organic pollutants (such as brominated methanes, chlorinated methanes, trihalomethanes, chlorinated benzenes, chlorinated ethenes, other polychlorinated hydrocarbons, dyes and pesticides) [26, 51–53, 56, 57]. The aforementioned metal nanoparticles are found to be very higher in reactivity, for example, self-ignition of nZVI is possible when exposed to air. Hence, oxidation is inhibited in order to preserve the nature of the chemical until they are exposed to the targeted contaminants.

Lin et al. [27] done a research using cation exchanger resin doped with nanoscale zero-valent copper (nZVC) which increases the accessible surface area of the catalyst by reducing aggregation of nZVC particles. During the reaction between CCl_4 and Cu^o, the Cu ion produced is recycled back by simultaneous cation exchange resin. The combination of sorption due to host resin and degradation by means of nZVC resulted in declination of the quantity of pollutant ($CCl_{4)}$ available in aqueous solution.

Dong et al. [8] researched a composite made by intercalating sodium carboxymethylcellulose (CMC) (serves as a stabilizer) into parallelized iron (Fe/Pd) nanoparticles. When comparing to CMC-stabilized nanoparticles, the pristine Fe/Pd particles, showed less stability against agglomeration, soil transport and chemical reactivity. It is concluded from batch dichlorination tests that CMC-stabilized nanoparticles photodegraded trichloroethene (TCE) which is found to be 17-folds faster than the non-stabilized nanoparticles.

5 Catalytic Reduction of 4-Nitrophenol

Recently, for the degradation of aromatic dye and nitro-compounds, various metal nanocatalysts has been developed in order to attain sustainable environment. Since 4-nitrophenol was found to be toxic, it has to be reduced to 4-aminophenol which is important in application aspect. Comparing to 4-nitrophenol, 4-aminophenol find its

Table 2 Summary of polymer-supported nanocomposites for photocatalysts

NPs	Polymer support	Pollutant	Preparation method	Results	References
TiO_2	Polyaniline (PAni)	Phenol	Aniline polymerization in presence of TiO_2	Under visible light, illumination phenol was degraded after 5 h in an aqueous solution	Li et al. [24]
TiO_2	Poly (3-hexylthiophene) (PHT)	Methyl orange (MO)	Dispersion of TiO_2 nanoparticles onto the polymer matrix	Under optimum molar ratio of TiO_2/PHT (75:1), the removal efficiency of MO reached 88.5% under 10 h illumination	Wang et al. [50]
TiO_2	Poly (1-naphthylamine) (PNA)	Methylene blue (MB)	In situ polymerization of 1-naphthylamine monomer with TiO_2	Under visible light illumination, MB was degraded nearly 60%	Ameen et al. [2]
Au/ TiO_2	Poly(methyl methacrylate) (PMMA)	Trypan blue (TB)	PMMA was first dissolved in tetrahydrofuran and TiO_2 powder suspended in a dissolved polymer	Degradation efficiency was 90% by Au-TiO_2/ PMMA thin film under sunlight	Elfeky and Al-Sherbini [9]
ZnO	PAni	Methylene blue (MB)	In situ polymerization method	% removal of MB after 1 h irradiation was 28% under UV and 82% under visible light	[38]
CdS	Chitosan	Congo red (CR)	Bio-mineralization simulation	85.9% of 0.02 g/L CR was degraded using 1.5 g/L composite catalyst under 180 min of illumination	Zhu et al. [67]
CdO	PAni	MB and Malachite green (MG)	Chemical oxidative polymerization	MB and MG dyes were photocatalytically degraded with removal efficiency of 99% under natural sunlight irradiation	Tadjarodi et al. [47]
Fe^0	Poly(methyl methacrylate) (PMMA)	Trichloroethene (TCE)	In situ synthesis with MMA and $FeSO_4.7H_20$ as a precursor	The observed degradation rate of TCE was 0.0034 h^{-1} and dechlorination efficiency was 62.3%	Wang et al. [51, 52]
Fe^0	Carboxymethyl cellulose (CMC)	Cr(VI)	In situ synthesis with $FeSO_4.7H_20$ as a precursor	After 1 h, higher removal efficiency was found in CMC/Fe^0 (94%) when compared with Fe^0 (22%)	Wang et al. [51, 52]

(continued)

Table 2 (continued)

NPs	Polymer support	Pollutant	Preparation method	Results	References
Cu^0	Chitosan	Cr(VI)	In situ synthesis with $CU(SO_4)_2.5H_20$ as a precursor	The concentration of Cr(VI) reduced from 50 to 2.21 mgL^{-1} in presence of 1 g composite after 24 h	Wu et al. [57]
Fe/Pd	Poly(vinylidene fluoride) (PVDF)	Trichloroacetic acid (TCAA)	Reduction deposition	Dechlorination of TCAA was observed within 180 min	Wang et al. [53]
Ni/Fe	Cellulose acetate	Trichloroethylene	Solvent-cast	Observed that reduction rate was proportional to the Ni content (0–14.3 wt%)	Wu and Ritchie [56]
Pd/Sn	Resin	Trichloroethylene	In situ reduction of Sn^{2+} to Sn^0 and the deposited Pb^0 via reduction of Pb^{4+}	When compared to commercial Sn, the rate of dechlorination in Resin-Pd/Sn was promoted by ~2 orders of magnitude	Lin et al. [26]

footprint as intermediate in the hair-dyeing agent, antipyretic drugs, anticorrosion lubricant and photographic developers etc. Hence, conversion of 4-nitrophenol (4-NP) by borohydride ions (BH_4^-) using metal nanocatalysts into 4-aminophenol (4-AP) via catalytic reduction became more significant. Various novel metal (such as Ag, Au, Pd, Pt etc.,), metal oxide (such as Cu_2O) and bimetal (such as Au/Ag, Au/Pt, Pd/Ag, Pd/ZnO_2 and Pt/CeO_2, etc.,) nanostructures were considered as catalyst for the reduction of 4-NP using $NaBH_4$, which is a pollutant in industrial wastewaters and non-biodegradable material, can be harmful to environment and human health [63].

For the enhancement of the activity of these type of catalysts, the particle morphology and size can be altered. Moreover, the catalytic performance is found higher while using bimetallic nanomaterials (such as Pt/Au alloy nanoparticles (NPs)) [62], Ag/Pd bimetallic NPs [19], and Au/Ag alloy nanoclusters [58]. Therefore, the important key variable for catalytic activity is considered as low-coordination sites and the elevated range of surface-to-volume ratio in nanocatalysts. In any case, these variables have the potential to reduce the surface energy of nanocatalyst and thereby increases the chance of agglomeration which influences the catalytic activity of nanocatalysts. In response to above problem, a promising technique has been proposed, that is anchoring of nanocatalyst on the support matrix, such as cellulose nanofibers [16], silica nanotubes [64], carbon nanotubes [22, 23], polymer-type matrices [13, 59], and graphene [22, 23]. Among these support materials, polymer-type carriers have received great interest for practical application because these materials contain various functional groups which act as anchor sites for loading nanoparticles (catalyst) and thereby improves the properties of nanoparticles, dispersibility and recyclability. Wu et al. [59] fabricated a composite particle of polystyrene/reduced graphene oxide@gold nanoparticles by a facile and controllable method and reported that as-prepared PS/RGO@AuNP composite has good dispersibility in water and additionally, showed excellent catalytic activity in the reduction of p-nitrophenol by sodium borohydride in aqueous solution. In Heidari [16] work, cost-effective biopolymer (nanofibrillated cellulose (NFC)) was used as support for silver nanoparticles and achieved shorter reduction period. Zhang et al. [65] immobilized silver nanoparticles in sulfhydryl functionalized poly(glycidyl methacrylate) microspheres for enhancing monodispersity and recyclability. Moreover, Ag NPs@PGMA-SH composite showed higher catalytic activity during the reduction of 4-NP, which was 1.3–1.32 times higher than reported in the literature. This catalyst also exhibited excellent reusability as a conversion higher than 92% (after 10 consecutive cycles).

6 Conclusions

Many researchers dedicated their work on CPs ever since the discovery of conducting polyacetylene in the 1970s and published more number of research results in this relevant field. The rapid development in the field of science and technology has led to the advancement in the CPs and their composite materials in nanoscale. The application of these polymer-based composites is found to be interesting. The well-known use of polymer-based composites in catalytic applications such as fuel cells, cross-coupling reaction, photocatalysis degradation and reduction of pollutants are discussed briefly. Among them, polymer composite as a catalyst in fuel cell and photocatalysis are often studied. Moreover, the interaction behaviour between the host polymers and the immobilized catalyst are highlighted. Others issues regarding the technology limitation and challenging tasks are also discussed.

References

1. Abdelhamid ME, O'Mullane AP, Snook GA (2015) Storing energy in plastics: a review on conducting polymers & their role in electrochemical energy storage. RSC Adv 5:11611–11626
2. Ameen S, Akhtar MS, Kim YS et al (2011) Nanocomposites of poly (1-naphthylamine)/SiO_2 and poly (1-naphthylamine)/TiO_2: Comparative photocatalytic activity evaluation towards methylene blue dye. Appl Catal B Environ 103(1–2):136–142
3. Antolini E (2009) Carbon supports for low-temperature fuel cell catalysts. Appl Catal B 88:1–24
4. Antolini E (2010) Composite materials: an emerging class of fuel cell catalyst supports. Appl Catal B-Environ 100:413–426
5. Baikeri S, Maimaitiyiming X (2017) Polypyrimidine/SWCNTS composite comprising Pt nanoparticles: possible electrocatalyst for fuel cell. Polym Sci Ser A 59(5):734–740
6. Bhattacharya S, Sengupta S (2004) Palladium catalyzed alkynylation of aryl halides (Sonogashira reaction) in water. Tetrahedron Lett 45(47):8733–8736
7. Costamagna P, Srinivasan S (2001) Quantum jumps in the PEMFC science and technology from the 1960s to the year 2000: part I. Fundamental scientific aspects. J Power Sources 102:242–252
8. Dong T, Luo H, Wang Y et al (2011) Stabilization of Fe–Pd bimetallic nanoparticles with sodium carboxymethyl cellulose for catalytic reduction of para-nitrochlorobenzene in water. Desalination 271(1–3):11–19
9. Elfeky SA, Al-Sherbini ASA (2011) Photocatalytic decomposition of trypan blue over nanocomposite thin films. Kinet Catal 52(3):391–396
10. Esmaeilpourv M, Javidi J, Dodeji FN, Hassannezhad H (2014) Fe3O4@SiO2-polymer-imid-Pd magnetic porous nanosphere as magnetically separable catalyst for Mizoroki-Heck and Suzuki-Miyaura coupling reactions. J Iran Chem Soc 11 (6):1703–1715
11. Garrett CE, Prasad K (2004) The art of meeting palladium specifications in active pharmaceutical ingredients produced by Pd-catalyzed reactions. Adv Synth Catal 346:889–900
12. Gomez-Romero P (2001) Hybrid organic–inorganic materials—in search of synergic activity. Adv Mater 13:163–174
13. Harish S, Mathiyarasu J, Phani KLN et al (2009) Synthesis of conducting polymer supported Pd nanoparticles in aqueous medium and catalytic activity towards 4-nitrophenol reduction. Catal Lett 128(1–2):197
14. Hasik M, Turek W, Nyczyk A et al (2009) Application of conjugated polymer–platinum group metal composites as heterogeneous catalysts. Catal Lett 127:304–311
15. Heck RF, Nolley JP Jr (1972) Palladium-catalyzed vinylic hydrogen substitution reactions with aryl, benzyl, and styryl halides. J Org Chem 37:2320–2322

16. Heidari H (2018) Ag nanoparticle/nanofibrillated cellulose composite as an effective and green catalyst for reduction of 4-nitrophenol. J Clust Sci:1–7
17. Heravi MM, Hashemi E, Beheshtiha YS et al (2014) PdCl2 on modified poly (styrene-co-maleic anhydride): A highly active and recyclable catalyst for the Suzuki-Miyaura and Sonogashira reactions. J Mol Catal A Chem 394:74–82
18. Islam M, Mondal P, Roy AS et al (2010) Use of a recyclable poly (N-vinyl carbazole) palladium (II) complex catalyst: Heck cross-coupling reaction under phosphine-free and aerobic conditions. Transition Met Chem 35(4):491–499
19. Jing H, Wang H (2015) Structural evolution of Ag–Pd bimetallic nanoparticles through controlled Galvanic replacement: effects of mild reducing agents. Chem Mater 27:2172–2180
20. Kumar A, Ramani VK (2013) RuO_2–SiO_2 mixed oxides as corrosion-resistant catalyst supports for polymer electrolyte fuel cells. Appl Catal B 138–139:43–50
21. Lauer MG, Thompson MK, Shaughnessy KH (2014) Controlling Olefin isomerization in the Heck reaction with neopentyl phosphine ligands. J Org Chem 79(22):10837–10848
22. Li H, Han L, Cooper-White J et al (2012) Palladium nanoparticles decorated carbon nanotubes: facile synthesis and their applications as highly efficient catalysts for the reduction of 4-nitrophenol. Green Chem 14(3):586–591
23. Li J, Liu CY, Liu Y (2012) Au/graphene hydrogel: synthesis, characterization and its use for catalytic reduction of 4-nitrophenol. J Mater Chem 22(17):8426–8430
24. Li X, Wang D, Cheng G et al (2008) Preparation of polyaniline modified TiO_2 nanoparticles and their photocatalytic activity under visible light illumination. Appl Catal B Environ 81(3–4):267–273
25. Liang Y, Xie YX, Li JH (2006) Modified palladium-catalyzed Sonogashira cross-coupling reactions under copper-, amine-, and solvent-free conditions. J Org Chem 71(1):379–381
26. Lin CJ, Liou YH, Lo SL (2009) Supported Pd/Sn bimetallic nanoparticles for reductive dechlorination of aqueous trichloroethylene. Chemosphere 74(2):314–319
27. Lin CJ, Lo SL, Liou YH (2005) Degradation of aqueous carbon tetrachloride by nanoscale zerovalent copper on a cation resin. Chemosphere 59(9):1299–1307
28. Lin Z, Wenya D, Nautiyal A et al (2018) Recent progress on nanostructured conducting polymers and composites: synthesis, application, and future aspects. Sci China Mater 61 (3):303–352
29. Luo C, Zhang Y, Wang Y (2005) Palladium nanoparticles in poly (ethyleneglycol): the efficient and recyclable catalyst for Heck reaction. J Mol Catal A: Chem 229(1–2):7–12
30. Maiyalagan T, Viswanathan B (2010) Synthesis, characterization and electrocatalytic activity of Pt supported on poly(3,4-ethylenedioxythiophene)–V_2O_5 nanocomposites electrodes for methanol oxidation. Mater Chem Phys 121:165–171
31. Mansor NB (2014) Development of catalysts and catalyst supports for polymer electrolyte fuel cells. Dissertation for the degree of Doctor of Philosophy, University College London, London
32. Marck G, Villiger A, Buchecker R (1994) Aryl couplings with heterogeneous palladium catalysts. Tetrahedron Lett 35:3277–3280
33. Mizoroki T, Mori K, Ozaki A (1971) Arylation of olefin with aryl iodide catalyzed by palladium. Bull Chem Soc Jpn 44:581–584
34. Mohsen E, Jaber J, Mehdi MA et al (2014) Synthesis and characterization of Fe3O4@ SiO2–polymer-imid–Pd magnetic porous nanospheres and their application as a novel recyclable catalyst for Sonogashira-Hagihara coupling reactions. J Iran Chem Soc 11(2):499–510
35. Mokrane S, Makhloufi L, Alonso-Vante N (2008) Electrochemistry of platinum nanoparticles supported in polypyrrole (PPy)/C composite materials. J Solid State Electrochem 12:569–574
36. Moniruzzaman M, Winey KI (2006) Polymer nanocomposites containing carbon nanotubes. Macromol 39(16):5194–5205
37. Muthuraman N, Guruvaiah PK, Agneeswara PG (2012) High performance carbon supported Pt–WO_3 nanocomposite electrocatalysts for polymer electrolyte membrane fuel cell. Mater Chem Phy 133:924–931

38. Olad A, Nosrati R (2012) Preparation, characterization, and photocatalytic activity of polyaniline/ZnO Nano composite. Res Chem Intermed 38:323–336
39. Pang H, Huang C, Chen J et al (2010) Preparation of polyaniline–tin dioxide composites and their application in methanol electro-oxidation. J Solid State Electrochem 14:169–174
40. Peng F, Zhou C, Wang H et al (2009) The role of RuO_2 in the electrocatalytic oxidation of methanol for direct methanol fuel cell. Catal Commun 10:533–537
41. Qu L, Dai L, Sun SS (2016) Conjugated polymers, fullerene C60, and carbon nanotubes for optoelectronic devices. In: Sun S-S, Dalton LR (eds) Introduction to organic electronic and optoelectronic materials and devices, 2nd edn. Taylor & Francis, CRC Press, Boca Raton
42. Roen LM, Paik CH, Jarvi TD (2004) Electrocatalytic corrosion of carbon support in PEMFC cathodes. Electrochem Sol State Lett 7:19–22
43. Sahoo NG, Rana S, Cho JW et al (2010) Polymer nanocomposites based on functionalized carbon nanotubes. Prog Polym Sci 35:837–867
44. Sarkar SM, Rahman ML, Chong KF et al (2017) Poly (hydroxamic acid) palladium catalyst for heck reactions and its application in the synthesis of Ozagrel. J Catal 350:103–110
45. Sharma S, Sarkar BR (2018) Efficient Mizoroki–Heck coupling reactions using phosphine-modified Pd (II)–picolinate complex. Synth Commun:1–9
46. Sun X, Zheng Y, Sun (2015) Pd nanoparticles immobilized on orange-like magnetic polymer-supported Fe_3O_4/PPy nanocomposites: a novel and highly active catalyst for suzuki reaction in water. Catal Lett 145(4):1047–1053
47. Tadjarodi A, Imani M, Kerdari H (2013) Experimental design to optimize the synthesis of CdO cauliflower-like nanostructure and high performance in photodegradation of toxic azo dyes. Mater Res Bull 48:935–942
48. Tamami B, Allahyari H, Farjadian F et al (2011) Synthesis and applications of poly (N-vinylimidazole) grafted silica-containing palladium nanoparticles as a new re-cyclable catalyst for Heck, Sonogashira and Suzuki coupling reactions. Iran Polym J 20:699–712
49. Taniguchi A, Akita T, Yasuda K et al (2004) Analysis of electrocatalyst degradation in PEMFC caused by cell reversal during fuel starvation. J Power Sources 130:42–49
50. Wang DS, Zhang J, Luo Q et al (2009) Characterization and photocatalytic activity of poly (3-hexylthiophene)-modified TiO_2 for degradation of methyl orange under visible light. J Hazard Mater 169(1–3):546–550
51. Wang Q, Qian H, Yang Y et al (2010) Reduction of hexavalent chromium by carboxymethyl cellulose-stabilized zero-valent iron nanoparticles. J Contam Hydrol 114(1–4):35–42
52. Wang W, Zhou M, Jin Z (2010) Reactivity characteristics of poly(methyl methacrylate) coated nanoscale iron particles for trichloroethylene remediation. J Hazard Mater 173(1–3):724–730
53. Wang X, Chen C, Liu H et al (2008) Preparation and characterization of PAA/PVDF membrane-immobilized Pd/Fe nanoparticles for dechlorination of trichloroacetic acid. Water Res 42(18):4656–4664
54. Wang X, Li W, Chen Z et al (2006) Durability investigation of carbon nanotube as catalyst support for proton exchange membrane fuel cell. J Power Sources 158:154–159
55. Wu G, Li L, Li JH et al (2005) Polyaniline-carbon composite films as supports of Pt and PtRu particles for methanol electrooxidation. Carbon 43:2579–2587
56. Wu LF, Ritchie SMC (2006) Removal of trichloroethylene from water by cellulose acetate supported bimetallic Ni/Fe nanoparticles. Chemosphere 63(2):285–292
57. Wu SJ, Liou TH, Mi FL (2009) Synthesis of zero-valent copper-chitosan nanocomposites and their application for treatment of hexavalent chromium. Bioresour Technol 100(19):4348–4353
58. Wu T, Zhang L, Gao J et al (2013) Fabrication of graphene oxide decorated with Au–Ag alloy nanoparticles and its superior catalytic performance for the reduction of 4-nitrophenol. J Mater Chem A 1(25):7384–7390
59. Wu Z, Wang L, Hu Y et al (2016) Facile synthesis of PS/RGO@ AuNP composite particles as highly active and reusable catalyst for catalytic reduction of p-nitrophenol. Colloid Polym Sci 294(7):1165–1172

60. Xie Z, Liu Z, Wang Y et al (2010) An overview of recent development in composite catalysts from porous materials for various reactions and processes. Int J Mol Sci 11(5):2152–2187
61. Xu Y, Peng X, Zeng H et al (2008) Study of an anti-poisoning catalyst for methanol electro-oxidation based on PAN-C composite carriers. C Chim 11:147–151
62. Ye W, Yu J, Zhou Y et al (2016) Green synthesis of Pt–Au dendrimer-like nanoparticles supported on polydopamine-functionalized graphene and their high performance toward 4-nitrophenol reduction. Appl Catal B Environ 181:371–378
63. You JG, Shanmugam C, Liu YW et al (2017) Boosting catalytic activity of metal nanoparticles for 4-nitrophenol reduction: modification of metal naoparticles with poly (diallyldimethylammonium chloride). J Hazard Mater 324:420–427
64. Zhang S, Gai S, He F et al (2014) In situ assembly of well-dispersed Ni nanoparticles on silica nanotubes and excellent catalytic activity in 4-nitrophenol reduction. Nanoscale 6(19):11181–11188
65. Zhang W, Sun Y, Zhang L (2015) In situ synthesis of monodisperse silver nanoparticles on sulfhydryl-functionalized poly (glycidyl methacrylate) microspheres for catalytic reduction of 4-nitrophenol. Ind Eng Chem Res 54(25):6480–6488
66. Zhao X, Lv L, Pan B et al (2011) Polymer-supported nanocomposites for environmental application: a review. Chem Eng J 170(2–3):381–394
67. Zhu H, Jiang R, Xiao L et al (2009) Photocatalytic decolorization and degradation of Congo red on innovative crosslinked chitosan/nano-CdS composite catalyst under visible light irradiation. J Hazard Mater 169(1–3):933–940
68. Zhu ZZ, Wang Z, Li HL (2008) Functional multi-walled carbon nanotube/polyaniline composite films as supports of platinum for formic acid electrooxidation. Appl Surf Sci 254 (10):2934–2940

Fabrication Methods of Sustainable Hydrogels

Cédric Delattre, Fiona Louis, Mitsuru Akashi, Michiya Matsusaki, Philippe Michaud and Guillaume Pierre

List of Abbreviations

2D	Two dimensions
3D	Three dimensions/three dimensional
4D	Four dimensional
AA	Acrylic acid
AD	Adipocytes
ADSC	Adipose-derived stem cells
BMP-2	Bone morphogenetic proteins 2
CC	Cell coating
CLSM	Confocal laser scanning microscopy
DMEM	Dulbecco's modified eagle medium
ECM(s)	Extracellular matrices
EDC	Ethyl carbodiimide
EG	Ethylene glycol
EGDMA	Ethylene glycol dimethacrylate

C. Delattre · P. Michaud · G. Pierre (✉)
Institut Pascal, Université Clermont Auvergne, CNRS, SIGMA Clermont,
63000 Clermont-Ferrand, France

F. Louis · M. Matsusaki
Joint Research Laboratory (TOPPAN) for Advanced Cell Regulatory Chemistry,
Graduate School of Engineering, Osaka University, 2-1 Yamadaoka, Suita,
Osaka 565-0871, Japan

M. Akashi
Department of Frontier Biosciences, Graduate School of Frontier Biosciences,
Osaka University, 1-3 Yamadaoaka, Suita, Osaka 565-0871, Japan

M. Matsusaki
Department of Applied Chemistry, Graduate School of Engineering,
Osaka University, 2-1 Yamadaoka, Suita, Osaka 565-0871, Japan

M. Matsusaki
JST-PRESTO, Kawaguchi, Japan

Inamuddin et al. (eds.), *Sustainable Polymer Composites and Nanocomposites*,
https://doi.org/10.1007/978-3-030-05399-4_13

FBS	Fetal bovine serum
FN	Fibronectin
G	Gelatin
HEMA	Hydroxyethyl methacrylate
HUVECs	Human umbilical vein endothelial cells
IPN	Interpenetrating polymeric
iPS-CMs	Induced pluripotent stem cell-derived cardiomyocytes
LbL	Layer-by-layer
LCA	Life-cycle assessment
LECs	Lymph epithelial cells
MBSCs	Bone marrow stromal cells
MCS	Maleic chitosan derivatives
MMT	Montmorillonite
NHDFs	Normal human dermal fibroblasts
PAAm	Polyacrylamide
PBS	Phosphate buffer saline
PEG	Poly(ethylene glycol)
PEGDA	Poly(ethylene glycol) diacrylate
PLGA	Poly(lactic-co-glycolic acid)
PPGs	Polyacrylamide particle gels
PPO-PEO	Poly(propylene oxide)-poly(ethylene oxide)
PVA	Poly (vinyl) alcohol
PVP	Poly (vinyl pyrrolidone)
PVSA	Poly-vinylsulfonic acid
RGD	Arginine-glycine-aspartic acid
TPVA	Thiol-terminated poly (vinyl alcohol)
VAc	Vinyl acetate

1 Introduction

What is a hydrogel? Answering this seemingly simple question could appear easy but this is not the case. Using the SciFinder portal (https://scifinder.cas.org), the word "hydrogels" led to 114,105 references (70% published between 2007 and 2017) including 28 book chapters, 6761 reviews, 24,324 patents and 83,187 articles (Fig. 1). Except for book chapters, the number of research articles, reviews and patents per year has doubled in less than 10 years. Basically, hydrogels are three-dimensional, smart and/or hungry polymer networks extensively swollen with water. Their sizes are variable, and they are named recently micro- and nanohydrogels when their sizes are reduced to 1 μm and 1 nm respectively [1]. Depending upon the pore size between polymer networks, the structure of hydrogel can be classified as nonporous, microporous, or superporous [2]. These complex structures

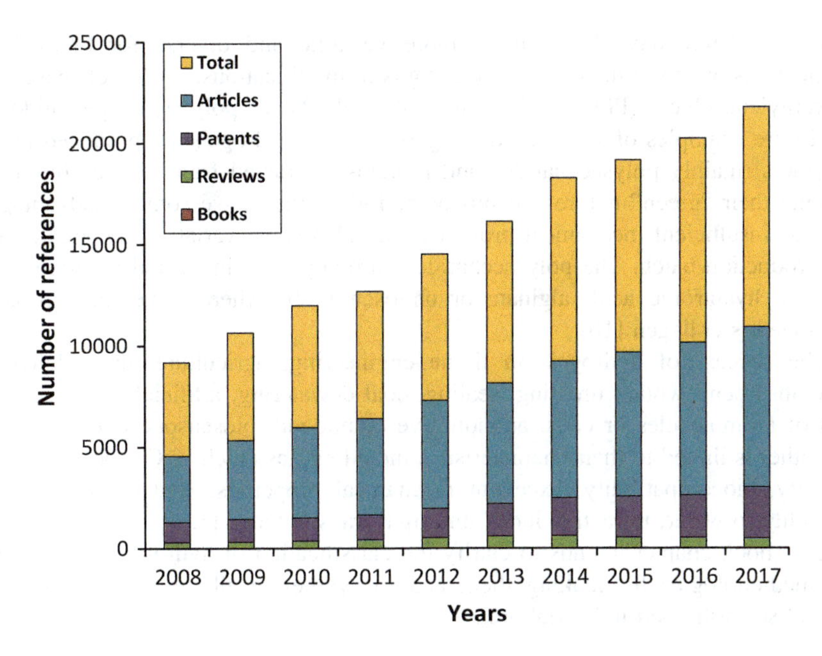

Fig. 1 Evolution of the scientific production relative to hydrogels between 2008 and 2017

are capable to absorb large quantities of water or other fluids such as biological liquids, but they do not dissolve in them. Owing to their varying compositions, hydrogels are classified into three categories for biomedical applications depending on their physical properties, i.e. liquid, semi-solid and solid [3]. They should be biocompatible, and non-toxic and can be biodegraded (or not) for specific applications. The most common hydrogels are cross-linked polymers generated by the polymerization of one or several monomers [4]. They can be also formed by the reticulation of synthetic or natural polymers [5]. Their hydrophilicity is mainly due to the presence of hydrophilic groups such as –OH, –$CONH_2$, –COOH, –CONH– or others [5]. Their ionic nature can be neutral, cationic or anionic, some of them having ampholytic behaviour. Hydrogels are sometimes named 'reversible' or 'physical' gels if molecular entanglements (ionic interactions, hydrophobic forces or hydrogen-bonding) play the main role in forming the network. It is sometimes possible to dissolve them by changing physicochemical environmental conditions, such as the ionic strength of the solution, light, magnetic field, pH, or temperature [6]. Hydrogels are non-reversible when a cross-linker leads to the formation of covalent bonds to build the network. All these families of hydrogels may have natural or synthetic origins. Novel hydrogels with specific properties made of natural and biodegradable polymers are in great demand. Nonetheless, during the 50 last years of scientific literature and patents speaking of hydrogels, the natural hydrogels have been gradually replaced by synthetic ones. Synthetic hydrogels are described in the literature as homopolymeric, copolymeric or multipolymeric hydrogels. They have been found more versatile and diverse for biomedical applications owing to their tailorable designs or modifications. Hydrogels based on polyethylene glycol (PEG), poly (vinyl) alcohol (PVA), poly (vinyl pyrrolidone) (PVP) are examples of synthetic hydrogels. Natural hydrogels are produced using polymers (mainly polysaccharides and proteins) extracted from various biomass. Despite their "green" nature and biocompatibility, they have some disadvantages such as insufficient mechanical properties but also some variations depending on the production batch. The polysaccharides currently used in natural hydrogels are dextran, hyaluronic acid, alginate or chitosan [7–9] whereas the main protein employed is collagen [10].

The success of hydrogels in tissue engineering, agriculture, drug delivery, superabsorbent, wound dressing, sealing, coal dewatering, artificial snow, separation of biomolecules or cells, antiadhesive compound, biosensors, contact lenses and other is linked to their characteristics including gas (such as O_2) permeability, stability, biocompatibility, excellent mechanical properties, wettability and permeability to water, refractive index and light transmittance [3, 4].

This book chapter intends to clarify the classification of hydrogels, to describe the methodologies for making them and to synthesize all their applications in several scientific and industrial areas.

2 Classifying Hydrogel: What's the Bottom Line?

Fabrication methodology of sustainable firstly requires a strong and comprehensive understanding of hydrogel products involved in the preparation. Recently, Varaprasad et al. [11] published a decent mini review describing an updated hydrogels classification, regarding their cross-linking and physical states but also some associated developments in miscellaneous applications. Based on different aspects (Fig. 2), hydrogels can be classified in many ways, depending on (i) source, which is of main importance for a sustainable point of view, (ii) polymeric composition, (iii) physicochemical composition, (iv) type of cross-linkers used, (v) network electrical charge and finally (vi) physical appearance [4]. Today, life-cycle assessment (LCA) of hydrogels should be also considered as a seventh item. In an eco-designing point of view, the good knowledge of the environmental effect and generated hydrogels is of first importance. LCA should be considered here as a comprehensive methodology to estimate and evaluate the environmental impact of the whole fabrication methodology of hydrogels, but also allover its whole life cycle according to ISO 14044. This notion implicated not only one single parameter but a succession of analyses throughout the product lifecycle. Thus, the best typical crop of the environmental impact of a product must consider the environmental courses all over the whole product's life, including the emission to land, water and air as well as the energy and material balance of product resources [12–14]. A complete LCA from the extraction of polymers/materials following by their modification/transformation/designing/manufacturing into a "sustainable"

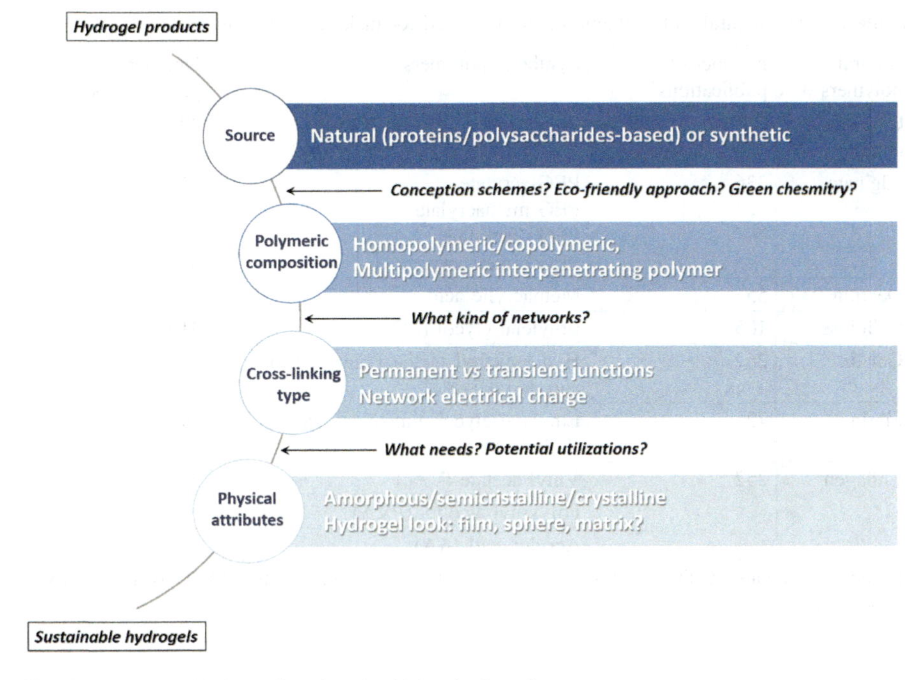

Fig. 2 The overall bottom line for classifying hydrogels

hydrogel, through the practical use of the product and finally the end of life circumstances, such as disposal or recycling, should be considered [15]. It was found that engineered bio-sourced materials should prospectively play an increasingly major role in our consumer society which is in perpetual search of sustainable and environmentally friendly materials. It was particularly reported that in comparison to petroleum-based materials, biopolymer materials could significantly reduce the energy/environmental balance impact. The use of biomass feedstocks and byproducts (polysaccharide-based composition for example) for making hydrogels is also not negligible and could contribute to trapping carbon thanks to this concept.

Hydrogels can be separated depending on their natural or synthetic origins. Thus, the hydrogel can contain natural polysaccharides and/or proteins such as chitosan, cellulose, alginate, starch, gelatin, or collagen. The list of synthetic polymers is longer, and the most common synthetic monomers used, especially in the pharmaceutical field, are probably PEG and its derivatives (PEG acrylate, methacrylate, diacrylate, dimethacrylate), ethylene glycol (EG), acrylic acid (AA) or hydroxyethyl methacrylate (HEMA). Table 1 gives an overview of these polymers commonly used for the preparation of hydrogels.

The type of polymer network generating during the fabrication method also greatly affects the classes of hydrogels [16]. From a single species of monomer, it is possible to obtain homopolymeric hydrogels which may have cross-linked skeletal structure, as reported by Takashi et al. [17]. The use of two or more different

Table 1 Some natural and synthetic monomers used for making hydrogels

Natural polymers	Number of publications[a]	Synthetic polymers	Number of publications
Chitosan/chitin	218/82	Poly(ethylene glycol) (PEG)	370
Alginate	336	PEG-acrylate PEG methacrylate PEG diacrylate PEG dimethacrylate	27 38 118 33
Dextran	55	Methacrylic acid	72
Cellulose	105	Ethylene glycol (EG)	319
Gelatin	262	Hydroxyethyl methacrylate (HEMA) and derivatives	59
Fibrin	72	Ethylene glycol dimethacrylate (EGDMA)	29
Collagen	252	Vinyl acetate (VAc)	2
		Polyvinyl alcohol (PVA)	101
		Acrylic acid (AA)	141

[a]Based on Scopus, with the keywords combination "polymer name AND hydrogel AND fabrication"

monomer species can be used to prepare copolymeric hydrogels, which needs at least a hydrophilic component and are characterized by a specific configuration in the chain of the polymer network [18]. Note that this kind of hydrogel can include a natural polymer in its structure [19]. The last type according to polymeric composition classification, i.e. the so-called multipolymer interpenetrating polymeric (IPN) hydrogels, is made of two independent natural and/or cross-linked synthetic polymers. The semi-ipn hydrogel is also reported in the literature and is composed of a cross-linked and a non-cross-linked polymer [20]. The kind of cross-linked interactions is also considered since it is possible to prepare hydrogels with chemically cross-linked networks (e.g. grafting, radical polymerization, condensation/enzymatic reaction, high-energy radiation) but also physically cross-linked networks. The cohesion of the last one is in general based on ionic and/or hydrophobic interactions, hydrogens bonds, polymer chain entanglements, stereo-complex formation, thermo-reversible gels, maturation due to heat-inducing aggregation, freeze-thawing [4, 11, 21]. The physical and chemical structure plus the physical appearance of hydrogels give birth to supplementary classes, e.g. non-crystalline, semi-crystalline, crystalline as well as gels (macro/micro/nano), matrix, film or microsphere. Generally speaking, some authors also report three main classes based on physical properties, i.e. solid, bio-adhesive and liquid-based hydrogels [11]. The preparation process is mainly involved for obtaining these looks (see Sect. 3). Finally, as reported by Ahmed [4], four groups describe the importance of electrical charge into hydrogel network on the cross-linked chains, i.e. neutral, ionic, amphoteric electrolyte and zwitterionic.

3 Methodology for Making Hydrogels and Sustainable Hydrogels

3.1 *Goals and Technical Features*

As described above, talking about hydrogels involves working on highly hydrophilic polymer networks presenting high swollen properties with water and aqueous media. Lot of works in the literature described hydrogels as a hydrocolloid gels material in which one, the dispersion medium is water [4, 22]. One of the most important applications of hydrogels largely described during the last decade is probably their uses as polymeric matrices for: (i) controlled releases of pharmaceutics drugs and and (ii) living cells entrapment/encapsulation [23–25]. Therefore, from the past 30 years, hydrogels continue to be a technology booming thank to their very high flexibility degree closely like human tissues [4, 26, 27]. A lot of studies was published for the optimization of hydrogels synthesis aiming to increase the technical features and improve their efficiency in pharmaceutical and biomedical applications [3, 4, 28]. In fact, the perpetual search for new biomaterials (e.g. artificial organs or tissues, filling materials, pharmaceutical release, diagnostic system) such as hydrogel is a fundamental notion in the biomedical sector. This requires a multidisciplinary approach involving physicians, biologists, chemists, and physicochemical. Hydrogels thus constitute a group of polymeric biomaterials having a hydrophilic structure allowing them to retain large quantities of water per area unit in a complex three-dimensional network. Moreover, thanks to the cross-links between the chains of the network, these biomaterials are highly resistant to dissolution in aqueous media. Consequently, as related by lot of authors [22, 23, 29, 30] we can easily list different functional features to characterize a most favorable hydrogel material for industrial applications such as: (i) highest biocompatibility, (ii) lowest solubility, (iii) highest swelling properties in water and in saline media, (iv) highest re-wetting ability depending on the desired applications, (v) highest biodegradability, (vi) highest stability and robustness in swelling surrounding, (vii) highest durability during storage before use, (viii) lowest releaser of toxic compound, (ix) odorless and/or colorlessness, (x) highest temperature, pH and photostability before and after swelling formation and, (xi) highest porosity to allow the fluid circulation and the best swelling properties. In general, it is well established that according to the use of the hydrogels, the main technical features to be developed must be optimized to consider a genius balance between the properties of the biomaterials and the targeted applications [3, 30]. Moreover, as very well related by Ahmed [4], it was clearly established that according to the nature, the intrinsic properties, the distribution and the density of compounds used (e.g. natural or synthetic polymers, cross-linkers, adhesives, adjuvants, extracellular matrices), hydrogel macrostructures could include diverse amount of water in swelling state in the application medium leading to higher water mass fraction in scaffold than polymers mass fraction itself. Therefore, if we focus particularly on the case of entrapment living cell application, another innovative technology is the cell coating

(CC). Generally, classic in vitro cultures in two dimensions (2D) can only present monolayer structures even after reaching cells confluency in the culture dish. One of the explanations might be the lack of extracellular matrices (ECMs) expressed by cells when they are maintained in 2D culture, which is necessary to switch into three-dimensional (3D) structure tissues. To overcome this issue, artificially addition of ECM surrounding the cells can be performed. This is called CC, which means nanometer- or micrometre-sized polymer thin films around cells, to induce their biological activities. This coating is performed using protein or polymer and can control for instance cell adhesion [31], growth direction [32, 33], or even killing specific bacteria [34]. Different coating methods exist, one of them is called "Layer-by-Layer" (LbL) formerly developed by Decher in 1981 to coat polymer and proteins onto substrate surfaces by dipping or spraying [35, 36]. This approach has been more and more used for cell surface modification, due to its easy preparation and tunable composition on cell surfaces under physiological conditions. The aim is to use specific ECM as coating components, known for their interactions with integrin receptor on the cell membranes. Many different types of ECM are found in our body, the most represented being collagen, laminin, hyaluronic acid, fibronectin or elastin. By selecting appropriate natural ECM components cytotoxicity is avoided while inducing specific cell-adhesive properties on molecules like RGD (arginine-glycine-aspartic acid) [37]. Among the cell membrane proteins, integrin receptors are important, determining the specificity for extracellular ligands as well as inducing intracellular signaling processes [38, 39]. Specifically, the $\alpha5\beta1$ integrin receptors can recognize the ECM component fibronectin (FN) since it contains the RGD sequence [40]. Gelatin (G) is a mixture of peptides and proteins produced from partial hydrolysis of collagen extracted from the skin, bones, and connective tissues. FN and G can interact with each other due to the collagen-binding domain found in FN, leading to nanometer-sized cell-adhesive surface films on cell surfaces, like the natural ECMs for multilayered structures without any cytotoxicity.

3.2 *Technologies Developed for Their Preparation*

A lot of strategies has been described in the literature to synthesize diverse forms of hydrogels. As recently mentioned in a very interesting review [30], hydrogels can be prepared by using several methodologies such as (i) polymerization grafting; (ii) chemical or physical cross-linking and (iii) radical cross-linking (Fig. 3). Conventionally, polymerization and radical processes are well defined as a good technology to generate varied hydrogels with controlled size, composition, particles distribution and morphologies [41–43]. Generally, the chemical method is the most common process related in literature for the preparation of hydrogel biomaterials with very good mechanical strength [23]. Nevertheless, the main drawback of chemical cross-linking is the use of toxic cross-linker such as for example

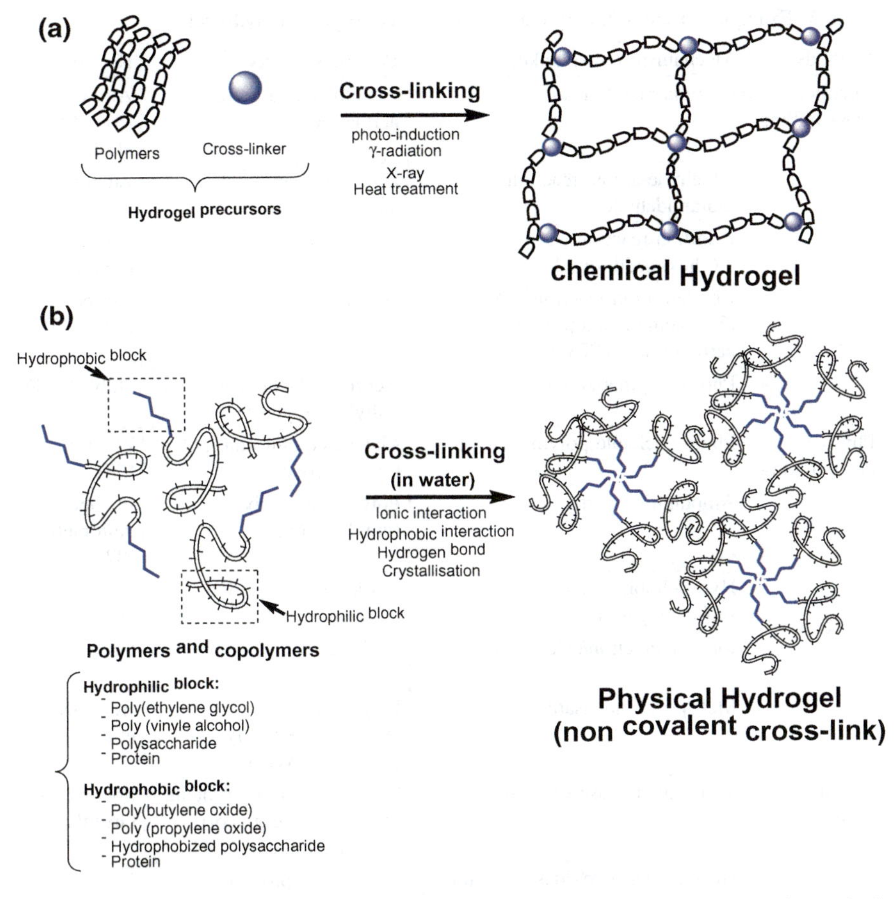

Fig. 3 The mains strategies to prepare hydrogels. **a** Chemical cross-linked polymers, **b** physically cross-linked polymers

glutaraldehyde and epichlorohydrin which must be removed for industrial and/or human applications. Consequently, a healthier alternative is the physical cross-link method [4, 22].

In a classic way, for the production of hydrogels (from macro- to nanogel 3D-networks) lot of cross-linking reactions have been performed (Table 2) such as: (1) the photo-induction reaction, (2) the Schiff-base reaction, (3) the thiol/disulfide reaction, (4) the carboxyl/amine reaction, (5) the amide/amine reaction and, (6) the click chemistry reaction [4, 23]. Furthermore, to perform the best hydrogel synthesis with controlled cross-link density and molecular sizes, studies traditionally recommend multi-steps procedure from polymerization to cross-link multifunctional mono/polymers having very reactive groups/functions [3, 4, 23]. As clearly mentioned by Mahinroosta et al. [30], the most important challenge for the preparation of hydrogels is the crucial control of the particle size distribution

Table 2 Examples of cross-linking reaction performed to prepare hydrogels

Methods	Mechanisms/cross-linker	Polymers/blocks	References
Chemical cross-linking	Gamma radiation	Cellulose, tara gum, acrylic acid	Alla et al. [81], Amin et al. [82]
	Aldehyde-amine reaction/ glutaraldehyde	Chitosan, polyvinyl alcohol	Zu et al. [83]
	Addition reaction/ 1,6-hexanedibromide	Scleroglucan	Coviello et al. [84]
	Condensation reaction/*N*,*N*-(3-dimethylaminopropyl)-*N*-ethyl carbodiimide (EDC)	Gelatin	Kuijpers et al. [85]
	Photo-polymerization	Acrylated lactic acid, poly (ethylene glycol)	Hubbell [86]
Physical cross-linking	Photo-clickable polymerization	Maleilated chitosan, poly (vinyl alcohol)	Zhou et al. [52]
	Amphiphilic graft	Polystyrene, poly (vinylpyridine)	Forster and Antonietti [87]
	Hydrophobic modification; self-aggregation	Cholesteryl, modified pullulan	Taniguchi et al. [88]
	Ionic interaction/calcium ions (Ca^{2+})	Alginate	Gacesa [89]
	Melt polycondensation	Poly(butylene terephthalate); Poly (ethylene glycol)	Bezemer et al. [90]
Enzymatic cross-linking	Transglutaminase reaction	Lysine-containing protein, poly-(ethylene glycol) glutaminamide	Sperinde and Griffith [44]
	Horseradish peroxidase reaction	Silk fibroin proteins, hyaluronic acid	Raia et al. [46]
	Alcalase reaction	Sucrose, acrylate, metacrylate	Chen et al. [45]
	Transglutaminase reaction	Hyaluronan	Broguiere et al. [54]
	Transglutaminase reaction	Gelatin	Yang et al. [91]

modulated by synthesis processes such as the regulation of monomer/polymer/ cross-linker ratio, the perfect adjustment of experimental conditions (e.g. pH, ionic strength, temperature). To accomplish this aim, other cross-linking alternative methods using enzymes were proposed to generate specific hydrogels [23]. As for example, Sperinde and Griffith [44] synthesized original hydrogels with cross-linking between a lysine-containing protein and a functionalized PEG by using transglutaminase (a human tissue enzyme). In their study, authors showed the catalyzed reaction between the amine group of lysine and the γ-carboxamide group

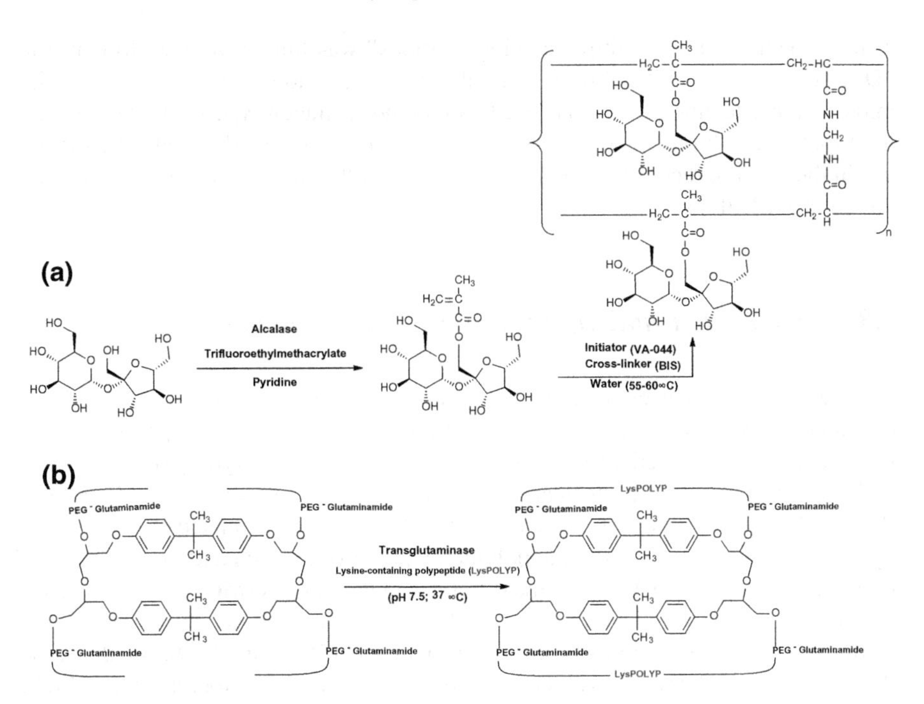

Fig. 4 Two mains examples of enzymatic cross-linking for the preparation of hydrogels. **a** Chemoenzymatic synthesis of poly-sucrose-methacrylate)-hydrogel (adapted from Chen et al. [45], **b** enzymatic cross-linking of glutaminamide polyethylene glycol (PEG) with lysine-containing polypeptide using transglutaminase

of PEG resulting in an attractive biomaterial allowing highly hydrated 3D-networks for living cells. In another enzymatic strategy, Chen et al. [45] developed chemo-enzymatic and enzymatic approaches to prepare a sugar-based hydrogel such as poly-sugar acrylate/methacrylate with highly water absorbents and drug delivery systems properties. As a final point, not to mention that recently, an enzymatic crosslinked silk fibroin proteins-hyaluronic acid hydrogel was produced by using horseradish peroxidase resulting in a highly elastic hydrogel with the application as scaffolds in tissue engineering [46]. The main enzymatic cross-linking way to prepare hydrogels are presented in Fig. 4.

Concerning the LbL coating (see Sect. 3.1), this "cell accumulation technique" technology provide around 6 nm thick FN-G films around cell surface, fabricated by FN and G coatings alternatively for 9 steps. Till now, over 100 μm thick 3D-tissue models with capillary networks [47, 48] were successfully constructed for further advanced studies as drug delivery systems, cancer cell invasion mechanism observation, or even manufacturing a biosensor chip [49]. Some tissues, for example, cartilage tissues [50], require a higher amount of ECMs, cell-cell distance in the tissues being in micrometre-sized level, and LbL FN-G coating methodology providing only nanometer-sized level. Another novel approach called "collagen coating method" and "multiple coating methods" was thus developed to construct 3D-tissue models with lower cell density and more ECM content using micro-coating technologies [51]. For this method, collagen type I, the most widely used material in biomedical and tissue engineering, was directly used. The method lies in the specific recognition abilities between another integrin receptor, α2β1, and the collagen I fibers.

3.3 Preparation and Optimization: Few Examples

In view of the above concerning the technologies used (chemical, physical and enzymatic cross-linking) to prepare hydrogels, a lot of optimization processes have been recently developed in order to synthesize improving hydrogel biomaterials with higher application performances. As for example, in their study, Zhou et al. [52] produce a very interesting maleilated chitosan/thiol-terminated PVA hydrogel by using photo-clickable thiol-ene polymerization process with and thiol-terminated poly (vinyl alcohol) (TPVA) and maleic chitosan derivatives (MCS) activated under UV light source (60 mW/cm^2) by a photo-initiator such as the 2-hydroxy-1-[4-(hydroxyethoxy) phenyl]-2-methyl-1-propanone (Darocur 2959). In this case, the author has clearly shown the potential of this new photocrosslinked MCS/ TPVA hydrogels as tissue engineering scaffolds biomaterial due to the efficient L929 cells attachment and proliferation.

Moreover, we can mention the work of Lu et al. [53] on the specific preparation of a new tissue adhesive phenolic glycol chitosan hydrogel using an optimized photo-cross-linkage process activated by blue-light illumination. In this study, author particularly showed that this biomaterial possesses hemostatic properties and very good tissue adhesiveness. Moreover, the encapsulation of antibiotic such as gentamycin into these hydrogels gave very advantageous antibacterial ability.

Recently, Broguiere et al. [54], proposed a new hyaluronan hydrogel synthesized by an optimized enzymatic cross-linking strategy using a transglutaminase activity from the activated blood coagulation factor XIII. Authors related that this hydrogel possesses higher significant ability for 3D neuronal network tissue engineering (strong synaptic connection, dendritic and axonal specification, faster neurite outgrowth) than classical hyaluronan gels. Last technological advances in the hydrogel biomaterial fields were performed lately (Fig. 5). First, we can cite Yan et al. [55] who developed a multiscale modeling approach to synthesize an extracellular matrix mimetic hydrogel with sequestered recombinant human bone morphogenetic protein-2 (rhBMP-2). This novel synthetic bone scaffold prepared by an optimized carbohydrazide/aldehyde cross-linking strategy was efficiently validated as by in vivo assay with rat ectopic model. Secondly, Kim et al. [56] designed a very smart heparin mimetic hydrogel to improve and stabilize bone morphogenetic proteins 2 (BMP-2) properties which are well known as one of the most important bone formation stimulators. In this study, authors prepared a hydrogel biomaterial surface model with an efficient photo-crosslink process using heparin and

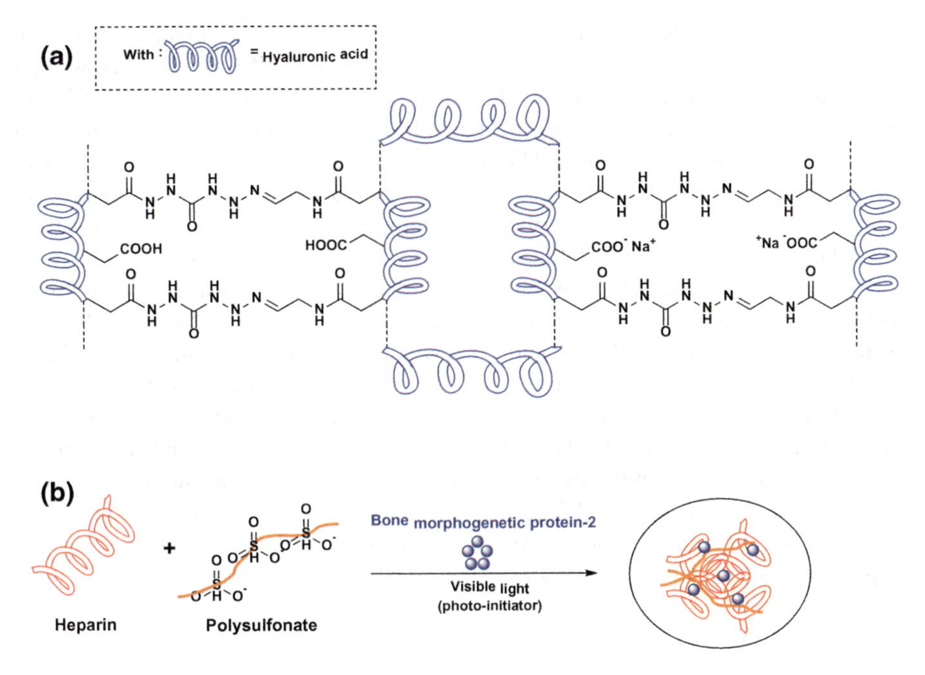

Fig. 5 Last technological advances in the hydrogel biomaterial fields. **a** An extracellular matrix mimetic hydrogel with cross-linked hyaluronic acid (adapted from Yan et al. [55]), **b** a new hydrogel biomaterial surface model with an efficient photo-crosslinking process using heparin and polysulfonate derivatives to encapsulate bone morphogenetic protein-2 (adapted from Kim et al. [56])

polysulfonate derivatives such as poly-4-styrenesulfonic acid (PSS) and poly-vinylsulfonic acid (PVSA). These sulfonated hydrogels were successful used to bind/encapsulate BMP-2 to increase osteogenesis and osteoinductive properties of bone marrow stromal cells (MBSCs) for bone tissue engineering.

Finally, other optimizations were done in the case of CC methodology, in order to evaluate coating effect on the construction of 3D-tissue models, cells without coating, cells coated with FN-G nanofilms or cells coated with collagen microfilms were seeded. Cells coated with collagen microfilms resulted at least 2-fold higher tissue than nano-coating (Fig. 6), with about 1/3 the cell number sufficient to fabricate equal-thick 3D-tissues. Collagen micro-coated cells led to tissues thickness up to 1871 μm. Also, cell-cell distances in these 3D-tissues were calculated as 15.6 ± 4.0 and 8.8 ± 3.1 μm, respectively. In comparison, cells without any coating resulted in compact distribution in some area and the very near distance between cells, about 3 times smaller than a cell with collagen micro-coating (data not shown). The difference in cell number also showed that the thickness of all samples increased according to the seeded cell number. The next step was the construction of functional 3D-tissue models. The development of 3D-vascularized thick tissues possessing high-density blood capillary networks is still an important issue for tissue engineering. To get a vascularized model, a sandwich culture with one layer of Human Umbilical Vein Endothelial Cells (HUVECs) seeded between 4 or 10-layers of FN-G nanofilm coated Normal Human Dermal Fibroblasts (NHDFs) was used. After 1 week, a highly developed homogenous capillary network with the tubular morphology of HUVECs was observed (Fig. 6B1, B2). With micrometre-sized collagen fiber-coated cells, vascularized network structures were also observed, suggesting that collagen coating method can be used to construct vascularized thick 3D-tissues (Fig. 6A1, A2). Other functional tissue models with FN-G nanofilm coated cells, as induced pluripotent stem cell-derived cardiomyocytes (iPS-CMs) or lymph epithelial cells (LECs), were also successfully constructed. Strong beating phenomenon was assessed during incubation [57]. When using collagen microfilm coating method, the constructed 3D tissues showed stable and strong beating (80 times/min) and high cell viability (>90%) after 4 days of incubation in 800 μm thick 3D tissues. Immunostaining assays such like actinin antibody, troponin T antibody, connexin antibody, and Azan stains were also performed to confirm the good functionality of cardiac myoblasts in the 3D iPS-CM tissues [24]. The collagen microfilm coating method can thus be applied to various types of cells and different purposes.

4 Innovative Sustainable Hydrogels: What's New?

4.1 Utilization of Current and Classical Hydrogel Products

Also known as smart and hungry three-dimensional networks, hydrogels are still subject of numerous papers and patents because of their high-tech potential for applications in a large range of fields, from the biomedical, biotechnology, agriculture to the pharmaceutical, microelectronics industry, oil recovery or cosmetic [30, 58]. Today, hydrogels must respond to physicochemical parameters (electric/magnetic field, solvent, pH, ionic strength, temperature) in their surroundings, change their physiochemical properties and be able to return to their initial states. Thus, the use of hydrogel products is obviously reliant on their technical features. Overall, the following items correctly characterize the functional properties that hydrogels should achieve [4], i.e. (i) the lowest soluble content and residual monomer, (ii) the lowest price, (iii) most eco-friendly approach (LCA concept), (iv) the highest stability and durability in the swelling environment, (v) the highest behavior against storage, (vi) the best potential for biodegradation without formation of toxic species, (vii) a light-stability without any color, odor and toxicity, (viii) a re-wetting capacity, that is, the ability to release or maintain solution trapped in the hydrogel (high flexibility), (ix) the highest absorbency under load, (x) the highest absorption capacity but also the possibility to correctly control a desired rate of absorption. Different items and levels in these features must be considered for making an "ideal" hydrogel. Nevertheless, authors recognize that no hydrogel can

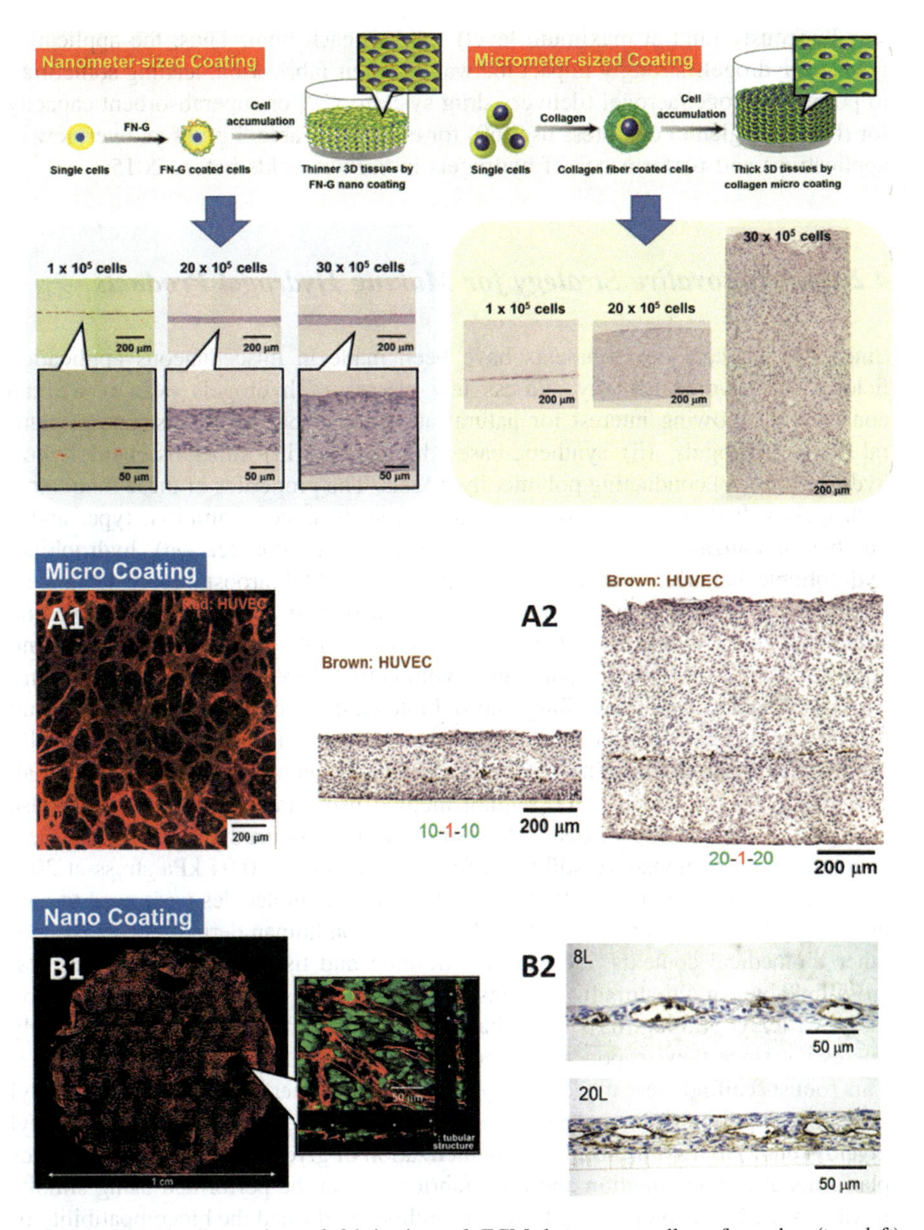

Fig. 6 Schematic illustration of fabrication of ECM layers on cell surfaces by (top left) nanometer-sized FN/G films and (top right) micrometre-sized collagen nanofiber matrices for construction of 3D-tissue models with higher and lower cell densities. Images of confocal laser scanning microscopy (CLSM) of vascularized 3D-tissue models constructed by **A1** nano- and **A2** micro-coatings by sandwich culture. Immunohistological staining images using anti-CD31 antibody of the 3D-tissue models by **B1** nano- and **B2** micro coating methods, adapted from Liu et al. [80]

simultaneously (and at maximum level) fulfil to each item. Thus, the applicative goals of hydrogels strongly impact the way for their fabrication, leading sometimes to porous hydrogel/aerogel (delivery drug system) [59] or superabsorbent capacity for (blood, hygienic) compress use [60] for example. Table 3 gives an overview of applications and performance of hydrogels in various fields before 2015.

4.2 An Innovative Strategy for Making Hydrogel Products

Since 2015, recent developments have been made in miscellaneous application fields. These works actually address few classes of hydrogels even if we can consider the growing interest for natural and sustainable hydrogels, i.e. (i) natural-based hydrogels, (ii) synthetic-based hydrogels, (iii) superabsorbent hybrid hydrogels or (iv) conducting polymer hydrogels. The properties of these innovative hydrogels will depend on several parameters such as concentration, type, and a number of ionizable groups, medium (and associated counter ion), hydrophilic/hydrophobic balance, charge, etc. As reported by Mahinroosta et al. [30], new intelligent hydrogels (Fig. 7) should be (i) sensitive to a wide range of external stimuli, such as temperature [61], enzymes [62], light [63] or pH (Aycan and Alemdar 2018), (ii) able to change their volumetric shape (expansion/contraction), (iii) swell and deswell (swelling ratio) biological fluids in particular for drug delivery applications and/or medical devices, etc. Recent developments especially involve the use of 3D printing for the fabrication of intelligent scaffolds. Tan et al. [64] published a cryogenic 3D printing method using the liquid to a solid phase change of a composite hydrogel. They successfully produced specific geometrical structures with compressive stiffness of O(1) kPa (0.49 ± 0.04 kPa stress at 30% compressive strain). Collagen type I, gelatin and other molecules were used to coat the 3D printed material before testing the systems on human dermal fibroblasts and other biomedical contexts, *e.g.* surgical training and tissue engineering. Besides, cross-linkable multi-stimuli responsive hydrogels were also prepared by direct-write 3D printing [65]. The behaviors of the new hydrogels were particularly interesting since they responded to shear-thinning, UV light but also temperature. This robust scaffold were made of poly(allyl glycidyl ether)-stat-poly(alkyl glycidyl ether)-block-poly(ethylene glycol)-block-poly(allyl glycidyl ether)-stat-poly(alkyl glycidyl ether) and synthesized by polymerization of glycidyl ethers. Bioelectronics platforms also gain attention and their fabrication can be performed using simple, flexible route by 3D bioprinting [66]. The authors confirmed the biocompatibility of the system against C2C12 murine myoblasts cell line. 3D printed hydrogels are in general programmable and responsive to environmental and fields signals, repeatable and stable over cycle/time, as reported by Lv et al. [67] for poly(ethylene glycol) diacrylate hydrogel microstructures which have excellent humidity responsiveness. New startups also try to take control of the cell microenvironment, e.g. Alvéole (http://alveolab.com, France) which develops innovative photopatterning solution (PRIMO) for 3D scaffold designing. This tool, using pseudo

Table 3 Overview of some hydrogel applications (before 2015)

Applications	Type of hydrogel	Properties/uses	References
Biomedical Biotechnology Pharmaceutical	Dextrin-based hydrogel/nanogel Dextrin-based hydrogel/urinary bladder matrix	Cell encapsulation and injectable nano/hydrogel	Silva et al. [92]
	Ionically cross-linkable (alginate) hyaluronate-based hydrogels	Drug (primary chondrocytes) delivery system for in vivo cartilage regeneration	Park et al. [93]
	Polyvinyl alcohol hydrogel/dextran copolymerization	Vascular grafting, high bio and hemocompatibility	Alexandre et al. [94]
	DNA hydrogel, DNA-capped gold nanoparticles	Enzymatic tool kit, adapter for protein detection, thermostable, plasmonics	Yang et al. [95]
	Fe3O4 nanoparticle/hydrogel magnetic nanocomposite	Excellent catalytic activity, sensitive toward H_2O_2 detection, environmental chelation	Gao et al. [96]
	Collagen/b-sodium glycerophosphate-carbodiimide hydrogel	Chitosan human-like assemblage, soft tissue defect filling, delivery system properties	Li et al. [97]
	TiO2/PEGDA hybrid hydrogel	Photodynamic therapy on tumour cells	Zhang et al. [98]
	Chitosan highly hydrated collagen hydrogel	Scaffold mimicking native extracellular matrix	Chicatun et al. [99]
	PVA/chitosan hydrogel	Swelling activity and antimicrobial potential, drug (sparfloxacin) delivery system	Abdel-Mohsen et al. [100]
	Poly(lactic acid)-poly(ethylene oxide) block copolymers	Nanoclusters as biosensors, enhance elastic modulus of the hydrogel, tissue engineering scaffold	Saffer et al. [101]
	PVA hydrogel	Arterial biomodelling	Kosukegawa et al. [102]
	Poly(*N*-isopropylacrylamide)/iron oxide magnetic hydrogel nanocomposites	Drug delivery system, hyperthermia for cancer treatment	Meenach et al. [103]
	Carboxy-methylated cellulose/chitosan hydrogel (irradiation)	Adsorption of heavy metal ions, water treatment	Zhao and Mitomo [104]
	Poly(acrylamide-co-acrylic acid)/chitosan nanostructured hydrogel	Drug (ascorbic acid) delivery system	Becerra-Bracamontes et al. [105]

(continued)

Table 3 (continued)

Applications	Type of hydrogel	Properties/uses	References
	Poly(ethylene glycol)/poly(e{open}-caprolactone)-based hydrogel	Cell delivery application, scaffold for cartilage formation (tissue collagen type II, aggrecan, SOX9, COMP gene expression)	Park et al. [106]
	PEG-PLGA graft copolymeric hydrogel	Molecular recognition for cancer treatment/imaging	Jeong and Gutowska [107]
	Methylcellulose-10%PEG Petrolatum-10%isopropyl myristate	Skin irritation (atopic eczema), a delivery system for allergens	Darsow et al. [108]
	PPO-PEO	Wound dressing and implant materials. Fill the spaces in a wound, isolate the area from bacterial infection	Corkhill and Hamilton [109]
Food packaging	Poly(*N*-isopropylacrylamide) nanohydrogel	Sensitive food packaging, heat capacity, delivery system, usable in w/o system	Fuciños et al. [110]
	Agar/κ-carrageenan/konjac glucomannan	Water and mechanical barrier properties, biohydrogel, antifogging, respiring hydrogel, antimicrobial property	Rhim and Wang [111]
	Poly(*N*-isopropylacrylamide) nanohydrogel Poly(*N*-isopropylacrylamide)(80%)/acrylic acid(20%)	Delivery system (pimaricin), antifungal, resistant to acidic conditions, thermal resistance	Fuciños et al. [112]
	Polyvinylpyrrolidone/ carboxymethylcellulose hydrogel	Biodegradable, eco-friendly material	Roy et al. [113]
	Carboxymethylcellulose/methacrylate hydrogel	Delivery system (by enzymes degradation), tool for detecting bacteria and fungi	Schneider et al. [114]
	Agar/silver nanoparticles hydrogel	Controlling microbial proliferation	Incoronato et al. [115]
	Collagen-based/dialdehyde starch hydrogel	Thermo-(ir)reversible gel, biodegradable, from waste proteins	Langmaier et al. [116]
Cosmetic	Carboxymethylcellulose hydrogel	Injectable, soft tissue filler, tunable hydrogel	Varma et al. [117]
	Poly(vinyl alcohol)/hyaluronic acid blends	Gelation time performance, high mechanical properties, blend hydrogel	Kodavaty and Deshpande [118]

(continued)

Table 3 (continued)

Applications	Type of hydrogel	Properties/uses	References
	Nanostructured lipid carriers-based hydrogel	Skin hydration, entrapment of oil (argan), easy formulated	Tichota et al. [119]
	Polyacrylamide-based hydrogel	Lip enhancement, widespread facial infections, potentially risky for health	Wang et al. [120, 121]
	Calcium hydroxylapatite-based hydrogel	Injectable, long-lasting un-permanent filler, highly biocompatible, dermal filler, potential body reactions (risk)	Pavicic [122]
	Palmitoyl glycol chitosan hydrogel	The delivery system (magnesium ascorbyl phosphate), UV protection, an inhibitor of tyrosinase	Wang et al. [120, 121]
	P(MAA-co-EGMA) hydrogel microparticles	Smart carrier, high skin permeability, pH-sensitive release behaviour	Lee and Kim [123]
	Xanthan/lignin hydrogel-epichlorohydrine	Thermal stability, hydrophobicity, decent biocompatibility	Raschip et al. [124]
	Xyloglucan-chitosan complex hydrogel	Non-toxic, antimicrobial and texture properties, thermal stability	Simi and Abraham [125]
Agriculture	Guar gum-g-polyacrylate-based hydrogel	Water retention and swelling behavior, application in moisture stress agriculture, pH-sensitive	Chandrika et al. [126]
	Fertilizing hydrogel	Work for establishing poplar	Böhlenius and Overgaard [127]
	ι/κ-carrageenan-agar/xanthan hydrogel blends	Coating seed, increase water holding capacity, increase gel strength	Hotta et al. [128]
	Chitin hydrogel	Biodegradable, plant growth regulator, excellent water uptake ability	Tang et al. [129]
	Nanocomposite PAAm/methyl cellulose/montmorillonite hydrogel	Nutrient carrier vehicle, increase water absorption speed	Bortolin et al. [130]
	Poly(acrylamide-co-acrylic acid)/$AlZnFe_2O_4$ hydrogel	Superabsorbent, improve wheat plant growth and establishment, enhance the moisture retention	Shahid et al. [131]
	Pectin-based polymer hydrogel	Remover of Cu^{2+} and Pb^{2+}, the release of fertilizers, help to conserve water, able to macromolecular relaxation	Guilherme et al. [132]

(continued)

Table 3 (continued)

Applications	Type of hydrogel	Properties/uses	References
Separation Oil recovery	Crosslinked dextran hydrogel-epichlorohydrin-toluene	Separation of liposome and drugs, water swelling property, high separation performance, decent sphericity and distribution	Bao et al. [133, 134]
	Dimethylamino ethyl methacrylate hydrogel	Thermo and pH-controllable, superhydrophilicity, superoleophobicity, o/w separation	Cao et al. [135]
	Clay nanosheets, *N,N*-dimethylacrylamide and 2-acrylamide-2-methylpropanesulfonic acid	The high-strength freestanding elastic hydrogel, selectively separation of Pb^{2+} and Cu^{2+}, recycle membrane	Bao et al. [133, 134]
	PEG-laponite XLG-acrylamide particle gels (PPGs)	3D dense network, mobility control, fracture-plugging applications, long-term thermal stability	Tongwa and Bai [136]
	Polyacrylamide/montmorillonite (Na-MMT)-chromium hydrogel	Enhance oil recovery, in-depth profile modifier, oil displacing agent, strong thermal stability	Zolfaghari et al. [137]
	Polyacrylamide/poly(vinyl alcohol)-chromium triacetate hydrogel	Gelation and swelling behaviour, enhance oil recovery	Aalaie et al. [138]
	PAM inverse emulsion/acrylamide	In-depth permeability control, fluid diverting, oil driving, increase oil productivity	Lei et al. [139]
	Xanthan/Zr^{4+} flowing gel	Enhance oil recovery, strongly resistant to shearing	Wang and Zhang [140]
	Conventional bilayer lipid membranes-hydrogel salt-bridge	pH and ion sensors, sensors for ligand-reception interactions, drug testing, enhance oil recovery	Tiena and Ottovaab [141]

hydrogel solution, allows the absorption of specific proteins on illuminated areas then cells to these proteins, respecting a defined micropattern, and thus 3D construction, for biomedical and tissue engineering for example. Finally, recent works highlighted 4D fabrication using shape-morphing hydrogel [68]. Alginate and hyaluronic acid were used as biopolymers for the conception of the hydrogel, and mouse bone marrow stromal cells for the biocompatibility tests. The authors were able to generate average internal tube diameters (20 μm), comparable to the smallest blood vessels without any loss of cell viability. Their statements are strong since this 4D (four-dimensional) biofabrication strategy aims to produce dynamically reconfigurable architectures, with tunable functionalities, as reported in Fig. 7. Wang et al. [69] recently published a comprehensive review concerning new development and biomedical applications of these hydrogels. This paper illustrates the work already done but mainly to perform in the fabrication processes which absolutely need to be (i) inexpensive, (ii) from and/or using nontoxic/non-hazardous materials and techniques and (iii) fine and easily-tuning possibilities. Thus, a wide range of papers are nowadays available in the literature and natural polymer (DNA, protein, polysaccharide) based-hydrogels take benefit from the situation [70–72].

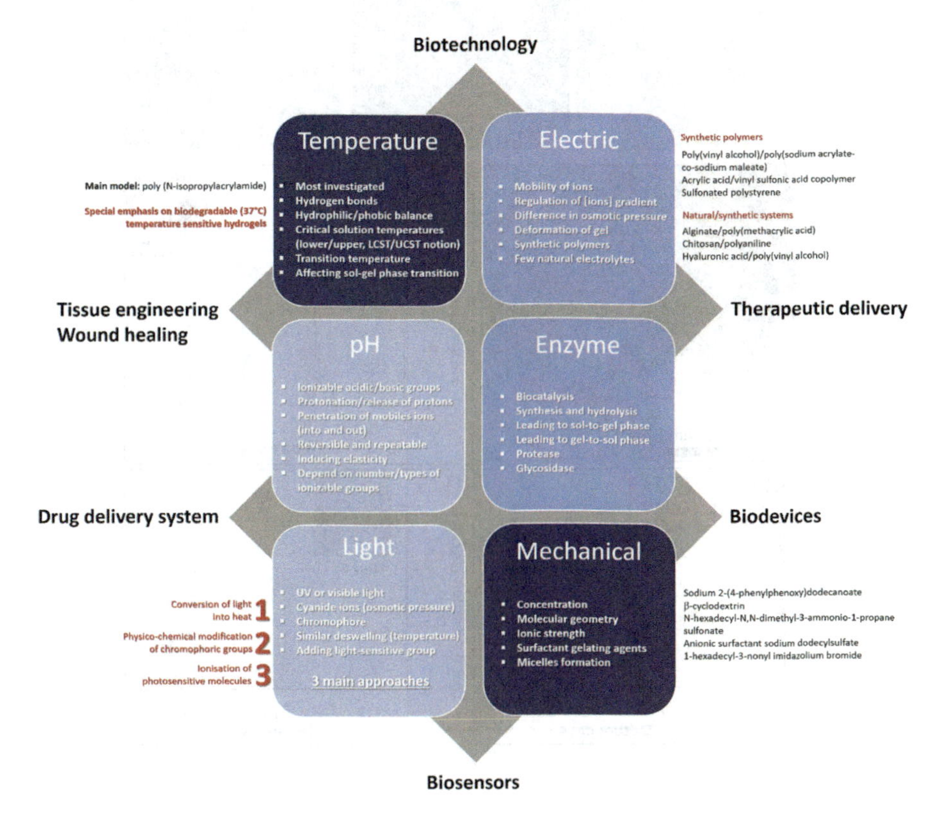

Fig. 7 Main strategies for the conception of new smart hydrogels and recently associated polymers

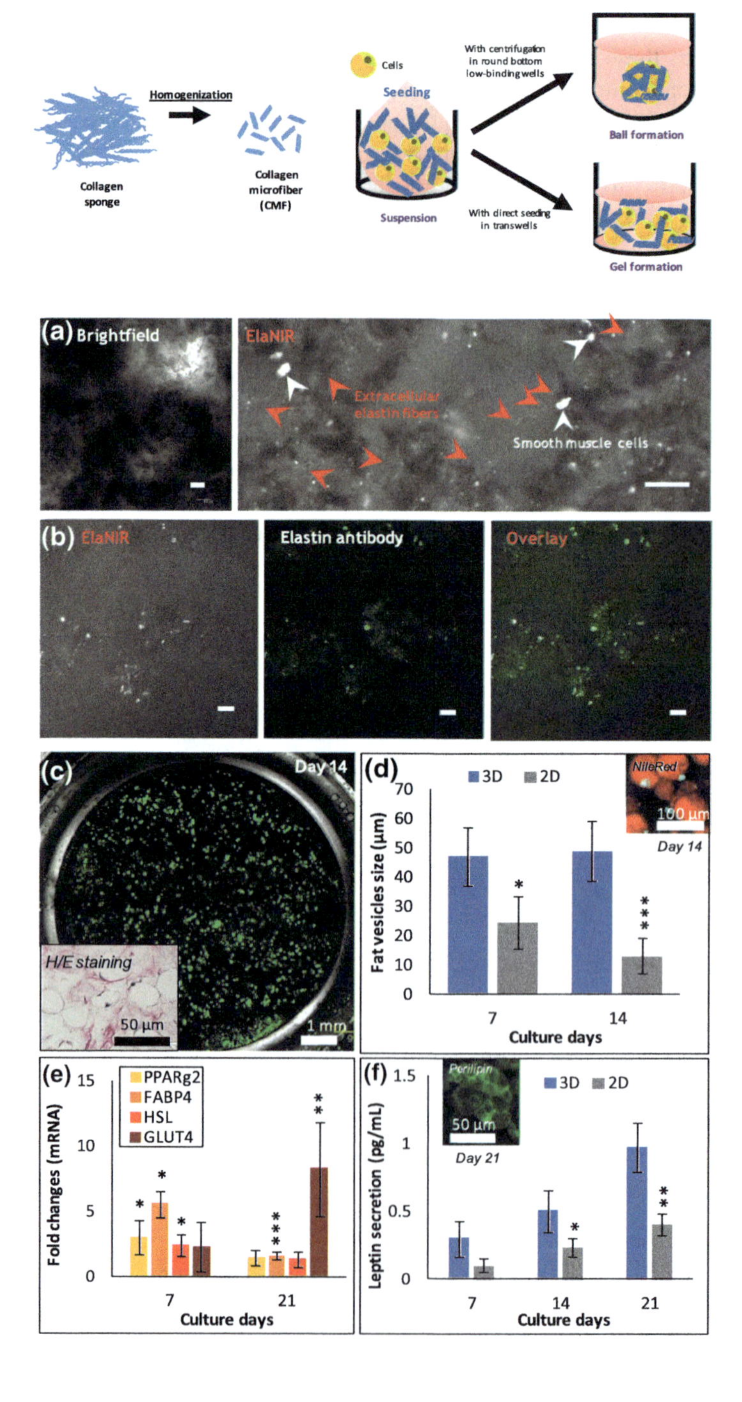
Homogenization
Collagen sponge
Collagen microfiber (CMF)
Cells
Seeding
Suspension
With centrifugation in round bottom low-binding wells
Ball formation
With direct seeding in transwells
Gel formation
(a) Brightfield
ElaNIR
Extracellular elastin fibers
Smooth muscle cells
(b) ElaNIR
Elastin antibody
Overlay
(c)
Day 14
H/E staining
50 µm
1 mm
(d)
3D
2D
NileRed
100 µm
Day 14
Fat vesicles size (µm)
70
60
50
40
30
20
10
0
7
14
Culture days
(e)
PPARg2
FABP4
HSL
GLUT4
Fold changes (mRNA)
15
10
5
0
7
21
Culture days
(f)
Perilipin
50 µm
Day 21
3D
2D
Leptin secretion (pg/mL)
1.5
1
0.5
0
7
14
21
Culture days

◂**Fig. 8** Schematic illustration of collagen microfiber hydrogels model construction. **a** Bright field (left) and ElaNIR signal in NIR channel (right). **b** ElaNIR signal was detected in NIR channel (left), the signal from an anti-elastin antibody (middle) and overlay imaging (right). Scale bar: 100 μm. **c** Live/dead assay attesting the good viability of mature adipocytes after 14 days in collagen microfibers hydrogels. **d** Adipose vesicles diameters measured each week using Nile red specific lipids staining. **e** Adipogenesis gene expression assessed by RT-qPCR on total RNA of ADSC (PPARγ2, FABP4, GLUT4 and HSL genes) expressed in fold changes regarding 2D condition. **f** Leptin secretion measurement by ELISA in the culture medium every week and normalized by DNA content. Perilipin immunostaining performed at day 21. Error bars represent SD. Tukey multiple comparison test (double-way ANOVA) was used with $*p < 0.05$, $**p < 0.01$, and $***p < 0.001$

4.3 Focus on the Nano to Micro ECM Gel Coating System

Classic collagen hydrogels are still widely used for tissue engineering despite its limited collagen content of maximum 0.3%, remaining far away from the in vivo conditions. In this context, the use of collagen microfibers instead of classic collagen dilution can achieve higher density (until 20–30 wt%). The method relies on the immersion of porcine type I collagen in ×10 PBS (Phosphate buffer saline), followed by its homogenization for 2 min to create the microfibers (VH-10 homogenizer, As One Corp., Osaka, Japan). After centrifugation, the microfibers were washed in DMEM (Dulbecco's modified eagle medium) without FBS (Fetal bovine serum) before being mixed with the cells suspension.

4.3.1 Live-Staining of Secreted Elastin by Smooth Muscle Cells in All Tissues

Elastin is one of the major components of the extracellular and thus often assessed during tissue regeneration. For example, the aorta wall is constructed with an arrangement of elastin in concentric lamellae presenting smooth muscle cells between them [73]. A new fluorescent ElaNIR probe was used to visualize specific elastin secretion in smooth muscle cells 3D-tissue models. Human umbilical artery smooth muscle cells were seeded in ball microfibers collagen tissues. These ball tissues were made by centrifuging the mix cells-collagen microfibers in round bottom low-binding plate wells. Smooth muscle cells ball tissues displayed an increase of extracellular elastin fibers as stained by ElaNIR probes (Fig. 8a), confirmed by consistent elastin antibody staining (Fig. 8b). Overall, the specific ElaNIR probe along with the in vivo like smooth muscle tissue can be used as a model for observation of elastin production during tissue regeneration or for the screening of elastin-enhancing chemicals in cosmetics products [74].

4.3.2 Adipose Tissue Regeneration Inducing and Maintaining the Functionality of Both Pre and Mature Adipocytes in Long-Term Cultures

Adipose tissue regeneration is currently a competitive challenge for either cosmetic/pharmaceutical assays or plastic surgery purposes. Conventional in vitro two-dimensional (2D) cell cultures using directly mature adipocytes (AD) showed limited culture time by quickly dedifferentiating [75] while getting sufficiency matured AD by differentiating adipose-derived stem cells (ADSC) usually required more than one month [76]. The existing three-dimensional (3D) models accelerated the ADSC adipogenesis, but the mature AD still cannot be maintained more than one-week in vitro cultures [77]. In this context, the construction of a biomimetic 3D-tissue is determinant. Using collagen microfibers, high-density collagen (until 20–30 wt%, see H/E staining (Fig. 8c), similar to in vivo [78] artificial adipose tissues were performed mixing the homogenized type I collagen with a mature AD or ADSC and seeding in 24 well transwells. These 3D-tissues ensured the long-term maintenance of unilocular mature AD with a good viability of 95% at day 14 (Live/Dead image Fig. 8c). On the contrary, the 2D mature AD showed significantly 4 times smaller multiple vesicles (Fig. 8d). Concerning ADSC, 3D adipogenic genes expression was found at least significantly doubled throughout the differentiation (even 8.3 times higher for GLUT4 at day 21, Fig. 8e), along with up to almost 4 times bigger fat vesicles observed at day 14 (data not shown). Perilipin immunostaining, the protein stabilizing the fat vesicles, and leptin secretion, the well-known safety protein, finally attested the up to the twice better functionality of 3D adipocytes (Fig. 5e). The obtained long-term functional maintenance and the faster adipogenesis made this model relevant for screening assays and reconstructive surgery.

5 Conclusions

More and more hydrogels have been published and patented these last years for their specific properties leading to various applications. Some of them are widely present in commercial products. Their success is linked to their ability to swell in aqueous solutions or suspensions. This book chapter has done the state of the art of the different natural or synthetic hydrogels available and described the obtaining processes for different uses in numerous industrial fields. The potential of these macromolecular networks has not been fully explored at this time notably in therapy for tissue engineering and drug delivery, two fields where only a few products are on the market. Covalently cross-linked hydrogels made in the absence of solvents are the more popular for this kind of applications. Undoubtedly, methods from supramolecular chemistry applied to the synthesis in aqueous environments of new hydrogels having modifiable properties could open the way of new possibilities. Supramolecular chemistry is a domain of chemistry that focuses on the building of macromolecular systems made up of non-covalently assembled molecular subunits. Current researches focusing on hydrogels for multidrug delivery by a single system and/or sequential delivery on demand with a high level of control by stimuli highlight also new opportunities. Another way of hydrogel improvement, notably for application in the human body, could be the control of their swelling capacity. Indeed, the swelling of hydrogels and/or their degradation after their implantation can alter surrounding tissues. This disadvantage can be limited by decreasing the polymer concentration but in this case, the time to form the hydrogel is too long. An interesting article recently published in Nature Biomedical Engineering proposed an original solution of a two-step gelation process of PEG hydrogels at low concentration [79]. In a first step, branched polymers clusters were generated but the crosslinking reaction was intentionally stopped just before the full gelation. The solution was then injected into the body at the required place and the clusters present in the co-crosslinked solution to form a gel in ten minutes. These hydrogels have low cytotoxicity and can act as an artificial vitreous body. So, the future of hydrogels in the therapeutic area will be probably linked to the development of innovative properties such as those described above but also to the reduction of costs for their obtaining.

Acknowledgements This research was supported by JST-PRESTO (15655131) and a Grant-in-Aid for Scientific Research (B) (26282138 and 17H02099). The authors also thank the program "Exploration Japon 2018" from Campus France, SST and SCAC (Ambassade de France au Japon).

References

1. Ganguly K, Chaturvedi K, More UA, Nadagouda MN, Aminabhavi TM (2014) Polysaccharide-based micro/nanohydrogels for delivering macromolecular therapeutics. J Control Release 193:162–173
2. Mastropietro DJ, Omidian H, Park K (2012) Drug delivery applications for superporous hydrogels. Expert Opin Drug Deliv 9:71–89
3. Calo E, Khutoryanskiy VV (2015) Biomedical applications of hydrogels: a review of patents and commercial products. Euro Polym J 65:252–267
4. Ahmed EM (2015) Hydrogels: preparation, characterizations and applications: a review. J Adv Res 6:105–121
5. Singhal R, Gupta K (2016) A review: tailor-made hydrogel structures (classifications and synthesis parameters). Polym Plast Technol Eng 55:54–70
6. Chai Q, Jiao Y, Yu X (2017) Hydrogels for biomedical applications: their characteristics and the mechanisms behind them. Gels 3, 6. https://doi.org/10.3390/gels3010006
7. Bescrades IG, Demirtas TT, Durukan MD et al (2015) Microwave-assisted fabrication of chitosan-hydroxyapatite superporous hydrogel composites as bone scaffold. J Tissue Eng Regen Med 9:1233–1246
8. Salimi-Kenari H, Mollaie F, Dashtimoghadam E et al (2018) Effects of chain length of the cross-linking agent on rheological and swelling characteristics of dextran hydrogels. Carbohydr Polym 181:141–149
9. Tavsanli B, Okay O (2017) Mechanically strong hyaluronic acid hydrogels with an interpenetrated network structure. Euro Polym J 94:185–195

10. Maisani M, Ziane S, Ehret C et al (2018) A new composite hydrogel combining the biological properties of collagen with the mechanical properties of a supramolecular scaffold for bone tissue engineering. J Tissue Eng Regen Med 12:1489–1500
11. Varaprasad K, Rahavendra GM, Jayaramudu T et al (2017) A mini review on hydrogels classification and recent developements in miscellaneous applucations. Mat Sci Eng C 79:958–971
12. Mati-Baouche N, Elchinger PH, de Baynast H, Pierre G, Delattre C, Michaud P (2014) Chitosan as an adhesive. Eur Polym J 60:198–213
13. Shonnard DR, Kicherer A, Saling P (2003) Industrial applications using BASF ecoefficiency analysis: perspectives on green engineering principles. Environ Sci Technol 37:5340–5348
14. Vink ETH, Rábago KR, Glassner DA, Gruber P (2003) Applications of life cycle assessment to nature works polylactide (PLA) production. Polym Degr Stab 80:403–19
15. Clark JH (2008) Green chemistry: today (and tomorrow). Green Chem 8:17–21
16. Tian H, Tang Z, Zhuang X et al (2012) Biodegradable synthetic polymers: preparation, functionalization and biomedical application. Prog Polym Sci 37:237–280
17. Takashi L, Hatsumi T, Makoto M et al (2007) Synthesis of porous poly (N-isopropylacrylamide) gel beads by sedimentation polymerization and their morphology. J Appl Polym Sci 104:842
18. Yang L, Chu JS, Fix JA (2002) Colon-specific drug delivery: new approaches and in vitro/in vivo evaluation. Int J Pharm 235:1–15
19. Zuluaga M, Gregnanin G, Cencetti C, Di Meo C, Gueguen V, Letourneur D, Meddahi-Pelle A, Pavon-Djavid G, Matricardi P (2018) PVA/dextran hydrogel patches as delivery system of antioxydant astaxanthin: a cardiovascular approach. Biomed Mater 13:1–13
20. Maolin Z, Jun L, Min Y et al (2000) The swelling behaviour of radiation prepared semi-interpenetrating polymer networks composed of polyNIPAAm and hydrophilic polymers. Radiat Phys Chem 58:397–400
21. Hacker MC, Mikos AG (2011) Synthetic polymers. In: Atala A, Lanza R, Thomson JA, Nerem RM (eds) Principles of regenerative medicine, 2nd edn. Academic Press, USA, pp 587–622
22. Ahmed EM, Aggor FS, Awad AM et al (2013) An innovative method for preparation of nanometal hydroxide superabsorbent hydrogel. Carbohydr Polym 91(2):693–698
23. Akhtar MF, Hanif M, Ranjha NM (2016) Methods of synthesis of hydrogels: a review. Saudi Pharm J 24(5):554–559
24. Liu CY, Matsusaki M, Akashi M (2015) Control of cell-cell distance and cell densities in millimeter-sized 3D tissues constructed by collagen nanofiber coating techniques. ACS Biomater Sci Eng 1:639–645
25. Yuangang Z, Ying Z, Xiuhua Z et al (2012) Preparation and characterization of chitosan–polyvinyl alcohol blend hydrogels for the controlled release of nano-insulin. Int J Biol Macromol 50(1):82–87
26. Brannon-Peppas L, Harland RS (1991) Absorbent polymer technology. J Control Release 17(3):297–298
27. Li Y, Huang G, Zhang X et al (2013) Magnetic hydrogels and their potential biomedical applications. Adv Funct Mater 23(6):660–672
28. Matsusaki M, Yoshida H, Akashi M (2007) The construction of 3D-engineered tissues composed of cells and extracellular matrices by hydrogel template approach. Biomaterials 28:2729–2737
29. Gobbi A, Whyte GP (2016) One-stage cartilage repair using a hyaluronic acid-based scaffold with activated bone marrow-derived mesenchymal stem cells compared with microfracture: five-year follow-up. Am J Sports Med 44(11):2846–2854
30. Mahinroosta M, Farsangi ZJ, Allahverdi A et al (2018) Hydrogels as intelligent materials: a brief review of synthesis, properties and applications. Mat Tod Chem 8:42–55
31. Pierschbacher MD, Ruoslahti E (1987) Influence of stereochemistry of the sequence Arg-Gly-Asp-Xaa on binding specificity in cell adhesion. J Biol Chem 262:17294–17298

32. Chang PC, Liu BY, Liu CM et al (2007) Bone tissue engineering with novel rhBMP2-PLLA composite scaffolds. J Biomed Mater Res A 81:771–780. https://doi.org/10.1002/jbm.a.31031
33. Recknor JB, Sakaguchi DS, Mallapragada SK (2006) Directed growth and selective differentiation of neural progenitor cells on micropatterned polymer substrates. Biomaterials 27:4098–4108. https://doi.org/10.1016/j.biomaterials.2006.03.029
34. Liss M, Petersen B, Wolf H et al (2002) An aptamer-based quartz crystal protein biosensor. Anal Chem 74:4488–4495
35. Decher G (1997) Fuzzy nanoassemblies: Toward layered polymeric multicomposites. Science 277:1232–1237. https://doi.org/10.1126/science.277.5330.1232
36. Decher G, Hong JD (2011) Buildup of ultrathin multilayer films by a self-assembly process, 1 consecutive adsorption of anionic and cationic bipolar amphiphiles on charged surfaces. Makromol Chem Macromol Symp 46:321–327. https://doi.org/10.1002/masy.19910460145
37. Ruoslahti E, Pierschbacher MD (1987) New perspectives in cell adhesion: RGD and integrins. Science 238:491–497
38. Arnaout MA, Mahalingam B, Xiong J-P (2005) Integrin structure, allostery, and bidirectional signaling. Annu Rev Cell Dev Biol 21:381–410
39. Campbell ID, Humphries MJ (2011) Integrin structure, activation, and interactions. Cold Spring Harb Perspect Biol 3:1–15. https://doi.org/10.1101/cshperspect.a004994
40. Nagae M, Re S, Mihara E et al (2012) Crystal structure of $\alpha5\beta1$ integrin ectodomain: atomic details of the fibronectin receptor. J Cell Biol 197:131–140. https://doi.org/10.1083/jcb.201111077
41. Liu P, Zhai M, Li J et al (2002) Radiation preparation and swelling behavior of sodium carboxymethyl cellulose hydrogels. Radiat Phys Chem 63(3–6):525–528
42. Said HM, Alla SGA, El-Naggar AWM (2004) Synthesis and characterization of novel gels based on carboxymethyl cellulose/acrylic acid prepared by electron beam irradiation. React Funct Polym 61(3):397–404
43. Sanson N, Rieger J (2010) Synthesis of nanogels/microgels by conventional and controlled radical crosslinking copolymerization. Polym Chem 1:965–977
44. Sperinde JJ, Griffith LG (1997) Synthesis and characterization of enzymatically-crosslinked-poly(ethylene glycol) hydrogels. Macromolecules 30:5255–5264
45. Chen X, Martin BD, Neubauer TK et al (1995) Enzymatic and chemoenzymatic approaches to synthesis of sugar-based polymer and hydrogels. Carbohydr Polym 28:15–21
46. Raia NR, Partlow BP, McGill M et al (2017) Enzymatically crosslinked silk-hyaluronic acid hydrogels. Biomaterials 131:58–67
47. Nishiguchi A, Matsusaki M, Asano Y et al (2014) Effects of angiogenic factors and 3D-microenvironments on vascularization within sandwich cultures. Biomaterials 35:4739–4748. https://doi.org/10.1016/j.biomaterials.2014.01.079
48. Nishiguchi A, Yoshida H, Matsusaki M et al (2011) Rapid construction of three-dimensional multilayered tissues with endothelial tube networks by the cell-accumulation technique. Adv Mater 23:3506–3510. https://doi.org/10.1002/adma.201101787
49. Matsusaki M, Akashi M (2014) Control of extracellular microenvironments using polymer/protein nanofilms for the development of three-dimensional human tissue chips. Polym J 46:524–536. https://doi.org/10.1038/pj.2014.20
50. Stockwell RA (1967) The cell density of human articular and costal cartilage. J Anat 101:753–763
51. Liu CY, Matsusaki M, Akashi M (2014) The construction of cell-density controlled three-dimensional tissues by coating micrometer-sized collagen fiber matrices on single cell surfaces. RSC Adv 4:46141–46144. https://doi.org/10.1039/C4RA09085C
52. Zhou Y, Zhao S, Zhang C et al (2018) Photopolymerized maleilated chitosan/thiol-terminated poly (vinyl alcohol) hydrogels as potential tissue engineering scaffolds. Carbohydr Polym 184:383–389
53. Lu M, Liu Y, Huang YC et al (2018) Fabrication of photo-crosslinkable glycol chitosan hydrogel as a tissue adhesive. Carbohydr Polym 181:668–674

54. Broguiere N, Isenmann L, Zenobi-Wong M (2016) Novel enzymatically cross-linked hyaluronan hydrogels support the formation of 3D neuronal networks. Biomaterials 99:47–55
55. Yan HJ, Casalini T, Hulsart-Billström G et al (2018) Synthetic design of growth factor sequestering extracellular matrix mimetic hydrogel for promoting in vivo bone formation. Biomaterials 161:190–202
56. Kim S, Cui ZK, Kim PJ, Jung LY, Lee M (2018) Design of hydrogels to stabilize and enhance bone morphogenetic protein activity by heparin mimetics. Acta Biomater (in press). https://doi.org/10.1016/j.actbio.2018.03.034
57. Amano Y, Nishiguchi A, Matsusaki M et al (2016) Development of vascularized iPSC derived 3D-cardiomyocyte tissues by filtration Layer-by-Layer technique and their application for pharmaceutical assays. Acta Biomater 33:110–121
58. Ullah F, Othman MBH, Javed F et al (2015) Classification, processing and application of hydrogels: a review. Mat Sci Eng C 57:414–433
59. Pierre G, Punta C, Delattre C et al (2017) TEMPO-mediated oxidation of polysaccharides: an ongoing story. Carbohydr Polym 165:71–85
60. Anisha S, Kumar SP, Kumar GV et al (2010) Hydrogels: a review. Int J Pharmaceut Sci Rev Res 4(2):97
61. Tylman M, Pieklarz K, Owczarz P et al (2018) Structure of chitosan thermosensitive gels containing graphene oxide. J Mol Struc 1161:530–535. https://doi.org/10.1016/j.molstruc.2018.02.065
62. Alshememry AK, El-Tokhy SS, Unsworth LD (2017) Using properties of tumor microenvironments for controlling local, on-demand delivery from biopolymer-based nanocarriers. Curr Pharm Des 23:5358–5391. https://doi.org/10.2174/1381612823666170522100545
63. Qin XH, Wang X, Rottmar M et al (2018) Near-infrared light-sensitive polyvinyl alcohol hydrogel photoresist for spatiotemporal control of cell-instructive 3D microenvironments. Adv Mat 30(1705564). https://doi.org/10.1002/adma.201705564
64. Tan Z, Parisi C, Di Silvio L et al (2017) Cryogenic 3D Printing of super soft hydrogels. Sci Reports 7(16668). https://doi.org/10.1038/s41598-017-16668-9
65. Karis DG, Ono RJ, Zhang M et al (2017) Cross-linkable multi-stimuli responsive hydrogel inks for direct-write 3D printing. Polym Chem 8:4199–4206
66. Agarwala S, Lee JM, Ng WL (2018) A novel 3D bioprinted flexible and biocompatible hydrogel bioelectronic platform. Biosens Bioelec 102:365–371
67. Lv C, Sun XC, Xia H et al (2018) Humidity-responsive actuation of programmable hydrogel microstructures based on 3D printing. Sens Act B: Chem 259:736–744. https://doi.org/10.1016/j.snb.2017.12.053
68. Kirillova A, Maxson R, Stoychev G et al (2017) 4D Biofabrication using shape-morphing hydrogels. Adv Mat 29(1703443). https://doi.org/10.1002/adma.201703443
69. Wang Y, Adokoh CK, Narain R (2018) Recent development and biomedical applications of self-healing hydrogels. Exp Opin Drug Deliv 15:77–91. https://doi.org/10.1080/17425247.2017.1360865
70. Aljohani W, Ullah MW, Li W et al (2018) Three-dimensional printing of alginate-gelatin-agar scaffolds using free-form motor assisted microsyringe extrusion system. J Polym Res 25 (62). https://doi.org/10.1007/s10965-018-1455-0
71. Azizi S, Mohamad R, Abdul Rahim R et al (2017) Hydrogel beads bio-nanocomposite based on Kappa-Carrageenan and green synthesized silver nanoparticles for biomedical applications. Int J Biol Macromol 104:423–431
72. Tedesco MT, Di Lisa D, Massobrio P (2018) Soft chitosan microbeads scaffold for 3D functional neuronal networks. Biomaterials 156:159–171
73. O'Connell MK, Murthy S, Phan S et al (2008) The three-dimensional micro- and nanostructure of the aortic medial lamellar unit measured using 3D confocal and electron microscopy imaging. Matrix Biol J Int Soc Matrix Biol 27:171–181. https://doi.org/10.1016/j.matbio.2007.10.008

74. Su D et al (2018) Elastin: a near-infrared zwitterionic fluorescent probe for in vivo elastin imaging. Chem J (accepted)
75. Lessard J, Pelletier M, Biertho L et al (2015) Characterization of dedifferentiating human mature adipocytes from the visceral and subcutaneous fat compartments: fibroblast-activation protein alpha and dipeptidyl peptidase 4 as major components of matrix remodeling. PLoS ONE 10:e0122065. https://doi.org/10.1371/journal.pone.0122065
76. Louis F, Pannetier P, Souguir Z et al (2017) A biomimetic hydrogel functionalized with adipose ECM components as a microenvironment for the 3D culture of human and murine adipocytes. Biotechnol Bioeng 114:1813–1824. https://doi.org/10.1002/bit.26306
77. Toda S, Uchihashi K, Aoki S et al (2009) Adipose tissue-organotypic culture system as a promising model for studying adipose tissue biology and regeneration. Organogenesis 5:50–56
78. Choi JS, Kim BS, Kim JY et al (2011) Decellularized extracellular matrix derived from human adipose tissue as a potential scaffold for allograft tissue engineering. J Biomed Mater Res A 97A:292–299. https://doi.org/10.1002/jbm.a.33056
79. Hayashi K, Okamoto F, Hoshi S et al (2017) Fast-forming hydrogel with ultralow polymeric content as an artificial vitreous body. Nat Biomed Eng 1:0044
80. Liu CY, Matsusaki M, Akashi M (2016) Three-dimensional tissue models constructed by cells with nanometer- or micrometer-sized films on the surfaces. Chem Rec 16:783–796
81. Alla SG, Sen M, El-Naggar AW (2012) Swelling and mechanical properties of superabsorbent hydrogels based on Tara gum/acrylic acid synthesized by gamma radiation. Carbohydr Polym 89(2):478–485
82. Amin MCI, Ahmad N, Halib N et al (2012) Synthesis and characterization of thermo- and pH-responsive bacterial cellulose/acrylic acid hydrogels for drug delivery. Carbohydr Polym 88(2):465–473
83. Zu Y, Zhang Y, Zhao X et al (2012) Preparation and characterization of chitosan polyvinyl alcohol blend hydrogels for the controlled release of nano-insulin. Int J Biol Macromol 50:82–87
84. Coviello T, Grassi M, Rambone G et al (1999) Novel hydrogel system from scleroglucan: synthesis and characterization. J Control Release 60(2–3):367–378
85. Kuijpers AJ, Van Wachem PB, Van Luyn MJ et al (2000) *In vivo* and *in vitro* release of lysozyme from cross-linked gelatin hydrogels: a model system for the delivery of antibacterial proteins from prosthetic heart valves. J Control Release 67:323–336
86. Hubbell JA (1996) Hydrogel systems for barriers and local drug delivery in the control of wound healing. J Control Release 39:305–313
87. Forster S, Antonietti M (1998) Amphiphilic block copolymers in structure-controlled nanomaterial hybrids. Adv Mater 10:195–217
88. Taniguchi I, Akiyoshi K, Sunamoto J (1999) Self-aggregate nanoparticles of cholesteryl and galactoside groups-substituted pullulan and their specific binding to galactose specific lectin, RCA120. Macromol Chem Phys 200:1555–1560
89. Gacesa P (1988) Alginates. Carbohydr Polym 8:161–182
90. Bezemer JM, Radersma R, Grijpma DW et al (2000) Zero-order release of lysozyme from (poly)ethylene glycol)/poly(butylene terephthalate) matrices. J Control Release 64(1–3):179–192
91. Yang G, Xiao Z, Ren X, Long H et al (2016) Enzymatically crosslinked gelatin hydrogel promotes the proliferation of adipose tissue-derived stromal cells. PeerJ 4:e2497
92. Silva DM, Nunes C, Pereira I et al (2014) Structural analysis of dextrins and characterization of dextrin-based biomedical hydrogels. Carbohydr Polym 114:458–466
93. Park H, Woo EK, Lee KY (2014) Ionically cross-linkable hyaluronate-based hydrogels for injectable cell delivery. J Control Release: Off J Control Release Soc 196:146–153
94. Alexandre N, Ribeiro J, Gärtner A et al (2014) Biocompatibility and hemocompatibility of polyvinyl alcohol hydrogel used for vascular grafting—in vitro and in vivo studies. J Biomed Mat Res 102:4262–4275
95. Yang D, Hartman MR, Derrien TL et al (2014) DNA materials: bridging nanotechnology and biotechnology. Acc Chem Res 47:1902–1911

96. Gao Y, Wei Z, Li F et al (2014) Synthesis of a morphology controllable Fe_3O_4 nanoparticle/ hydrogel magnetic nanocomposite inspired by magnetotactic bacteria and its application in H_2O_2 detection. Green Chem 16:1255–1261
97. Li X, Fan DD, Deng JJ et al (2013) Synthesis and characterization of chitosan human like collagen/β-sodium glycerophosphate-carbodiimide hydrogel. Asian J Chem 25:9613–9616
98. Zhang H, Shi R, Xie A et al (2013) Novel TiO_2/PEGDA hybrid hydrogel prepared in situ on tumor cells for effective photodynamic therapy. ACS Appl Mater Interfaces 5:12317–12322
99. Chicatun F, Muja N, Serpooshan V et al (2013) Effect of chitosan incorporation on the consolidation process of highly hydrated collagen hydrogel scaffolds. Soft Matter 9:10811–10821
100. Abdel-Mohsen AM, Aly AS, Hrdina R et al (2011) Eco-synthesis of PVA/chitosan hydrogels for biomedical application. J Polym Env 19:1005–1012
101. Saffer EM, Tew GN, Bhatia SR (2011) Poly(lactic acid)-poly(ethylene oxide) block copolymers: new directions in self-assembly and biomedical applications. Curr Med Chem 18:5676–5686
102. Kosukegawa H, Mamada K, Kuroki K et al (2009) Evaluation of compliance of poly (vinyl alcohol) hydrogel for development of arterial biomodeling. In: Proceedings of the 13th international conference on biomedical engineering, IFMBE, pp 1993–1995
103. Meenach S, Anderson AA, Suthar M et al (2009) Biocompatibility analysis of magnetic hydrogel nanocomposites based on poly(N-isopropylacrylamide) and iron oxide. J Biomed Mar Res Part A 91A:903–909
104. Zhao L, Mitomo H (2008) Adsorption of heavy metal ions from aqueous solution onto chitosan entrapped CM-cellulose hydrogels synthesized by irradiation. J Appl Polym Sci 110:1388–1395
105. Becerra-bracamontes F, Sanchez-Diaz JC, Gonzalez-Alvarez A et al (2007) Design of a drug delivery system based on poly(acrylamide-co-acrylic acid)/chitosan nanostructured hydrogels. J Appl Polym Sci 106:3939–3944
106. Park JS, Woo DG, Sun BK et al (2007) In vitro and in vivo test of PEG/PCL-based hydrogel scaffold for cell delivery application. J Control Release 124:51–59
107. Jeong B, Gutowska A (2002) Lessons from nature: stimuli-responsive polymers and their biomedical applications. Trends Biotechnol 20:305–311
108. Darsow U, Vieluf D, Ring J (1995) Atopy patch test with different vehicles and allergen concentrations: an approach to standardization. J Allergy Clin Immunol 95:677–684
109. Corkhill PH, Hamilton CJ, Tighe BJ (1989) Synthetic hydrogels. VI. Hydrogel composites as wound dressings and implant materials. Biomaterials 10:3–10
110. Fuciños C, Fuciños P, Miguez M et al (2014) Temperature- and pH-sensitive nanohydrogels of poly(N-Isopropylacrylamide) for food packaging applications: modelling the swelling-collapse behavior. PLoS ONE 9:e87190. https://doi.org/10.1371/journal.pone.0087190
111. Rhim JW, Wang LF (2013) Mechanical and water barrier properties of agar/κ-carrageenan/ konjac glucomannan ternary blend biohydrogel films. Carbohydr Polym 96:71–81
112. Fuciños C, Guerra NP, Teijon JM et al (2012) Use of poly(N-isopropylacrylamide) nanohydrogels for the controlled release of pimaricin in active packaging. J Food Sci 77: N21–28. https://doi.org/10.1111/j.1750-3841.2012.02781.x
113. Roy N, Saha N, Kitano T et al (2012) Biodegradation of PVP-CMC hydrogel film: a useful food packaging material. Carbohydr Polym 89:346–353
114. Schneider KP, Gewessler U, Flock T et al (2012) Signal enhancement in polysaccharide-based sensors for infections by incorporation of chemically modified laccase. N Biotechnol 29:502–509
115. Incoronato AL, Conte A, Buonocore GG et al (2011) Agar hydrogel with silver nanoparticles to prolong the shelf life of Fior di Latte cheese. J Dairy Sci 94:1697–1704
116. Langmaier F, Mokrejs P, Kolomaznik K et al (2008) Biodegradable packing materials from hydrolysates of collagen waste proteins. Waste Manag 28:549–556

117. Varma DM, Gold GT, Taub PJ et al (2014) Injectable carboxymethylcellulose hydrogels for soft tissue filler applications. Acta Biomater 10:4996–5004
118. Kodavaty J, Deshpande AP (2014) Regimes of microstructural evolution as observed from rheology and surface morphology of crosslinked poly(vinyl alcohol) and hyaluronic acid blends during gelation. J Appl Polym Sci 131:1–10. https://doi.org/10.1002/APP.41081
119. Tichota DM, Silva AC, Sousa Lobo JM et al (2014) Design, characterization, and clinical evaluation of argan oil nanostructured lipid carriers to improve skin hydration. Int J Nanomed 9:3855–3864
120. Wang PC, Huang YL, Hou SS et al (2013a) Lauroyl/palmitoyl glycol chitosan gels enhance skin delivery of magnesium ascorbyl phosphate. J Cosmet Sci 64:273–286
121. Wang Y, Du R, Yu T (2013b) Systematical method for polyacrylamide and residual acrylamide detection in cosmetic surgery products and example application. Sci Justice 53:350–357
122. Pavicic T (2013) Calcium hydroxylapatite filler: an overview of safety and tolerability. J Drugs Dermatol 12:996–1002
123. Lee E, Kim B (2011) Smart delivery system for cosmetic ingredients using pH-sensitive polymer hydrogel particles. Korean J Chem Eng 28:1347. https://doi.org/10.1007/s11814-010-0509-8
124. Raschip IE, Hitruc EG, Vasile C (2011) Semi-interpenetrating polymer networks containing polysaccharides. II. Xanthan/lignin networks: a spectral and thermal characterization. High Perf Polym 23:219–229. https://doi.org/10.1177/0954008311399112
125. Simi CK, Abraham TE (2010) Transparent xyloglucan–chitosan complex hydrogels for different applications. Food Hydrocoll 24:72–80
126. Chandrika KSVP, Singh A, Sarkar DJ et al (2014) pH-sensitive crosslinked guar gum-based superabsorbent hydrogels: swelling response in simulated environments and water retention behavior in plant growth media. J Appl Polym Sci 131:1–12. https://doi.org/10.1002/APP.41060
127. Böhlenius H, Overgaard R (2014) Effects of direct application of fertilizers and hydrogel on the establishment of poplar cuttings. Forests 5:2967–2979
128. Hotta M, Kennedy J, Higginbotham CL et al (2014) Synthesis and characterisation of novel ι-Carrageenan hydrogel blends for agricultural seed coating application. Appl Mech Mat 679:81–91
129. Tang H, Zhang L, Hu L et al (2014) Application of chitin hydrogels for seed germination, seedling growth of rapeseed. J Plant Growth Regul 33:195–201
130. Bortolin A, Aouada FA, Mattoso LH et al (2013) Nanocomposite PAAm/methyl cellulose/ montmorillonite hydrogel: evidence of synergistic effects for the slow release of fertilizers. J Agric Food Chem 61:7431–7439
131. Shahid SA, Qidwai AA, Anwar F et al (2012) Improvement in the water retention characteristics of sandy loam soil using a newly synthesized poly(acrylamide-co-acrylic acid)/ $AlZnFe_2O_4$ superabsorbent hydrogel nanocomposite material. Molecules 17:9397–9412
132. Guilherme MR, Reis AV, Paulino AT et al (2010) Pectin-based polymer hydrogel as a carrier for release of agricultural nutrients and removal of heavy metals from wastewater. J Appl Polym Sci 117:3146–3154
133. Bao JM, Wang YZ, Li YX (2014a) Preparation of crosslinked dextran hydrogel microspheres by inverse suspension polymerization and its application in separation of liposome and drug. Xiandai Huagong/Modern Chem Indus 34:55–58, 60. ISSN: 02534320
134. Bao S, Wu D, Wang Q et al (2014b) Functional elastic hydrogel as recyclable membrane for the adsorption and degradation of methylene blue. PLoS One 9(2):e88802. https://doi.org/10.1371/journal.pone.0088802
135. Cao Y, Lui N, Fu C et al (2014) Thermo and pH dual-responsive materials for controllable oil/water separation. ACS Appl Mater Interfaces 6:2026–2030
136. Tongwa P, Bai B (2014) Degradable nanocomposite preformed particle gel for chemical enhanced oil recovery applications. J Petrol Sci Eng 124:35–45

137. Zolfaghari R, Katlab AA, Nabavizadeh J et al (2006) Preparation and characterization of nanocomposite hydrogels based on polyacrylamide for enhanced oil recovery applications. J Appl Polym Sci 100:2096–2103
138. Aalaie J, Vasheghani-Farahani E, Semsarzadeh MA et al (2008) Gelation and swelling behavior of semi-interpenetrating polymer network hydrogels based on polyacrylamide and poly(vinyl alcohol). J Macromol Sci Part B 47:1017–1027
139. Lei ZX, Chen YM, Chen YW et al (2006) Preliminary results of pilot test on indepth permeability profile control/emulsion flood by using PAM inverse emulsion. Oilfield Chem 23:81–84
140. Wang XM, Zhang DS (2003) A preliminary study on xanthan/zirconium flowable gel as flooding fluid. Oildfield Chem 20:157–159
141. Tiena HT, Ottovaab AL (1998) Supported planar lipid bilayers (s-BLMs) as electrochemical biosensors. Electrochim Acta 43:3587–3610

Application of Sustainable Nanocomposites for Water Purification Process

Hayelom Dargo Beyene and Tekilt Gebregiorgs Ambaye

1 Introduction

Water is one of the most vital bases for the living system and is used in daily life activities. Due to rapid industrial growth, natural water resources are affected by several water pollutants. The World Health Organization (WHO) 2014 report on water supply and sanitation estimated that 748 million people still lack safe drinking water, 2.5 billion peoples without access sanitation and 3900 children die every day due to poor quality water and communicable diseases [1]. These statistics indicated that water pollution by numerous pollutants becomes an alarming issue worldwide. Consequently, competent water treatment technologies have been established to raise the potential of water resources and to decline the challenges and concerns associated with water pollution. In this regard, nanocomposite has to play a significant role in the water purification technology including potable water treatment, wastewater desalination, and treatment in order to deliver the real technology to clean water at a lower price using less energy by decreasing further ecological impacts.

Nanomaterials are materials which have the structural components sized from 1 to 100 nm [2]. They have unique properties when compared with other conventional materials, such as mechanical, electrical, optical, and magnetic properties due to their the small size and higher specific surface area, nanomaterials [2]. In recent years, nanomaterials have been effectively applied to numerous perspectives as catalysis [3], medicine [3], sensing, and biology [4]. They have extensive applications to prevent several environmental problems like water and wastewater

H. D. Beyene (✉)
Department of Chemistry, Adigrat University, P.O. Box: 50, Adigrat, Ethiopia
e-mail: hayeda21@gmail.com

T. G. Ambaye
Department of Chemistry, Mekelle University, Mekelle, Ethiopia

Inamuddin et al. (eds.), *Sustainable Polymer Composites and Nanocomposites*,
https://doi.org/10.1007/978-3-030-05399-4_14

treatment. Because, nanomaterials have the potential to eliminate different toxins, for instance, heavy metals, organic pollutants, inorganic anions, and pathogens [5]. Zero-valent metal nanoparticles (nZVI), metal oxides nanoparticles, carbon nanotubes (CNTs) and nanocomposites are the most recent appropriate nanomaterials for water and wastewater treatment [6].

The nZVI is one of the most useful nanomaterials for water purification [7–9]. The nZVI has a role in water purification as an electron subscriber which encourages the conversion toxic metals to safe forms (the reduction of chromium from hexavalent into trivalent form), adsorption, co-precipitation processes and strong reducing ability [10]. The nZVI has discovered real application for eliminating various organic and inorganic pollutants such as polychlorinated compounds [11, 12], Nitrates, phosphates and perchlorates [13, 14], nitroaromatic compounds [15], organic dyes [16], phenols [17], heavy metals [18], metalloids [19], and radio elements [6, 20].

Other nanoparticles like silver (Ag), titanium oxide (TiO_2), zinc oxide (ZnO), iron oxides and CNT are applied in water treatment technology. Silver nanoparticles (AgNPs) are very noxious to microbes and hence have solid antibacterial effects for an extensive variety of microorganisms (viruses, bacteria, and fungi) [21]. AgNPs are the promising antimicrobial agents, which have been extensively used for water disinfection [21]. AgNPs have the removal potential for bacteria's like methicillin-resistant *Staphylococcus aureus*, ampicillin resistant *E. coli*, a common water contaminant, erythromycin resistant *Streptococcus pyogenes* and vancomycin-resistant *Staphylococcus aureus* [22], *Pseudomonas aeruginosa, Vibrio cholera* [23], *Bacillus subtilis* [24]. There are different ways of Nanosilver disinfection mechanisms such as the interaction of AgNPs with DNA, altering the membrane and altering the enzymatic activity and thus destroy it [25–28], the dissolution of AgNPs that able to react through the thiol sets of enzymes disable, and interrupt usual services the cell [29].

In nanocomposites (NCs), there is no a previous documented review of their application in water and wastewater treatment perspectives. NCs are formed through the combination of more than two materials having various physical and chemical properties and unique interface [28], [30]. Composites have many advantages than other compounds due to their unique characteristics such as high durability, high rigidity, high strength, gas-barrier features, corrosion resistance, low density, and heat resistance. The combination of the matrix (continuous phase) and the reinforced materials (dispersed) is knowns as composite materials. They are materials of the 21st century which are multiple phase materials a minimum one of the phase's displays sizes from the range 10–100 nm [31]. Todays, NC materials have developed as appropriate choices to overwhelmed restrictions of various manufacturing tools. NCs have wide practice in various fields such as life sciences, drug distribution schemes, and wastewater treatment. In NCs, the nanoparticles were merged within diverse functionalized materials like multiwall CNT, activated carbon, cheap graphene oxide, and polymeric media. NCs have a number of application in the area of food packaging [32–34], anti-corrosion barrier protection [31], biomedical [31, 35] and coating [36]. This chapter focuses on the exciting NC types and their current application in water purifications. Besides, the future perspective of nanocomposites in water treatment also addressed.

2 Conventional Water Purifications Technologies

Surface water (spring, rivers, and lakes) and unconventional water resources (which are not available for direct use. For example, wastewater, seawater and brackish water) are the major universal water resources potentials [37]. Globally, the upsurge in industrialization and urbanization with a quick population growth and weather change contributes to the pollution of freshwater resources [38–40]. Table 1 shows the available conventional water purification technologies such as coagulation and flocculation, air flotation and advanced oxidation processes. These methods are very quiet in removing the contaminants efficiently. However, these methods possess several challenges related to the formation of either secondary pollution or higher energy requirement. Therefore, a massive attention should be given to the improvement/innovation of technologies having ecologically friendly, low energy consumption and economical feasible treatments perspectives applicable to the feasible water sanitization systems. To meet the demand for clean water standards, many authors have been focused on the suitable and economically viable water purification approaches including water remediation, reclamation, and desalination [41].

Table 1 Water purification methods [24, 41]

Water purification technologies	Contaminate removed
Coagulation and flocculation	Turbidity, dissolved organic carbon, bacteria and chemical contaminants such as cyanide compounds, phosphorus, fluorides, arsenic etc
Boiling	Kill the bacterial cultures
Distillation	To destroy microbial cells and unwanted chemicals such as calcium, lead, magnesium
Ultraviolent treatment	Can achieve disinfection of about 99.99%
Ultrasound	Damage cellular structures of bacteria
Ozone	It is effective in eradicating tastes, odour, colour, iron, and manganese; and not affected by pH and temperature
Chlorine	Kills several waterborne pathogens
Catalytic process	Applied to breakdown down an extensive diversity of organic materials like organic acids, estrogens, pesticides, dyes, crude oil and microbes
Bioremediation	Eradicating heavy metals, organic toxins, pesticides and dyes by plant extracts and microbes.

3 Types of Nanocomposites and Its Application in Water Purification

The use of nanoparticles in water management has associated with some practical problems, such as accumulation, tough separation, drainage into the contact water, possess environmental and human health [30]. One capable approach to improve the application of nano-particulate materials is to develop NC materials that take advantages of both the hosts and impregnated nanoparticles (Fig. 1) [42]. NCs have the potential to mitigate the discharge of nanoparticles into the environment, and improves the suitability of nanotechnology with current infrastructures. The NCs are essentially multiphase solid material, including porous media, colloids, gels, and copolymers in a broad sense. The selection of hosts for nanocomposites is of great significance, and even dominates the performance of the resultant nanocomposites. Compared with free nanomaterials, the performance and usability of nanocomposites were significantly improved, in terms of nanoparticle dispersion, stability, and recyclability. Hence, nanocomposite materials could bond the gap between nanoscopic and mesoscopic scale. Till now, nanocomposites were believed to be the most likely way to forward water nanotechnology from laboratory up to the large-scale applications [41].

3.1 *Metal Nanocomposite*

Polymer-supported nanosilver has recognized antibacterial properties of polyurethane and cellulose acetate impregnated nanosilver-fiber composites have good inhibition activity for Gram-positive and negative bacteria. The dispersion

Fig. 1 Application of nanocomposite for water purification [42]

nanoparticle in polyurethane foam has gained effective antibacterial filters [43]. Once announced in polymeric membranes, a decrease of biofouling as well as good pathogen eradication efficiency was perceived. Nanosilver was also used in the making of economicall y feasible microfilters for handling drinking water which is mainly preferred in unreachable regions [44].

Silver-alginate composite beads were effectively prepared using three different methods. Specifically, the adsorption-reduction (AR), hydraulic retention time (HRT) and simultaneous gelation-reduction (SGR) composite beads were talented to succeed a disinfection effectiveness for portable water purifying. Those Composite beads equipped using diverse methods were established effective cleaning in the *E. coli* to various degrees. Both SGR and the AR beads confirmed equivalent disinfection efficiency but, the SGR beads released knowingly more Ag than the AR beads fix, indicating that the SGR beads may have a higher lifespan than the AR beads without losing sterilization success. These results weight the significance of improving the synthesis method in yielding material configurations that lead to the essential physical properties of numerous aspects [45].

The synthesized novel NC containing AgNPs and mesoporous alumina have been used for the elimination of dye compounds like methyl orange, bromothymol blue, and reactive yellow from synthetic waste. The results display that the silver/mesoporous alumina nanocomposite (Ag/OMA NC) was noble adsorbent for the elimination of anionic dyes from aqueous solution, and also this NCs had an antibacterial activity against both Gram-negative and Gram-positive bacteria [46].

The addition of AgNPs and *Moringa oleifera* seed powder were improved graphene structure which improves the removal efficiency of pollutants from liquid industrial waste like textile, tannery, and paper mill. The adsorption study of the adsorbents clearly revealed that the graphene loaded with AgNPs and seed powder of *Moringa. oleifera* composite (GAM) designated superior results compared to normal adsorbents due to the configuration of GAM sorbent which is recognized by the high surface area, biocidal action, adsorption activity AgNPs, and coagulation property of *Moringa oleifera.* Thus motivated the composite to be novel, economically feasible, and environmental suitable and promising adsorbent for water treatment [47].

Bimetallic nanocomposites supported on carbon are of great interest. Carbon supported bimetallic nanoparticles have reduced surface area which enhances their properties to a large extent. Nowadays, Water pollution is crucial problem happen due to existence contaminates like chemicals, microbes (fungi, bacteria, and virus) in water by human activities. Nanotechnology offers an alternative to the water purification. The bimetallic nanoparticles like ruthenium-palladium are used as reinforce to develop NC on the surface carbon matrix which had successfully helped in wastewater treatment having perchlorate as the main pollutant [31]. Others like, NCs of Au/Pd nanoparticles reinforced on TO_2 have been synthesis by microemulsion means and being used as an efficient photocatalyst due to their high light absorption ability. Bimetallic NCs of Fe/Ni-K have the capacity to remove DBG from the wastewater. The degree of eliminating DBG in the NC (Ni/Fe-K) is greater than that of separate kaolin and the bimetallic nanoparticles (Ni/Fe) [31, 48].

3.2 Metal Oxide Nanocomposite

The Metal oxide nanocomposite (MONC) are often used as adsorbents, photocatalyst, and devices to challenge environmental pollution problems. MONC are used merging with graphene, silica, other oxides, carbon nanotube (CNT), polymers for the removal of various organic and inorganic of pollutants [49]. Currently, removal of organic pollutants from wastewater is one of the most significant alarms in water pollution control. In last decade, the interest in solving global water pollution by means of photocatalysis is increased rapidly using metal oxide nanoparticles (TiO_2 and ZnO). However, the use of basic TiO_2 and ZnO nanoparticles are limited because of their extensive band gap and the high recombination rate of photo produced charges. Coupling is developing an approach to increase the destroying degree of organic contaminants under visible light conditions. MONC provide a current technique to modify the properties of semiconductor metal oxide photocatalyst through encouraging charge transfer processes and improving charge separation [50].

The alumina composite reinforced by CNTs was produced by rising CNT above Fe and Ni-doped energetic alumina. The composite was influenced by numerous factors able to initiated high capacity synthesis which is factors includes activated alumina, CNT, amorphous carbon and various surface functional groups such as carboxyl, carbonyl and hydroxyl present in the clusters [51]. Ihsanullah et al. [52] deliberated that the consequence CNT/Al_2O_3 for actual elimination of chlorophenol and phenol from aqueous solutions. Alumina ornamented onto the exterior of multi-well carbon nanotube (MWCNT) was an inspiring adsorbent for immediate removal of Cd^{+2} and trichloroethylene (TCE) from poisoned groundwater. Electrostatic interactions, the hydrogen bond interactions and the protonation or hydroxylation of Al_2O_3 are the adsorption mechanism of Al_2O_3/MWCNTs to remove Cd^{+2} and TCE from the polluted water Fig. 2 [53, 54].

TiO_2 is the new greater type of composites based Metal oxide. TiO_2 nanocomposite has received more attention in water purification due to its nontoxicity, and the ability for the photo-oxidative degradation contaminates such as MB [55], benzene derivatives [56], and carbamazepine [57] were powerfully photodegraded by CNT/TiO_2 composites. Researchers described that the bond of carbon-oxygen-titanium can enlarge the light absorption to longer wavelengths and hence potentially improvement of the photocatalytic action [53]. Senusi et al. [40] also indicated that synthesized TiO_2-zeolite NCs for the innovative water treatment of industrial dyes. The results indicated that the nanocomposite followed an adsorption concerned with photocatalytic degradation, which is mainly effective for eradicating trace dye compounds [40]. A novel Cu–TiO_2–SiO_2 NCs synthesized by a sol-gel method and used to degrade Rhodamine Blue in water modelling the dyes wastewater under both UV and visible light irradiation. Studies revealed that the Cu–TiO_2–SiO_2 nanocomposite has smaller crystalline size, higher surface area, and slight agglomeration by judging from the characteristic analysis. The Cu–TiO_2–SiO_2 nanocomposite exhibited higher photocatalytic activity than TiO_2 for the

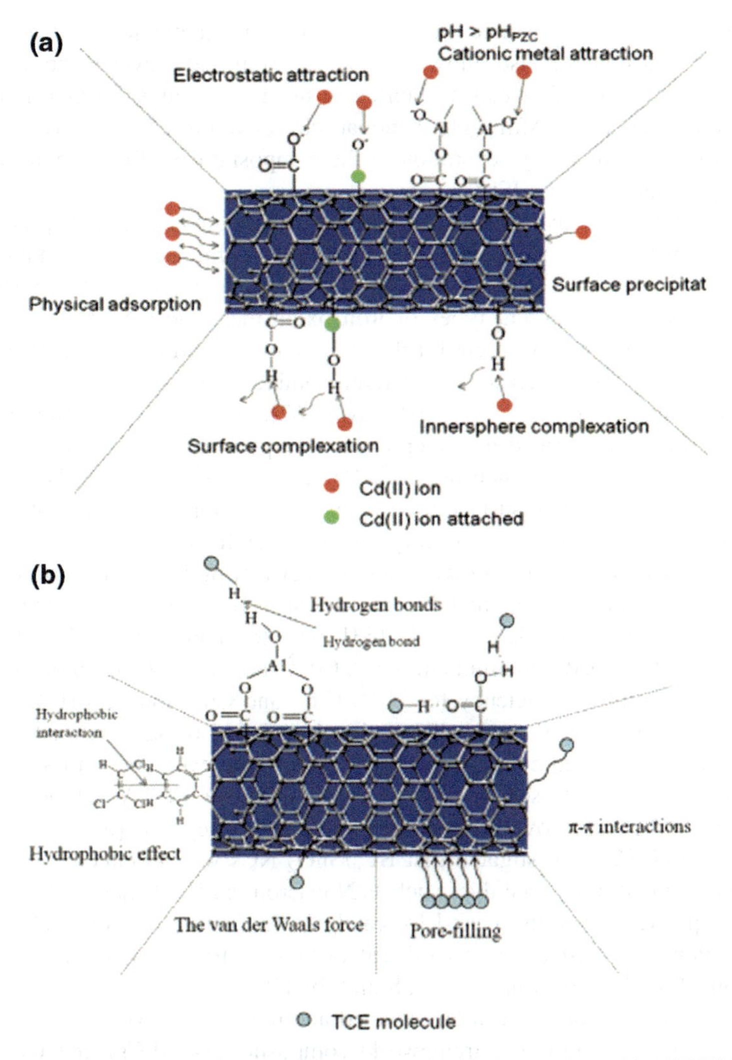

Fig. 2 The diagram representation of Cd(II) ion (**a**) and TCE (**b**) interface with Al2O3/MWCNTs [53]

degradation of Rhodamine Blue under both UV and visible light irradiation. The increase in the photocatalytic activity may be due to the lower recombination rate of electron-hole and the high dispersion of SiO_2 [58].

Iron oxides (i.e. Fe_2O_3 and Fe_3O_4) are unique and talented magnetic constituents which create a new composite with CNTs, and graphene. This is one of the greatest smart magnetic metallic oxides and has established extensive consideration due to its exceptional physical and chemical properties and several benefits such as high reversible capacity, rich abundance, cheap, and environmentally friend [54]. Magnetic nanoparticles are highly advantageous than nonmagnetic nanoparticles since they can simply isolate from water via a magnetic field. Magnetic field separation is a practice also allows simple isolation and recycled the adsorbents. Magnetic nanocomposites can be fabricated using magnetite (Fe_3O_4), maghemite (Fe_2O_3), and jacobsite ($MnFe_2O_4$) nanoparticles as reinforcer filling on a polymer matrix which permits easy separation of the composite from the aqueous solutions after the sorption process [29].

Researchers were investigated series of magnetic alginate polymers prepared and batch trials were shown to examine their capacity to eliminate heavy metal ions such as Co^{+2}, Cr^{+6}, Ni^{+4}, Pb^{+2}, Cu^{+2}, Mn^{+2}, La^{+3} and organic dyes (MB and MO) from aqueous solutions. Different types of iron oxide magnetic composites have been positively useful as an adsorbent for the elimination of various targets of impurities from water and wastewater such as naphthylamine [59], metals [59], phenol [59], and tetracycline [60], As^{+3}, As^{+5} [61], dyes [62]. Moreover, graphene-based iron oxide NCs have confirmed an exceptional adsorption capacity to fix extra heavy metals and organic dyes such as Cr^{+6}, Pb^{+2}, Co^{+2}, neutral red, MB etc. due to magnetic properties, high surface to volume ratio and rapid diffusion rate [59].

In addition to the above, Currently, many researchers have studied also on the practice of metal oxide NCs for water and wastewater purification. Currently, many scientists focus on the heavy metals removal due to their strong influence on health and environment. The Saad et al. [63] was to manufacture ZnO@Chitosan nanocomposite (ZONC) to eliminate Pb^{+2}, Cd^{+2} and Cu^{+2} ions from unclean water with optimal removal efficiency for Pb^{+2}, Cd^{+2} and Cu^{+2} ions at pH 4, 6 and 6.5 with adsorption capacity were 476.1, 135.1 and 117.6 mg/g, respectively. The researchers also studied nonstop adsorption-desorption cyclic outcomes established that ZONC can be reused after recovery of ions by EDTA solution, and the regenerate ZOCS used over without significant efficiency loss [63].

Singh et al. [64] investigated that BC_4/SnO_2 NCs was an effective catalyst for the degradation of industrial dyes such as Novacron red Huntsman (NRH) and MB. This composite is also discovered for catalysis destruction of industrial dyes. The Degradation study displays that 1 g/L catalyst concentration of BC_4/SnO_2 destroys NRH and MB dye up to nearly 97.38 and 79.41%, respectively, in 20 min using sun radiation. The catalyst can be recycled and recovered [64].

Zr-magnetic metal-organic frameworks composites (Zr-MFCs) are an amino-rich prepared by a facile and efficient strategy. The achieved Zr-MFCs were confirmed to be effective adsorbents with feasible adsorption ability and fast adsorption kinetics for metal ions and organic dyes removal from water. The amine-decorated MFCs were very efficient for metal ions and dyes elimination than row MFC-O. MFC-N confirmed the maximum ability for Pb^{2+} (102 mg g^{-1}) and MB (128 mg g^{-1}), while MFC-O revealed the maximum ability for MB (219 mg g^{-1}). Furthermore, Zr-MFCs have also good removal efficiency for anionic and cationic dyes from the miscellaneous solution by adjusting pH. Zr-MFC adsorbents can be simply improved by removing metal ions and/or organic dyes from the adsorbents with appropriate reagents without change adsorption capacity up to 6 generations. The attained results confirmed the prepared MFCs have the great application perspective as interesting adsorbents for water treatment [65].

Saad et al. [63] investigated a facile method for in situ fabrication of ZnO@Chitosan nanocomposite (ZONS), and the attained composite demonstrated noble ability and rapid kinetics for Pb^{+2}, Cd^{+2} and Cu^{+2} ions adsorption. The main advantage of this product is the recovery of metal ions and the significant ability for adsorption after many series of recycling. The ZONC demonstrations important feasibility in ecological remediation for wastewater treatment and can attain the increasing need for the purification of water resources [63].

3.3 Carbon Nanocomposite

A magnetic multi-wall carbon nanotube (MMWCNT) nanocomposite was used as an adsorbent for removal of cationic dyes from aqueous solutions. The MMWCNT nanocomposite was composed of viable multi-wall CNT and IONPs. The elimination of MB, neutral red and brilliant cresyl blue was deliberate using MMWCNT nanocomposite adsorbent. Investigations were carried out to study adsorption kinetics, the adsorption capacity of the sorbent and the effect of sorption dosage and pH values on the elimination of cationic dyes [66].

Mesoporous carbon with entrenched iron carbide nanoparticles (ICNPs) was effectively synthesized via a facile impregnation-carbonization method. Biomass was used as a carbon basis and an iron pioneer was rooted to create mesopores through a catalytic graphitization reaction. The pore conformation of the NCs structured by the iron pioneer loadings and the immovable ICNPs support as a dynamic component of magnetic isolation next sorption. The newly produced mesopores were established as a critical feature to increase the adsorption capacity of organic dyes while immovable ICNPs are responsible for the careful removal of heavy metal ions (Zn^{2+}, Cu^{2+}, Ni^{2+}, $Cr^{6+,}$ and Pb^{2+}). Composed with the desirable elimination of extra noxious heavy metal species (Cr^{6+} and Pb^{2+}), these mesoporous NCs show favourable applications in impurity removal from water. The facile material preparation permits appropriate scale-up production with economical feasible and lowest ecological impact [67].

Advanced technologies integrating with engineered nanoparticles into biochar fabrication schemes might increase the roles of biochar for numerous uses comprising soil fertility upgrading, carbon sequestration, and wastewater treatment. Inyang et al. [68], investigated that removal ability MB was evaluated in batch sorption using untreated hickory biochars (HC), bagasse biochars (BC) and CNT-biochar composites (HC-CNT and BC-CNT, respectively). The addition of CNTs considerably enriched the physiochemical properties of HC-CNT and BC-CNT such as extreme thermal stabilities, surface areas, and pore volumes. These results recommend that electrostatic magnetism was the principal devices for the removal of MB onto the biochar-nanocomposite. Hybridized CNT-biochar NC can be considered as capable, cheap adsorbent material for eliminating dyes and organic contaminants from water [68].

Carbon-nanocomposites (CNCs) are constituents that have two or more elements prepared to form a composite mixture with CNTs as the primary host synthesized the poly 1,8-diaminonaphthalene/MWCNTs COOH hybrid material which could be used as an active sorbent for the separation Cd^{+2} and Pb^{+2} at trace levels [40]. Muneeb et al. [69] was primed a new NCs from biomass used for the removal of selected heavy metals (As, Cr, Cu, Pd and Zn) from the wastewater. With the increase in pH, there was a decline in percentage adsorption of the metals [70].

Tian et al. [70] stated an eco-friendly, effective and synergistic nanocomposite development for new antibacterial agents using both iron oxide nanoparticles (IONPs) and AgNPs on the surface of graphene oxide (GO), resulted in novel GO-IONP-Ag nanocomposite. When associated with pure AgNPs, GO-IONP-Ag offers deliberately improved bacteriocidal action to both Gram-negative bacteria and Gram-positive bacteria. GO has the beautiful benefit through GO-IONP-Ag composite to kill Gram-positive bacteria at small agent concentration. Moreover, GO-IONP-Ag nanocomposite can simply reprocess by magnetic separation, low cost, and environmentally. In the account of those exceptional benefits, the developed GO-IONP-Ag nanocomposite can use for prospective requests as a multifunctional sterile agent in the diverse area [71].

3.4 Polymer Nanocomposite

Polymer nanocomposites (PNCs) are a superior type of tools which nanoparticles spread in a polymer matrix resulting in novel materials having unique physical and chemical properties [70]. Polymers are special supports for nanomaterials as they usually possess tunable porous structures, excellent mechanical properties, and chemically bounded functional groups. PNCs are prospecting materials for their sound performance in water and wastewater treatment. Adsorption of contaminant through PNC is among various treatment technologies, which is considered as an advanced tool in water treatment technology. PNCs often integrate the essential advantages of both the nanoparticles and the polymeric matrix [72]. PNCs could be synthesis by either joining nanoparticles into polymer structures or by fixing polymers to nanoparticles. Direct compounding and in situ synthesis are two leading approaches used in the manufacture of several PNCs as shown in Fig. 3 [41]. PNC has of great potential for pollutants removal including heavy metals (Cu, Pb, Cr (III), Ni), As, F, and P. The pollutants were often removed through multiple mechanisms including surface complexation, electrostatic attraction and co-precipitation [73].

These PNCs are avoided challenging issues such as nanoparticle dissolution, which is common when using free nanoparticles [72]. Some of the nanocomposites were also responsive to regeneration and recycle without significant capacity loss, which is critical from economic outlook. Since of the large size of the PNCs, they could be simply isolated from treated water.

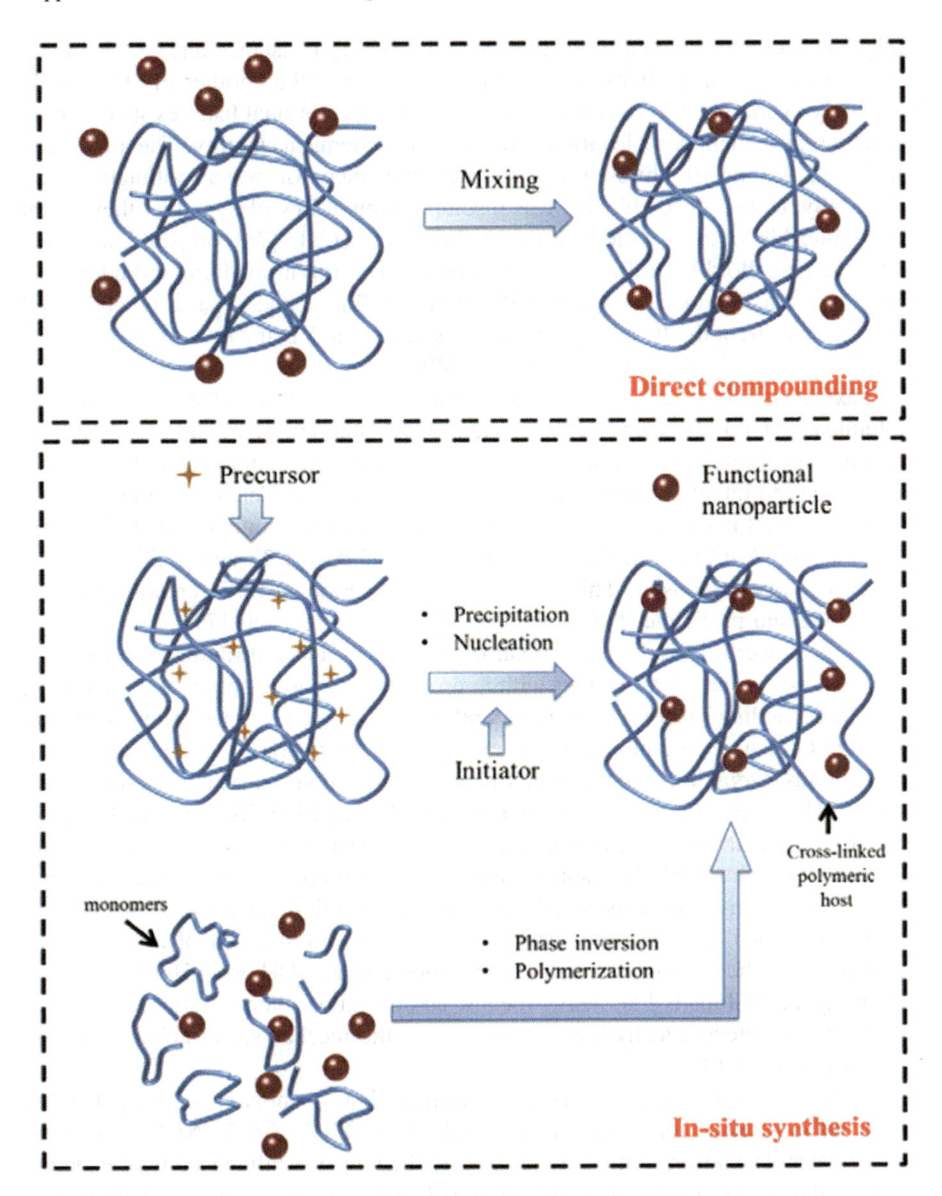

Fig. 3 Graphic of fusion methods for PNCs. Adapted with permission [41]

Alginate [74], macromolecule (polypyrrole) [75] polyaniline [76], porous resins [77] and ion-exchangers [41] are most extensively used polymeric hosts. New types of polymeric hosts are essentially bio-polymers such as chitosan and cellulose. They are plentiful in nature and eco-friendly. However, they could suffer a serious biodegradation problem in the long-term application. cellulose showed good chemical stability and mechanical strength, due to its densely and systematic aligned, hydrogen-bonded molecules, sound swelling resistance and its characteristics such as hydrophilicity and chirality. Chitosan is the another most naturally rich polysaccharide next to cellulose. Chitosan has exceptional features such as high reactivity, excellent complexation behaviour, and chemical stability. The amino and hydroxyl groups of chitosan aid as energetic sites for water pollutants [78]. Generally, cross-linked chitosan was insoluble even at low pHs, so that they might be applicable over a wide pH range. Djerahov et al. [78] prepared a steady CS-AgNPs colloid by diffusing the AgNPs sol in chitosan medium and additional recycled it to attain a cast film with high steadiness under packing and good mechanical strength. It showed efficient isolation and extraction of Al^{+3}, Cd^{+2}, Cu^{+2}, Co^{+2}, Fe^{+3}, Ni^{+2}, Pb + ^{2} and Zn^{+2} [40, 78].

Saxena and Saxena [79] developed Bimetal oxide fixed PNC by means of Alumina and IONPs with Nylon-6,6 and Poly (sodium-4-,styrenesulphonate) as polymer medium for pollutants elimination from the water. The prepared NCs have maximum pollutant removal capacities for all factors. The exclusion of total alkalinity, total hardness, calcium, magnesium, chloride, nitrate, fluoride, TDS and EC was 66.67, 42.85, 66.67, 25, 58.66, 34.78, 63.85, 41.27 and 41.37% respectively by this composite. This is an indication period towards emerging multifunctional and profitable PNCs for water remediation requests [79].

CNTs powerfully sorb varied polar organic compounds attributable to the stuff miscellaneous interfaces together with hydrophobic impact, peppiness interactions, covalent bonding, valence bonding, and electrical connections. The п-electron wealthy CNT apparent allows energy exchanges with carbon-based molecules with C=C bonds. Organic compounds that have used functional groups like –COOH, –OH, $-NH_2$ might additionally kind a bond with the graphitic CNT exterior that pays electrons. Electricity magnetism enables the surface assimilation of exciting carbon-based chemicals like some antibiotics at appropriate pH range. PNCs are sorbents tailored adsorbents which are talented for eliminating different types of pollutants. Their internal shells can be hydrophobic for sorption of organic compounds while the exterior channels can be tailored (e.g., –OH or $-NH_2$) for sorption of inorganic pollutants like heavy metals. complexation, electrostatic interactions, hydrophobic effect, and hydrogen bonding are the mechanism established during sorption process [80].

Carboxymethyl-cyclodextrin polymer adapted Fe_3O_4 nanoparticles (Copolymers) was manufactured for selective elimination of Pb^{2+}, Cd^{2+}, Ni^{2+} ions from wastewater. The adsorption efficiency of metallic ions was influenced by the factors like contact time, a dose of copolymers pH, ionic strength, and temperature. At equilibrium condition in single sorption way, the optimum uptakes of the adsorbent for Pb^{2+}, Cd^{2+}, and Ni^{2+} were 64.5, 27.7 and 13.2 mg g^{-1} respectively at 45 min and 25 °C. The PNC improved the sorption capacity since of the chelating abilities of the several hydroxyl and carboxyl sets in polymer support with metal ions. In mixed adsorption experiments, CDpoly-MNPs might favorably high sorption of Pb^{2+} ions with an attraction order of $Pb^{2+} \gg Cd^{2+} > Ni^{2+}$ [81].

Khaydarov. et al. [82] studies a new technique for emerging nanocarbon-conjugated polymer nanocomposites (NCPC) by means of carbon colloids as nanoparticle and polyethyleneimine as a matrix for metal ions removal from water. The researchers have been examined the efficiency of NCPC depends on size carbon colloids, synthesis NCPC and its chemical features, the ratio of carbon colloids and polyethyleneimine, the speed of coagulation NCPC, interaction mechanism, removal potential NCPC against pH. The bonding capacity adsorbent was 4.0–5.7 mmol/g with divalent metal ions at pH 6 which sorption has above 99% removal efficiency for Zn^{2+}, Cd^{2+}, Cu^{2+}, Hg^{2+}, Ni^{2+}, Cr^{6+} [82].

Clay can found the suitable matrix for varnish of polyaniline. The characterization outcomes of NC established that the clay sheet was develop layered in the synthesis NC. Parameters like contact time, pH, and concentration were determined the adsorption capacity of modified adsorbent. The researchers were announced new clay NC which use of polyaniline improved clay nanocomposite as an adsorbent for water purification of lead ions. It can be used as talented sorption scheme incoming water and wastewater treatment in order to eliminate lead ion [83].

Nithya and Sudha [84] studied using chitosan-g-poly(butyl acrylate)/bentonite NC as an adsorbent for chromium, lead and other significant physicochemical water quality parameters such as total solids (TS), biological oxygen demand (BOD), chemical oxygen demand (COD), total hardness, salinity, turbidity and conductivity from the tannery wastewater. The effect of some parameters, such as contact time, pH and dose adsorbent was assessed. The outcomes showed that NC can be used tannery wastewater treatment containing heavy metals powerfully [84].

3.5 Membranes Nanocomposite

In membrane technology, porous materials are plays capturing role to trap pollutants. Inclusive, numerous forms of membranes with diverse pore sizes engaged in water treatment process including microfiltration, ultrafiltration, reverse osmosis and nanofiltration membranes which depend on their shared materials that would be clean out through each process as shown Fig. 4 [42]. The existing membranes have numerous challenges for water purification, such as the exchange link between permeability, selectivity and low resistance to fouling. Recent progress in nanotechnology have offered the growth of the new generation membrane for water purification [40]. Nanocomposite membrane (NCM) has a great role in water purification and reuses for several bases of water such as drinking water, brackish, seawater, and wastewater treatment. NCM is an innovative type of membranes prepared by merging complex constituents with nanomaterials that are developing as a promising tool to answer membrane separation problems. The innovative NCM can be deliberate to fulfil exact water purification uses by calibration their assembly and chemical characters (e.g., water-heating, porousness, charge density, and thermal and mechanical stability) and announcing distinctive functionalities (e.g., medicinal drug, photocatalytic or adsorbent capabilities). The advance of membranes with high permeability, rejection and smart protective property is way

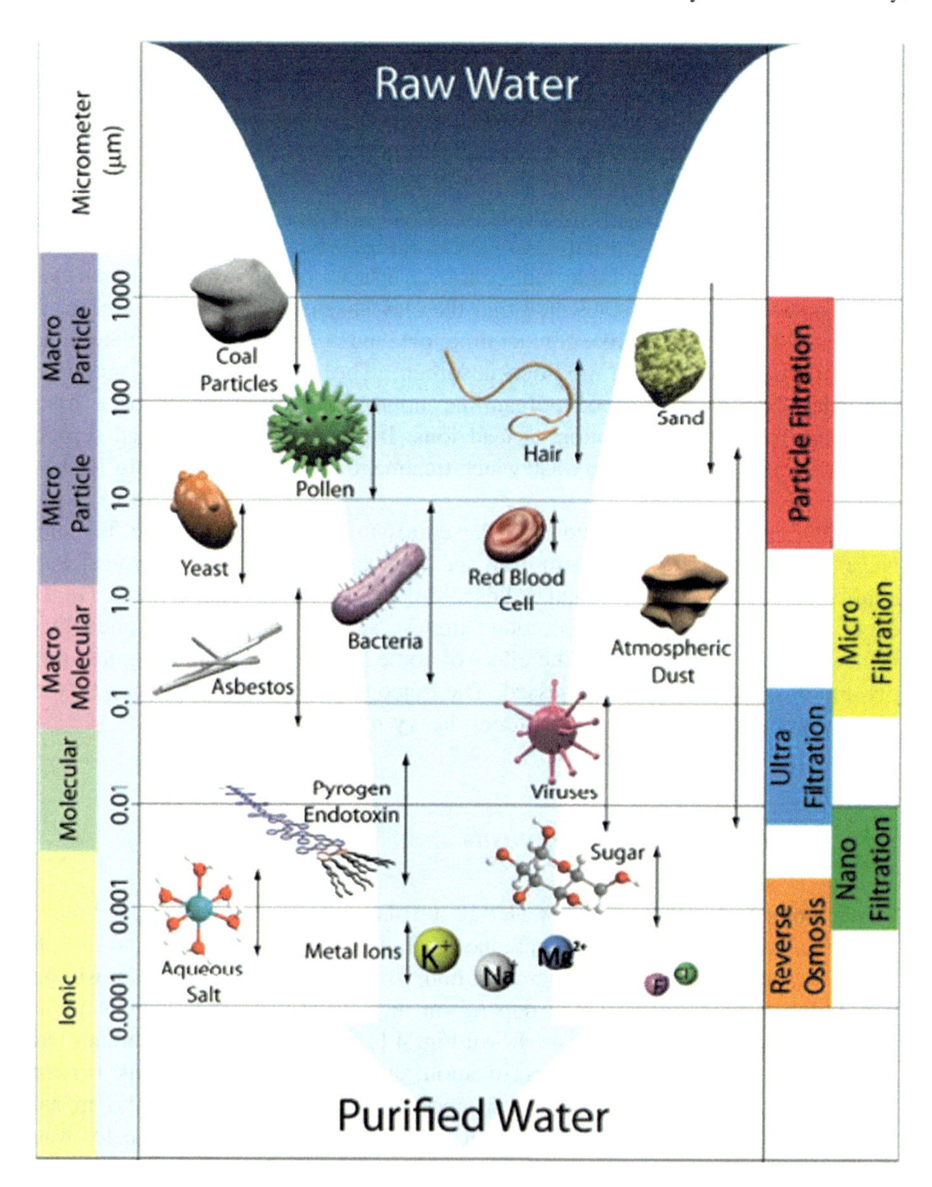

Fig. 4 Schematic illustration of membrane filtration [42]

required for water purification beneath the context of energy potency and cost-effectiveness. According to membrane assembly and position of nanomaterials, they can be classified into four groups: (1) conventional nanocomposite, (2) thin-film nanocomposite (TFN), (3) thin-film composite (TFC) with nanocomposite substrate, and (4) surface-located nanocomposite [85].

In water treatment applications, membranes have to significantly determine hydrophilicity, surface structure, and high toughness with respect to physicochemical and mechanical stability. Pore size and porosity have also strong significant in membrane separation practices. NCM is a mixture of material that can have nanoscale inorganic and/or organic solid phases in a porous structure. These nanoscale constituents enrich membrane assets that would other not be fulfilled by the polymer only [85]. Nanomaterials can improve numerous characteristics of mechanical strength, thermal stability, antifouling properties, permeability, and selectivity which have enhanced membrane separations process. Various constituents such as CNTs, graphene and GO, silica and zeolites, metal and metal oxides, polymers, dendrimers and biological nanomaterials are used in NCM to improve water purification performance [86, 87].

NCM able to reflect as a novel class of filtration tools containing hybrid medium membranes and surface active membranes. Hybrid medium membranes use nanofillers, which are auxiliary to a medium material. In most cases, the nanofillers are inanimate and fixed in a polymeric or inorganic oxide medium. These nanofibers article has larger specific surface area leading to a higher surface-to-mass ratio [88]. NCMs are materials which have no single application of separating pollutants from water. They are also introducing new functionalities such as adsorption [89], photocatalysis [90], antimicrobial activity [91] and surface modification [91] which promoted adsorbing, degrading, and/or deactivating contaminates.

Most of the researchers have confirmed that the integration of nanomaterials into polymers besides to adjust assembly and physicochemical assets like hydrophilicity, porosity, and charge density, chemical, the thermal and mechanical stability of membranes, they are also announced exceptional characteristics such as bactericidal and photocatalytic features into the membranes. The effects of nanofiller on the performance of on the 3 type's NCMs are explained as follows.

3.5.1 Conventional Nanocomposite Membranes

Synthesis of CNM is commonly built on phase inversion (PI) technique in which nanofibers are discrete in polymer solution previous to the PI method as shown Fig. 5 [85]. It can be synthesised in either flat area or deep fiber arrangements. CNM is mostly applied in microfiltration or Ultrafiltration methods because it's typical porous arrangement.

It is known to join nanoparticles inside the polymer medium to create efficient membranes with an exact ability to adsorb heavy metals from water. For instance, incorporated PANI/Fe_3O_4 NPs inside polyethersulfone (PES) [92] and chitosan drops inside ethylene vinyl alcohol (EVAL) medium [93] had to remove Cu (II) water. Both outlooks have confirmed the opportunity of making CNM for the adsorptive elimination of impurities from water.

In the CNM research area, TiO_2 has also merged into numerous membrane mediums to deliver membrane with photocatalytic actions. TiO_2 has been extensively used for water treatment since its exceptional photocatalytic action, solidity,

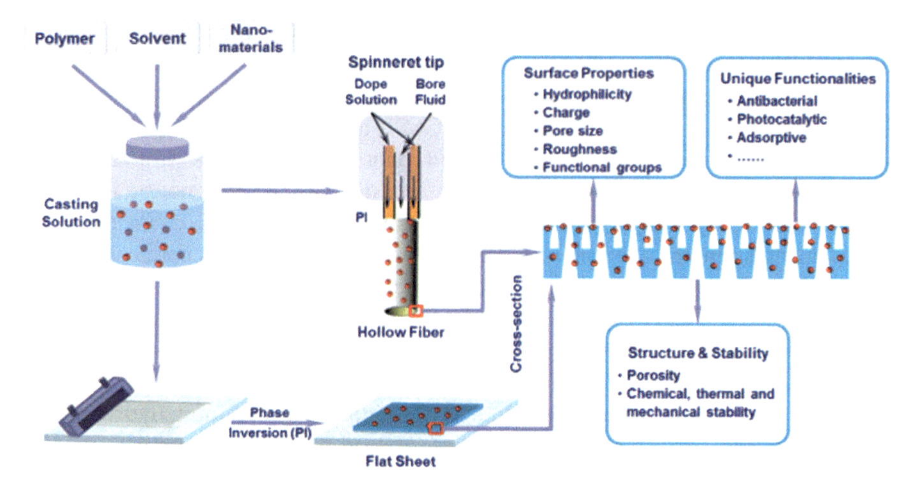

Fig. 5 Production of conventional CNM through the PI process [85]

and simplicity for its fabrication [84]. Evolving antimicrobial membranes will be expected to increase membrane efficiency and lifespan meaningfully which benefits to deliver microbes free clean water. For example, Ag is an excellent biocidal that usually used as an antimicrobial agent in CNM [85, 94]. AgNPs introduced into various metrics such as cellulose acetate [84], PSU [84], and PES [95] enhanced the membrane anti-bacterial activity, virus removal, and biofouling resistance respectively. The efficiency of CNM can be improved by the role reinforcement as indicated in Table 2.

Table 2 Type and role of reinforcement for conventional nanocomposite membrane [84, 91, 96–101]

Type of Reinforcement	Role of Reinforcement
Carbon Nanotube	Incorporation for improved properties such as anti-biofouling and good strength
metal oxide (TiO_2, ZnO, SiO_2, Al_2O_3, Fe_3O_4)	Adjusts the assembly and physicochemical assets, such as hydrophilicity, Porosity, charge density, and chemical, thermal, and mechanical stability of membranes. Introduces the unique characteristics such as antifouling, and photocatalytic action into the membranes.
Metals (Ag, Cu, Se)	Antimicrobial functionality
Nano clay	Improvement in abrasion resistance
Organic Material	Enhance in hydrophilicity, upgrading sorption capacity, and anti-compaction, the antifouling performance of resultant membranes.
AgNPs	Reduce biofouling
Zeolite	Improvement hydrophilicity, advance cross-linking property and increase membrane inflexibility
Biomaterial	Water-channel membrane proteins
Hybrid material	Synergistic effect

3.5.2 Thin-Film Nanocomposites

Thin film nanocomposites (TFNs) membrane contains an extreme tinny wall sheet above a more permeable assistant material. TFN is interfacially synthesized by reverse osmosis or nanofiltration membrane which is extensively applied to remove heavy metals, desalinate seawater/brackish water, hardness causing salts, organic contaminants like pesticides, insecticides and disinfection intermediates. Researchers have been focused to advance water flux, toxin elimination, and antifouling characteristics of TFC l (1) to adapt the auxiliary film thus the linkage among the wall layer and the second layer might be improved, and (2) to enhance the wall layer by changing the IP settings, i.e. exchanging monomers, applying physical layering [102]. Materials like zeolites, CNTs, silica, Ag, and TiO_2 used for CNM synthesis have also been discovered to make TFN membranes [85, 101].

In general technologies yield NCs, a novel theory has been projected centred on diffusing nanomaterials into the extremely tinny wall to increase membrane efficiency for water purification [84]. The known production method is done the in situ IP course among aqueous phenylenediamine (MPD) and trimesoylchloride (TMC) organic solution as shown in Fig. 6. The nanofiller able to spread either in aqueous or an organic phase.

The additions nanoparticle make ready the thin films membranes to yield benefit of the properties of the nanomaterials. Adding of nanoparticles to in between polymerization routes or exterior accessory by self-assembly has announced the concept of TFN, which offer possible profits of improved separation efficiency, reduction fouling, antimicrobial action, and other novel properties. Like TFC membranes, TFN membrane performance can be achieved with nanoparticle

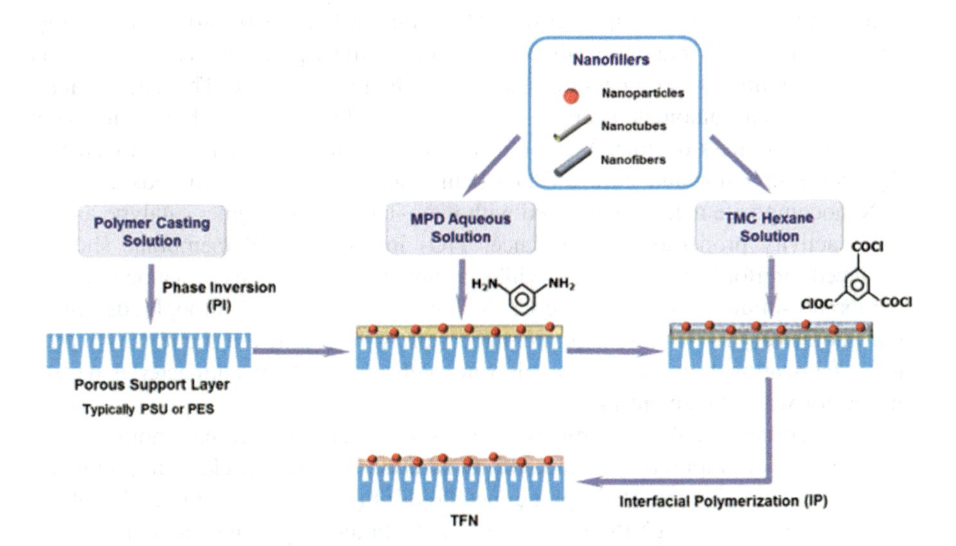

Fig. 6 Production of TFN membranes through the IP method [85]

Table 3 Summary of TFN membranes with nanocomposite substrate [85, 103]

Type of reinforcement	Role of reinforcement
Carboxylic MWNTs	Better antifouling and anti-oxidative properties
Zeolite	Salt elimination; Fighting to physical compaction
MWNTs	Increase the flexible strength of substrate and salt elimination
Ag-zeolite/PA-PSf	Improved water penetrability, Reduced tendency for biofouling
Titania/PA-PES	Reduced porousness and improved elimination at small unit additions, Improved permeability and reduced salt refusal beyond 5 wt%
Zeolite/PA-PSf	Improved interaction with water and superficial charge, Reduced superficial irregularity, Improved water penetrability by 80%
Zeolite/PA-PSf	Improved interaction with water, Increased water penetrability, Improved salt removal in RO testing

Note PA Polyamide, *PSf* Polysulfone, *PES* polyethersulfone

additions to the preserve membrane the coating film, or both. Like CNM, The efficiency of TFN can be improved by the role reinforcement as indicated in Table 3.

3.5.3 TFC with Nanocomposite Substrate

This membrane has been established to look at the consequences of nanofiller on membrane compassion manners. During this category, oxide nanoparticles were entrenched into the postscript substrate [104] that utilized at IP process to arrange TFC film. The ready membrane displays a better primary porousness and minor flux failure throughout the compassion related with the first TFC one. The nanoparticles deliver necessary automated care to moderate the failure of permeable arrangement and resist thickness decline. Membranes with NC substrate tolerate so much less physical compassion and show a vital role in sustaining high water porousness [96].

Nanocomposite membrane coated with nano-TiO_2 shown higher catalytic and/or photo activity properties. For instance, TiO_2 imbedded PES membrane showed enhanced antifouling capability while a novel anatase/titanate nanocomposite membrane simultaneously remove Cr (VI) and 4-chlorophenol through adsorption and photocatalytic oxidation. Impregnation of AgNPs into the membrane would allow fabricating thin-film nanocomposite with an excellent antibacterial performance for water treatment [41].

An important number of articles on membrane nanoscience has motivated on production of multipurpose membranes by addition of nanoparticles into polymeric or inorganic membranes. Hydrophilic metal oxides (e.g., Al_2O_3, TiO_2, and zeolite), antimicrobials (e.g., AgNPs and CNTs), and photocatalytic nanomaterials (e.g., bi-metallic nanomaterials, TiO_2) are the nanomaterials used this application.

The additional water loving metal oxide nanoparticles played a great role to decline fouling by improving the membrane hydrophilicity while the adding of metal oxide nanoparticles such as alumina [105], silica, zeolite [79] and TiO_2 has contributed to increasing membrane surface hydrophilicity, water permeability, or fouling resistance to polymeric ultrafiltration membranes. Besides to this, this metal and/or metal oxide nanoparticles also aid to improve the mechanical and thermal solidity of polymeric membranes, decreasing the destructive influence of compassion and heat on membrane porousness [79, 106].

Qin et al. [107] investigated that handling the wastewater effluent generated from oil refinery and shell gas was difficult since this type of waste effluent was contaminated by contents of oils and salts. This type of wastewater was difficult to treat using conventional membranes because the membrane was severe fouling or failure by salts. The researchers developed another NCFO membrane for succeeding direct oil/water isolation and desalination. This NCFO membrane was accumulated an oil preventing and salt eliminating hydrogel separating layer on surface GO nanosheets imparted polymeric sustenance layer. The hydrogel separating layer governs strong water heating that leads to superior antifouling competency under several oil/, water emulsions, and the imparted GO in support layer can considerably moderate interior concentration polarization by decreasing FO membrane. Compared with viable FO membrane, the new membrane establishes triple water flux, higher eliminations for oil (>99.9%) and salts (>99.7%) and pointedly worse fouling attraction when examined with replicated shale gas wastewater as shown the Fig. 7. These combined benefits will validate this new NCFO membrane with wide requests in handling highly salty and oily effluents [107].

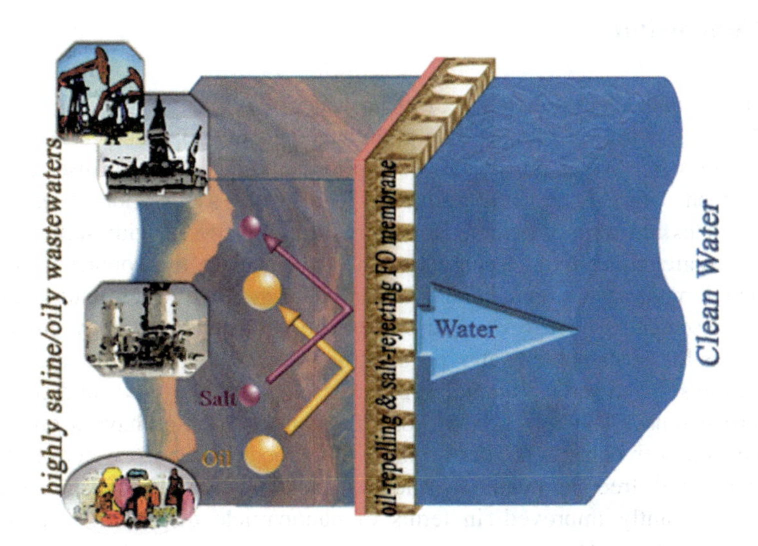

Fig. 7 Illustration of immediate oil/water separation and desalination by Hydrogel/GOFO membrane [107]

4 Future Outlooks

In this chapter, the most extensively studied nanocomposite, Metal nanocomposite, nanocomposite zero-valent metals (Ag, Pb, and Zn), MONCs (TiO_2, ZnO, and iron oxides), PNCs and MNCs were highlighted. Moreover, their applications in water purification were discussed in detail. Since the current rapid water demand development and sustainable application, NC look exceptionally favourable materials for water purification.

However, more studies are quiet required to solve the NC encounters. Still, now, insufficient types of nanocomposite are available commercially. Meanwhile, less production price is critical to confirm their extensive requests for water purification, future research has to devote to developing the commercial competence of NCs. Moreover, with progressively widespread applications of the NCs in water treatment, there are increasing alarms about their potential noxiousness to the environment and human health. Existing evidence in the literature has discovered that numerous NCs. However, principles for evaluating the noxiousness of NCs are somewhat inadequate at present-day. Hence, widespread assessment of the toxicity of NC is the crucial necessity to confirm their real applications. What is more, the assessment and contrast of the performance of numerous NC in water purification are recognized standards. It is hard to relate the performances of diverse nanoparticles and figure out talented NC that merits extra application. Consequently, the performance assessment tool of the NC in water purification ought to be perfected in the future.

5 Conclusion

Growing demand and deficiency of clean water as a result of rapid urbanization, population growth, and climate disruption have become unparalleled urgent global issues. Globally, Water purification is a priority issue for human use, ecosystem management, agriculture, and industry. The water sanitization process using nanoparticles are quite efficient. However, these are linked with some weakness such as aggregation, tough separation, and leakage into the contact water, environmental impact and human health. Therefore, to improve water treatment process system, researchers have been paid to develop eco-friendly, energy efficient and low price for sustainable water purification. The nanocomposites are basically multiphase solid materials, including porous media, colloids, gels, and copolymers in a broad sense. The selection of hosts for nanocomposites have a great consequence, and even controls the performance of nanocomposites in water purification. Compared with free nanomaterials, the efficiency and usability of nanocomposites were significantly improved, in terms of nanoparticle dispersion, stability, and recyclability. Nowadays, nanocomposites were supposed to be the supreme likely way of advancing water nanotechnology from lab study to large-scale application.

A number of the researcher was investigated nanocomposites synthesis from metal, metal oxide, carbon, polymer and membrane are the common materials used for water purification. Polymer nanocomposites (PNCs) are a superior class of materials which nanoparticles (NPs) dispersed in a polymer matrix resulting in novel materials having unique physical and chemical properties [74]. Polymers are special supports for nanomaterials as they usually possess tunable porous structures, excellent mechanical properties, and chemically bounded functional groups. Polymer-based nanocomposites (PNCs) are prospecting materials for their sound performance in water and wastewater treatment. Nanocomposite membrane has a great role in water purification and reuses for various sources of water such as drinking water, brackish, seawater, and wastewater treatment. Nanocomposite membranes is an innovative type of membranes prepared by merging polymeric materials with nanoparticles are developing as an encouraging solution to above challenges.

References

1. WHO/UNICEF (2014) Progress on drinking water and sanitation. Monitoring Programme update, WHO report, pp 1–18
2. Dargo H, Ayaliew A, Kassa H (2017) Synthesis paradigm and applications of silver nanoparticles (AgNPs), a review. Sustain Mater Technol 13:18–23
3. Liang XJ, Kumar A, Shi D, Cui D (2012) Nanostructures for medicine and pharmaceuticals. J Nanomaterials 2012:2012–2014
4. Kusior A, Klich-Kafel J, Trenczek-Zajac A, Swierczek K, Radecka M, Zakrzewska K (2013) TiO2-SnO2 nanomaterials for gas sensing and photocatalysis. J Eur Ceram Soc 33(12): 2285–2290
5. Diana S, Luigi R, Vincenzo V (2017) Progress in Nanomaterials Applications for Water Purification, In: Lofrano, Gi, Libralato, Giovanni, Brown, Jeanette (Eds) Nanotechnologies for Environmental Remediation, Applications and Implications, 1st ed, pp 1–24. Springer International Publishing AG
6. Lu H et al (2014) An overveiw of nanomaterials for water and wastewater treatment. J Environ Anal Chem 2016(2):10–12
7. Mueller NC et al (2012) Application of nanoscale zero valent iron (NZVI) for groundwater remediation in Europe. Environ Sci Pollut Res 19(2):550–558
8. Karn B, Kuiken T, Otto M (2009) Nanotechnology and in situ remediation: a review of the benefits and potential risks. Environ Health Perspect 117(12):1823–1831
9. Kumar D, Parashar A, Chandrasekaran N, Mukherjee A (2017) The stability and fate of synthesized zero-valent iron nanoparticles in freshwater microcosm system. 3 Biotech 7(3): 1–9
10. Fu F, Dionysiou DD, Liu H (2014) The use of zero-valent iron for groundwater remediation and wastewater treatment: a review. J Hazard Mater 267:194–205
11. Amin MT, Alazba AA, Manzoor U (2014) A review on removal of pollutants from water/ wastewater using different types of nanomaterials. Adv Mater Sci Eng vol 2014:ID 825910
12. Ghasemzadeh G, Momenpour M, Omidi F, Hosseini MR, Ahani M, Barzegari A (2014) Applications of nanomaterials in water treatment and environmental remediation. Front Environ Sci Eng 8(4):471–482
13. Marková Z et al (2013) Air stable magnetic bimetallic Fe-Ag nanoparticles for advanced antimicrobial treatment and phosphorus removal. Environ Sci Technol 47(10):5285–5293

14. Muradova GG, Gadjieva SR, Di L, Vilardi G (2016) Nitrates removal by bimetallic nanoparticles in water. Chem Eng Trans 47:205–210
15. Xiong Z, Lai B, Yang P, Zhou Y, Wang J, Fang S (2015) Comparative study on the reactivity of Fe/Cu bimetallic particles and zero valent iron (ZVI) under different conditions of N < inf > 2</inf > air or without aeration. J Hazard Mater 297:261–268
16. Hoag GE, Collins JB, Holcomb JL, Hoag JR, Nadagouda MN, Varma RS (2009) Degradation of bromothymol blue by 'greener' nano-scale zero-valent iron synthesized using tea polyphenols. J Mater Chem 19(45):8671–8677
17. Sun Z, Song G, Du R, Hu X (2017) Modification of a Pd-loaded electrode with a carbon nanotubes-polypyrrole interlayer and its dechlorination performance for 2,3-dichlorophenol. RSC Adv 7(36):22054–22062
18. Arancibia-Miranda N et al (2016) Nanoscale zero valent supported by zeolite and montmorillonite: template effect of the removal of lead ion from an aqueous solution. J Hazard Mater 301:371–380
19. Ling L, Pan B, Zhang WX (2014) Removal of selenium from water with nanoscale zero-valent iron: mechanisms of intraparticle reduction of Se (IV). Water Res 71(34): 274–281
20. Ling L, Zhang WX (2015) Enrichment and encapsulation of uranium with iron nanoparticle. J Am Chem Soc 137(8):2788–2791
21. Mahmoudi M, Serpooshan V (2012) Silver-coated engineered magnetic nanoparticles are promising for the success in the fight against antibacterial resistance threat. ACS Nano 6 (3):2656–2664
22. Lara HH, Romero-Urbina DG, Pierce C, Lopez-Ribot JL, Arellano-Jiménez MJ, Jose-Yacaman M (2015) Effect of silver nanoparticles on Candida albicans biofilms: an ultrastructural study. J Nanobiotechnol 13(1):1–12
23. Morones JR et al (2005) The bactericidal effect of silver nanoparticles. Nanotechnology 16(10):2346–2353
24. Surendhiran D, Sirajunnisa A, Tamilselvam K (2017) Silver–magnetic nanocomposites for water purification. Environ Chem Lett 15(3):367–386
25. Kim JS et al (2007) Antimicrobial effects of silver nanoparticles. Nanomed Nanotechnol Biol Med 3(1):95–101
26. Xiu Z-M, Ma J, Alvarez PJJ (2011) Differential effect of common ligands and molecular oxygen on antimicrobial activity of silver nanoparticles versus silver ions. Environ Sci Technol 45(20):9003–9008
27. Mlalila NG, Swai HS, Hilonga A, Kadam DM (2017) Antimicrobial dependence of silver nanoparticles on surface plasmon resonance bands against Escherichia coli. Nanotechnol Sci Appl 10:1–9
28. Ishida H, Campbell S, Blackwell J (2000) General approach to nanocomposite preparation. Chem Mater 12(5):1260–1267
29. Tapas RS (2017) Polymer Nanocomposites for Environmental Applications. In: Deba KT, Bibhu PS (Eds) Properties and Applications of Polymer Nanocomposites, Clay and Carbon Based Polymer Nanocomposites, 1st ed, pp 77-99. Springer-Verlag GmbH Germany
30. Gehrke I, Geiser A, Somborn-Schulz A (2015) Innovations in nanotechnology for water treatment. Nanotechnol Sci Appl 8:1–17
31. Sharma G, Amit K, Shweta, Mu N, Ram PD, Zeid AA, Genene TM (2017) Novel development of nanoparticles to bimetallic nanoparticles and their composites: a review. J King Saud Univ Sci. https://doi.org/10.1016/j.jksus.2017.06.012
32. Rhim JW, Park HM, Ha CS (2013) Bio-nanocomposites for food packaging applications. Prog Polym Sci 38(10–11):1629–1652
33. de Azeredo HMC (2009) Nanocomposites for food packaging applications. Food Res Int 42(9):1240–1253
34. Othman SH (2014) Bio-nanocomposite materials for food packaging applications: types of biopolymer and nano-sized filler. Agric Agric Sci Procedia 2:296–303

35. Zare Y, Shabani I (2016) Polymer/metal nanocomposites for biomedical applications. Mater Sci Eng C 60:195–203
36. Veprek S, Veprek-Heijman MJG (2008) Industrial applications of superhard nanocomposite coatings. Surf Coat Technol 202(21):5063–5073
37. Zhang R et al (2016) Antifouling membranes for sustainable water purification: strategies and mechanisms. Chem Soc Rev 45(21):5888–5924
38. Galiano F et al (2015) A step forward to a more efficient wastewater treatment by membrane surface modification via polymerizable bicontinuous microemulsion. J Membr Sci 482: 103–114
39. Manawi Y, Kochkodan V, Hussein MA, Khaleel MA, Khraisheh M, Hilal N (2016) Can carbon-based nanomaterials revolutionize membrane fabrication for water treatment and desalination? Desalination 391:69–88
40. Senusi F, Shahadat M, Ismail S, Hamid SA (2018) Recent advancement in membrane technology for water purification, In : Oves M (ed) Modern age environmental problems and their remediation, Recent Advancement, 1st edn. Springer International Publishing AG, pp 1–237
41. Zhang Y et al (2016) Nanomaterials-enabled water and wastewater treatment. NanoImpact 3–4:22–39
42. Lee A, Elam JW, Darling SB (2016) Membrane materials for water purification: design, development, and application. Environ Sci Water Res Technol 2(1):17–42
43. Botes M, Cloete TE (2010) The potential of nanofibers and nanobiocides in water purification. Crit Rev Microbiol 36(1):68–81
44. Peter-Varbanets M, Zurbrügg C, Swartz C, Pronk W (2009) Decentralized systems for potable water and the potential of membrane technology. Water Res 43(2):245–265
45. Lin S, Huang R, Cheng Y, Liu J, Lau BLT, Wiesner MR (2013) Silver nanoparticle-alginate composite beads for point-of-use drinking water disinfection. Water Res 47(12):3959–3965
46. Yahyaei B, Azizian S, Mohammadzadeh A, Pajohi-Alamoti M (2015) Chemical and biological treatment of waste water with a novel silver/ordered mesoporous alumina nanocomposite. J Iran Chem Soc 12(1):167–174
47. Firdhouse MJ, Lalitha P (2016) Nanosilver-decorated nanographene and their adsorption performance in waste water treatment. Bioresour Bioprocess 3(1):12
48. Liu X, Chen Z, Chen Z, Megharaj M, Naidu R (2013) Remediation of direct black G in wastewater using kaolin-supported bimetallic Fe/Ni nanoparticles. Chem Eng J 223:764–771
49. Lateef A, Nazir R (2017) Metal nanocomposites : synthesis, characterization and their applications, In: P. DS, (ed) Science and applications of tailored nanostructures, 1st edn. One central press, Italy, pp 239–240
50. Ray C, Pal T (2017) Recent advances of metal-metal oxide nanocomposites and their tailored nanostructures in numerous catalytic applications. J Mater Chem A 5(20):9465–9487
51. Sankararamakrishnan N, Jaiswal M, Verma N (2014) Composite nanofloral clusters of carbon nanotubes and activated alumina: an efficient sorbent for heavy metal removal. Chem Eng J 235:1–9
52. Ihsanullah, Asmaly HA, Saleh TA, Laoui T, Gupta VK, Atieh MA (2015) Enhanced adsorption of phenols from liquids by aluminum oxide/carbon nanotubes: comprehensive study from synthesis to surface properties. J Mol Liq 206(February):176–182
53. Liang J et al (2015) Facile synthesis of alumina-decorated multi-walled carbon nanotubes for simultaneous adsorption of cadmium ion and trichloroethylene. Chem Eng J 273:101–110
54. Mallakpour S, Khadem E (2016) Carbon nanotube–metal oxide nanocomposites: fabrication, properties and applications. Chem Eng J 302(May):344–367
55. Ming-Zheng G, Chun-Yan C, Jian-Ying H, Shu-Hui L, Song-Nan Z, Shu D, Qing-Song L, Ke-Qin Z, Yue-Kun L (2016) Synthesis, modification, and photo/photoelectrocatalytic degradation applications of TiO2 nanotube arrays: a review. Nanotechnol Rev 5(1). https://doi.org/10.1515/ntrev-2015-0049

56. Silva CG, Faria JL (2010) Photocatalytic oxidation of benzene derivatives in aqueous suspensions: synergic effect induced by the introduction of carbon nanotubes in a TiO2 matrix. Appl Catal B Environ 101(1–2):81–89
57. Martínez C, Canle LM, Fernández MI, Santaballa JA, Faria J (2011) Kinetics and mechanism of aqueous degradation of carbamazepine by heterogeneous photocatalysis using nanocrystalline TiO2, ZnO and multi-walled carbon nanotubes-anatase composites. Appl Catal B Environ 102(3–4):563–571
58. Li J, Zhen D, Sui G, Zhang C, Deng Q, Jia L (2012) Nanocomposite of Cu–TiO < SUB > 2</SUB > –SiO < SUB > 2</SUB > with high photoactive performance for degradation of rhodamine B dye in aqueous wastewater. J Nanosci Nanotechnol 12(8): 6265–6270
59. Khan M et al (2015) Graphene based metal and metal oxide nanocomposites: synthesis, properties and their applications. J Mater Chem A 3(37):18753–18808
60. Ma J, Zhang J, Xiong Z, Yong Y, Zhao XS (2011) Preparation, characterization and antibacterial properties of silver-modified graphene oxide. J Mater Chem 21(10):3350–3352
61. Chandra V, Park J, Chun Y, Lee JW, Hwang IC, Kim KS (2010) Water-dispersible magnetite-reduced graphene oxide composites for arsenic removal. ACS Nano 4(7):3979–3986
62. Geng Z et al (2012) Highly efficient dye adsorption and removal: a functional hybrid of reduced graphene oxide-Fe3O4 nanoparticles as an easily regenerative adsorbent. J Mater Chem 22(8):3527–3535
63. Saad AHA, Azzam AM, El-Wakeel ST, Mostafa BB, Abd El-latif MB (2018) Removal of toxic metal ions from wastewater using ZnO@Chitosan core-shell nanocomposite. Environ Nanotechnol Monit Manag 9(August):67–75
64. Singh P et al (2018) Specially designed B4C/SnO2 nanocomposite for photocatalysis: traditional ceramic with unique properties. Appl Nanosci 8(1–2):1–9
65. Huang L, He M, Chen B, Hu B (2018) Magnetic Zr-MOFs nanocomposites for rapid removal of heavy metal ions and dyes from water. Chemosphere 199:435–444
66. Gong JL et al (2009) Removal of cationic dyes from aqueous solution using magnetic multi-wall carbon nanotube nanocomposite as adsorbent. J Hazard Mater 164(2–3):1517–1522
67. Chen L et al (2016) Facile synthesis of mesoporous carbon nanocomposites from natural biomass for efficient dye adsorption and selective heavy metal removal. RSC Adv 6(3): 2259–2269
68. Inyang M, Gao B, Zimmerman A, Zhang M, Chen H (2014) Synthesis, characterization, and dye sorption ability of carbon nanotube-biochar nanocomposites. Chem Eng J 236:39–46
69. Muneeb M, Zahoor M, Muhammad B, AliKhan F, Ullah R, AbdEI-Salam NM (2017) Removal of heavy metals from drinking water by magnetic carbon nanostructures prepared from biomass. J Nanomater 2017:10
70. Tian T et al (2014) Graphene-based nanocomposite as an effective, multifunctional, and recyclable antibacterial agent. ACS Appl Mater Interfaces 6(11):8542–8548
71. Zarei M (2017) Application of nanocomposite polymer hydrogels for ultra-sensitive fluorescence detection of proteins in gel electrophoresis. TrAC - Trends Anal Chem 93:7–22
72. Zhao S et al (2012) Performance improvement of polysulfone ultrafiltration membrane using well-dispersed polyaniline-poly(vinylpyrrolidone) nanocomposite as the additive. Ind Eng Chem Res 51(12):4661–4672
73. Pan B, Xu J, Wu B, Li Z, Liu X (2013) Enhanced removal of fluoride by polystyrene anion exchanger supported hydrous zirconium oxide nanoparticles. Environ Sci Technol 47(16): 9347–9354
74. Settanni, G, Zhou, J, Suo, T, Schöttler, S, Landfester, K, Schmid, F, Mailänder, V (2017) Protein corona composition of poly (ethylene glycol)- and poly (phosphoester)-coated nanoparticles correlates strongly with the amino acid composition of the protein surface. Nanoscale 9(6):2138–2144

75. Kelta B, Taddesse AM, Yadav OP, Diaz I, Mayoral Á (2017) Nano-crystalline titanium (IV) tungstomolybdate cation exchanger: Synthesis, characterization and ion exchange properties. J Environ Chem Eng 5(1):1004–1014
76. Zhang L, Liu J, Guo X (2018) Investigation on mechanism of phosphate removal on carbonized sludge adsorbent. J Environ Sci (China) 64:335–344
77. Vunain E, Mishra AK, Mamba BB (2016) Dendrimers, mesoporous silicas and chitosan-based nanosorbents for the removal of heavy-metal ions: a review. Int J Biol Macromol 86:570–586
78. Djerahov L, Vasileva P, Karadjova I, Kurakalva RM, Aradhi KK (2016) Chitosan film loaded with silver nanoparticles - Sorbent for solid phase extraction of Al (III), Cd (II), Cu (II), Co (II), Fe (III), Ni (II), Pb (II) and Zn (II). Carbohydr Polym 147(March):45–52
79. Saxena S, Saxena U (2016) Development of bimetal oxide doped multifunctional polymer nanocomposite for water treatment. Int Nano Lett 6(4):223–234
80. Qu X, Alvarez PJJ, Li Q (2013) Applications of nanotechnology in water and wastewater treatment. Water Res 47(12):3931–3946
81. Zayed A et al (2013) Fe3O4/cyclodextrin polymer nanocomposites for selective heavy metals removal from industrial wastewater. Carbohydr Polym 91(1):322–332
82. Khaydarov RA, Khaydarov RR, Gapurova O (2010) Water purification from metal ions using carbon nanoparticle-conjugated polymer nanocomposites. Water Res 44(6):1927–1933
83. Piri S, Zanjani ZA, Piri F, Zamani A, Yaftian M, Davari M (2016) Potential of polyaniline modified clay nanocomposite as a selective decontamination adsorbent for Pb (II) ions from contaminated waters; kinetics and thermodynamic study. J Environ Health Sci Eng 14(1): 1–10
84. Nithya R, Sudha PN (2017) Removal of heavy metals from tannery effluent using chitosan-g-poly (butyl acrylate)/bentonite nanocomposite as an adsorbent. Text Cloth Sustain 2(1):7
85. Yin J, Deng B (2015) Polymer-matrix nanocomposite membranes for water treatment. J Membr Sci 479:256–275
86. Shen YX, Saboe PO, Sines IT, Erbakan M, Kumar M (2014) Biomimetic membranes: a review. J Memb Sci 454:359–381
87. Hernández S, Saad A, Ormsbee L, Bhattacharyya D (2016) Nanocomposite and responsive membranes for water treatment, In: Hankins NP, Singh R (ed) Emerging membrane technology for sustainable water treatment, 1st edn. Elsevier B.V., USA, pp 389–431
88. Nasreen SAAN, Sundarrajan S, Nizar SAS, Balamurugan R, Ramakrishna S (2013) Advancement in electrospun nanofibrous membranes modification and their application in water treatment. Membr (Basel) 3(4):266–284
89. Fard AK et al (2018) Inorganic membranes: preparation and application for water treatment and desalination. Mater (Basel) 11(1):74
90. Razzaq H, Nawaz H, Siddiqa A, Siddiq M, Qaisar S (2016) Madridge a brief review on nanocomposites based on PVDF with nanostructured TiO2 as filler. J Nanotechnol 1(1): 29–35
91. Pant HR et al (2014) One-step fabrication of multifunctional composite polyurethane spider-web-like nanofibrous membrane for water purification. J Hazard Mater 264:25–33
92. Daraei P et al (2012) Novel polyethersulfone nanocomposite membrane prepared by PANI/ Fe 3O 4 nanoparticles with enhanced performance for Cu (II) removal from water. J Membr Sci 415–416:250–259
93. Tetala KKR, Stamatialis DF (2013) Mixed matrix membranes for efficient adsorption of copper ions from aqueous solutions. Sep Purif Technol 104:214–220
94. Lopez Goerne TM (2011) Study of Bacterial Sensitivity to Ag-TiO2 Nanoparticles. J Nanomed Nanotechnol s5(01):2
95. Liu S, Fang F, Wu J, Zhang K (2015) The anti-biofouling properties of thin-film composite nanofiltration membranes grafted with biogenic silver nanoparticles. Desalination 375(November):121–128

96. Tewari PK (2016) Nanocomposite membrane technology, 1st edn. CRC Press Taylor & Francis Group, Boca Raton
97. Ladewig B, Al-Shaeli MNZ (2017) Fundamental of membrane process. In: Ladewig B, Al-Shaeli MNZ (eds) Fundamentals of membrane bioreactors, 1st edn. Springer Nature Singapore, Singapore, pp 13–38
98. Jamshidi Gohari R, Halakoo E, Nazri NAM, Lau WJ, Matsuura T, Ismail AF (2014) Improving performance and antifouling capability of PES UF membranes via blending with highly hydrophilic hydrous manganese dioxide nanoparticles. Desalination 335(1):87–95
99. Jamshidi Gohari R, Lau WJ, Matsuura T, Ismail AF (2013) Fabrication and characterization of novel PES/Fe-Mn binary oxide UF mixed matrix membrane for adsorptive removal of as (III) from contaminated water solution. Sep Purif Technol 118:64–72
100. Akar N, Asar B, Dizge N, Koyuncu I (2013) Investigation of characterization and biofouling properties of PES membrane containing selenium and copper nanoparticles. J Membr Sci 437:216–226
101. Manjarrez Nevárez L et al (2011) Biopolymers-based nanocomposites: membranes from propionated lignin and cellulose for water purification. Carbohydr Polym 86(2):732–741
102. Jeong BH et al (2007) Interfacial polymerization of thin film nanocomposites: a new concept for reverse osmosis membranes. J Membr Sci 294(1–2):1–7
103. Pendergast MM, Hoek EMV (2011) A review of water treatment membrane nanotechnologies. Energy Environ Sci 4(6):1946–1971
104. Lind ML, Suk DE, Nguyen TV, Hoek EMV (2010) Tailoring the structure of thin film nanocomposite membranes to achieve seawater RO membrane performance. Environ Sci Technol 44(21):8230–8235
105. Maximous N, Nakhla G, Wong K, Wan W (2010) Optimization of Al2O3/PES membranes for wastewater filtration. Sep Purif Technol 73(2):294–301
106. Pendergast MTM, Nygaard JM, Ghosh AK, Hoek EMV (2010) Using nanocomposite materials technology to understand and control reverse osmosis membrane compaction. Desalination 261(3):255–263
107. Qin D, Liu Z, Delai Sun D, Song X, Bai H (2015) A new nanocomposite forward osmosis membrane custom-designed for treating shale gas wastewater. Sci Rep 5(January):1–14

Sustainable Nanocomposites in Food Packaging

H. Anuar, F. B. Ali, Y. F. Buys, M. A. Siti Nur E'zzati, A. R. Siti Munirah Salimah, M. S. Mahmud, N. Mohd Nordin and S. A. Adli

List of Abbreviations

APs	Alkyl phenols
CNCs	Cellulose nanocrystals
CNT	Carbon nanotube
CNW	Cellulose nanowhisker
CO_2PC	Carbon dioxide permeability coefficient
CO_2TR	Carbon dioxide transmission rate
DSC	Differential scanning calorimetry
DMF	N, N-dimethylformamide
EVOH	Ethylene vinyl alcohol copolymer
HDPE	High-density polyethylene
HPMC	Hydroxyl propyl methyl cellulose
HT	Hydroxytyrosol
HV	Hydroxyl-valerate
LDPE	Low-density polyethylene
MgO	Magnesium oxide
MMT	Montmorillonite
MWNT	Multi-walled carbon nanotube
OPC	Oxygen permeability coefficient
OTR	Oxygen transmission rate
PANI	Polyaniline
PCL	Poly(e-caprolactone)
PEG	Polyethylene glycol
PEGME	Polyethylene glycol methyl ether

H. Anuar (✉) · Y. F. Buys · M. A. Siti Nur E'zzati · A. R. Siti Munirah Salimah · M. S. Mahmud · N. Mohd Nordin
Department of Manufacturing and Materials Engineering, Faculty of Engineering, International Islamic University Malaysia, Jalan Gombak, 53100 Kuala Lumpur, Malaysia
e-mail: hazleen@iium.edu.my

F. B. Ali · S. A. Adli
Department of Biotechnology Engineering, Faculty of Engineering, International Islamic University Malaysia, Jalan Gombak, 53100 Kuala Lumpur, Malaysia

Inamuddin et al. (eds.), *Sustainable Polymer Composites and Nanocomposites*,
https://doi.org/10.1007/978-3-030-05399-4_15

PET	Polyethylene terephthalate
PHA	Poly hydroxyalkanoate
PHB	Polyhydroxy butyrate
PHBV	Polyhydroxybutyrate-co-hydroxyvalerate
PLA	Polylactic acid
PP	Polypropylene
PVA	Polyvinyl alcohol
ROP	Ring-opening polymerization
SEM	Scanning electron microscopy
TGA	Thermogravimetric analysis
WHO	World health organization
WSC	Water-soluble chitosan
WVPC	Water vapour permeability coefficient
WVTR	Water vapour transmission rate

1 Introduction

Human activities have already brought about changes to the ecological system worldwide, thereby giving rise to global warming and pollution. These problems have led to increased environmental degradation, which is destroying many things. Many plastic packaging industries are trying to minimize their reliance on synthetic materials in order to reduce toxicity levels [47]. Therefore, new environmental guidelines are forcing the exploration of novel packaging materials that are compatible with the environment. Such materials are currently being developed from various natural resources, among which biopolymers are the most popular. Biopolymers have many advantages compared to petroleum-based plastics in terms of performance, processability, and biodegradability [82]. There are several types of biopolymer matrices including polylactic acid (PLA), polyvinyl alcohol (PVA), polyhydroxy butyrate (PHB) and starch-based polymers. The poor gas and water barrier properties, unstable mechanical properties and weak resistivity of plastics have limited their use in various applications [55]. Therefore, to improve their mechanical and barrier properties, and to impart novel features to them, biopolymers are reinforced with various nanofillers for the development of nanocomposites.

1.1 Nanocomposites

Various terms have been used for advanced multifunctional packaging such as smart, active, and intelligent packaging. The idea behind using such terms for this technology is to describe the selected reaction of a certain type of packaging when located in different environments. This reaction allows the smart or active packaging to sense changes in certain properties of food products and report them to the customer [40]. Active food packaging materials encompassing nanocomposites play several roles, including those of protecting food products from the outside environment, as well as increasing their shelf life by preserving their value for a longer period of time [79]. Other than that, smart packaging materials prevent oxidation of the food item from taking place, hence delaying its deterioration. They also function as a moisture controller, anti-microbial agent and freshness indicator [40]. Thus, active and intelligent packaging systems also deal with nanocomposite materials when they show ideal properties of smart packaging through their features.

Nanocomposites have been used in recent years in many industries such as in textiles, home decorations, packaging, furnishings, and agriculture. The application of natural fibers in nanocomposite materials is not a serious concern due to new technologies on the fabrication of synthetic fibers [79]. These synthetic fibers are mass produced for application in various industries. However, during the fabrication of these fibers, pollutants are produced that have an effect on the environment. The structure of biopolymer plastics tends to limit their application when there is a lack of reinforcement such as nanofillers or nanofibers [55]. Synthetic fibers are improved by reinforcement with nanofibers to enable them to be used in packaging applications. During this phase, the load is transferred by the matrix to the nanofibers. Apart from the mechanical and thermal properties, permeability is considered to be the most important parameter in the selection of materials for food packaging applications [40]. A high resistance to vapour and moisture is important when choosing a suitable polymer-based plastic. Bio-nanocomposites provide an excellent balance in mechanical, thermal and barrier properties.

The incorporation of nanofillers from inorganic and organic natural fibers can lead to green food packaging [43]. Natural fibers are more compatible with the environment than other fibers because they are biodegradable and renewable. In addition, these materials are cheap and have a low level of toxicity [49]. Nanofibers are measured in nanometres (10^{-9} m), and the term ‘nano’ relates to a range of nanoparticles. The measurement of the diameter of nanoparticles changes to nanometres following the shrinkage of micrometre materials. Deepa et al. [13] mentioned that the nanometre measurement of constituents is equivalent to being 80,000 times thinner than that of the hair of a person. However, this depends on whether the nanocomposite materials originated from material or fiber sources. Sometimes, they may become tiny particles but do not reach the nano-size range. The unique properties of nanocomposite materials are that they enhance the properties of composites, they have more attractive features in terms of their mechanical properties, they have flexible surfaces, and high surface area ratios [75]. Their surface area ratios are 10^3 times more than those of micrometre materials [48].

Nanofillers play a potential role in improving or altering the properties of the material in nanocomposites. Nanocomposite materials have a more acceptable demand due to the specific advantages that they have over non-biodegradable products. Being more environmentally friendly, nanomaterials are used in many applications like the aerospace, automotive, electronics and biotechnology sectors [63]. Nanofibers and their composites offer a highly attractive contemporary to the research line. Nanocomposites show an outstanding improvement due to the nano-reinforcement. Basically, nanofibers are extracted for the stiffness that they provide to the reinforcement in nanocomposite materials [48]. Unlike natural fibers, synthetic fibers are produced from petrochemicals, while natural fibers originate from farm or agricultural crops. Due to environmental processes, biopolymer nanocomposite packaging is subject to degradation when exposed to environmental conditions such as moisture, sunlight, temperature or different pH solutions [77].

In nanoclay composites, the barrier properties depend on the orientation of the nanofiller, and it dispersion condition in the polymer matrix. The clay layers able to create a barrier to delay the diffusion of molecules through the food packaging. According to Majeed et al. [40], nanoclay composites have gained a lot of attention from packaging manufacturers due to their cost-effectiveness, wide availability and simple processability. Nanoclay-based composites are usually reinforced with a polymeric starch-based material. Talegaonkar et al. [79] and his group identified the reason behind the successful use of nanoclay composites as a nanofiller in food packaging materials. According to them, the nanoclay that is used as a nanofiller improves the barrier properties of the biocomposite, possibly by increasing the tortuosity of the diffusive path for permeants, such as gas, moisture, and vapour. Li et al. [36] also added that the addition of 60 wt% of nanoclay in agarose starch increases the elastic modulus by five times than that of neat agarose starch at 2.4 GPa. It has been proven that the mechanical behaviour of bio-nanocomposites can be improved with the presence of nanoclay in it.

Smart or active packaging plays the most important role in the security of food products by preserving their integrity throughout their shelf life. The use of nanocomposite materials in food packaging industries is a new green technology aimed at attaining markedly enhanced packaging properties like improved mechanical properties and water resistance, increased thermal stability and barrier properties, and decreased migration activity. In addition, nanocomposites also offer an environmental approach, where they can be degraded within a few months. There are various types of nanofiller-based natural fibers that are suitable for preparation as nanocomposite materials. Thus, the nanocomposite materials are remarkably incorporated as packaging materials and also provide smart properties to the packaging system, thereby ultimately increasing the shelf life of food products and improving product quality.

2 Nanocomposite Preparation

The properties of nanocomposites heavily depend on the dispersion states of the reinforcement phase in the matrix. In order to optimize the properties of nanocomposites, it is necessary to disperse the particles in homogeneous "nano-level" dispersion. For the cases of clay reinforced nanocomposites, there are three possible polymer/clay nanocomposite structures, i.e. flocculated, intercalated, and exfoliated [71], with exfoliation structure, followed by intercalation are the preferred ones. It also should be noted that in preparing nanocomposites specimens, degradation of the matrices should be avoided.

Various procedures to disperse the nanoparticles inside the polymeric matrices have been reported, and mainly can be classified into three classes, i.e. solution casting, melt mixing and in situ polymerization.

2.1 Polymer Solution Casting

Solvent casting technology, for the production of high-quality flexible films, has attracted widespread interest from plastic manufacturers. Driven by the requirements of the photographic industry well over a century ago, the solvent casting method has gradually declined due to the development of new film extrusion technology. However, due to the existence of certain limitations which constrain the applicability of conventional polymer processing methods, researchers' have adopted several measures to develop and modify polymer processing methods to suit natural polymers [53]. Solution casting is known as solvent casting or the wet processing method and is used for forming polymer thin films. Polymer films have many technological applications and advantages in industries such as packaging, medical devices, and photography tools and are also used in the coating industry. Notably, uniformity is the main characteristic for any of these applications.

Solution casting is a unique process where it is not reliant upon conventional extrusion and injection moulding machines, yet it incorporates well-mixed constituents like those produced by more traditional methods [30]. Solution casting is a manufacturing process that involves mixing of the solubilized polymer matrix and filler under continuous agitation via stirring, which is then followed by casting into a flat mould and solution drying. Moreover, it is often referred to as the evaporation method where the film is left for one day [58]. Salehifar et al. [65] stated in their findings that the solvent casting process is used in industrial food packaging for a broad range of applications. Accordingly, each polymer matrix has its own co-solvent to assist in dissolving the polymer. Notably, alcohol, chemical solvent, water or dissolving solutions are used in this kind of process as co-solvent [38]. Moreover, the dissolved polymer process can be improved using applied heat or via a readjusted PH condition which will enhance the formation of properties of the thin film matrix.

For the raw materials in the solvent casting method, several prerequisites are required. First, the polymer matrix should be soluble in the inorganic solvent or PH water. Also, the solution needs to be stable with suitable minimum solid substance and viscosity to achieve satisfactory performance [57]. The second prerequisite is the opportunity to produce a homogeneous thin film and removing the film after the drying process [66]. Figure 1 illustrates the solvent casting process to produce

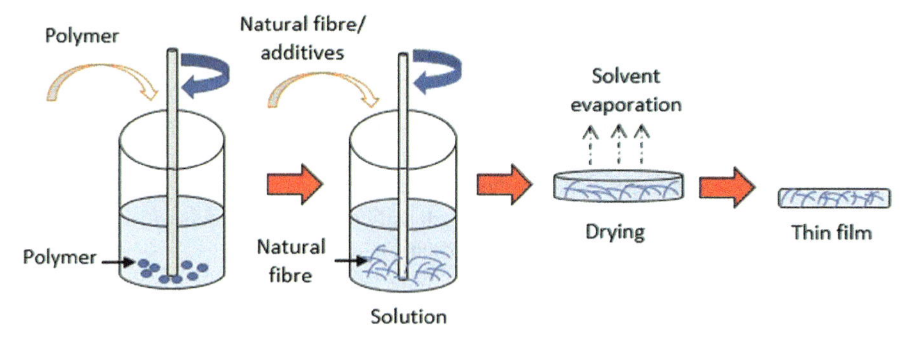

Fig. 1 Process to produce polymer nanocomposites film

biocomposite film. Notably, basic film applications are produced using different polymer and solvent combinations. During the casting process, the solid polymer matrix of various shapes (i.e. granules, powder or pellet size) is gradually dissolved in pure co-solvent. As a precautionary measure, the material type and geometry of the paddle or magnetic stirrer needs to be carefully selected due to the significant difference in viscosity between the polymer and co-solvent which may affect the mixture solvent [69]. Therefore, the temperature and mechanical stirring rate need to be carefully controlled given it affects the quality of solvent evaporation, polymer chain and skin formation of the polymer film. Typically, the induced temperature is varied between room temperature and the boiling point of the solvent [30]. Also, the dissolution time usually takes several iterations which are dependent on the type of solvent and method used.

The advantages of the manufacturing process of polymer solution casting over traditional film extrusion methods are mainly due to the unique approach and without the need to apply thermal or mechanical stress. Therefore, the degradation process or adverse side reaction is considered to be significant. Accordingly, solution casting is becoming an attractive process where the production of film offers uniform thickness distribution, maximum optical purity and extremely low toxicity [69]. Additionally, the process is conducted at a low temperature where it is relevant for thermally activated films although, in some circumstances, it may be temperature sensitive with active constituents. According to Salit et al. [66], due to the process being conducted at low temperature, the thin film shelf life can be extended to incorporate a much longer period. Furthermore, the non-melting process from soluble raw materials will possibility produce high-temperature resistant films. Notably, the main advantages of solvent casting are primarily due to the total cost of manufacturing prototypes and production volumes, which are much less than in extrusion blending. Also, this includes an inexpensive mould with a standard manufacturing/production line [57].

Next, a case study is used to examine nanocellulose reinforced with biodegradable polymer matrix where the focus is on the polymer process and to achieve outcomes. Jiang et al. [27] reported on cellulose nanowhisker

(CNW) reinforced poly(3-hydroxybutyrate) (PHB) by solvent casting and extrusion blending. Polyethylene glycol (PEG) acts as a compatibiliser or dispersion agent, and N, N-dimethylformamide (DMF) is co-solvent for this composite. Firstly, CNW/PEG was dissolved in a DMF stable suspension for 4 h of sonication at 258 °C. PHBV was also dissolved in DMF, and the mixture continued to be sonicated further for 2 h for complete homogeneity, followed by the solvent mixture casted on a glass plate. The film PHBV/CNW composite was formed following the complete evaporation of DMF. Finally, the sample was removed from the glass plate and kept in a desiccator to maintain a constant humidity. The homogeneous dispersion of CNW posed a significant challenge in preparing the nanocomposite, due to the possibility of hydrogen bonding inducing agglomeration on the sample.

The solvent casting technique demonstrated better dispersion of CNW in the nanocomposite compared to the extrusion method in which good dispersion significantly improved the properties of the PHBV nanocomposite [27]. PEG was also added to increase the dispersion in the nanocomposite. The addition of CNW also enhanced the PLA molecular mobility. Whisker content from 0 to 30 wt% also increased the tensile strength and Young's modulus obtained. Therefore, the enhanced tensile properties indicate good dispersion and strong interfacial adhesion between the fiber and the polymer matrix. The water resistance of the composite also increases with whisker content. By comparing the preparation method, the result for the composite prepared via extrusion showed lower tensile properties as compared to solution casting. Agglomeration of cellulose nanowhisker was reported to be observed under the scanning electron microscopy (SEM) micrograph and no intimate contact between cellulose nanowhisker and polymer matrix even though with the presence of a compatibilizer. Notably, the influences of agglomeration decreased the performance of the composite.

In conclusion, solution casting has numerous benefits compared to other fabrication methods due to lower production temperature required and no mechanical stress. Medical manufacturers are producing materials such as silicone urethane, via solution casting to produce breakthrough products subsequently excel in manufacturing performance and shorter lead times in supplying the demands. Solution casting as a manufacturing process depends only on the reaction between the co-solvent and polymer or the natural fiber mixture [53]. One of the primary reasons is due to solvent casting requiring lower energy, thereby reducing costs in recovering the solvent, and only a small investment is necessary for installing the facilities needed in handling the solvents and dope solutions which differs from other manufacturing methods [30]. Therefore, high quality of the thin film can only be achieved using the solvent casting methods and not using other methods.

2.2 *Polymerization*

Polymers are basically made of a large number of repeating unit of monomers. There are various methods that can be used to produce the polymers; step-growth polymerization method (condensation polymer) and chain growth polymerization method (cationic polymerization and anionic polymerization). Depends on the constituents of the monomer, homopolymer and copolymers can be prepared. For an example, homopolymer polypropylene (PP) has been used extensively in food packaging industry as this polymer is known for its' exceptional properties [8]. PP is polymerized from its monomer which is propylene through Ziegler-Natta polymerization and by metallocene catalysis polymerization. Despite its low impact strength at low temperatures and high gas permeability, PP is reinforced with other components such as fillers, inorganic materials and many more in order to improve its properties. Previous studies show significant improvement in gas barrier properties of PP when blended with montmorillonite (MMT) by a twin-screw extruder [11, 93] and coating with corn zein nanocomposite [34].

Another polymer extensively used as packaging material is polyethylene terephthalate (PET). PET is produced by a step-growth polymerization of ethylene glycol and terephthalic acid or dimethyl terephthalate [20]. PET exhibit exceptional properties such as strength, permeability, chemical resistance and high transparency making it suitable for many applications. PET is also blended with other components such as clay by stretch blow molding machine in order to study the migration of aluminium and silicon from the composite into acidic food simulant bottle [20]. The study shows that the migration of aluminium and silicon is dependent on storage time and temperature. Another study had produced a nanocomposite by blending PET with layered double hydroxide by using high energy ball milling process and the results obtained showed that the oxygen diffusion and permeability are lower than that of neat PET [80].

Both PP and PET are known as non-degradable polymers. Despite its non-degradability, they are extensively used as food packaging material as they possess suitable properties for application in packaging. Nowadays, the biopolymer is emerging to substitute this conventional plastic as they can degrade naturally into the environment [20]. Biopolymers are categorized into their synthesis process namely from biomass product, microorganism, biotechnology and oil-products. Starch can be used to form a biodegradable film and different sources of starch can be obtained from plant biomass such as corn or sugarcane [81]. A study done by Heydari et al. [23] prepared corn starch film with glycerol by casting method where with increasing glycerol content, the tensile strength decreases. Chitosan is another polymer synthesize from chitin which is the most abundant agro-polymer in nature next to cellulose. Chitosan nanocellulose is developed in previous studies in order to improve its anti-microbial properties in food packaging, for extending the shelf life of meat in particular [14].

Poly hydroxyalkanoate (PHA) is a biopolymer synthesize from a microorganism. The homopolymer of PHA, poly(3-hydroxybutyrate) (PHB) and copolymer hydroxyl-valerate (HV) have been extensively studied to improve their properties [81]. From a previous study, PHA is combined with nano keratin by using two methods, direct melt compounding and pre-incorporated into an electrospun masterbatch of PHA and also solution casting [31]. Barrier properties are enhanced by both methods while sample prepared by solvent casting shows good adhesion.

Another study had been done by incorporating polyhydroxybutyrate-co-hydroxyvalerate (PHBV) with cellulose nanowhiskers using electrospinning techniques and showed improve barrier properties despite the extreme brittleness [31]. In another study, PHB had also been fabricated with cellulose nanocrystal by using solution casting technique and the results from the study show enhance gas barrier and migration properties [67]. Previous studies had also blended Poly (3-hydroxybutyrate-co-3-hydroxyvalerate) with other components such as clay [31], zinc oxide [15] and cellulose nanocrystal [92].

PLA is a polymer formed by ring opening polymerization of lactic acid [81]. PLA is considered a renewable material as it is manufactured by fermentation of renewable agriculture resources. Despite the attention towards PLA for its commercialization towards substituting conventional plastic, PLA brittleness and high cost limit its application. Many research had been done to modify PLA in order to widen its applications. PLA/zinc oxide film prepared by twin-screw extruder proven to improve PLA mechanical properties [44]. Other than that, PLA had also been blended with MMT by using a twin-screw extruder to form a nanocomposite. Comparing to neat PLA, the study shows that PLA/MMT exhibit good mechanical and oxygen barrier properties. Another study had been done by Pinto et al. [59] where PLA is incorporated with graphene by using solvent casting method. This study also shows improvement in mechanical and oxygen barrier properties.

Poly(e-caprolactone) (PCL) is a synthetic biopolymer that can be synthesized either by ring opening polymerization from ε-caprolactone (monomer) or by free radical ring opening polymerization of 2-methylene-1-3-dioxepane [6]. Because of its high cost and brittleness, PCL applications is limited to medical purpose. However, many researchers have attempted to modify PCL by blending it with other components in order to widen its applications to other fields such as in food packaging sector. Beltrán et al. [6] had done a study on blending PCL with hydroxytyrosol (HT) and a commercial montmorillonite by melt blending method. The presence of montmorillonite decrease oxygen permeability but enhance the elasticity PCL. Another study had developed an antimicrobial PCL/clay nanocomposite films by melt blending method [90]. Results obtained from the study shows that mechanical water vapour barrier properties and antimicrobial properties improved for the nanocomposite, showing that it is suitable for the application in food packaging.

PVA is another type of synthetic biopolymer. Unlike other polymers, PVA is not synthesized from its monomer, instead, PVA is prepared by partial or complete hydrolysis of polyvinyl acetate to remove acetate groups [10]. According to Butnaru et al. [10], for PVA to be used in applications such as food sector and pharmaceutical, PVA need to be cross-linked first by the freeze-thawing method. Butnaru et al. [10] also had done a study to prepare a nanocomposite by blending PVA with chitosan and clay. The study shows enhanced properties in terms of thermal stability, mechanical and antimicrobial properties. A previous study had also been done where PVA had been blended with cellulose nanocrystal by a solvent casting method and the result proven to improve its mechanical properties [18] (Table 1).

Table 1 Summary of nanocomposites preparation in food packaging industry

No.	Nanocomposite	Processing method	References	Remarks
1	Polypropylene blended with montmorillonite (MMT) (PP/MMT)	Polypropylene (PP): Ziegler-Natta polymerization and by metallocene catalysis polymerization	Zehetmeyer et al. [93]	Improve gas barrier properties
2	Polypropylene/Organoclay Nanocomposites	Polypropylene (PP): in situ polymerization supported Ziegler-Natta catalysts	Almeida et al. [1]	Improve thermal degradation
3	PET blended with clay	Polyethylene terephthalate (PET): step-growth polymerization of ethylene glycol and terephthalic acid or dimethyl terephthalate	Galotto and Ulloa [20]	Migration of aluminium and silicon is dependent on storage time and temperature
4	PET blended with layered double hydroxide (PET/LDH)		Tammaro et al. [80]	Oxygen diffusion and permeability is lower than that of neat PET
5	Corn starch film with glycerol	Casting method	Heydari et al. [23]	With increasing glycerol content, the tensile strength decreases
6	Chitosan nanocellulose	Casting method	Dehnad et al. [14]	Improve its anti-microbial properties in food packaging (extending the shelf life of meat)
7	PHA combined with nano keratin	Poly hydroxyalkanoate (PHA): synthesized from a microorganism (fermentation)	Lagarón et al. [31]	Barrier properties are enhanced by both methods while sample prepared by solvent casting shows good adhesion
8	PHB fabricated with cellulose nanocrystal	Poly(3-hydroxybutyrate) (PHB): synthesize from a microorganism (fermentation)	Sengupta et al. [67]	Enhance gas barrier and migration properties

(continued)

Table 1 (continued)

No.	Nanocomposite	Processing method	References	Remarks
9	Polyhydroxybutyrate-co-hydroxyvalerate (PHBV) with cellulose nanowhiskers	Polyhydroxybutyrate-co-hydroxyvalerate (PHBV): synthesized from mixed microbial cultures	Lagarón et al. [31]	Improve barrier properties despite the extreme brittleness
10	Poly (3-hydroxybutyrate-co-3-hydroxyvalerate) blended with clay		Lagarón et al. [31]	Improve thermal stability
11	Poly (3-hydroxybutyrate-co-3-hydroxyvalerate) blended with zinc oxide		Diez-Pascual and Diez-Vicente [15]	Improve stiffness, strength, toughness, and glass transition Temperature as well as a reduction in water uptake and oxygen and water vapour permeability
12	Poly (3-hydroxybutyrate-co-3-hydroxyvalerate) blended with cellulose nanocrystal		Yu et al. [92]	Improve mechanical performance, thermal stability, barrier, and migration properties, improve PLA mechanical properties
13	PLA/ZnO film	Polylactic acid (PLA): Ring-opening polymerization (ROP) of lactic acid	Marra et al. [44]	
14	PLA blended with montmorillonite (MMT) (PLA/MMT) PLA blended with graphene		Pinto et al. [59]	PLA/MMT exhibit good mechanical and oxygen barrier properties Improvement in mechanical and oxygen barrier properties
	PCL blended with hydroxytyrosol (HT) and a commercial montmorillonite PCL/clay nanocomposite films	Poly(e-caprolactone) (PCL): ring opening polymerization from ε-caprolactone (monomer) or by free radical ring opening polymerization of 2-methylene-1-3-dioxepane	Beltrán et al. [6]	Presence of montmorillonite decrease oxygen permeability but enhance the elasticity PCL
			Yahiaoui et al. [90]	Mechanical, water vapour barrier properties and antimicrobial properties improved
15	PVA blended with cellulose nanocrystal	Poly(vinyl alcohol) (PVA): partial or complete hydrolysis of polyvinyl acetate to remove acetate groups	Fortunati et al. [18]	Improve its mechanical properties
16	Polyaniline/polyvinyl alcohol/Ag (PANI/PVA/Ag)	Poly(vinyl alcohol) (PVA): in situ chemical oxidation polymerization of aniline monomer	Ghaffari-Moghaddam and Eslah [21]	Improve antibacterial properties

2.3 Melt Mixing

Amongst the three methods, melt mixing is the most economically attractive and scalable method for dispersing nanoparticles into polymers. To some extent, this method is also preferable over other two methods because it aligns well with the currently established industrial processing routes. This method also can be considered as environmentally benign, due to the unnecessity to use organic solvents. Melt mixing process involves melting of the thermoplastic polymer matrix to form viscous liquid followed by incorporation of nanoparticles.

In order to ensure proper dispersion of the nanoparticles during melt processing, two key factors need to be considered, i.e. (i) favourable interaction between the polymer matrix and the nanoparticle, and (ii) suitable processing conditions [50]. The favourable enthalpic interaction between particles and the matrix is necessary, because, in the absence of such favourable interactions, the dispersion of the nanoparticles within the matrix becomes difficult, and may only result in the formation of micro-composites instead of nanocomposites. Meanwhile, the processing conditions are also needed to be controlled carefully. Although high shear mixing in principle can improve the dispersion states of the particles, too high mechanical shearing, combined with high temperature and possible oxidation, may degrade certain types of polymers, especially biopolymers [50].

There are numerous works that reported successful attempts in preparing PLA based nanocomposites by melt mixing method. PLA nanocomposites reinforced with various types of particles such as silica [32, 86, 94], clay [33, 72], carbon nanotube [87], graphene [28, 46], titanium dioxide (TiO_2) [17] and nanocellulose [2, 22] have been successfully prepared by melt mixing process. In most cases, nanoparticles were surface modified to facilitate favourable interaction with the polymer matrix, which leads to better dispersion states. For example, –COOH surface modified multi-walled carbon nanotube (MWNTs) displayed better dispersion in the PLA matrix than the unmodified MWNTs [87], or polyethylene glycol methyl ether (PEGME)-modified nanosilica exhibited better dispersion states in the PLA matrix compared to pristine nanosilica.

Beside surface modification of nanoparticles, processing conditions in melt mixing method also affect the dispersion state of the particles in PLA nanocomposites. Villmow et al. [83] showed that higher rotating speed (500 rpm compared to 100 rpm), more mixing elements in extruder screw configuration, and temperature profile with raising value towards the extruder die contributed to better dispersion of CNT inside PLA matrix. Okubo et al. [52] exhibited that dispersion of cellulosenano- or microfibers was improved with decreased gap distance in roll-milling, while Oksman et al. [51] also pointed out that the type of screw in extruder influence the dispersion states of nanoparticles inside the matrix, i.e. in case of nanocellulose processing, the co-rotating twin-screw extruders are preferred over counter-rotating, because they are better for mixing and dispersing. In principle, higher mechanical shearing may lead to better dispersion of nanoparticles inside the matrix.

Preparation of PHA family based nanocomposites by melt mixing method has also been reported. The examples include PHBV reinforced with carbon nanotube (CNT) [68], PHBV reinforced with hydroxyapatite [54], PHB/clay nanocomposites [9, 39], etc. Similar to PLA based nanocomposites, surface chemistry of nanoparticles play a critical role in helping good dispersion. Organo-modified MMT dispersed better compared pristine MMT inside PHB matrix [9]. In PHBV matrix, hydroxyapatite treated with silane coupling agent exhibited better dispersion states and mechanical properties compared to untreated hydroxyapatite particles [54]. However, since many of PHA are sensitive to thermo-mechanical degradation, the melt-processing of PHA nanocomposites occurred within a short window and caution should be taken be taken to avoid degradation during processing [9, 50].

3 Characterization of Nanocomposites

3.1 Mechanical Property

Mechanical characteristic is one of the crucial factors in nanocomposite especially in designing food packaging. Chemical resistance of the product can be affected by its process-ability and mechanical properties such as tensile strength, tear strength, elongation, toughness, softness and puncture resistance [5]. So, it is mandatory to perform the suitability test for nanocomposite food packaging stored with food as a function of duration. For example, the tensile test is conducted to determine the tensile strength, Young's modulus, yield stress and elongation at break of the nanocomposite. Generally, this testing is carried out according to the ASTM D88 standard testing method for tensile properties of thin plastic sheeting and ASTM D638 for testing materials up to 14 mm of thickness.

Moreover, impact properties test is also conducted to determine the amount of energy needed for the plastic deformation at certain condition according to ASTM D1709 for plastic film. Impact test consists of Charpy and IZOD specimen configuration according to ASTM D256 and ASTM 6110, respectively. The flexural test measures the force required to bend the sample under three-point loading conditions according to ASTM D790. This test is conducted as an indication of a material's stiffness when flexed. The sample lies on a support span and the load is applied to the centre by the loading nose producing three-point bending.

The modification of polymer content leads to increase in tensile performance up to 50% of the original value [5]. Nanocomposite film with the addition of water-soluble chitosan (WSC) increased the tensile strength of the polymer [78]. Relative elongation is also an important mechanical property as it could affect the biodegradable rate during the degradation process. As studied by Mostafa et al. [45], the percentage of elongation values for starch blend film was reduced in inoculated soil faster than in non-inoculated soil. Elongation of starch blend film dropped rapidly by 56% in inoculated soil compared to 12% decrement in elongation within the fixed time basis. Therefore, mechanical stability properties help to improve the new innovations in biobased food packaging for commercial use.

3.2 *Thermal Property*

Addition of nanoparticles into polymeric materials also may enhance the thermal properties. Higher thermal stability is important in food packaging application to withstand a wide range of temperature exposure during food processing, transportation, and storage [55]. A recent report by De Silva et al. [12] showed that incorporation of magnesium oxide (MgO) into chitosan exhibited higher thermal stability and fire retardancy, due to the high thermal stability of MgO along with the homogeneous distribution of MgO nanoparticles in chitosan films. Kim and Cha [29] also reported that addition of organically modified MMT nanoclays into ethylene vinyl alcohol copolymer (EVOH) resulted in enhancement of thermal stability as indicated by thermogravimetric analysis (TGA), as well as shifting the crystallization temperature of EVOH as displayed by differential scanning calorimetry (DSC) result.

3.3 *Degradation Behaviour*

Development of new food packaging materials that capable to be degraded in a controlled manner has been a focus of interest of researchers nowadays. Incorporation of nanoparticles into degradable polymeric matrix may affect the degradation behaviour of the composites. Addition of cellulose nanocrystals (CNCs) into PLA helps to enhance the disintegrability rate of PLA since this nanocellulose is equipped with hydrophilic nature [3, 19]. However, different observation has been found when PLA is added with coated CNC with surface surfactant (s-CNC). Even though s-CNC accelerate the disintegration process of PLA, however, water diffusion is slightly restricted due to the improvement in barrier properties and therefore slowdowns the hydrolysis process and delays the degradation process [19]. It can be concluded that the addition of appropriate plasticizers will help to improve the disintegration of PLA in composting conditions [3, 35], which indirectly speed up the disintegration of PLA-plasticizer nanocomposites.

In other literature reported by Pandey et al. [56], the addition of addition of nano-layered silicate into PLA matrix increased the biodegradation rate of the materials. The increment was predicted due to the presence of terminal hydroxylated edge groups in the silicate layers. Addition of 4% of nano-layered silicate lead to the good dispersion of silicate layer in PLA matrix and these hydroxy groups begin the heterogeneous hydrolysis of PLA matrix by absorbing water from compost. Up to one month, the weight loss and hydrolysis of PLA and PLA with 4% filler is about the same, however, after one month, the degradation of nanocomposites increased significantly.

3.4 Migration Testing

Migration is the process where chemicals are transferred from nanocomposites to food as they come in contact with each other. The term "migration", according to Icoz and Eker [26], is used to define the diffusion of substances from high concentration regions to low concentration regions within the food and the packaging material. The chemical substances that are added to polymeric materials to enhance the characteristics of a composite may interact with the food components and move into the food during the processing, storage, and distribution. The quality, such as taste and odour, as well as safety of the food will be affected and cause significant health problems to the consumer. There are various factors that influence the rate of migration, such as the contact time, temperature, structure of the nanocomposite (thickness for plastics), chemical properties (molecular size, polarity, vapor pressure, etc.) of the migrants, and the types of materials that come in contact with the food [7, 26]. Migration can occur in different ways such as contact migration, gas phase migration, penetration migration, set-off migration, and condensation or distillation migration [26].

Generally, this testing is carried out according to the EN 1186 standard testing method, and several types of food simulants are used to analyze the overall migration testing. These food simulants are chosen as they are less chemically complex than foods [7] according to the regulations by EU 10/2011. The recommended simulants, as stated by Bhunia et al. [7], are: ethanol (10% v/v) (simulant A) to simulate aqueous foods (pH > 4.5); acetic acid (3% v/v) (simulant B) to simulate acidic foods; ethanol (20% v/v) (simulant C) to simulate alcoholic products; ethanol (50% v/v) (simulant D1) and vegetable oil (simulant D2) to simulate fatty foods, and lastly, Tenax (PPPO) (simulant E) to simulate dry food. The overall migration is limited to 10 mg/dm^2 on a contact area basis or 60 mg/kg in the stimulant or food (for plastics).

The food contact material or migration testing is important especially in producing food packaging as it affects the quality and shelf life of food and has a bad impact on human health. The substances that migrate into food can transform to become toxic and hazardous to consumers, especially babies and children. The composition of the materials that come in contact with food, such as additives and plasticizers, may affect the safety and quality of food. They affect the human reproductive system and are carcinogenic as both of them are described as endocrine disruptors. Phthalate esters, alkyl phenols (APs) and 2,2-bis (4-hydroxyphenyl) propane, also known as bisphenol A, are examples of hazardous chemicals that have serious toxic effects if exposed to the human body even at low concentrations [85].

Therefore, migration testing is very important in food packaging applications in order to maintain the safety and quality of the food and also consumer health. It is necessary to choose carefully the materials and their properties to protect the quality and safety of food.

3.5 Antimicrobial Testing

Nanomaterial has revealed many benefits in various fields. As the uses of nanomaterials have been increasing, it has been found to be a promising advancement for the food packaging productions in the global market. However, the use of nanomaterials in food packaging could be challenging due to the method in reducing particle size and the fact that characteristic of nanoparticles material is quite different from microparticles [24]. Food packaging offers to maintain the quality of food taste, prolong shell life and protect food from contamination. However, food security is an excessive problem due to increasing disease from contamination adverse from food packaging materials. World Health Organization (WHO) stated the spoilage of food arise at any time during production of food to consume. Pathogens bacterial are to blame in many cases because it easily grows on food surface [84]. This leads to the introduction of nanocomposite as an active antimicrobial packaging and particularly designed to control the bacteria from adversely affect health. By using active packaging, the internal environment is modified via constant interaction with the food over the specified shell life [41].

Nanoscale materials consist of a high surface to volume ratio than micro-size particles which make them easier to attach with a numerous number of the molecule and increased the efficiency [88]. Nanoparticles such as copper, silver and zinc oxide usually used in the retarded growth of microorganism due to strong antibacterial or antifungal activities. The reinforcement of this kind nanoparticles in biocomposite may overcome the problem of food spoilage. So, the main goal of active packaging enhances the packaging products quality by guarantee food security and improve sensory properties of food while sustaining food quality [61]. Active antimicrobial packaging may be described as a structure that alters the environment inside the plastic package by modifying the condition of the packaged food system. Food spoilage can be formed when it is exposed to air via processing, production, and packaging of food [70]. Generally, active food packaging is divided into two categories; biodegradables and non-biodegradable materials. Petroleum-based packaging has particularly benefit of high mechanical strength, low cost of raw material, availability, ease production process, good barrier properties but not biodegradable [84]. They cause environmental pollution as they can take hundreds of years to completely degrade.

Besides, petroleum-based packaging material can also preserve food over the particular period but, currently, consumers do prefer products made of biodegradable resources [76]. Besides, those packaging have chemical additives diffused through the plastic barrier into food products thus provides a negative effect on the health. Biopolymer materials have attracted widespread industries because they can be decomposed easily. Therefore, utilization of antimicrobial material in biopolymer packaging can provide significant enhancement of property which can help to increase health security and shelf life of food. It is one of the methods that effectively kill or inhibit pathogenic microorganism from growth in infected food [62]. Hence, a good food packaging able to reduce contamination over the food [44, 64].

The improvement of biopolymer-based nanocomposites with antibacterial agent presents interesting potentials because the bio-based polymer can be varied according to specific technology requirements and also fulfil the nanostructures with envisaged applications [60]. The antimicrobial elements most commonly introduced into packaging products can be characterized into many types such as plant oil extract, enzymes, bacteriocin, preservation, chemical substance and others [76, 84]. Previous literature have recognized the potential application of antimicrobial nanocomposite for wound dressing [89], bactericides [60] and medical devices [16].

Nanoparticles have different activities depending on pathogenic and spoilage microorganism species [44]. Pinto et al. [60] reported based on their finding in antimicrobial activity test, the presence of nanoparticle copper (0.93–4.95%) can against *K. pneumoniae* (Gram-negative bacteria) in cellulose nanocomposite specimens. The inhibition of bacteria depends on the nanofiller content in the specimens, where increasing copper content gave significant inhibited bacteria growth. This study believed that Gram-negative bacteria can be affected by copper-based material as they have a strong antimicrobial action against peptidoglycan layer. The interaction of nanoparticles could cause in changing its permeability and the structures promote the membrane degradation which ultimately causing the death of bacteria [25]. According to Lai et al. [33], the incorporated nanoclay into PVA film showed strong antimicrobial activity against Gram-positive bacteria (*L. moncytogenes* and *S.aureus*). Nanoclay was known as an active agent due to its highly dispersed in biocomposite and increased their exposure toward microorganism [37]. These may result in deteriorated bacterial cell membranes and cause cell lysis. Current food technology used antimicrobial in covering substances because this method will make sure food product are safe and lead to extend shell life [61]. Therefore, selection of removal of any microorganism should be identified for selection of an appropriate antimicrobial agent.

The main function of nanomaterials to act as an antioxidant agent is, it consists of microbial cidal effects and microbial static effects on the microorganism. Microbial static effects related to the active function of substance to enhance the concentration below minimal inhibitory during preserved or storage process of food product [41]. This state may reduce perishable to occur by contamination. The main concept permitted in this packaging are wrapping system with presence of antimicrobial agent either natural or synthetic agents. The existence of an antimicrobial agent in packaging product is more significant than adding antimicrobial agent directly onto food, the fact of covered product that consists antimicrobial agent could not diffuse into food product as well as can be thrown away after its' usage [25]. The major challenge in food packaging is the design of nanocomposite material with active antimicrobial products. Lately, many customers concern about potential health risks when chemical additives are added to packaging products. The achievement by scientists with the new formulation of reinforced nanoparticles verified the potential of bio-nanocomposites as an active food packaging, can be considered as an alternative to conventional composites used in the packaging material.

3.6 Optical Behaviour

Greater lightness and transparency in food packaging are vital as they are in high demand by consumers. Most non-composite foods packaging are colourless and transparent. These characteristics, however, are slightly affected by the addition of other materials, which in the end influence the aesthetic value of the food packaging. Arrieta et al. [4] revealed that the lightness of PLA is 94.08 ± 0.07, while that of PLA/limonene is 93.86 ± 0.10. Both PLA and PLA/limonene have high brightness characteristics, but their lightness is reduced once additives are added. In other researches, the lightness and transparency of coated films for food packaging were affected in the presence of proteins and plasticizers.

Lee et al. [34] stated that the transparency of protein-coated films was increased from 16.9 to 19.7 for corn zein, 13.3 to 30.3 for soy protein isolate and 17.8 to 28.6 for whey protein isolate. Moreover, the transparency of protein-coated films also increased with the addition of plasticizers. Yoo and Krochta [91] compared the transparencies of biopolymer, biopolymer blended and synthetic polymer films. The results indicated that the percentage of transparency of biopolymers like hydroxyl propyl methyl cellulose (HPMC) films was higher compared to synthetic polymers such as low-density polyethylene (LDPE), high-density polyethylene (HDPE) and polypropylene (PP) films.

Generally, the optical properties of nanocomposites are evaluated according to the *CIELAB* colour system. In this system, the colour coordinates L^*, a^* and b^* represent lightness, red-green and yellow-blue, respectively. An increase in the L^* value indicates that the nanocomposite specimen is lightening. The colour is switching to red if the* value is increasing, while it is turning green if the value is decreasing. The increase in the value of b^* shows that the colour is switching towards yellow, and a decreasing value indicates that it is turning to blue. Therefore, the total colour difference (ΔE) can be determined by using Eq. 1.

$$\Delta E = (\Delta L^2 + \Delta a^2 + \Delta b^2)^{1/2} \tag{1}$$

Transparency and opacity are essential properties in food packaging application as they improve the appearance of a product. Opacity is the degree to which light is not allowed to travel through. Products with high transparency, especially in packaging, have become more popular in the market as consumers prefer see-through rather than opaque packaging. Thus, nanocomposites with high transparency and lightness are important in food packaging applications as there is a high demand for them from consumers.

3.7 Permeability and Barrier Properties

The determination of the barrier properties of the nanocomposite is important as it determines the shelf-life of the product. Various factors may affect the barrier properties to gases and vapours on the nanocomposite such as environmental conditions like temperature and relative humidity. Permeability according to Siracusa [73], stated that the ability of the gas or vapour transmits through a resisting material. Permeates diffusion is influenced by the structure of the nanocomposite, permeability to specific gases or vapour, area, thickness, temperature, the difference in pressure or concentration gradient across the nanocomposite.

Oxygen transmission rate (OTR), water vapour transmission rate (WVTR) and carbon dioxide transmission rate (CO_2TR) are main examples of the barrier properties that used in food packaging application [74]. The oxygen permeability coefficient (OPC) exhibits the amount of oxygen that permeates into packaging materials in term of per unit area and time ($\mathrm{kg\,m/m^2\,s\,Pa}$) [73, 74]. When the nanocomposite film packaging has lower oxygen permeability coefficients, it prolongs the product's lifespan as the pressure of the oxygen inside the packaging decrease to the point where it slows down the oxidation. The OPC is correlated to the OTR by using Eq. (2)

$$OPC = \frac{OTR \cdot l}{\Delta P} \tag{2}$$

For water vapour permeability coefficient (WVPC) and carbon dioxide permeability coefficient (CO_2PC), both indicate the amount of water vapour and carbon dioxide, respectively that permeates per unit area and time as well ($\mathrm{kg\,m/m^2\,s\,Pa}$). The WVPC and CO_2PC are correlated to the WVTR and CO_2TR by using Eqs. (3) and (4), respectively.

$$WVPC = \frac{WVTR \cdot l}{\Delta P} \tag{3}$$

$$CO_2PC = \frac{CO_2TR \cdot l}{\Delta P} \tag{4}$$

where l is the thickness of the nanocomposite film and ΔP is P_1–P_2, P_1 is the gas partial pressure at the temperature test on the test side and P_2 is equal to zero on detector side.

In order to improve barrier properties to gasses, vapour and aromas due to the sensitivity of various food products to oxygen degradation, microbial growth stimulated by moisture and aroma for keeping the food quality, nano or microfilters can be incorporated into biopolymers food packaging. Mali et al. [42] stated that the oxygen permeability was decreased while mechanical and heal seal properties were improved in bio-nanocomposite films from sago starch and bovine gelatin with nanorod-rich zinc oxide as the nanofillers.

Thus, oxygen and water vapour permeabilities in nanocomposite are essential properties as they give a major impact on the shelf-life of fresh and processed foods. It is important to know if there is an interaction between food and packaging and factors that influence the transport mechanism through the material.

Acknowledgements The authors wish to thank Ministry of Education Malaysia for the Fundamental Research Grant Scheme, FRGS14-105-0346, FRGS14-108-0349, FRGS16-003-0502 and RIGS16-085-0249 for the financial support and International Islamic University Malaysia for the facilities and equipment in making these studies a success.

References

1. Almeida LA, Marques MDFV, Dahmouche K (2015) Synthesis of polypropylene/organoclay nanocomposites via in situ polymerization with improved thermal and dynamic-mechanical properties. J Nanosci Nanotechnol 15(3):2514–2522
2. Arias A, Heuzey MC, Huneault MA et al (2015) Enhanced dispersion of cellulose nanocrystals in melt-processed polylactide-based nanocomposites. Cellulose 22(1):483–498
3. Arrieta MP, Fortunati E, Dominici F et al (2014) PLA-PHB/cellulose based films: mechanical, barrier and disintegration properties. Polym Degrad Stab 107:139–149
4. Arrieta MP, Lopez J, Ferrandiz S et al (2013) Characterization of PLA-limonene blends for food packaging applications. Polym Testing 32(4):760–768
5. Badmus AA, Gauri S, Ali NI et al (2015) Mechanical stability of biobased food packaging materials. Food Sci Q Manag 39:41–47
6. Beltrán A, Valente AJM, Jiménez A et al (2014) Characterization of poly(ε-caprolactone)-based nanocomposites containing hydroxytyrosol for active food packaging. J Agric Food Chem 62(10):2244–2252
7. Bhunia K, Sablani S, Tang J et al (2013) Migration of chemical compounds from packaging polymers during microwave, conventional heat treatment and storage. Compr Rev Food Sci Food Saf 12:523–545
8. Boone Lox F, Pottie S (1993) Deficiencies of polypropylene in its use as a food-packaging material—a review. Packag Technol Sci 6(June):277–281
9. Bordes P, Pollet E, Bourbigot S et al (2008) Structure and properties of PHA/clay nano-biocomposites prepared by melt intercalation. Macromol Chem Phys 209(14):1473–1484
10. Butnaru E, Cheaburu CN, Yilmaz O et al (2016) Poly(vinyl alcohol)/chitosan/montmorillonite nanocomposites for food packaging applications: influence of montmorillonite content. High Perform Polym 28(10):1124–1138
11. Choi RN, Cheigh CI, Lee SY et al (2011) Preparation and properties of polypropylene/clay nanocomposites for food packaging. J Food Sci 76(8):62–67
12. De Silva RT, Mantilaka MMMGPG, Ratnayake SP et al (2017) Nano-MgO reinforced chitosan nanocomposites for high performance packaging applications with improved mechanical, thermal and barrier properties. Carbohydr Polym 157:739–747
13. Deepa B, Abraham E, Cherian BM et al (2011) Structure, morphology and thermal characteristics of banana nanofibers obtained by steam explosion. Bioresour Technol 102(2): 1988–1997
14. Dehnad D, Mirzaei H, Emam-Djomeh Z et al (2014) Thermal and antimicrobial properties of chitosan-nanocellulose films for extending shelf life of ground meat. Carbohydr Polym 109:148–154

15. Diez-Pascual AM, Diez-Vicente, Angel L (2014) ZnO-reinforced poly (3-hydroxybutyrate-co -3-hydroxyvalerate) bionanocomposites with antimicrobial function for food packaging. Appl Mater Interf 9822–9834
16. Feldman D (2016) Review-polymer nanocomposites in medicine. J Macromol Sci, Part A: Pure and Appl Chem 53(1):55–62
17. Fonseca C, Ochoa A, Ulloa MT et al (2015) Poly (lactic acid)/TiO2 nanocomposites as alternative biocidal and antifungal materials. Mater Sci Eng 57:314–320
18. Fortunati E, Puglia D, Monti M et al (2012) Cellulose nanocrystals extracted from okra fibers in PVA nanocomposites. J Appl Polym Sci 128(5):3220–3230
19. Fortunati E, Rinaldi S, Peltzer M et al (2014) Nano-biocomposite films with modified cellulose nanocrystals and synthesized silver nanoparticles. Carbohydr Polym 101:1122–1133
20. Galotto M, Ulloa P (2010) Effect of high-pressure food processing on the mass transfer properties of selected packaging materials. Packag Technol Sci 23(May):253–266
21. Ghaffari-Moghaddam M, Eslahi H (2014) Synthesis, characterization and antibacterial properties of a novel nanocomposite based on polyaniline/polyvinyl alcohol/Ag. Arab J Chem 7(5):846–855
22. Herrera N, Mathew AP, Oksman K (2015) Plasticized polylactic acid/cellulose nanocomposites prepared using melt-extrusion and liquid feeding: mechanical, thermal and optical properties. Compos Sci Technol 106:149–155
23. Heydari A, Alemzadeh I, Vossoughi M (2013) Functional properties of biodegradable corn starch nanocomposites for food packaging applications. Mater Des 50:954–961
24. Honarvar Z, Hadian Z, Mashayekh M (2016) Nanocomposites in food packaging applications and their risk assessment for health. Electr Phys 8(6):2531–2538
25. Hu W, Chen S, Yang J et al (2014) Functionalized bacterial cellulose derivatives and nanocomposites. Carbohydr Polym 101(1):1043–1060
26. Icoz A, Eker B (2016) Selection of food packaging material, migration and its effects on food quality. In: 1st international conference on quality of life, Center for Quality, Faculty of Engineering, Unviersity of Kragujevac, June 2016
27. Jiang L, Morelius E, Zhang J et al (2008) Study of the Poly(3-hyroxybutyrate-co-3-hydroxyvalerate)/Cellulose Nanowhisker composites prepared by solution casting and melt processing. J Compos Mater 42(24):2629–2645
28. Kim IH, Jeong YG (2010) Polylactide/exfoliated graphite nanocomposites with enhanced thermal stability, mechanical modulus, and electrical conductivity. J Polym Sci, Part B: Polym Phys 48(8):850–858
29. Kim SW, Cha SH (2014) Thermal, mechanical, and gas barrier properties of ethylene–vinyl alcohol copolymer-based nanocomposites for food packaging films: Effects of nanoclay loading. J Appl Polym Sci 131(11):40289
30. Kong I, Tshai KY, Hoque ME (2015) Manufacturing of natural fibre-reinforced polymer composites by solvent casting method. In: Salit MS (ed) Manufacturing of natural fibre reinforced polymer composites. Springer International Publishing, Switzerland, pp 331–349
31. Lagarón JM, López-Rubio A, José Fabra M (2015) Bio-based packaging. J Appl Polym Sci 133(2):42971
32. Lai SM, Hsieh YT (2016) Preparation and properties of polylactic acid (PLA)/silica nanocomposites. J Macromol Sci, Part B 55(3):211–228
33. Lai SM, Wu SH, Lin GG et al (2014) Unusual mechanical properties of melt-blended poly (lactic acid)(PLA)/clay nanocomposites. Eur Polym J 52:193–206
34. Lee JW, Son SM, Hong SI (2008) Characterization of protein-coated polypropylene films as a novel composite structure for active food packaging application. J Food Eng 86(4):484–493
35. Lemmouchi Y, Murariu M, Santos DAM et al (2009) Plasticization of poly(lactide) with blends of tributyl citrate and low molecular weight poly(D, L-lactide)-B-poly(ethylene glycol) copolymers. Eur Polym J 45(10):2839–2848
36. Li X, Gao H, Scrivens WA et al (2005) Structural and mechanical characterization of nanoclay-reinforced agarose nanocomposites. Nanotechnology 16(10):2020–2029

37. Liu G, Song Y, Wang J et al (2014) Effects of nanoclay type on the physical and antimicrobial properties of PVOH-based nanocomposite films. LWT-Food Sci Technol 57(2):562–568
38. Lunineau G, Rahaman A (2012) A review of strategies for improving the degradation properties of laminated continuous-fiber/epoxy composites with carbon-basednanoreinforcements. Carbon 50:2377–2395
39. Maiti P, Batt CA, Giannelis EP (2007) New biodegradable polyhydroxybutyrate/layered silicate nanocomposites. Biomacromol 8(11):3393–3400
40. Majeed K, Jawaid M, Hassan A et al (2013) Potential materials for food packaging from nanoclay/natural fibres filled hybrid composites. Mater Des 46:341–410
41. Malhotra B, Anu K, Harsha K (2015) Review-antimicrobial food packaging: potential and pitfalls. Front Microbiol 6(611):1–9
42. Mali S, Victoria EM, Garcia M et al (2004) Barrier, mechanical and optical properties of plasticized Yam starch films. Carbohydr Polym 56:129–135
43. Marina R, Alfonso J, Mercedes P et al (2014) Development of novel nano-biocomposite antioxidant films based on poly (lactic acid) and thymol for active packaging. Food Chem 1–31
44. Marra A, Silvestre C, Duraccio D et al (2016) Polylactic acid/zinc oxide biocomposite films for food packaging application. Int J Biol Macromol 88:254–262
45. Mostafa HM, Sourell H, Bockisch FJ (2010) The mechanical properties of some bioplastics under different soil types for use as a biodegradable drip tubes. CIGR Ejournal 12:1–8
46. Murariu M, Dechief AL, Bonnaud L et al (2010) The production and properties of polylactide composites filled with expanded graphite. Polym Degrad Stab 95(5):889–900
47. Mushi NE, Utsel S, Berglund LA (2014) Nanostructured biocomposite films of high toughness based on native chitin nanofibers and chitosan. Front Chem 2:99
48. Nascimento P, Marim R, Carvalho G et al (2016) Nanocellulose produced from rice hulls and its effect on the properties of biodegradable starch films. Mater Res 19(1):167–174
49. Neelamana IK, Thomas S, Parameswaranpillai J (2013) Characteristics of banana fibers and banana fiber reinforced phenol formaldehyde composites-macroscale to nanoscale. J Appl Polym Sci 130(2):1239–1246
50. Ojijo V, Sinha Ray S (2013) Processing strategies in bionanocomposites. Prog Polym Sci 38 (10–11):1543–1589
51. Oksman K, Aitomäki Y, Mathew AP et al (2016) Review of the recent developments in cellulose nanocomposite processing. Compos Part A: Appl Sci Manuf 83:2–18
52. Okubo K, Fujii T, Thostenson ET (2009) Multi-scale hybrid biocomposite: processing and mechanical characterization of bamboo fiber reinforced PLA with microfibrillated cellulose. Compos Part A: Appl Sci Manuf 40(4):469–475
53. Olatunji O, Richard O (2016) Processing and characterization of natural polymers. In: Olatunji O (ed) Natural polymers. Springer International Publishing, Switzerland, pp 19–34
54. Öner M, İlhan B (2016) Fabrication of poly (3-hydroxybutyrate-co-3-hydroxyvalerate) biocomposites with reinforcement by hydroxyapatite using extrusion processing. Mater Sci Eng, C 65:19–26
55. Othman SH (2014) Bio-nanocomposite materials for food packaging applications: types of biopolymer and nano-sized filler. Agric Agric Sci Proc 2:296–303
56. Pandey JK, Reddy KR, Kumar AP et al (2005) An overview on the degradability of polymer nanocomposites. Polym Degrad Stab 88:234–250
57. Pazourkova L, Martynkova G, Placha D (2015) Preparation and mechanical properties of polymeric nanocomposites with hydroxyapatite and hydroxyapatite/clay mineral fillers—review. J Nanotechnol: Nanomed Nanobiotechnol 2(007):1–8
58. Petersson L, Kvien I, Oksman K (2007) Structure and thermal properties of poly (lactic acid)/ cellulose Whiskers nanocomposites. Compos Sci Technol 67:2535–2544
59. Pinto AM, Cabral J, Tanaka DAP et al (2012) Effect of incorporation of graphene oxide and graphene nanoplatelets on mechanical and gas permeability properties of poly(lactic acid) films. Polym Int 62(1):33–40
60. Pinto RJB, Daina S, Sadocco P et al (2013) Antibacterial activity of nanocomposites of copper and cellulose. BioMed Res Int, 1–6

61. Qin Y, Yang J, Xue J (2015) Characterization of antimicrobial poly(lactic acid)/ poly (trimethylene carbonate) films with cinnamaldehyde. J Mater Sci V50(3):1150–1158
62. Ramos M, Jiménez A, Peltzer M et al (2014) Development of novel nano-composite antioxidants films based on poly (lactic acid) and tymol for active packaging. Food Chem 162:149–155
63. Saba N, Paridah MT, Mohammad J (2014) Potentiality of nano filler/natural fiber filled polymer hybrid composites. A review. Polymers 2014(6): 2247–2273
64. Saeedeh SA, Hosseini H, Mohammadifar MA et al (2014) Characterization of k-carrageenan films incorporated plant essentialoils with improved antimicrobial activity. Carbohydr Polym 101:582–591
65. Salehifar M, Mohammad Hadi BN, Reza A et al (2013) Effect of LDPE/MWCNT films on the shelf life of Iranian Lavash bread. Eur J Exp Biol 3(6):183–188
66. Salit MS, Jawaid M, Yusoff NB, Hoque ME (2015) Manufacturing of natural fibre reinforced polymer composites. Springer, Switzerland
67. Sengupta R, Chakraborty S, Bandyopadhyay S et al (2007) A short review on rubber/ clay nanocomposites with emphasis on mechanical properties. Engineering 47:21–25
68. Shan GF, Gong X, Chen WP et al (2011) Effect of multi-walled carbon nanotubes on crystallization behavior of poly (3-hydroxybutyrate-co-3-hydroxyvalerate). Coll Polym Sci 289(9):1005–1014
69. Siemann U (2005) Solvent cast technology—a versatile tool for thin film production. Prog Coll Polym Sci 130:1–14
70. Singh P, Ali AW, Sven S (2011) Active packaging of food products: recent trends. Nutr Food Sci 41(4):249–260
71. Sinha RS, Bousmina M (2005) Biodegradable polymers and their layered silicate nanocomposites: in greening the 21st century materials world. Prog Mater Sci 50:962–1079
72. Sinha RS, Maiti P, Okamoto M et al (2002) New polylactide/layered silicate nanocomposites. 1. Preparation, characterization, and properties. Macromolecules 35(8):3104–3110
73. Siracusa V (2012) Food packaging permeability behaviour: a report. Int J Polym Sci 2012:1–11
74. SiracusaV Rocculi P, Romani S et al (2008) Biodegradable polymers for food packaging: a review. Trends Food Sci Technol 19(12):634–643
75. Siti Nur E'zzati MA, Anuar H, Siti Munirah Salimah, AR (2018) Effect of coupling agent on durian skin fibre nanocomposite reinforced polypropylene. In: IOP conference series: materials science and engineering, University Islamic University Malaysia, Malaysia, 8–9 August 2017
76. Souza AC, Souza C, Dias AMA et al (2014) Impregnation of cinnamaldehyde into cassava starch biocomposite films using supercritical fluid technology for the development of food active packaging. Carbohydr Polym 102:830–837
77. Sriupayoa J, Supaphola P, Blackwell J et al (2005) Preparation and characterization of alpha-Chitin Whisker-reinforced poly (vinyl alcohol) nanocomposite films with or without heat treatment. Polymer 46(15):5637–5644
78. Suzuki S, Shimahashi K, Takahara J et al (2005) Effect of addition of water-soluble chitin on amylose film. Biomacromol 6:3238–3242
79. Talegaonkar S, Sharma H, Pandey S et al (2017) Chapter 3: Bionanocomposites: smart biodegradable packaging material for food preservation. In: Grumezescu AM (ed) Food Packaging, Nanotechnology in the Agri-Food Industry 7. Elsevier, United Kingdom, pp 79–110
80. Tammaro L, Vittoria V, Bugatti V (2014) Dispersion of modified layered double hydroxides in poly(ethylene terephthalate) by high energy ball milling for food packaging applications. Eur Polym J 52(1):172–180
81. Tang XZ, Kumar P, Alavi S et al (2012) Recent advances in biopolymers and biopolymer-based nanocomposites for food packaging materials. Cri Rev Food Sci Nut 52(5):426–442
82. Valdés A, Mellinas AC, Ramos M et al (2014) Natural additives and agricultural wastes in biopolymer formulations for food packaging. Front Chem 2(6):10

83. Villmow T, Pötschke P, Pegel S et al (2008) Influence of twin-screw extrusion conditions on the dispersion of multi-walled carbon nanotubes in a poly (lactic acid) matrix. Polymer 49 (16):3500–3509
84. Vodnar DV, Oana LP, Francisc VD (2015) Antimicrobial efficiency of edible films in food industry. Notule Botanicae Horti Agrobotamici 3(2):302–312
85. Wagner M, Oehlmann J (2009) Endocrine disruptors in bottled mineral water: total estrogenic burden and migration from plastic bottles. Environ Sci Pollut Res 16(3):278–286
86. Wen X, Lin Y, Han C et al (2009) Thermomechanical and optical properties ofbiodegradable poly($_L$-lactide)/silica nanocomposites by melt compounding. J Appl Polym Sci 114:3379–3388
87. Wu D, Wu L, Zhang M et al (2008) Viscoelasticity and thermal stability of polylactide composites with various functionalized carbon nanotubes. Polym Degrad Stab 93(8):1577–1584
88. Wu J, Yu C, Li Q (2017) Novel regenerable antimicrobial nanocomposite membranes: effect of silver loading and valence state. J Membr Sci 531:68–76
89. Wu J, Zheng Y, Song W et al (2014) In situ synthesis of silver-nanoparticles/bacterial cellulose composites for slow-released antimicrobial wound dressing. Carbohydr Polym 104(1):762–771
90. Yahiaoui F, Benhacine F, Ferfera-Harrar H et al (2014) Development of antimicrobial PCL/ nanoclay nanocomposite films with enhanced mechanical and water vapor barrier properties for packaging applications. Polym Bull 72(2): 235–254
91. Yoo S, Krochta JM (2011) Whey protein–polysaccharide blended edible film formation and barrier, tensile, thermal and transparency properties. J Sci Food Agric 91(14):2628–2636
92. Yu H, Yan C, Yao J (2014) Fully biodegradable food packaging materials based on functionalized cellulose nanocrystals/poly(3-hydroxybutyrate-co-3-hydroxyvalerate) nanocomposites. RSC Adv 4(104):59792–59802
93. Zehetmeyer G, Soares RMD, Brandelli A et al (2012) Evaluation of polypropylene/ montmorillonite nanocomposites as food packaging material. Polym Bull 68(8):2199–2217
94. Zheng Y, Yan K, Zhao Y et al (2016) Preparation and characterization of poly (L-lactic acid)/ hollow silica nanospheres nanocomposites. Fibers Polym 17(12):2020–2026

Mechanical Techniques for Enhanced Dispersion of Cellulose Nanocrystals in Polymer Matrices

Jamileh Shojaeiarani, Dilpreet S. Bajwa and Kerry Hartman

1 Introduction

The growing concern regarding the impact of petroleum-based polymers on the environment and the development of biopolymers call for a transition from petroleum-based polymers to sustainable biopolymers. However, such a transition possess a substantial challenge for scientists and industries, requiring innovative materials and the applicable methods for improving the potential application of new materials.

Among all different types of biopolymers, cellulose is one of the most abundant organic compound available on the earth and it is the primary structural component of the cell wall of various plants, many forms of algae and the oomycetes. It is also present in different species of fungi, bacteria, and some sea animals such as tunicates [1]. Cellulose crystallites contain highly ordered, crystalline portions along with some disordered (amorphous) domains [2]; crystallinity index of the celluloses indicates the ratio between the area of the crystalline regions and the total area. Cellulose can be transformed into micro or nano-scale products with different shape and crystallinity using different methods such as acid hydrolysis, combined with mechanical shearing and enzymatic hydrolysis [3–7]. During the process, the amorphous or disordered regions of cellulose are hydrolyzed, and the crystalline regions with higher resistance to acid attack remain intact [8–10]. The resulting crystalline segments with the dimensions of nanometer is called nanocellulose (NC); generally, the family of NC can be classified as cellulose nanocrystals (CNCs), cellulose nanofiber (CNF), and cellulose nanowhisker (CNW) [11].

J. Shojaeiarani (✉) · D. S. Bajwa
Department of Mechanical Engineering, North Dakota State University,
Fargo, ND 58102, USA
e-mail: jamileh.shojaeiarani@ndsu.edu

K. Hartman
Department of sciences, Nueta Hidatsa Sahnish College, New Town, ND 58763, USA
e-mail: khartm@nhsc.edu

Inamuddin et al. (eds.), *Sustainable Polymer Composites and Nanocomposites*,
https://doi.org/10.1007/978-3-030-05399-4_16

In general, CNCs with a strength over 10 GPa and the elastic modulus of 150 GPa [12] has attracted wide attention as a reinforcing agent and have been employed in the nanocomposite, soft-tissue replacement, and food packaging industry for several decades [7, 13, 14]. The specific structure of CNCs contains several free hydroxyl groups on the surface (Fig. 1). The strong hydrophilic character of CNCs due to the presence of free hydroxyl groups on the surface of CNCs restricts the application of different solvents processing as the medium for solution blending [15].

The hydroxyl group located at C_6 is primary alcohol and at C_2 and C_3 are secondary alcohols. It has been reported that the hydroxyl group located at 6 positions has reactivity ten times higher than the other hydroxyl group [16]. The high reactivity of the hydroxyl groups to form hydrogen bonds strongly influence the overall properties of cellulose such as the reactivity of the hydroxyl groups, hierarchical organization, crystallinity, and limited solubility in most solvents [17, 18].

There are plenty of studied highlighting the optimum characteristics of CNCs as reinforcing agent. It is reported that the homogeneous dispersion of CNCs within the polymeric matrix is an essential step for achieving the superior properties of CNCs in composite materials [19, 20]. In addition, the dispersion of CNCs into the hydrophobic polymer with water-insoluble nature is a big issue. Therefore, several techniques have been experimented to decrease the affinity for moisture of the CNCs and improve the compatibility with a nonpolar polymer. The presence of hydroxyl groups on the surface of CNCs provides an opportunity for application of different surface modification techniques to alter the hydrophilicity and improve the compatibility with different nonpolar polymer matrices [21, 22]. Much research has been devoted to moderate the hydrophilicity of cellulose nanocrystals using physical and chemical modifications [23].

The application of different chemical-oriented surface modification methods is the most common method to alter the hydrophobicity nature of CNCs and enhance the compatibility between CNCs and nonpolar polymer. However, the need for developing new manufacturing processes capable of scaling up motivated the academia to find out innovative processing techniques. In the literature, two innovative manufacturing processes can be found: the application of liquid feeding and the application of masterbatch approach. This chapter contains contributions to the field of cellulose nanocomposites in the area of mechanical processing reporting new advances of the emerging ideas about manufacturing processes, which mainly focus on the achieved mechanical improvement.

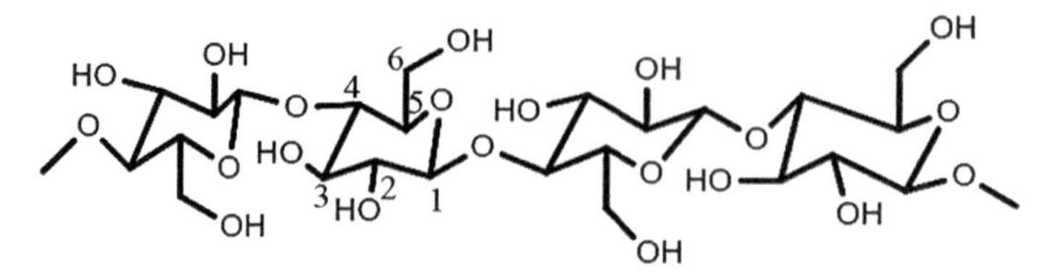

Fig. 1 Schematic representation of a cellulose molecule

2 Liquid Feeding

The application of extruder to shape thermoplastic materials dates back to 1935 when the first extruder machine was built by Paul Troester [24]. Since then, it has become the most broadly used processing technique through the development of different types of extruders capable of serving in different fields. The dramatic growth in plastic processing industry makes it essential to feed solid and liquid phases into extruders. In solid feeding extruders, the forces generated from rotating the screw and the stationary barrel move the materials down in the screw channel. In Liquid feeding extruders, the liquid can be fed into an extruder through a liquid injection nozzle.

It is reported that the drying process of cellulose nanocrystals results in the formation of irreversible aggregates which cannot be re-dispersed through an extrusion process. The application of liquid feeding seems to be a possible option to limit the formation of cellulose nanocrystal agglomerates. The incorporation of liquid and solid phase in an extruder could be difficult and the liquid feed rate, as well as liquid temperature, need to be monitored carefully, since the liquid temperature can strongly influence the viscosity and the change in liquid viscosity can result in pellet slippage on the barrel wall and consequently form undesirable product [25].

The first report of liquid feeding application of cellulose nanofillers into a polymer was by Oksman [26]. The extrusion process was implemented using an extruder equipped with a peridtalic peristaltic pump which controlled the liquid feeding rate. Two different feeding methods were used: the dry materials were fed into the extruder from a top mounted hopper into the barrel taking advantage from gravimetric feeding and the aqueous cellulose nanowhisker suspension was fed into the extruder using a vacuum pump to ensure the constant liquid feeding rate. In the extrusion process, the existing solvent in the liquid phase was removed by atmospheric venting (Zones 7 and 8) as well as vacuum venting (zone 10) (Fig. 2).

The elaboration of achieving uniformly dispersed cellulose nanowhisker in this work resulted in the generation of the high amount of solvent vapour during the extrusion process. TEM analysis of the composite samples exhibited partly dispersed cellulose nanowhisker into the matrix as well as thermal degradation of cellulose nanowhisker [26].

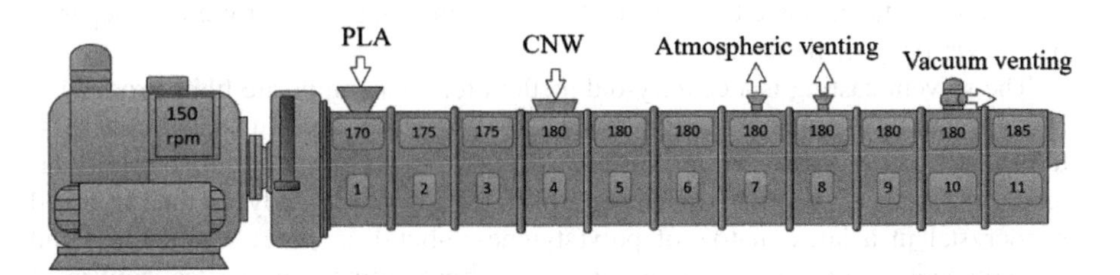

Fig. 2 Schematic image of extrusion process with liquid feeding [26]

In another work similar to cellulose nanowhisker liquid feeding, cellulose nanofibers were fed into PLA in a liquid phase. The high viscosity of cellulose nanofiber suspension reduced the uniform dispersion thorough composite samples [27]. The need for a specific extruder capable of feeding liquid and dry matter was reported as an essential need for incorporating liquid cellulose nanofiber into the polymer matrix.

3 Masterbatch Approach

The incorporation of cellulose nanocrystals into the different polymer matrix in a step-wise manner is one of the most commonly used preprocessing techniques in nanocomposites preparation. It is reported that the application of masterbatch can maximize the dispersion of cellulose nanocrystals in a polymer matrix, however, the time-consuming nature is the main weakness of the masterbatch approach [28, 29]. In masterbatch approach, a selective polymer is employed as a carrier for cellulose nanocrystals. The polymer can be either the same or different than the host polymer in the nanocomposite [30–34]. The highly concentrated masterbatches can be diluted in the extrusion process by adding polymer using the let-down ratio or mixing ratio (CNCs: polymer, generally between 1:14 and 1:20). The let-down ratio is of paramount importance since high mixing ratio might limit the uniform dispersion of CNCs in the polymer matrix [35]. Solvent casting and spin-coating are two methods employed in preparing CNCs masterbatches in literature.

3.1 Solvent Casting

Solvent casting has a widespread use in different applications owing to its simplicity and low-cost processing [36–39]. Solvent casting technique contains solubilization, casting, and solvent evaporation steps [40–45]. In solvent casting method, a polymer melt or polymer solution is applied on a flat surface, the solvent is then evaporated leaving a solid film. The evaporation rate of the solvent depends on the boiling point of the solvent, the viscosity of the solution, the pressure and the ambient temperature [46]. The rheological properties of the polymeric solution are of huge importance since the film thickness and the roughness of the film depends on the viscosity of the solution.

The solvent casting is a century-old method for nanocomposite films production [47–49] and is the most common method for preparing highly concentrated masterbatches. The application of solvent casting in composites manufacturing was reported for the first time by Favier et al. [50]. In that study, a tunicin-based nanocrystal in a latex matrix of poly(styrene-*co*-butyl acrylate) was studied and the competitive mechanical properties in corresponding composites confirmed the capability of solvent casting technique in composite films preparation.

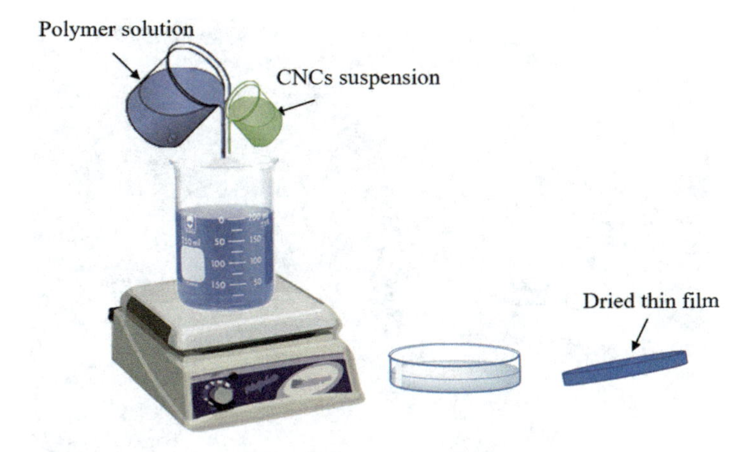

Fig. 3 Schematic depicting of solvent casting method

The preparation of thin films with uniform thickness, maximum optical clarity, and low haze were some advantages reported for solvent casting technology [51–53]. In general, the literature regarding the preparation of cellulose nanoparticles masterbatches involves the solvent casting as the main technique [20, 30, 54].

In solvent casting method, a polymer is first dissolved in a selective solvent either at room temperature or at elevated temperatures [55, 56]. The nanocelluloses are dispersed in either same or different solvent separately. The application of sonication and homogenization techniques can be used to increase the dispersion of nanoparticles through the solvent prior to the addition to the polymer solution [57]. The solution of polymer and CNCs suspension are then mixed together using magnetic stirrer and then poured into a flat-bottomed glass Petri-dishes and the solvent is evaporated and consolidate the films (Fig. 3).

3.1.1 Formation of Aggregates in Masterbatch Films

The morphology of thin film masterbatches generally affects the mechanical properties of final composites since the high concentration of CNCs in masterbatches tends to form CNCs' aggregates [35]. The time-intensive drying process in solvent casting method can intensify the formation of micro-sized cellulose aggregates in the masterbatch. The slow evaporation rate in solvent casting permits solvent molecules to exclude the nanoparticles and pushing them into closer proximity and leading to the formation of unavoidable CNC aggregates [35, 58]. In a recently published work, the formation of CNC aggregates in the masterbatch films was explored. In the aforementioned work, chloroform was used as the solvent and the percentage of CNCs in masterbatch films was 15% and the ultimate thickness was kept constant at 1.12 mm. Figure 4 illustrated the SEM image from a cross-section of the masterbatch.

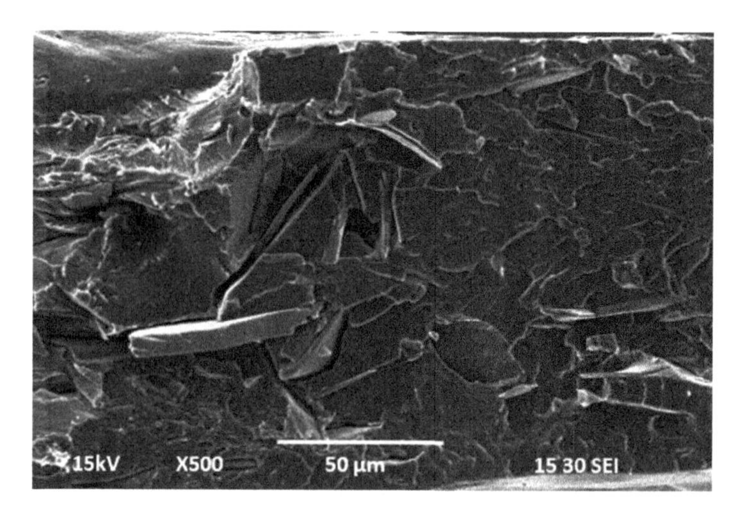

Fig. 4 Cross-sectional SEM image of solvent cast masterbatch film [35]

The formation of permanent CNC aggregates with relatively large effective size suggested strong aggregation in solvent cast masterbatches. It was reported that CNC aggregates formed during masterbatch preparation are difficult to separate during the extrusion process and those aggregates might result in poor adhesion between cellulose nanocrystals and polymer matrix [59].

3.2 *Spin-Coating*

Spin-coating is a common method employed to prepare thin films with thickness in the order of micrometre to nanometers. In this method, a liquid is deposited on a substrate, which can either be static or rotate at a specific angular velocity [60–62]. The deposited liquid generally consists of volatile solvents and non-volatile solute, and the non-volatile solute forms a thin film after solvent evaporation. Spin-coating involves four consecutive stages: deposition, spin-up, spin-off, and evaporation with some overlap in spin-off and evaporation steps [63]. During spin-off state, a film of liquid tends to spread with a uniform thickness, and after reaching a uniform thickness it tends to remain the uniform thickness. This behaviour suggests that mixture viscosity does not depend on shear and it would be constant throughout the substrate [64]. The equilibrium between centrifugal force generated from rotating substrate and the hydrodynamic (viscous) force evolving from the viscosity of mixture governs the efficiency of the formation of thin films with desirable thickness [35, 65]. In another word, the solution viscosity and the spinning speed mainly influence the film forming procedure. In general, the uniformity of thin film depends on the spinning speed, the concentration of mixture, and the volatility of the solvent [66]. The desired film thickness can be achieved by adjusting the spinning time and speed [67].

The most common application of spin-coating method is in the field of microelectronic thin films preparation. This method was first used by Emil et al., who studied the thin film formation of Newtonian fluid on a rotating substrate [68]. The application of this method in polymer films has been investigated in several theoretical and experimental studies [65, 69, 70].

A recently published work introduced an efficient spin-coating method for masterbatch preparation [35]. In this study, thin films with the thickness of the order of micrometres were spread evenly over the glass substrate using a combination of centrifugal force and the surface tension of the solution (Fig. 5).

In this work, the PLA-CNCs mixture was loaded in a syringe and injected through a needle (diameter = 500 μm) onto the centre of the rotating glass substrate with 100 mm diameter. The solvent evaporates simultaneously as the solution is applied to the substrate [62]. The spinning speed was kept constant at 400 rpm. The PLA-CNC mixture was loaded onto the centre of rotating substrate for 180 s.

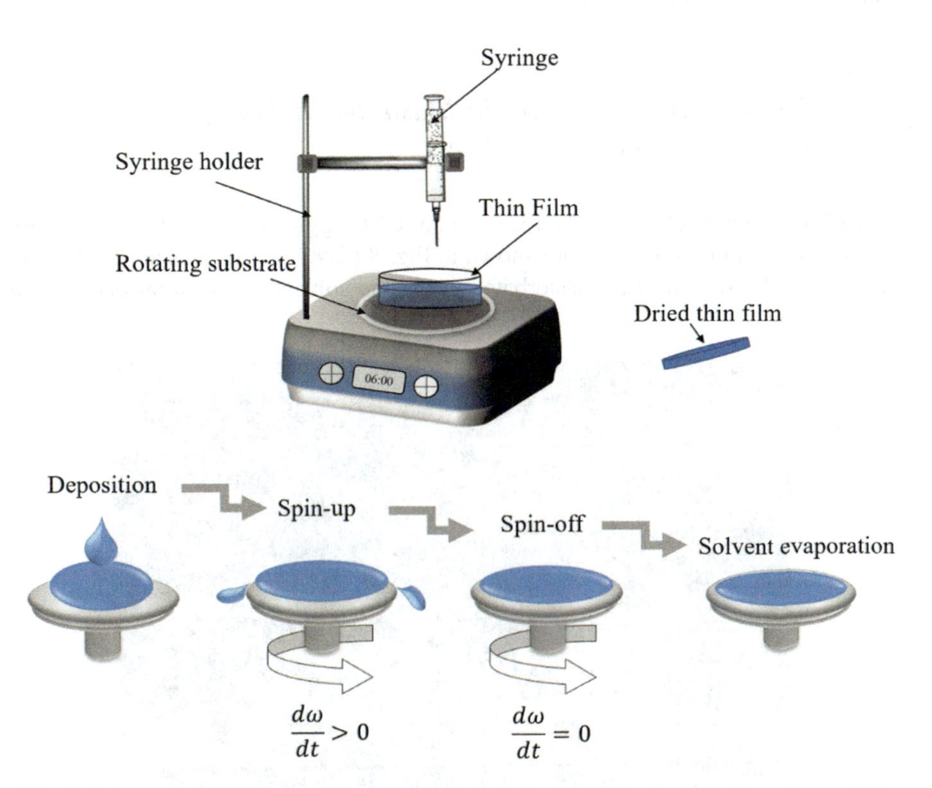

Fig. 5 Schematic illustration of the spin-coating method and four consecutive stages of the spin-coating method

3.2.1 Formation of Aggregates in Masterbatch Films

The volatile solvents used in spin-coating technique with a high evaporation rate can influence the formation of CNC aggregates through the polymer matrix. In fact, the high evaporation rate and low vaporizing time of the solvent from thin films in spin-coating method limit the movement of CNCs through the matrix and inhibits their assembly into micro-sized aggregates [35]. In fact, the rapid increase in the viscosity of solution as a result of the high evaporation rate of solvent kinetically traps the CNCs in the polymer matrix and hinder their movement for making more CNCs bundles. In the spin- coating method, thin film masterbatches are effectively dried out during spinning step. It is reported that in the spin-coating process as the solution is injected onto rotating substrate, the solvent evaporates and this, in turn, results in trapping the individual cellulose nanocrystals from forming big aggregates (Fig. 6).

The SEM micrograph of the free surface of spin-coated masterbatch exhibits that the high evaporation rate results in the formation of widespread voids on the free surface of masterbatches (Fig. 7).

3.3 Variation of Aggregates in Masterbatch Along the Cross-Sectional Thickness

The CNCs concentration and the formation of CNC aggregate through the thickness of the thin film masterbatches are shown in Fig. 8 (CNCs aggregates are pointed by arrow). In the solvent cast masterbatch, since the solvent evaporation occurs at a

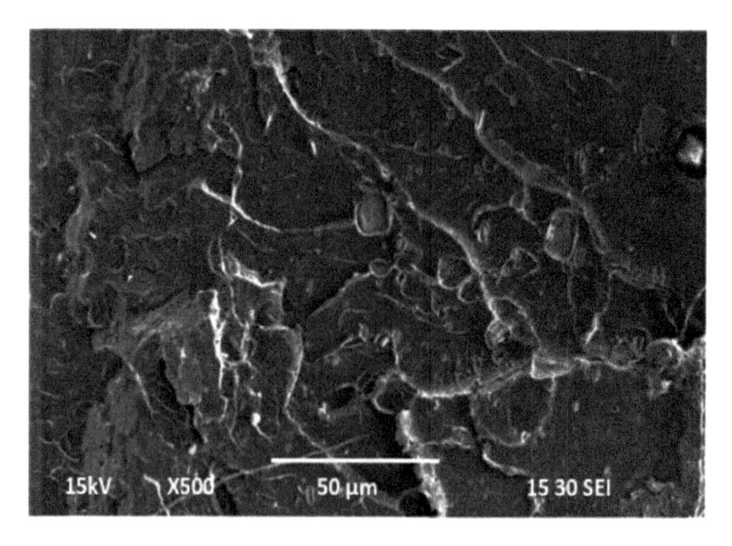

Fig. 6 Cross-sectional SEM image of solvent cast masterbatch film [35]

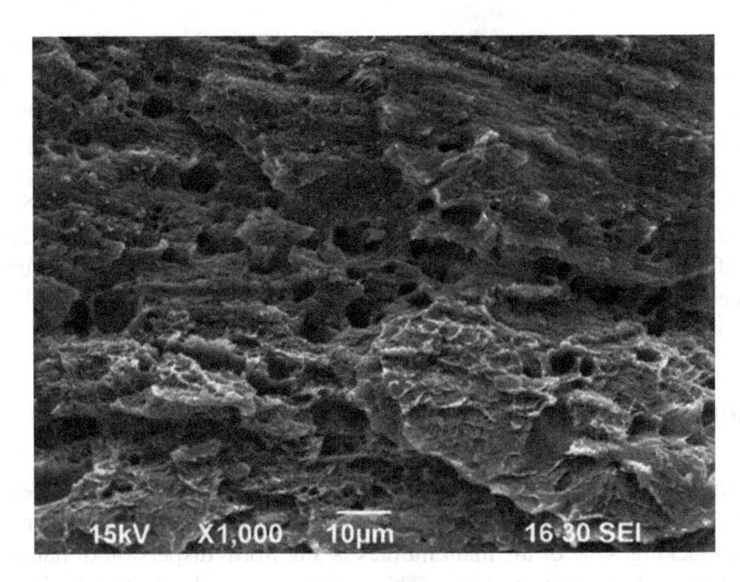

Fig. 7 The presence of voids on cross-sectional SEM image of spin-coat masterbatch film

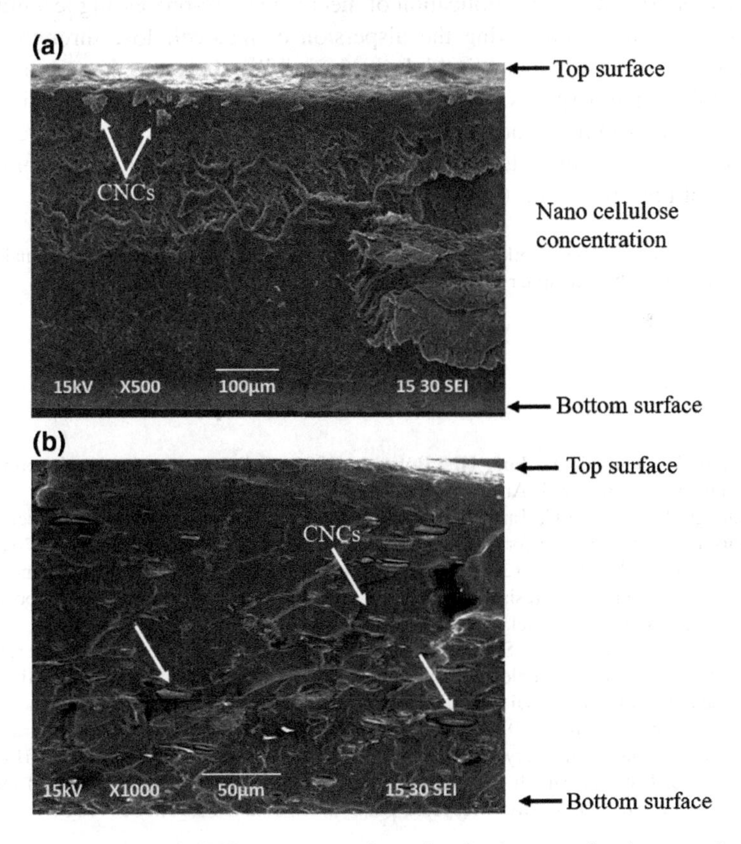

Fig. 8 The comparison of CNCs aggregates formation in the masterbatches: **a** solvent cast, **b** spin-coated

relatively low rate from the free surface of the film, the concentration of CNCs varies vertically along with the thickness of the thin film and the highest solute concentration happens close to the free surface. The increasing CNCs concentration can lead to the formation of CNC aggregates (Fig. 5a), however, in the spin-coated masterbatch, the CNC aggregates with perpendicular orientation with respect to the film thickness are scattered throughout the masterbatch thickness (Fig. 5b). These observations confirm the lower CNCs mobility in the spin-coated masterbatch as a result of the high evaporation rate of the solvent as well as the centrifugal force generated from rotating substrate [35].

4 Conclusion

The potential of nanocellulose to improve material properties has been widely accepted, however, application of nanocellulose in commercial polymer products has been lacking. One of the main hurdles is a uniform dispersion of nanocellulose material through polymer matrices. This review shows clearly that besides surface modification treatment, the application of mechanical pre-processing techniques has a great potential for improving the dispersion of nanocellulose through polymer matrices. This, in turn, results in higher compatibility between the polymer matrix and cellulose nanocrystals. The different techniques discussed in this chapter demonstrated an improvement in the performance characteristics of corresponding nanocomposites. All these techniques are expected to widen the domain of different mechanical pre-processing techniques for using nanocellulose materials.

Acknowledgements This work is based upon works supported by the National Science Foundation, ND EPSCoR under grant No. 11A1355466.

References

1. Klemm D, Heublein B, Fink HP, Bohn A (2005) Cellulose: fascinating biopolymer and sustainable raw material. Angew Chem Int Ed 44(22):3358–3393
2. Akhlaghi SP, Berry RC, Tam KC (2013) Surface modification of cellulose nanocrystal with chitosan oligosaccharide for drug delivery applications. Cellulose 20(4):1747–1764
3. Gozdecki C, Wilczyn A (2015) Effects of wood particle size and test specimen size on mechanical and water resistance properties of injected wood–high density polyethylene composite. Wood Fiber Sci 47(4):365–374
4. Siqueira G, Tapin-Lingua S, Bras J, da Silva Perez D, Dufresne A (2010) Morphological investigation of nanoparticles obtained from combined mechanical shearing, and enzymatic and acid hydrolysis of sisal fibers. Cellulose 17(6):1147–1158
5. Wang Q, Zhao X, Zhu J (2014) Kinetics of strong acid hydrolysis of a bleached kraft pulp for producing cellulose nanocrystals (CNCs). Ind Eng Chem Res 53(27):11007–11014
6. Filson PB, Dawson-Andoh BE, Schwegler-Berry D (2009) Enzymatic-mediated production of cellulose nanocrystals from recycled pulp. Green Chem 11(11):1808–1814

7. Ahola S, Turon X, Osterberg M, Laine J, Rojas O (2008) Enzymatic hydrolysis of native cellulose nanofibrils and other cellulose model films: effect of surface structure. Langmuir 24(20):11592–11599
8. Angles MN, Dufresne A (2001) Plasticized starch/tunicin whiskers nanocomposite materials. 2. Mechanical behavior. Macromolecules 34(9):2921–2931
9. Turbak AF, Snyder FW, Sandberg KR (1983) Microfibrillated cellulose, a new cellulose product: properties, uses, and commercial potential. J Appl Polym Sci: Appl Polym Symp (United States), vol 37, No. CONF-8205234-Vol. 2. ITT Rayonier Inc., Shelton, WA
10. Dong XM, Revol J-F, Gray DG (1998) Effect of microcrystallite preparation conditions on the formation of colloid crystals of cellulose. Cellulose 5(1):19–32
11. Hindi SS (2017) Differentiation and synonyms standardization of amorphous and crystalline cellulosic products. Nanosci Nanotechnol 4(3):73–85
12. Iwamoto S, Kai W, Isogai A, Iwata T (2009) Elastic modulus of single cellulose microfibrils from tunicate measured by atomic force microscopy. Biomacromol 10(9):2571–2576
13. Miao C, Hamad WY (2013) Cellulose reinforced polymer composites and nanocomposites: a critical review. Cellulose 20(5):2221–2262
14. Abitbol T, Rivkin A, Cao Y, Nevo Y, Abraham E, Ben-Shalom T, Lapidot S, Shoseyov O (2016) Nanocellulose, a tiny fiber with huge applications. Curr Opin Biotechnol 39:76–88
15. Nagalakshmaiah M (2016) Melt processing of cellulose nanocrystals: thermal, mechanical and rheological properties of polymer nanocomposites. Grenoble Alpes
16. Hebeish A, Guthrie J (2012) The chemistry and technology of cellulosic copolymers, vol 4. Springer Science & Business Media
17. Habibi Y, Lucia LA, Rojas OJ (2010) Cellulose nanocrystals: chemistry, self-assembly, and applications. Chem Rev 110(6):3479–3500
18. Popa V (2011) Polysaccharides in medicinal and pharmaceutical applications. Smithers Rapra
19. Sokolova Y, Shubanov S, Kandyrin L, Kalugina E (2009) Polymer nanocomposites and their structure and properties. A review. Plast Massy 3:18–23
20. Shojaeiarani J, Bajwa DS, Stark NM (2018) Green esterification: A new approach to improve thermal and mechanical properties of poly(lactic acid) composites reinforced by cellulose nanocrystals. J Appl Polym Sci
21. Eyley S, Thielemans W (2014) Surface modification of cellulose nanocrystals. Nanoscale 6(14):7764–7779
22. Lucia LA, Rojas O (2009) The nanoscience and technology of renewable biomaterials. Wiley
23. Thakur VK (2014) Nanocellulose polymer nanocomposites: fundamentals and applications. Wiley
24. Rauwendaal C (2014) Polymer extrusion: Carl Hanser Verlag GmbH Co KG
25. Giles Jr HF, Mount III EM, Wagner Jr JR (2004) Extrusion: the definitive processing guide and handbook. William Andrew
26. Oksman K, Mathew AP, Bondeson D, Kvien I (2006) Manufacturing process of cellulose whiskers/polylactic acid nanocomposites. Compos sci technol 66(15):2776–2784
27. Herrera N, Mathew AP, Oksman K (2015) Plasticized polylactic acid/cellulose nanocomposites prepared using melt-extrusion and liquid feeding: mechanical, thermal and optical properties. Compos Sci Technol 106:149–155
28. Pracella M, Haque MM-U, Puglia D (2014) Morphology and properties tuning of PLA/cellulose nanocrystals bio-nanocomposites by means of reactive functionalization and blending with PVAc. Polymer 55(16):3720–3728
29. Mariano M, El Kissi N, Dufresne A (2015) Melt processing of cellulose nanocrystal reinforced polycarbonate from a masterbatch process. Eur Polym J 69:208–223
30. Jonoobi M, Harun J, Mathew AP, Oksman K (2010) Mechanical properties of cellulose nanofiber (CNF) reinforced polylactic acid (PLA) prepared by twin screw extrusion. Compos Sci Technol 70(12):1742–1747
31. Gong G, Mathew AP, Oksman K (2011) Toughening effect of cellulose nanowhiskers on polyvinyl acetate: fracture toughness and viscoelastic analysis. Polym Compos 32(10):1492–1498

32. Corrêa AC, de Morais Teixeira E, Carmona VB, Teodoro KBR, Ribeiro C, Mattoso LHC, Marconcini JM (2014) Obtaining nanocomposites of polyamide 6 and cellulose whiskers via extrusion and injection molding. Cellulose 21(1):311–322
33. Lee S-H, Teramoto Y, Endo T (2011) Cellulose nanofiber-reinforced polycaprolactone/polypropylene hybrid nanocomposite. Compos A Appl Sci Manuf 42(2):151–156
34. Yang W, Fortunati E, Dominici F, Giovanale G, Mazzaglia A, Balestra G, Kenny J, Puglia D (2016) Synergic effect of cellulose and lignin nanostructures in PLA based systems for food antibacterial packaging. Eur Polymer J 79:1–12
35. Shojaeiarani J, Bajwa D, Stark N (2018) Spin-coating: a new approach for improving dispersion of cellulose nanocrytals and mechanical properties of poly(lactic acid) composites. Carbohyd polym
36. Rezakazemi M, Sadrzadeh M, Mohammadi T, Matsuura T (2017) Methods for the preparation of organic-inorganic nanocomposite polymer electrolyte membranes for fuel cells. In: Inamuddin D, Mohammad A, Asiri AM (eds) Organic-inorganic composite polymer electrolyte membranes. Springer International Publishing, Cham, pp 311–325
37. Rezakazemi M, Ebadi Amooghin A, Montazer-Rahmati MM, Ismail AF, Matsuura T (2014) State-of-the-art membrane based CO_2 separation using mixed matrix membranes (MMMs): an overview on current status and future directions. Prog Polym Sci 39(5):817–861
38. Baheri B, Shahverdi M, Rezakazemi M, Motaee E, Mohammadi T (2014) Performance of PVA/NaA mixed matrix membrane for removal of water from Ethylene Glycol solutions by pervaporation. Chem Eng Commun 202(3):316–321
39. Shahverdi M, Baheri B, Rezakazemi M, Motaee E, Mohammadi T (2013) Pervaporation study of ethylene glycol dehydration through synthesized (PVA-4A)/polypropylene mixed matrix composite membranes. Polym Eng Sci 53(7):1487–1493
40. Dashti A, Harami HR, Rezakazemi M (2018) Accurate prediction of solubility of gases within H_2-selective nanocomposite membranes using committee machine intelligent system. Int J Hydrogen Energy 43(13):6614–6624
41. Rezakazemi M, Dashti A, Asghari M, Shirazian S (2017) H_2—selective mixed matrix membranes modeling using ANFIS, PSO-ANFIS. GA-ANFIS. Int J Hydrogen Energy 42(22):15211–15225
42. Rostamizadeh M, Rezakazemi M, Shahidi K, Mohammadi T (2013) Gas permeation through H_2-selective mixed matrix membranes: experimental and neural network modeling. Int J Hydrogen Energy 38(2):1128–1135
43. Rezakazemi M, Mohammadi T (2013) Gas sorption in H_2-selective mixed matrix membranes: experimental and neural network modeling. Int J Hydrogen Energy 38(32):14035–14041
44. Rezakazemi M, Shahidi K, Mohammadi T (2012) Sorption properties of hydrogen-selective PDMS/zeolite 4A mixed matrix membrane. Int J Hydrogen Energy 37(22):17275–17284
45. Rezakazemi M, Shahidi K, Mohammadi T (2012) Hydrogen separation and purification using crosslinkable PDMS/zeolite A nanoparticles mixed matrix membranes. Int J Hydrogen Energy 37(19):14576–14589
46. Chinaglia DL, Gregorio R, Stefanello JC, Pisani Altafim RA, Wirges W, Wang F, Gerhard R (2010) Influence of the solvent evaporation rate on the crystalline phases of solution-cast poly (vinylidene fluoride) films. J Appl Polym Sci 116(2):785–791
47. Rezakazemi M, Vatani A, Mohammadi T (2016) Synthesis and gas transport properties of crosslinked poly(dimethylsiloxane) nanocomposite membranes using octatrimethylsiloxy POSS nanoparticles. J Nat Gas Sci Eng 30:10–18
48. Rezakazemi M, Vatani A, Mohammadi T (2015) Synergistic interactions between POSS and fumed silica and their effect on the properties of crosslinked PDMS nanocomposite membranes. RSC Adv 5(100):82460–82470
49. Farno E, Rezakazemi M, Mohammadi T, Kasiri N (2014) Ternary gas permeation through synthesized pdms membranes: experimental and CFD simulation basedon sorption-dependent system using neural network model. Polym Eng Sci 54(1):215–226
50. Favier V, Canova G, Cavaillé J, Chanzy H, Dufresne A, Gauthier C (1995) Nanocomposite materials from latex and cellulose whiskers. Polym Adv Technol 6(5):351–355

51. Siemann U (2005) Solvent cast technology—A versatile tool for thin film production. In: Scattering methods and the properties of polym mater, pp 307–316
52. Anbukarasu P, Sauvageau D, Elias A (2015) Tuning the properties of polyhydroxybutyrate films using acetic acid via solvent casting. Sci Rep 5:17884
53. Hsu S-T, Yao YL (2014) Effect of film formation method and annealing on morphology and crystal structure of Poly (L-Lactic Acid) films. J Manuf Sci Eng 136(2):021006
54. Jonoobi M, Mathew AP, Abdi MM, Makinejad MD, Oksman K (2012) A comparison of modified and unmodified cellulose nanofiber reinforced polylactic acid (PLA) prepared by twin screw extrusion. J Polym Environ 20(4):991–997
55. Rezakazemi M, Sadrzadeh M, Matsuura T (2018) Thermally stable polymers for advanced high-performance gas separation membranes. Progr Energy Combust Sci 66:1–41
56. Sadeghi A, Nazem H, Rezakazemi M, Shirazian S (2018) Predictive construction of phase diagram of ternary solutions containing polymer/solvent/nonsolvent using modified Flory-Huggins model. J Mol Liq 263:282–287
57. Dufresne A (2013) Nanocellulose: a new ageless bionanomaterial. Mater Today 16(6): 220–227
58. Bruckner JR, Kuhnhold A, Honorato-Rios C, Schilling T, Lagerwall JP (2016) Enhancing self-assembly in cellulose nanocrystal suspensions using high-permittivity solvents. Langmuir 32(38):9854–9862
59. Mathew AP, Oksman K, Sain M (2005) Mechanical properties of biodegradable composites from poly lactic acid (PLA) and microcrystalline cellulose (MCC). J Appl Polym Sci 97(5): 2014–2025
60. Mellbring O, Kihlman Øiseth S, Krozer A, Lausmaa J, Hjertberg T (2001) Spin coating and characterization of thin high-density polyethylene films. Macromolecules 34(21):7496–7503
61. Norrman K, Ghanbari-Siahkali A, Larsen N (2005) 6 Studies of spin-coated polymer films. Annu Rep Sect "C" (Physical Chemistry) 101:174–201
62. Hall DB, Underhill P, Torkelson JM (1998) Spin coating of thin and ultrathin polymer films. Polym Eng Sci 38(12):2039–2045
63. Syed JA, Lu H, Tang S, Meng X (2015) Enhanced corrosion protective PANI-PAA/PEI multilayer composite coatings for 316SS by spin coating technique. Appl Surf Sci 325:160–169
64. Brinker C, Hurd A, Schunk P, Frye G, Ashley C (1992) Review of sol-gel thin film formation. J Non-Cryst Solids 147:424–436
65. Danglad-Flores J, Eickelmann S, Riegler H (2018) Deposition of polymer films by spin casting: a quantitative analysis. Chem Eng Sci
66. Sahu N, Parija B, Panigrahi S (2009) Fundamental understanding and modeling of spin coating process: a review. Indian J Phys 83(4):493–502
67. Lien S-Y, Wuu D-S, Yeh W-C, Liu J-C (2006) Tri-layer antireflection coatings (SiO_2/SiO_2–TiO_2/TiO_2) for silicon solar cells using a sol–gel technique. Sol Energy Mater Sol Cells 90(16):2710–2719
68. Emslie AG, Bonner FT, Peck LG (1958) Flow of a viscous liquid on a rotating disk. J Appl Phys 29(5):858–862
69. Herrera MA, Sirviö JA, Mathew AP, Oksman K (2016) Environmental friendly and sustainable gas barrier on porous materials: nanocellulose coatings prepared using spin-and dip-coating. Mater Des 93:19–25
70. Zabihi F, Xie Y, Gao S, Eslamian M (2015) Morphology, conductivity, and wetting characteristics of PEDOT: PSS thin films deposited by spin and spray coating. Appl Surf Sci 338:163–177

Processing and Industrial Applications of Sustainable Nanocomposites Containing Nanofillers

Khadija Zadeh, Sadiya Waseem, Kishor Kumar Sadasivuni, Kalim Deshmukh, Aqib Muzaffar, M. Basheer Ahamed and Mariam Al-Ali AlMaadeed

List of Abbreviations

3D	Three dimensional
Ag NWs	Silver nanowires
BNCs	Bionanocomposites
BOEA	Battery operated portable handheld electrospinning apparatus
CS	Chitosan
$CaCO_3$	Calcium carbonate
CNTs	Carbon nanotubes
ESO	Epoxidized soybean oil
FTIR	Fourier transform infrared spectroscopy
HNTs	Halloysite nanotubes
GO	Graphene oxide
GNFs	Graphite nanoflakes
GNPs	Graphene nanoplatelets
GBR	Guide bone regeneration
GTR	Guide tissue regeneration
HA	Hyalumeric acid
LBL	Layer By layer
MWCNTs	Multiwalled carbon nanotubes
NPs	Nanoparticles
NFs	Nanofillers

K. Zadeh · K. K. Sadasivuni (✉)
Center for Advanced Materials, Qatar University, PO Box 2713, Doha, Qatar
e-mail: kishor_kumars@yahoo.com

S. Waseem
Advanced Carbon Products, CSIR-NPL, New Delhi 110012, India

K. Deshmukh · A. Muzaffar · M. Basheer Ahamed
Department of Physics, B.S. Abdur Rahman Crescent Institute of Science and Technology, Chennai 600048, Tamil Nadu, India

M. A.-A. AlMaadeed
Materials Science and Technology Program, Qatar University, PO Box 2713, Doha, Qatar

Inamuddin et al. (eds.), *Sustainable Polymer Composites and Nanocomposites*,
https://doi.org/10.1007/978-3-030-05399-4_17

NF	Nanofiller
NMs	Nanomaterials
NCs	Nanocomposites
PNCs	Polymer nanocomposites
PVA	Polyvinylalcohol
PEG	Polyethylene glycol
PANI	Polyaniline
PCO	Poly (cyclooctene)
P3HT	Poly (3-hexylthiophene)
PLA	Poly (lactic acid)
POMA	Poly (O-methoxyaniline)
PCL	Poly (caprolactone)
PTAA	Poly (3-thiophene acetic acid)
PLGA	Poly (lactic-co-glycolic acid)
SWCNTs	Single-walled carbon nanotubes
SiO_2	Silicon dioxide
SMPs	Shape memory polymers
TiO_2	Titanium dioxide
WVP	Water vapour permeability
ZnO	Zinc oxide
Zn	Zinc

1 Introduction

In the scientific research field, nanotechnology is a hugely popular area which covers polymer science and technology, micro and nanoelectronics, biomaterials etc. [1–3]. Polymer nanocomposites (PNCs) for industrial applications belong to a certain category of reinforced polymers having a lesser amount of effectively dispersed nanoparticles (NPs) [4, 5]. These PNCs are more beneficial than micro composites due to their size and interaction capability at the low filler loadings [6, 7]. The NPs in the form of fillers possess large specific surface area and higher surface energy when added to polymeric matrix which leads to changes in the morphological surfaces of the overall nanocomposites (NCs) [8–10]. The interaction of nanofillers (NFs) with the polymer matrix alters the chain mobility of the polymer and generates new trap centres in the NCs thereby changing desirable properties [11–13]. The analysis based on the use of materials in nanosize range provides an opportunity to design and produce new materials with enhanced bending, flexibility and improved physical properties for various interdisciplinary fields [14–16]. Since the NCs consist of different constituents having different structures compositions and properties, therefore, it leads to the development of materials with multi-functionality [17, 18]. The evolution of technology regarding the synthesis of new materials to be served for various applications has been diverted towards the development of synthetic strategies for production of NCs [19, 20]. The technologically advanced synthetic strategies provide various advantages over conventional procedures like the technique used to procure homogenous large grained materials [21]. The evolution of NCs in the recent times is solely attributed to their new, desirable and advanced properties compared to the conventional materials [22].

PNCs comprise an essentially important class of commercial materials with applications including electrical insulators, thermal conductors, damping, and aerospace. In the composites having micrometre-scale dimensions, the properties are limited due to lack of optimization of micrometre-scale composite fillers [23–26]. The fillers provide optimization of one property at the expense of other like stiffness at expense of toughness and toughness is achieved at the expense of optical clarity etc. [27]. The presence of defects adds to the limitations of the conventional composites arising as a result of regions of the high or low volume fraction of micrometer-scale fillers which in turn ultimately leads to the failure of the composite [28, 29]. These limitations associated with such types of composites led to the development of NFs based PNCs. In NCs, unprecedented advantages of combined properties were observed like the insertion of equiaxed NPs in thermoplastics leads to increase in the tensile strength, stress yield and Young's modulus in contrast to the pure polymer [30–32]. The development of chemical and in situ processing pertaining to NPs and NCs respectively has provided control over the morphology of such materials [33–37]. In addition to that, the advancement in the development process has provided the capability to manage the control over the interface between the matrix and the filler [35, 38]. Besides, the NCs are exciting materials due to their unique feature to attain a combination of properties pertaining to their potential as industrial and commercial materials [36]. The technical community has made substantial progress in the processing of NCs [39, 40]. However, the progress is still under the initial or basic phase embracing the understanding of NCs, their tailoring and optimization of properties [41]. The optimization achieved so far yields the ability to change the size, shape, volume fraction and degree of dispersion.

Most commercial materials used nowadays, are based on petroleum or their derived products and plastics. Both of these are used in various fields like packaging industry, petroleum industries which produce a negative impact on the environment. Their constant use has led to issues like pollution of air, water, and soil which ultimately leads to the global warming. Their toxic behaviour and non-degrading or long degrading periods lead to an enormous rise in the concentration of CO_2 [42]. As such there arises a need for alternative materials which are degradable and sustainable. The examples of such sustainable alternatives include biopolymers and clay NCs as renewable sources to develop inventive green materials like polymer blends containing NFs and bio-nanocomposites (BNCs) [31]. Pectin, a sustainable polymer is used to develop smart green materials for specific applications. A blend of pectin and Chitosan (CS) are predicted to be used as carriers in pharmaceutical industry [43]. In another study, it was reported that the NCs developed using pectin and starch is expected to serve for the conservative cause in food packaging industry owing to the enhanced mechanical properties and oxygen barrier tendencies [44]. The bio-films based on gelatin and pectin as sustainable and biocompatible NCs displayed improved tensile strength and water-resistant characteristics presenting an alternative material for fragile and watertight packing [45]. In a similar study, graphene oxide (GO) reinforced polyvinyl alcohol (PVA)/polyethylene glycol (PEG) blend composites were reported as high-performance dielectric material [46]. PEG is biocompatible, non-toxic and exhibits enriched water solubility. It acts as an efficient plasticizer for many biopolymers and NCs. The addition of a suitable amount of PEG to CS as per reports enhances the mechanical properties of the overall NCs [47]. The amount of PEG is significant due to the dependence of effectiveness of plasticization, variation in glass transition temperature on concentration [48]. The sustainability of polymers filled with nanoclay is well investigated for unique properties from both physical and chemical point of view [49]. The most promising NFs among nanoclay materials are the halloysite nanotubes (HNTs) due to their size and biocompatibility which makes it an ideal material for use in the biotechnology field, for water decontamination, as anti-corrosive coatings on metals, for humidity control and packaging [49–52].

The polymers sometimes do not display the desired properties. In such circumstances, NFs are added to the polymer matrix or blended with another suitable polymer to achieve the desired properties [53, 54]. The addition of small quantity of NPs is intended to facilitate new properties of the composites. However, the attainment of new properties is proportional to the surface treatment of the NFs and the processing technique applied. In polymer nanocomposites, the presence of NFs does not create large stress concentrations or do not alter ductility of the polymer [55]. Generally, the NFs with optical clarity (not scattering light) are added to the polymer matrix which subsequently changes the electrical and mechanical properties. For example, the addition of carbon nanotubes (CNTs) to polymer matrix greatly enhances the strength of the overall NCs. This type of interaction has been reported in the polymer matrix and single-walled carbon nanotubes (SWCNTs) wherein filler displays higher optical gain and large interfacial area [56]. In the formation of the NCs, the prime concern remains the credibility of NFs integration with polymers to display desired properties pertaining to the particular application.

In the industrial field, there is an urgency to seek new functional materials acquiring unique properties which can meet the difficult challenges concerning this field. As far as nanotechnology is concerned, there is no guideline pertaining to the mixing of NPs with polymers in composite structure to achieve the required properties [57]. The main challenges for NCs are to display and maintain the unique and multiple properties on a large scale through the use of conventional chemicals and materials [58]. For an industrial purpose, it is essential to establish a method or model to determine which nanofiller (NF) can be incorporated effectively for the formation of NCs and what new and improved properties can be achieved by following such procedures [59]. This can be achieved by determining the extent of effectiveness of dispersion of NPs in the matrix and its impact on the structure of polymer to yield desired properties [60]. Once the basic model is established, it is imperative to determine how the mixing of various NPs in polymer matrix influences the structure and properties of the NCs [61, 62]. In addition to the basic processing models, the interaction between NFs and polymer matrix also requires prime focus due to which the consequences on the overall NCs can be attained. Therefore, it requires a combination and optimum utilization of numerical modelling, characterization, and informatics towards the formation of particular NCs encompassing desired properties [63, 64]. It is noteworthy, that the nanocomposite exhibits properties which are significantly different from the NFs and the polymer.

The processing technique has a huge impact on the electrical, thermal and mechanical properties of NCs to a great extent [65]. In case of NCs, just like the composites at the macroscale, the properties are dictated by the distribution, orientation, and interaction between the polymer matrix and NF [66]. At the nanoscale, the biggest challenge is presented by the dispersion of NFs like CNTs in the polymer matrices to attain considerable interfacial adhesion between the NF and polymer matrix [67, 68]. This is because the propensity of NFs like CNT's for the reinforcement to aggregate endures unless high shear forces are applied for homogeneous dispersion of the polymer and NF. However, it is important to control the intensity of the mixing, as over mixing often leads to structural damage of NF [69]. In addition to that, the viscosity parameter requires prime attention due to vicious nature of polymer and NF solution and as such, there arises difficulty in the moulding of NCs [70]. The other issue in the processing of NCs is the compatibility of NFs with polymer and solvents. To overcome these issues, several approaches have been devised and implemented to achieve desired mixing of NFs and polymers in solvents [71]. These approaches include dry powder mixing; melt mixing and surfactant-assisted mixing [72–77]. However, there is still room for more creative and advanced processing techniques for better results.

Nanocomposite technology has implications of exceptionally wide range like in medical field for the surgical purpose, in the agriculture sector for effective crop treatment, in construction for high tensile strength materials [74–78]. The development of sustainable nanocomposite technology can lead to processing of advanced materials and devices having applications devoted towards better living, safety and eco-friendly in nature [79]. As such the development of novel NCs with suitable NFs has seen enormous growth in the recent times. The rise of NCs can be attributed to their value-added properties which are obtained without altering the basic properties of their components [80, 81]. The integration of graphene with the polymer leads to enhancement of electrical, physical, mechanical and barrier properties of the polymer composite at exceptionally low filler loadings.

The NCs for industrial applications are still under developing stage although a lot of success has been achieved in understanding their properties. The conceptual analysis and proofs of concepts have been consistently made, however, the optimization of nanocomposite materials is still obscure [82]. The optimization of nanocomposite materials is achievable only when the physics behind the control of improved properties are appropriately and thoroughly understood along with the establishment of the routes of processing to get the desired structures is obtained [79, 83]. This is because the fabrication of the new NCs, the optimization through modelling provides the material response and structural property linkages [84].

After the establishment of the nanocomposite structure through modelling, assistance in the development of material processing is provided. This chapter provides an insight of processing of sustainable NCs containing NFs used in industrial applications.

2 Fabrication Techniques of Nanocomposites

In the industrial field, the fabrication techniques regarding the development of composites based on nanomaterials (NMs) play a pivotal role in determining the properties and applications of the nanocomposite material. The composite material is composed of at least two different types of material phases which are interconnected by means of inherent interfaces. The formation of stable NCs requires the compatibility and reduction of interfacial tension involving different material phases [85]. When it comes to nanocomposite materials, they contain NPs of nanoscale dimensions dispersed with even small interfacial separations. In contrast to that, the NCs are inherited with a large surface area due to which there is a substantial enhancement in many of its properties [86]. Therefore, it is important to consider a proper fabrication technique to form stable NCs without aggregation and phase separation [87]. For fabrication of NCs containing NFs, there are several conventional techniques like intercalation method including in situ polymerization, solution mixing, melt compounding. Additionally, other techniques like sol-gel, molecular composite and direct dispersion are also quite vividly used. In order to get a nanocomposite which exhibits unique properties, it is essential that the interaction between the components is strong in nature [88]. By using proper fabrication, the intended interaction can be achieved and therefore the nanocomposite will exhibit unique characteristics. In general, for stable nanocomposite formation containing NFs, the interfacial tension should be lower than 5×10^{-4} N/m [89]. In the above-listed techniques, the most exploited method used in the fabrication of PNCs includes layer exfoliation and intercalation [90]. The other commonly used fabricating methods comprise the flexible polymerization process and cost friendly melt compound process. For industrial manufacture line, it is essential to develop techniques which produce sustainable, cost-effective and eco-friendly NCs. In this chapter, some of the technical features of the fabrication methods developed for the industrial purpose are described.

2.1 Intercalation Method

Considering the fabrication of NCs containing NFs, intercalation method is the most popular method. This method comprises of a top-down approach in which the fillers are downsized to nano-dimensions [90]. In this method, the polymer or monomers layers are intercalated between layers of inorganic layered substances

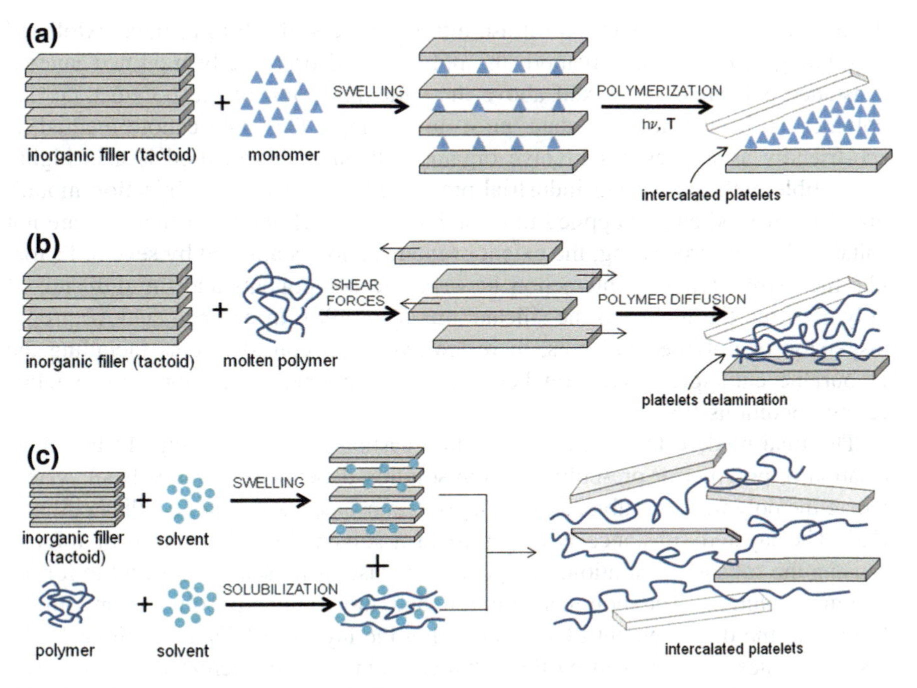

Fig. 1 Schematic representation of different intercalation methods: **a** in situ polymerization, **b** melt processing, and **c** solution casting [93]. Copyright 2010. Reproduced with permission from The Royal Society of Chemistry (RSC)

[91]. Within the polymer matrix, the inorganic substance is dispersed during the process of polymerization or melt-compounding thereby exfoliating the layered inorganic substance through each layer causing uniform dispersion of NFs [92]. The intercalation consist of three methods as shown schematically in Fig. 1 comprising of in situ polymerization, melt compounding and solution casting [93].

In the first method called as in situ polymerization, the layered substance like nanoclay is inflated within a liquid monomer or monomer solution as shown in Fig. 1a for the formation of polymer in between the intercalated layers. The polymerization commences on the application of either heat, an organic initiator or via the use of a catalyst [93]. The catalyst is fixed by means of cation exchange within the layered substance prior to inflation. However, the disadvantage of this method is the quick affinity of the inorganic substance towards sediment from the organic polymer and how rapid the phase separation occurs [94]. The interaction between solvent and NFs at the solvent-filler interface can be enhanced by incorporation of specific groups which are linked with the interface in order to stabilize the NPs dispersion [95].

The second method called melt intercalation method or melt processing as shown in Fig. 1b comprises annealing occurring statically or under application of shear [96]. The shear is necessary for delamination of the layered substance in order to allow the polymer chains to diffuse into interlayers of NF to achieve exfoliated morphology [97]. In this method, the polymer and organically modified layered substance (silicates) are mixed above the softening point of the polymer i.e. the polymer is in the molten state prior to mixing. This fabrication method is eco-friendly as it does not involve organic solvents. This method is also highly compatible with the current industrial process like extrusion and injection moulding. This method can be applied to those NCs for which the other methods are not suitable. In melt processing, the extent of intercalation is affected by several factors like the thermodynamic interaction between the components and the diffusion of polymer from the melt into the silicate interlayers [98]. In order to achieve proper dispersion of polymer and NFs, there are two main considerations including the favourable enthalpic interaction between the components and appropriate fabricating conditions [99].

The final method termed, is the solution casting as shown in Fig. 1c the intercalation of polymer or pre-polymer from solution dependent on the solvent system where the polymer is soluble, and the layered substance is swellable (silicate fillers) [73]. The layered substance is dispersed in a solvent like chloroform or toluene causing the substance inflation. The polymer is also dispersed in a solvent to form a polymer solution. The polymer solution and inorganic NF solution are mixed leading to the displacement of solvent within the layers of NFs and ultimately the displaced sites are occupied by the polymer chains due to intercalation [100]. The mixing of the polymer solution with the delaminated nanoparticles causes strong interaction between the components of the composite [81]. The intercalation is attributed to the entropy resulting from desorption of the solvent molecules thereby balancing the decrease in entropy of the confined intercalated chains [101]. From the intercalated structure, when the solvent is removed, a nanocomposite comprising NF layer with the polymer is formed. The advantage of this method is quick rapid exfoliation of the stacked layers by application of an appropriate solvent [102]. On the other hand, the usage of organic solvents is strongly discarded as it makes the method unsafe and hazardous to the environment [103].

2.2 *Sol-Gel Method*

The sol-gel method comprises of a bottom-up approach in which NCs are formed using combined in situ NF formation and polymerization. In the sol-gel method, the NCs are made at relatively low temperature comprising of hydrolysis of the constituent molecular precursors followed by polycondensation to a glass-like form [104]. The sol-gel method allocates incorporation of organic and inorganic additives during the glassy network formation taking place at room temperature [105, 106]. Due to this reason, sol-gel has been conventionally used in the fabrication of glasses, polycrystals, porous composites, organic and inorganic NCs and ceramic materials. The process is initiated using metal alkoxide, melted using water, alcohol, ammonia or an acid for homogeneous dispersion. The metal alkoxide

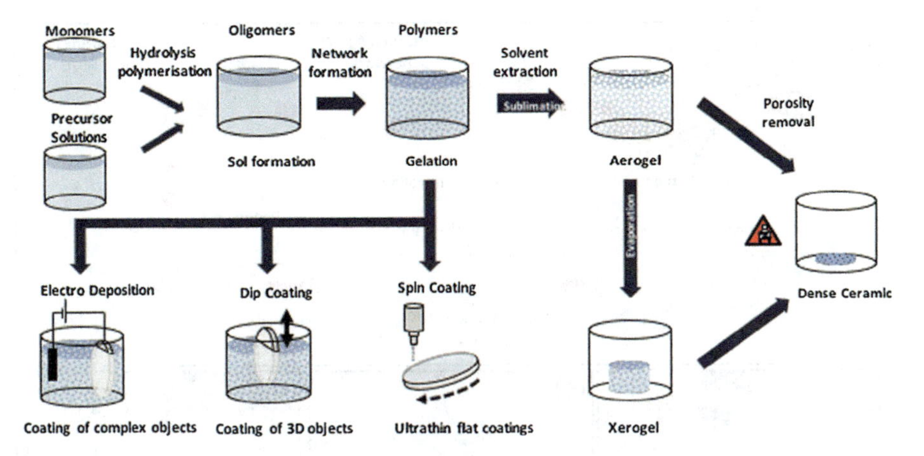

Fig. 2 Schematic representation of sol-gel synthesis routes [108]. Reproduced with permission from The Elsevier Ltd.

undergoes hydrolysis on reaction with water to form metal hydroxide and alcohol [107]. The sol-gel method is expected to be one of the key technologies in future for fabrication of NMs. The sol-gel method for processing of composite films is shown in Fig. 2. The process is initiated with the selection of appropriate precursors reacting at various steps to form colloidal particles or polymeric gels [108]. However, the prime requirements for this method for thin film deposition by spin coating or drop casting is stable sol and this sol is converted into a polymeric gel.

The metal ions bordered by ligands are mainly used as precursors for sol-gel reaction due to radial reaction mode. For the synthesis of zinc oxide (ZnO) thin film, the precursors used were zinc acetate dihydrate; $Zn(CH_3COO)_2$. $(H_2O)_2$ and ethanol were used as solvent [96]. The zinc (Zn) precursor and the dopant element were first dissolved in an appropriate solvent. This was followed by the addition of stabilizing agent (Monoethanolamine) to avoid premature precipitation and quick conversion of the sol into the gel. The solution of precursor material and the stabilizer is constantly kept on magnetic hot plate bath under constant stirring while maintaining the temperature at 25–80 °C in an oil bath leading to the formation of sol. The sol after attaining stability is aged for 24 h at room temperature. This is followed by post heat treatment where the parameters including drying time and temperature are altered. The procedure is continued several times to yield homogeneous, crystalline and single phase ZnO films after annealing.

2.3 *Direct Dispersion Method*

This method of nanocomposite fabrication is a top-down approach and comprises of surface modification of NFs aiding to the usage of chemicals to improve their

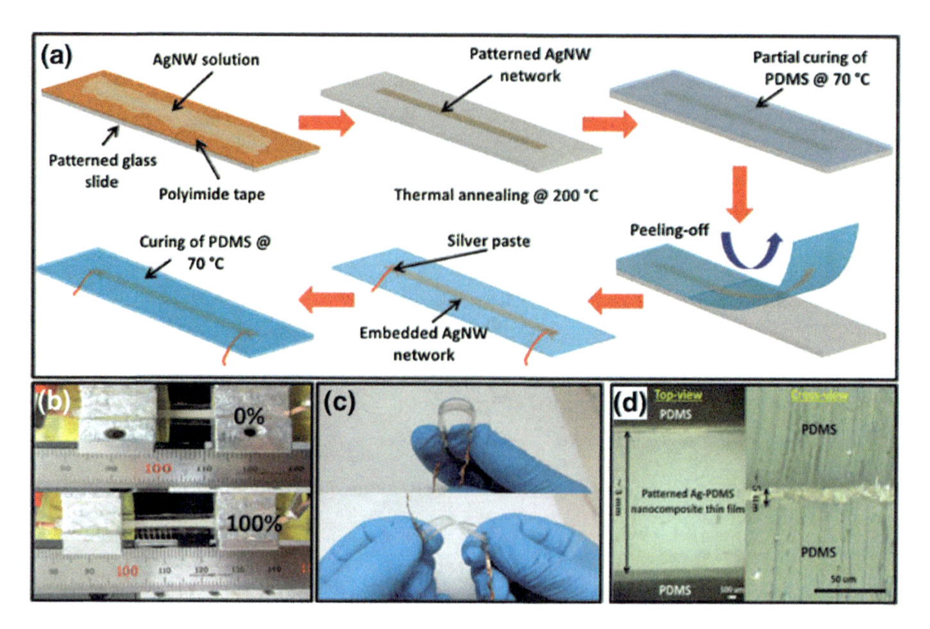

Fig. 3 **a** Fabrication process of sandwich-structured PDMS/AgNW/PDMS based strain sensors, **b** Strain sensor before and after 100% stretching, **c** strain sensor under bending and twisting, **d** optical microscopy images on top and cross-section of the strain sensor [111]. Reproduced with permission from. The American Chemical Society

compatibility with polymers [109]. The surface modified NFs are dispersed homogeneously in the polymer solution. The NCs fabricated by this method include fabrication of stretchable sensors, coating agents which are photo-hardened using functionalized silica NPs, comb-shaped block copolymers which are silver (Ag) protected, polyamide nanocomposite with silica nano-particles surface treated aminobutyric acid etc. [110–113]. The schematic representation for fabrication of stretchable sensors [111] is shown in Fig. 3. It can be seen that the fabricated sandwich-structured strain sensors exhibit excellent flexibility, stretching and bending ability. The top and cross-sectional optical images of the samples demonstrated the fabrication of well-patterned silver nanowires (AgNWs)-PNCs film with an average thickness of 5 μm.

Direct dispersion method is suitable for fabrication of polymer-based composites containing NFs or micro-fillers [42]. Generally, the polymer and the fillers are mixed in two ways i.e. either in presence of a solvent or without solvent [43]. The first case involves melting of the constituents while the second case involves mixing of the constituents in a solution [45]. Besides the above methods, there are several other methods used in the fabrication of NCs, which are listed in Table 1.

Table 1 Various techniques for fabrication of nanocomposites

Fabrication technique	Nanocomposites	Properties	References
Intercalation	• Polyethylene/ clay nanocomposites • Conducting polyaniline (PANI) and montmorillonite (Mmt)	• Reduction in heat release rate by 32% • Enhanced thermal stability	[114, 115]
Solvent casting	• Poly (3-hexylthiophene) (P3HT) • Gelatin and bioactive glass	• Higher mechanical strength and conductivity • Improved biocompatibility and porosity	[116, 117]
Melt compounding	• Cellulose whiskers-Polylactic acid • Polyethylene erephthalate/ graphene	• Improvement in material elongation • Low percolation threshold and superior electrical conductivity	[118, 119]
In-situ polymerization	• $LiFePo_4$-carbon • Graphene oxide-epoxy	• High power and long cycling life • Increased storage modulus, electrical conductivity and thermal stability	[120, 121]
Doctor blade	• Graphene nanoplatelet-Liquid crystalline polymer • TiO_2 -SiO_2 nanocomposite	• Better electrical conductivity and improvement shear modulus • Superior crystallinity, photocurrent density and photovoltaic performance	[122, 123]
Compression molding	• Exfoliated graphite–polypropylene • Polyethylene-layered silicate	• Enhanced flexibility and lower percolation threshold • Thermodynamic stability and improvement in tensile strength	[124, 125]
Sol-gel	• Ag-TiO_2 nanocomposite • Carbon nanotubes—TiO_2	• Exhibition of antimicrobial properties • Improvement in photo-catalytic properties	[126, 127]
Direct dispersion	• TiO_2-graphene nanocomposite • Graphene metal nanocomposite films	• Decrease in lateral resistance and increase in photo-activity • Enhanced optical, electrical and mechanical properties	[109, 128]

3 Applications of Sustainable Nanocomposites

Consistent efforts have been made in the last two decades in the area of nanotechnology in order to get NMs with determined functionality and promising prospects of sustainable materials [129]. Tailoring the properties of sustainable

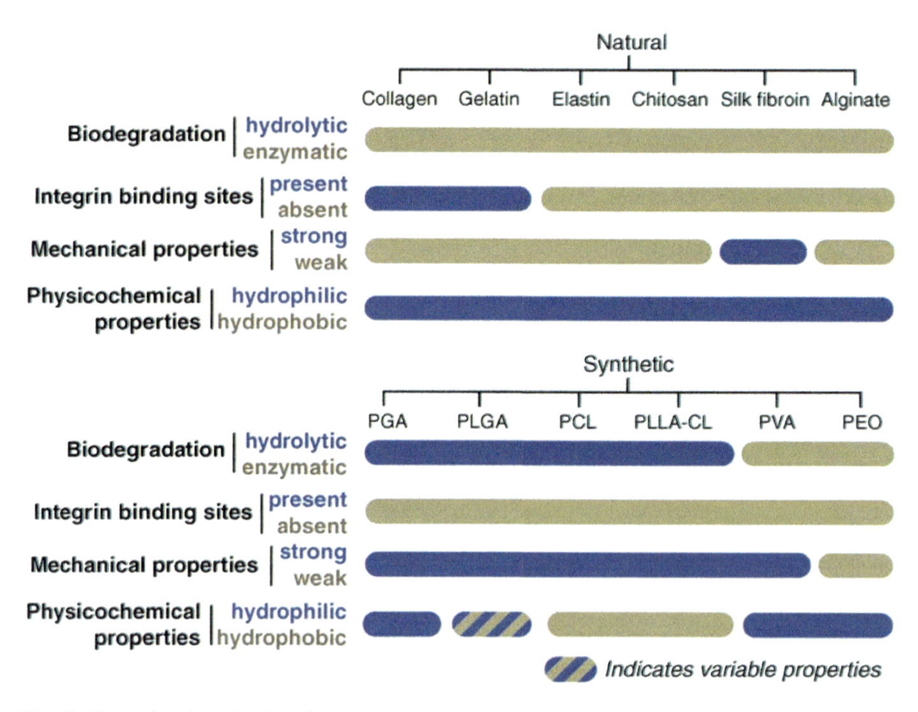

Fig. 4 Natural and synthetic polymers commonly used in the preparation of polyblend nanofibers with their key biological, mechanical and physicochemical properties [131]. Reproduced with permission from The Elsevier Ltd.

Table 2 Several polymer nanocomposites with their specific applications

Nanocomposites	Application	Properties	References
PCL/CNC/PEG	Shape memory polymer	Nanocomposites possessed good cytocompatibility and enhance mechanical properties	[132]
PLA/MWCNT	Electronics	Highly efficient nanocomposites as anti-static agents and can be applied for coating of electronic devices	[133]
PLA/CNC	Food packaging	Enhancement of stiffness, strength, and it is also easily processed to make a film for use in industrial packaging	[134]
PCL/lignin	Biomedical "tissue engineering"	Electrospun PCL-Lig scaffold enhanced the biological response	[135]

nanocomposite required biopolymer matrices. Biopolymers are polymers that biodegrade with different properties including renewability, sustainability, non-toxicity, and biodegradability [130]. Figure 4 illustrates classes of biodegradable polymers from different resources as sustainable matrices for the NFs [131].

The selection of NFs depends entirely on the application field of biological, mechanical and physicochemical properties. There is tremendous interest in the development of sustainable NCs for a wide range of commercial applications and many of these applications are already sought after for industrial applications. Table 2 shows several applications of sustainable NCs.

3.1 Electronic Applications

The behaviour of NCs in terms of electrical conductivity by large depends on the properties of NMs. For instance, some polymeric materials are considered poor electrical conductor due to the existence of high bandgap but, in the nano revolution researchers focused on sustainable polymeric nanomaterials having many advantages such as flexibility, inexpensive, non-corrosive and conductive materials in order to fabricate biodegradable and transient electronic devices for biomedical and other important applications [136]. Mai et al. [137] developed an intelligent sustainable nanocomposite sensor system comprising of Polylactic acid (PLA)/CNTs that can sense the degrading levels of the biopolymer which is correlated with the changes in electrical resistivity of the PLA/CNT NCs. Today the researchers mainly concentrate on the synthesis and development of new smart materials. These materials react with externally applied force like, thermal, light or magnetic. Piezoelectric is an example of such smart materials.

Piezoelectric materials convert mechanical energy (force) to an electrical signal. Thus, it can be seen that indeed the smart materials act as sensor component. Poly (vinylidene fluoride) is an example of smart polymer matrix; however, the addition of nanomaterials such as (SWNTs) is used as NF to tailor the sensing and power harvesting with enhanced performance [138, 139]. The use of smart materials as sensing and shape changing devices has been enhanced due to increase emphasis on nanocomposite materials. A typical smart material assembly contains:

- Sensor components: containing smart materials that monitor changes in pressure, temperature, light current or magnetic field.
- A communication network that relays changes detected by sensor component
- Actuator part that react to the command, it could be also smart materials such as piezoelectric.

Biopolymers and biopolymer NCs are desired materials for supercapacitor applications. These materials should possess low cost, high conductivity, high voltage window and high storage capacity in order to find application as a supercapacitor electrode. Higher conductivity can be achieved by using suitable electrode material through negative or positive "doping" with ions either by oxidation or reduction process. Christinelli et al. [140] reported the fabrication of supercapacitors with layer-by-layer (LBL) technique using poly (o-methoxyaniline) (POMA) and poly (3-thiophene acetic acid) (PTAA). The results were compared with POMA casting film. The film behaviour with increasing bilayer numbers through a self-doping process which initiated each layer to act as a capacitor in parallel resulting in an increase in the overall capacitance of the film and the surface area. Hence, in this proposition, it is reasonable to expect an increase in the active area of the film with an increase in the number of bilayers [140].

In recent years, highly flexible and planar supercapacitor electrodes have been prepared using graphite nanoflakes (GNFs) on polymer lapping films as a flexible substrate. Botta et al. [141] reported the effect of the addition of GNFs into PLA matrix and evaluated the effect of reprocessing on the properties of PLA filled with graphene nanoplatelets (GNPs). In particular, the morphological analyses, intrinsic viscosity measurements, thermal, rheological and mechanical tests were carried out on the materials reprocessed using five subsequent extrusion cycles. It was found that addition of GNPs in the PLA matrix resulted in the decrease of the degradation rate as a function of the reprocessing cycles. Biodegradable polymers with high crystallinity, high hydrophobicity, and facile processing are used to maintain the functionality of electronic devices that are exposed to aqueous solutions, high salinity environments, and elevated temperature. Alam et al. [142] reported a technically benign procedure to combine vermiculite nanoplatelets with nanocellulose fiber dispersions to form functional biohybrid films. The unique combination of excellent oxygen-barrier properties and optical transparency of these biohybrid materials suggests their potential as an alternative in the flexible packaging of oxygen-sensitive devices such as in the displays of a light emitting diode, gas-storage and as barrier coatings in large volume-packaging.

3.2 *Shape Memory and Biomedical Applications*

Shape-memory polymers (SMPs) have attracted significant attention due to their fascinating applications. CNTs possess attractive properties such as high thermal and electrical conductivity which encourages towards utilizing them as the functional filler in PNCs. Wang et al. [143] reported the development of a chemically cross-linked polycyclooctene (PCO), multiwall carbon nanotubes (MWCNTs) and polyethylene based shape memory NCs with co-continuous structure and selective distribution of MWCNTs in the PCO matrix. It was noted that the selective distribution of MWCNTs in the PCO matrix reduces the conductivity. The shape memory composites with thermally responsive thicknesses could be employed as an artificial tendon [144] as shown in Fig. 5. The biodegradable polymers are proposed to provide temporary assisting support for cell growth which otherwise gets degraded with time in a controlled way into nontoxic products [145]. In tissue engineering, the methods from materials engineering and life science are employed to create artificial constructs for regeneration of new tissue.

One of the most promising approaches toward cells growth is Scaffold, carried via application of temporary support for regrowth of the targeted tissues without the loss of the three-dimensional stable structure [146]. In the biomedical application,

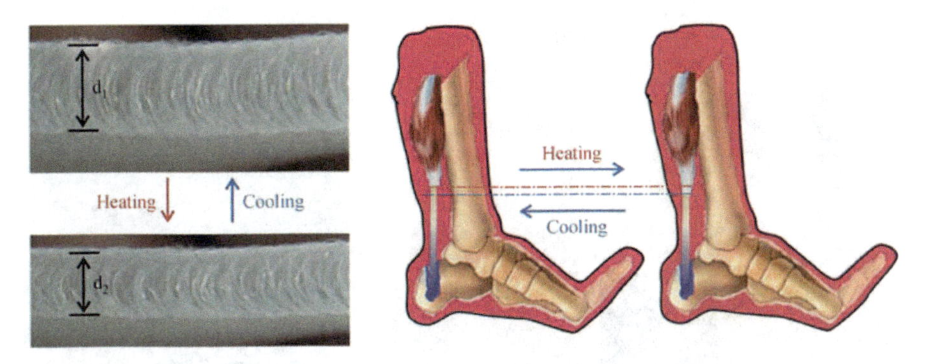

Fig. 5 Shape memory application sustainable polymer composite in biomedical field [144]. Reproduced with permission from The Springer Nature

the tissue engineering scaffold is one of the massive fields of research in recent years because of its potential in the repair or replacement of damaged tissues and organs. Tissue engineering scaffolds are three dimensional (3D) backbones that can be used for cell proliferation migration and differentiation which ultimately aid in the formation of the extracellular matrix [147].

Biopolymer scaffolds is a major part of tissue engineering which becomes helpful through cell seeding, proliferation, and formation of new tissues providing great scopes in the research field of engineering. The diversity of tissues depends majorly on 3 factors such as pore size, porosity and surface area which are widely recognized in the field of tissue engineering scaffold [148]. For example, PLA/CNTs have been extensively investigated for biomedical and other applications. It has a potential use in biomedical scaffolds for tissue engineering. The electrical conductivity of the carbon-based nanostructures plays a vital role in direct cell growth as they are capable of conducting and stimulating an electric field in the process of healing of tissues [149].

Fujihara et al. [150] developed guided bone regeneration (GBR) membranes by reinforcing calcium carbonate ($CaCO_3$) NPs into polycaprolactone (PCL) matrix to synthesize nanofibers with the help of electrospinning technique. The nanofibrous membrane was fabricated by developing two layers. These layers are called mechanical support layer and PCL/$CaCO_3$ nanocomposite as a functional layer. Figure 6 shows the membrane supported osteoblast attachment and proliferation of a battery operated portable handheld electrospinning apparatus (BOEA) which has potential application in rapid hemostatic treatments [151].

Biodegradable NCs has fascinating application in dental field Such as periodontal and alveolar bone regeneration. Both Guided tissue regeneration (GTR) and GBR works on the principle of placement of a barrier membrane with the motive of preventing epithelial migration into the defective area, thus allowing sufficient and prolonged time for periodontal ligament, cementum and bone regeneration. Park et al. [152] developed a membrane of biodegradable poly (lactic-co-glycolic acid) (PLGA) and hyaluronic acid (HA) to obtain HA-PLGA NPs used for bone and

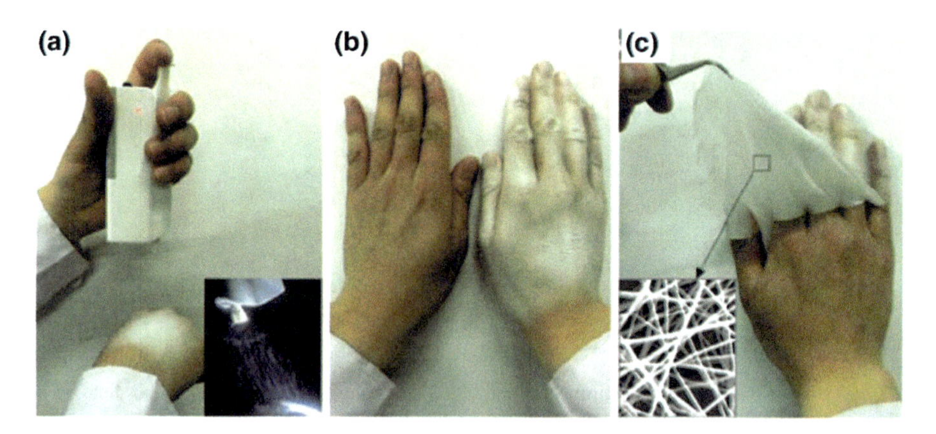

Fig. 6 **a** The BOEA spinning process on hand, **b** fabricated PLA fibrous membrane on another hand in 2 min, **c** fibrous membrane showing good flexibility and compactness [151]. Reproduced with permission from The Elsevier Ltd.

periodontal regeneration as shown in Fig. 7. Several biomedical applications of CS composites [154] have been reported as shown in Fig. 8. CS is the potential material to be used as artificial kidney membrane, hypocholesterolemic agents, drug delivery systems, absorbable sutures and supports for immobilized enzymes. CS has some unique characteristic advantages due to its nontoxicity and biodegradability which doesn't damage the environment. Being a biocompatible material, CS breaks down slowly into harmless products that are absorbed completely by the body.

3.3 Mechanical Applications

NFs are used to improve the mechanical properties of biopolymers as reinforcement materials. Example of such NMs is nanocellulose which is one of the most abundant biomass materials extracted from cellulose. It has been showed to be an environmentally friendly material with excellent mechanical properties owing to its unique nano-structure. Nanocellulose has been extensively used as functional materials in a variety of applications [155]. To induce a wide range of bending and geometries in the surfaces of such materials, the prototype can be laminated with a wood veneer surface from either side. The flexible fireboard can be fixed with desired geometry by fixing another veneer layer on the other side of it which produces sandwich panels as shown in Fig. 9. Similarly, flat or curved sandwich panels can be fabricated from the same materials which are developed following the same production process with different post-production processes. When the veneer is applied from both the sides of the panel, the form is fixed and no geometrical variations are required [156].

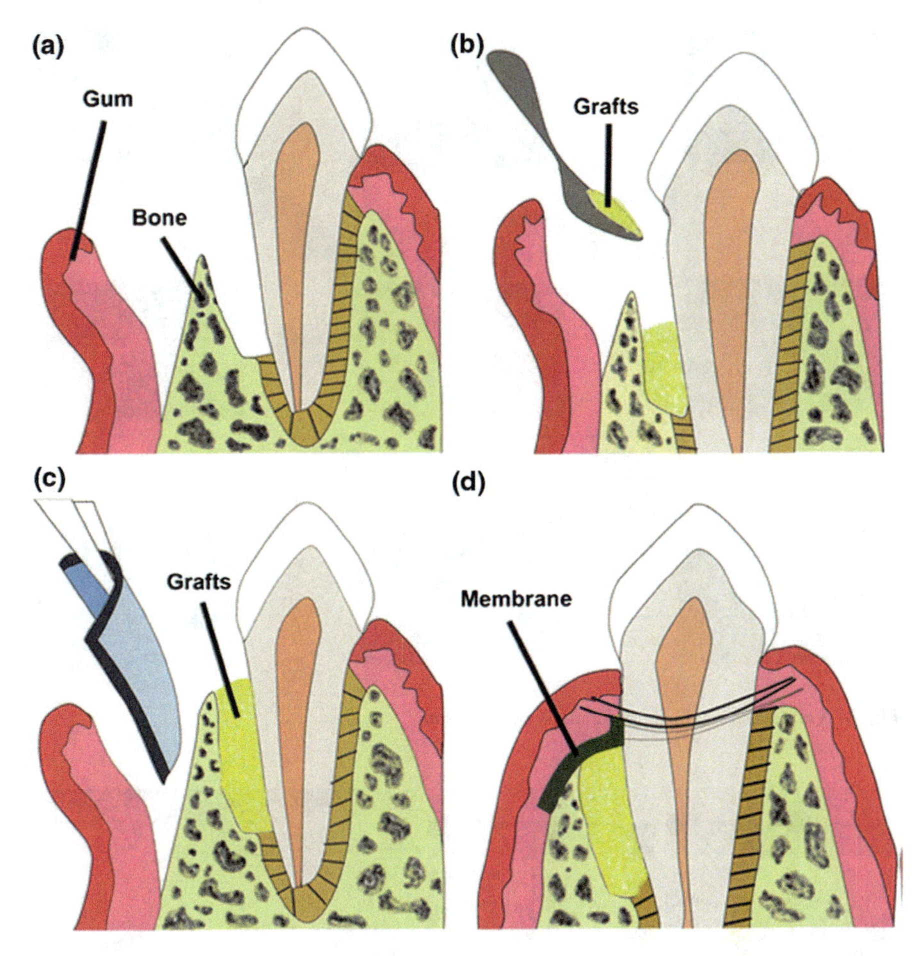

Fig. 7 Schematic representation of a combination therapy of bone grafts and GTR membrane for periodontal regeneration. Reprinted from Ref. [153]. Reproduced with permission from The Elsevier Ltd.

The great biocompatibility and biodegradability, in particular, render nanocellulose is seen as promising sustainable NF. Bulota et al. [157] introduced acetylated micro-fibrillo cellulose to PLA with fiber contents 2–20% using solution casting method. The maximum tensile strength was observed with fiber content over 10% whereas Young's modulus was increased by approximately 15%. However, no improvement was observed in case of tensile strength. The strain recorded in the materials at the time of fracture improved from 8.4% to 76.1% with 5% fiber loading. Miao et al. [158] reported preparation of epoxidized soybean oil (ESO) based paper cellulose composites. The study demonstrated the good compatibility between ESO and cellulose paper. Boron trifluoride diethyl etherate was used as catalyst and ESO was in situ polymerized on the morphological surface of

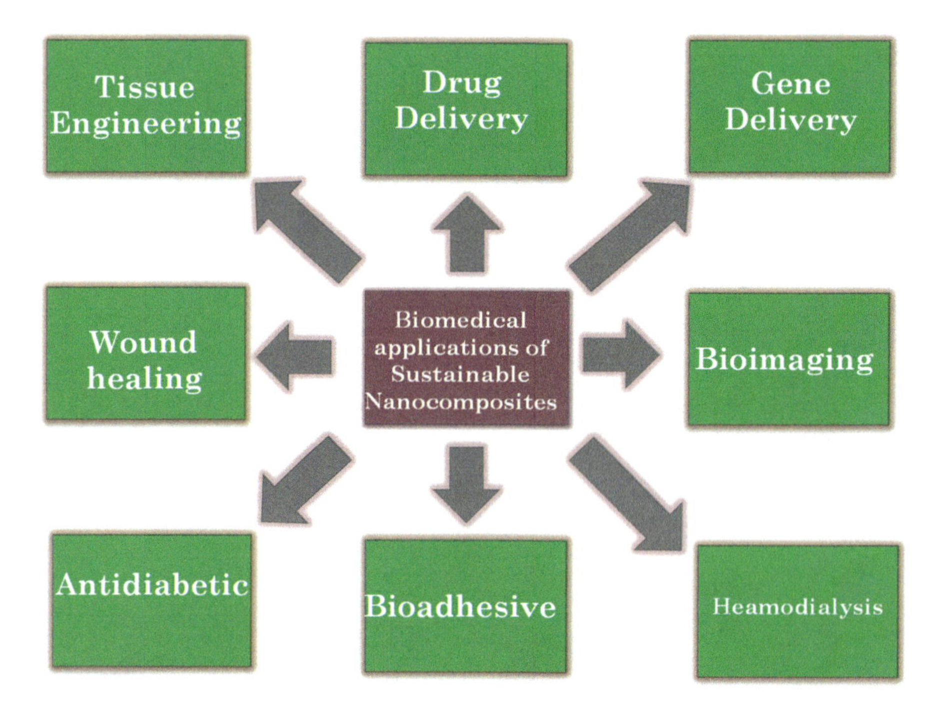

Fig. 8 Various biomedical applications of Chitosan

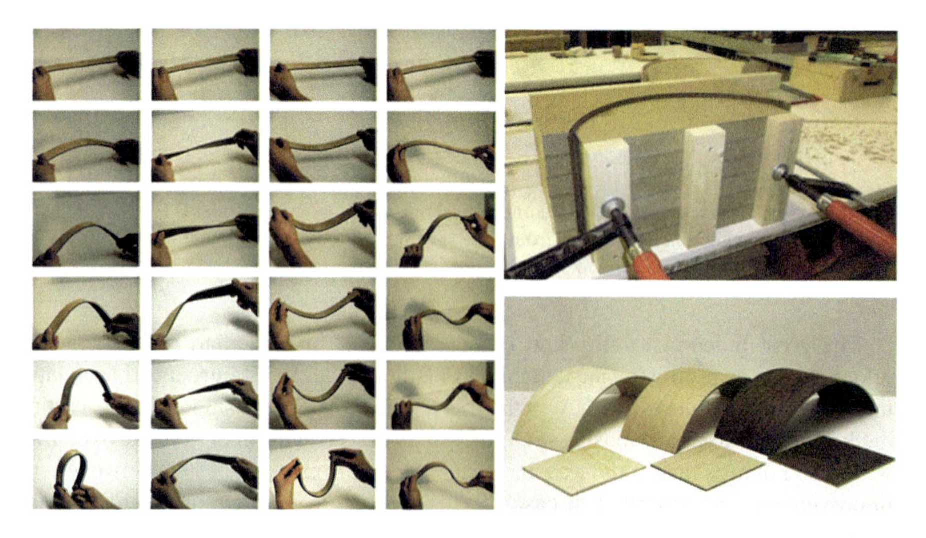

Fig. 9 Illustration of a wood veneer surface with excellent flexibility and a wide range of geometries [156]. Reproduced with permission from The Elsevier Ltd.

the cellulose. The long alkyl chain of the ESO is assumed to improve fiber dispersion in PLA matrix. The polymerization of ESO was confirmed using Fourier transform infrared spectroscopy (FTIR). Water vapour permeability (WVP) measurements revealed that these composites can be used as the potential water resistant material. Majeed et al. [159] reviewed a work on natural fiber/nanoclay reinforced polymeric materials for food packaging application. The biodegradable hybrid material obtained by mixing natural fibers with nanoclay exhibits improved barrier properties. At some optimum proportion, this combination can show excellent dispersion along with good compatibility with the matrix. This combination of hybrid materials has excellent mechanical strength at very low cost. This material has most suitable application as vapour sensitive materials to be used in electronic goods and pharmaceutical packaging.

4 Conclusions

This chapter discussed the significant progress in the processing of sustainable NCs containing NFs. The sustainable NCs formation requires polymer which is as bio-based polymer matrix along with NFs. In this chapter, some of the technical features of the fabrication methods were described for the industrial purposes. The processing technique greatly influences the electrical, biological and mechanical properties of sustainable NCs to a great extent. Biodegradable polymers enhanced by NFs are attractive and befitting candidates for applications in several fields such as biomedical, especially as drug delivery, tissue engineering and in, electronic such as supercapacitors, piezoelectric materials, and biosensors. Mechanical and barrier properties of the Biopolymer NCs can be greatly improved by reinforcing them with various NFs. The resultant BNCs maintains their vital essence of biodegradability even after infusion. Biopolymer composites with anticipated properties have the ample potential to replace conventional materials presently used in various field of applications.

References

1. Mrlik M, Sobolciak P, Krupa I, Kasak P (2018) Light-controllable viscoelastic properties of a photolabile carboxybetaine ester-based polymer with mucus and cellulose sulfate. Emergent Mater 1(1–2):1–1
2. Meng T, Yi C, Liu L, Karim A, Gong X (2018) Enhanced thermoelectric properties of two-dimensional conjugated polymers. Emergent Mater 1(1–2):1
3. Popelka A, Sobolciak P, Mrlík M, Nogellova Z, Chodák I, Ouederni M, Al-Maadeed MA, Krupa I (2018) Foamy phase change materials based on linear low-density polyethylene and paraffin wax blends. Emergent Mater 1(1–2):1–8

4. Deshmukh K, Ahamed MB, Deshmukh RR, Pasha SKK, Bhagat PR, Chidambaram K (2016) Biopolymer composites with high dielectric performance: interface engineering. In: Sadasivuni KK, Ponnamma D, Kim J, Cabibihan JJ, AlMaadeed MAA (eds) Biopolymer composites in electronics. Elsevier, Amsterdam, pp 27–128
5. Deshmukh K, Sankaran S, Ahamed MB, Sadasivuni KK, Pasha SKK, Ponnamma D, Sreekanth PSR, Chidambaram K (2017) Dielectric spectroscopy. In: Thomas S, Mishra RK, Thomas R, Zachariah AK (eds) Instrumental techniques to the characterizations of nanomaterials. Elsevier, Amsterdam, pp 237–299
6. Thangamani GJ, Deshmukh K, Chidambaram K, Ahamed MB, Sadasivuni KK, Ponnamma D, Faisal M, Nambiraj NA, Pasha SKK (2018) Influence of CuO nanoparticles and graphene nanoplatelets on the sensing behavior of poly (vinylalcohol) nanocomposites for the detection of ethanol and propanol vapors. J Mater Sci Mater Electron 29(6):5186–5205
7. Badgayan ND, Samanta S, Sahu SK, Venkata Siva SB, Sadasivuni KK, Sahu D, Rama Sreekanth PS (2017) Tribological behaviour of 1D and 2D nanofiller based high density polyethylene nanocomposites: a run in and steady state phase analysis. Wear 376–377:1379–1390
8. Thangamani GJ, Deshmukh K, Sadasivuni KK, Chidambaram K, Ahamed MB, Ponnamma D, AlMaadeed MAA, Pasha SKK (2017) Recent advances in electrochemical biosensors and gas sensors based on graphene and carbon nanotubes (CNT): a review. Ad Mater Lett 8(3):196–205
9. Sathapathy KD, Deshmukh K, Ahamed MB, Sadasivuni KK, Ponnamma D, Pasha SKK, AlMaadeed MAA, Ahmad J (2017) High quality factor poly (vinylidenefluoride) based novel nanocomposites filled with graphene nanoplatelets and vanadium pentoxide for high-Q capacitor applications. Ad Mater Lett 8(3):288–294
10. Mohanapriya MK, Deshmukh K, Chidambaram K, Ahamed MB, Sadasivuni KK, Ponnamma D, AlMaadeed MAA, Deshmukh RR, Pasha SKK (2017) Polyvinyl alcohol (PVA)/Polystyrene sulfonic acid (PSSA)/carbon black nanocomposites for flexible energy storage device applications. J Mater Sci Mater Electron 28(8):6099–6111
11. Abdullah N, Yusof N, Ismail AF, Othman FE, Jaafar J, Jye LW, Salleh WN, Aziz F, Misdan N (2018) Effects of manganese (VI) oxide on polyacrylonitrile-based activated carbon nanofibers (ACNFs) and its preliminary study for adsorption of lead (II) ions. Emergent Mater 1(1–2):1–6
12. Mohanapriya MK, Deshmukh K, Ahamed MB, Chidambaram K, Pasha SKK (2016) Zeolite 4A filled poly (3, 4-ethylenedioxythiophene): (polystyrenesulfonate) and polyvinyl alcohol blend nanocomposites as high-k dielectric materials for embedded capacitor applications. Ad Mater Lett 7(12):996–1002
13. Muzaffar A, Ahamed MB, Deshmukh K, Faisal M, Pasha SKK (2018) Enhanced electromagnetic absorption in NiO and $BaTiO_3$ based polyvinylidene fluoride nanocomposites. Mater Lett 218:217–220
14. Ponnamma D, Sadasivuni KK, Strankowski M, Moldenaers P, Thomas S, Grohens Y (2013) Interrelated shape memory and Payne effect in polyurethane/graphene oxide nanocomposites. RSC Adv 3(36):16068–16079
15. Ponnamma D, Sadasivuni KK, Strankowski M, Guo Q, Thomas S (2013) Synergistic effect of multiwalled carbon nanotubes and reduced graphene oxide in natural rubber for sensing applications. Soft Matter 9(43):10343–10353
16. Sadasivuni KK, Castro M, Saiter A, Delbreilh L, Feller JF, Thomas S, Grohens Y (2013) Development of poly(isobutylene-co-isoprene)/reduced graphene oxide nanocomposites for barrier, dielectric and sensing applications. Mater Lett 96:109–112
17. Mohanapriya MK, Deshmukh K, Ahamed MB, Chidambaram K, Pasha SKK (2016) Influence of cerium oxide (CeO_2) nanoparticles on the structural, morphological, mechanical and dielectric properties of PVA/PPy blend nanocomposites. Mater Today Proc 3(6):1864–1873
18. Sadasivuni KK, Saiter A, Gautier N, Thomas S, Grohens Y (2013) Effect of molecular interaction on the performance of poly (isobutylene-co-isoprene)/graphene and clay nanocomposites. Colloids Polymer Sci 291(7):1729–1740

19. Fayyad EM, Abdullah AM, Hassan MK, Mohamed AM, Jarjoura G, Farhat Z (2018) Recent advances in electroless-plated Ni-P and its composites for erosion and corrosion applications: a review. Emergent Mater 1(1–2):1–22
20. Illa MP, Khandelwal M, Sharma CS (2018) Bacterial cellulose-derived carbon nanofibers as anode for lithium-ion batteries. Emergent Mater 1(3–4):1–6
21. Nisar U, Amin R, Shakoor A, Essehli R, Al-Qaradawi S, Kahraman R, Belharouak I (2018) Synthesis and electrochemical characterization of Cr-doped lithium-rich Li 1.2 Ni 0.16 Mn 0.56 Co 0.08-x Cr x O 2 cathodes. Emergent Mater 1(3–4):1–0
22. Reddy YG, Awasthi AM, Chary AS, Reddy SN (2018) Characterization and ion transport studies through impedance spectroscopy on (1-x) Pb (NO 3) 2: xAl 2 O 3 composite solid electrolytes. Emergent Mater 1(3–4):1–0
23. Fadiran OO, Girouard N, Meredith JC (2018) Pollen fillers for reinforcing and strengthening of epoxy composites. Emergent Mater 1(1–2):95–103
24. Selmy AE, Soliman M, Allam NK (2018) Refractory plasmonics boost the performance of thin-film solar cells. Emergent Mater 1(3–4):1–7
25. Ponnamma D, Erturk A, Parangusan H, Deshmukh K, Ahamed MB, Al-Maadeed MA (2018) Stretchable quaternary phasic PVDF-HFP nanocomposite films containing graphene-titania-$SrTiO_3$ for mechanical energy harvesting. Emergent Mater 1(1–2):55–65
26. Ponnamma D, Sadasivuni KK, Grohens Y, Guo Q, Thomas S (2014) Carbon nanotube based elastomer composites-an approach towards multifunctional materials. J Mater Chem C 2(40):8446–8485
27. Fujiyama-Novak JH, Rufino V, Amaral RA, Habert AC, Borges CP, Mano B (2016) Oxygen permeability of nanocomposite-based polyolefin films. Macromol Symp 368(1):19–23
28. Sadasivuni KK, Ponnamma D, Kumar B, Strankowski M, Cardinaels R, Moldenaers P, Thomas S, Grohens Y (2014) Dielectric properties of modified graphene oxide filled polyurethane nanocomposites and its correlation with rheology. Compos Sci Technol 104:18–25
29. Deshmukh K, Ahamed MB, Sadasivuni KK, Ponnamma D, AlMaadeed MAA, Pasha SKK, Deshmukh RR, Chidambaram K (2017) Graphene oxide reinforced poly (4-styrenesulfonic acid)/polyvinyl alcohol blend composites with enhanced dielectric properties for portable and flexible electronics. Mater Chem Phys 186:188–201
30. Cooke KO, Khan TI (2018) Effect of thermal processing on the tribology of nanocrystalline Ni/TiO_2 coatings. Emergent Mater 1(3–4):1–9
31. Rahman M, Hamdan S, Hashim DM, Islam M, Takagi H (2015) Bamboo fiber polypropylene composites: effect of fiber treatment and nano clay on mechanical and thermal properties. J Vinyl Add Tech 21(4):253–258
32. Chen T, Xie Y, Wei Q, Wang XA, Hagman O, Karlsson O, Liu J, Lin M (2016) Improving the mechanical properties of ultra-low density plant fiber composite (ULD_PFC) by refining treatment. BioResources 11(4):8558–8569
33. Chen RS, Ahmad S (2017) Mechanical performance and flame retardancy of rice husk/ organoclay-reinforced blend of recycled plastics. Mater Chem Phys 198:57–65
34. Tasdemir M (2017) Effects of olive pit and almond shell powder on polypropylene. Key Eng Mater Trans Tech 733:65–68
35. Arjmandi R, Hassan A, Majeed K, Zakaria Z (2015) Rice husk filled polymer composites. Int J Polymer Sci 32. Article ID 501471
36. Majeed K, Hassan A, Bakar AA, Jawaid M (2016) Effect of montmorillonite (MMT) content on the mechanical, oxygen barrier, and thermal properties of rice husk/MMT hybrid filler-filled low-density polyethylene nanocomposite blown films. J Thermoplast Compos Mater 29(7):1003–1019
37. Ahmad J, Deshmukh K, Habib M, Hägg MB (2013) Influence of TiO_2 on the chemical, mechanical and gas separation properties of polyvinylalcohol-titanium dioxide (PVA/TiO_2) nanocomposite membrane. Int J Polym Anal Charact 18(4):287–296

38. Mohanapriya MK, Deshmukh K, Ahamed MB, Chidambaram K, Pasha SKK (2015) Structural, morphological and dielectric properties of multiphase nanocomposites consisting of polycarbonate, barium titanate and carbon black nanoparticles. Int J Chem Tech Res 8(5): 32–41
39. Feller JF, Sadasivuni KK, Castro M, Bellegou H, Pillin I, Thomas S, Grohens Y (2015) Gas barrier efficiency of clay and graphene-poly(isobutylene-co-isoprene) nanocomposite membranes evidenced by a quantum resistive vapour sensor cell. Nanocomposites 1(4): 96–105
40. Kafy A, Sadasivuni KK, Akther A, Min SK, Kim J (2015) Cellulose/graphene nanocomposites as multifunctional electronic and solvent sensor material. Mater Lett 159:20–23
41. Akhtar MN, Sulong AB, Nazir MS, Majeed K, Radzi MK, Ismail NF, Raza MR (2017) Kenaf-biocomposites: manufacturing, characterization, and applications. In: Green biocomposites. Springer International Publication, Berlin, pp 225–254
42. Cavallaro G, Lazzara G, Milioto S (2013) Sustainable nanocomposites based on halloysite nanotubes and pectin/polyethylene glycol blend. Polym Degrad Stab 98(12):2529–2536
43. Ghaffari A, Navaee K, Oskoui M, Bayati K, Rafiee-Tehrani M (2007) Preparation and characterization of free mixed-film of pectin/chitosan/Eudragit® RS intended for sigmoidal drug delivery. Eur J Pharm Biopharm 67(1):175–186
44. Miyamoto H, Yamane C, Seguchi M, Okajima K (2010) Comparison between cellulose blend films prepared from aqueous sodium hydroxide solution and edible films of biopolymers with possible application for new food materials. Food Sci Technol Res 17(1): 21–30
45. Mishra RK, Majeed AB, Banthia AK (2011) Development and characterization of pectin/gelatin hydrogel membranes for wound dressing. Int J Plas Technol 15(1):82–95
46. Deshmukh K, Ahamed MB, Sadasivuni KK, Ponnamma D, Deshmukh RR, Pasha SK, AlMaadeed MA, Chidambaram K (2016) Graphene oxide reinforced polyvinyl alcohol/polyethylene glycol blend composites as high-performance dielectric material. J Polym Res 23:159
47. Gunbas ID, Aydemir SU, Gülceİz S, Delioğlu Gürhan I, Hasirci N (2012) Semi-IPN chitosan/PEG microspheres and films for biomedical applications: characterization and sustained release optimization. Ind Eng Chem Res 51(37):11946–11954
48. Altinisik A, Yurdakoc K (2011) Synthesis, characterization, and enzymatic degradation of chitosan/PEG hydrogel films. J Appl Polym Sci 122(3):1556–1563
49. Ruiz-Hitzky E, Sobral MM, Gómez-Avilés A, Nunes C, Ruiz-García C, Ferreira P, Aranda P (2016) Clay-graphene nanoplatelets functional conducting composites. Adv Func Mater 26(41):7394–7405
50. Liu M, Wu C, Jiao Y, Xiong S, Zhou C (2013) Chitosan–halloysite nanotubes nanocomposite scaffolds for tissue engineering. J Mater Chem B 1:2078–2089
51. Abdullayev E, Lvov Y (2010) Clay nanotubes for corrosion inhibitor encapsulation: release control with end stoppers. J Mater Chem 20:6681–6687
52. Lvov Y, Abdullayev E (2013) Functional polymer–clay nanotube composites with sustained release of chemical agents. Prog Polym Sci 38(10–11):1690–1719
53. Deshmukh K, Ahamed MB, Sadasivuni KK, Ponnamma D, Deshmukh RR, Trimukhe AM, Pasha SK, Polu AR, AlMaadeed MA, Chidambaram K (2017) Solution-processed white graphene-reinforced ferroelectric polymer nanocomposites with improved thermal conductivity and dielectric properties for electronic encapsulation. J Polym Res 24:27
54. Deshmukh K, Ahmad J, Hägg MB (2014) Fabrication and characterization of polymer blends consisting of cationic polyallylamine and anionic polyvinyl alcohol. Ionics 20: 957–967
55. Deshmukh K, Ahamed MB, Deshmukh RR, Sadasivuni KK, Ponnamma D, Pasha SK, AlMaadeed MA, Polu AR, Chidambaram K (2017) Eeonomer 200F®: a high-performance nanofiller for polymer reinforcement-Investigation of the structure, morphology and dielectric properties of polyvinyl alcohol/Eeonomer-200F® nanocomposites for embedded capacitor applications. J Electron Mater 46(4):2406–2418

56. Spitalsky Z, Tasis D, Papagelis K, Galiotis C (2010) Carbon nanotube–polymer composites: chemistry, processing, mechanical and electrical properties. Prog Polym Sci 35(3):357–401
57. Deshmukh K, Ahamed MB, Deshmukh RR, Pasha SK, Sadasivuni KK, Ponnamma D, Chidambaram K (2016) Synergistic effect of vanadium pentoxide and graphene oxide in polyvinyl alcohol for energy storage application. Eur Polymer J 76:14–27
58. Lau WJ, Gray S, Matsuura T, Emadzadeh D, Chen JP, Ismail AF (2015) A review on polyamide thin film nanocomposite (TFN) membranes: history, applications, challenges and approaches. Water Res 80:306–824
59. Janson A, Minier-Matar J, Al-Shamari E, Hussain A, Sharma R, Rowley D, Adham S (2018) Evaluation of new ion exchange resins for hardness removal from boiler feedwater. Emergent Mater 1(1–2):1–1
60. Nagaraj A, Govindaraj D, Rajan M (2018) Magnesium oxide entrapped Polypyrrole hybrid nanocomposite as an efficient selective scavenger for fluoride ion in drinking water. Emergent Mater 1(1–2):1–9
61. Hegab HM, Zou L (2015) Graphene oxide-assisted membranes: fabrication and potential applications in desalination and water purification. J Membr Sci 484:95–106
62. Deshmukh K, Ahamed MB, Sadasivuni KK, Ponnamma D, AlMaadeed MAA, Deshmukh RR, Pasha SKK, Polu AR, Chidambaram K (2017) Fumed SiO_2 nanoparticle reinforced biopolymer blend nanocomposites with high dielectric constant and low dielectric loss for flexible organic electronics. J Appl Polym Sci 134(5):44427
63. Basavaiah K, Kahsay MH, Rama Devi D (2018) Green synthesis of magnetite nanoparticles using aqueous pod extract of Dolichos lablab L for an efficient adsorption of crystal violet. Emergent Mater 1(3–4):1–2
64. Deshmukh K, Ahamed MB, Deshmukh RR, Pasha SKK, Sadasivuni KK, Polu AR, Ponnamma D, AlMaadeed MAA, Chidambaram K (2017) Newly developed biodegradable polymer nanocomposites of cellulose acetate and Al_2O_3 nanoparticles with enhanced dielectric performance for embedded passive applications. J Mater Sci Mater Electron 28(1):973–986
65. Parambath SV, Ponnamma D, Sadasivuni KK, Thomas S, Stephen R (2017) Effect of polyhedral oligomeric siliseuioxane on the physical properties of polyvinyl alcohol. J Appl Polym Sci 134(43):45447
66. Ponnamma D, Chamakh MM, Deshmukh K, Ahamed MB, Alper E, Sharma P, AlMaadeed MAA (2017) Ceramic based polymer nanocomposites as piezoelectric materials. In: Ponnamma D, Sadasivuni KK, Cabibihan JJ, AlMaadeed MAA (eds) the Book "Smart polymer nanocomposites. Springer Publications, Berlin, pp 77–94
67. Kim H, Macosko CW (2008) Morphology and properties of polyester/exfoliated graphite nanocomposites. Macromolecules 41(9):3317–3327
68. Andrews R, Jacques D, Minot M, Rantell T (2002) Fabrication of carbon multiwall nanotube/polymer composites by shear mixing. Macromol Mater Eng 287(6):395–403
69. Hou Y, Cheng Y, Hobson T, Liu J (2010) Design and synthesis of hierarchical MnO_2 nanospheres/carbon nanotubes/conducting polymer ternary composite for high performance electrochemical electrodes. Nano Lett 10(7):2727–2733
70. Ajayan PM, Tour JM (2007) Materials science: nanotube composites. Nature 447(7148): 1066–1068
71. Thakur VK, Kessler MR (2015) Self-healing polymer nanocomposite materials: a review. Polymer 69:369–383
72. Deshmukh K, Ahamed MB, Pasha SK, Deshmukh RR, Bhagat PR (2015) Highly dispersible graphene oxide reinforced polypyrrole/polyvinyl alcohol blend nanocomposites with high dielectric constant and low dielectric loss. RSC Adv 5:61933–61945
73. Potts JR, Dreyer DR, Bielawski CW, Ruoff RS (2011) Graphene-based polymer nanocomposites. Polymer 52:5–25
74. El Achaby M, Arrakhiz FE, Vaudreuil S, Kacem Qaiss A, Bousmina M, Fassi-Fehri O (2012) Mechanical, thermal, and rheological properties of graphene-based polypropylene nanocomposites prepared by melt mixing. Polym Compos 33(5):733–744

75. Tang QY, Chan YC, Wong NB, Cheung R (2010) Surfactant-assisted processing of polyimide/multiwall carbon nanotube nanocomposites for microelectronics applications. Polym Int 59(9):1240–1245
76. Inam F, Heaton A, Brown P, Peijs T, Reece MJ (2014) Effects of dispersion surfactants on the properties of ceramic–carbon nanotube (CNT) nanocomposites. Ceram Int 40(1): 511–516
77. Tkalya EE, Ghislandi M, de With G, Koning CE (2012) The use of surfactants for dispersing carbon nanotubes and graphene to make conductive nanocomposites. Curr Opin Colloid Inter Sci 17(4):225–232
78. Veprek S, Veprek-Heijman MJ (2008) Industrial applications of superhard nanocomposite coatings. Surf Coat Technol 202(21):5063–5073
79. Fukushima K, Wu MH, Bocchini S, Rasyida A, Yang MC (2012) PBAT based nanocomposites for medical and industrial applications. Mater Sci Eng C 32:1331–1351
80. Ponnamma D, Saiter A, Saiter JM, Thomas S, Grohens Y, AlMaadeed MAA, Sadasivuni KK (2016) Influence of temperature on the confinement effect of micro and nanolevel graphite filled poly(isoprene-co-isobutylene) composites. J Polym Res 23:125
81. Stankovich S, Dikin DA, Dommett GH, Kohlhaas KM, Zimney EJ, Stach EA, Piner RD, Nguyen ST, Ruoff RS (2006) Graphene-based composite materials. Nature 442:282–286
82. Deshmukh K, Ahamed MB, Deshmukh RR, Pasha SKK, Chidambaram K, Sadasivuni KK, Ponnamma D, AlMaadeed MA (2016) Ecofriendly synthesis of graphene oxide reinforced hydroxypropyl methyl cellulose/polyvinylalcohol blend nanocomposites filled with zinc oxide nanoparticles for high-k capacitor applications. Polymer-Plastics Technol Eng 55(12): 1240–1253
83. Deshmukh K, Ahamed MB, Polu AR, Sadasivuni KK, Pasha SK, Ponnamma D, AlMaadeed MA, Deshmukh RR, Chidambaram K (2016) Impedance spectroscopy, ionic conductivity and dielectric studies of new Li + ion conducting polymer blend electrolytes based on biodegradable polymers for solid state battery applications. J Mater Sci Mater Electron 27(11):11410–11424
84. Stephenson T, Li Z, Olsen B, Mitlin D (2014) Lithium ion battery applications of molybdenum disulfide (MoS 2) nanocomposites. Energy Environ Sci 7:209–231
85. Pfaendner R (2010) Nanocomposites: industrial opportunity or challenge? Polym Degrad Stab 95(3):369–373
86. Ahmad R, Griffete N, Lamouri A, Felidj N, Chehimi MM, Mangeney C (2015) Nanocomposites of gold nanoparticles@ molecularly imprinted polymers: chemistry, processing, and applications in sensors. Chem Mater 27(16):5464–5478
87. Siqueira G, Mathew AP, Oksman K (2011) Processing of cellulose nanowhiskers/cellulose acetate butyrate nanocomposites using sol–gel process to facilitate dispersion. Compos Sci Technol 71(16):1886–1892
88. Biswas M, Ray SS (2001) Recent progress in synthesis and evaluation of polymer-montmorillonite nanocomposites. Adv Polym Sci 155:167–222
89. Deshmukh K, Ahamed MB, Deshmukh RR, Pasha SK, Sadasivuni KK, Ponnamma D, AlMaadeed MA (2017) Striking multiple synergies in novel three-phase fluoropolymer nanocomposites by combining titanium dioxide and graphene oxide as hybrid fillers. J Mater Sci Mater Electron 28(1):559–575
90. Deshmukh K, Ahamed MB, Deshmukh RR, Bhagat PR, Pasha SK, Bhagat A, Shirbhate R, Telare F, Lakhani C (2016) Influence of K_2CrO_4 doping on the structural, optical and dielectric properties of polyvinyl alcohol/K_2CrO_4 composite films. Polymer-Plastics Technol Eng 55(3):231–241
91. Parry S, Pancoast J, Mildenhall S (2015) Chemical and bonding effects of exposing uncured PBI-NBR insulation to ambient conditions. J Appl Polym Sci 132(40):42636
92. Balachandran M, Devanathan S, Muraleekrishnan R, Bhagawan SS (2012) Optimizing properties of nanoclay–nitrile rubber (NBR) composites using face centred central composite design. Mater Des 35:854–862

93. Unalan IU, Cerri G, Marcuzzo E, Cozzolino CA, Farris S (2014) Nanocomposite films and coatings using inorganic nanobuilding blocks (NBB): current applications and future opportunities in the food packaging sector. RSC Adv 4(56):29393–29428
94. Liu L, Jia D, Luo Y, Guo B (2006) Preparation, structure and properties of nitrile–butadiene rubber–organoclay nanocomposites by reactive mixing intercalation method. J Appl Polym Sci 100(3):1905–1913
95. Fuentes-Alventosa JM, Introzzi L, Santo N, Cerri G, Brundu A, Farris S (2013) Self-assembled nanostructured biohybrid coatings by an integrated sol–gel/intercalation' approach. RSC Adv 3(47):25086–25096
96. Kim H, Abdala AA, Macosko CW (2010) Graphene/polymer nanocomposites. Macromolecules 43(16):6515–6530
97. Ramanathan T, Abdala AA, Stankovich S, Dikin DA, Herrera-Alonso M, Piner RD, Adamson DH, Schniepp HC, Chen XR, Ruoff RS, Nguyen ST (2008) Functionalized graphene sheets for polymer nanocomposites. Nat Nanotechnol 3:327–331
98. Dennis H, Hunter DL, Chang D, Kim S, White JL, Cho JW, Paul DR (2001) Effect of melt processing conditions on the extent of exfoliation in organoclay-based nanocomposites. Polymer 42(23):9513–9522
99. Njuguna J, Pielichowski K, Desai S (2008) Nanofiller-reinforced polymer nanocomposites. Polym Adv Technol 19(8):947–959
100. Capadona JR, Van Den Berg O, Capadona LA, Schroeter M, Rowan SJ, Tyler DJ, Weder C (2007) A versatile approach for the processing of polymer nanocomposites with self-assembled nanofibre templates. Nat Nanotechnol 2:765–769
101. Moniruzzaman M, Winey KI (2006) Polymer nanocomposites containing carbon nanotubes. Macromolecules 39(16):5194–5205
102. Roth SV, Herzog G, Körstgens V, Buffet A, Schwartzkopf M, Perlich J, Kashem MA, Döhrmann R, Gehrke R, Rothkirch A, Stassig K (2011) In situ observation of cluster formation during nanoparticle solution casting on a colloidal film. J Phys Condens Matter 23 (25):254208
103. Al-Hussein M, Schindler M, Ruderer MA, Perlich J, Schwartzkopf M, Herzog G, Heidmann B, Buffet A, Roth SV, Müller-Buschbaum P (2013) In situ X-ray study of the structural evolution of gold nano-domains by spray deposition on thin conductive P3HT films. Langmuir 29(8):2490–2497
104. Klein LC (2013) Sol-gel optics: processing and applications. Springer Publications, Berlin
105. Zhang J, Zhang M, Lin L, Wang X (2015) Sol processing of conjugated carbon nitride powders for thin-film fabrication. Angew Chem Int Ed 54(21):6297–6301
106. Neena D, Shah AH, Deshmukh K, Ahmad H, Fu DJ, Kondamareddy KK, Kumar P, Dwivedi RK, Sing V (2016) Influence of (Co-Mn)co- doping on the microstructures, optical properties of sol gel derived ZnO nanoparticles. Eur Phys J D 70:53
107. Zhang J, Chen Y, Wang X (2015) Two-dimensional covalent carbon nitride nanosheets: synthesis, functionalization, and applications. Energy Environ Sci 8(11):3092–3108
108. Owens Gareth J, Singh Rajendra K, Foroutan Farzad, Alqaysi Mustafa, Han Cheol-Min, Mahapatra Chinmaya, Kim Hae-Won, Knowles Jonathan C (2016) Sol–gel based materials for biomedical applications. Prog Mater Sci 77:1–79
109. Williams G, Seger B, Kamat PV (2008) TiO_2-graphene nanocomposites. UV-assisted photocatalytic reduction of graphene oxide. ACS Nano 2(7):1487–1491
110. Vatani M, Lu Y, Lee KS, Kim HC, Choi JW (2013) Direct-write stretchable sensors using single-walled carbon nanotube/polymer matrix. J Electron Packag 135(1):011009
111. Morteza A, Aekachan P, Sangjun L, Seunghwa R, Park I (2014) Highly stretchable and sensitive strain sensor based on silver nanowire elastomer nanocomposite. ACS Nano 8(5): 5154–5163
112. Lu Y, Vatani M, Choi JW (2013) Direct-write/cure conductive polymer nanocomposites for 3D structural electronics. J Mech Sci Technol 27(10):2929–2934

113. Vatani M, Engeberg ED, Choi JW (2014) Detection of the position, direction and speed of sliding contact with a multi-layer compliant tactile sensor fabricated using direct-print technology. Smart Mater Struct 23(9):095008
114. Wang S, Hu Y, Zhongkai Q, Wang Z, Chen Z, Fan W (2003) Preparation and flammability properties of polyethylene/clay nanocomposites by melt intercalation method from Na+ montmorillonite. Mater Lett 57:2675–2678
115. Yoshimoto S, Ohashi F, Ohnishi Y, Nonami T (2004) Synthesis of polyaniline–montmorillonite nanocomposites by the mechanochemical intercalation method. Synth Met 145(2–3):265–270
116. Kuila BK, Nandi AK (2004) Physical, mechanical, and conductivity properties of poly (3-hexylthiophene)-montmorillonite clay nanocomposites produced by the solvent casting method. Macromolecules 37(23):8577–8584
117. Mozafari M, Moztarzadeh F, Rabiee M, Azami M, Maleknia S, Tahriri M, Moztarzadeh Z, Nezafati N (2010) Development of macroporous nanocomposite scaffolds of gelatin/bioactive glass prepared through layer solvent casting combined with lamination technique for bone tissue engineering. Ceram Int 36(8):2431–2439
118. Oksman K, Mathew AP, Bondeson D, Kvien I (2006) Manufacturing process of cellulose whiskers/polylactic acid nanocomposites. Compos Sci Technol 66(15):2776–2784
119. Zhang HB, Zheng WG, Yan Q, Yang Y, Wang JW, Lu ZH, Ji GY, Yu ZZ (2010) Electrically conductive polyethylene terephthalate/graphene nanocomposites prepared by melt compounding. Polymer 51(5):1191–1196
120. Wang Y, Wang Y, Hosono E, Wang K, Zhou H (2008) The design of a LiFePO4/carbon nanocomposite with a core–shell structure and its synthesis by an in situ polymerization restriction method. Angew Chem Int Ed 47(39):7461–7465
121. Bao C, Guo Y, Song L, Kan Y, Qian X, Hu Y (2011) In situ preparation of functionalized graphene oxide/epoxy nanocomposites with effective reinforcements. J Mater Chem 21(35):13290–13298
122. Biswas S, Fukushima H, Drzal LT (2011) Mechanical and electrical property enhancement in exfoliated graphene nanoplatelet/liquid crystalline polymer nanocomposites. Compos A Appl Sci Manuf 42(4):371–375
123. Hussain I, Tran HP, Jaksik J, Moore J, Islam N, Uddin MJ (2018) Functional materials, device architecture, and flexibility of perovskite solar cell. Emergent Mater 1(3–4):1–22
124. Kalaitzidou K, Fukushima H, Drzal LT (2007) A new compounding method for exfoliated graphite–polypropylene nanocomposites with enhanced flexural properties and lower percolation threshold. Compos Sci Technol 67(10):2045–2051
125. Alexandre M, Dubois P, Sun T, Garces JM, Jérôme R (2002) Polyethylene-layered silicate nanocomposites prepared by the polymerization-filling technique: synthesis and mechanical properties. Polymer 43(8):2123–2132
126. Zhang H, Chen G (2009) Potent antibacterial activities of Ag/TiO_2 nanocomposite powders synthesized by a one-pot sol-gel method. Environ Sci Technol 43(8):2905–2910
127. Jitianu A, Cacciaguerra T, Benoit R, Delpeux S, Beguin F, Bonnamy S (2004) Synthesis and characterization of carbon nanotubes–TiO_2 nanocomposites. Carbon 42(5–6):1147–1151
128. Liu C, Wang K, Luo S, Tang Y, Chen L (2011) Direct electrodeposition of graphene enabling the one step synthesis of graphene–metal nanocomposite films. Small 7(9):1203–1206
129. Belgacem MN, Gandini A (2008) Monomers, polymers and composites from renewable resources. Elsevier Publications, Amsterdam
130. Abdul Khalil HPS, Bhat AH, Ireana Yusra AF (2012) Green composites from sustainable cellulose nanofibrils: a review. Carbohyd Polym 87(2):963–979
131. Jonathan G, Zhang M (2010) Polyblend nanofibers for biomedical applications: perspectives and challenges. Trends Biotechnol 28(4):189–197
132. Liu Y, Li Y, Yang G, Zheng X, Zhou S (2015) Multi-stimulus-responsive shape-memory polymer nanocomposite network cross-linked by cellulose nanocrystals. ACS Appl Mater Inter 7(7):4118–4126

133. Yeom J, Oh EJ, Reddy M (2009) Guided bone regeneration by poly (lactic-*co*-glycolic acid) grafted hyaluronic acid bi-layer films for periodontal barrier applications. Acta Biomaterialia 5(9):3394–3403
134. Reddy MM, Vivekanandhan S, Misra M, Bhatia SK, Mohanty AK (2013) Biobased plastics and bionanocomposites: current status and future opportunities. Progress Polymer Sci 38 (10):1653–1689
135. Salami M, Kaveian F, Rafienia M, Saber-Samandari S, Khandan A, Naeimi M (2017) Electrospun polycaprolactone/lignin-based nanocomposite as a novel tissue scaffold for biomedical applications. J Med Signals Sens 7(4):228–238
136. Sadasivuni KK, Ponnamma D, Kim J, Cabibihan JJ, AlMaadeed MAA (2017) Introduction of biopolymer composites. In: Ponnamma D (ed) Biopolymer Composites in electronics. Elsevier, Amsterdam
137. Mai F, Habibi Y, Jean-Marie R, Philippe D, Feller JF, Ton P, Emiliano B (2013) Poly (lactic acid)/carbon nanotube nanocomposites with integrated degradation sensing. Polymer 54(25): 6818–6823
138. Sadasivuni KK Ponnamma D, Cabibihan JJ, AlMaadeed MAA (2016) Electronic applications of polydimethylsiloxane and its composites. In: Ponnamma D, Sadasivuni KK, Wan C, Thomas S, AlMaadeed MAA (eds) Flexible and stretchable electronic composites. Springer Publication, Berlin, pp 199–228
139. Okonkwo PC, Collins E, Okonkwo E (2017) Application of biopolymer composites in super capacitor. In: Sadasivuni KK Cabibihan JJ, Ponnamma D, AlMaadeed MAA (eds) Biopolymer composites in electronics. Springer Publication, Berlin, pp 487–503
140. Christinelli WA, Gonçalves R, Pereira EC (2016) A new generation of electrochemical supercapacitors based on layer-by-layer polymer films. J Power Sources 303:73–80
141. Botta L, Scaffaro R, Sutera F, Mistretta MC (2018) Reprocessing of PLA/graphene nanoplatelets nanocomposites. Polymers 10:18
142. Alam J, Alam M, Raja M, Abduljaleel Z, Dass LA (2014) MWCNTs-reinforced epoxidized linseed oil plasticized polylactic acid nanocomposite and its electroactive shape memory behaviour. Int J Mol Sci 15(11):19924–19937
143. Wang ZW, Zhao J, Chen M, Yang MH, Tang LY, Dang ZM, Chen FH, Huang MM, Dong X (2014) Dually actuated triple shape memory polymers of cross-linked polycyclooctene-carbon nanotube/polyethylene nanocomposites. ACS Appl Mater Interfaces 6(22):20051–20059
144. Wang K, Strandman S, Zhu XX (2017) A mini review: shape memory polymers for biomedical applications. Front Chem Sci Technol 11(2):143–153
145. Ikada Y (2006) Scope of tissue engineering. Tissue engineering: fundamentals and applications. Inter Sci Technol 8:1–90
146. Wei G, Ma PX (2004) Structure and properties of nano-hydroxyapatite/polymer composite scaffolds for bone tissue engineering. Biomaterials 25(19):4749–4757
147. Salgado AJ, Coutinho OP, Reis RL (2004) Bone tissue engineering: state of the art and future trends. Macromol Biosci 4(8):743–765
148. McCullen SD, Stevens DR, Roberts WA, Clarke LI, Bernacki SH, Gorga RE, Loboa EG (2007) Characterization of electrospun nanocomposite scaffolds and biocompatibility with adipose-derived human mesenchymal stem cells. Int J Nanomed 2(2):253–263
149. Sowmya S, Bumgardener JD, Chennazhi KP, Nair SV, Jayakumara R (2013) Role of nanostructured biopolymers and bioceramics in enamel, dentin and periodontal tissue regeneration. Progress Polymer Sci 38(10–11):1748–1772
150. Fujihara K, Kotaki M, Ramakrishna S (2005) Guided bone regeneration membrane made of polycaprolactone/calcium carbonate composite nano-fibers. Biomaterials 26(19):4139–4147
151. Minghuan L, Xiao-Peng D, Ye-Ming L, Da-Peng Y, Yun-Ze L (2017) Electrospun nanofibers for wound healing. Mater Sci Eng C 76:1413–1423

152. Park JK, Yeom J, Oh EJ, Reddy M, Kim JY, Cho DW, Lim HP, Kim NS, Park SW, Shin HI, Yang DJ, Park KB, Hahn SK (2009) Guided bone regeneration by poly(lactic-co-glycolic acid) grafted hyaluronic acid bi-layer films for periodontal barrier applications. Acta Biomaterialia 5(9):3394–3403
153. Chen FM, Zhang J, Zhang M, An Y, Chen F, Wu ZF (2010) A review on endogenous regenerative technology in periodontal regenerative medicine. Biomaterials 31(31): 7892–7927
154. Shakeel A, Saiqa I (2016) Chitosan based scaffolds and their applications in wound healing. Achieve Life Sci 10(1):27–37
155. Zadeh KM, Ponnamma D, Al-Maadeed MAA (2017) Date palm fibre filled recycled ternary polymer blend composites with enhanced flame retardancy. Polymer Test 61:341–348
156. Dahy H (2017) Biocomposite materials based on annual natural fibres and biopolymers—design, fabrication and customized applications in architecture. Construct Build Mater 147:212–220
157. Bulota M, Kreitsmann K, Hughes M, Paltakari J (2012) Acetylated mcrofibrillated cellulose as a toughening agent in poly (lactic acid). J Appl Polymer Sci 126(S1):E448–E457
158. Miao S, Liu K, Wang P, Su Z, Zhang S (2015) Preparation and characterization of epoxidized soybean oil-based paper composite as potential water-resistant materials. J Appl Polymer Sci 132(10):41575
159. Majeed K, Jawaid M, Hassan A, Abu Bakar A, Khalil HPSA, Salema AA, Inuwa I (2013) Potential materials for food packaging from nanoclay/natural fibers filled hybrid composites. Mater Des 46:391–410

Recent Advances in Paper-Based Analytical Devices: A Pivotal Step Forward in Building Next-Generation Sensor Technology

Charu Agarwal and Levente Csóka

List of Abbreviations

μPADs	Microfluidic paper-based analytical devices
Ab	Antibody
ABTS	2,2′-azino-bis(3-ethylbenzothiazoline-6-sulfonic acid)
ACV	Alternating current voltammetry
AEC	3-amino-9-ethylcarbazole
AFP	α-fetoprotein
AKD	Alkyl ketene dimer
AP	Absorbent pad
BA	Biogenic amines
BDDE	Boron-doped diamond electrode
BHB	*β*-hydroxybutyrate
BPA	Bisphenol A
CA125	Carcinoma antigen 125
CA199	Carcinoma antigen 199
CB	Carbon black
CBM	Carbohydrate binding molecule
CEA	Carcinoembryonic antigen
CFU	Colony forming units
CNCs	Carbon nanocrystals
CPRG	Chlorophenol red *β*-galactopyranoside
Cy3	Cyanine 3
DAB	3,3′-diaminobenzidine
DAP	1,8-diaminonaphthalene
ECL	Electrochemiluminescent
eGFP	Enhanced green fluorescent protein

C. Agarwal · L. Csóka (✉)
Institute of Wood Based Products and Technologies,
University of Sopron, Sopron 9400, Hungary
e-mail: levente.csoka@skk.nyme.hu

C. Agarwal
e-mail: charu.agarwal3@gmail.com

Inamuddin et al. (eds.), *Sustainable Polymer Composites and Nanocomposites*,
https://doi.org/10.1007/978-3-030-05399-4_18

ELISA	Enzyme linked immunosorbent assay
FRET	Förster resonance energy transfer
GA	Glucoamylase
GNPs	Gold nanoparticles
GNRs	Gold nanorods
GO	Graphene oxide
GOx	Glucose oxidase
GQDs	Graphene quantum dots
HPV	Human papillomavirus
HRP	Horseradish peroxidase
IgG	Immunoglobulin G
LFIA	Lateral flow immunoassay
LRET	Luminescence resonance energy transfer
LSPR	Localized surface plasmon resonance
MEMS	Micro-electro-mechanical systems
MIP	Molecularly imprinted polymers
MNPs	Magnetic nanoparticles
NPs	Nanoparticles
NQS	Sodium 1,2-naphthoquinone-4-sulfonate
OTB	*O*-toluidine blue
PATP	*P*-aminothiophenol
PB	Prussian blue
PBS	Phosphate buffered saline
PC	Phycocyanin
PC-paper	Parylene C-coated paper
PEC	Photoelectrochemical
PEDOT:PSS	Poly(3,4-ethylenedioxythiophene):poly(styrenesulfonate)
PL	Photoluminescence
POC	Point-of-care
poly(DAB)	Poly-3,3′-diaminobenzidine
PVP	Polyvinylpyrrolidone
QDs	Quantum dots
R6G	Rhodamine-6G
RDX	1,3,5-trinitroperhydro-1,3,5-triazine
RGO	Reduced graphene oxide
SCO	Spin-crossover
SEM	Scanning electron micrographs
SERS	Surface-enhanced Raman spectroscopy
SP	Sample pad
SPCE	Screen-printed carbon electrode
SWV	Square wave voltammetry
TA	Thioctic acid
TBPB	Tetrabromophenol blue
TG	Thioguanine

Thi	Thionine
TMB	3,3′,5,5′-tetramethylbenzidine
TMPyP	5,10,15,20-tetrakis(1-methyl-4-pyridinio)porphyrin tetraiodide
TNT	2,4,6-trinitrotoluene
TPA	Tri-*n*-propylamine
ubi	Ubiquitin
UV-Vis	Ultraviolet-visible
WHO	World health organization

1 Introduction

The term "point-of-care (POC) testing" has gained immense popularity since the last decade especially in connection to the diagnostic applications. It has offered the fascinating possibility of providing rapid results in places with limited availability of resources [1]. Lately, the paper has attracted the significant attention of the global research community as a substrate for the development of dip-sticks, lateral flow immunoassays (LFIAs) and microfluidic paper-based analytical devices (μPADs) [2–4]. The interest in making use of paper stemmed from the belief that the conventional devices were too complicated and expensive to be used on a large scale in developing countries. The low cost of paper and its ability to wick fluids by capillary action due to hydrophilicity, thus eliminating the need for pumps or external power to transfer fluids, were the primary motivating factors inspiring researchers to work with paper as a new generation material for sensing devices [5]. Additionally, the porosity and high surface-to-volume ratio of paper are advantageous for assays wherein the reagents are bound on the paper surface and its flammable nature allows for easy disposal of paper devices via incineration. The μPADs are extremely promising for POC diagnostic devices for use in resource-limited settings owing to their ease of operation, low cost and ability to work without external power supply or supporting equipment [6]. In addition, μPADs are generally easy to fabricate, user-friendly and offer simple readouts of analytical assay results [7]. Further, paper substrates are easily compatible with printing processes like ink-jet printing, screen printing and flexographic printing [8]. Finally, being environmentally friendly, the paper is a sustainable material, which enables it to stand out amongst its peers as a pathway for "green" sensors [9].

Since the last decade, the focus has shifted towards quantitative sensings such as that required for glucose and tumour markers than mere qualitative analysis as in case of pH and pregnancy test strips. Following the seminal work on microfluidic paper-based analytical devices by Whitesides' group in 2007 [10], a myriad of works have been published on sensing devices using paper as a substrate for the detection of a plethora of analytes. Paper substrates have been functionalized with biomolecules [11], metal and metal oxide nanoparticles [12], quantum dots [13], aptamers [14], spin-crossover particles [15] and metamaterials [9] to develop a range of sensors. These sensors find major applications in clinical diagnostics and therapeutics [16–18], environmental monitoring and analysis [19], food and water quality [20, 21] and forensics [22]. For instance, visual detection of DNA on paper chips has been reported which have the ability to identify and distinguish dog and human genomic and mitochondrial samples for forensic purposes [23]. Similarly, paper-based electrochemical sensors have shown potential for the detection of K-562 cells, one of the most aggressive human chronic myelogenous leukaemia cell lines, based on the release of hydrogen peroxide (H_2O_2) from cells [24]. This development has been made possible by the multiple attributes of paper-based sensors as per the "ASSURED" criteria of the World Health Organization (WHO) for effective POC testing devices that stands for affordable, sensitive, specific, user-friendly, rapid and robust, equipment-free, delivered to end-users [25, 26].

This chapter presents a concise review of paper-based analytical devices used for sensing in various sectors comprising of biomedical diagnostics, environmental monitoring, food and water safety as well as forensics and security, in the ongoing decade with a focus on the discussion of sensing principle or the mechanism of detection (Fig. 1). The various analytes have been broadly grouped based on the classical detection principles of colorimetry, electrochemistry, and luminescence-based sensing.

2 Sensing in the Physical World

Paper-based sensors have been used to measure some of the most fundamental parameters in basic and applied sciences such as temperature [27], pH [28] and humidity [29, 30]. Precise measurement of these physical quantities is required in areas as diverse as medicine, biotechnology, environmental science, and meteorology to name a few. An optical temperature sensor was fabricated by soaking the paper in a temperature sensitive luminescent indicator dichlorotris(1,10-phenanthroline)ruthenium(II) hydrate ($Ru(phen)_3$) [31]. The dried paper was subsequently laminated to eliminate oxygen cross-sensitivity by preventing its diffusion. A linear response to temperature was obtained with phase fluorimetry and ratio imaging. In another study, paper thermometers with an ultrafast response and high stability were developed using an ionic liquid (1-ethyl-3-methyl imidazolium bis(trifluoromethylsulfonyl)imide) deposited on paper by means of pen writing or inkjet printing [27]. The low viscosity and hydrophobic nature of the ionic ink facilitated easy writing while resisting the hydration by moisture in the atmosphere (Fig. 2a). The sensing ability of the paper chip was quantified by the relative change in conductivity against temperature change. The conductivity was contributed solely by the ionic liquid since paper itself is electrically insulated with a very high resistance. The thermal response of the paper thermometer reached over 60% by raising the ambient temperature and remained the

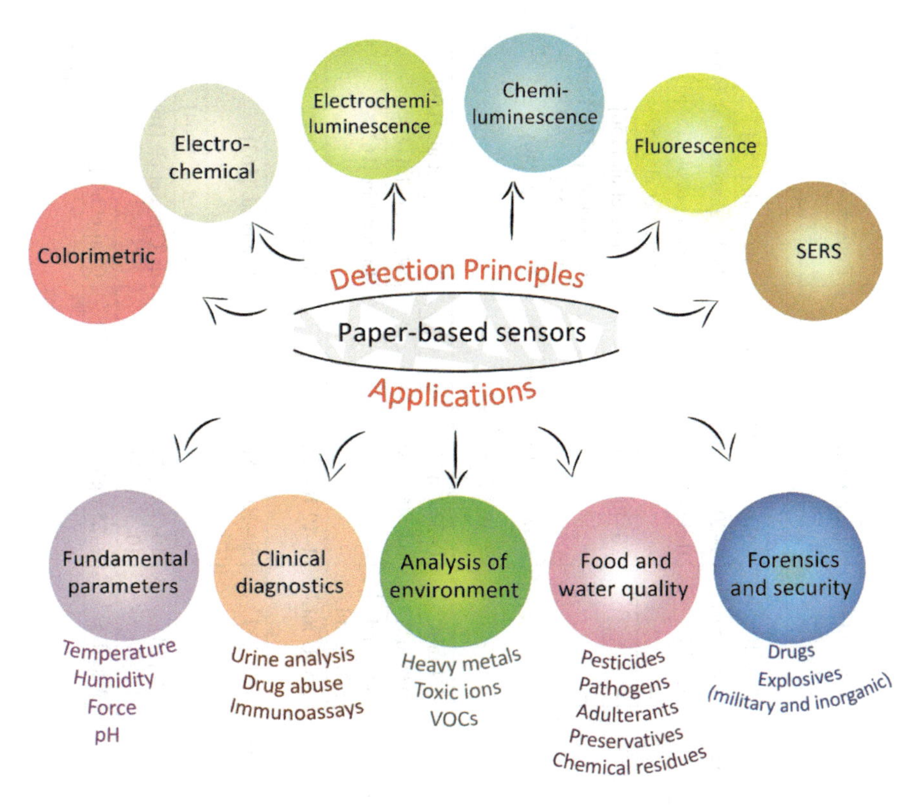

Fig. 1 Sensing approaches of paper-based analytical devices for various applications

same after multiple heating and cooling cycles (Fig. 2b) [27]. Cellulose nanocomposites showing spin-crossover phenomena and having thermochromic properties have been fabricated by adsorption of spin-crossover (SCO) nanoparticles onto linter fibres [32, 33]. The $[Fe(hptrz)_3](OTs)_2$ (hptrz = 4-heptyl-1,2,4-triazole, OTs = *p*-toluenesulfonyl) SCO particles are known to display an abrupt spin transition with a hysteresis loop close to room temperature, a characteristic which is highly relevant for potential application in paper thermometry. The curves for the temperature dependence on the optical reflectance revealed that the reflectance changes occurred rather abruptly in the heating and cooling modes, thus unambiguously relating to the spin transition process of the SCO particle complex [32].

Taking advantage of the hydrophilicity of paper, humidity sensors with good stability and reproducibility were developed by simply "writing" the carbon electrodes by pencils and ink marker consisting of oxidized multi-walled carbon nanotubes [34]. Similarly, pH sensors were fabricated using phenol red and chlorophenol red indicators on paper. The indicators produced selective colour changes, which were captured by a smartphone camera and processed to extract the hue from the colour space [35]. Similarly, a graphene-based paper sensor fabricated by vacuum filtration process was reported for sensing the pH of the analyte by

Table 1 Detection of various biomedical analytes

Target	Sensing agents/sensor components	Limit of detection	Linear range	References
Colorimetric sensing				
Glucose	GOx, HRP	3×10^{-4} M	1.0×10^{-3}–11.0×10^{-3} M	Zhu et al. [152]
Urease enzyme	Hexamethyldisilazane, tetra-ethylorthosilicate	1 unit mL^{-1}	1–20 units mL^{-1}	Malekghasemi et al. [153]
Lactate dehydrogenase	Pullulan coating to immobilize reagents	13 U L^{-1}	0–450 U L^{-1}	Kannan et al. [154]
Zika virus RNA genome	CRISPR-based tool	–	–	Pardee et al. [155]
Antioxidants	Ceria nanoparticles	–	20–400 μM	Sharpe et al. [156]
Blood typing	Antibodies	–	–	Guan et al. [157], Khatri et al. [158]
Procaine	Reagents	0.9 μmol L^{-1}	5–60 μmol L^{-1}	Silva et al. [39]
Electrochemical sensing				
Glucose	Plasma isolation membranes	3.4 mM	0–33.1 mM	Noiphung et al. [159]
Adenosine triphosphate	Paper-based electrode	0.08 μM	0.3–450 μM	Wang et al. [160]
Rabbit immunoglobulin G HIV p24 antigen	Protein probes immobilized on ZnO nanowires grown on paper	60 fg mL^{-1} 300 fg mL^{-1}	– –	Li and Liu [161]
Butyrylcholinesterase activity in serum	Carbon black/Prussian blue	0.5 IU mL^{-1}	1–12 IU mL^{-1}	Scordo et al. [162]
Acetaminophen	AuNPs-polyglutamic acid/single-walled carbon nanotube	15.0 μM	50–300 μM	Lee et al. [163]
p-aminophenol	Au & Pt microwire electrodes	31 μM	–	Adkins et al. [164]
Fe^{2+} Dopamine	Carbon nanotube electrodes	10 μM 10 μM	10–200 μM 10–100 μM	da Costa et al. [165]
Paraoxon (nerve agent)	Carbon black/Prussian blue	3 μg/L	3–25 μg/L	Cinti et al. [166]

(continued)

Table 1 (continued)

Target	Sensing agents/sensor components	Limit of detection	Linear range	References
Norepinephrine Serotonin	Boron doped diamond paste electrodes	2.5 μM 0.5 μM	2.5–100 μM 0.5–7.5 μM	Nantaphol et al. [167]
Luminescence-based sensing				
Polynucleotide kinase activity	λ exonuclease assisted fluorescence quenching	0.0001 U mL^{-1}	–	Zhang et al. [168]
Immunoglobulin E (IgE)	Upconversion nanoparticles & carbon nanoparticles leading to LRET	–	0.5–80 ng mL^{-1}	Jiang et al. [169]
Other types				
Cancer cells: MCF-7 HL-60 K562	Graphene oxide & aptamer leading to FRET	62 cells mL^{-1} 70 cells mL^{-1} 65 cells mL^{-1}	180–8 × 10^7 cells mL^{-1} 210–7 × 10^7 cells mL^{-1} 200–7 × 10^7 cells mL^{-1}	Liang et al. [170]
Pulse and motion of body	Strain/deforming angle measurement using RGO	–	–	Saha et al. [171]
Hepatitis B virus DNA	Nafion-coated paper for ion concentration polarization	150 copies mL^{-1}	–	Gong et al. [172]
Glucose/DNA/protein	Resistive temperature detector (RTD) for calorimetric determination	–	–	Davaji et al. [173]

Table 2 Detection of various environmental analytes

Target	Sensing agents/sensor components	Limit of detection	Linear range	References
Colorimetric sensing				
Hg^{2+} ion	AuNPs with oligonucleotide sequences	50 nM	25–100 nM	Chen et al. [174]
Organophosphate pesticides	Acetylcholinesterase	–	–	Sicard et al. [175]
Nitrite ion	Mixed indicator (*N*-(1-naphthyl)-ethylenediamine & p-amino benzenesulfonamide	–	0.156–1.25 mmol L^{-1}	Wang et al. [176]
Metals	Reagents and buffer	0.75 μg (Fe) 0.75 μg (Ni) 0.75 μg (Cu) 0.12 μg (Cr)	1.5–15 μg (Fe) 1.5–15 μg (Ni) 3.0–15 μg (Cu) 0.38–6.0 μg (Cr)	Rattanarat et al. [177]
Dissolved ammonia in water	Reagents for modified Berthelot's reaction	10 mg L^{-1}	10–200 mg L^{-1}	Cho et al. [178]
Electrochemical sensing				
Ferricyanide	UV curable screen printing & ink-jet inks	8 μM	0.05–1 mM	Lamas-Ardisana et al. [179]
Chlorine	poly(3,4-ethylenedioxythiophene): poly(styrenesulfonate)	0.5 ppm	0.5–500 ppm	Qin et al. [180]
Phosphate	Reagents screen-printed on electrodes	4 μM	0–300 μM	Cinti et al. [181]
Bisphenol A	AuNPs with multi-walled carbon nanotubes	0.03 mg L^{-1}	0.2–20 mg L^{-1}	Li et al. [182]
Pb^{2+} Cd^{2+}	Boron doped diamond paste electrodes	1 ppb 25 ppb	1–200 ppb 25–200 ppb	Nantaphol et al. [167]
Formaldehyde	Microbial fuel cell-based paper sensor	–	–	Chouler et al. [183]
Luminescence-based sensing				
Formaldehyde	Fluorescence quenching using acetoacetanilide reagent	0.2 ppm	0–8 ppm	Liu et al. [132]
Hydrazine	Fluorescence quenching of fluorogens with aggregation-induced emission	143 ppb	0–60 μM	Zhang et al. [184]
Other types				
Volatile organic compounds	Measurement of mechanical deflection angle of thin polymers adhered to paper	–	–	Fraiwan et al. [185]
Titania (TiO_2) nanoparticles	Photocatalytic effect of TiO_2 on methylene blue	–	0–5000 ppm	Bulbul et al. [186]

Fig. 2 **a** Fabrication process showing ionic liquid (IL) paper chip written with a pen, **b** thermal response with on-off cycles of the paper chip between 45 and 25 °C. Adapted with permission from Ref. [27] © 2017 ACS publisher

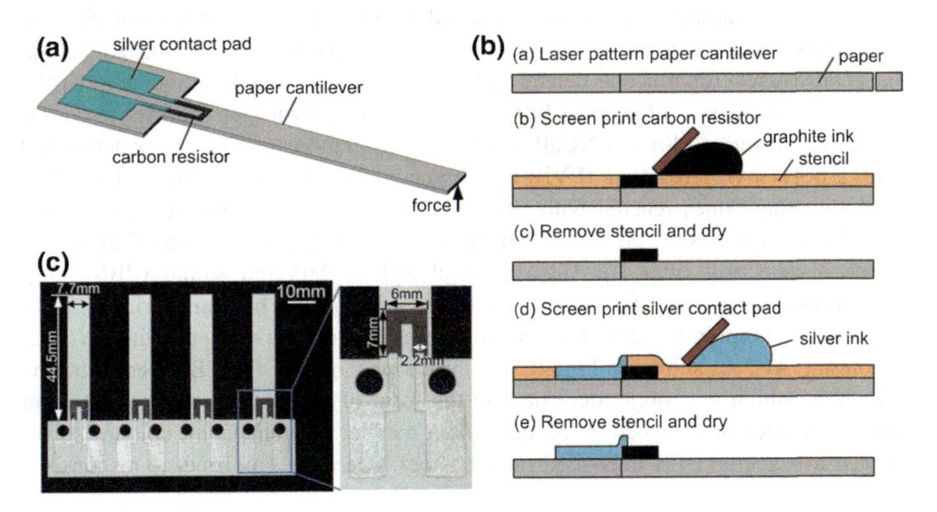

Fig. 3 **a** Schematic showing a paper-based piezoresistive force sensor with a carbon resistor as the sensing component, **b** fabrication of the sensor by laser cutting of paper and screen printing of carbon and silver inks, **c** picture of an array of four devices. Adapted with permission from Ref. [37] © 2011 RSC publisher

directly measuring the resistance across the sensor [36]. The sensor showed a sensitivity of 30.8 Ω/pH and a high linearity. Another study developed a MEMS (micro-electro-mechanical systems) sensor for the measurement of forces based on the piezoresistive effect of conducting materials patterned on paper [37]. Paper was preferred over commonly used silicon for the construction of the sensor primarily due to its low cost, lightweight, disposability, and ease of fabrication. As shown in Fig. 3a, the carbon resistor experienced a mechanical strain/stress when a force was applied to the cantilever beam. This induced a change in resistance of the resistor, which allowed quantification of the applied force. The carbon resistors used high-resistivity graphite ink, while the contact pads used low-resistivity silver ink (Fig. 3b, c). The sensor had a resolution of 120 μN, a measurement range of ±16 mN and a sensitivity of 0.84 mV mN^{-1} [37].

3 Sensing in Biomedical Health Care and Clinical Diagnostics

3.1 Colorimetric Sensing

Colorimetric spot tests using paper are rapid, inexpensive and can be done in locations with limited infrastructure. Further, the white paper provides a strong contrast against a coloured substrate that allows direct checking of the results with the naked eye [38, 39]. This has immensely facilitated the use of paper in a number of colorimetric assays for diagnostics and therapeutics. For example, colorimetric test strips prepared using a chromogen (2,4,6-tribromo-3-hydroxybenzoic acid) have been used for glucose detection by measuring their colour intensity as the differential diffusive reflectance [40]. The sensing of hydrogen sulfide (H_2S) gas from live cancer cell lines (LNCaP and PC-3) was demonstrated using paper coated with a polyvinylpyrrolidone (PVP) membrane containing silver/Nafion. The silver in the coating zones reacted with sulfide, giving a brown colour of silver sulfide (Ag_2S). The assay showed a high sensitivity, selectivity and reproducibility with a limit of detection of 1.4 μM Na_2S in phosphate buffered saline (PBS) [38]. A text-displaying assay for urinary protein was fabricated by using printing techniques and combining the classical colorimetric indicator system with an inert colourant [41]. As shown in Fig. 4a, tetrabromophenol blue (TBPB), a colorimetric indicator, which was inkjet-deposited in the form of symbols on paper underwent colour changes in a concentration-dependent manner. A transparent coloured layer served to screen TBPB deposited on the paper. After the colorimetric response, a

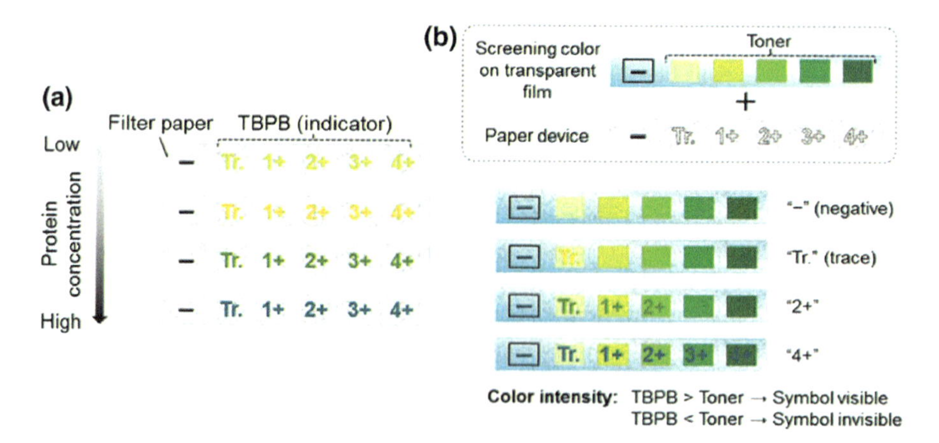

Fig. 4 Schematic showing the principle of the paper device utilizing a TBPB-based colorimetric indicator system for protein: **a** inkjet-deposited TBPB symbols exhibiting colorimetric response depending on the sample concentration, **b** response after the colorimetric reaction, where transparent film with screening color is overlaid on the paper device. Adapted with permission from Ref. [41] © 2017 ACS publisher

series of screening colours served to shield the indicator symbols with weaker colour intensity than that of the respective screening colour, making the symbols invisible to the human eye, as shown in Fig. 4b. In order to improve the sensitivity of bioassays on μPADs, a bienzyme system was combined with a drying method to achieve signal amplification and reduction of the background signal, respectively for the detection of glucose and uric acid [42].

Paper-based immunoassays with antibody-antigen interaction are gaining popularity in resource-limited settings due to their simplicity and affordability. In a paper-based enzyme-linked immunosorbent assay (ELISA), paper modified with chitosan and glutaraldehyde to enhance antigen (ubiquitin or ubi & enhanced green fluorescent protein or eGFP) immobilization was used to detect targeted antibodies (anti-ubi & anti-eGFP). The cationic chitosan was bonded to anionic cellulose forming a layer on the surface of the paper, while glutaraldehyde served as a cross-linker to facilitate the covalent attachment of chitosan with the protein groups. The assay used a drinking straw for washing and incubation to avoid the need of pipettes and shakers. A visible green colour resulted on catalysis of the 2,2′-azino-bis(3-ethylbenzothiazoline-6-sulfonic acid) (ABTS) substrate by protein L HRP in the presence of targeted antibodies with a detection limit of 0.5 nM [43]. Similar immunoassay for the detection of anti-*Leishmania* antibodies was achieved using a paper-based 96-well ELISA [44].

Aptamers, single-stranded oligonucleotides, have been widely used in μPADs for the selective and sensitive detection of a large number of analytes such as proteins, cells ions and microbes due to their ability to specifically recognize and interact with targets. The advantage of aptamer-based assays is that by simply changing the aptamer, sensing of different targets can be done by synthesizing the respective target-responsive hydrogels [45]. An aptamer-crosslinked hydrogel was used for detection of cocaine in urine taking advantage of their ability to bind to a specific target molecule [46]. As shown in Fig. 5, the hydrogel collapsed in presence of the target due to aptamer sequence (L-Apt) dissociating from polymer chains (P-SA and P-SB), thereby releasing the trapped glucoamylase (GA) and generating glucose. The capillary action caused glucose to flow in solution along the channel, where it was subsequently converted into H_2O_2 by glucose oxidase (GOx) already present on the substrate. Similarly, 3,3′-diaminobenzidine (DAB) was converted into a brown-coloured bar of poly-3,3′-diaminobenzidine (poly(DAB)) by horseradish peroxidase (HRP), whose length could be positively correlated to the amount of target to achieve visual distance-base quantitative detection. The cascade of enzymatic steps for signal amplification made it possible to achieve highly sensitive detection of targets [46]. Earlier, the same group reported the simultaneous detection of multiple analytes such as cocaine, adenosine and Pb^{2+} in urine using target-induced phase-transformation of the aptamer hydrogel to mediate fluid flow and signal readout in the μPAD [47]. Another group used DNA-triggered hybridization chain reaction to capture hairpin probes and bind GOx tags via biotin-streptavidin interactions, where its quantity could be positively related to the adenosine analyte [48].

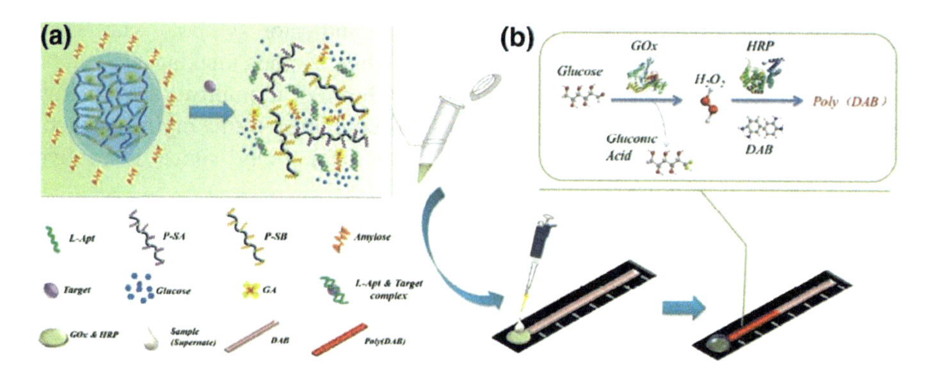

Fig. 5 Schematic of aptamer-based μPAD for distance-based visual quantitative analysis of cocaine in urine: **a** target-induced dissolution of hydrogel releasing GA to catalyse the production of glucose, **b** conversion of glucose to gluconic acid catalysed by GOx generating H_2O_2, which reacts with DAB catalysed by HRP to yield a brown stripe of poly(DAB) for signal readout. Adapted with permission from Ref. [46] © 2016 ACS publisher

3.2 *Electrochemical Sensing*

The detection of simple and complex analytes of clinical significance has been achieved based on the concept of electrochemistry including glucose [49], uric acid [50], drugs such as diazepam [51] and ketamine [52], tumor markers [53] and DNA viruses such as human papillomavirus (HPV) [54]. Detection of carcinoembryonic antigen was performed by a label-free electrochemical immunosensor fabricated by coating nanocomposites of amino functional graphene (NH_2-G)/thionine (Thi)/gold nanoparticles (AuNPs) on the screen-printed working electrode [55]. The concept was based on the fact that the decreased response currents of Thi were proportional to the concentrations of corresponding antigens due to the formation of antibody-antigen immunocomplex. The cyclic voltammetry and differential pulse voltammetry results revealed the stability of peak currents, thus indicating that the electroactive material was tightly bound to the electrode. The determination of antigen solutions showed linear working ranges of 50 pg mL^{-1}–500 ng mL^{-1} with the limit of detection as 10 pg mL^{-1} [55]. Another study reported the use of conducting paper modified by poly(3,4-ethylenedioxythiophene):poly(styrenesulfonate) (PEDOT:PSS) and reduced graphene oxide (RGO) for the detection of carcinoembryonic antigen. There was a significant increase in the electrical conductivity of the paper due to conformational rearrangement in the polymer and strong non-covalent interactions between PEDOT and cellulose [56].

In order to eliminate the need for enzymatic amplification and to improve the sensitivity, the analyte (biotin/streptavidin) was labelled with silver nanoparticles (AgNPs) and magnetic microbeads. Detection limit as low as 767 fM could be achieved by magnetic preconcentration of AgNP labels followed by their oxidation to Ag^+ by slipping a piece of paper to deliver the oxidizing agent at a specific time and

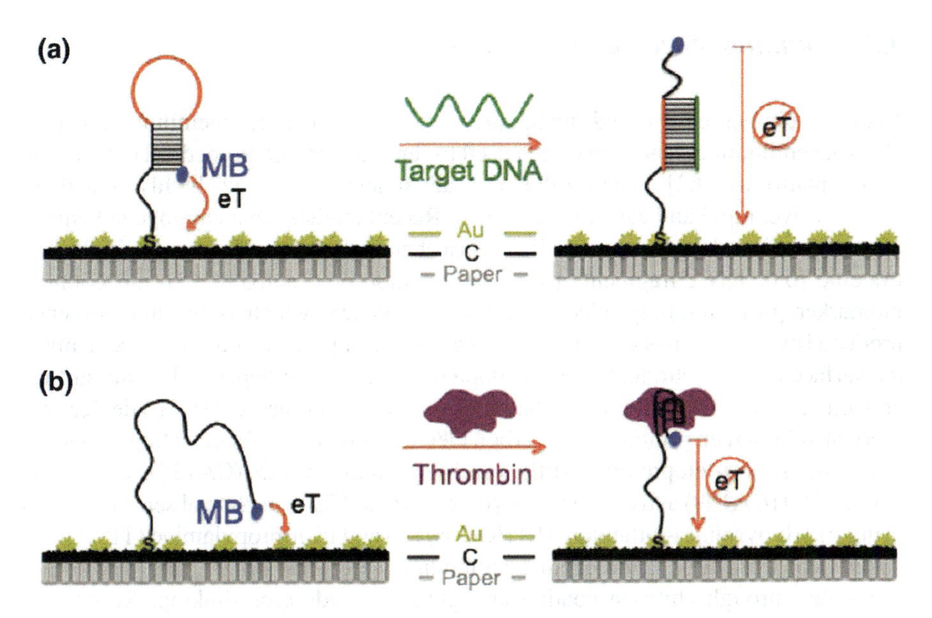

Fig. 6 Sensing mechanism of DNA and thrombin on a PAD. Adapted with permission from Ref. [61] © 2014 ACS publisher

point in the device [57]. In another work, a "pop-up" sensor fabricated from paper folded into a 3D structure enabled electrochemical detection of β-hydroxybutyrate (BHB), a biomarker for diabetic ketoacidosis, using a glucometer [58]. Similarly, amperometric detection of troponin, a cardiovascular biomarker, was done using conducting paper electrodes formed by coating a layer of polyaniline. The change in the oxidation current of polyaniline was proportional to the analyte concentration. The assay showed a sensitivity of 5.5 μA/ng mL^{-1} cm^{-2} over a wide physiological range of 1–100 ng mL^{-1} [59]. Parylene C-coated newspaper acted as a sensing electrode for the detection of pathogenic *E. coli* based on DNA hybridization, showing excellent performance in the cyclic voltammetry and electrochemical impedance spectroscopy experiments with a detection limit of 0.16 nM [60].

Based on target-induced folding or unfolding of an aptamer linked to an electrochemical label, the detection of DNA and thrombin down to limits of 30 and 16 nM, respectively was achieved [61]. The oligonucleotide probes had a pendant redox reporter (methylene blue) at the distal end and a thiol at the proximal end for easy attachment to a gold electrode. On binding of the analyte, the probe underwent a conformational change that altered the location of the redox reporter relative to the electrode, as shown in Fig. 6. The sensor was "on" if the redox reporter moved closer to the electrode and "off" if, it moved away. This conformational change resulted in a change in faradaic current that was easily detected using alternating current voltammetry (ACV) or square wave voltammetry (SWV).

3.3 Luminescence-Based Sensing

Various luminescence-based techniques viz. fluorescence, chemiluminescence, electrochemiluminescence have been used to sense a range of biomedical analytes on paper platforms [62]. They offer the advantages of being highly sensitive, non-invasive, rapid and easy to implement. Recent studies have employed chemiluminescence, i.e., the generation of light via a chemical reaction, for the detection of L-cysteine [63], DNA fragments [64], carcinoembryonic antigen [65] and cotinine biomarker [66]. Similarly, electrochemiluminescence, which is the luminescence produced by relaxation of excited state molecules during electron-transfer occurring at the surface of an electrode, has been adopted for sensing of hepatitis B virus surface antigen in serum [67]. A 3D paper-based electrochemiluminescent (ECL) device was fabricated by screen printing eight carbon electrodes on paper for detecting a panel of tumor markers-α-fetoprotein (AFP), carcinoma antigen 125 (CA125), carcinoma antigen 199 (CA199) and carcinoembryonic antigen (CEA) in clinical serum samples using tris-(bipyridine)-ruthenium (II) ($Ru(bpy)_3^{2+}$) and tri-*n*-propylamine (TPA) [68]. As evident from Fig. 7, the capture antibodies were immobilized on the working electrodes through chitosan coating and glutaraldehyde cross-linking. $Ru(bpy)_3^{2+}$-labeled signal antibodies were added to corresponding electrodes to carry out the

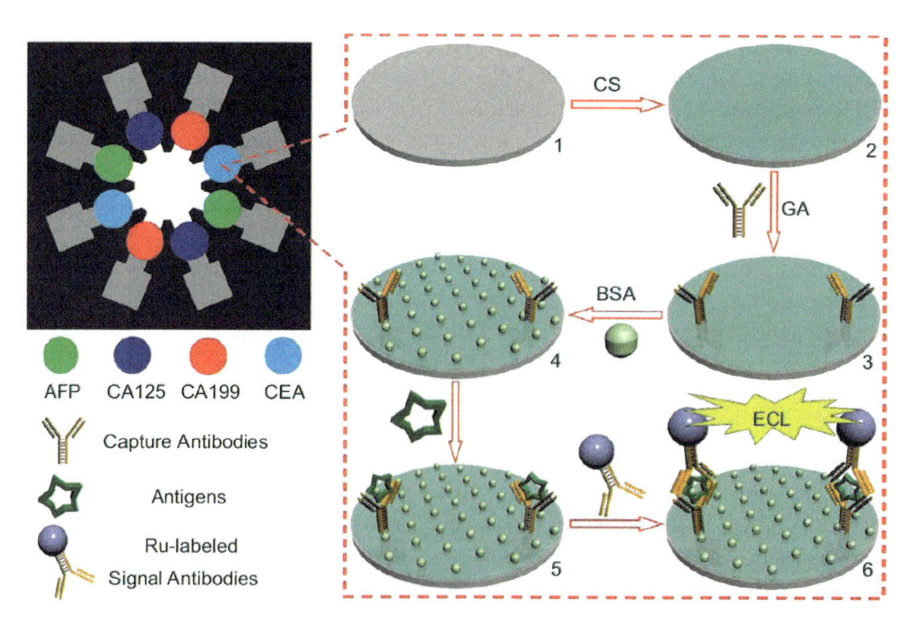

Fig. 7 Fabrication and detection principle for paper-based ECL device: (1) screen-printed carbon working electrode; (2) after chitosan modification; (3) after immobilization of capture antibodies; (4) after blocking & washing; (5) after capturing and washing; (6) after incubation with signal antibodies, washing & ECL reaction. Adapted with permission from Ref. [68] © 2012 Elsevier publisher

immunoreactions and ECL detections. The sandwich immunocomplexes were found to give much higher ECL response than the nonspecific adsorption of signal antibodies without antigen. Moreover, since the ECL intensity rose with antigen concentration, the immune device could be for the determination of sensitive antigens. The limits of detection for the four markers were 0.15 ng mL^{-1}, 0.6 U mL^{-1}, 0.17 U mL^{-1} and 0.5 ng mL^{-1}, respectively [68]. Another study fabricated a photoelectrochemical (PEC) immunosensor using CdS quantum dots (QDs) deposited on a paper working electrode modified with zinc oxide nanorods grown on reduced graphene oxide as the photoactive matrix and chemiluminescence reagent/enzyme/antibody bioconjugate as the label for sensitive detection of cancer antigen 125 [69].

Fluorescence, which is by far the most widely employed optical detection technique, has been used for sensing numerous targets including aluminium (Al^{3+}) detection in living cells [70], Cu^{2+} in human urine [71], sulfur dioxide derivative (SO_3^{2-}) in mitochondria [72], β-D-galactosidase enzyme [73], peptide, protein and DNA [74]. In order to overcome the limitations of conventional methods of attaching biomolecules onto the cellulose surface, a chemo-enzymatic method for activating cellulose with functional groups by click chemistry was proposed using propargylated xyloglucan as a molecular anchor [75]. They demonstrated the detection of esterase with a biosensor fabricated by tethering a chromogenic moiety/ fluorophore to the cellulose substrate, instead of a biomolecular entity, thus simultaneously addressing the issues of signal attenuation and potential toxicity due to chromophore diffusion after substrate cleavage. The sensitivity and selectivity of fluorescence-based sensing with a quick response time makes them preferred over other detection techniques in a number of bioassays [72]. Fluorescent probes have successfully been employed to detect antibiotic-resistant genes from various bacteria [76]. Another study used fluorescence emission of fluorescein-labeled DNA probe to detect hybridization of DNA strands captured on antibodies via CBM-ZZ (carbohydrate binding molecule with high affinity for cellulose and double Z fragment of staphylococcal protein A, which recognizes IgG antibodies). The antibodies anchored via CBM-ZZ fusions were combined with wax printed μPADs for capturing and detecting DNA hybrids [77]. Other studies reporting nucleic acid assays have employed chemiluminescence [78] and luminescence resonance energy transfer (LRET) [79].

3.4 Other Sensor Types

Surface-enhanced Raman spectroscopy (SERS) is a non-destructive and sensitive technique used in molecular detection, which combines the specificity of Raman spectroscopy with the metal nanostructure-induced sensitivity provided by plasmon assisted scattering [80]. Detection of rhodamine-6G at ppb level was achieved using office paper as a SERS substrate taking advantage of its low porosity and the better ink retaining the capacity to obtain a uniform distribution of silver nanostars along with higher values for the enhancement factor [80]. A SERS paper strip fabricated

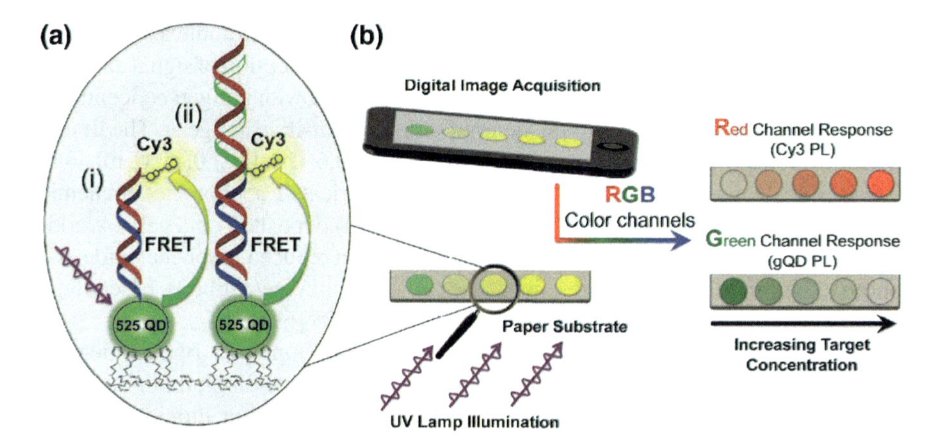

Fig. 8 **a** Design of paper-based QD-FRET nucleic acid hybridization assay. Modified paper-substrates immobilized with QD-probe conjugates, and hybridization assays in (i) a direct format or (ii) a sandwich format provided the proximity for FRET sensitized emission from the acceptor dye when the substrate was illuminated with a UV lamp, **b** digital imaging of paper substrates showing increase in Cy3 PL with corresponding decrease in gQD PL as target concentration increases, associated with R and G channels, respectively. Adapted with permission from Ref. [88] © 2014 ACS publisher

by in situ synthesis of AuNPs on cellulose fibers could detect a cancer marker in a whole blood sample [81]. The nanoparticles not only generated condensed hot spots on the fibres but also enhanced the size exclusion effect of paper. Several other studies have reported SERS-based paper sensors [82–86].

Similarly, localized surface plasmon resonance (LSPR)-based biosensors, where the bulk refractive index sensitivity and the electromagnetic decay length of the metal nanostructures are employed as optical transducers, have been used to detect contagious biomolecules [8] and cardiac biomarkers [87]. The strong electric fields thus created around the nanoparticles improved the signal intensity dramatically and allowed for extremely sensitive detection with only a small amount of analyte required. The Au and Ag NPs deposited on a paper substrate by the laser-induced annealing technique showed a rapid colour change on binding with cysteine, which could be observed by the naked eye and measured spectroscopically [8].

A nucleic acid hybridization assay was developed using green-emitting QDs (gQDs) immobilized on imidazole-modified paper, as donors in Förster resonance energy transfer (FRET) (Fig. 8a). A hybridization event brought the Cyanine 3 (Cy3) acceptor dye in close proximity to the immobilized gQDs and was responsible for a FRET sensitized emission from the dye, which served as an analytical signal. The photoluminescence (PL) intensities of gQDs and Cy3 were associated with the green (G) and red (R) imaging channels of an iPad camera after R-G-B splitting of the acquired images, as shown in Fig. 8b. The use of dry paper substrates for data acquisition offered 10-fold higher sensitivity and 10-fold lower limit of detection for the assay as compared to the hydrated paper substrates [88].

Table 1 lists some of the major biomedical analytes detected using different sensing mechanisms.

4 Sensing for Environmental Monitoring

4.1 Colorimetric Sensing

There has been a growing interest in the colorimetric detection of heavy metals and toxic compounds, which is an instrumental part of the rapid on-site environmental analysis and its real-time monitoring. The detection of toxic ions such as cobalt (Co^{2+}) [89], silver (Ag^{+}) [90], chloride (Cl^{-}) [91] and copper (Cu^{2+}) [92] in aquatic environments such as wastewater or groundwater has been reported on paper-based sensing platforms. A multiplexed patterned sensor for the detection of heavy metals was fabricated by ink-jet printing of sol-gel based bio-inks that allowed colorimetric visualization of the enzymatic activity of β-galactosidase [93]. As shown in Fig. 9a, the chromogenic substrate, chlorophenol red β-galactopyranoside (CPRG), is hydrolyzed by the β-galactosidase enzyme to give a red-magenta product. It is printed as a substrate zone and is moved to the sensing zone by lateral flow of a sample along the paper sensor. The presence of heavy metals in the sample causes a loss of the red-magenta colour in a concentration-dependent manner (Fig. 9b). In order to identify the different metals such as Hg^{2+}, Cu^{2+}, Cr^{4+} and Ni^{2+} in the mixture, a multiplexed sensor was made with two assay arms as controls, one for testing a mixture of metal ions using the β-galactosidase assay, and four additional assay arms that had different colorimetric reagents (Fig. 9c). Interestingly, time-programmable assays for metal ion determination have been developed, where the acceleration or delay of fluid transport can be controlled by adjusting capillary and laminar flow on paper without active pumping [94]. This was achieved by razor-crafting open channels on the paper using a cutting blade in a perpendicular or longitudinal directions to the direction of flow.

In order to eliminate the need for on-spot calibration using standard solutions and reduce the influence of paper inhomogeneity, matrix effects and environmental conditions on the results, a generic approach using calibrant-loaded paper was developed for multiple-point standard addition calibration [95]. The sensing areas were pre-loaded with an excess of colorimetric reagents and known amounts of the analyte; thus, a coloured product was developed before analysis. During analysis, the excess of reagents present in the sensing zone reacted with the analyte in the sample generating a combined colorimetric signal corresponding to the total concentration of the analyte in the sensing area. The combined analytical signals were used to generate a multiple standard additions curve and calculate the analyte concentration [95].

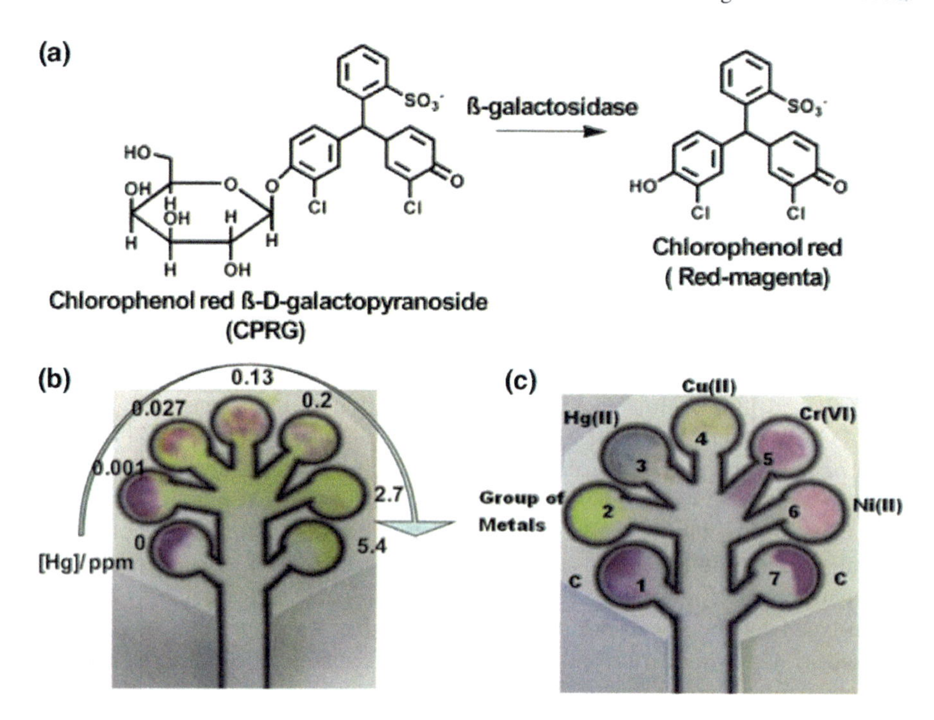

Fig. 9 **a** Detection principle for sensing of heavy metal ions, **b** dose-dependent color intensity of Hg^{2+} on the β-galactosidase-immobilized paper sensor, **c** detection of individual metals from a mixture on the multiplexed patterned sensor. Adapted with permission from Ref. [93] © 2011 ACS publisher

4.2 Electrochemical Sensing

Coulometric sensors were developed for the detection of halides in water samples using thin-layer coulometry, where the target ions are preferentially and exhaustively transported through an ion-selective membrane [96]. The sequential oxidation/plating of the halide at the silver wire and its subsequent regeneration with an inverted potential was monitored by cyclic voltammetry. The paper served the dual function of transporting the sample by capillarity as well as the making of an exhaustive electrochemical process. Confinement of the sample between a silver element and a Nafion membrane facilitated the resolution of a mixture of halides in a wide concentration range- from $10^{-4.8}$ to 0.1 M for iodide and bromide and from $10^{-4.5}$ to 0.6 M for chloride, with a detection limit of 10^{-5} M [96].

There have been studies based on the principle of potentiometric sensing of ions such as sodium and potassium [97]. Potentiometric ion sensors were fabricated using a newspaper and coating it with parylene C (PC-paper) to impart hydrophobicity along with improved mechanical properties and chemical stability [98]. A two-electrode configuration containing ion-selective and reference electrodes was achieved by depositing polyaniline and chloride on pre-patterned gold

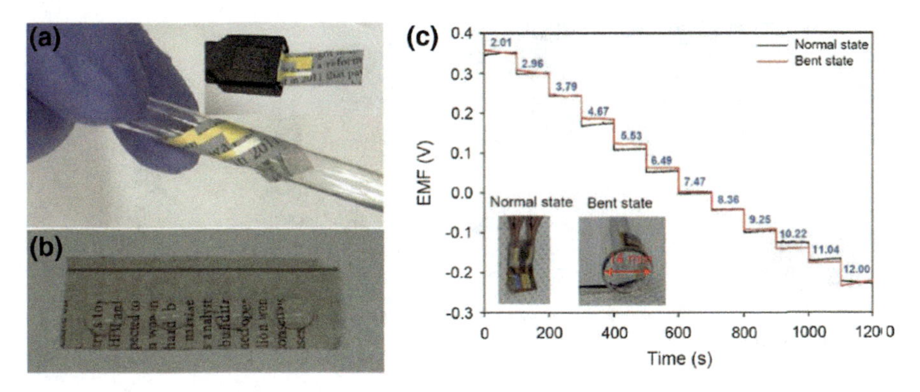

Fig. 10 **a** Image showing USB-type sensing platform and flexibility of ion sensor, **b** water droplets on PC-paper, **c** response of pH sensors with increasing pH in normal and bent states, while inset shows pH sensors in normal and bent states. Adapted with permission from Ref. [98] © 2017 Elsevier publisher

and silver layers on the sensors, respectively. These pH sensors were tested by increasing the pH levels from 2 to 12, as shown in Fig. 10c. A plot of EMF signals against the pH generated a line indicating behaviour according to Nernst equation. Moreover, this response remained the unaffected on sing bent electrodes, thus confirming the mechanical resistance of the sensors. Further, they showed good repeatability and a low potential drift [98].

4.3 Luminescence-Based Sensing

Recently, a photoluminescent nanopaper was fabricated by embedding nitrogen-doped carbon quantum dots (N-CQDs) into bacterial cellulose nanofibers for the selective detection of iodide (I^-) down to 6.1 ppm [99]. The blue emission of N-CQDs embedded paper remarkably quenched in the presence of I^-. On the contrary, hardly any change could be observed for other tested anions like chloride, fluoride, bromide, acetate, sulfate and nitrate. The quenching effect resulted from the photo-induced electron transfer between N-CQDs in the nanopaper and iodide ions in the sample [99]. Another work demonstrated the use of photoluminescent copper nanoclusters for sensing of H_2S in spring-water down to 650 nM [100]. The sensing was based on the strong Cu–S interaction resulting in the formation of non-luminescent CuS particles and subsequent photoluminescence quenching effect. The simultaneous detection of toxic heavy metal pollutants using a paper-based oligonucleotide ECL sensor was developed with carbon nanocrystals (CNCs) capped silica nanoparticles (Si@CNCs) and $Ru(bpy)_3^{2+}$-AuNPs clusters (Ru@AuNPs) as ECL labels (Fig. 11) [101]. The immobilized Si@CNCs-tagged DNA showed an ECL signal in the cathodic potential for Pb^{2+} detection, while the Ru@AuNPs modified one exhibited the signal in the anodic potential for Hg^{2+}

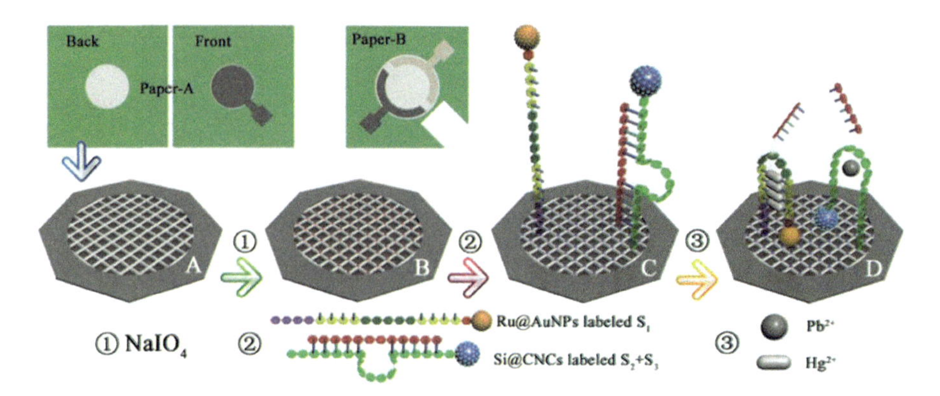

Fig. 11 Schematic illustration of a paper-based ECL device. The immunoassay was on the back of wax-patterned paper working zone: (1) the paper was spotted with $NaIO_4$; (2) after immobilized with Ru@AuNPs labeled DNA strands for Hg^{2+} and Si@CNCs labeled DNA strands for Pb^{2+}; (3) after capturing with Pb^{2+} and Hg^{2+}. Adapted with permission from Ref. [101] © 2013 Elsevier publisher

detection. The Pb^{2+} and Hg^{2+} ions induced a conformational change of DNA strands through the formation of G-quadruplex and T–Hg–T complex with detection limits of 10 pM and 0.2 nM, respectively.

Very recently, a "turn-off" fluorescence sensor was fabricated by immobilizing a cationic porphyrin, 5,10,15,20-tetrakis(1-methyl-4-pyridinio)porphyrin tetraiodide (TMPyP), onto a pre-patterned paper for the detection of Cu^{2+} [102]. The fluorescence signal of TMPyP (with two distinct maxima at 674 and 711 nm), which was easily visible to naked eyes under illumination, showed selective and substantial quenching effect in the presence of 20 mM Cu^{2+} against other metal ions. In contrast, a "turn-on" fluorescence sensor was developed to detect cyanide ion (CN^-) in water with a detection limit of 39.9 nM [103]. The sensing was based on the nucleophilic addition of CN^- to- the β-conjugated position of the barbituric acid moiety of the sensor, thus disrupting the π-conjugation and hampering the intramolecular charge transfer to produce a colorimetric and fluorescent response. Similar fluorescence-based sensing of phycocyanin (PC), a pigment-protein occurring in water bodies, has been demonstrated using QDs with molecularly imprinted polymers (MIP) deposited on paper [104]. As shown in the Fig. 12, in the absence of template molecule on the surface of QDs, the electrons of QDs absorb the UV energy and get excited. Subsequently, the QDs emit green fluorescence when the excited electrons return to the ground state. If the energy level of phycocyanin is higher than that of QDs, when the excited electrons of QDs return to the ground state, they will not generate fluorescence. Thus, adsorption of more phycocyanin onto the surface imprinting sites results in greater fluorescence quenching. Similarly, a ratiometric oxygen sensor based on fluorescence-phosphorescence was designed to detect as low as 0.01% of oxygen [105].

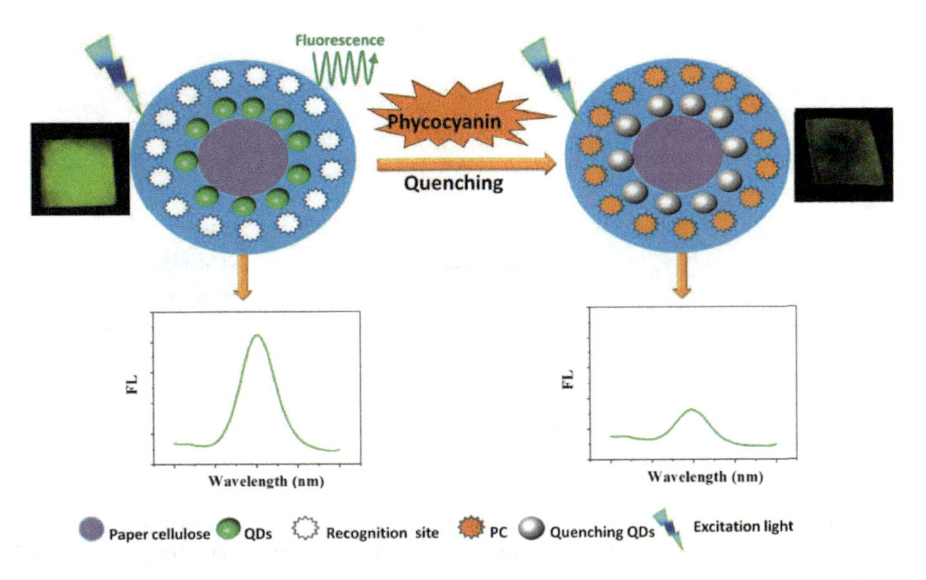

Fig. 12 Sensing principle of paper@QDs@PC-MIPs. Adapted with permission from Ref. [104] © 2017 ACS publisher

4.4 *Other Sensor Types*

Very recently, graphene quantum dots (GQDs) were immobilized on a paper substrate as a screening tool for organic pollutants such as 4-nitrophenol and paraoxon in seawater [13]. The luminescent blue GQDs were quenched in the presence of the target based on the phenomenon of FRET. Likewise, FRET was exploited on a paper-based sensor strip containing carbon nanodots (C-dots) conjugated to rhodamine for the detection of Al^{3+} ions [106]. As evident from Fig. 13, the hybrid system showed blue emission at 410 nm from C-dots since the unopened rhodamine could not absorb the energy from the blue emission of C-dots, leading to no FRET. On exposure to Al^{3+}, the metal ions induced ring-opening of rhodamine on C-dots through the chelation of the rhodamine-6G (R6G) moiety with metal ions, leading to a spectral overlap of the absorption of the donor (C-dots) and the emission of the ring-opened acceptor (rhodamine), finally emitting yellow. The addition of Al^{3+} to C-dots-R6G solution led to a gradual decrease in emission intensity at 410 nm with concomitant increase in emission at 560 nm, thus exhibiting a change in emission from blue to yellow (inset of Fig. 13).

SERS was exploited on a filter paper for the sensing of thiram and ferbam as model pesticides in trace amounts with detection limits of 0.46 and 0.49 mM, respectively [107]. In order to improve the retention of AgNPs on paper, the hydrophilic hydroxyl groups were converted to hydrophobic alkyl groups by the application of alkyl ketene dimer (AKD). This led to an increased contact angle with a reduced contact area and longer retention time of the aqueous dispersion due

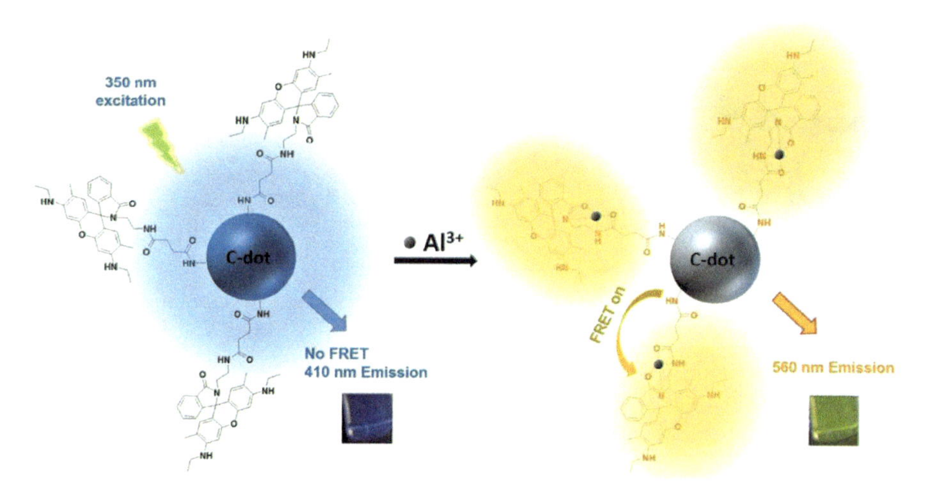

Fig. 13 FRET-based sensing mechanism of the C-dots-R6G in response to Al^{3+}. Photographs in the inset are under 365 nm UV light. Adapted with permission from Ref. [106] © 2015 ACS publisher

to lower absorption, thus concentrating the nanoparticles on paper. The SERS signal was strongly intensified by the increased number of SERS hot spots owing to the increased density of the nanoparticles on a small contact area of the paper surface [107]. In a similar fashion, SERS was implemented on a paper biosensor to enhance the low intensity of Raman signals for detecting wastewater components [108] and organic pollutants [109]. Table 2 lists the various environmental analytes detected by different sensing principles.

5 Sensing for Food and Water Quality

5.1 *Colorimetric Sensing*

Paper-based sensors have shown immense potential for monitoring and detecting various contaminants such as bacteria and chemical residues present in food, which holds utmost significance to ensure "food safety". Foodborne pathogens like *E. coli, Salmonella typhimurium*, and *L. monocytogenes* have been detected in food samples by measuring the colour change when an enzyme associated with the pathogen of interest reacts with a chromogenic substrate [110]. A colorimetric assay for the detection of *L. monocytogenes* consisted of magnetic nanoparticles conjugated with protease specific substrate to be selectively cleaved by the *L. monocytogenes* proteases with a limit of detection of 2.17×10^2 CFU/mL [111]. The ability of the protease to break the peptide bonds in a protein substrate led to a colour change from black to golden. Gas sensors using diatomite on paper for detecting volatile

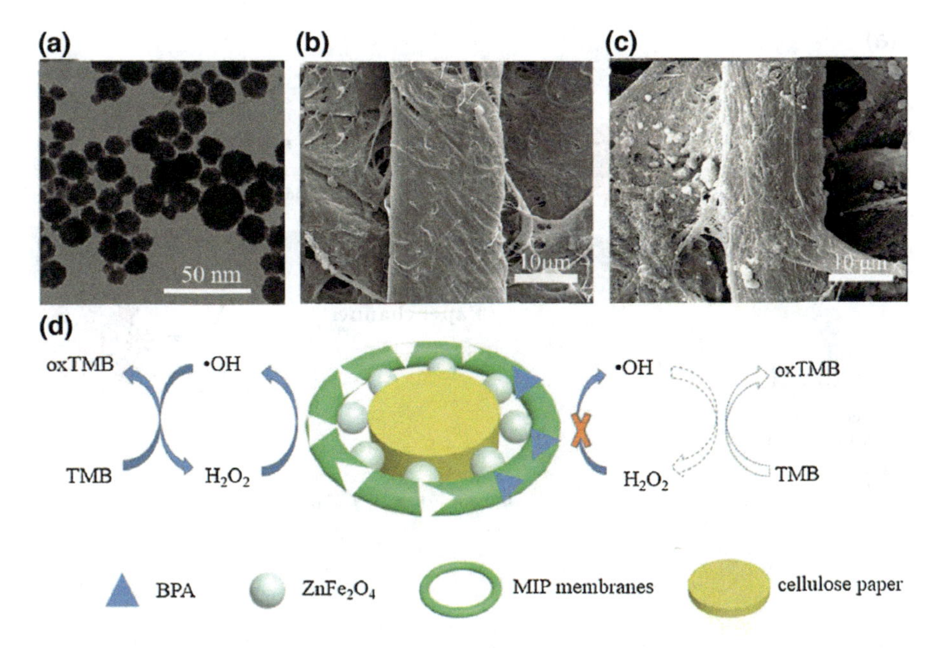

Fig. 14 **a** TEM of the synthesized $ZnFe_2O_4$ MNPs, SEM images of **b** synthesized MIP membranes and **c** MIP membranes@$ZnFe_2O_4$, **d** mechanism for the detection of BPA. Adapted with permission from Ref. [114] © 2017 Elsevier publisher

nitrogenous compounds such as ammonia as an indicating factor for meat spoilage [112]. The colour change of a pH-sensitive dye like bromophenol blue coated on paper occurs in presence of ammonia, which acts as a Lewis base and induces the colour change due to the release of hydrogen. Another work used the metachromasy of a dye, *o*-toluidine blue (OTB) to detect charged macromolecules, indicated by a characteristic colour change from blue to pink. On coming in contact with ionic molecules, the dye molecules align with the charges on the molecules, resulting in a shift in the wavelength of maximum absorbance of the dye [113].

The colorimetric sensing technique has been used to detect several toxic chemical contaminants in food such as bisphenol A [114], benzoic acid [115], clenbuterol [116], melamine [117], residual antibiotics [118] and estrogens [119]. Bisphenol A (BPA) has been widely used in food packaging such as the plastic beverage bottles and causes contamination of food as well as the environment [114]. A sensor was developed for the detection of BPA combining the intrinsic peroxidase activity and the colorimetric potential of $ZnFe_2O_4$ magnetic nanoparticles ($ZnFe_2O_4$ MNPs) and the adsorption capacity of molecularly imprinted polymer (MIP) membranes. Figure 14 shows the morphology of the MNPs and their binding on the membranes. As can be seen from Fig. 14d, in the absence of BPA, H_2O_2 was adsorbed on the $ZnFe_2O_4$ MNPs and activated by the ferric ions to generate ·OH, which subsequently oxidized 3,3′,5,5′-tetramethylbenzidine (TMB) to form a blue complex due to the transfer of charge. In contrast, no ·OH

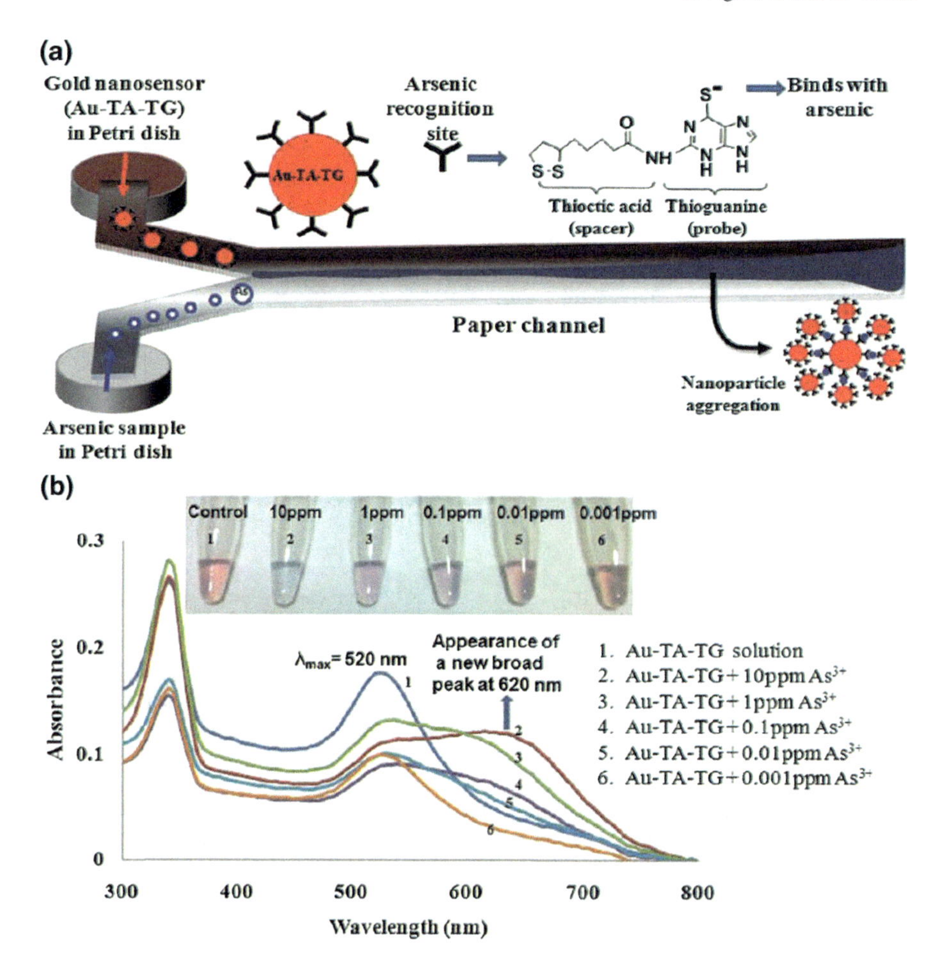

Fig. 15 **a** Schematic representation of gold nanosensor (Au–TA–TG) μPAD for arsenic detection, **b** UV-Vis absorption spectra of Au–TA–TG treated with different concentrations of arsenic. Inset: visual colour changes of the Au–TA–TG. Adapted with permission from Ref. [124] © 2014 RSC publisher

could be generated when BPA was anchored on the membrane surface. The assay was linear over the range 10–1000 nM with a limit of detection of 6.18 nM [114]. Other studies have also reported the detection of pesticides [120], food preservatives such as sulfur dioxide [121] and foodborne carcinogens such as aflatoxin B_1 [122]. Similarly, studies have reported the detection of cadmium (Cd^{2+}) [123] and nitrite [35] toxic ions in drinking water.

The very low concentration of arsenic (As) has been detected in water, 1 ppb which is lower than the WHO's reference standard for drinking water, using a microfluidic paper nanosensor fabricated by anchoring gold nanoparticles (Au) to thioctic acid (TA) and thioguanine (TG). As shown in Fig. 15a, when the two extended arms of the Y-channel are touched into the Au–TA–TG and As solutions, the fluids started flowing into the channel due to capillary action and rapidly interacts to form a visible bluish-black precipitate. TA acts as a spacer arm to increase the detection ability by reducing the steric hindrance. During detection, TG probes interact with As^{3+} ions leading to aggregation of nanoparticles. This causes a visual colour change due to inter-particles coupled plasmon resonance from aggregated Au nanoparticles. Figure 15b shows the shift in UV-Vis spectra wavelength (from λ_{max} 520 nm to λ_{max} 620 nm) indicating As binding with Au–TA–TG and formation of aggregates in various concentrations of As solutions [124].

5.2 Electrochemical Sensing

Electrochemical detection of ethanol in beer has been reported using screen-printed electrodes on office paper modified with Carbon Black and Prussian Blue (CB/PBNPs) nanocomposite [125]. Ethanol was quantified indirectly from the concentration of H_2O_2, which was produced by the enzymatic reaction between alcohol oxidase (bio-recognition element) and ethanol. A quick response was obtained in just 40 s with a sensitivity of 9.13 μA/mM cm^2 and a detection limit of 0.52 mM. The presence of lead and cadmium ions in rice and fish was sensed with bismuth-modified, boron-doped diamond electrode (Bi-BDDE) [126]. The Bi-BDDE increased the sensitivity of ion detection; exhibiting sharp and well-defined peaks in the stripping voltammograms, which were significantly higher than the stripping peak current observed in case of bismuth-modified screen-printed carbon electrode (Bi-SPCE). The superior performance of BDDE was attributed to the fact that bismuth could form "fused" alloys with ions, which rendered them ready to be reduced.

The detection of cadmium in rice was performed by a double-sided conductive carbon tape coated with a thin layer of gold for stripping analysis of Cd^{2+} coupled with in situ electrodeposition of bismuth [127]. Likewise, resistive sensors on paper with copper acetate were developed for monitoring the quality of raw broiler meat where H_2S was detected as an end product of microbial metabolism by replicating the aerobic and anaerobic conditions of modified atmosphere packaging in the laboratory [128]. Similarly, capacitive humidity sensors built for smart packaging showed a logarithmic response on paper [129].

5.3 Luminescence-Based Sensing

The photoluminescent quenching ability of graphene oxide (GO) was demonstrated for the detection of the pathogen (*E. coli*) using an LFIA with a limit of detection as low as 10 CFU mL^{-1} [130]. As shown in Fig. 16a, test, and control lines were

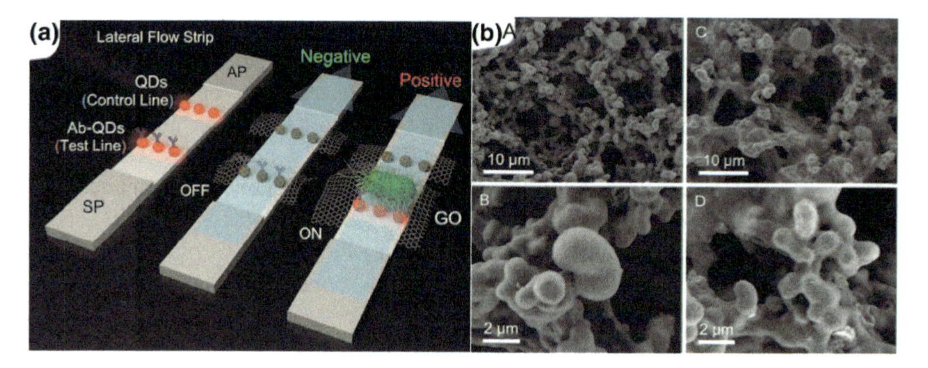

Fig. 16 **a** Photoluminescent lateral flow assay revealed by graphene oxide (GO) for *E. coli* detection, **b** SEM images of the paper-based sensing platform: (A & B) Bare detection line, (C & D) Graphene oxide coated detection line. Adapted with permission from Ref. [130] © 2015 ACS publisher

printed on a nitrocellulose substrate with antibody decorated CdSe@ZnS quantum dots (Ab-QDs) and bare QDs, respectively. Capillary forces caused the sample flow from the sample pad (SP) to the absorbent pad (AP). In the absence of target pathogen, the test line is quenched by GO. On the contrary, when the target is present, it is selectively captured by the Ab-QDs and the test line is not significantly quenched compared to the control line due to hindered resonance energy transfer caused by the increased distance between the donor and the acceptor. The control line, with no pathogen-binding molecules, is always quenched thus ensuring the correctness of the assay. The scanning electron micrographs (SEM) of the biosensing platform are shown in Fig. 16b [130]. Other studies have employed the fluorescence-based sensing approach for the determination of chemical contaminants in food such as formaldehyde [131, 132], Hg^{2+}, Ag^{+} and aminoglycoside antibiotic residues [133].

Another immunoassay developed for the detection of *Salmonella*, a common foodborne pathogen, was based on chemiluminescent reaction indicated by the quantification of adenosine triphosphate (ATP) [134]. The wax-printed μPAD was Z-folded for liquid handling without external driving force. As shown in Fig. 17, the μPAD was divided into two separate functional zones- reagent zone A and testing zone B by the folding area. Each functional zone possessed humped circles that were opposite. For testing, the reagent and tested samples were put into the corresponding zones; then the hump of the two circles was connected in order to allow the liquid to flow through the Z-folding. The ATP aptamer, immobilized on the reagent zone, bonded to the target and released the HRP tagged DNA associated with the aptamer. The released HRP tagged DNA then went through the micro-channel humps and arrived at the testing zone B via Z-folding of the μPAD. Eventually, H_2O_2 was added to lead to a colour change (since 3-amino-9-ethylcarbazole (AEC) coated on the testing zone was oxidized by HRP/H_2O_2), thus determining the concentration of ATP. The detection limit was 1 μM for ATP and around 2.6×10^7 CFU/mL for *Salmonella* [134]. Another work

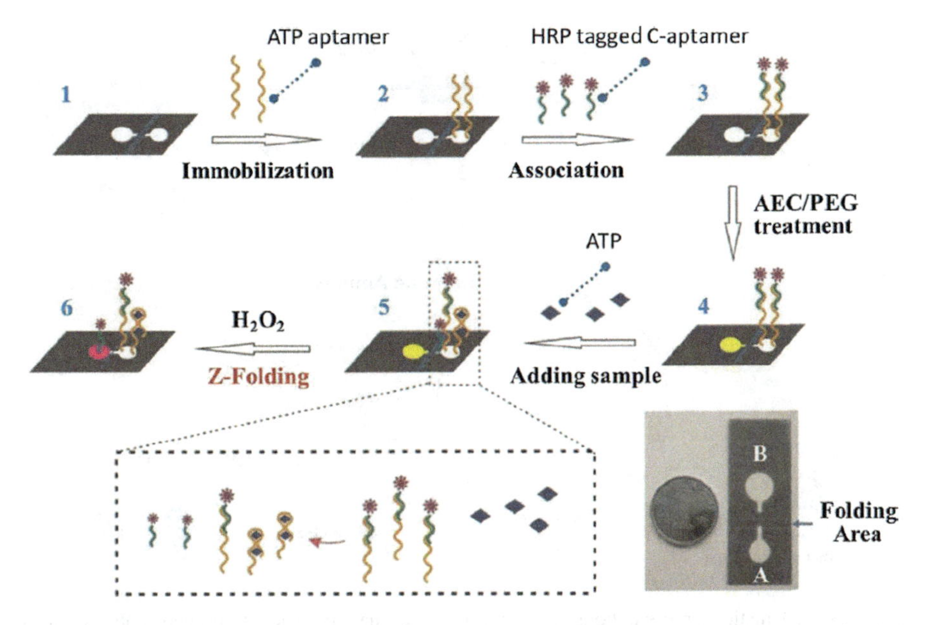

Fig. 17 Schematic showing the fabrication of Z-folded μPAD for *Salmonella* detection via ATP quantification. Adapted with permission from Ref. [134] © 2015 Elsevier publisher

developed an LFIA based on enzyme-catalyzed chemiluminescence for the detection of fumonisins, which is produced by *Fusarium* mould species, in maize with a detection limit of 2.5 μg L^{-1} [135]. The immunological and chemiluminescence reactions were conducted sequentially in the LFIA strip and the HRP-labeled tracer antibody was revealed by chemiluminescence to achieve the quantification. In a similar way, chemiluminescence-based sensing has been exploited on paper platforms for the detection of dichlorvos pesticides in vegetables [136, 137].

5.4 Other Sensor Types

A triple-mode colorimetric, fluorescent, and SERS sensor was developed for detecting nitrite ion, a widely used meat preservative, based on the hybrid assembly of gold nanorods (GNRs)-azo-gold nanoparticles (GNPs) (GNRs-Azo-GNPs) on a paper strip [138]. A quantitative colorimetric response resulted on the addition of nitrite by the Griess reaction between *p*-aminothiophenol (PATP) and 1,8-diaminonaphthalene (DAP), causing a drop in the fluorescence intensities due to FRET from DAP to azo-moiety. The sensor with a limit of detection of 0.01 μM also served as a SERS substrate for quantitative analysis of nitrite [138]. A filter paper with AgNPs served as a rag to collect pesticide residues from the peels of apples, bananas, and tomatoes [139]. The thiram and paraoxon residues could be collected by simply swabbing the paper substrate across a wide surface area of the peels. The

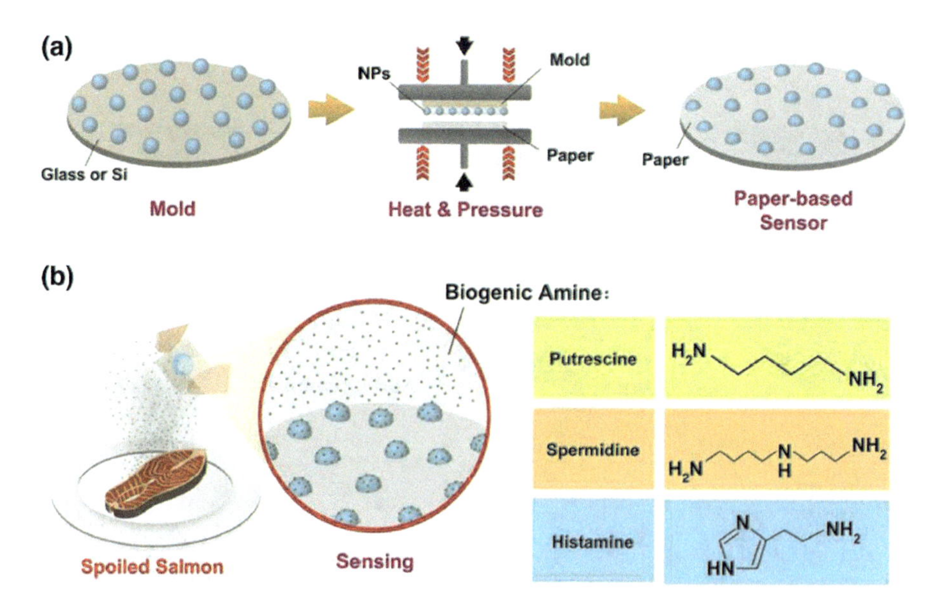

Fig. 18 Schematic representations showing **a** metal nanoparticles imprinted onto a paper, **b** nanoparticle-embedded paper to detect volatile biogenic amines released from spoiled salmon. Adapted with permission from Ref. [140] © 2017 ACS publisher

AgNP-decorated filter paper with 'dynamic SERS' provided not only the hot spots from nanostructures and concentrated target molecules but also a high sensitivity and reproducibility pesticide detection. Based on plasmonic refractometric sensing, gas sensors were fabricated by imprinting nanoparticles on inkjet paper [140]. Refractometric sensing is based on the adsorption of the analyte on the surface of the nanoparticles resulting in a red-shift in the wavelength and an extinction increase of the LSPR peak. Plasmonic nanoparticles such as gold and silver were deposited on paper using reversal lithography (Fig. 18a). The high reflectance and smoothness of inkjet papers were exploited for sensing of biogenic amines (BA) released from spoiled fish. The levels of BA, generated through microbial action, were used as an indicator of food freshness. During detection, the volatile BAs attached spontaneously onto the nanoparticle surface via their amino groups forming a dense layer (Fig. 18b). This led to a change in the local refractive index around the nanoparticles, resulting in the red-shift of the LSPR wavelength [140].

6 Sensing for Forensics and Security

Lately, sensing high-energy materials has gained immense importance in security and forensics in view of their potential toxicity and environmental risks. The modern methods for their detection such as gas chromatography coupled with mass spectrometry, ion mobility spectrometry and energy dispersive X-ray spectroscopy are all infrastructure-intensive and cannot be implemented outside a laboratory setting [141, 142]. This motivated the development of complementary techniques for rapid, straightforward and on-site detection of explosives. Several sensing approaches including colorimetry [143], luminescence quenching [141], fluorescence quenching [144], SERS [145] and others [146] have been implemented on paper platforms to achieve these goals. The detection of improvised explosives has been demonstrated by μPADs containing reagents capable of reacting with explosives and resulting in a specific colorimetric reaction [143, 147]. A single eluent reservoir was used to extract the explosives and transport them to the testing zones by capillary action. Inorganic explosives with nitrate, nitrite, ammonium, chlorate and perchlorate oxidizers, as well as military explosives such as 2,4,6-trinitrotoluene (TNT) and 1,3,5-trinitroperhydro-1,3,5-triazine (RDX), along with urea nitrate and organic peroxides could be detected with limits of detection between 0.39 and 19.8 μg [143]. Another study detected TNT down to 14 ng/cm^2 based on the photothermal effect of the complex formed by charge transfer between an electron-rich group in polyaniline and an electron-deficient nitro group in TNT, which led to a near infra-red absorption at 800 nm, inducing a temperature rise proportional to the concentration of TNT upon irradiation [148].

Detection of drugs is of paramount significance for crime scene investigations. Using an office paper, a simple colorimetric sensor was fabricated for sensing of phenacetin, an analgesic and antipyretic drug, widely used as a cutting agent by traffickers for illicit drugs such as cocaine to increase profits [149]. The sensing mechanism involved hydrolysis of the adulterant in the presence of an acid followed by reaction of the formed compound with sodium 1,2-naphthoquinone-4-sulfonate (NQS) to yield a reddish-orange product, which could be read by a smartphone camera. Likewise, the electrochemical detection of alprazolam, a psychotropic drug, was performed by methylene blue doped silver core-shell palladium nano-hybrids (Ag@Pd nano-hybrids) on a chip. The cyclic voltammetry tests showed that Ag@Pd nano-hybrids generated a strong electrochemical signal in PBS. Due to the redox transition of methylene blue, the drug was reduced to its corresponding dihydro derivative on the modified surface of the biosensor chip [150].

7 Summary, Challenges and Future Perspectives

The present chapter has reviewed the developments in the paper-based sensing devices with an emphasis on the fundamental sensing approaches. Considering the multitude of striking features of paper, it is no surprise that there has been a dramatic increase in the number of research groups working on analytical devices using paper as a substrate. The hydrophilicity and low-cost of paper are the two key attributes that have made it a versatile sensing platform in virtually every sector from healthcare to security. The paper-based sensing technology has come a long way from simple qualitative test strips to the 3D complex assays that have the potential to carry out multiple chemical reactions in a simple format without the need for any external equipment. Although paper-based sensing technology has seen immense development since the last decade and is undoubtedly the dawn of a new day in building a next-generation sensor technology, there are still a number of concerns that need to be addressed. The clinical performance of the existing devices needs to be improved as varying specificity and sensitivity may lead to false-positive or false-negative results [26]. Since different detection methods could also cause varying performance, it would be imperative to assess them to reach an optimal sensitivity and specificity as well as to reduce cost. Especially in case of colorimetric assays, there exists the possibility of subjective judgement among the users in interpreting the results with the naked eye in different illumination environments, which could lead to controversial readouts. The reagents used in the assay, particularly the biomolecules such as enzymes and antibodies, being highly susceptible to the surrounding environment, may degrade in extreme conditions thus affecting the robustness of the assay and having a detrimental influence on its performance. Additionally, the changes in temperature and humidity may affect the long-term stability of the immobilized reagents, leading to varying signals. Furthermore, batch-to-batch variation may also have an impact on the reproducibility of testing [151]. Eventually, the extent of commercialization into the market, outside the research laboratories will be the deciding factor for the ultimate success of paper-based sensing devices.

Acknowledgements CA is grateful to the Tempus Public Foundation (TPF) for providing financial assistance under the *Stipendium Hungaricum Programme*. This chapter was also made in frame of the "EFOP-3.6.1-16-2016-00018—Improving the role of research + development + innovation in the higher education through institutional developments assisting intelligent specialization in Sopron and Szombathely."

References

1. Nayak S, Blumenfeld NR, Laksanasopin T, Sia SK (2017) Point-of-care diagnostics: recent developments in a connected age. Anal Chem 89:102–123
2. Gong MM, Sinton D (2017) Turning the page: advancing paper-based microfluidics for broad diagnostic application. Chem Rev 117:8447–8480
3. Sher M, Zhuang R, Demirci U, Asghar W (2017) Paper-based analytical devices for clinical diagnosis: recent advances in the fabrication techniques and sensing mechanisms. Expert Rev Mol Diagn 17:351–366
4. Zarei M (2017) Portable biosensing devices for point-of-care diagnostics: recent developments and applications. Trends Anal Chem 91:26–41
5. Martinez AW (2011) Microfluidic paper-based analytical devices: from POCKET to paper-based ELISA. Bioanalysis 3:2589–2592
6. Martinez AW, Phillips ST, Whitesides GM, Carrilho E (2010) Diagnostics for the developing world: Microfluidic paper-based analytical devices. Anal Chem 82:3–10
7. Morbioli GG, Mazzu-Nascimento T, Stockton AM, Carrilho E (2017) Technical aspects and challenges of colorimetric detection with microfluidic paper-based analytical devices (mPADs)—a review. Anal Chim Acta 970:1–22

8. Tseng SC et al (2012) Eco-friendly plasmonic sensors: using the photothermal effect to prepare metal nanoparticle-containing test papers for highly sensitive colorimetric detection. Anal Chem 84:5140–5145
9. Tao H et al (2011) Metamaterials on paper as a sensing platform. Adv Mater 23:3197–3201
10. Martinez AW, Phillips ST, Butte MJ, Whitesides GM (2007) Patterned paper as a platform for inexpensive, low-volume, portable bioassays. Angew Chem Int Ed 46:1318–1320
11. Parolo C, Merkoci A (2013) Paper-based nanobiosensors for diagnostics. Chem Soc Rev 42:450–457
12. Ge S, Zhang L, Zhang Y, Lan F, Yan M, Yu J (2017) Nanomaterials-modified cellulose paper as a platform for biosensing applications. Nanoscale 9:4366–4382
13. Alvarez-Diduk R, Orozco J, Merkoci A (2017) Paper strip-embedded graphene quantum dots: a screening device with a smartphone readout. Sci Rep 7:976–984
14. Carrasquilla C, Little JR, Li Y, Brennan JD (2015) Patterned paper sensors printed with long-chain DNA aptamers. Chemistry 21:7369–7373
15. Rat S et al (2016) Elastic coupling between spin-crossover particles and cellulose fibers. Chem Commun 52:11267–11269
16. Lim WY, Goh BT, Khor SM (2017) Microfluidic paper-based analytical devices for potential use in quantitative and direct detection of disease biomarkers in clinical analysis. J Chromatogr B 1060:424–442
17. Mahato K, Srivastava A, Chandra P (2017) Paper based diagnostics for personalized health care: emerging technologies and commercial aspects. Biosens Bioelectron 96:246–259
18. Wang S, Chinnasamy T, Lifson MA, Inci F, Demirci U (2016) Flexible substrate-based devices for point-of-care diagnostics. Trends Biotechnol 34:909–921
19. Meredith NA, Quinn C, Cate DM, Reilly TH 3rd, Volckens J, Henry CS (2016) Paper-based analytical devices for environmental analysis. Analyst 141:1874–1887
20. Bulbul G, Hayat A, Andreescu S (2015) Portable nanoparticle-based sensors for food safety assessment. Sensors 15:30736–30758
21. Hua M, Li S, Wang S, Lu X (2018) Detecting chemical hazards in foods using microfluidic paper-based analytical devices (μPADs): The real-world application. Micromachines 9:32–45
22. Ansari N, Lodha A, Pandya A, Menon SK (2017) Determination of cause of death using paper-based microfluidic device as a colorimetric probe. Anal Methods 9:5632–5639
23. Song Y, Gyarmati P, Araujo AC, Lundeberg J, Brumer H 3rd, Stahl PL (2014) Visual detection of DNA on paper chips. Anal Chem 86:1575–1582
24. Ge S, Zhang L, Zhang Y, Liu H, Huang J, Yan M, Yu J (2015) Electrochemical K-562 cells sensor based on origami paper device for point-of-care testing. Talanta 145:12–19
25. Nilghaz A, Guan L, Tan W, Shen W (2016) Advances of paper-based microfluidics for diagnostics—the original motivation and current status. ACS Sens 1:1382–1393
26. Yamada K, Shibata H, Suzuki K, Citterio D (2017) Toward practical application of paper-based microfluidics for medical diagnostics: state-of-the-art and challenges. Lab Chip 17:1206–1249
27. Tao X, Jia H, He Y, Liao S, Wang Y (2017) Ultrafast paper thermometers based on a green sensing ink. ACS Sens 2:449–454
28. Yang J, Kwak TJ, Zhang X, McClain R, Chang WJ, Gunasekaran S (2016) Iridium oxide-reduced graphene oxide nanohybrid thin film modified screen-printed electrodes as disposable electrochemical paper microfluidic pH sensors. J Vis Exp 117:e53339–e53345
29. Balde M, Vena A, Sorli B (2015) Fabrication of porous anodic aluminium oxide layers on paper for humidity sensors. Sens Actuators B Chem 220:829–839
30. Yuan Y, Zhang Y, Liu R, Liu J, Li Z, Liu X (2016) Humidity sensor fabricated by inkjet-printing photosensitive conductive inks PEDOT:PVMA on a paper substrate. RSC Adv 6:47498–47508
31. Koren K, Kühl M (2015) A simple laminated paper-based sensor for temperature sensing and imaging. Sens Actuators B Chem 210:124–128

32. Nagy V et al (2014) Cellulose fiber nanocomposites displaying spin-crossover properties. Colloids Surf A Physicochem Eng Asp 456:35–40
33. Nagy V, Suleimanov I, Molnár G, Salmon L, Bousseksou A, Csóka L (2015) Cellulose–spin crossover particle composite papers with reverse printing performance: a proof of concept. J Mater Chem C 3:7897–7905
34. Zhao H, Zhang T, Qi R, Dai J, Liu S, Fei T (2017) Drawn on paper: a reproducible humidity sensitive device by handwriting. ACS Appl Mater Interfaces 9:28002–28009
35. Lopez-Ruiz N, Curto VF, Erenas MM, Benito-Lopez F, Diamond D, Palma AJ, Capitan-Vallvey LF (2014) Smartphone-based simultaneous pH and nitrite colorimetric determination for paper microfluidic devices. Anal Chem 86:9554–9562
36. Lee C-Y, Lei KF, Tsai S-W, Tsang N-M (2016) Development of graphene-based sensors on paper substrate for the measurement of pH value of analyte. BioChip J 10:182–188
37. Liu X, Mwangi M, Li X, O'Brien M, Whitesides GM (2011) Paper-based piezoresistive MEMS sensors. Lab Chip 11:2189–2196
38. Lee J, Lee YJ, Ahn YJ, Choi S, Lee G-J (2018) A simple and facile paper-based colorimetric assay for detection of free hydrogen sulfide in prostate cancer cells. Sens Actuators B Chem 256:828–834
39. Silva TG, de Araujo WR, Munoz RA, Richter EM, Santana MH, Coltro WK, Paixao TR (2016) Simple and sensitive paper-based device coupling electrochemical sample pretreatment and colorimetric detection. Anal Chem 88:5145–5151
40. Cha R, Wang D, He Z, Ni Y (2012) Development of cellulose paper testing strips for quick measurement of glucose using chromogen agent. Carbohydr Polym 88:1414–1419
41. Yamada K, Suzuki K, Citterio D (2017) Text-displaying colorimetric paper-based analytical device. ACS Sens 2:1247–1254
42. Chen X, Chen J, Wang F, Xiang X, Luo M, Ji X, He Z (2012) Determination of glucose and uric acid with bienzyme colorimetry on microfluidic paper-based analysis devices. Biosens Bioelectron 35:363–368
43. Chan SK, Lim TS (2016) A straw-housed paper-based colorimetric antibody–antigen sensor. Anal Methods 8:1431–1436
44. Costa MN et al (2014) A low cost, safe, disposable, rapid and self-sustainable paper-based platform for diagnostic testing: lab-on-paper. Nanotechnology 25:094006–094017
45. Tian T et al (2016) Integration of target responsive hydrogel with cascaded enzymatic reactions and microfluidic paper-based analytic devices (mPADs) for point-of-care testing (POCT). Biosens Bioelectron 77:537–542
46. Wei X et al (2016) Microfluidic distance readout sweet hydrogel integrated paper-based analytical device (μDiSH-PAD) for visual quantitative point-of-care testing. Anal Chem 88:2345–2352
47. Wei X et al (2015) Target-responsive DNA hydrogel mediated "stop-flow" microfluidic paper-based analytic device for rapid, portable and visual detection of multiple targets. Anal Chem 87:4275–4282
48. Zhang Y et al (2016) Naked-eye quantitative aptamer-based assay on paper device. Biosens Bioelectron 78:538–546
49. Santhiago M, Kubota LT (2013) A new approach for paper-based analytical devices with electrochemical detection based on graphite pencil electrodes. Sens Actuators B Chem 177:224–230
50. Yao Y, Zhang C (2016) A novel screen-printed microfluidic paper-based electrochemical device for detection of glucose and uric acid in urine. Biomed Microdevices 18:92–100
51. Narang J et al (2017) Lab on paper chip integrated with Si@GNRs for electroanalysis of diazepam. Anal Chim Acta 980:50–57
52. Narang J, Malhotra N, Singhal C, Mathur A, Chakraborty D, Ingle A, Pundir CS (2017) Point of care with micro fluidic paper based device incorporated with nanocrys of Zeolite–GO for electrochemical sensing of date rape drug. Proc Technol 27:91–93

53. Wang P, Ge L, Yan M, Song X, Ge S, Yu J (2012) Paper-based three-dimensional electrochemical immunodevice based on multi-walled carbon nanotubes functionalized paper for sensitive point-of-care testing. Biosens Bioelectron 32:238–243
54. Teengam P, Siangproh W, Tuantranont A, Henry CS, Vilaivan T, Chailapakul O (2017) Electrochemical paper-based peptide nucleic acid biosensor for detecting human papillomavirus. Anal Chim Acta 952:32–40
55. Wang Y et al (2016) A novel label-free microfluidic paper-based immunosensor for highly sensitive electrochemical detection of carcinoembryonic antigen. Biosens Bioelectron 83:319–326
56. Kumar S et al (2015) Reduced graphene oxide modified smart conducting paper for cancer biosensor. Biosens Bioelectron 73:114–122
57. Scida K, Cunningham JC, Renault C, Richards I, Crooks RM (2014) Simple, sensitive, and quantitative electrochemical detection method for paper analytical devices. Anal Chem 86:6501–6507
58. Wang CC et al (2016) A paper-based “pop-up” electrochemical device for analysis of beta-hydroxybutyrate. Anal Chem 88:6326–6333
59. Jagadeesan KK, Kumar S, Sumana G (2012) Application of conducting paper for selective detection of troponin. Electrochem Commun 20:71–74
60. Yang M et al (2016) Flexible and disposable sensing platforms based on newspaper. ACS Appl Mater Interfaces 8:34978–34984
61. Cunningham JC, Brenes NJ, Crooks RM (2014) Paper electrochemical device for detection of DNA and thrombin by target-induced conformational switching. Anal Chem 86:6166–6170
62. Mirasoli M, Guardigli M, Michelini E, Roda A (2014) Recent advancements in chemical luminescence-based lab-on-chip and microfluidic platforms for bioanalysis. J Pharm Biomed Anal 87:36–52
63. Liu W, Luo J, Guo Y, Kou J, Li B, Zhang Z (2014) Nanoparticle coated paper-based chemiluminescence device for the determination of L-cysteine. Talanta 120:336–341
64. Liu F, Zhang C (2015) A novel paper-based microfluidic enhanced chemiluminescence biosensor for facile, reliable and highly-sensitive gene detection of *Listeria monocytogenes*. Sens Actuators B Chem 209:399–406
65. Zhao M, Li H, Liu W, Guo Y, Chu W (2016) Plasma treatment of paper for protein immobilization on paper-based chemiluminescence immunodevice. Biosens Bioelectron 79:581–588
66. Liu W, Cassano CL, Xu X, Fan ZH (2013) Laminated paper-based analytical devices (LPAD) with origami-enabled chemiluminescence immunoassay for cotinine detection in mouse serum. Anal Chem 85:10270–10276
67. Chen Y, Wang J, Liu Z, Wang X, Li X, Shan G (2018) A simple and versatile paper-based electrochemiluminescence biosensing platform for hepatitis B virus surface antigen detection. Biochem Eng J 129:1–6
68. Ge L, Yan J, Song X, Yan M, Ge S, Yu J (2012) Three-dimensional paper-based electrochemiluminescence immunodevice for multiplexed measurement of biomarkers and point-of-care testing. Biomaterials 33:1024–1031
69. Ge S, Liang L, Lan F, Zhang Y, Wang Y, Yan M, Yu J (2016) Photoelectrochemical immunoassay based on chemiluminescence as internal excited light source. Sens Actuators B Chem 234:324–331
70. Shi L, Li L, Li X, Zhang G, Zhang Y, Dong C, Shuang S (2017) Excitation-independent yellow-fluorescent nitrogen-doped carbon nanodots for biological imaging and paper-based sensing. Sens Actuators B Chem 251:234–241
71. Cai Y, You J, You Z, Dong F, Du S, Zhang L (2018) Profuse color-evolution-based fluorescent test paper sensor for rapid and visual monitoring of endogenous Cu^{2+} in human urine. Biosens Bioelectron 99:332–337

72. Samanta S, Halder S, Dey P, Manna U, Ramesh A, Das G (2018) A ratiometric fluorogenic probe for the real-time detection of SO_3^{2-} in aqueous medium: Application in a cellulose paper based device and potential to sense SO_3^{2-} in mitochondria. Analyst 143:250–257
73. Thom NK, Lewis GG, Yeung K, Phillips ST (2014) Quantitative fluorescence assays using a self-powered paper-based microfluidic device and a camera-equipped cellular phone. RSC Adv 4:1334–1340
74. Mei Q, Zhang Z (2012) Photoluminescent graphene oxide ink to print sensors onto microporous membranes for versatile visualization bioassays. Angew Chem Int Ed 51:5602–5606
75. Derikvand F, Yin DT, Barrett R, Brumer H (2016) Cellulose-based biosensors for esterase detection. Anal Chem 88:2989–2993
76. Li B, Zhou X, Liu H, Deng H, Huang R, Xing D (2018a) Simultaneous detection of antibiotic resistance genes on paper-based chip using $[Ru(phen)_2dppz]^{2+}$ turn-on fluorescence probe. ACS Appl Mater Interfaces. https://doi.org/10.1021/acsami.7b17653
77. Rosa AM, Louro AF, Martins SA, Inacio J, Azevedo AM, Prazeres DM (2014) Capture and detection of DNA hybrids on paper via the anchoring of antibodies with fusions of carbohydrate binding modules and ZZ-domains. Anal Chem 86:4340–4347
78. Wang Y, Wang S, Ge S, Wang S, Yan M, Zang D, Yu J (2013) Facile and sensitive paper-based chemiluminescence DNA biosensor using carbon dots dotted nanoporous gold signal amplification label. Anal Methods 5:1328–1336
79. Zhou F, Noor MO, Krull UJ (2014) Luminescence resonance energy transfer-based nucleic acid hybridization assay on cellulose paper with upconverting phosphor as donors. Anal Chem 86:2719–2726
80. Oliveira MJ et al (2017) Office paper decorated with silver nanostars—an alternative cost effective platform for trace analyte detection by SERS. Sci Rep 7:2480–2493
81. Hu SW, Qiao S, Pan JB, Kang B, Xu JJ, Chen HY (2018) A paper-based SERS test strip for quantitative detection of Mucin-1 in whole blood. Talanta 179:9–14
82. Abbas A, Brimer A, Slocik JM, Tian L, Naik RR, Singamaneni S (2013) Multifunctional analytical platform on a paper strip: separation, preconcentration, and subattomolar detection. Anal Chem 85:3977–3983
83. Banaei N, Foley A, Houghton JM, Sun Y, Kim B (2017) Multiplex detection of pancreatic cancer biomarkers using a SERS-based immunoassay. Nanotechnology 28:455101–455111
84. Kim W-S, Shin J-H, Park H-K, Choi S (2016) A low-cost, monometallic, surface-enhanced Raman scattering-functionalized paper platform for spot-on bioassays. Sens Actuators B Chem 222:1112–1118
85. Liu Q et al (2014) Paper-based plasmonic platform for sensitive, noninvasive, and rapid cancer screening. Biosens Bioelectron 54:128–134
86. Ngo YH, Li D, Simon GP, Garnier G (2012) Gold nanoparticle-paper as a three-dimensional surface enhanced Raman scattering substrate. Langmuir 28:8782–8790
87. Tadepalli S et al (2015) Peptide functionalized gold nanorods for the sensitive detection of a cardiac biomarker using plasmonic paper devices. Sci Rep 5:16206–16216
88. Noor MO, Krull UJ (2014) Camera-based ratiometric fluorescence transduction of nucleic acid hybridization with reagentless signal amplification on a paper-based platform using immobilized quantum dots as donors. Anal Chem 86:10331–10339
89. Liu X, Yang Y, Li Q, Wang Z, Xing X, Wang Y (2018c) Portably colorimetric paper sensor based on ZnS quantum dots for semi-quantitative detection of Co^{2+} through the measurement of grey level. Sens Actuators B Chem. https://doi.org/10.1016/j.snb.2018.01.121
90. Dhavamani J, Mujawar LH, El-Shahawi MS (2018) Hand drawn paper-based optical assay plate for rapid and trace level determination of Ag^+ in water. Sens Actuators B Chem 258:321–330
91. Yakoh A, Rattanarat P, Siangproh W, Chailapakul O (2018) Simple and selective paper-based colorimetric sensor for determination of chloride ion in environmental samples using label-free silver nanoprisms. Talanta 178:134–140

92. Chaiyo S, Siangproh W, Apilux A, Chailapakul O (2015) Highly selective and sensitive paper-based colorimetric sensor using thiosulfate catalytic etching of silver nanoplates for trace determination of copper ions. Anal Chim Acta 866:75–83
93. Hossain SM, Brennan JD (2011) *b*-Galactosidase-based colorimetric paper sensor for determination of heavy metals. Anal Chem 83:8772–8778
94. Giokas DL, Tsogas GZ, Vlessidis AG (2014) Programming fluid transport in paper-based microfluidic devices using razor-crafted open channels. Anal Chem 86:6202–6207
95. Kappi FA, Tsogas GZ, Christodouleas DC, Giokas DL (2017) Calibrant-loaded paper-based analytical devices for standard addition quantitative assays. Sens Actuators B Chem 253:860–867
96. Cuartero M, Crespo GA, Bakker E (2015) Paper-based thin-layer coulometric sensor for halide determination. Anal Chem 87:1981–1990
97. Ruecha N, Chailapakul O, Suzuki K, Citterio D (2017) Fully inkjet-printed paper-based potentiometric ion-sensing devices. Anal Chem 89:10608–10616
98. Yoon JH et al (2017) Fabrication of newspaper-based potentiometric platforms for flexible and disposable ion sensors. J Colloid Interface Sci 508:167–173
99. Zor E, Alpaydin S, Arici A, Saglam ME, Bingol H (2018) Photoluminescent nanopaper-based microcuvette for iodide detection in seawater. Sens Actuators B Chem 254:1216–1224
100. Chen PC, Li YC, Ma JY, Huang JY, Chen CF, Chang HT (2016) Size-tunable copper nanocluster aggregates and their application in hydrogen sulfide sensing on paper-based devices. Sci Rep 6:24882–24890
101. Zhang M, Ge L, Ge S, Yan M, Yu J, Huang J, Liu S (2013) Three-dimensional paper-based electrochemiluminescence device for simultaneous detection of Pb^{2+} and Hg^{2+} based on potential-control technique. Biosens Bioelectron 41:544–550
102. Prabphal J, Vilaivan T, Praneenararat T (2018) Fabrication of a paper-based turn-off fluorescence sensor for Cu^{2+} ion from a pyridinium porphyrin. ChemistrySelect 3:894–899
103. Sun T, Niu Q, Li Y, Li T, Hu T, Wang E, Liu H (2018) A novel oligothiophene-based colorimetric and fluorescent "turn on" sensor for highly selective and sensitive detection of cyanide in aqueous media and its practical applications in water and food samples. Sens Actuators B Chem 258:64–71
104. Li B, Zhang Z, Qi J, Zhou N, Qin S, Choo J, Chen L (2017) Quantum dot-based molecularly imprinted polymers on three-dimensional origami paper microfluidic chip for fluorescence detection of phycocyanin. ACS Sens 2:243–250
105. Zhao H, Zang L, Wang L, Qin F, Zhang Z, Cao W (2015) Luminescence ratiometric oxygen sensor based on gadolinium labeled porphyrin and filter paper. Sens Actuators B Chem 215:405–411
106. Kim Y, Jang G, Lee TS (2015) New fluorescent metal-ion detection using a paper-based sensor strip containing tethered rhodamine carbon nanodots. ACS Appl Mater Interfaces 7:15649–15657
107. Lee M, Oh K, Choi HK, Lee SG, Youn HJ, Lee HL, Jeong DH (2018) Subnanomolar sensitivity of filter paper-based SERS sensor for pesticide detection by hydrophobicity change of paper surface. ACS Sens 3:151–159
108. Lee J-C, Kim W, Choi S (2017) Fabrication of a SERS-encoded microfluidic paper-based analytical chip for the point-of-assay of wastewater. Int J Precis Eng Manuf-Green Tech 4:221–226
109. Bharadwaj S, Pandey A, Yagci B, Ozguz V, Qureshi A (2018) Graphene nano-mesh-Ag-ZnO hybrid paper for sensitive SERS sensing and self-cleaning of organic pollutants. Chem Eng J 336:445–455
110. Jokerst JC, Adkins JA, Bisha B, Mentele MM, Goodridge LD, Henry CS (2012) Development of a paper-based analytical device for colorimetric detection of select foodborne pathogens. Anal Chem 84:2900–2907

111. Alhogail S, Suaifan G, Zourob M (2016) Rapid colorimetric sensing platform for the detection of *Listeria monocytogenes* foodborne pathogen. Biosens Bioelectron 86:1061–1066
112. Hakovirta M, Aksoy B, Hakovirta J (2015) Self-assembled micro-structured sensors for food safety in paper based food packaging. Mater Sci Eng, C 53:331–335
113. Swerin A, Mira I (2014) Ink-jettable paper-based sensor for charged macromolecules and surfactants. Sens Actuators B Chem 195:389–395
114. Kong Q, Wang Y, Zhang L, Ge S, Yu J (2017) A novel microfluidic paper-based colorimetric sensor based on molecularly imprinted polymer membranes for highly selective and sensitive detection of bisphenol A. Sens Actuators B Chem 243:130–136
115. Liu C-C, Wang Y-N, Fu L-M, Chen K-L (2018) Microfluidic paper-based chip platform for benzoic acid detection in food. Food Chem 249:162–167
116. Ma L, Nilghaz A, Choi JR, Liu X, Lu X (2018) Rapid detection of clenbuterol in milk using microfluidic paper-based ELISA. Food Chem 246:437–441
117. Gao N, Huang P, Wu F (2018) Colorimetric detection of melamine in milk based on Triton X-100 modified gold nanoparticles and its paper-based application. Spectrochim Acta A Mol Biomol Spectrosc 192:174–180
118. Ha N-R, Jung I-P, Kim S-H, Kim AR, Yoon M-Y (2017) Paper chip-based colorimetric sensing assay for ultra-sensitive detection of residual kanamycin. Process Biochem 62:161–168
119. Xiao L, Zhang Z, Wu C, Han L, Zhang H (2017) Molecularly imprinted polymer grafted paper-based method for the detection of 17b-Estradiol. Food Chem 221:82–86
120. Nouanthavong S, Nacapricha D, Henry CS, Sameenoi Y (2016) Pesticide analysis using nanoceria-coated paper-based devices as a detection platform. Analyst 141(5):1837–1846
121. Liu C-C, Wang Y-N, Fu L-M, Yang D-Y (2017) Rapid integrated microfluidic paper-based system for sulfur dioxide detection. Chem Eng J 316:790–796
122. Busa LS, Mohammadi S, Maeki M, Ishida A, Tani H, Tokeshi M (2016) A competitive immunoassay system for microfluidic paper-based analytical detection of small size molecules. Analyst 141:6598–6603
123. Lopez Marzo AM, Pons J, Blake DA, Merkoci A (2013) All-integrated and highly sensitive paper based device with sample treatment platform for Cd^{2+} immunodetection in drinking/tap waters. Anal Chem 85:3532–3538
124. Nath P, Arun RK, Chanda N (2014) A paper based microfluidic device for the detection of arsenic using a gold nanosensor. RSC Adv 4:59558–59561
125. Cinti S, Basso M, Moscone D, Arduini F (2017) A paper-based nanomodified electrochemical biosensor for ethanol detection in beers. Anal Chim Acta 960:123–130
126. Chaiyo S, Apiluk A, Siangproh W, Chailapakul O (2016) High sensitivity and specificity simultaneous determination of lead, cadmium and copper using μPAD with dual electrochemical and colorimetric detection. Sens Actuators B Chem 233:540–549
127. Bi X-M, Wang H-R, Ge L-Q, Zhou D-M, Xu J-Z, Gu H-Y, Bao N (2018) Gold-coated nanostructured carbon tape for rapid electrochemical detection of cadmium in rice with in situ electrodeposition of bismuth in paper-based analytical devices. Sens Actuators B Chem. https://doi.org/10.1016/j.snb.2018.01.007
128. Koskela J et al (2015) Monitoring the quality of raw poultry by detecting hydrogen sulfide with printed sensors. Sens Actuators B Chem 218:89–96
129. Mraovic M, Muck T, Pivar M, Trontelj J, Pletersek A (2014) Humidity sensors printed on recycled paper and cardboard. Sensors 14:13628–13643
130. Morales-Narváez E, Naghdi T, Zor E, Merkoçi A (2015) Photoluminescent lateral-flow immunoassay revealed by graphene oxide: highly sensitive paper-based pathogen detection. Anal Chem 87:8573–8577
131. Guzman JMCC, Tayo LL, Liu C-C, Wang Y-N, Fu L-M (2018) Rapid microfluidic paper-based platform for low concentration formaldehyde detection. Sens Actuators B Chem 255:3623–3629

132. Liu C-C, Wang Y-N, Fu L-M, Huang Y-H (2018) Microfluidic paper-based chip platform for formaldehyde concentration detection. Chem Eng J 332:695–701
133. Zhang Y, Zuo P, Ye BC (2015) A low-cost and simple paper-based microfluidic device for simultaneous multiplex determination of different types of chemical contaminants in food. Biosens Bioelectron 68:14–19
134. Jin SQ, Guo SM, Zuo P, Ye BC (2015) A cost-effective Z-folding controlled liquid handling microfluidic paper analysis device for pathogen detection via ATP quantification. Biosens Bioelectron 63:379–383
135. Mirasoli M, Buragina A, Dolci LS, Simoni P, Anfossi L, Giraudi G, Roda A (2012) Chemiluminescence-based biosensor for fumonisins quantitative detection in maize samples. Biosens Bioelectron 32:283–287
136. Liu W, Guo Y, Luo J, Kou J, Zheng H, Li B, Zhang Z (2015) A molecularly imprinted polymer based a lab-on-paper chemiluminescence device for the detection of dichlorvos. Spectrochim Acta A Mol Biomol Spectrosc 141:51–57
137. Liu W, Kou J, Xing H, Li B (2014) Paper-based chromatographic chemiluminescence chip for the detection of dichlorvos in vegetables. Biosens Bioelectron 52:76–81
138. Li D, Ma Y, Duan H, Deng W, Li D (2018) Griess reaction-based paper strip for colorimetric/fluorescent/SERS triple sensing of nitrite. Biosens Bioelectron 99:389–398
139. Zhu Y, Li M, Yu D, Yang L (2014) A novel paper rag as 'D-SERS' substrate for detection of pesticide residues at various peels. Talanta 128:117–124
140. Tseng SY, Li SY, Yi SY, Sun AY, Gao DY, Wan D (2017) Food quality monitor: Paper-based plasmonic sensors prepared through reversal nanoimprinting for rapid detection of biogenic amine odorants. ACS Appl Mater Interfaces 9:17307–17317
141. Gonzalez CM, Iqbal M, Dasog M, Piercey DG, Lockwood R, Klapotke TM, Veinot JG (2014) Detection of high-energy compounds using photoluminescent silicon nanocrystal paper based sensors. Nanoscale 6:2608–2612
142. Sablok K, Bhalla V, Sharma P, Kaushal R, Chaudhary S, Suri CR (2013) Amine functionalized graphene oxide/CNT nanocomposite for ultrasensitive electrochemical detection of trinitrotoluene. J Hazard Mater 248–249:322–328
143. Peters KL, Corbin I, Kaufman LM, Zreibe K, Blanes L, McCord BR (2015) Simultaneous colorimetric detection of improvised explosive compounds using microfluidic paper-based analytical devices (μPADs). Anal Methods 7:63–70
144. Aparna RS, Anjali Devi JS, Sachidanandan P, George S (2018) Polyethylene imine capped copper nanoclusters-fluorescent and colorimetric onsite sensor for the trace level detection of TNT. Sens Actuators B Chem 254:811–819
145. Nergiz SZ, Gandra N, Farrell ME, Tian L, Pellegrino PM, Singamaneni S (2013) Biomimetic SERS substrate: peptide recognition elements for highly selective chemical detection in chemically complex media. J Mater Chem A 1:6543–6549
146. Hughes S, Dasary SS, Begum S, Williams N, Yu H (2015) Meisenheimer complex between 2,4,6-trinitrotoluene and 3-aminopropyltriethoxysilane and its use for a paper-based sensor. Sens Biosensing Res 5:37–41
147. Pesenti A, Taudte RV, McCord B, Doble P, Roux C, Blanes L (2014) Coupling paper-based microfluidics and lab on a chip technologies for confirmatory analysis of trinitro aromatic explosives. Anal Chem 86:4707–4714
148. Huang S, He Q, Xu S, Wang L (2015) Polyaniline-based photothermal paper sensor for sensitive and selective detection of 2,4,6-trinitrotoluene. Anal Chem 87:5451–5456
149. da Silva GO, de Araujo WR, Paixao T (2018) Portable and low-cost colorimetric office paper-based device for phenacetin detection in seized cocaine samples. Talanta 176:674–678
150. Narang J, Malhotra N, Singhal C, Mathur A, Pn AK, Pundir CS (2017) Detection of alprazolam with a lab on paper economical device integrated with urchin like Ag@ Pd shell nano-hybrids. Mater Sci Eng, C 80:728–735
151. Hu J, Wang S, Wang L, Li F, Pingguan-Murphy B, Lu TJ, Xu F (2014) Advances in paper-based point-of-care diagnostics. Biosens Bioelectron 54:585–597

152. Zhu W-J, Feng D-Q, Chen M, Chen Z-D, Zhu R, Fang H-L, Wang W (2014) Bienzyme colorimetric detection of glucose with self-calibration based on tree-shaped paper strip. Sens Actuators B Chem 190:414–418
153. Malekghasemi S, Kahveci E, Duman M (2016) Rapid and alternative fabrication method for microfluidic paper based analytical devices. Talanta 159:401–411
154. Kannan B, Jahanshahi-Anbuhi S, Pelton RH, Li Y, Filipe CD, Brennan JD (2015) Printed paper sensors for serum lactate dehydrogenase using pullulan-based inks to immobilize reagents. Anal Chem 87:9288–9293
155. Pardee K et al (2016) Rapid, low-cost detection of Zika virus using programmable biomolecular components. Cell 165:1255–1266
156. Sharpe E, Frasco T, Andreescu D, Andreescu S (2013) Portable ceria nanoparticle-based assay for rapid detection of food antioxidants (NanoCerac). Analyst 138:249–262
157. Guan L, Tian J, Cao R, Li M, Cai Z, Shen W (2014) Barcode-like paper sensor for smartphone diagnostics: an application of blood typing. Anal Chem 86:11362–11367
158. Khatri V, Halasz K, Trandafilovic LV, Dimitrijevic-Brankovic S, Mohanty P, Djokovic V, Csoka L (2014) ZnO-modified cellulose fiber sheets for antibody immobilization. Carbohydr Polym 109:139–147
159. Noiphung J, Songjaroen T, Dungchai W, Henry CS, Chailapakul O, Laiwattanapaisal W (2013) Electrochemical detection of glucose from whole blood using paper-based microfluidic devices. Anal Chim Acta 788:39–45
160. Wang P, Cheng Z, Chen Q, Qu L, Miao X, Feng Q (2018) Construction of a paper-based electrochemical biosensing platform for rapid and accurate detection of adenosine triphosphate (ATP). Sens Actuators B Chem 256:931–937
161. Li X, Liu X (2016) A microfluidic paper-based origami nanobiosensor for label-free, ultrasensitive immunoassays. Adv Healthc Mater 5:1326–1335
162. Scordo G, Moscone D, Palleschi G, Arduini F (2018) A reagent-free paper-based sensor embedded in a 3D printing device for cholinesterase activity measurement in serum. Sens Actuators B Chem 258:1015–1021
163. Lee SH, Lee JH, Tran V-K, Ko E, Park CH, Chung WS, Seong GH (2016) Determination of acetaminophen using functional paper-based electrochemical devices. Sens Actuators B Chem 232:514–522
164. Adkins JA, Noviana E, Henry CS (2016) Development of a quasi-steady flow electrochemical paper-based analytical device. Anal Chem 88:10639–10647
165. da Costa TH, Song E, Tortorich RP, Choi J-W (2015) A paper-based electrochemical sensor using inkjet-printed carbon nanotube electrodes. ECS J Solid State Sci Technol 4:S3044–S3047
166. Cinti S, Minotti C, Moscone D, Palleschi G, Arduini F (2017) Fully integrated ready-to-use paper-based electrochemical biosensor to detect nerve agents. Biosens Bioelectron 93:46–51
167. Nantaphol S, Channon RB, Kondo T, Siangproh W, Chailapakul O, Henry CS (2017) Boron doped diamond paste electrodes for microfluidic paper-based analytical devices. Anal Chem 89:4100–4107
168. Zhang H, Zhao Z, Lei Z, Wang Z (2016) Sensitive detection of polynucleotide kinase activity by paper-based fluorescence assay with l exonuclease-assistance. Anal Chem 88:11358–11363
169. Jiang P, He M, Shen L, Shi A, Liu Z (2017) A paper-supported aptasensor for total IgE based on luminescence resonance energy transfer from upconversion nanoparticles to carbon nanoparticles. Sens Actuators B Chem 239:319–324
170. Liang L et al (2016) Aptamer-based fluorescent and visual biosensor for multiplexed monitoring of cancer cells in microfluidic paper-based analytical devices. Sens Actuators B Chem 229:347–354
171. Saha B, Baek S, Lee J (2017) Highly sensitive bendable and foldable paper sensors based on reduced graphene oxide. ACS Appl Mater Interfaces 9:4658–4666
172. Gong MM, Nosrati R, San Gabriel MC, Zini A, Sinton D (2015) Direct DNA analysis with paper-based ion concentration polarization. J Am Chem Soc 137:13913–13919

173. Davaji B, Lee CH (2014) A paper-based calorimetric microfluidics platform for bio-chemical sensing. Biosens Bioelectron 59:120–126
174. Chen GH, Chen WY, Yen YC, Wang CW, Chang HT, Chen CF (2014) Detection of mercury (II) ions using colorimetric gold nanoparticles on paper-based analytical devices. Anal Chem 86:6843–6849
175. Sicard C et al (2015) Tools for water quality monitoring and mapping using paper-based sensors and cell phones. Water Res 70:360–369
176. Wang B, Lin Z, Wang M (2015) Fabrication of a paper-based microfluidic device to readily determine nitrite ion concentration by simple colorimetric assay. J Chem Educ 92:733–736
177. Rattanarat P, Dungchai W, Cate D, Volckens J, Chailapakul O, Henry CS (2014) Multilayer paper-based device for colorimetric and electrochemical quantification of metals. Anal Chem 86:3555–3562
178. Cho YB, Jeong SH, Chun H, Kim YS (2018) Selective colorimetric detection of dissolved ammonia in water via modified Berthelot's reaction on porous paper. Sens Actuators B Chem 256:167–175
179. Lamas-Ardisana PJ et al (2017) Disposable electrochemical paper-based devices fully fabricated by screen-printing technique. Electrochem Commun 75:25–28
180. Qin Y, Pan S, Howlader MM, Ghosh R, Hu NX, Deen MJ (2016) Paper-based, hand-drawn free chlorine sensor with poly(3,4-ethylenedioxythiophene):poly(styrenesulfonate). Anal Chem 88:10384–10389
181. Cinti S, Talarico D, Palleschi G, Moscone D, Arduini F (2016) Novel reagentless paper-based screen-printed electrochemical sensor to detect phosphate. Anal Chim Acta 919:78–84
182. Li H, Wang W, Lv Q, Xi G, Bai H, Zhang Q (2016) Disposable paper-based electrochemical sensor based on stacked gold nanoparticles supported carbon nanotubes for the determination of bisphenol A. Electrochem Commun 68:104–107
183. Chouler J, Cruz-Izquierdo A, Rengaraj S, Scott JL, Di Lorenzo M (2018) A screen-printed paper microbial fuel cell biosensor for detection of toxic compounds in water. Biosens Bioelectron 102:49–56
184. Zhang R, Zhang C-J, Song Z, Liang J, Kwok RTK, Tang BZ, Liu B (2016) AIEgens for real-time naked-eye sensing of hydrazine in solution and on a paper substrate: structure-dependent signal output and selectivity. J Mater Chem C 4:2834–2842
185. Fraiwan A, Lee H, Choi S (2016) A paper-based cantilever array sensor: monitoring volatile organic compounds with naked eye. Talanta 158:57–62
186. Bulbul G, Eskandarloo H, Abbaspourrad A (2018) A novel paper based colorimetric assay for the detection of TiO_2 nanoparticles. Anal Methods 10:275–280

Polymers and Polymer Composites for Adsorptive Removal of Dyes in Water Treatment

Weiya Huang, Shuhong Wang and Dan Li

List of Abbreviations

AB 25	Acid blue 25
AF	Acid fuchsin
AB	Amido black 10B
AS	Almond shell waste
APT	Attapulgite
BY 28	Basic yellow 28
BF	Basic fuchsin
BV 14	Basic violet 14
CFA	Coal fly ash
CNT	Carbon nanotube
CMC	Carboxymethyl cellulose
mCS/CNT	Magnetic chitosan-decorated carbon nanotube
CR	Congo red
CV	Crystal violet
Bent/CMC-g-P (DMAEMA)	Carboxymethyl cellulose grafted by poly(2-(dimethylamino) ethyl methacrylate) modified bentonite
DB	Direct blue 199
ES	Emeraldine salt
EB	Emeraldine base
EY	Eosin Y
GV	Gentian violet
HA-Am-PAA-B	Humic acid-immobilized amine modified polyacrylamide/ bentonite composite
TRGO/PVA	Hydrothermally reduced graphene oxide/poly (vinyl alcohol)

W. Huang · S. Wang
School of Metallurgy and Chemical Engineering, Jiangxi University of Science and Technology, Ganzhou, China

D. Li (✉)
School of Engineering and Information Technology, Murdoch University, Murdoch, Australia
e-mail: L.Li@murdoch.edu.au

Inamuddin et al. (eds.), *Sustainable Polymer Composites and Nanocomposites*,
https://doi.org/10.1007/978-3-030-05399-4_19

hPEI-CE	Hyperbranched polyethyleneimine functionalized cellulose
IC	Indigo carmine
MNPs	Magnetic nanoparticles
MS	Mesoporous silica
MB	Methylene blue
MG	Methylene green
MO	Methyl orange
NR	Neutral red
OVS	Octavinylsilsesquioxane
OR	Oil red O
PAM	Polyacrylamide
PVI	Poly(1-vinyl imidazole)
PSSMA	Poly(4-styrenesulfonic acid-co-maleic acid) sodium
PSSMA/M-rGO	Poly(4-styrenesulfonic acid-co-maleic acid) sodium modified magnetic reduced graphene oxide nanocomposite
Mt	Montmorillonite
Pal	Palygorskite
poly(AN-co-ST)	Poly(acrylonitrile-co-styrene)
PAMAM	Polyamidoamine
PANI	Polyaniline
PANI -FP	Polyaniline-coated filter papers
PANI–MS@Fe_3O_4	Polyaniline functionalized magnetic mesoporous silica composite
PEI	Polyethyleneimine
PmPD	Poly(m-phenylenediamine)
PGA	Poly(γ-glutamic acid)
PVA	Poly (vinyl alcohol)
PV	Proflavine
r-GO-PIL	PVI polymer functionalized reduced graphene oxide
RB5	Reactive black 5
RR228	Reactive red 228
RB	Rhodamine B
RB2	Rose Bengal
TC	Tetracycline
TPE	Tetraphenylethylene
PAmABAmPD	Terpolymer of aniline/m-aminobenzoic acid/m-phenylenediamine
TCAS	Thiacalix[4]arene tetrasulfonate
β-CD	β-cyclodextrin
20CMC-Bent	20% of CMC in the total amount of CMC + Bent composite
RR2	Reactive red 2
MDI	4,4′-diphenylmethane diisocyanate
TAT	1,3,5-triacryloylhexahydro-1,3,5-triazine

1 Introduction

Natural and synthetic dyes have been commonly used to colour substrates, including food, drugs, cosmetics, paper, leather, plastics and textile products. Synthetic dyes play the dominating role particularly in the fabric and textile industry as compared with natural dyes, which insufficiently meet the industrial demand and therefore are more often used in the food industry [1]. It is estimated that more than 10,000 types of dyes are being used in various industrial applications. According to the literature, more than 8×10^5 tons of synthetic dyes are produced annually worldwide, in which 10–15% is released into natural water environment [2, 3]. Synthetic dyes often exhibit resistance to biodegradation and their persistence in the environment results in pollution, which has become a severe problem worldwide [4]. In particular, many of the synthetic dyes are toxic, carcinogenic and mutagenic. Therefore, dyes in wastewater, even in a small amount, are undesirable and should be properly removed before they enter the environment [5].

According to the applications of dyes, they are classified into groups of acidic, basic, reactive, direct, dispersed, and sulphur dyes; whilst based on the chemical structure of dyes, they are classified as nitro, azo, indigoid, anthraquinone, triarylmethane and nitroso dyes [6]. The most common way is to classify them as anionic (e.g. acid, reactive, and direct dyes), cationic (basic dyes), and nonionic dyes (dispersed dyes), depending on the ionic charge of dye molecules [6]. Cationic dyes were reported with greater toxicity than anionic dyes, because of their easy interaction with negatively charged cell membrane surfaces and entry into cells [5]. The commonly used dyes with their molecular structure, molecular weight, classification and λ_{max} were summarized in Table 4.

In order to remove dyes from wastewater, several biological, chemical and physical methods have been developed, including coagulation, adsorption, membrane separation, precipitation, chemical oxidation, and aerobic or anaerobic treatment [6–8]. Among them, adsorption seems to be the favourite technique because of low cost and diversity of adsorbents, easy operation, environmental and economic sustainability [7, 8]. Various adsorbents have been investigated for dye removal, such as activated carbon, carbon nanotubes, graphene, clay minerals, metal oxides, polymers, non-conventional low-cost adsorbents (e.g. agricultural and industrial waste or by-products), etc. [1, 4, 6, 9–11]. Several parameters, including high adsorption capacity, fast adsorption rate, good selectivity, wide availability, low cost, easy regeneration and feasible reusability, are intensively concerned when designing and selecting a suitable adsorbent to target dye removal. In general, hydrophilic adsorbents are preferable for use, which attract the dye molecules via electrostatic interaction, van der Waals force, π–π interaction, and hydrogen bonding [12].

Recently, polymers and polymer composites have attracted tremendous attention as adsorbents for dye treatments [4, 12–14]. Polymers exhibit a number of excellent properties [13], such as high strength, good flexibility, chemical inertness, hydrophilic surface chemistry, which are able to act as matrix materials to fabricate polymer composites with inorganics. Their properties can be tuned via functionalization with chemical groups, crosslinking or blending with organic materials which are capable to interact with dye pollutants. Therefore, polymers and polymer composites serve as promising adsorbent materials for high-performance dye removal from water.

The current development of polymers or polymer composites as adsorbents concentrates on the use of nanostructures, such as nanosized particles, which provided great external surface area and exhibited high adsorption to dyes. For example, the nano-sized poly(m-phenylenediamine) (PmPD) exhibited enhanced adsorption capacity, which was calculated by the Langmuir model (387.6 mg/g) towards Orange G (OG), as compared with that of micro-sized PmPD (163.9 mg/g) [15]. However, the nanoparticles might suffer from aggregation, affecting adsorption capacity and kinetics. To improve application potential, polymers and their composites were synthesized in the form of nanofibers, which can be successfully produced by an electrospinning method. These one-dimensional (1D) nanoscale adsorbents, due to their great surface-to-volume ratios, were widely studied for the enhanced properties [16–19]. In addition, porous structures are highly desirable, because they can facilitate the diffusion of dyes into the adsorbents, resulting in fast and efficient adsorption [20]. The porous structure is also favourable to increase the swelling rate of the polymers; however, their mechanic strength would be reduced with increasing porosity. This can be solved by fabricating composite with other inorganic materials, e.g. bentonite [20]. Polymeric adsorbents, in the forms of hydrogels or xerogels which are three-dimensionally crosslinked hydrophilic polymers, were synthesized, functionalized and used as super-adsorbents to remove dyes from aqueous solution, ascribed to their high physicochemical stability and good regeneration property [21–24].

2 Modified or Functionalized Polymers and Polymer Composites

The surface chemistry of adsorbents, such as surface acidity/basicity and points of zero charge, significantly affects the adsorption of dyes. For targeting the removal of anionic dyes, surface properties of adsorbents were modified to minimize negative charge and increase positive charge on surfaces. Since that adsorption takes place on the surfaces of polymer adsorbents, another important factor for consideration is the density of functional groups on surfaces [14, 25]. For example, Qiu et al. synthesized three polystyrene resins with significantly different surface functionality and studied their adsorption performance for an anionic dye Reactive Black 5 dye (RB5), as shown in Fig. 1a [14]. As compared with commercial polymer XAD-4 of a low-degree functionality, NG-8, which was synthesized in the laboratory, had primarily acidic functional groups; and its aminated product MN-8

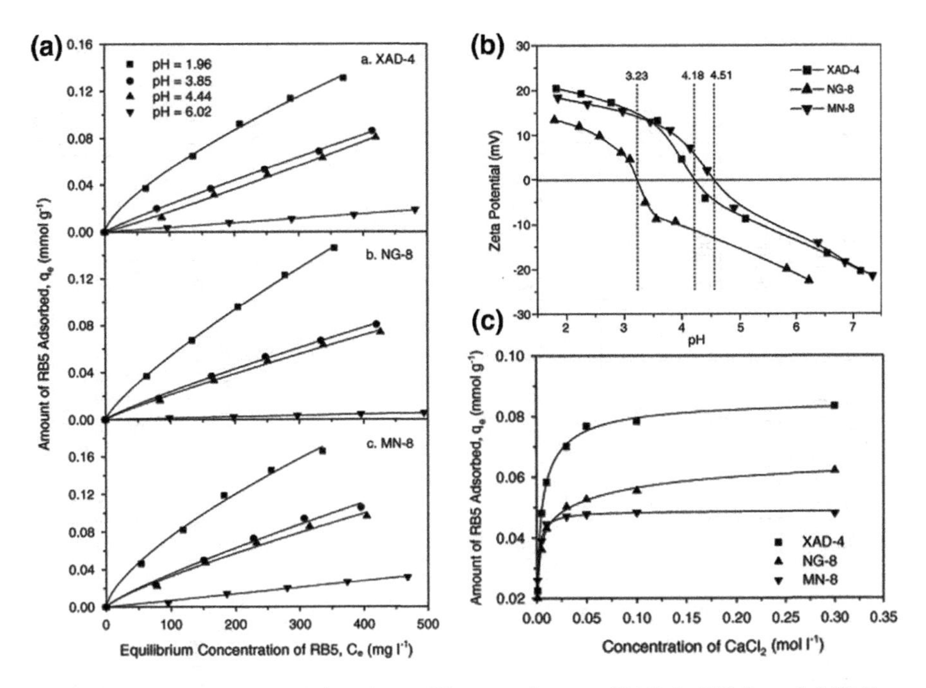

Fig. 1 **a** Adsorption of RB5 by three different polymers (XAD-4, NG-8, and MN-8) at pH = 1.96, 3.85, 4.44, and 6.02; **b** zeta potentials of XAD-4, NG-8, and MN-8 as a function of solution pH and determined points of zero charge; **c** effect of $CaCl_2$ concentration with initial RB5 concentration of 100 mg/L (reprinted from [14] with permission from Elsevier)

possessed mainly basic amino groups. It was found that XAD-4 exhibited non-polar nature; however moderate polarity was seen for NG-8 and MN-8. Their corresponding point of zero charge was 4.18, 3.23, and 4.51, respectively, as shown in Fig. 1b. The equilibrium adsorption capacity of RB5 for XAD-4, NG-8 and MN-8 was recorded as 115.05, 129.94 and 158.69 mg/g, accordingly, at pH = 1.90 and RB5 equilibrium concentration of 300 mg/L (Table 1). Moreover, the adsorption decreased at high solution pH for all adsorbents. MN-8 was the most effective adsorbent in RB5 removal at all tested pH values; which might probably lie in its increasing basicity caused by the protonation of amine groups and in turn creation of more positive charges on the surface. It was noted that the enhanced adsorption was observed in the presence of $CaCl_2$, as shown in Fig. 1c, possibly ascribed to the neutralization of negative surface charge by Ca^{2+} and RB5-Ca^{2+} pairings [14].

Some biopolymers, such as chitosan, have a large number of functional groups in their molecules, such as hydroxyl (–OH) and primary amine (–NH_2), which can be utilized as functional groups for treatment of dyes due to the electrostatic interaction or hydrogen bonding force [26, 27]. Due to the presence of some groups, such as sulfate (–SO_3H) or carboxyl (–COOH), some biopolymers are negatively charged and have potential application for adsorption of cationic dyes [5, 28]. For example, κ-carrageenan is a highly negatively charged natural polysaccharide and was used

Table 1 Dye removal by modified or functionalized polymer composites

Adsorbent	Dye	Adsorption capacity (mg g^{-1})	pH	Temperature (°C)	Initial concentration (mg L^{-1})	References
EDTA-EDA-PAN nanofibers[h]	Methyl orange (MO)	99.15[a]	4	25	5–300	[18]
	Reactive red (RR)	110.0[a]	4	25	5–300	
Ti(IV) functionalized chitosan molecularly imprinted polymer (Ti-CSMIP)	Active brilliant red X-3B	161.1[a]	6.0–7.0	20	/	[27]
Carboxylated functionalized acrylonitrile-styrene co-polymer nanofibers	Basic violet 14 (BV 14)	67.11[a]	6.2	25	1–100	[17]
Silsesquioxane-based tetraphenylethene-linked nanoporous polymers	Rhodamine B (RB)	1666[a]	/	/	/	[34]
	Congo red (CR)	1040[a]				
	Crystal violet (CV)	862[a]				
	Methylene blue (MB)	144[a]				
	Methyl orange (MO)	67[a]				
Mesoporous silica functionalized with polypeptides	Methylene blue (MB)	4.537[b]	/	room temperature	11	[66]
XAD-4[c]	Reactive black 5 (RB5)	115.05[b]/153.73[b]	1.90/ 6.05[f]	30 ± 0.5	300	[14]
NG-8[d]		129.94[b]	1.90	30 ± 0.5	300	
MN-8[e]		158.69[b]	1.90	30 ± 0.5	300	
Crosslinked chitosan with TAT[g]	Acid orange 7	858.28[b]	5.0	40	/	[31]

(continued)

Table 1 (continued)

Adsorbent	Dye	Adsorption capacity (mg g^{-1})	pH	Temperature (°C)	Initial concentration (mg L^{-1})	References
	Acid red 88	640.61[b]	6.0	40	/	
Hydrothermally reduced graphene oxide/poly (vinyl alcohol) (TRGO/PVA) aerogels	Neutral red (NR)	306.2[a]	7	30	/	[12]
	Indigo carmine (IC)	250.0[a]	2	30	/	
Functional poly(m-phenylenediamine) nanoparticles	Orange G (OG)	387.6[a]	/	30	/	[15]
β-cyclodextrin functionalized poly (styrene-alt-maleic anhydride)	Basic fuchsin (BF)	295[b]	6.5 ± 0.1	25	20–350	[32]
	Methylene blue (MB)	528[b]	6.5 ± 0.1	25	20–350	
Poly(1-vinylimidazole) (PVI) -modified graphene sheets	Methyl blue	1910[a]	/	room temperature	100–700	[36]
Cellulose functionalized with hyperbranched Polyethylenimine	Congo red (CR)	2107[a]/2100[b]	5.0	25	100–1000	[25]
	Basic yellow 28 (BY28)	1865[a]/1860[b]	9.0	25	100–1000	

[a]Maximum adsorption capacities calculated by Langmuir model
[b]Equilibrium adsorption capacity
[c]A representative St-DVB resin developed by Rohm & Haas, displays an excellent adsorptive affinity for small organic compounds (e.g., phenols), was purchased from Rohm & Haas (Philadelphia, PA)
[d]Was synthesized using the Friedel-Crafts reaction through self-crosslinking of a chloromethylated copolymer of St-DVB
[e]Was obtained by aminating swollen NG-8 using 40% dimethylamine solution at 45 °C for 10 h
[f]pH = 6.05 solution with 0.01 mol l^{-1} $CaCl_2$
[g]Chitosan crosslinked with 1,3,5-triacryloylhexahydro-1,3,5-triazine (TAT)
[h]Ethylenediaminetetraacetic acid (EDTA) and ethylenediamine (EDA) crosslinker modified electrospun polyacrylonitrile (PAN) nanofibers

to modify carbon nanotubes to improve the adsorption performance for cationic dye methylene blue (MB) [5]. The magnetite nanoparticles decorated with poly (γ-glutamic acid) (PGA), which is an anionic polypeptide with α-carboxyl groups and synthesized by Bacillus species in a fermentation process, had the Langmuir maximum adsorption capacity of 78.67 mg/g for MB [28].

To enhance their adsorption performance for dye removal, it is an effective way by introducing new functional groups or increasing the density of surface functionality. For example, Wang et al. chemically modified biopolymer chitosan via carboxymethylation, which introduced different amounts of active –OH, –COOH, and $-NH_2$ groups onto the chitosan by controlling the degree of substitution. The resulting N, O-carboxymethyl chitosan showed enhanced adsorption capacity to remove cationic dye MB, with the Langmuir maximum adsorption capacity of 351 mg/g [29]. It is also a good method to fabricate bifunctional materials, which show applications in not only dye decolourization, but also other areas, such as heavy metal ion removal and bacterial capturing [30, 31]. Functional groups, such as amine, can be protonated to be positively charged and adsorb negatively charged dye molecules via electrostatic attraction. For example, the cellulose functionalized with quaternary ammonium groups had an enhanced adsorption of 190 mg/g at pH = 3 for reactive red 228 (RR228), due to the electrostatic attraction (cellulose-R-N^+ $(C_2H_5)_3 \cdots SO_3^-$) between the positive quaternary ammonium group and SO_3H group from negatively charged dye molecules RR228 [3]. Other functional groups, e.g. carboxylic, can improve the polarity of adsorbent to enhance its sorption affinity to cationic dyes [17, 32]. For example, the carboxylated poly (acrylonitrile-co-styrene) nanofiber showed the Langmuir-derived maximum dye adsorption capacity, 67.11 mg/g, when removing basic violet 14 dye (BV 14) (as shown in Table 1). Its adsorption equilibrium was achieved within 30 min and the dye adsorption followed the pseudo-second-order model, suggesting chemisorption [17]. In addition to carboxyl functional groups, the internal hydrophobic cavity in the functional molecule β-cyclodextrin (β-CD), which is a torus-shaped cyclic oligosaccharide made up of seven α-1,4-linked-D-glucopyranose units, formed inclusion complexes with dye organic molecules through host-guest interactions. As a result, the β-CD functionalized poly (styrene-alt-maleic anhydride) adsorbent showed a high equilibrium adsorption quantity of 272.56 and 366.35 mg/g for basic fuchsin (BF) and MB (at pH = 6.5, and initial BF and MB concentration of 95.7 and 93.6 mg/L) (Table 1), respectively; which was one magnitude greater than that of the unfunctionalized adsorbent [32]. In addition, the adsorption capacities increased with increasing initial dyes concentrations from 20 to 350 mg/L, and their Langmuir maximum adsorption capacity for BF and MB reached 298.5 and 531.9 mg/g, respectively [32].

The fabrication of polymers consisting of different organic segments can improve their physical and chemical properties [4, 22]. It would endow new properties to the resulting polymeric materials for dye decolourization applications [17, 33]. Chitosan is a copolymer of D-glucosamine and N-acetyl-D-glucosamine units with a large number of –OH and $-NH_2$ groups; it is prepared by the N-deacetylation of chitin in an aqueous alkaline solution [4]. These functional groups on chitosan chains work as sites for electrostatic interaction and coordination in dye adsorptive removal. However, the disadvantage of ready dissolution in an acidic medium makes the phase separation very difficult, thus limiting its application for dye removal. Chemical modification by crosslinking, grafting and/or other methods can improve its stability in acidic solution and enhance its mechanical strength [4]. After introducing a hexahydrotriazine ring into the crosslinked structure of chitosan with 1,3,5-triacryloylhexahydro-1,3,5-triazine (TAT), the crosslinked chitosan showed the promising adsorption of 858.28 mg/g towards CI Acid Orange 7 at pH 5.0, while that of 640.61 mg/g to Acid Red 88 at pH 6.0, as shown in Table 1 [31]. Besides, the flexibility of the structure of the modified polymer materials might have some certain effect on the adsorption process [31]. Poly(acrylonitrile-co-styrene) (poly(AN-co-ST)), as an important random copolymer of acrylonitrile and styrene, exhibits high resistance to heat, chemicals and oil, accompanied with other features, e.g. high rigidity and superior transparency [17]. The carboxylated functionalized poly(AN-co-ST) nanofibers exhibited a good adsorption capacity to BV 14, with the maximum capacity of 67.11 mg/g, as shown in Table 1. Liu et al. fabricated a series of hybrid polymers after the Friedel–Crafts reaction of tetraphenylethylene (TPE) and octavinylsilsesquioxane (OVS) (Fig. 2a, b) [34]. In addition to the high luminescence properties ascribed to TPE, the resulting silsesquioxane based TPE-bridged polymers had found adsorption use in gas, dye and metal ion detection. Particularly, the adsorption capacity of hybrid polymer for cationic dye rhodamine B (RB), anionic dye congo red (CR) and crystal violet (CV) was 1666, 1042 and 862 mg/g, respectively, as shown in Table 1. The adsorption process was mainly governed by size-selective mechanism ascribed to the unique bimodal pore structure in the hybrid polymers with micropores centring at $\sim$1.4 nm and mesopores centring at $\sim$4.5 nm, as well as the high surface area of up to 1910 cm^2/g [34]. Zhu et al. modified cellulose with cationic hyperbranched polyethyleneimine (hPEI) via forming Schiff base structure between the amino groups of hPEI and the aldehyde groups on the chemically oxidized cellulose surface (Fig. 2c, d). Because of a large number of amino groups, the resulting functionalized hPEI-CE copolymers displayed the high Langmuir-derived adsorption capacity of 2107 mg/g to anionic dye CR at pH 5.0 and that of 1865 mg/g to cationic basic yellow 28 (BY28) at pH 9.0, since the pH_{pzc} of the copolymer was recorded at pH 8.6 [25].

Recently, the combination of organic and inorganic materials has become an important strategy to modify properties of adsorbents. The resulting adsorbents may not only take advantage of the characteristic of each material, but also improve adsorption performance [27, 35]. Several inorganic compounds or materials, such as mesoporous silica or carbon, carbon nanotubes, graphene, clay, fly ash, metal oxides, and inorganic salts, etc, have been widely used to fabricate inorganic-organic materials as adsorbents. Among them, graphene has drawn significant research interest [12, 36], which can form covalent bonds with a polymer, e.g. as shown in Fig. 3a, by utilizing a diazonium addition reaction and the subsequent grafting of poly(1-vinyl imidazole) (PVI) onto the graphene via a quaternarization reaction. The resulting PVI polymer functionalized reduced graphene

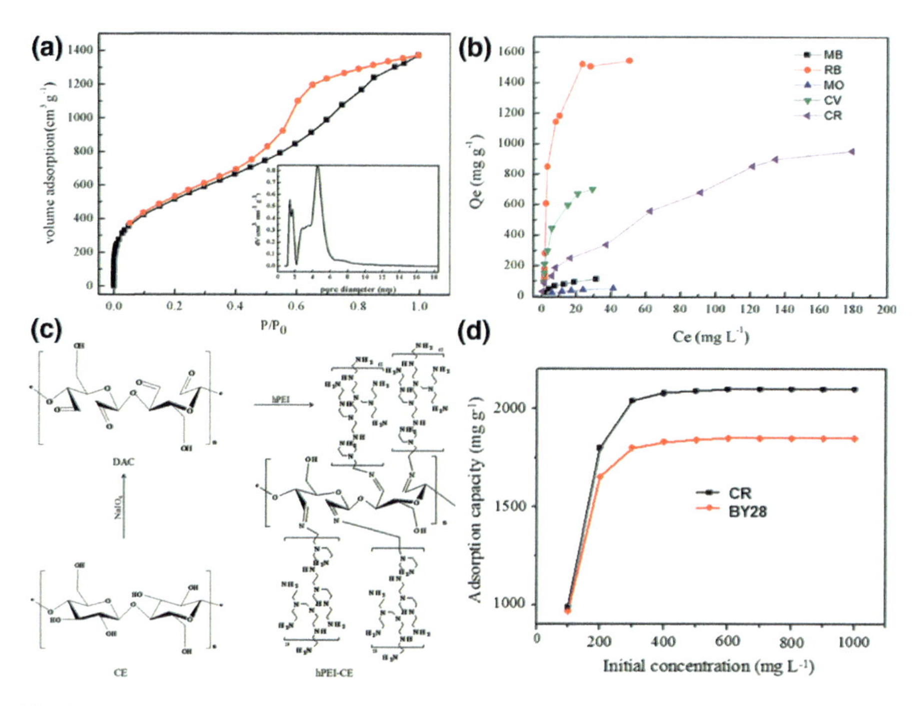

Fig. 2 **a** N_2 adsorption-desorption isotherm (inset:pore-size distribution) and **b** equilibrium adsorption capacity of silsesquioxane based TPE-bridged polymers (reproduced from [34] with permission of The Royal Society of Chemistry); **c** synthesis of cationic hyperbranched polyethyleneimine (hPEI) functionalized cellulose (hPEI-CE) and **d** initial concentration on the adsorption capacity of hPEI-CE for CR and BY28 (reproduced from [25])

oxide (r-GO-PIL) displayed enhanced adsorption efficient towards anionic dyes methyl blue as compared to the unmodified graphene (r-GO), as shown in Fig. 3b. The adsorption equilibrium was almost reached after 1120 min, as shown in Fig. 3c. The adsorption process fitted well with the Langmuir isotherm model and the maximum adsorption capacity was 1910 mg/g; such highly effective absorption was explained by the van der Waals forces and electrostatic interactions between r-GO-PIL and dye [36]. Xiao and co-workers fabricated hydrothermally reduced graphene oxide/poly (vinyl alcohol) (TRGO/PVA) in the form of aerogels via an in situ hydrothermal reduction followed by direct sol-aerogel transformation strategy (Fig. 4a–d). The resulting materials showed an attractive adsorption of various cationic, anionic and nonionic dyes. Their Langmuir-derived maximum adsorption capacity for cationic neutral red (NR) and anionic indigo carmine (IC) dye was 306.2 and 250.0 mg/g (Table 1), respectively (Fig. 4e). Noticeably, other common cationic, anionic or nonionic dyes were also able to be removed by the TRGO/PVA aerogel (Fig. 4f). This attractive adsorption capacity was mostly due to the $\pi - \pi$ interactions between aromatic or heterocyclic structures of dyes and graphene sheets; whilst the additional electrostatic attraction contributes to

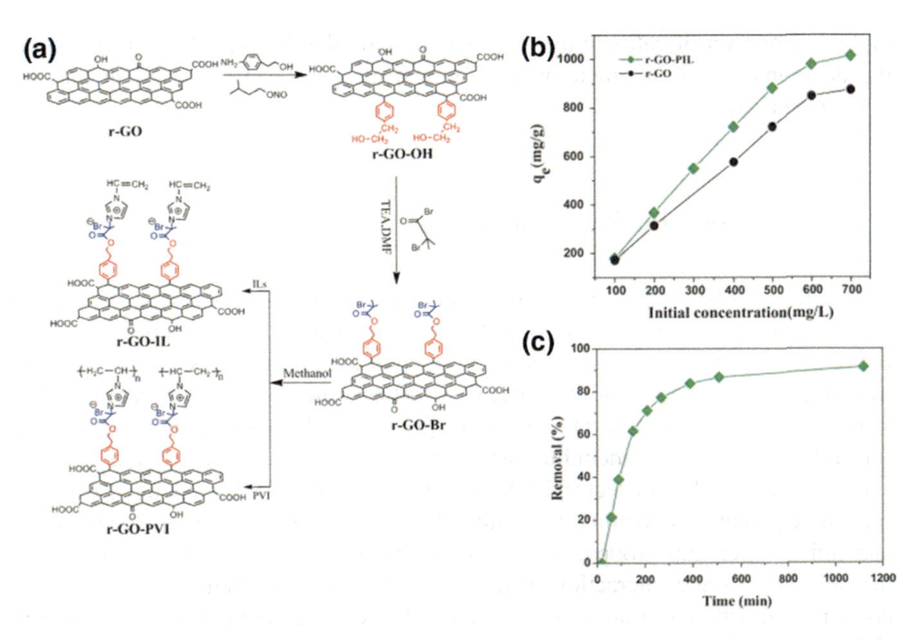

Fig. 3 **a** Grafting of poly(1-vinyl imidazole) onto the surface of the chemically reduced graphene oxide; **b** adsorption capacity of r-GO and r-GO-PIL for the methyl blue at room temperature; **c** Removal of MB by r-GO-PIL as a function of time (initial concentration of dye: 350 mg/L, solution volume: 30 mL, amount of r-GO-PIL: 5 mg) (reprinted from [36] with permission from Elsevier)

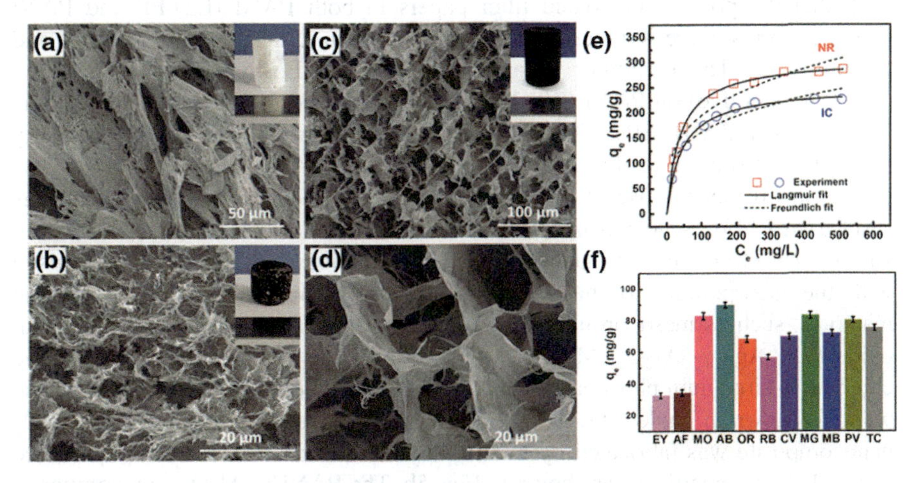

Fig. 4 SEM images of poly (vinyl alcohol) (PVA) **a**, TRGO **b**, and TRGO/PVA (0.5 wt%/1.0 wt %) (**c**: low magnification, **d**: high magnification) aerogels (insets: optical photographs of aerogels); **e** adsorption isotherms using TRGO/PVA (0.5 wt%/1.0 wt%) aerogel fitted by the Langmuir and Freundlich models; **f** adsorption of anionic dyes, eosin Y(EY), acid fuchsin (AF), methyl orange (MO) and amido black 10B (AB) (at pH = 2), as well as the adsorption of nonionic dye, oil red O (OR), cationic dyes, RB, CV, methylene green (MG) and MB, and aromatic drugs, proflavine (PV) and tetracycline (TC) (at pH = 7) (reprinted from [12] with permission from Elsevier)

higher adsorption of cationic dye NR than anionic dye IC [12]. More progress on the development of inorganic-organic adsorbents has been reviewed in the following sections.

3 Polyaniline and Its Composites

Polyaniline (PANI) is one of the amine-containing conjugated polymers, which has been extensively investigated during the last two decades. Many merits have been realized in related to its properties, such as good environmental stability, low-cost monomers for synthesis, high conductivity, instinct redox property, air and moisture stability, water insolubility and flexibility [37–39]. It has been found applications for different purposes, including anti-corrosion coating, batteries and sensors, supercapacitor and optoelectronic devices [15, 37, 39, 40]. Especially, PANI has incredible potential in treating dye-polluted water, because of the presence of amine and imine functional groups, which can act as the chelating and adsorbing sites through electrostatic interaction or hydrogen bond [8, 24]. When these nitrogen-containing functional groups are protonated, PANI is present in its emeraldine salt (ES) state; whilst those are deprotonated, PANI exhibits in its form of emeraldine base (EB). This conversion can be easily achieved by the treatment using acid or base [41]. It is notable that the deprotonated PANI-EB favours the selective adsorption of cationic dyes; whilst the PANI-ES preferentially adsorbs anionic dyes due to the electrostatic interactions. This was supported by Majumdar's research [42], that the polyaniline-coated filter papers in both PANI (ES)-FP and PANI (EB)-FP forms were synthesized to remove seven different anionic dyes and cationic dyes. The Langmuir adsorption capacity of Eosin yellow (EY) (as an anionic model dye) on PANI (ES)-FP was 4.3 mg/g and that of MB (as a cationic model dye) on PANI (EB)-FP was 1.3 mg/g at neutral pH. Figure 5a schematically shows the adsorption of dyes by PANI (ES)-FP and PANI (EB)-FP.

It is noted that the use of PANI powders as adsorbent could be limited by its surface area [43]. Research interest was attracted at the fabrication of PANI into various nanostructures, i.e. nanoparticles and nanotubes [38, 44], or its composite with the incorporation of nanosized inorganics [8, 40, 45]. Various inorganic materials, such as mesoporous silica, carbon nanotubes, nanosized metal or metal oxide (e.g. Ag, γ- Al_2O_3, MgO, and ZrO_2), as well as metal salts (e.g. cupric chloride, α-zirconium phosphate, et al.), have been explored; the summary of those works is shown in Table 2 [8, 37, 38, 41, 43, 45–47]. For example, PANI/γ- Al_2O_3 nanocomposite was fabricated by in situ polymerization of aniline in the presence of γ- Al_2O_3 nanoparticles, as shown in Fig. 5b. The PANI/γ- Al_2O_3 nanocomposite showed a high adsorption capacity of 1000 mg/g for cationic dye Direct blue 199 (DB) at pH 2, calculated by the Langmuir isotherm model. The adsorption process was proceeded via electrostatic attraction between the protonated ammonia groups in PANI ($R\text{-}NH_3^+$) and sulfonic groups from anionic dye ions ($D\text{-}SO_3^-$), in the form of ($R\text{-}NH_3 \cdots O_3S\text{-}D$) [8].

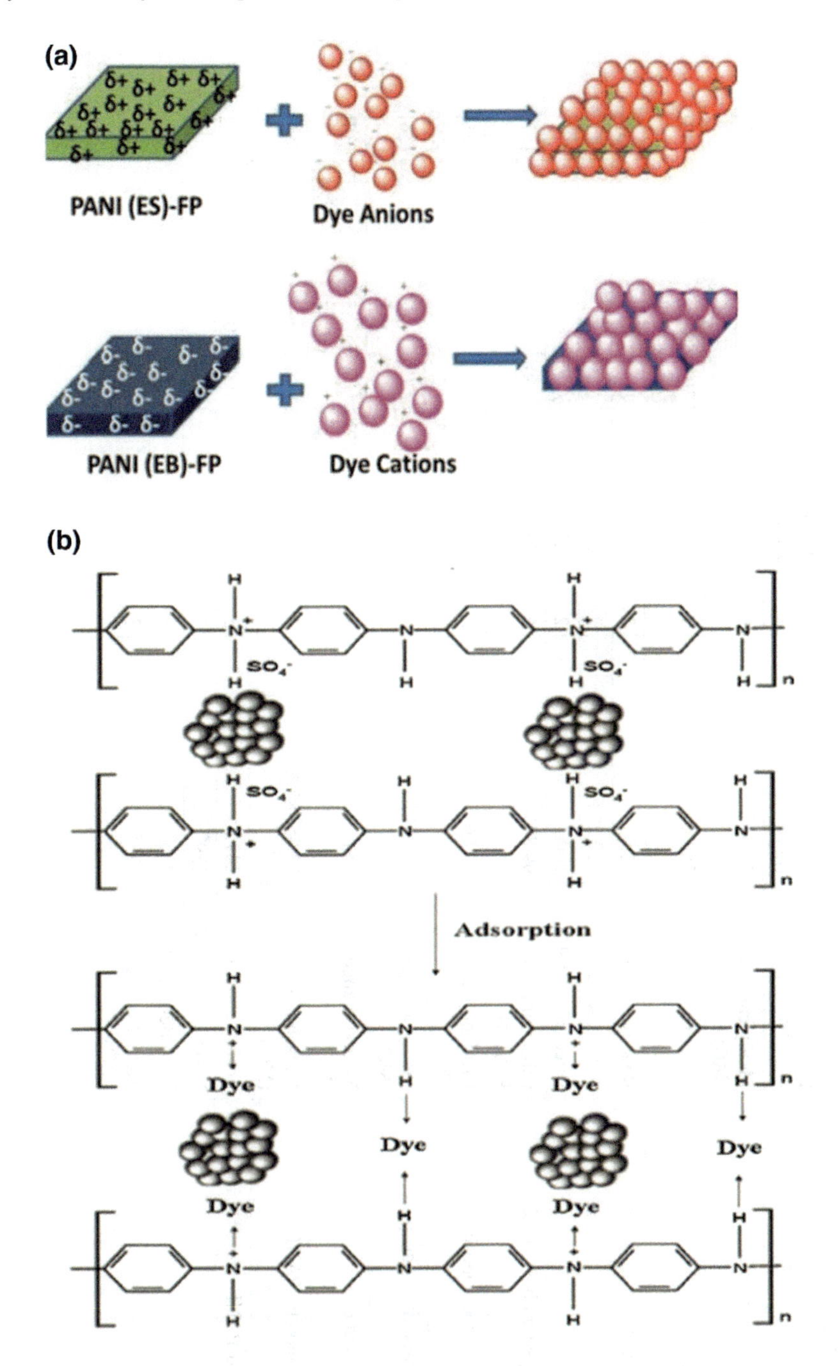

Fig. 5 **a** Schematic of adsorption of anionic and cationic dyes by PANI (ES)-FP and PANI (EB)-FP; **b** adsorption of anionic dyes by PANI/γ- Al_2O_3 nanocomposite (modified & reprinted with permission from [42, 8])

Table 2 Dye removal by polyaniline (PANI) composites

Adsorbent	Dye	Adsorption capacity (mg g^{-1})	pH	Temperature (°C)	Initial concentration (mg L^{-1})	References
PANI/zirconium oxide	Methylene blue (MB)	77.51[a]	/	26 ± 1	10–40	[37]
PANI/MgO	Reactive orange 16 (RO)	558.4[a]	7	30	50–250	[43]
PANI/γ- Al_2O_3	Reactive red 194	71.9[a]	2	25	/	[8]
	Acid blue 62	222.2[a]	2	25	/	
	Direct blue 199	1000[a]	2	25	/	
PANI doped with 8% $CuCl_2$	Cibacron navy P-2R-01	99.83%[b]	6	25	60	[47]
PANI/Ag nanocomposite	Brilliant green (BG)	23.66[a]	/	30	/	[45]
PANI/α-zirconium phosphate	Methyl orange (MO)	377.46[a]	4	/	/	[38]
PANI nanotube	Methyl orange (MO)	254.15[a]	4	/	/	[38]
Carbon nanotube/PANI	Malachite green (MG)	13.95[e]/88%[b]	7	20	16	[41]
Folic Acid-PANI	Eosin yellow (EY)	247.5[a]	3	25	100–230	[24]
	Eosin yellow (EY)	239[e]	3	25	200	
	Rose Bengal (RB2)	206[e]	3	25	200	
	Methyl orange (MO)	173[e]	3	25	200	

(continued)

Table 2 (continued)

Adsorbent	Dye	Adsorption capacity (mg g^{-1})	pH	Temperature (°C)	Initial concentration (mg L^{-1})	References
Fe_3O_4@AmABAmPD-TCAS[c]	Methylene blue (MB)	31.64[a]	8	30	5–50	[33]
	Malachite green (MG)	29.07[a]	8	30	5–50	
PANI-MS@Fe_3O_4[d]	Methyl orange (MO)	55.74[a]	4	0	4–32	[48]
PANI emeraldine salt (ES)-coated filter paper	Eosin yellow (EY)	4.26[a]	7	/	1–25	[42]
	Methylene blue (MB)	1.26[a]	7	/	1–25	

[a]Maximum adsorption capacity calculated by Langmuir isotherm model
[b]Removal rate %
[c]The adsorbent is composed of a Fe_3O_4 core and polyaniline–aminobenzoic acid–phenylenediamine terpolymer shell functionalized with thiacalix(4)arene tetrasulfonate as the internal dopant
[d]Polyaniline functionalized magnetic mesoporous silica iron oxide (MS@Fe3O4) nanoparticles
[e]Equilibrium adsorption capacity

A synergistic effect of PANI and inorganic materials was seen on improving the adsorptive removal of dyes from the literature [37, 40, 43]. For example, the carbon nanotube (CNT)/PANI composites fabricated by Zeng et al. had an equilibrium adsorption capacity of 13.95 mg/g for cationic dye malachite green (MG) at an initial MG concentration of 16 mg/L, which was 15% higher than that of neat PANI. This was probably due to the strong interaction of CNT-PANI, as well as its high porosity and large surface area [41]. Wang and co-workers synthesized PANI and α-zirconium phosphate composite (PANI/α-ZrP) via in situ oxidative polymerization reaction. The resulting PANI/α-ZrP exhibited BET surface area of 30.40 m^2/g, showing plate-like α-ZrP structures decorated by PANI thin layer of small and uniform fibrillar nanostructure (Fig. 6b); this differed from the structure of PANI nanotube (Fig. 6a). Thanks to this unique structure, the PANI/α-ZrP composites showed the Langmuir-derived maximum methyl orange (MO) adsorption of 377.46 mg/g, which was greater than that of PANI nanotubes (254.15 mg/g), as shown in Fig. 6c. In the kinetic study (Fig. 6d), the fast adsorption in first 60 min was probably ascribed to the electrostatic interactions between sorption sites of PANI/α-ZrP, i.e. amine and imine functional groups,

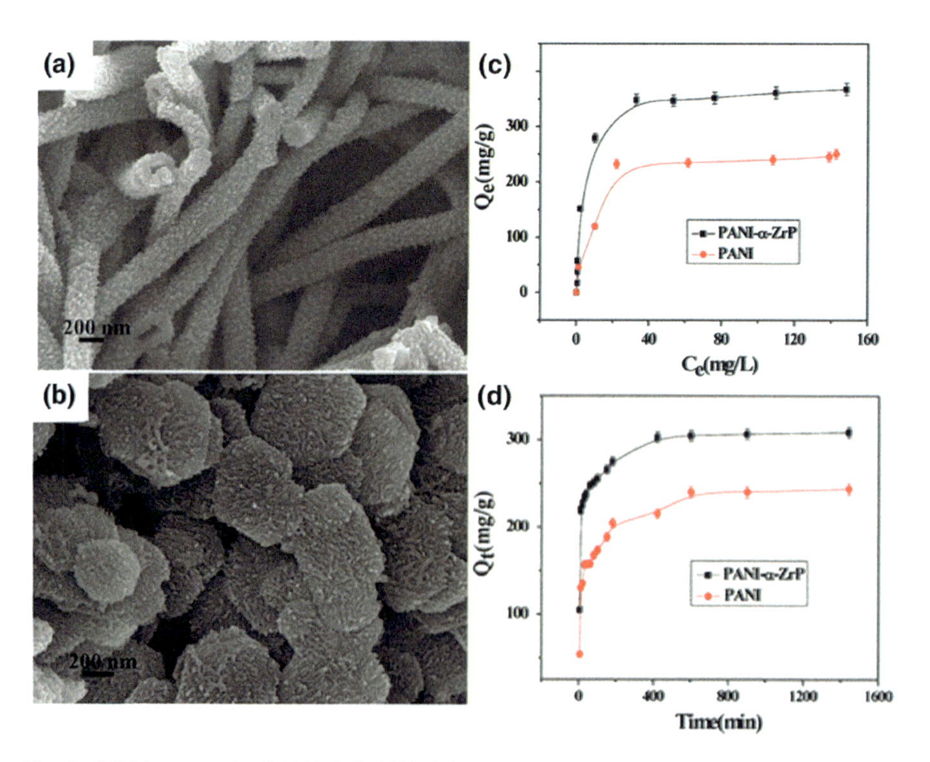

Fig. 6 SEM images of **a** PANI, **b** PANI/α-ZrP (α-ZrP/aniline mol ratio = 1:30); **c** adsorption isotherms of MO on PANI nanotubes and PANI/α-ZrP nanocomposites; **d** adsorption kinetics curves for the adsorption of MO with initial concentration of 100.0 mg/L by PANI nanotubes and PANI/α-ZrP nanocomposites (reprinted with permission from [38])

and MO molecule; after that, the adsorption process slowed down and reached equilibrium [38].

Organic functional materials, such as folic acid, aminobenzoic acid, phenylenediamine, etc., and other materials, e.g. low-cost filter paper, have been used to improve the properties of PANI for dye-containing water purification, as shown in Table 2 [24, 33, 42]. For example, Das et al. fabricated a novel porous folic acid/PANI hydrogel, in which folic acid was used as a cross-linker to PANI. The as-synthesized folic acid/PANI hydrogel exhibited high specific surface area with 3D interconnected pores. The folic acid/PANI xerogels were found efficient in removing anionic dyes, such as EY, Rose Bengal (RB2), MO, ascribed to the electrostatic attraction between anionic dyes and positively charged PANI, which was presented in emeraldine salt (ES) state [24]. Moreover, magnetic nanoparticles, such as Fe_3O_4, were introduced to PANI composites for efficient magnetic separation and reuse [33, 48]; this has been covered in the following section.

4 Magnetic Polymer Composites

The development of magnetic polymer composites is of importance for recycling and reusing in practical application. The magnetic separation offers the ability to recover and reuse the suspended adsorbents after multiple cycles of adsorption, which is of great significance for sustainable process management. In particular, the magnetic separation is believed to be more efficient than conventional centrifugation and filtration, which might be subject to the risk of blockage of filters and loss of adsorbents [5, 49].

In the past decades, the adsorbents combined with magnetic nanoparticles (MNPs) have attracted great research interest, ascribed to their good adsorption capacity, high adsorption rate and convenient recycling of solids by employing an external magnetic field. Fe_3O_4 MNPs are most commonly used as magnetic particles due to their unique property such as low toxicity, biocompatibility and easy handling of magnetic separation [5, 50, 51]. Fe_3O_4 MNPs can be synthesized by precipitation in a fine mixture of Fe(II) and Fe(III) solution with ammonium hydroxide under nitrogen atmosphere [28], or via a solvothermal route under heating [26]. However, the practical application of bare Fe_3O_4 MNPs has seen several limitations, such as the leaching of iron in strong acid solution and a high tendency to aggregate [28]. Therefore, an effective surface coating of Fe_3O_4 MNPs with appropriate coating materials, such as inorganic silica or polymer is necessary to enhance the stabilization of MNPs.

The combination of polymer with MNPs resulted in magnetic polymer composite; it benefits from both abundant functional groups of the polymer, which endows high affinity towards dyes, and magnetic properties of MNPs. So far, such materials have been widely studied on the removal of dyes from aqueous solution, as shown in Table 3. The presence of nitrogen-containing groups in the polymers, i.e. polyethyleneimine, poly 1, 4-phenylenediamine, and polyaniline, etc., makes them

Table 3 Dye removal by magnetic polymer composites

Polymers and composites	Magnetic particles	Dye	Adsorption capacity (mg g^{-1})	pH	Temperature (°C)	References
Poly 1, 4-phenylenediamine	Fe_3O_4 nanoparticles	Direct red 81 (DR81)	144.92[a]	4.0	/	[52]
Cross-linked polyethylenimine	Fe_3O_4-NH_2 cubic crystalline	Alizarin red S (ARS)	256.1[a]	3.0	30	[51]
		Methyl orange (MO)	244.4[a]	3.0	30	
		Methyl blue	172.1[a]	3.0	30	
		Nuclear fast red (NFR)	138.8[a]	3.0	30	
		Sunset yellow FCF (SY)	145.8[a]	3.0	30	
		Alizarin green (AG)	134.6[a]	3.0	30	
Chitosan/poly(vinyl alcohol) hydrogel beads	Fe_3O_4 nanoparticles	Congo red (CR)	467.3[a]	/	25 ± 0.2	[22]
Oxidized multiwalled carbon nanotube (OMWCNT)- κ-carrageenan	Fe_3O_4 (10–25 nm)	Methylene blue (MB)	396.63[a]	6.5	25	[5]
Poly(γ-glutamic acid) (PGA)	Fe_3O_4 (8.3 nm)	Methylene blue (MB)	78.67[a]	6.0	28	[28]
Terpolymer of aniline/m-aminobenzoic acid/ m-phenylenediamine	Fe_3O_4 (39 nm)	Methylene blue (MB)	31.64[a]	8.0	30	[33]
		Malachite green (MG)	29.07[a]	8.0	30	

(continued)

Table 3 (continued)

Polymers and composites	Magnetic particles	Dye	Adsorption capacity (mg g^{-1})	pH	Temperature (°C)	References
Carboxymethyl chitosan-g-poly(acrylamide) (CMC-g-PAAm).	Fe_3O_4 (<3 nm)/ laponite RD	Crystal violet (CV)	120[a]	/	24	[55]
Polyaniline (PANI)/mesoporous silica	Fe_3O_4 (110–130 nm)	Methyl orange (MO)	55.74[a]	4.0	0	[48]
Poly(4-styrenesulfonic acid-co-maleic acid) sodium (PSSMA)	Fe_3O_4 (~50 nm)	Basic fuchsin (BF)	588.2[a]	7.0	25	[54]
		Crystal violet (CV)	384.6[a]	7.0	25	
		Methylene blue (MB)	270.3[a]	7.0	25	
Chitosan/carbon nanotube	Fe_3O_4 (~65 nm)	Acid red 18 (AR18)	691.0[a]	3.0	30	[26]
Titania-polyaniline	$CoFe_2O_4$	Methyl orange (MO)	168.57[a]	/	/	[40]
Polysaccharide resin	$CuFe_2O_4$	Methylene blue (MB)	366.6[a]	8.0	/	[30]
Polyacrylamide microspheres	γ-Fe_2O_3	Methylene blue (MB)	1990[a]	8	30	[53]
		Neutral red	1937[b]	/	30	
		Gentian violet	1850[b]	/	30	

[a]Maximum adsorption capacity calculated by Langmuir isotherm model

[b]At C_0 of 100 mg/L, the equilibrium adsorption capacities for GV and NR are 1850 and 1937 mg/g

easily be positively charged in an acidic medium and in turn, facilitates the adsorption of negatively charged dyes, such as Direct red 81, and Acid Red 18 (AR18), etc., via strong electrostatic attraction, which is considered to be the dominant adsorption mechanism [48, 50–52]. The maximum adsorption capacity calculated by the Langmuir model was in the range of 55.7–256.1 mg/g at pH = 3.0 or 4.0, as shown in Table 3 [48, 51, 52]. With anionic functional groups (e.g. $-COO^-$, $-SO_3^-$), the polymers such as natural polysaccharides, poly(γ-glutamic acid) (PGA) and poly(4-styrenesulfonic acid-co-maleic acid) sodium (PSSMA) are highly negatively charged at pH 6.0–8.0 [5, 28, 33]; therefore, electrostatic interaction formed between polymers and cationic dyes, as shown in Fig. 7a [28]. In addition, other mechanisms, such as π–π interaction, hydrogen bonding, ion exchange and hydrophobic interaction are recommended in the adsorption process of dyes onto some magnetic polymer composites [30, 33, 39]. For example, in addition to electrostatic interactions between the functional groups ($-SO_3$, −OH and −NH) in the polymer with the cationic dyes (MB and MG), dye was captured via π–π interactions between hydrophobic residues of dyes and aromatic cavity from the polymer backbone of magnetic polymer nanocomposite, as shown in Fig. 7b [33]. More importantly, magnetic polymer nanocomposite with porous structure or surface is believed to have extraordinary adsorption performance, i.e. a high sorption capacity as well as a fast adsorption process. For example, the porous magnetic polyacrylamide (PAM) microspheres reached an equilibrium adsorption for MB in about 200 min, and the maximum adsorption capacity calculated by Langmuir model was 1990 mg/g. Their equilibrium adsorption capacities for gentian violet (GV) and neutral red (NR) at an initial concentration of 100 mg/L, were 1850 and 1937 mg/g, respectively, as shown in Table 3 [53]. The equilibrium adsorption capacity of MB onto porous magnetic PAM microspheres increased from 263 to 1977 mg/g, when initial dye concentration was changed from 5 to 300 mg/L [53].

The internal architectures of these magnetic polymer nanocomposites were examined by TEM. Two were distinguished, MNPs embedded in cross-linked polymer and core-shell structured polymer@MNPs, as shown in Fig. 8 [33, 54]. Several methods were used to fabricate magnetic polymer composites with the former structure. For example, a two-step strategy included the synthesis of MNPs

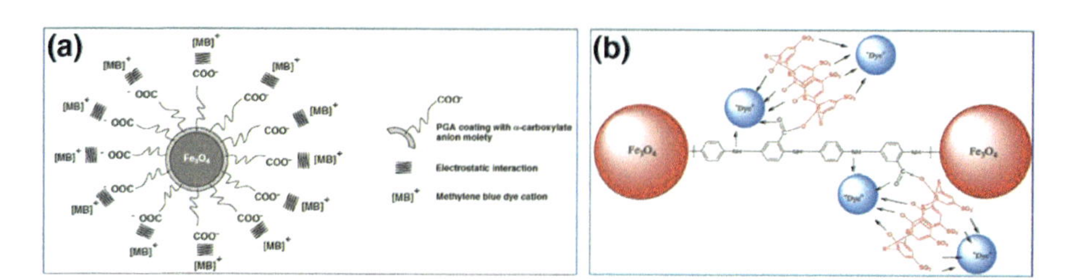

Fig. 7 Schematic diagram of proposed mechanism of **a** MB dye adsorption on PGA-MNPs (MNPs, magnetite nanoparticles; PGA, poly(γ-glutamic acid);); and **b** cationic dye adsorption on to Fe_3O_4@PAmABAmPD-TCAS nano adsorbent (PAmABAmPD, terpolymer of aniline/m-aminobenzoic acid/m-phenylenediamine; TCAS, Thiacalix(4)arene tetrasulfonate) (reprinted with permission from [28, 33])

and their dispersion in monomer-containing solution, followed by polymerization. In addition, a simple and facile one-pot solvothermal method was developed [54] and a dispersing route, in which porous PAM microspheres were dispersed in Fe(II) and Fe(III) solution and then in NaOH solution at 100 °C [53], was adopted. On the other side, core-shell structured polymer@ MNPs could be fabricated by firstly coating a dense silica and subsequently mesoporous silica, which generated pores on the surface of magnetic nanoparticles (Fe_3O_4) and allowed the PANI conjugated into the pores of mesoporous silica (MS), and finally to obtain PANI-MS@Fe_3O_4 nanocomposites [48]. Other ways to fabricate the core-shell structure were by directly coating the MNPs with a water-soluble polymer in deionized water [28], or via in situ coprecipitation method in which the synthesis started from a fine mixture of Fe(II) and Fe(III) salts and doped copolymer [33].

The reusability is considered as a key performance for investigation. The desorption of dye-loaded adsorbent was conducted by adjusting the pH of the aqueous solution since the tendency of maximum dye recovery was in general inversely proportional to the trend observed for the dye adsorption at different pHs. The adsorption of anionic dyes onto magnetic polymer composites is unfavourable at the alkaline medium, thus, NaOH or ammonium solution could be selected as effective dye desorption eluent to regenerate the spent adsorbents [51]. On the other side,

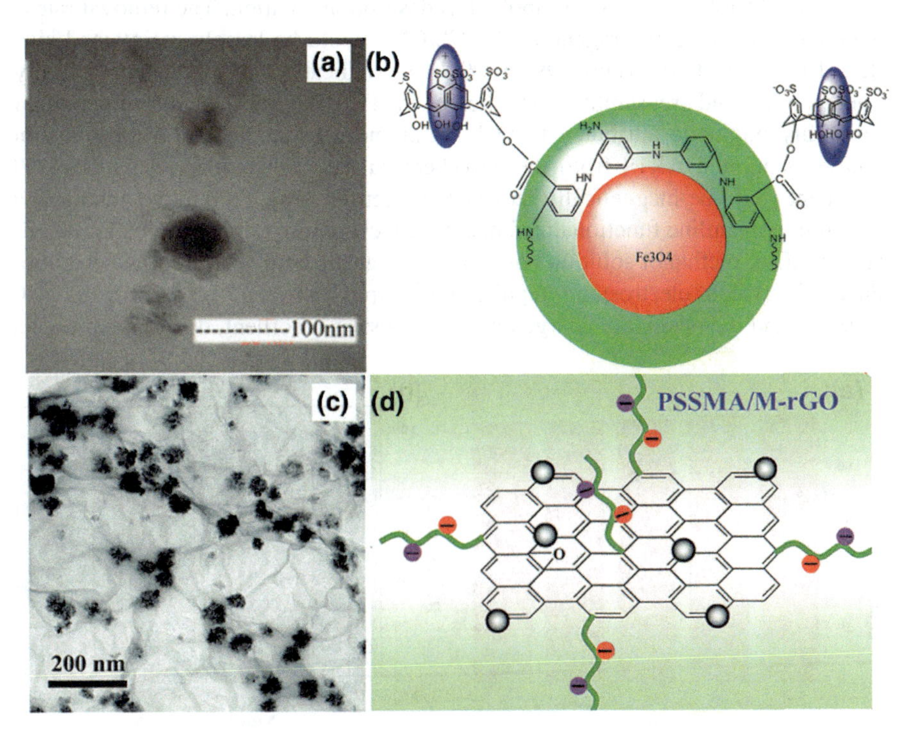

Fig. 8 TEM image and structural schematic of (**a**, **b**) Fe_3O_4@PAmABAmPD-TCAS and (**c**, **d**) poly(4-styrenesulfonic acid-co-maleic acid) sodium (PSSMA) modified magnetic reduced graphene oxide nanocomposite (PSSMA/M-rGO) (reprinted with permission from [33, 54])

desorption of cationic dyes was usually favoured in acid solution, e.g. 100–91% of the MB recovered from PGA-MNPs at pH = 1–3 [28]. The MB adsorbed onto montmorillonite/polyaniline/Fe_3O_4 (Mt/PANI/Fe_3O_4) nanocomposite could be successfully desorbed using 0.5 M HCl as the desorbing agent, and almost no decrease in the adsorption ratio was observed upon five cycles, as shown in Fig. 9a [39].

With the aid of a solvent, such as methanol or ethanol, solvent desorption technique was used to enhance the regeneration of the exhausted magnetic adsorbents [26, 48, 53]. The pH of solvent solution affected the surface charge and functional groups of adsorbent, the properties of dye molecule and in turn the desorption process [26, 48, 53]. The cationic dye MB saturated magnetic PAM microspheres could be completely regenerated (100% desorption) with acid methanol-water solution (50 v/v%, 10 mL) of pH 2 as the desorption solvent; while another cationic dye neutral red (NR) adsorbed magnetic PAM microspheres could be regenerated by neutral methanol-water solution as desorption solvent for three cycles of repeating washing process [53]. This is probably due to the difference in the charge of dye molecules, that the molecule of NR is less positively charged than MB [53]. To desorb anionic dye AR18 from the saturated magnetic chitosan-decorated carbon nanotube (mCS/CNT), which formed by compositing MNPs in chitosan-decorated carbon nanotube (CS/CNT), basic solvent solution, i.e. ammonia/ethanol (v/v, 2:3) was applied as desorption solution. The removal rate of anionic dye AR18 using regenerated mCS/CNT could be largely maintained even after 10 consecutive cycles (99.11–99.76%), indicating the excellent stability, regeneration, and reusability, as shown in Fig. 9b [26]. Such an excellent reusability might be attributed to OH^- in ammonia/ethanol elution enabling the release of anionic AR18 from the adsorbent surface; the regenerated mCS/CNT adsorbent could be magnetically recycled and reused [26]. To remove anionic dye MO from polyaniline functionalized magnetic mesoporous silica composite (PANI–MS@Fe_3O_4), methanol solution containing 4% acetic acid was used as desorption eluent, in which electrostatic repulsion occurred between the protonated MO molecule and the positively charged nanocomposites adsorbent, due to the presence

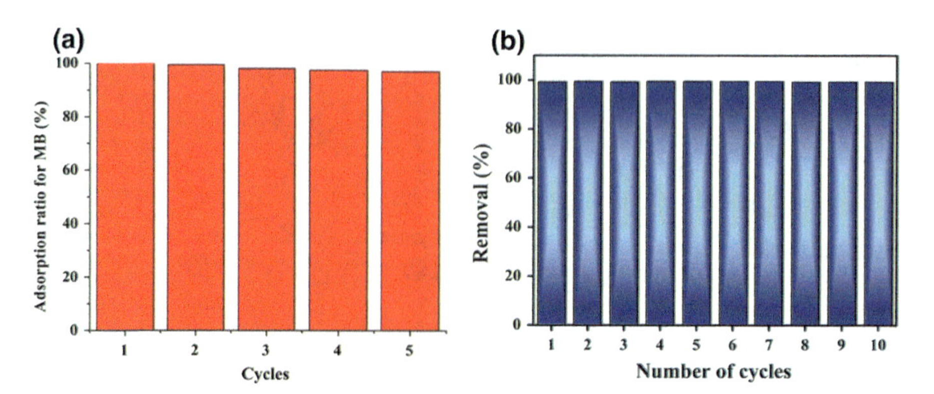

Fig. 9 **a** Adsorption cycles of MB onto Mt/PANI/Fe_3O_4 composites; **b** AR18 onto magnetic chitosan-decorated carbon nanotube (mCS/CNT) (reprinted with permission from [26, 39])

of nitrogen-containing functionalities (imine and amine groups) in PANI. Thus, there was still 80.25% of the anionic dye MO adsorbed onto the regenerated PANI–MS@Fe_3O_4 after three cycles of adsorption-desorption [48]. Other desorption solution, e.g. ethanol/water with 0.5 M KCl, was used to desorb cationic dye CV efficiently (desorption rate >97%) [55]. Taking the merit of magnetic separation, the spent magnetic polymer composites can be easily recycled and potentially reused.

5 Polymer/Clay Composites

The economic efficiency, as well as adsorption performance, are of extra importance when selecting a suitable adsorbent for practical application. Natural clays are hydrated layered aluminosilicates, which are widely available and low-cost, thus may be more viable and inexpensive for use as adsorbents to remove dyes from wastewater. Because of abundant silicon hydroxyls, negative surface charges and exchangeable cations (usually Na^+ and Ca^{2+}), clay minerals, e.g. attapulgite, are capable to adsorb cation dyes (e.g. MB) from water [56]. However, the relatively low adsorption capacity especially for anionic dyes due to the negatively charged surface limited their application. Organo-modification of clays with polymers is an essential way to fabricate effective adsorbents with enhanced adsorption capacity toward cationic or anionic dyes pollutants [7, 56]. Due to the presence of reactive –OH groups on their surfaces, clays can interact with reactive sites of polymers and monomers leading to the generation of polymer/clay composites [57]. In the last decades, significant research interest has been attracted on the development of polymer composites as adsorbents by incorporating inorganic clays such as attapulgite, montmorillonite, vermiculite, palygorskite and bentonite into polymeric matrices [20, 39, 57, 58]. The blending of clays not only potentially reduced fabrication cost, but also improved some properties, e.g. mechanical and thermal stability or swelling ability [20].

Among different clay minerals, bentonite is known for its good specific surface area and source abundance. The carboxymethyl cellulose (CMC) grafted by poly(2-(dimethylamino) ethyl methacrylate) modified bentonite (Bent/CMC-g-P(DMAEMA)) showed a maximum adsorption of 110.7 mg/g toward anionic MO (operation conditions: pH 6.86, 298 K and 50 min), and its removal rate was almost doubled in compared with bentonite [7]. In addition to ionic interaction, the hydrogen bonding interactions between the hydroxyl group (Si-OH, Al-OH, Fe-OH, and Mg-OH) in parent Bent and –O, –N in dye molecular greatly contributed to the observed good adsorption performance. By changing weight % of Bent in the polymer/clay composites, the surface characteristic of Bent/CMC-g-P(DMAEMA) was tuned; the resulting adsorption capacity was optimized when prepared using 20% CMC in the total amount of CMC + Bent [7]. Besides, humic acid-immobilized amine modified polyacrylamide/bentonite composite (HA-Am-PAA-B) was capable to remove cationic dyes (MG, MB, and CV) from their single and binary component solutions, as a result of the negative surface charge of the adsorbent [59]. The carrageenan-graft-poly (acrylamide)/bentonite exhibited a maximum MB adsorption capacity up to 156.25 mg/g and its adsorption was well fitted by the Langmuir isotherm model [20].

Attapulgite (APT) and palygorskite (Pal) are another two available clays, which exist in fibrous/rod morphology [56, 58]. Featured with one-dimensional nanoscale and a large number of silanol groups, APT and Pal are good inorganic candidate materials for composite fabrication. The APT/Fe_3O_4/PANI nanocomposites showed a much higher adsorption ratio, 96.0%, for CR at pH of 7, as compared with that of APT, 14.5% (298 K, contact time: 60 min) [56]. The adsorption capacity of Reactive Red 3BS onto the polyamidoamine (PAMAM) dendrimer-functionalized Pal adsorbents markedly increased from 34.2 to 322.6 mg/g (293 K, contact time: 20 min) by increasing addition of PAMAM, which contains amino-terminated groups. The anionic dyes could be trapped and then stabilized in the cavities of PAMAM dendrimers due to the host-guest affinity [58].

Temperature is one of the factors affecting equilibrium capacity of adsorbents, which correlates with the exothermicity or endothermicity nature of adsorption. Mostly, the adsorption capacity increased by increasing of temperature during the endothermic adsorption of dyes onto polymer/clay composites, which could be confirmed by the positive values of enthalpy change ($\Delta H°$) [57, 60]. The diffusion rate of dye molecules from solution to the adsorbent surface was also enhanced with increasing temperature, thus, adsorption capacity was improved [20, 57]. In the case of MO adsorption onto 20CMC-Bent (20% of CMC in the total amount of CMC + Bent composite), the negative $\Delta H°$ value indicated an exothermic adsorption process [7].

The adsorption capacity of cationic dye onto polymer/clay composites was observed to increase at higher pH [20, 57]; this was explained by the stronger electrostatic attraction formed between deprotonated functional groups of the adsorbents and positively-charged dye molecules. The maximum removal rate % of MG, MB, and CV onto the HA-Am-PAA-B was 99.7, 99.3, and 98.8%, respectively, at pH 6.0 with an initial dye concentration of 200 μmol/L [59]. The optimized adsorption pH was observed in the range of 6.0–8.0, under which the carboxylic and phenolic groups were deprotonated and the surface charge of adsorbent was negative (pH_{pzc} 4.8) [59].

Interestingly, at pH = 6.3, the Mt/PANI/Fe_3O_4 composite, which was protonated via treating with 0.1 M HCl, displayed a maximum adsorption ratio of 98.1% toward anionic dye CR; whilst the dedoped Mt/PANI/Fe_3O_4 showed excellent adsorption ratio of 99.6 and 96.2% to cationic dyes MB and Brilliant green (BG), respectively (298 K, contact time: 60 min) [39]. This difference resulted from the variation on the surface charge, that the dedoped Mt/PANI/Fe_3O_4 was negatively charged, as compared to positively charged Mt/PANI/Fe_3O_4 after acid treatment. The resulting clay/polymer composite showed excellent adsorption capacity to both cationic dyes and anionic dyes, in relative to the original Mt adsorbent [39]. Moreover, by introducing magnetic particles, such as Fe_3O_4 and $CoFe_2O_4$, the clay/polymer composites were able to be recovered by magnetic separation for reuse [39, 56, 60]. For example, via magnetic separation, no decrease in the adsorption ratio of

MB onto the dedoped Mt/PANI/Fe_3O_4 composite was observed upon five successive cycles [39].

6 Polymer/by-Products or Waste Composites

Recently, there has been increasing research interest in exploring polymer composites synthesized with industrial or agricultural by-products or waste, such as coir pith, fly ash, flax shrive, sawdust, sugarcane molasses, and almond shell waste, etc., due to the biodegradability, low cost and great availability [2, 3, 61–65]. By-products or waste materials may have a certain affinity to cationic dyes, because of the presence of surface hydroxyl groups; however, the adsorption capacity of anionic dyes was usually found to be low [63]. For example, the use of flax shrive cellulose exhibited only ~0.5 mg/g adsorption towards RR228 [3]; whilst that of almond shell waste (AS) to acid blue 25 (AB25) was less than 2.4 mg/g [63]. Benefited from functional groups of polymers, their incorporated polymer composites exhibited enhanced dye adsorption capacity.

In order to achieve good adsorption property, the physical or chemical modification is usually adopted to increase surface groups of waste/by-products for functionalization with polymers and in turn improve their adhesion with the polymer matrix. For example, after heat and alkali treatment, the surface area of coal fly ash (CFA), a coal combustion by-product, increased from 66.78 to 102.89 m^2/g, accompanied with greater total pore volume and average pore size; these features were believed to contribute in enhancing adsorption capacity. Moreover, the modified CFA possessed a greater amount of surface silanol groups (Si–OH), which could be functionalized with polyethyleneimine (PEI) to fabricate composite adsorbents. Thanks to the amine group at the end of the PEI chain as well as the high amine density, the Langmuir maximum adsorption capacity of CFA/PEI composite was found to be 316.75 mg/g for reactive red 2 (RR2) (pH = 3, 313 K) and 174.83 mg/g for MG (pH = 8, 308 K), respectively [61]. Goes and co-workers chemically modified cellulose with 4,4′-diphenylmethane diisocyanate (MDI); the resulting material showed enhanced adsorption capacity [2]. The equilibrium adsorption capacities of MB onto polyurethane foams with unmodified cellulose was 1.57 mg/g; that on polyurethane with chemically modified cellulose increased to 1.83 mg/g under the same operating conditions. This difference might result from the structural changes caused by the reaction of MDI on cellulose surface [2].

7 Conclusions

Adsorption has been investigated as an effective approach for the treatment of dye-containing wastewater. Polymers and polymer composites exhibited attractive features of high strength, good flexibility, and ease of modification or functionalization, thus as potential adsorbents with high adsorption capacity, fast adsorption rate and good reusability. Recent advances in polymers and polymer composites, such as PANI composites, magnetic polymer composites, polymer/clay composites and polymer/by-products or waste composites, were reviewed in terms of their properties and adsorption. The regeneration of spent polymers and polymer composites and subsequent reuse was also discussed, which potentially improved the cost efficiency of adsorbents.

Both the properties of adsorbents and dyes were considered as the key factors tuning adsorptive dye removal from aqueous solution. Generally, polymers and polymer composites attract the dye molecules via electrostatic interaction, van der Waals forces, π–π interaction, and hydrogen bonding. Electrostatic interaction and hydrogen bonding are two of the most commonly underlying mechanisms governing the sorption process, which were found to be affected by the complex physicochemical natures of dyes (e.g. cationic, anionic and nonionic dyes) and the properties of adsorbents (such as surface acidity or charge, zeta potential, pore size and its distribution, and surface area). The polymer sorbents with enhanced positively charged surface sites via modification or functionalization were found to favour the removal of anionic dyes and vice versa. By combining with inorganic materials, e.g. graphene, the resulting polymer composites exhibited a broad-spectrum adsorption performance, which was explained by the π–π interactions between aromatic or heterocyclic structures of dyes and graphene sheets.

Polymers and polymer composites with special characteristics, such as the presence of an internal hydrophobic cavity, in the forms of nanosized particles or 1D nanofibers, and the introduction of 3D porous structure (i.e. nanopores or mesopores), could facilitate the adsorption. The selectivity during adsorption could be achieved via tuning the pore size in the porous polymers or polymer composites, which was suggested as being governed by the size-selective mechanism.

Note that the adsorption performance was significantly affected by a number of operating conditions, including solution pH, temperature, initial dye concentration, and equilibrium time. For example, the solution pH would vary both solution chemistry and surface binding sites of adsorbents. Therefore, it is necessary for optimization of operation conditions towards enhanced adsorptive dye removal from water.

Acknowledgements The authors gratefully acknowledge funding from the National Natural Science Foundation of China (No.21607064, No.21263005 and No.21567008), and Qingjiang youth Talent program of Jiangxi University of Science and Technology (No. JXUSTQJYX20170005).

Appendix

See Table 4.

Table 4 Molecular structure/weight, abbreviation, species and λ_{max} of dyes in reports

Dye	Abbreviation	Molecular weight (g mol^{-1})	λ_{max} (nm)	Ionic species	Molecular structure	References
Eosin yellow	EY	645.89	517	Anionic	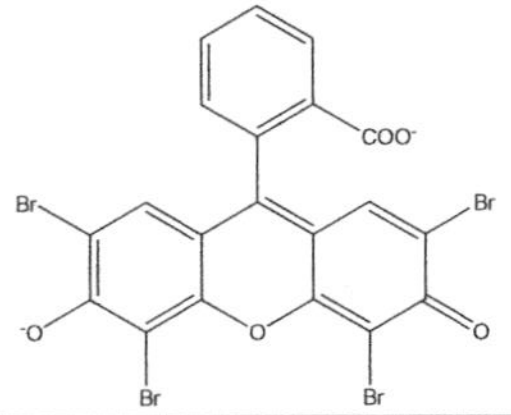	[42]
Eosin blue	EB	578.09	516	Anionic	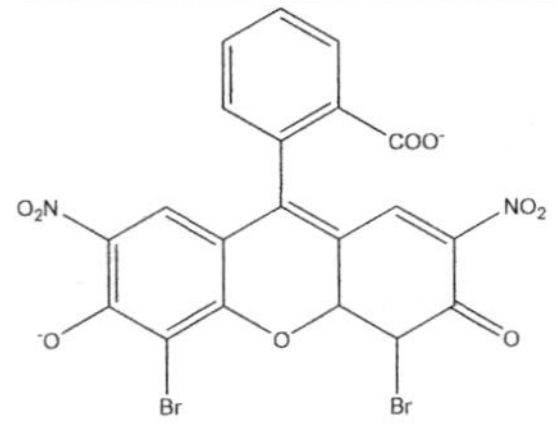	[42]
Orange G	OG	452.37	480	Anionic	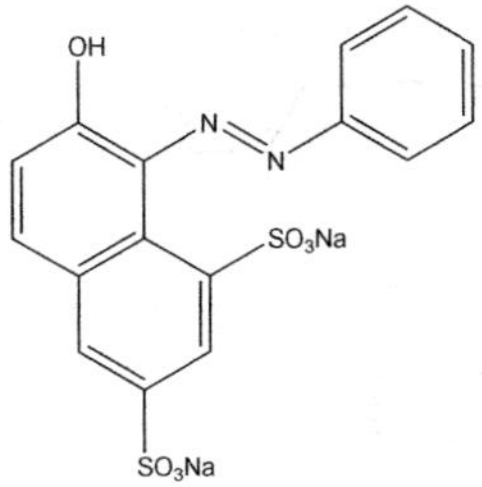	[42]

(continued)

Table 4 (continued)

Dye	Abbreviation	Molecular weight (g mol^{-1})	λ_{max} (nm)	Ionic species	Molecular structure	References
Reactive orange-14	RO	631.383	420	Anionic	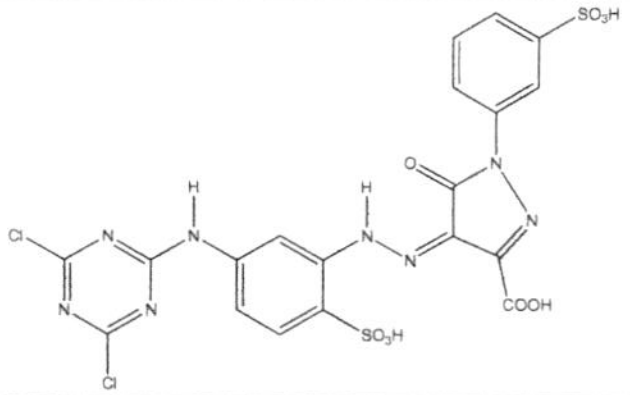	[42]
OrangeII (C I Acid orange 7)	/	350.32	483	Anionic	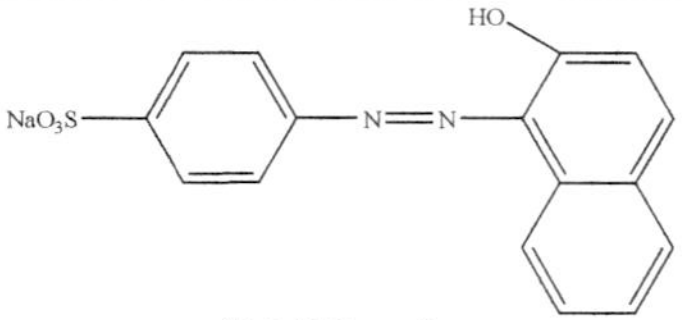 Cl Acid Orange 7	[31, 42]
Rhodanmine B	RHB	479.01	554	Cationic	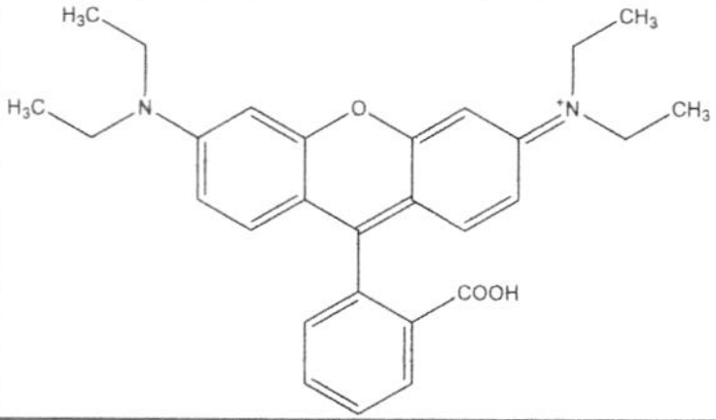	[42]

(continued)

Table 4 (continued)

Dye	Abbreviation	Molecular weight (g mol^{-1})	λ_{max} (nm)	Ionic species	Molecular structure	References
Methylene blue	MB	373.9	663	Cationic	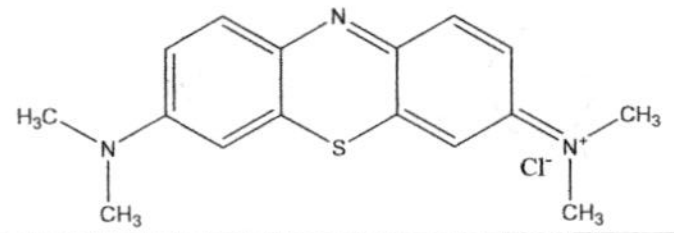	[28]
Basic violet 14	BV	337.85	545	Cationic	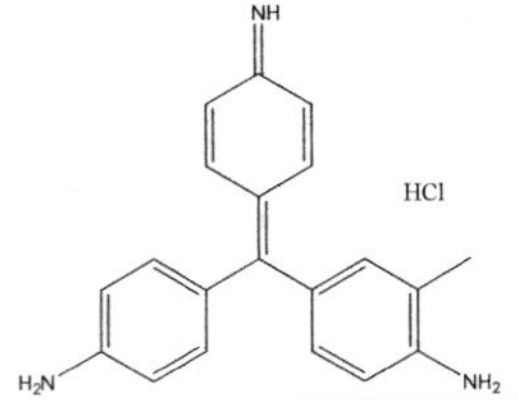	[17]
Reactive black 5	RB5	991.82	597	Anionic	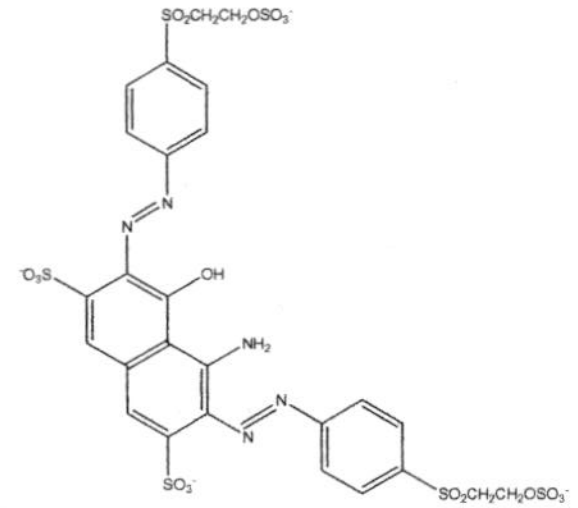	[14]
Brilliant green	BG	482.63	625	Cationic	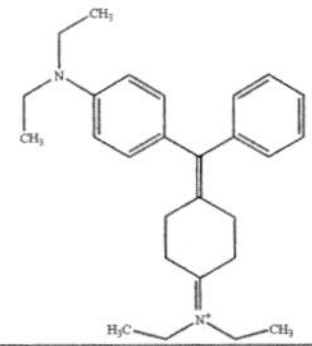	[45]

(continued)

Table 4 (continued)

Dye	Abbreviation	Molecular weight (g mol^{-1})	λ_{max} (nm)	Ionic species	Molecular structure	References
Reactive red 194	RR-194	984.2062	542	Anionic	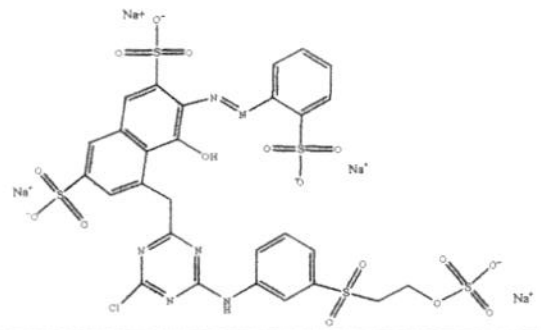	[8]
Direct blue 199	DB-199	775.21	608	Anionic	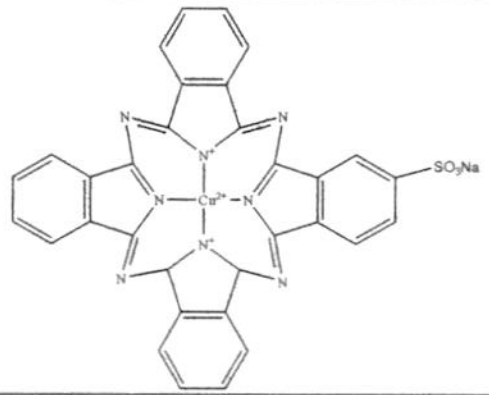	[8]
Acid blue 62	AB-62	400.45	620	Anionic	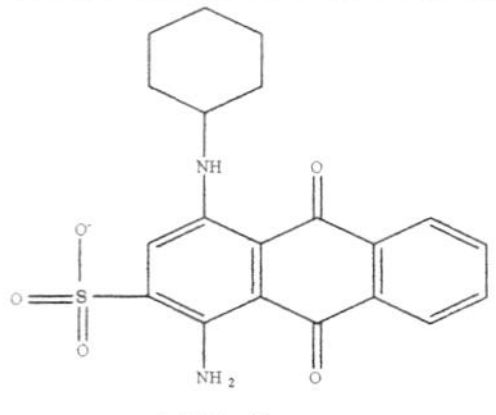 Acid Blue 62	[8]
Cl Acid red 88	AR 88	400.38	505	Anionic	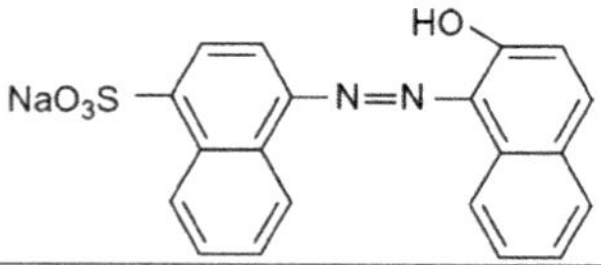	[31]

(continued)

Table 4 (continued)

Dye	Abbreviation	Molecular weight (g mol^{-1})	λ_{max} (nm)	Ionic species	Molecular structure	References
Alizarin red S	ARS	342.26	423	Anionic	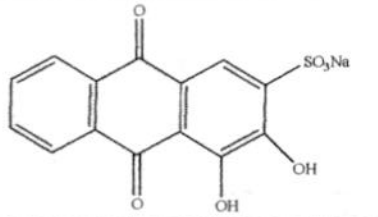	[51]
Nuclear fast red	NFR	357.28	518	Anionic	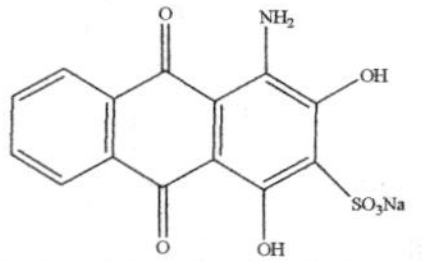	[51]
Alizarin green	AG	622.58	642	Anionic	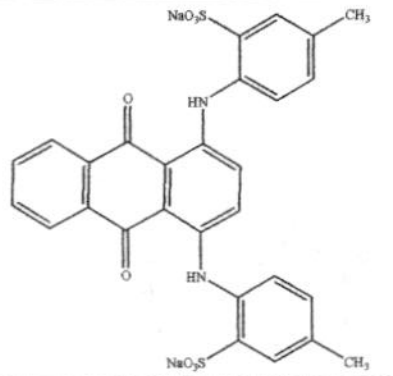	[51]
Methyl orange	MO	327.33	463	Anionic	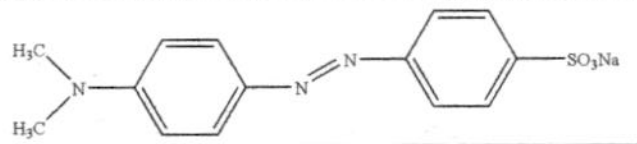	[51]
Sunset yellow FCF	SY	452.37	482	Anionic	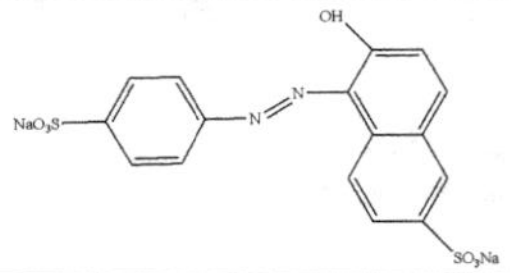	[51]

(continued)

Table 4 (continued)

Dye	Abbreviation	Molecular weight (g mol^{-1})	λ_{max} (nm)	Ionic species	Molecular structure	References
Methyl blue	/	799.80	600	Anionic	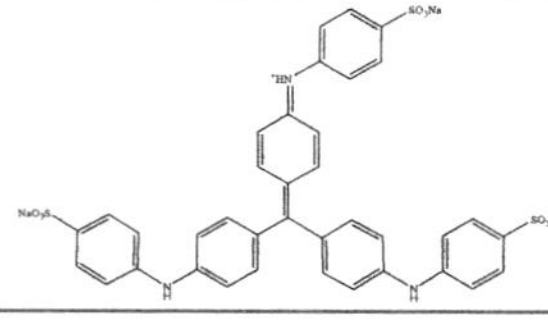 SO_3Na, HN, NaO_3S, SO_3Na, H, H	[51]
Congo red	CR	696.68	497/500	Anionic	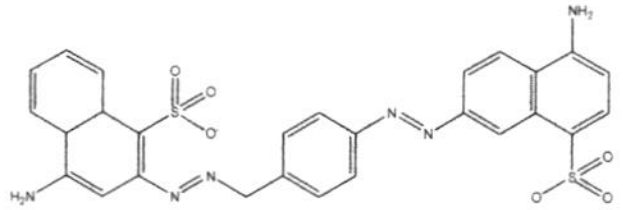 NH_2, O, S, O, O^-, N=N, N=N, H_2N, O^-, S, O, O	[14, 34]
Malachite green	MG	364.92	617	Cationic	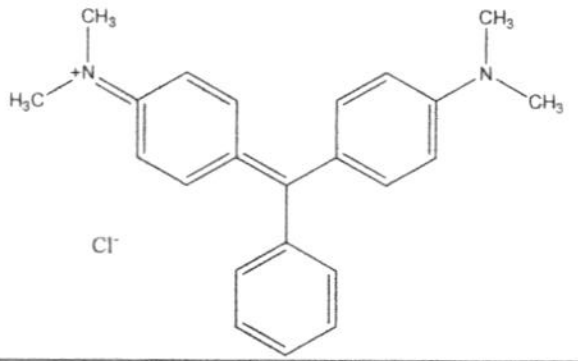 CH_3, ^+N, H_3C, CH_3, N, CH_3, Cl^-	[41]
Acid red 18	AR18	604.47	507	Anionic	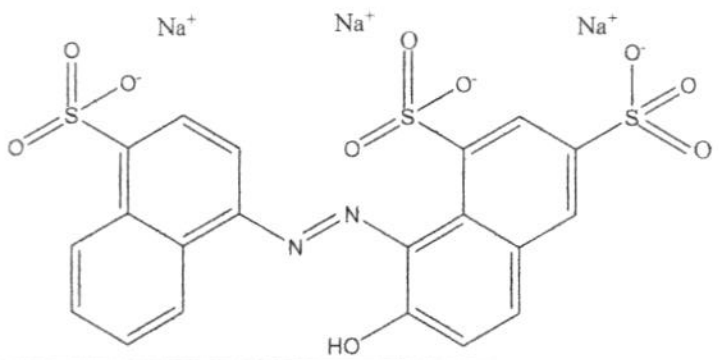 Na^+, Na^+, Na^+, O, O^-, S, O, O, O^-, S, O, O^-, S, O, O, N=N, HO	[26, 65]

(continued)

Table 4 (continued)

Dye	Abbreviation	Molecular weight (g mol^{-1})	λ_{max} (nm)	Ionic species	Molecular structure	References
Reactive red 3BS (C.I. reactive red 195)	/	1118.83	546	Anionic	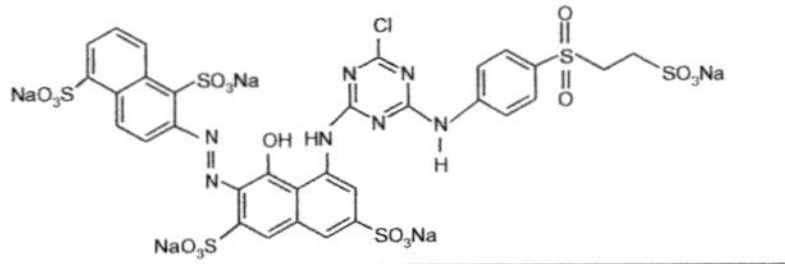 NaO_3S, SO_3Na, Cl, N, N, N, N, H, O, S, O, SO_3Na, N, N, OH, HN, NaO_3S, SO_3Na	[58]
Reactive red	RR	613.92	547	Anionic	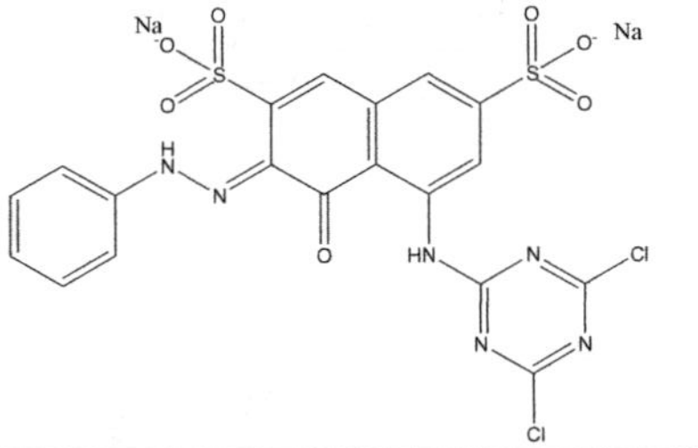 Na, ^-O, O, S, O, O, S, O^-, Na, O, H, N, N, O, HN, N, Cl, N, N, Cl	[25]
Active brilliant red X-3B	/	615.33	540.5	Anionic	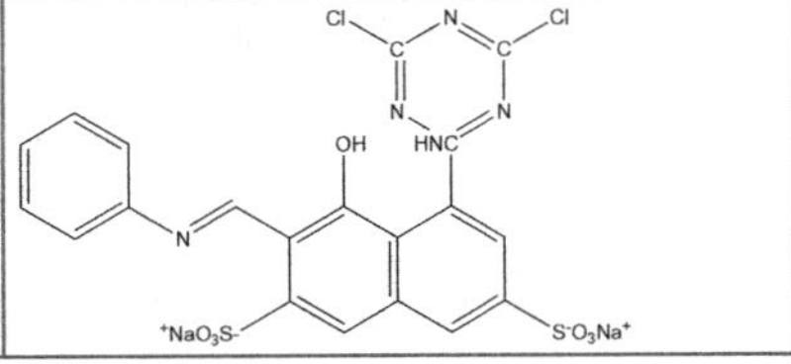 Cl, C, N, C, Cl, N, N, OH, HNC, N, $^+NaO_3S^-$, $S^-O_3Na^+$	[27]

(continued)

Table 4 (continued)

Dye	Abbreviation	Molecular weight (g mol^{-1})	λ_{max} (nm)	Ionic species	Molecular structure	References
Crystal violet (methyl violet 10B)	CV	407.99	595	Cationic	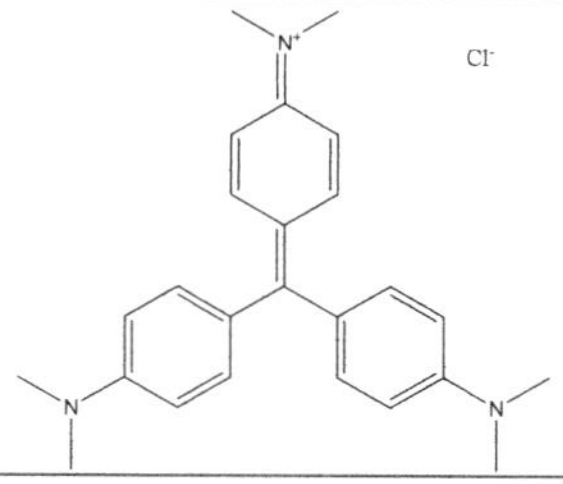	[55]
Basic yellow 28	BY28	433.52	420	Cationic	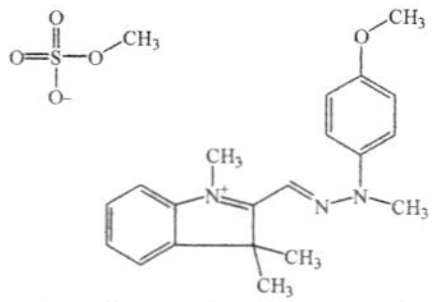	[25]
Brilliant blue 133	BB133	792.85	609.5	Anionic	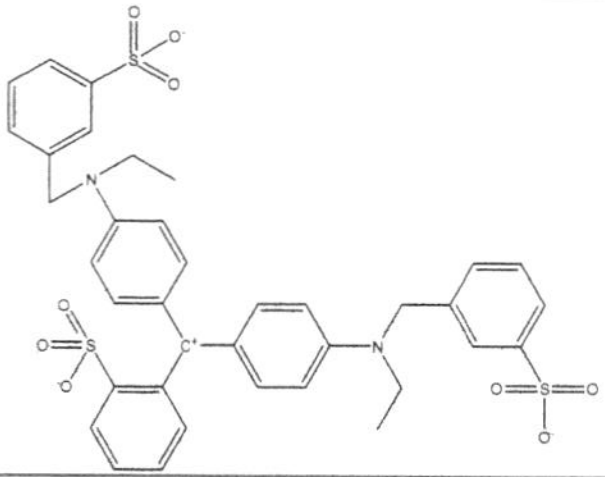	[25]

References

1. Ngulube T et al (2017) An update on synthetic dyes adsorption onto clay based minerals: a state-of-art review. J Environ Manag 191:35–57
2. Goes MM et al (2016) Polyurethane foams synthesized from cellulose-based wastes: kinetics studies of dye adsorption. Ind Crops Prod 85:149–158
3. Wang L, Li J (2013) Adsorption of C.I. reactive red 228 dye from aqueous solution by modified cellulose from flax shive: kinetics, equilibrium, and thermodynamics. Ind Crops Prod 42(Supplement C):153–158
4. Kanmani P et al (2017) Environmental applications of chitosan and cellulosic biopolymers: a comprehensive outlook. Bioresour Technol 242:295–303
5. Duman O et al (2016) Synthesis of magnetic oxidized multiwalled carbon nanotube-kappa-carrageenan-Fe_3O_4 nanocomposite adsorbent and its application in cationic methylene blue dye adsorption. Carbohydr Polym 147:79–88
6. Rajabi M, Mahanpoor K, Moradi O (2017) Removal of dye molecules from aqueous solution by carbon nanotubes and carbon nanotube functional groups: critical review. Rsc Adv 7(74): 47083–47090
7. Li W et al (2017) Tunable adsorption properties of bentonite/carboxymethyl cellulose-g-poly (2-(dimethylamino) ethyl methacrylate) composites toward anionic dyes. Chem Eng Res Des 124:260–270
8. Javadian H, Angaji MT, Naushad M (2014) Synthesis and characterization of polyaniline/gamma-alumina nanocomposite: a comparative study for the adsorption of three different anionic dyes. J Ind Eng Chem 20(5):3890–3900
9. Ezzatahmadi N et al (2017) Clay-supported nanoscale zero-valent iron composite materials for the remediation of contaminated aqueous solutions: a review. Chem Eng J 312:336–350
10. Pandey S (2017) A comprehensive review on recent developments in bentonite-based materials used as adsorbents for wastewater treatment. J Mol Liq 241:1091–1113
11. Sulyman M, Namiesnik J, Gierak A (2017) Low-cost adsorbents derived from agricultural by-products/wastes for enhancing contaminant uptakes from wastewater: a review. Polish J Environ Stud 26(2):479–510
12. Xiao J et al (2017) Multifunctional graphene/poly(vinyl alcohol) aerogels: in situ hydrothermal preparation and applications in broad-spectrum adsorption for dyes and oils. Carbon 123:354–363
13. Voisin H et al (2017) Nanocellulose-based materials for water purification. Nanomaterials 7(3):19
14. Qiu Y, Ling F (2006) Role of surface functionality in the adsorption of anionic dyes on modified polymeric sorbents. Chemosphere 64(6):963–971
15. Zhang L et al (2012) Facile and large-scale synthesis of functional poly(m-phenylenediamine) nanoparticles by Cu2 + -assisted method with superior ability for dye adsorption. J Mater Chem 22(35):18244–18251
16. Gezici O et al (2016) Humic-makeup approach for simultaneous functionalization of polyacrylonitrile nanofibers during electrospinning process, and dye adsorption study. Soft Mater 14(4):278–287
17. Elkady MF, El-Aassar MR, Hassan HS (2016) Adsorption profile of basic dye onto novel fabricated carboxylated functionalized co-polymer nanofibers. Polymers 8(5):177
18. Chauque EFC et al (2017) Electrospun polyacrylonitrile nanofibers functionalized with EDTA for adsorption of ionic dyes. Phys Chem Earth 100:201–211
19. Almasian A, Olya ME, Mahmoodi NM (2015) Preparation and adsorption behavior of diethylenetriamine/polyacrylonitrile composite nanofibers for a direct dye removal. Fibers Polym 16(9):1925–1934
20. Pourjavadi A et al (2016) Porous Carrageenan-g-polyacrylamide/bentonite superabsorbent composites: swelling and dye adsorption behavior. J Polym Res **23**(3)

21. Ekici S, Isikver Y, Saraydin D (2006) Poly(acrylamide-sepiolite) composite hydrogels: preparation, swelling and dye adsorption properties. Polym Bull 57(2):231–241
22. Zhu HY et al (2012) Novel magnetic chitosan/poly(vinyl alcohol) hydrogel beads: preparation, characterization and application for adsorption of dye from aqueous solution. Bioresour Technol 105:24–30
23. Zhu L et al (2017) Adsorption of dyes onto sodium alginate graft poly(acrylic acid-co-2-acrylamide-2-methyl propane sulfonic acid)/kaolin hydrogel composite. Polym Polymer Compos 25(8):627–634
24. Das S et al (2017) Folic acid-polyaniline hybrid hydrogel for adsorption/reduction of chromium(VI) and selective adsorption of anionic dye from water. Acs Sustain Chem Eng 5(10):9325–9337
25. Zhu W et al (2016) Functionalization of cellulose with hyperbranched polyethylenimine for selective dye adsorption and separation. Cellulose 23(6):3785–3797
26. Wang S et al (2014) Highly efficient removal of acid Red 18 from aqueous solution by magnetically retrievable Chitosan/carbon nanotube: batch study, isotherms, kinetics, and thermodynamics. J Chem Eng Data 59(1):39–51
27. Deng H, Wei Z, Wang X (2017) Enhanced adsorption of active brilliant red X-3B dye on chitosan molecularly imprinted polymer functionalized with Ti(IV) as Lewis acid. Carbohydr Polym 157:1190–1197
28. Inbaraj BS, Chen BH (2011) Dye adsorption characteristics of magnetite nanoparticles coated with a biopolymer poly(gamma-glutamic acid). Bioresour Technol 102(19):8868–8876
29. Wang L, Li Q (2010) Wang A Adsorption of cationic dye on N,O-carboxymethyl-chitosan from aqueous solutions: equilibrium, kinetics, and adsorption mechanism. Polym Bull 65(9): 961–975
30. Beyki MH et al (2017) Clean approach to synthesis of graphene like CuFe2O4@polysaccharide resin nanohybrid: bifunctional compound for dye adsorption and bacterial capturing. Carbohydr Polym 174:128–136
31. Shimizu Y, Saito Y, Nakamura T (2006) Crosslinking of chitosan with a trifunctional crosslinker and the adsorption of acid dyes and metal ions onto the resulting polymer. Adsorpt Sci Technol 24(1):29–39
32. Zhang X et al (2015) Adsorption of basic dyes on beta-cyclodextrin functionalized poly (styrene-alt-maleic anhydride). Sep Sci Technol 50(7):947–957
33. Lakouraj MM, Norouzian R-S, Balo S (2015) Preparation and cationic dye adsorption of novel Fe_3O_4 supermagnetic/thiacalix 4 arene tetrasulfonate self-doped/polyaniline nanocomposite: kinetics, isotherms, and thermodynamic study. J Chem Eng Data 60(8):2262–2272
34. Liu H, Liu H (2017) Selective dye adsorption and metal ion detection using multifunctional silsesquioxane-based tetraphenylethene-linked nanoporous polymers. J Mater Chem A 5(19): 9156–9162
35. Liu J et al (2017) Preparation of polyhedral oligomeric silsesquioxane based cross-linked inorganic-organic nanohybrid as adsorbent for selective removal of acidic dyes from aqueous solution. J Coll Interf Sci 497:402–412
36. Zhao W et al (2015) Functionalized graphene sheets with poly(ionic liquid)s and high adsorption capacity of anionic dyes. Appl Surf Sci 326:276–284
37. Agarwal S et al (2016) Synthesis and characteristics of polyaniline/zirconium oxide conductive nanocomposite for dye adsorption application. J Mol Liq 218:494–498
38. Wang L et al (2012) Stable organic-inorganic hybrid of polyaniline/α-zirconium phosphate for efficient removal of organic pollutants in water environment. ACS Appl Mater Interf 4(5):2686–2692
39. Mu B et al (2016) Preparation, characterization and application on dye adsorption of a well-defined two-dimensional superparamagnetic clay/polyaniline/Fe3O4 nanocomposite. Appl Clay Sci 132:7–16
40. Xiong P et al (2013) Ternary titania-cobalt ferrite-polyaniline nanocomposite: a magnetically recyclable hybrid for adsorption and photodegradation of dyes under visible light. Ind Eng Chem Res 52(30):10105–10113

41. Zeng Y et al (2013) Enhanced adsorption of malachite green onto carbon nanotube/polyaniline composites. J Appl Polym Sci 127(4):2475–2482
42. Majumdar S, Saikia U, Mahanta D (2015) Polyaniline-coated filter papers: cost effective hybrid materials for adsorption of dyes. J Chem Eng Data 60(11):3382–3391
43. Pandiselvi K, Manikumar A, Thambidurai S (2014) Synthesis of novel polyaniline/MgO composite for enhanced adsorption of reactive dye. J Appl Polym Sci 131(9)
44. Saad M et al (2017) Synthesis of polyaniline nanoparticles and their application for the removal of crystal violet dye by ultrasonicated adsorption process based on response surface methodology. Ultrason Sonochem 34:600–608
45. Salem MA, Elsharkawy RG, Hablas MF (2016) Adsorption of brilliant green dye by polyaniline/silver nanocomposite: kinetic, equilibrium, and thermodynamic studies. Eur Polym J 75:577–590
46. Pandiselvi K, Thambidurai S (2016) Synthesis of adsorption cum photocatalytic nature of polyaniline-ZnO/chitosan composite for removal of textile dyes. Desalin Water Treat 57(18): 8343–8357
47. Bingol D et al (2012) Analysis of adsorption of reactive azo dye onto $CuCl_2$ doped polyaniline using Box-Behnken design approach. Synth Metals 162(17–18):1566–1571
48. Mahto TK et al (2015) Kinetic and thermodynamic study of polyaniline functionalized magnetic mesoporous silica for magnetic field guided dye adsorption. Rsc Adv 5(59):47909–47919
49. Anuradha Jabasingh S, Ravi T, Yimam A (2017) Magnetic hetero-structures as prospective sorbents to aid arsenic elimination from life water streams. Water Sci
50. Zhou L, He B, Huang J (2013) One-step synthesis of robust amine- and vinyl-capped magnetic iron oxide nanoparticles for polymer grafting, dye adsorption, and catalysis. Acs Appl Mater Interf 5(17):8678–8685
51. Chen B et al (2016) Magnetically recoverable cross-linked polyethylenimine as a novel adsorbent for removal of anionic dyes with different structures from aqueous solution. J Taiwan Inst Chem Eng 67:191–201
52. Beyki MH et al (2016) Green synthesized Fe_3O_4 nanoparticles as a magnetic core to prepare poly 1, 4 phenylenediamine nanocomposite: employment for fast adsorption of lead ions and azo dye. Desalin Water Treat 57(59):28875–28886
53. Yao T et al (2015) Investigation on efficient adsorption of cationic dyes on porous magnetic polyacrylamide microspheres. J Hazard Mater 292:90–97
54. Song Y-B et al (2015) Poly(4-styrenesulfonic acid-co-maleic acid)-sodium-modified magnetic reduced graphene oxide for enhanced adsorption performance toward cationic dyes. Rsc Adv 5(106):87030–87042
55. Mahdavinia GR, Karami S (2015) Synthesis of magnetic carboxymethyl chitosan-g-poly (acrylamide)/laponite RD nanocomposites with enhanced dye adsorption capacity. Polym Bull 72(9):2241–2262
56. Mu B, Wang A (2015) One-pot fabrication of multifunctional superparamagnetic attapulgite/Fe_3O_4/polyaniline nanocomposites served as an adsorbent and catalyst support. J Mater Chem A 3(1):281–289
57. Dong K et al (2013) Polyurethane-attapulgite porous material: preparation, characterization, and application for dye adsorption. J Appl Polym Sci 129(4):1697–1706
58. Zhou S et al (2015) Novel polyamidoamine dendrimer-functionalized palygorskite adsorbents with high adsorption capacity for $Pb2^+$ and reactive dyes. Appl Clay Sci 107:220–229
59. Anirudhan TS, Suchithra PS (2009) Adsorption characteristics of humic acid-immobilized amine modified polyacrylamide/bentonite composite for cationic dyes in aqueous solutions. J Environ Sci 21(7):884–891
60. Ai L, Zhou Y, Jiang J (2011) Removal of methylene blue from aqueous solution by montmorillonite/$CoFe_2O_4$ composite with magnetic separation performance. Desalination 266(1):72–77

61. Dash S et al (2016) Fabrication of inexpensive polyethylenimine-functionalized fly ash for highly enhanced adsorption of both cationic and anionic toxic dyes from water. Energy Fuels 30(8):6646–6653
62. Gonte RR, Shelar G, Balasubramanian K (2014) Polymer-agro-waste composites for removal of Congo red dye from wastewater: adsorption isotherms and kinetics. Desalin Water Treat 52(40–42):7797–7811
63. Jabli M et al (2017) Almond shell waste (Prunus dulcis): Functionalization with dimethy-diallyl-ammonium-chloride-diallylamin-co-polymer and chitosan polymer and its investigation in dye adsorption. J Mol Liq 240:35–44
64. Namasivayam C, Sureshkumar MV (2006) Anionic dye adsorption characteristics of surfactant-modified coir pith, a 'waste' lignocellulosic polymer. J Appl Polym Sci 100(2): 1538–1546
65. Shabandokht M, Binaeian E, Tayebi H-A (2016) Adsorption of food dye acid red 18 onto polyaniline-modified rice husk composite: isotherm and kinetic analysis. Desalin Water Treat 57(57):27638–27650
66. Lu Y-S et al (2016) Direct assembly of mesoporous silica functionalized with polypeptides for efficient dye adsorption. Chem-a Eur J 22(3):1159–1164

Current Scenario of Nanocomposite Materials for Fuel Cell Applications

Raveendra M. Hegde, Mahaveer D. Kurkuri and Madhuprasad Kigga

List of Abbreviations

FC	Fuel cell
PEM	Proton exchange membrane
PEMFC	Proton exchange membrane fuel cell
AFC	Alkaline fuel cell
DMFC	Direct methanol fuel cell
PAFC	Phosphoric acid fuel cell
MCFC	Molten carbonate fuel cell
SOFC	Solid oxide fuel cell
MEA	Membrane electrode assembly
GDL	Gas diffusion layers
RH	Relative humidity
PEEK	Poly (ether ether ketone)
SPEEK	Sulfonated poly (ether ether ketone)
PBI	Polybenzimidazole
PVA	Polyvinyl alcohol
ORR	Oxygen reduction reaction
CV	Cyclic voltammetry
XRD	X-ray diffraction
DFT	Density functional theory
PECVD	Plasma enhanced chemical vapor deposition
CNT	Carbon nanotubes
SWCNT	Single-walled carbon nanotubes
MWCNT	Multi-walled carbon nanotubes
IEC	Ion exchange capacity

R. M. Hegde · M. D. Kurkuri (✉) · M. Kigga (✉)
Centre for Nano and Material Sciences, Jain University, Jain Global Campus,
Bengaluru 562 112, Karnataka, India
e-mail: mahaveer.kurkuri@jainuniversity.ac.in

M. Kigga
e-mail: madhuprasad@jainuniversity.ac.in

Inamuddin et al. (eds.), *Sustainable Polymer Composites and Nanocomposites*,
https://doi.org/10.1007/978-3-030-05399-4_20

1 Introduction

The increasing population and rapid urbanization are leading to a sudden rise in the energy consumption worldwide. To fulfil this huge demand, fossil fuels have been extensively exploited as a major source of energy until date. However, this source of energy consumption is creating irreversible damage to the environment by producing huge and rapid carbon footprint and hence constant depletion of this source is eventually leading to the energy crisis in most of the countries today [1]. In addition, it has been predicted that there will be more than 50% increase in energy demand by next 25 years, which cannot be met just by relying majorly on fossil fuel source. Therefore, exploring the alternative energy sources have become inevitable in fulfilling the energy demand of future generation. Consequently, renewable energy sources, on the other hand, are promising sources of energy and are very fast developing technology as it provides the energy with significantly less pollution. Though several alternative technologies such as solar, wind, tidal, biomass have been developed, they all have been restricted for complete commercial exploitation because of their drawbacks such as storage issues, energy conversion, efficiency, transportation etc. Consequently, fuel cells (FCs) are considered to be a future source of energy to meet the rising energy demand and thus, have been attracting the wide scientific community to develop sustainable technology for efficient energy generation [2]. It is an emerging field of research in energy materials as it provides clean and green energy source with a conversion efficiency as high as 60%. Contrasting to other electrochemical sources like batteries, FCs requires the constant supply of fuel and oxygen to retain the continuous electrochemical reaction process, whereas, in a battery, the previously loaded chemicals in a batch process format attain the energy. However, FCs obtain the energy sources continuously from the exterior of the cell i.e., hydrogen from the fuel and oxygen from the environment [3]. Also, the possibilities of rapid recharging, off-grid operations, a significant reduction in weight, noise-free operations are the added advantages with FCs. More importantly, FCs reduce the greenhouse gas emission and produce only water as the main by-product. Therefore, it is significantly applied in portable energy devices, automobiles, transportation vehicles [4]. However, the conditions like high-temperature operation, availability of fuel source and infrastructures are hindering the further development of FCs and their usage in large scale.

1.1 Working Principle of FC

FCs are the galvanic cells, which work on the principle of oxidation and reduction reactions at the anode and cathode respectively. They use hydrogen or hydrogen-rich fuels for electrochemical reactions as hydrogen contains high energy per unit weight than any other fuels. A conventional FC is illustrated in Fig. 1

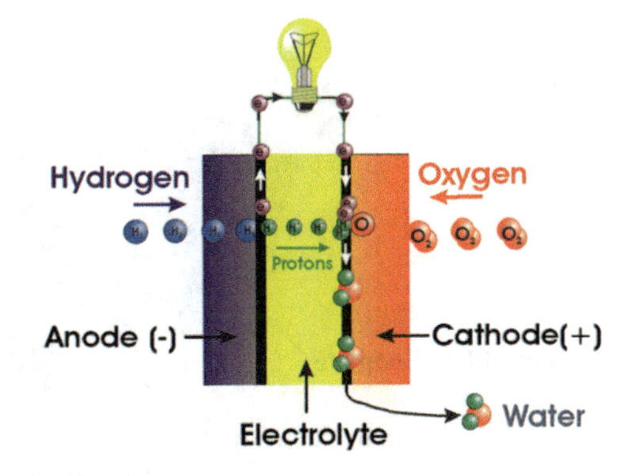

Fig. 1 Schematic representation of a conventional fuel cell. Reproduced with permission from Ref. [5], Copyright 2010, Elsevier

which consists of a sandwiched membrane between the anode, cathode, and electrolyte by forming a membrane electrode assembly (MEA) along with gas diffusion layers (GDL).

The electrochemical reactions are involving the continuous oxidation of hydrogen at anode and reduction of oxygen at the cathode. This chemical reaction produces the electrons, which are opposed by the electrolytic membrane and they move through an external circuit to produce direct current. The reactions involved in FCs are given below;

$$\text{At anode: } 2H_2 \rightarrow 4H^+ + 4e^-$$
$$\text{At cathode: } O_2 + 4H^+ + 4e^- \rightarrow 2H_2O$$
$$\text{Overall reaction: } 2H_2 + O_2 \rightarrow 2H_2O$$

To dissociate the hydrogen fuel into constituent ions, the FCs are equipped with the catalyst at the anode. Generally, platinum is most widely used as anode catalyst as it exhibits highest electrocatalytic activities in organic fuel redox reactions. The outer layer of catalyst is constructed with GDL, which promotes the transfer of reactants into catalyst layer and helps in the removal of by-product, water. GDL is composed of a thick porous array of carbon fibers, which provides a conductive pathway for current collection. Further, GDL also helps in the electronic connection between bipolar plate and electrode through the channel of MEA. In addition, GDL enhances the mechanical strength of MEA and protects the catalyst layer from corrosion.

The amount of total current produced by the FC depends on cell size, type of the cell, operating temperature and extent of gaseous pressure applied to it. During the electrochemical reactions, FC produces only a small amount of current in the range of 0.6–0.87 V. Therefore, in order to obtain a high voltage, a parallel or series of single cells are constructed as illustrated in Fig. 2. Further, each FC in a stack is separated by a bipolar plate which assists in the uniform distribution of fuel and oxygen in MEA. Polymer-based gaskets are inserted around the edges of MEA to

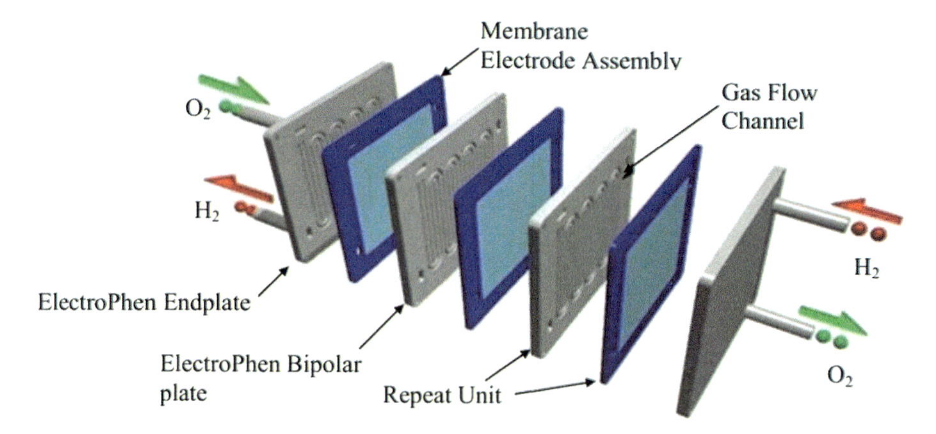

Fig. 2 Schematic representation for the construction of fuel cell stack. Reproduced from Ref. [6], Copyright 2013, Bentham Open

strengthen the FC stack as it is much effective in holding the MEA and adjusting the required pressure inside the system. The produced current is collected by a collector plate which is presented at the terminal part of MEA.

1.2 Proton Conduction Mechanism in FC

Proton conduction across the electrolyte membrane is the prime requirement to attain high current density and it is, therefore, efficiency determining a factor for the FC. This proton conduction depends on the modification made to the membrane, extent of sulfonation, relative humidity (RH), and temperature. The proton conduction phenomena generally follow either Grotthuss mechanism or Vehicle mechanism [7]. In Grotthuss mechanism, proton jumps from one ionic site to another through hydrogen bond network (Fig. 3a). For example, in Nafion® membranes, a proton hops from sulfonic acid (SO_3H) site to the nearby acceptor site i.e., water molecule which has potential for proton movement throughout the membrane.

On the other hand, according to the vehicle mechanism (or en masse diffusion mechanism), proton transfer takes place by the diffusion of carrier species in the form of hydrated ions in the electrolyte (Fig. 3b). Here the protons attach itself to the acceptor molecule and thus acceptor molecule moves from one end to another leading to the proton movement across the membrane. Both the above mechanisms depend on the nature and properties of nanocomposites membrane used in the FC. Moreover, the proton conductivity increases with RH of the system due to the presence of more hydrated protons in the system. Therefore, the proposed models can be used to design the suitable organic and inorganic fillers for polymeric membranes compared to pure polymer membranes [9].

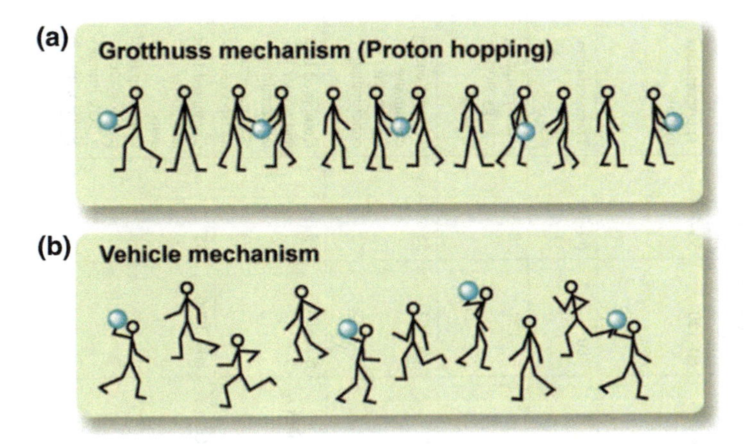

Fig. 3 Schematic representation of **a** Grotthuss mechanism and **b** Vehicle mechanism. Reproduced with permission from Ref. [8], Copyright 2008, American Chemical Society

FCs are generally classified based on the nature of electrolyte used in the cell, operating temperature and types of fuel/oxidants used. Different types of FCs with their properties, efficiency and operating temperature are listed in Table 1.

2 Nanocomposites in FC

To overcome the problems associated with low proton conduction, low current density, fuel crossover, carbon monoxide (CO) poisoning in conventional FCs, in the recent years a new variety of organic, inorganic and polymer-based nanocomposites have been developed. Nanocomposites are hybrid materials with conventional components in it [10, 11] which would be the solution for some of the challenges by providing improved water retention capacity, high energy conversion, and suppression of fuel crossover. Modification of FC composites with organic and inorganic materials is a growing technology in energy material development. Nanomaterials have been integrated with polymers to enhance their original characteristics such as thermal and chemical stabilities [12, 13]. This is attributed to the strong interfacial interaction between the polymers and inserted material. Preparation method of the nanocomposite is one of the important factors in the improvement of FCs as it alters the microstructure of the membranes [14]. Nanomaterials with conventional composite materials have specific properties such as high surface area, specific functional groups, interaction capacity [15–22] that would increase the catalytic performance of the electrodes in FC in terms of rigidity and thermal stability. On the other hand, organic nanocomposites such as sulfonated poly (ether ether ketone) (SPEEK) and polybenzimidazole (PBI) have provided the high flexibility, durability, and processability to the components of the FC [23]. The different types of potential nanocomposites and its applications in FC are described

Table 1 Types of important fuel cells and comparison of their properties

Type of fuel cell	Electrolyte	Charge carrier	Anode reaction	Cathode reaction	Overall reaction	Working temperature (°C)	Efficiency (per cell) [%]	Applications
Proton Exchange Membrane Fuel Cell (PEMFC)	Polymer membrane	H^+	$2H_2 \rightarrow 4H^+ + 4e^-$	½ $O_2 + 4H^+ + 4e^- \rightarrow 2H_2O$	$2H_2 + ½\ O_2 \rightarrow 2H_2O$	60–120	50–70	Transportation, military
Alkaline Fuel Cell (AFC)	Aqueous alkaline solution	OH^-	$2H_2 + 4OH^- \rightarrow 4H_2O + 4e^-$	$O_2 + 4e^- + 2H_2O \rightarrow 4OH^-$	$2H_2 + O_2 \rightarrow 2H_2O$	<80	60–70	Transportation, space
Direct Methanol Fuel Cell (DMFC)	Polymer membrane	H^+	$CH_3OH + H_2O \rightarrow CO_2 + 6H^+ + 6e^-$	3/2 $O_2 + 6H^+ + 6e^- \rightarrow 3H_3O$	CH_2OH + 3/2 $O_2 \rightarrow CO_2 + 2H_2O$	60–120	20–30	Transportation, energy storage systems
Phosphoric Acid Fuel Cell (PAFC)	Molten phosphoric acid	H^+	$2H_2 \rightarrow 4H^+ + 4e^-$	O_2 (g) $+ 4H^+ + 4e^- \rightarrow 2H_2O$	$2H_2 + O_2 \rightarrow 2H_2O$	150–200	55	Thermal energy consuming system, air conditioning system
Molten Carbonate Fuel Cell (MCFC)	Molten alkaline carbonate	CO_3^{2-}	$H_2 + CO_3^{2-} \rightarrow H_2O + CO_2 + 2e^-$	½ $O_2 + CO_2 + 2e^- \rightarrow CO_3^{2-}$	H_2 + ½ $O_2 + CO_2 \rightarrow H_2O + CO_2$	600–650	55	Combined heat and power for decentralized systems and transportation
Solid Oxide Fuel Cell (SOFC)	Solid ceramic electrolyte (Yttrium stabilized zirconia)	O^{2-}	$H_2 + O_2^- \rightarrow H_2O + 2e^-$	½ $O_2 + 2e^- \rightarrow O_2^-$	$H_2 + ½\ O_2 \rightarrow 2H_2O$	800–1000	55	Combined heat and power units, uninterruptible power systems (UPS), Primary power units

in the following section. Furthermore, the efficient working of FC by the integration of various nanocomposite materials for the anode, cathode, and hybrid membranes are summarized below.

2.1 *Nafion®- Metal Oxide-Based Nanocomposite*

Nafion® is a fluoropolymer which is comprised of sulfonated tetrafluoroethylene backbone and it has been extensively using in the development of FC especially the PEMFC due to its excellent proton conductivity. The proton on sulfonic acid can jump from one site to another through Grotthuss mechanism which makes them conduct protons easily and thus, prevents the electron conduction. However, the property of becoming dehydrated at high temperature, high cost and high fuel crossover of Nafion® made its applications in FC limited to a certain extent. Therefore, the modification of Nafion® with other nanomaterials to overcome these drawbacks has gained significant importance.

Recently, Nafion® modified with phosphonate and sulfonate silica nanoparticles (NIM_PO_3 and NIM_SO_3) for high proton conduction was reported [24]. The synthesized nanocomposite membrane (Fig. 4) revealed the excellent proton conducting property even at a low relative humidity and elevated temperature greater than 80 °C. The conductivity of sulfonate-based nanocomposites at 130 °C and 30% RH was 50 mS/cm. The NIM_SO_3 membranes also exhibited the promotion of water retention capacity by increasing the water uptake 32 (± 2) wt% and better mechanical stiffness even above the temperature of 200 °C.

Mohammadi et al. fabricated the Nafion® with metal oxide nanoparticles for PEMFC [25]. They recast the commercial Nafion® with 75 nm sized TiO_2/ZrO_2 nanoparticles by sol-gel and blending method respectively. Nafion®/ZrO_2 nanocomposite membrane offered good proton conduction and with an increase in the concentration of ZrO_2 and Nafion®/TiO_2 membrane displayed a better water retention capacity than Nafion® membranes modified by other conventional methods. Further, these membranes also unveiled the highest PEMFC performance

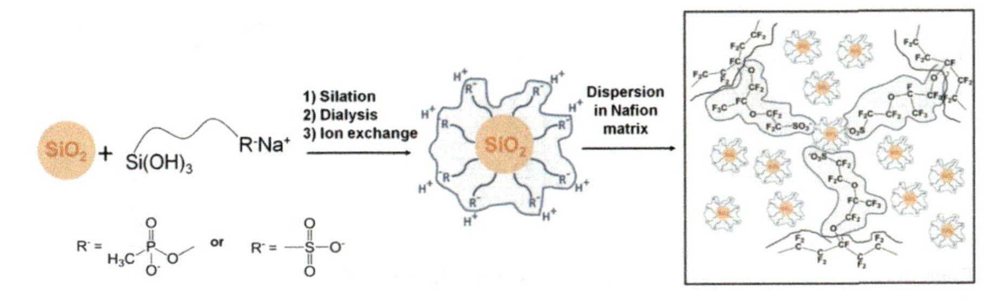

Fig. 4 Schematic representation of the synthesis of Nafion® membranes functionalized with phosphonate and sulfonate nanoparticles. Reproduced with permission from Ref. [24], Copyright 2016, Elsevier

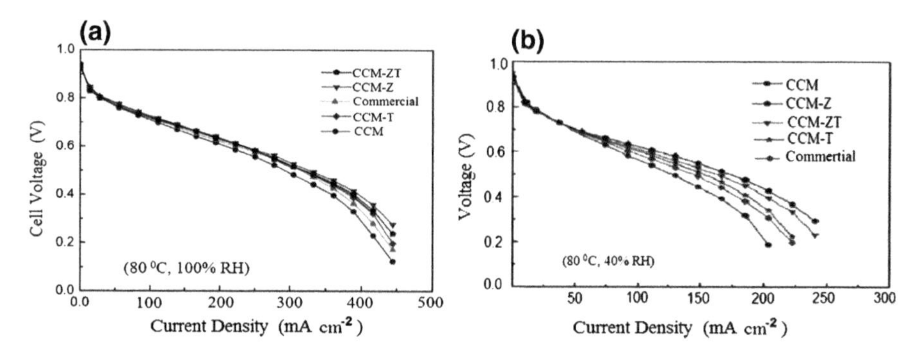

Fig. 5 Comparison of the polarization curves for modified composite membranes and Nafion® membranes at 110 °C, **a** 100% RH, **b** 30% RH. Reproduced with permission from Ref. [25], Copyright 2013, Elsevier

with respect to 1–5 polarization under 110 °C, 0.6 V and 30% RH at 1 atm (Fig. 5). Thus, they proved to be the best-modified membranes for PEMFC.

Integration of Fe_2TiO_5 in Nafion® membranes prepared in water, ethanol and water-ethanol solvent increased proton exchange capacity in FCs [26]. The comparison of the efficiency of membranes revealed that the modified membranes prepared in water solvents are superior to the membranes prepared in either of the solvents. This was due to the fact that water being a polar solvent led the nanoparticles into microscopic swelling and offered strong hydrogen bond whereas, the less polar solvent ethanol couldn't make that to happen. The proton conductivity of 226 mS/cm was obtained by the insertion of 2% Fe_2TiO_5 in commercial Nafion® membrane using water, ethanol and water-ethanol solvents. In addition, the membrane also showed a better water uptake capacity due to the hydrophilic nature of nanoparticles summarized (Table 2).

Table 2 Properties of solvent uptake and proton conduction in Nafion® and modified nanocomposite membranes at 25 °C (95% RH) and 110 °C (70% RH)

Sample code	Thickness (μm)	Solvent uptake (%)	Proton conductivity (mS/cm)	
			25 °C	110 °C
NH-2[a]	231	33	226	240
NHE-2[a]	226	25	87	93
NE-2[a]	240	35	72	80
N[a]	224	22	21	24
N[b]	230	29	6	–
N[c]	238	31	1	–

Reproduced with permission from Ref. [26], Copyright 2014, Elsevier
[a]Membranes stored in water solvent
[b]Membranes stored in a mixture of water-ethanol solvent
[c]Membranes stored in pure ethanol solvent

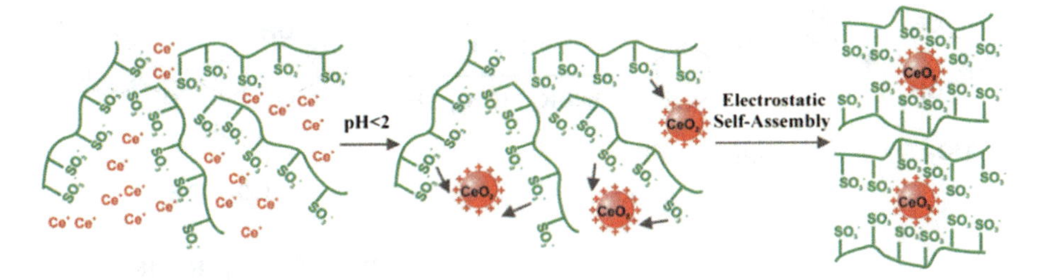

Fig. 6 Graphical representation of self-assembled Nafion®/CeO_2 nanocomposites. Reproduced with permission from Ref. [27], Copyright 2012, Elsevier

The Nafion®/CeO_2 membranes are proved to be the better nanocomposite materials for FC applications. Zhao and co-workers prepared a self-assembly of Nafion®/CeO_2 for electrolyte membrane between positively charged CeO_2 nanoparticle with negatively charged SO_3^- group of Nafion® as illustrated in Fig. 6 [27]. The hybrid nanocomposite membranes were displayed the superior proton exchange capacity and dimensional stability than pristine Nafion® membrane below the RH of 75% and low fluoride emission rate. In addition, the prepared nanocomposite membrane showed a very low fluoride emission rate of 43.05, 8.67, 6.01, and 4.47 mg/h for 1, 3, 5 and 10 wt% the CeO_2, respectively. However, the pristine Nafion® membrane showed 55.78 mg/h and 11.64 mg/h by the same Nafion® membrane synthesized by a sol-gel process. The material also exhibited the irreversible open reduction rate of 1.13×10^{-4} mV/s which was much lower than pristine Nafion® membrane (5.78×10^{-4} mV/s) and the Nafion® membrane prepared through sol-gel method (5.78×10^{-4} mV/s).

Cozzi and co-workers [28] modified the pristine Nafion® with propyl sulfonic (RSO_3H) acid on TiO_2 nanoparticles (TiO_2–RSO_3H) to achieve a higher efficiency in IEC and proton conductivity. The synthesis of TiO_2–RSO_3H is illustrated in Fig. 7. The covalently grafted hybrid materials as a nanocomposite with Nafion® promoted the efficiency of DMFC. The conductivity value of this material obtained was 80 mS/cm at 140 °C at the composition of 10 wt% TiO_2–RSO_3H in a single

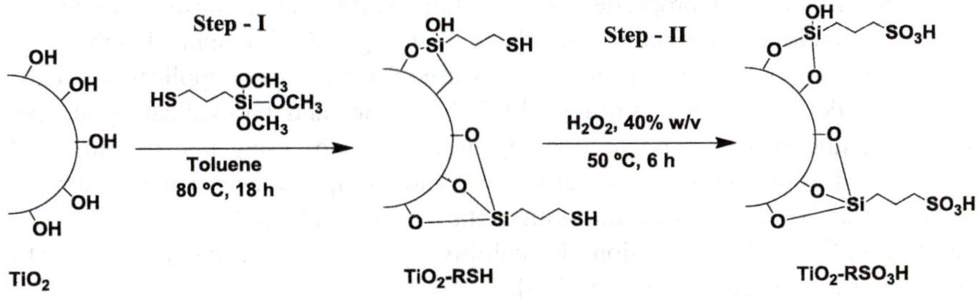

Fig. 7 Functionalization of TiO_2 nanoparticles with propyl sulfonic acid. Reproduced with permission from Ref. [28], Copyright 2014, Elsevier

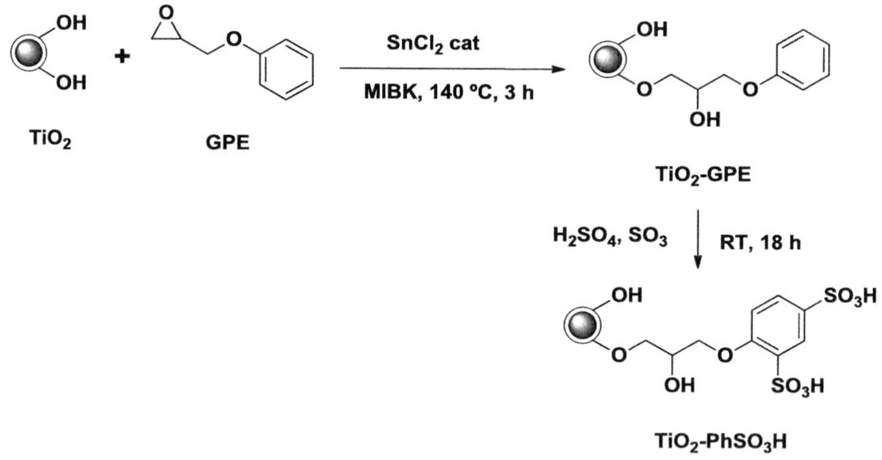

Fig. 8 Functionalization of TiO_2 nanoparticles with phenyl sulfonic acid. Reproduced with permission from Ref. [29], Copyright 2014, Elsevier

cell of DMFC. Also, they showed a supreme power density of 64 mW/cm^2 which is about 40% more than the pristine Nafion® composite membrane.

The organic functionalization of TiO_2 with phenyl sulfonic groups was able to overcome the less proton conduction in FC [29]. Further, the grafting of glycidyl phenyl ether group on the oxide surface was confirmed (Fig. 8) and the IEC was increased due to the covalently bound phenyl sulfonic group. The experimental observation showed that the prepared hybrid membranes reached the highest conductivity of 110 mS/cm at 140 °C with the concentration of 10 wt% TiO_2–$PhSO_3H$. The material also showed the better properties such as reduced methanol crossover (up to 20%) compared to the unfilled Nafion® membrane.

2.2 *Graphene-Based Nanocomposites*

Graphene has typical properties such as high surface area, surface active sites, electrical conductivity, excellent mechanical strength, high chemical stability and low metal loading capacity. Therefore, it is used for varieties of applications such as electronic devices, energy storage [30–32], sensors and biomedical applications [33]. In addition, the graphene is being used as a supporting material due to the presence of epoxy groups and carboxylic acid groups enhances the proton conducting capacity of the material with metal electrocatalysts for oxygen reduction reaction in FCs [34]. In addition, the stability of nano-catalysts can be increased by dispersing the metal on graphene [35].

The synthetic methods for the preparation of doped graphene and graphene supported nano electrocatalysts with respect to their structure-dependent properties and further developments were discussed by Liu et al. [36]. They have summarized the synthesis and characterization of graphene nanocomposites with various metals electrocatalysts for cathode and anode materials of FC. They elucidated the components into various types i.e. (1) graphene supported metal-free electrocatalysts for high oxygen reduction reaction (ORR) in acid and alkaline electrolytic medium. (2) graphene supported non-noble metals for efficient electrocatalysts (3) graphene-based Pt-free electrocatalysts and alloy nanomaterial for low-cost FC material (4) graphene-supported Pt-based nano-catalysts for increased ORR in electrolytes. These materials were found to participate in the electrooxidation of organic molecules at the anode of Direct Methanol Fuel Cells (DMFC) and ORR at the cathode. By the experimental observation, graphene or co-doped graphene with N, S, B, and P are found to be an excellent cost-effective cathode catalyst for FC applications.

Considering the advantages of graphene, Huang and co-workers developed a novel graphene nanoplate-Pt (GNP/Pt) composite electrocatalysts to obtain the high-performance DMFC [37]. They synthesized a series of graphene nanoplate (GNP/Pt), reduced graphene oxide nanoplate (RGO/Pt) and Vulcan XC-72 Pt (XC-72/Pt) with 0.17 mL of 0.45 M $Pt(NO_3)_2$ as demonstrated in Fig. 9. Further, its electrochemical activity was measured through cyclic voltammetry (CV) and the X-ray diffraction (XRD) patterns which revealed that the Pt was uniformly dispersed over the graphene nanoplates and confirmed the formation of an intact composite. This reduced the probable catalytic poisoning due to methanol oxidation and thus electrocatalytic activity was increased. The time required to increase the electrode potential was significantly decreased in the order of GNP/Pt ($\sim$130 s) followed by RGO/Pt (50 s) and XC-72/Pt ($\sim$30 s). Thus, they concluded that the GNP/Pt was superior for electrocatalytic activity than RGO/Pt and XC-72/Pt. Also, it is worth in mentioning that GNP can be used as the best catalyst supports for DMFC.

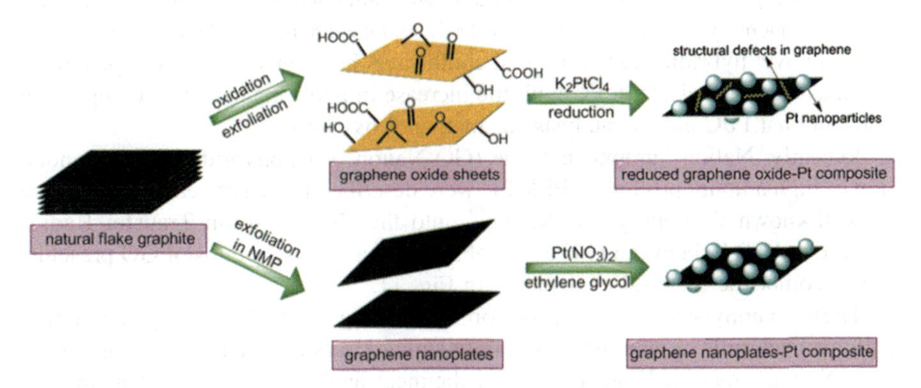

Fig. 9 Illustrations of the synthesis of reduced graphene oxide-Pt composite and graphene nanoplate-Pt composite by the traditional oxidation-reduction method and soft chemical method, respectively. Reproduced with permission from Ref. [37], Copyright 2012, Elsevier

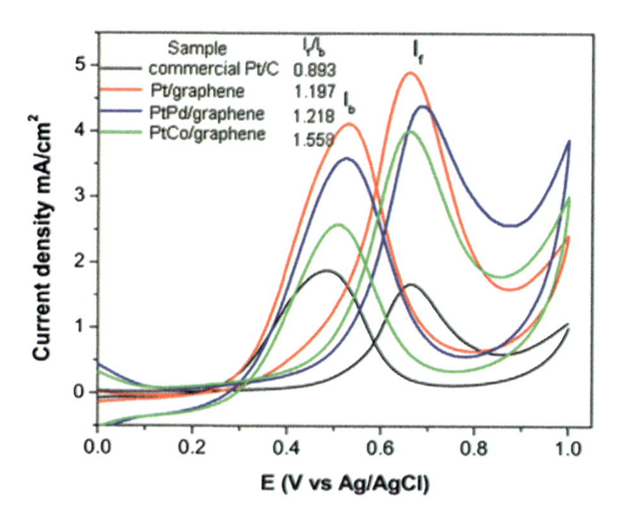

Fig. 10 A comparison of cyclic voltammograms in 0.5 M CH_3OH/0.5 M $HClO_4$ electrolyte at a scanning rate of 50 mV/s displayed by modified materials to evaluate the methanol oxidation and tolerance to CO poisoning. Reproduced with permission from Ref. [38], Copyright 2013, Elsevier

The application of exfoliated graphene-supported Pt and Pt-based alloys as electrocatalysts were elucidated to enhance the performance of DMFC [38]. A low cost and environment-friendly method of "Thermal Expansion and Liquid Exfoliation Solvothermal Reaction (TELESR)" was used to hybridize the exfoliated graphene sheets to load the Pt metal and alloys such as Pt/Pd, Pt/Co nano-clusters. This method improved the methanol oxidation in FC by increasing the electro-catalytic activity. The methanol oxidation was found to be $I_f/I_b = 1.218$ and 1.558 in PtPd/graphene and PtCo/graphene respectively during an electrochemical analysis. Also, the modified material showed high conductivity and tolerant to carbon monoxide poisoning (Pt/graphene, $I_f/I_b = 1.197$) compared to commercial Pt/C catalyst ($I_f/I_b = 0.893$) as shown in Fig. 10. This was due to the interaction of graphene with Pt electronic environment occurred and graphene played a major role in controlling the electronic environment with attached Pt atom which was confirmed by Density Functional Theory (DFT) studies. Therefore, the reported process gives a novel hybridized material that could save the extensive use of expensive Pt metal as electrocatalysts in FC with the increase in performance when compared to conventional Pt/C electrocatalysts for methanol oxidation.

Recently, Nafion®-graphene oxide (GO-Nafion®) nanohybrids for the conduction of high amount protons in PEMFC were described [39]. The reaction of a chain of well-known fluoropolymer, Nafion® onto the GO via Atom Transfer Radical Addition (ATRA) between C-F group of Nafion® and C=C groups of GO presented a nanocomposite material as depicted in Fig. 11.

These nanohybrids form the proton conducting fields by the aggregation of sulfonic acid units of Nafion® material which increases the interfacial compatibility with Nafion® matrix. From the electrochemical analysis, it was evident that the developed nanohybrids showed 1.6 folds high performance in conducting the protons compared to the commercially available Nafion® 112 membranes which

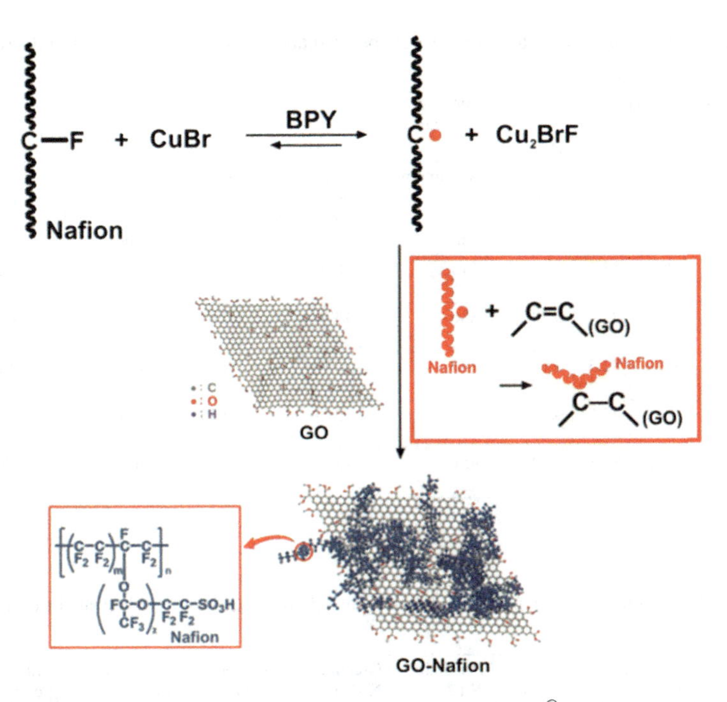

Fig. 11 Schematic representation of the reaction between Nafion® chains onto GO surfaces through an atom transfer radical addition reaction to prepare GO-Nafion® hybrid material. Reproduced with permission from Ref. [39], Copyright 2016, Elsevier

have been summarized in Table 3. Thus, a new variety of material which can be used as nano additive in the fabrication of Nafion® based nanocomposite for the PEM for FC was displayed.

A transition metal hierarchical porous N-doped graphene foams (HPGFs) were prepared by using silica nanoparticles as a template were reported by Zhou and co-workers [40]. The material exhibited the excellent property of ORR in 0.1 M KOH solution with high onset potential of 1.03 V and the limiting current of ~9 mA/cm^2 which was 1.7 times higher than the commercial Pt/C catalyst. Also, an excellent catalytic performance in acidic medium with an onset potential of 0.81 V and the limiting current was up to ~10 mA/cm^2 (Table 4) was observed. Further, the material also showed good methanol tolerance, long-term durability in both acidic and basic conditions. Such excellent material is a model for the development of applied energy systems such as FCs and metal-air batteries.

Nitrogen-doped graphene sheets prepared through plasma enhanced chemical vapour deposition (PECVD) method were studied as anode material for a Microbial Fuel Cell (MFC) [41]. The doping of nitrogen affected the electronic conductivity and catalytic activity due to the formation of structural defects. The material performed with excellent electrocatalytic activity towards glucose oxidation mediated via *Escherichia coli* due to the adjacent contact between microorganism and

Table 3 Proton conduction and single cell performance among various modified GO-Nafion® nanocomposite [39]

Membrane (with % of GO)	Proton conductivity (mS/cm)		Single cell tests					
			H_2/O_2			H_2/air		
	20 °C	95 °C	Open cell voltage (V)	Maximum power density (mW/cm^2)	Current density at 0.6 V (mS/cm^2)	Open cell voltage (V)	Maximum power density (mW/cm^2)	Current density at 0.6 V (mS/cm^2)
Recast Nafion®	25.8	53.3	0.97	713	1018	0.98	417	619
NM/GO-0.05	36.7	72.4	0.98	886	1376	0.94	586	826
NM/GO-0.10	40.8	82.3	0.98	743	1059	0.95	450	595
NM/GO-0.15	22.3	47.5	0.96	836	1215	0.95	509	652
Nafion® 212	40.8	88.3	0.99	951	1347	0.98	563	740

Table 4 Electrochemical properties exhibited by HPGFs and Pt/C catalyst at different physical parameters.

Catalysts	Surface area (m^2/g)	N content (%)	Pyridinic N (%)	Graphtic N (%)	Oxidized N (%)	Onset potential (V)		Limiting current density (mA/cm^2)	
						KOH	$HClO_4$	KOH	$HClO_4$
HPGF-1	918.7	3.15	37.9	51.9	10.2	1.03	0.81	9.08	9.90
HPGF-2	325.9	5.49	41.9	52.0	6.1	0.99	0.81	5.87	5.05
HPGF-3	567.7	3.09	36.8	50.6	12.6	0.97	0.80	7.33	5.27
Pt/C (20 wt%)	–	–	–	–	–	1.04	1.00	5.51	5.40

electrode (Fig. 12). The doped nanocomposite material showed the power density of 1008 mW/cm^2 at a current density of 6300 mA/m^2. This is a good metal free nanocomposite material which showed better performance in MFC.

2.3 *Carbon Nanotubes and Its Hybrid Nanocomposites*

Carbon nanotubes (CNT) are an important class of materials which significantly used in the field of material science, nanotechnology, optics, electronics, sensors, energy materials [42] due to their excellent mechanical, electrical and optical properties. Due to their light weight and high electrical conductivity, both Single-Walled Carbon Nanotubes (SWCNT) and Multi-Walled Carbon Nanotubes (MWCNT) [43] have obtained substantial consideration in FC application. They tend to increase the catalytic performance, steadiness, corrosion resistance, electron

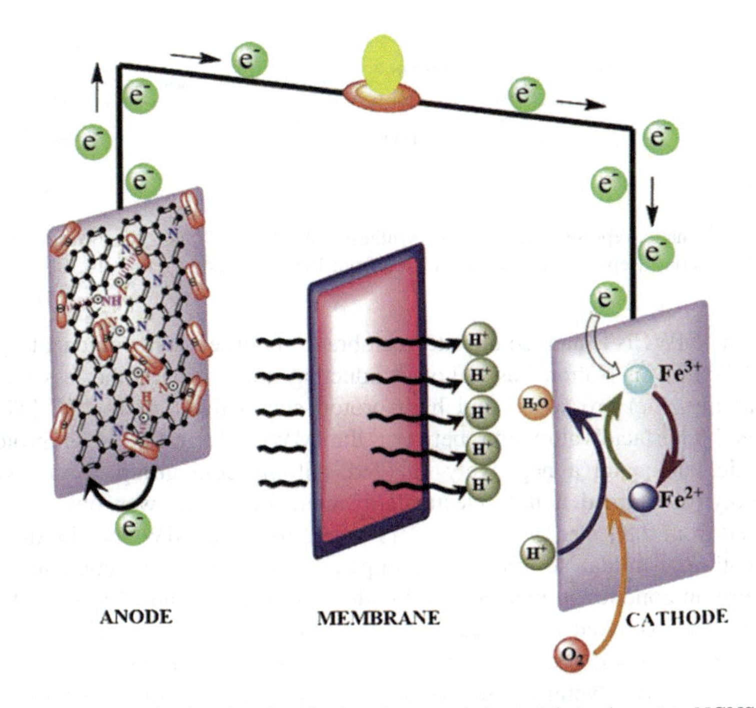

Fig. 12 Predicted mechanism involved in the electrogenic bacterial attachment to NGNS/cathode catalyst. Reproduced with permission from Ref. [41], Copyright 2015, Elsevier

transmission capacity and decreases the overall fuel cell cost [44]. Therefore, their usage as nanocomposite materials has extensively been studied in FC applications, especially in MFC.

CNT as an alternative cathode support and catalyst in MFCs was demonstrated by preparing a hybrid material containing CNT/Pt enriched with Palm Oil Mill Effluent. This material showed a better catalytic activity than undecorated Pt metal [45]. Incorporation of only 25% of Pt in the hybrid material reduced the use of precious Pt metal and thus helped in reducing the cost of FC. The material increased the MFC output voltage from 31.8 to 169.7 mW/m^2 at the chemical oxygen demand of 100 mg/L and 2000 mg/L, respectively which becomes a novel material as a catalyst in FC.

In 2014, Mehdinia et al. [46] compared the electrochemical performance of MFC using MWCNT-SnO_2/Glassy Carbon Electrode (GCE), MWCNT/GCE and bare GCE as anode material. From this experiment, it was concluded that the fabricated MWCNT-SnO_2/GCE (Fig. 13) produced a high electrochemical activity owing to the insertion of SnO_2 and high electron conductivity, high surface area properties of MWCNT into GCE. The power densities of MWCNT-SnO_2/(GCE), MWCNT/GCE, and bare GCE anode were found to be 1421, 699, 457 mW/m^2 respectively. Finally, the MWCNT-SnO_2/(GCE) nanocomposite represented as the best anode material for MFC due to its clean and green preparation and high electrochemical activity.

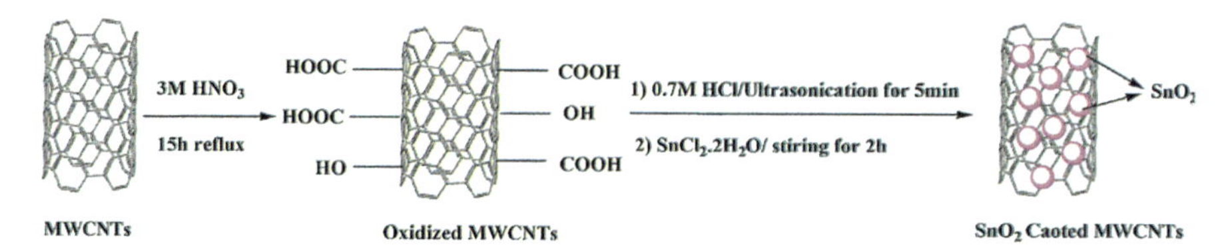

Fig. 13 Schematic representation of the synthesis of MWCNT/SnO_2 nanocomposite as an electrode material. Reproduced with permission from Ref. [46], Copyright 2014, Elsevier

A new MWCNT nanocomposite membrane modified with imidazole groups (MWCNT-Im) on Nafion® as proton conducting units showed the less methanol permeability, fuel crossover, and high proton conductivity in DMFC [47]. The combined interfacial attraction between the MWCNT grafted with protonated imidazole units, and a negatively charged sulfonic acid group of the Nafion® fluoropolymer formed a new electrostatic interaction. IEC was observed to be increased from 0.89 meq/g of Nafion® 117–0.92 meq/g of MWCNT-Im due to the participation of imidazole functional groups as new proton conduction sites added to the proton conduction mechanism. In addition, Nafion® modified with 0.5 wt% MWCNT-Im exhibited decreased methanol permeability of 1×10^{-6} cm^2/s with increasing in temperature by holding methanol molecules through a formation of complex structure. Overall, compared to neat Nafion®, the modified nanocomposite material showed excellent performance and accepted as a promising material in the application of FC.

In other work, a versatile metal-free catalyst for oxygen reduction reaction in FC was designed by Zhong et al. [48] They reformed the nitrogen, iron, and cobalt functionalized CNT (FeCoN-CNTs) with N-doped carbon foams (NCFs) with a 3D structure which provides a strong porous structure and large catalytically active sites. The composite material exhibited a synergetic effect of Fe/Co and the N species by forming the Fe/Co–N_x complex in the carbon material. In comparison with the commercial Pt/C catalyst, the newly fabricated material showed better performance in terms of resistance for fuel crossover and electrocatalysis in alkaline medium.

Recently, MWCNT functionalized with manganese oxide/polypyrrole (MWCNT-MnO_2/PPy) as an anode material in MFC was successfully demonstrated to produce the electricity from sewage water [49]. They electrochemically deposited the MWCNT-MnO_2/PPy on the surface of carbon cloth electrode as shown in Fig. 14. The fabricated electrode displayed electrical conductivity of 0.1185 S/m along with the band gap value of 0.8 eV and power density of 112.5.4 mW/m^2. Hence, it is a good example of MWCNT nanocomposite materials for generating the electricity from sewage source.

Mirzaei et al. [50] investigated a new catalyst support that made up of Pt/MWCNT nanocomposite for PEMFC. This hydrothermally prepared catalyst support material achieved more activity even after 4000 cycles, whereas the Pt/C catalyst showed no activity after 2000 cycles. The nanocomposite material revealed

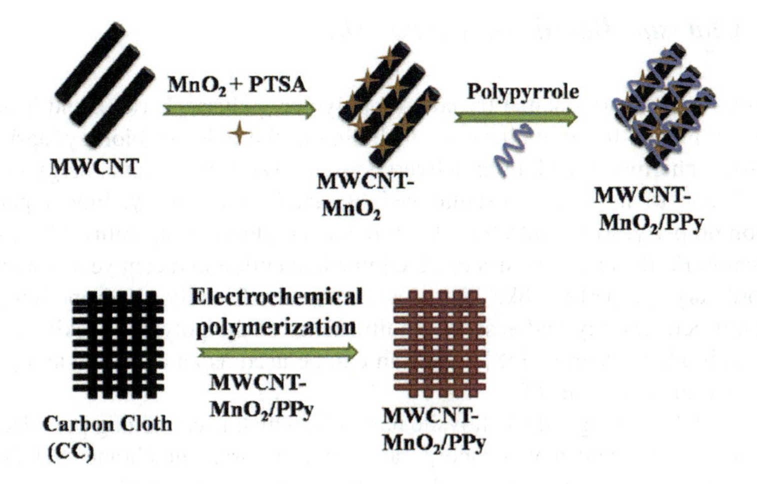

Fig. 14 Schematic representation of the synthesis of MWCNT-MnO_2/Ppy and proposed electrochemical polymerization on CC. Reproduced with permission from Ref. [49], Copyright 2016, Elsevier

Table 5 Comparison of the electrochemical and physical characteristics of Pt/MWCNTs and Pt/C.

Catalyst	Q_H (μc)	Specific surface area (m^2/g)	Particle size (nm)	Specific surface area (m^2/g) after 4000 cycle	OCV (V)
Pt/MWCNTs	0.178	36.46	7	5.37	+0.9
Pt/C	0.700	52.08	3.9	0.58	+0.96

Reproduced with permission from Ref. [50], Copyright 2017, Elsevier

Electrochemical Specific Surface Areas (ECSA) were initially 52.08 m^2/g was reduced to 0.58 m^2/g after 4000 cycles. Also showed good catalyst stability by achieving the ECSA value from 36.46 to 5.37 m^2/g (Table 5).

The HNO_3–H_2SO_4 functionalized CNT nanocomposite with MnO_2 demonstrated good catalyst for ORR in MFC maintained in neutral solution [51]. The ORR was increased by the unique interaction between MnO_2 and CNT which was fabricated through hydrothermal method. From the electrochemical measurements, it was observed that the MFC incorporated with the present material reached a power density of 520 mW/m^2 which is higher than the pristine CNT (275 mW/m^2) and MnO_2 with HNO_3–H_2SO_4 functionalized CNT (fCNT) (440 mW/m^2). Furthermore, the columbic efficiency was found to be 28.65% which was higher than the three mentioned material. Finally, it was concluded that the material is an excellent replacement for Pt/C catalyst material in MFC.

2.4 Chitosan-Based Nanocomposites

Chitosan is an environmental friendly biopolymer (polysaccharide), and hence was employed in significant applications including in the fields of biology, agriculture, industries, pharmaceutical, drug delivery system, dye removal and energy materials [52–57] due to its high molecular weight, antifungal activity, biocompatibility, gelation property, well-controlled structure and conduction capability. The usage of chitosan in FC development has received much attention in recent years owing to its extraordinary properties like that low cost, eco-friendly, hydrophilicity, low methanol permeability and ease of modifications of the polymer backbone of chitosan with other materials [58]. Chitosan can be used as an electrode and polymer electrolyte membrane in FC.

Bai et al. [59] designed a halloysite nanotube which is containing polyelectrolyte brushes (SHNTs) and it was incorporated into chitosan membrane. SHNTs generated a strong electrostatic interaction with the chitosan chain which improved the thermal and chemical properties by inhibiting the chain mobility. The nanocomposite membrane also overcome the problem of proton conduction in DMFC in an effective way by showing the highest conductivity of 18.6 mS/cm and IEC value of 0.204 mmol/g (Fig. 15) with an increase in the concentration of SHNT.

The phosphate and triphosphate salt complex membranes were inserted into chitosan membrane using chitosan hydrogel as electrode binder for increasing the proton conducting property in borohydride FCs [60]. This modified membrane was reached the highest power density of 685 mW/cm^2 at 60 °C which is almost 50% higher than the commercial Nafion® membrane. Also, the modified nanocomposite membrane showed a highest thermal stability of 200 °C as shown in Fig. 16.

A triple layer chitosan nanocomposite membrane having high efficiency in terms of power output, methanol permeability and proton exchange were demonstrated by Sadrabadi and group [61]. They coated two thin layers of chitosan on both the sides of Nafion® 105, chitosan acts as methanol barrier layer due to the presence of amino and hydroxyl group. Further, proton conductivity, methanol permeability, open circuit voltage measurements were proved too superior for multilayer Nafion®117 membranes with the thickness of 150–170 μm. In addition, power output was found to be 68.10 mW/cm^2 by feeding 5 M methanol which is 72% more than that of Nafion®117 membranes. Moreover, the lesser methanol crossover, ease of preparation and low cost will be the advantages to use as polyelectrolyte for DMFC (Table 6).

Chitosan was modified with polymeric reactive dyes which are containing quaternary ammonium groups (PRDQA) through blending followed by dyeing processes and studied for OH^- conduction in AFC [62]. The combined framework of CTS/PRDQA (1:0.5 by mass) nanocomposite membrane was exhibited an excellent OH^- conductivity of 8.17 mS/cm at room temperature (Table 7). The highest power density of 29.1 mW/cm^2 at a current density of 57.4 mA/cm^2 and open circuit voltage of 991.6 mV in an H_2/O_2 system was achieved which is appreciably better performance than pristine CTS membrane.

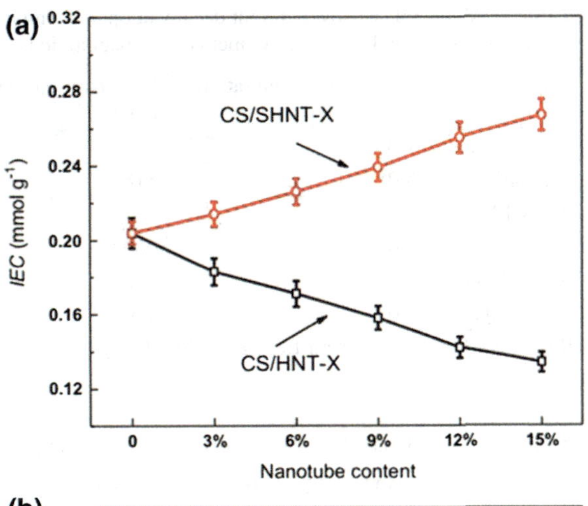

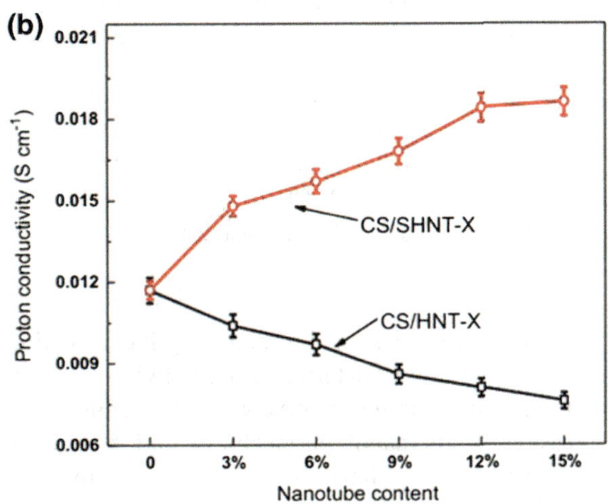

Fig. 15 A comparison of chitosan control and nanohybrid membranes at 25 °C at 100% RH in **a** IEC and **b** proton conductivity. Reproduced with permission from Ref. [59], Copyright 2014, Elsevier

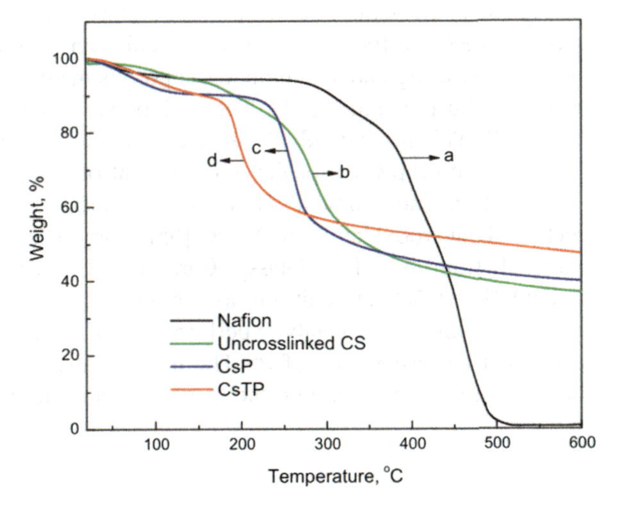

Fig. 16 TGA thermograms to demonstrate the stabilities of **a** Nafion®, **b** pristine chitosan, **c** chitosan modified with phosphate (CsP) and **d** chitosan modified with triphosphate (CsTP) membranes. Reproduced with permission from Ref. [60], Copyright 2012, Elsevier

Table 6 Methanol crossover current density at open circuit condition and limiting current density of various MEAs at 1 M and 5 M methanol concentrations at 70 °C

Sample	Methanol concentration (M)	Crossover current (mA/cm^2)	Limiting current (mA/cm^2)
Nafion®117	1.0	156	530
Nafion®117	5.0	518	260
CGS-12	1.0	142	445
CGS-12	5.0	460	285
Multi-layer	1.0	136	575
Multi-layer	5.0	420	385

Reproduced with permission from Ref. [61], Copyright 2012, Elsevier

Table 7 Physical and chemical properties of CTS/PRDQA membrane at room temperature

CTS/PRDQA (by mass)	WU (g/g)	Σ ($\times$ 10^{-3} S/cm)	Swelling ratio		IEC (meqiv./g)
			Δ S/S (%)	Δ V/V (%)	
1:0.125	0.82	2.15	41.65	81.17	0.41
1:0.25	0.73	3.92	33.93	73.89	0.65
1:05	0.67	8.17	24.44	65.42	1.08
1:0.75	0.62	9.09	21.52	49.69	1.24

Reproduced from Ref. [62], Copyright 2016, ESG

Li et al. [63] designed a matrix by incorporating a varying amount of chitosan nanoparticles into quaternized polyvinyl alcohol (QPVA) (Fig. 17) for FC applications. The modified matrix with 10% chitosan nanoparticle showed better suppression of methanol permeability and higher ionic conductivity than pristine QPVA. Also, they have prepared glutaraldehyde cross-linked nanocomposite film which has exhibited superior peak power density of 67 mW/cm^2 in DMFC.

Free-standing Chitosan/Phosphotungstic acid membranes for H_2O_2 FCs were reported recently [64]. To induce the chitosan crosslinking, the membrane was prepared by using anodic alumina as a porous medium to liberate the oxo-metallate anions as illustrated in Fig. 18. The peak power density of the membrane was found to be 350 mW/cm^2 and polyelectrolyte conductivity was 18 mS/cm which provided the excellent properties for the development of FC.

Pt-chitosan incorporated to $LaFeO_3$ nanoparticles with CNT has been studied for methanol electrooxidation in DMFC [65]. They prepared a layer by layer electrode material from modified Glassy Carbon (GC) with Pt nanoparticles (PtNPs), $LaFeO_3$NPs, CNT and chitosan as a binder. Integration of LaFeO3NPs and CNT material promoted the catalytic performance of the cell due to the presence of CNT decreased the dissolution of Pt. The method also reported that the methanol oxidation can be improved with the loading of a small amount of Pt.

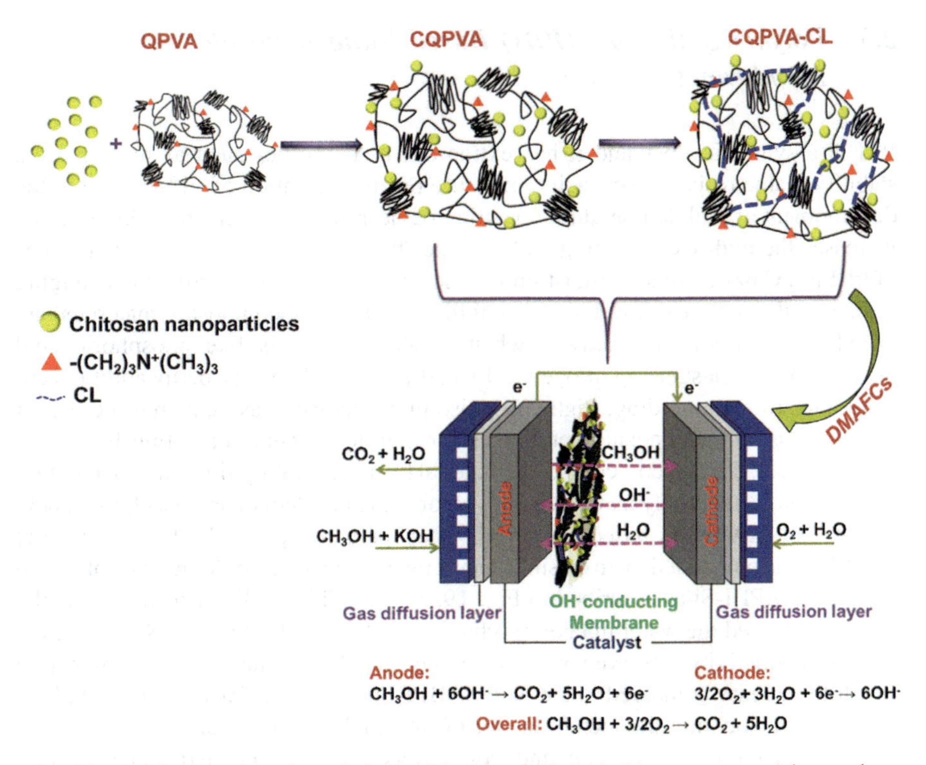

Fig. 17 Pictorial representation of glutaraldehyde cross-linked chitosan nanoparticle membrane for DMFC. Reproduced with permission from Ref. [63], Copyright 2016, Elsevier

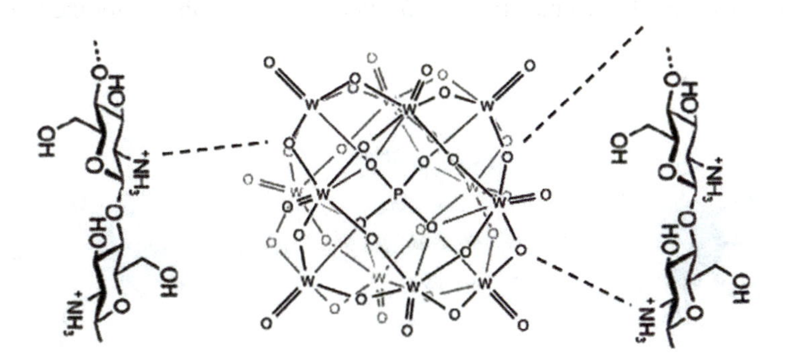

Fig. 18 Schematic representation of crosslinking the chitosan to the anionic species PTA^{3-}. Reproduced with permission from Ref. [64], Copyright 2015, Elsevier

2.5 Polybenzimidazole (PBI) Based Nanocomposite Membranes

PBI is a synthetic fiber and it has extraordinary properties such as high thermal stability, flame resistance capacity, moisture regains, retention of stiffness led to the development of high temperature operating devices. The close chain packing in PBI increases the hydrogen bonding and provides the rigidity to the materials. The use of PBI in FC has obtained utmost importance because of its high proton exchanging capacity, thermal and mechanical stabilities. PBI nanocomposite materials are excellent in proton conductivity when doped with acids like phosphoric acid because of its plasticizing property. In addition, the low gas permeability, zero water-electro osmotic drag, high CO poisoning tolerance has delivered the use of PBI units as nanocomposite membranes for high-temperature operating FCs.

Linlin et al. developed a composite material by incorporating silica in PBI to obtain high proton conductivity as well as to overcome the problem of acid leaching in acid doped PBI (dABPBI) based FCs [66]. They prepared poly (2,5-benzimidazole) (ABPBI) and embedded them on silica membrane in a methanesulfonic acid solvent to fabricate ABPBI-Si as described in Fig. 19. This modified sulfonated silica-ABPBI matrix improved the water uptake, mechanical and thermal stabilities. Similarly, the proton conductivity observed was 38 mS/cm at 140 °C and 1% RH which is a two-fold increase compared to the bare dABPBI membrane. Thus, it proved to be a promising nanocomposite membrane for FC at a higher temperature.

In a related work, Suryani and co-workers [67] incorporated silica on PBI membrane using N-(p-carboxyphenyl)maleimide (pCPM). This functionalized silica (SNP-pCPM) increased the interfacial compatibility between PBI and silica nanoparticles (SNPs) in PBI/SNP as described in Fig. 20 which created a proton conducting channel. This material was further used as nanofillers for the preparation

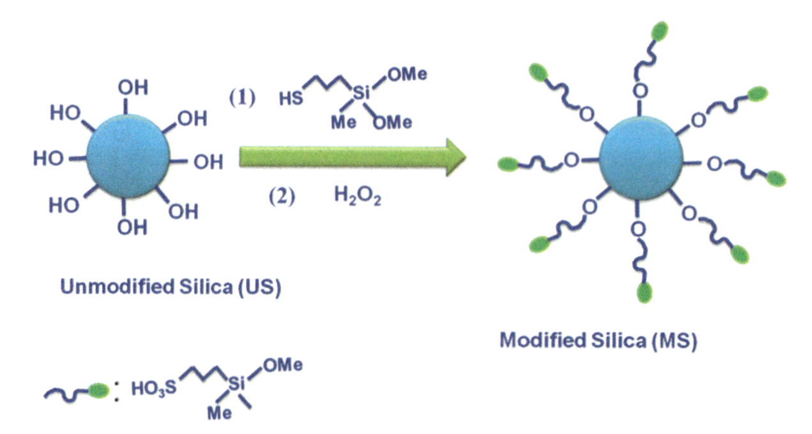

Fig. 19 Schematic representation of silica nanoparticles with methanesulfonic acid. Reproduced with permission from Ref. [66], Copyright 2012, Elsevier

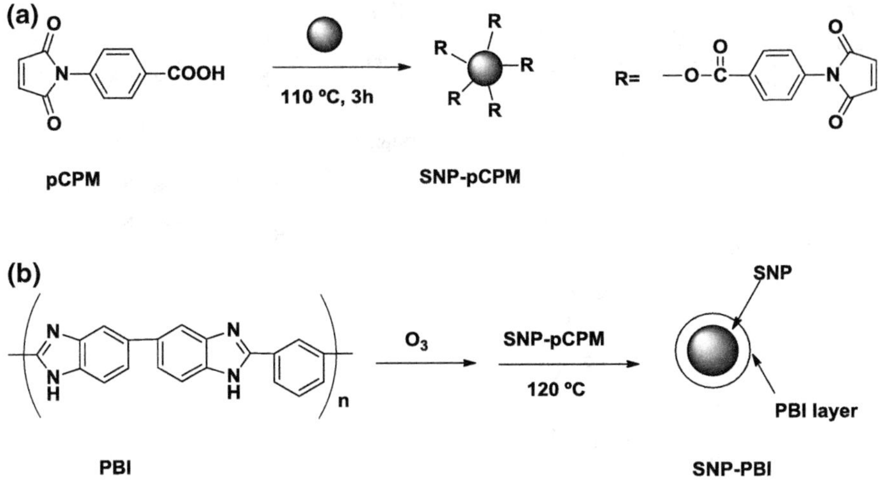

Fig. 20 **a** Functionalization of pCPM with silica nanoparticles **b** PBI functionalization with silica modified pCPM for proton conducting electrolyte membrane in FCs

of PBI/SNP-PBI membrane. With this modification, FCs displayed an excellent proton conductivity of 50 mS/cm at 160 °C by the use of 10 wt% composition of SNP-PBI. Also, a high power density of 650 mW/cm^2 was achieved in a single cell test which is higher than that of pristine PBI membrane.

The proton conductivity of PBI membrane was increased by functionalizing with Barium Zirconate ($BaZrO_3$) [68]. This modification increased the proton conduction to 125 mS/cm at 180 °C and 5% RH. At the same condition, the power density of 56 mW/cm^2 and a current density of 1120 mA/cm^2 were attained by the use of 4 wt% $BaZrO_3$ as nanofillers in PBI-$BaZrO_3$ (PBZ) membrane. This material proved to be a promising polyelectrolyte in FC membranes to improve proton conduction.

PBI/SiO_2 nanocomposite membranes were used to enhance the proton conduction membranes in PEMFC [69]. 5 wt% of SiO_2 as inorganic nanofiller was cast with PBI in dimethylacetamide solvent. The presence of SiO_2 facilitated proton conduction and acid retention properties. The nanohybrid membrane attained a conductivity of 102.7 mS/cm at 180 °C. Further, maximum cell voltage attained by this nanohybrid was found to be 0.6 V as 240 mA/cm^2.

A new material consisting of PBI functionalized with CNT was found to be very useful in FC applications. Wu et al. [70] doped this PBI/CNT and PBI membranes with KOH solution to make it more hydroxide conductive in ADMFC. They supplied the FC with 2 M methanol in 6 M KOH as anode fuel and humidified oxygen as oxidant at the cathode. This system attained a power density of 104.7 mW/cm^2 at 90 °C. In addition to this, thermal stability was enhanced and methanol permeability was reduced after the loading of 0.05–1% PBI functionalized CNT into PBI matrix. Therefore, it is considered to be a potential nanocomposite membrane for ADMFC.

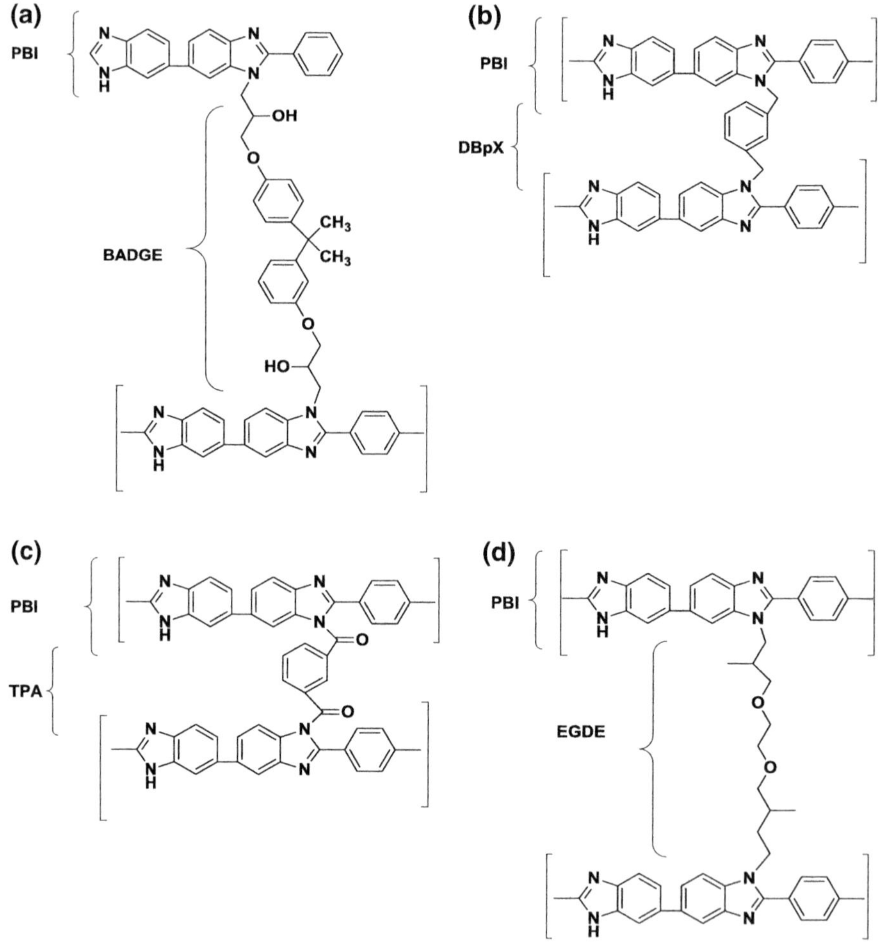

Fig. 21 Mechanism showing the crosslinking of **a** PBI/BADGE, **b** PBI/DBpX, **c** PBI/TPA and **d** PBI/EGDE. Reproduced with permission from Ref. [71], Copyright 2017 Elsevier

Recently, Özdemir et al. [71] studied the efficiency of PBI membranes fabricated with various cross-linkers for PEMFC. They cross-linked the PBI membrane with bisphenol A diglycidyl ether (BADGE), ethylene glycol diglycidyl ether (EGDE), α-α′-dibromo-p-xylene (DBpX) and terephthaldehyde (TPA) as illustrated in Fig. 21. From the electrochemical analysis, it was evident that PBI/BADGE nanocomposite membrane presented the superior acid retention property, however poor proton conduction. On the other hand, PBI/DBpX membrane was reached the maximum proton conductivity of 151 mS/cm at 180 °C. Meanwhile, PBI/BADGE composite materials exhibited maximum power density of 0.123 W. Therefore, they concluded that the PBI/BADGE and PBI/DBpX nanocomposite membranes are well suited for the polyelectrolyte membranes in FCs.

2.6 Poly (ether ether ketone) Based Nanocomposite

Poly (ether ether ketone) (PEEK) are the organic thermoplastics which have been used extensively for the modification of PEM in FC due to their high thermal and mechanical stabilities. High proton conductivity was noticed when they are cast in organic solvents as they permit the direct electrophilic sulfonation [72]. There are few reports on the preparation of sulfonated PEEK (SPEEK) membrane to increase the efficiency of PEM in FC [73]. However, there are some limitations like methanol crossover and less stability in FC caused by the high sulfonation in these membranes. Therefore, the modifications of SPEEK membranes require good maintenance and skills in preparation methods.

Modified PEM was prepared by incorporating SPEEK in poly (ether sulfone) (PES) in MFC at different concentration [74]. The conductivity of PES membranes was increased with 3–5% of SPEEK addition as summarized in Table 8. At 5% of SPEEK addition, the membrane achieved a high power density of 17,000 mW/cm^2. Further, the conductivity was reduced from 0.000615 to 0.0693 mS/cm and capacitance was reduced from 3.0×10^{-7} to 1.56×10^{-3} F. Such performance of nanocomposite materials are due to the sulfonated groups present which enhance the hydrophilic nature of the SPEEK membrane. Eventually, the modified PES membranes were considered as the best replacing PEM material to Nafion® 112 and 117 for MFC applications.

Later in 2013, the functionalization of SPEEK membranes with organically functionalized GO (SSi-GO) for DMFC applications was reported [75]. The synthesized material made up of SPEEK and sulfonated GO with further oxidation was carried out as shown in Fig. 22.

Integration of this material into the membrane increased the proton conductivity and lower the methanol permeability when compared to pristine Nafion® 112 and Nafion® 115 membranes (Table 9). Similarly, the condensed matrix of SSi-GO membrane established high mechanical and chemical stabilities.

The incorporation of sulfonated GO (s-GO) in SPEEK was investigated for the efficient FC functions [76]. The embedded SPEEK/s-GO composite membrane

Table 8 The modified nanocomposites with high performance to increase the efficiency of MFC

Type of membrane	Max. power density (mW/m)	Current density (mA m^{-2})	COD removal (%)	E_c (%)
N-112	0.003	0.124	55 ± 2	1.0 ± 0.4
N-117	3.630	9.842	60 ± 2	14.9 ± 5.5
PES	0.030	0.011	64 ± 4	0.5 ± 0.2
PES/SPEEK 3%	0.065	0.181	66 ± 6	5.4 ± 1.4
PES/SPEEK 5%	6.665	17.527	68 ± 6	26.3 ± 13.3

Reproduced with permission from Ref. [74] Copyright 2012 Elsevier

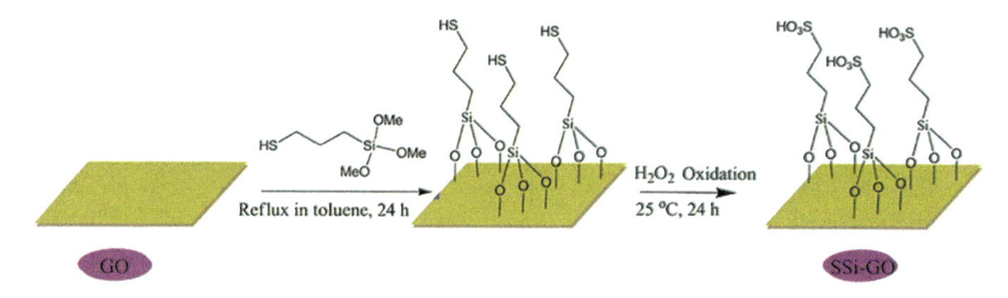

Fig. 22 Schematic illustration for the synthetic route of SSi-GO membrane. Reproduced with permission from Ref. [75], Copyright 2013, Elsevier

Table 9 Comparison of the efficiency of SPEEK membrane and modified SPEEK membranes [75]

Membrane	IEC (meq/g)	WU (%)	SW (%)	σ (mS/cm)	P (× 10^6 cm^2/s)
SPEEK	1.710	40.1	15.5	88.1	1.15
SPEEK/GO (5 wt%)	1.704	33.1	13.8	98.3	0.59
SPEEK/SSi-GO (3 wt%)	1.792	38.5	16.2	146.7	0.72
SPEEK/SSi-GO (5 wt%)	1.864	49.9	17.0	160.2	0.83
SPEEK/SSi-GO (8 wt%)	1.872	50.9	18.3	162.6	1.36
Nafion® 112	0.941	37.1	13.9	125.4	1.53

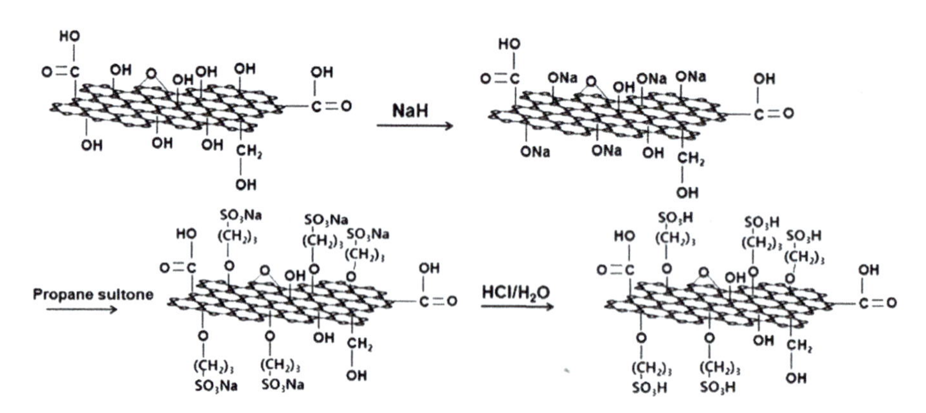

Fig. 23 A synthetic pathway for the sulfonation of GO to improve proton conduction. Reproduced with permission from Ref. [76], Copyright 2013, Elsevier

exhibited the improved proton conductivity of 8.41 mS/cm owing to the presence of high amount of sulfonic acid groups on the membrane (Fig. 23). The increased mechanical property of the membrane was attributed to the presence of strong hydrogen bonds between the s-GO and SPEEK. With the increase of GO content in the membrane, the elastic modulus and intrinsic strength were also increased. Thus, lower methanol permeability of 2.6388×10^{-7} cm^2/s was attained and therefore, it is considered to be the best nanocomposite material for the applications in DMFC.

Wang et al. prepared a novel anhydrous membrane for PEM by embedding the dopamine-modified silica nanoparticles ($DSiO_2$) into SPEEK polymer [77]. The enhancement of thermal and mechanical stabilities are due to uniformly dispersed $DSiO_2$ in SPEEK which is increased interfacial electrostatic attraction. There was an occurrence of high proton conduction at the acid-base site of SPEEK/$DSiO_2$ via Grotthuss mechanism due to the formation of small aggregation of sulfonic acid ionic channels. By the experimentation, it was revealed that the prepared hybrid membrane with 15% $DSiO_2$ achieved a conductivity of 4.52 mS/cm at 120 °C under anhydrous conditions. The membrane also attained a power density of 111.7 mW/cm^2 and open cell potential of 0.98 V which is considered to be the superior performance in PEM under anhydrous conditions.

Incorporation of novel nanomaterial as graphitic carbon nitride (g-C_3N_4) sheets into SPEEK membrane illustrated in Fig. 24 offers the superior efficiency compared to the pristine SPEEK membrane [78]. The fabricated SPEEK membrane with 5 wt % of g-C_3N_4, proton conductivity up to 786 mS/cm at 20 °C was achieved. As previously mentioned, the Grotthuss effect played a major role in the proton conduction due to the presence of acid-base pair at the composite site. Further, it favoured the resistance of methanol crossover and maintained the power density up to 39% due to the periodic vacancies in the lattice of the g-C_3N_4 membrane. Therefore, this is accepted to be the best nanocomposite materials for FC applications.

Sulfated zirconia nanoparticles with SPEEK modified nanocomposites (Fig. 25) were found to be the best composite materials for FCs [79]. The combination of

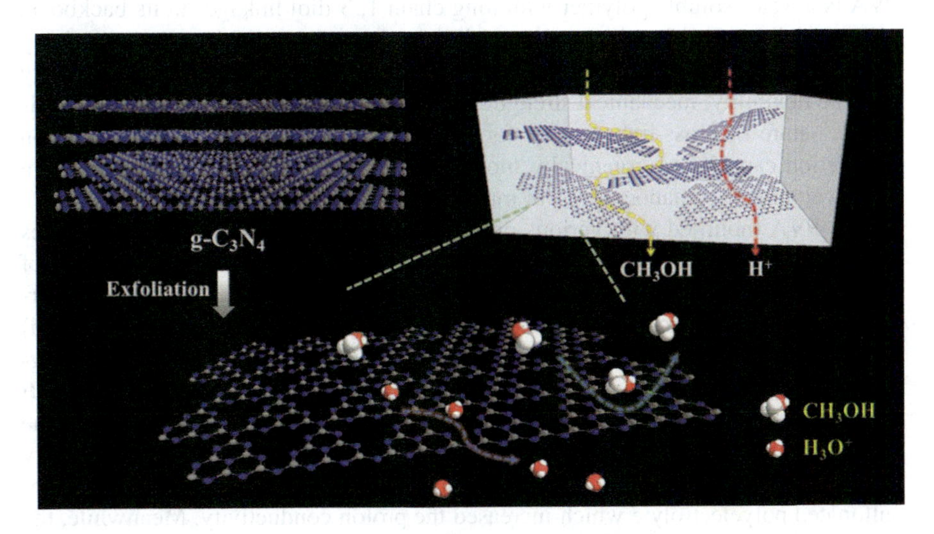

Fig. 24 Schematic illustration of the incorporation of g-C_3N_4 into SPEEK nanocomposite membrane. Reproduced with permission from Ref. [78], Copyright 2016, Elsevier

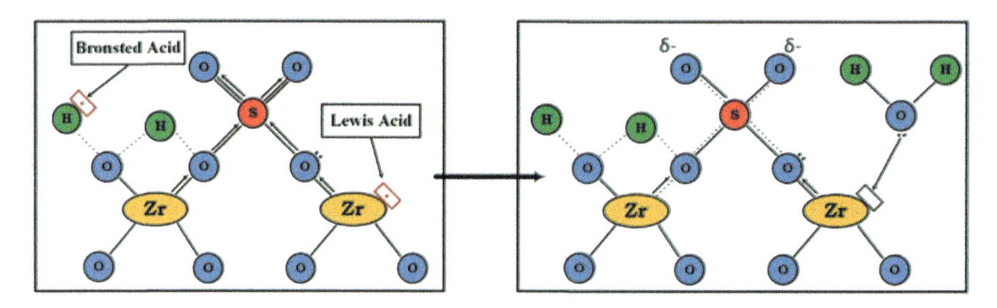

Fig. 25 Representation of the surface modification of SPEEK with sulfated zirconia nanoparticles. Reproduced with permission from Ref. [79], Copyright 2016, Elsevier

sulfonation time (6.9 h) and inorganic additives (5.94 wt%) produced the superior membranes. The proton conductivity observed was 3.88 mS/cm and oxidative stability of 102 min. In addition, the membrane also obtained the mechanical strength of 898 MPa.

SPEEK as an electrode ionomer material showed good catalytic activity for PEMFC [80]. The electrodes were prepared with 15, 25, 35 wt% of SPEEK and Pt content of 0.3, 0.4 and 0.5 mg/cm^2 were investigated for electrochemical reactions. The experiments demonstrated that the 25% of ionomer with 0.5 mg/cm^2 was found to be suitable for cathode catalyst for PEMFC.

2.7 Polyvinyl Alcohol (PVA) Based Nanocomposites

PVA is a water-soluble polymer with long chain 1, 3-diol linkages in its backbone. It has excellent film forming and emulsifying properties and hence is extensively used in membrane technology, adhesive materials development, and drug delivery etc. The flexibility, mechanical strength, thermal stability of PVA materials depends on the nature of cross-linkers used for material preparation. The excellent proton conduction capacity of organically modified PVA material attained importance in the development of nanocomposite materials in FC material applications.

The PVA modified with various poly(sulfonic acid) grafted silica nanoparticles were studied by Salarizadeh and co-workers [81]. First, they altered the surface of nanoparticles by APTES followed by preparing a nanocomposite membrane by various concentration from 0 to 20% of SNP with poly (styrene sulfonic acid) (PSSA-G-SNP), poly (2-acrylamido-2methyl-1-propane sulfonic acid) (PAMPS-g-SN) by using glutardialdehyde as a cross-linking agent as demonstrated in Fig. 26. It was observed that the membranes consisting of 5 wt% of PAMPS-g-SNP exhibited the better performance for FCs applications. This material showed the proton conductivity of 10.4 mS/cm by signifying that sulfonation of silica nanoparticles by grafting of sulfonated polyelectrolyte which increased the proton conductivity. Meanwhile, the

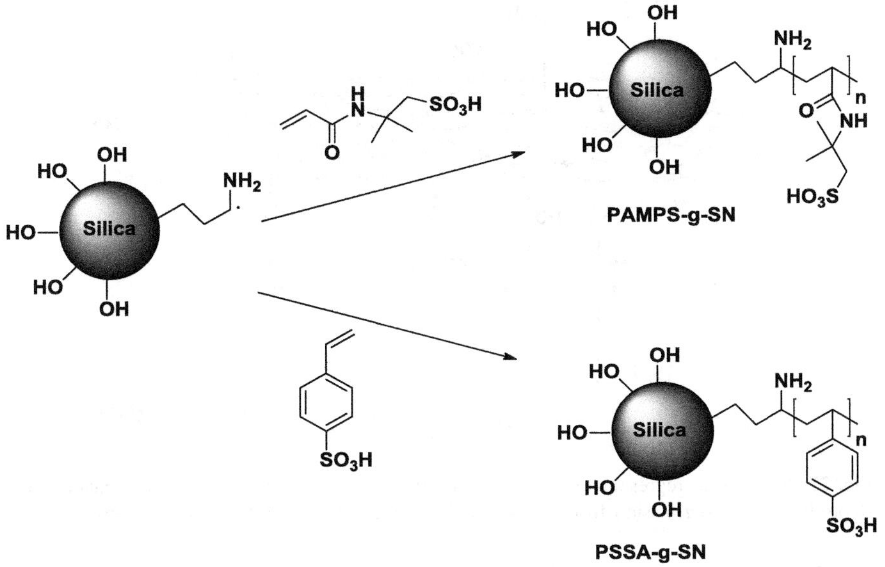

Fig. 26 Mechanism of surface-initiated free radical polymerization of AMPS and SSA onto APTES modified silica nanoparticles initiated by Ce(IV)-based redox initiation system. Reproduced with permission from Ref. [81], Copyright 2013, Elsevier

water uptake capacity was also increased for the same concentration, then diminished rapidly. Therefore, it was concluded that the prepared membrane can be effectively used for high proton conduction in FC.

A new variety of PEM for FCs prepared by PVA and the ionic liquid was reported by Liew et al. [82]. The prepared nanocomposite membrane with PVA/ammonium acetate/1-butyl-3-methylimidazolium chloride (BmImCl) reveals that increased ionic conductivity with ionic liquid mass loading. After the addition of 50 wt% of BmImCl, the PEM reached the highest ionic conductivity 5.74 ± 0.01 mS/cm. This was due to the fact that the addition of ionic liquid enhanced the plasticizing effect to the membrane. In addition, the nanocomposites also exhibited the thermal stability of up to 250 °C and achieved a maximum power density of 18 mW/cm^2 at room temperature with an operational current of 750 mA.

In 2014, a nanocomposite membrane was prepared by cross-linking of PVA with aryl sulfonated graphene oxide (SGO) for versatile applications in FC material [83]. The modification of the GO surface with aryl diazonium salt of the sulfanilic acid by further introducing 5 wt% SGO into PVA matrix by using glutaraldehyde as a cross-linking agent. The hybrid membrane showed high thermal stability with melting temperature of 223 °C, the tensile strength of 67.8 MPa and proton conductivity of 50 mS/cm. The increased proton conductivity owing to the interaction between—SO_3H of SGO and –OH group of GO via Grotthuss mechanism and Vehicle mechanism was explained in Fig. 27. Furthermore, the power density of 16.15 mW/cm^2 was attained. Overall, the modified PVA/SGO nanocomposite

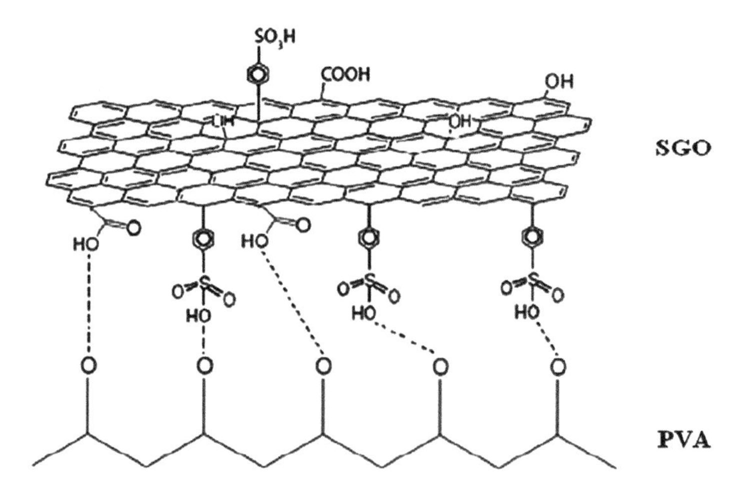

Fig. 27 Schematic representation of the reaction between PVA and SGO nanoparticles. Reproduced with permission from Copyright 2014, [83], American Chemical Society

membrane showed higher efficiency compared to unmodified GO and Nafion® membranes in terms of thermal stability and proton conductivity.

In other work, a novel hybrid membrane based on the infiltration of a blend of SPEEK with PVA in water and SPEEK blended with polyvinyl butyral (PVB) into electrospun nanofibres were reported [84]. A covering of hydrated SPEEK-30% PVB showed better proton conductivity. Further, the nanofibre incorporation into SPEEK-PVA matrix improved the mechanical stability, and methanol barrier capacity as explained in Table 10. Eventually, the prepared membrane was found useful for the applications in DMFC.

Yang et al. [85] prepared a blended membrane consisting of PVA and sodium alginate (PVASA) polymers of various concentration for alkaline solid PEM. This was further treated with glutaraldehyde to form a cross-linked membrane of PVASA (PVASA-GA). From the electrochemical measurements, it was observed that the ionic conductivity was found to be 91 mS/cm at 25 °C for the membrane containing 60:40 of PVA and SA. In addition, the methanol permeability was achieved in the order of 10^{-7} cm^2/s. On the other hand, the membrane prepared with 80:20 of PVA and SA cross-linked with GA up to 60 min, exhibited the selectivity value of 21.50×10^3 S/cm^3s^1. Also, the PVASA with 80:20 composition showed a maximum power density of 20.7 mW/cm^2 which was achieved at $E_{p.max}$ with a peak current density of 89.20 mA/cm^2 at 30 °C. Therefore, the membrane was found to be extensively used in DMFC application.

Recently, the application of PVA modified with functionalized CNT (m-CNT) in ADEFC was reported [86]. Addition of m-CNT into PVA membrane increased the alkaline uptake and ionic conductivity of KOH doped electrolyte membrane. At the same time, a lower swelling ratio and ethanol permeability were suppressed in comparison with pristine PVA membrane. Also, the highest power density of 65 mW/cm^2 at 60 °C was achieved.

Table 10 Comparison of the performance of nanocomposite and SPEEK-35% PVA

Membrane	Crosslinking temperature (°C)	Metahnol permeability(cm^2/s)	Proton conductivity(S/cm)	Modified characteristic factor (S^2 s cm^{-4})
SPEEK-35% PVA	110	$5.81 \pm 0.20 \times 10^{-6}$	$1.11 \pm 0.08) \cdot 10^{-2}$	21.2 ± 3.2
Nanocomposite	110	$4.43 \pm 0.21 \times 10^{-6}$	$1.35 \pm 0.11) \cdot 10^{-2}$	41.1 ± 5.8
SPEEK-35% PVA	120	$4.70 \pm 0.13 \times 10^{-6}$	$1.10 \pm 0.05) \cdot 10^{-2}$	25.7 ± 1.9
Nanocomposite	120	$3.82 \pm 0.18 \times 10^{-6}$	$1.03 \pm 0.08) \cdot 10^{-2}$	27.8 ± 3.6
SPEEK-35% PVA	130	$2.18 \pm 0.07 \times 10^{-6}$	$5.84 \pm 0.32) \cdot 10^{-3}$	15.6 ± 1.5
Nanocomposite	130	$2.02 \pm 0.11 \times 10^{-6}$	$2.50 \pm 0.18) \cdot 10^{-3}$	3.1 ± 0.3
SPEEK-35% PVA	140	$1.19 \pm 0.06 \times 10^{-6}$	$3.53 \pm 0.13) \cdot 10^{-3}$	10.5 ± 0.3
Nanocomposite	140	$1.34 \pm 0.09 \times 10^{-6}$	$1.63 \pm 0.10) \cdot 10^{-3}$	2.0 ± 0.1
Nafion® 115	_	$3.71 \pm 0.05 \times 10^{-6}$	$3.64 \pm 0.11) \cdot 10^{-2}$	357 ± 24

3 Conclusion

In this chapter, we summarized the current scenario of FC research and development around the world. The modification of composites with various nanomaterials is the developing technology to overcome the problems such as fuel crossover, low proton conduction, expensive raw materials, higher cost, and low membrane stability. The advantages of hybrid nanocomposites such as Nafion®, CNT, GO, SPEEK, PBI, PVA in FCs are comprehensively discussed. A variety of improved materials as anode, cathode, and membranes are elucidated. The improved efficiency of FCs in terms of high proton conduction, better water retention, suppression of fuel crossover, high chemical and thermal stabilities are precisely explained.

Acknowledgements The authors acknowledge the financial support from DST Nanomission, India (SR/NM/NS-20/2014), DST, India (DST-TM-WTI-2K14-213) and SERB-DST, India (YSS/2015/000013) for financial support. We also thank Jain University, India for providing facilities.

References

1. Hajilary N, Shahi A, Rezakazemi M (2018) Evaluation of socio-economic factors on CO_2 emissions in Iran: factorial design and multivariable methods. J Clean Prod 189:108–115
2. Hashemi F, Rowshanzamir S, Rezakazemi M (2012) CFD simulation of PEM fuel cell performance: effect of straight and serpentine flow fields. Math Comput Model 55(3–4):1540–1557
3. Winter M, Brodd RJ (2004) What are batteries, fuel cells, and supercapacitors? Chem Rev 104:4245–4269
4. Lemons RA (1990) Fuel cells for transportation. J Power Sources 29(1–2):251–264
5. Peighambardoust S, Rowshanzamir S, Amjadi M (2010) Review of the proton exchange membranes for fuel cell applications. Int J Hydrogen Energy 35(17):9349–9384
6. Giorgi L, Leccese F (2013) Fuel cells: technologies and applications. Open Fuel Cells J 6: 1–20
7. Agmon N (1995) The grotthuss mechanism. Chem Phys Lett 244(5):456–462
8. Ueki T, Watanabe M (2008) Macromolecules in ionic liquids: progress, challenges, and opportunities. Macromolecules 41(11):3739–3749
9. Rezakazemi M, Amooghin AE, Montazer Rashmati MM, Ismail AF, Matsuura T (2014) State-of-the-art membrane based CO_2 separation using mixed matrix membranes (MMMs): an overview on current status and future directions. Prog Polym Sci 39(5):817–861
10. Rezakazemi M, Razavi S, Mohammadi T, Nazari GA (2011) Simulation and determination of optimum conditions of pervaporative dehydration of isopropanol process using synthesized PVA–APTEOS/TEOS nanocomposite membranes by means of expert systems. J Membr Sci 379(1–2):224–232
11. Dashti A, Harami HR, Rezakazemi M (2018) Accurate prediction of solubility of gases within H_2-selective nanocomposite membranes using committee machine intelligent system. Int J Hydrogen Energy 43(13):6614–6624
12. Rezakazemi M, Vatani A, Mohammadi T (2016) Synthesis and gas transport properties of crosslinked poly(dimethylsiloxane) nanocomposite membranes using octatrimethylsiloxy POSS nanoparticles. J Nat Gas Sci Eng 30:10–18
13. Rezakazemi M, Vatani A, Mohammadi T (2015) Synergistic interactions between POSS and fumed silica and their effect on the properties of crosslinked PDMS nanocomposite membranes. RSC Adv 5(100):82460–82470
14. Rezakazemi M, Sadrzadeh M, Mohammadi T, Matsuura T (2017) Methods for the preparation of organic–inorganic nanocomposite polymer electrolyte membranes for fuel cells. In: Inamuddin D, Mohammad A, Asiri AM (eds) Organic-inorganic composite polymer electrolyte membranes. Springer International Publishing, Cham, pp 311–325
15. Sodeifian G, Mojtaba R, Asghari M, Rezakzemi M (2018) Polyurethane-SAPO-34 mixed matrix membrane for CO_2/CH_4 and CO_2/N_2 separation. Chin J Chem Eng. https://doi.org/10.1016/j.cjche.2018.03.012
16. Rezakazemi M, Dashti A, Asghari M, Saeed S (2017) H_2-selective mixed matrix membranes modeling using ANFIS, PSO-ANFIS GA-ANFIS. Int J Hydrogen Energy 42(22):15211–15225
17. Baheri B, Mahnaz S, Razakazemi M, Elahe M, Mohammadi T (2014) Performance of PVA/NaA mixed matrix membrane for removal of water from ethylene glycol solutions by pervaporation. Chem Eng Commun 202(3):316–321
18. Shahverdi M, Baheri B, Razakazemi M, Elahe M, Mohammadi T (2013) Pervaporation study of ethylene glycol dehydration through synthesized (PVA-4A)/polypropylene mixed matrix composite membranes. Polym Eng Sci 53(7):1487–1493
19. Rostamizadeh M, Rezakazemi M, Shahidi K, Mohammadi T (2013) Gas permeation through H_2-selective mixed matrix membranes: Experimental and neural network modeling. Int J Hydrogen Energy 38(2):1128–1135

20. Rezakazemi M, Mohammadi T (2013) Gas sorption in H_2-selective mixed matrix membranes: experimental and neural network modeling. Int J Hydrogen Energy 38(32):14035–14041
21. Rezakazemi M, Shahidi K, Mohammadi T (2012) Sorption properties of hydrogen-selective PDMS/zeolite 4A mixed matrix membrane. Int J Hydrogen Energy 37(22):17275–17284
22. Rezakazemi M, Shahidi K, Mohammadi T (2012) Hydrogen separation and purification using crosslinkable PDMS/zeolite a nanoparticles mixed matrix membranes. Int J Hydrogen Energy 37(19):14576–14589
23. Rezakazemi M, Sadrzadeh M, Matsuura T (2018) Thermally stable polymers for advanced high-performance gas separation membranes. Prog Energy Combust Sci 66:1–41
24. Boutsika LG, Enotiadis A, Nicotera I, Simari C, Charalambopoulou G, Giannelis EP, Steriotis T (2016) Nafion® nanocomposite membranes with enhanced properties at high temperature and low humidity environments. Int J Hydrogen Energy 41(47):22406–22414
25. Mohammadi G, Jahanshahi M, Rahimpour A (2013) Fabrication and evaluation of Nafion nanocomposite membrane based on ZrO_2–TiO_2 binary nanoparticles as fuel cell MEA. Int J Hydrogen Energy 38(22):9387–9394
26. Hooshyari K, Javanbhakt M, Naji L, Enhessari M (2014) Nanocomposite proton exchange membranes based on Nafion containing Fe_2 TiO_5 nanoparticles in water and alcohol environments for PEMFC. J Membr Sci 454:74–81
27. Wang Z, Tang H, Zhang H, Lei M, Chen R, Xiao P, Pan M (2012) Synthesis of Nafion/CeO_2 hybrid for chemically durable proton exchange membrane of fuel cell. J Membr Sci 421: 201–210
28. Cozzi D, de Bonis C, D'Epifanio A, Mecheri B, Tavares AC, Licoccia S (2014) Organically functionalized titanium oxide/Nafion composite proton exchange membranes for fuel cells applications. J Power Sources 248:1127–1132
29. de Bonis C, Cozzi D, Mecheri B, D'Epifanio A, Rainer A, De Porcellenis D, Licoccia S (2014) Effect of filler surface functionalization on the performance of Nafion/Titanium oxide composite membranes. Electrochim Acta 147:418–425
30. Yang Y, Cuiping H, Beibei J, James I, Chengen H, Dean S, Tao J, Zhiqun L (2016) Graphene-based materials with tailored nanostructures for energy conversion and storage. Mater Sci Eng R Rep 102:1–72
31. Farooqui U, Ahmad A, Hamid N (2018) Graphene oxide: a promising membrane material for fuel cells. Renew Sust Energy Rev 82:714–733
32. Tsang AC, Kwok HY, Leung DY (2017) The use of graphene based materials for fuel cell, photovoltaics, and supercapacitor electrode materials. Solid State Sci 67:A1–A14
33. Das TK, Prusty S (2013) Graphene-based polymer composites and their applications. Polym Plast Technol Eng 52(4):319–331
34. Zhu C, Dong S (2013) Recent progress in graphene-based nanomaterials as advanced electrocatalysts towards oxygen reduction reaction. Nanoscale 5(5):1753–1767
35. Fampiou I, Ramasubramaniam A (2012) Binding of Pt nanoclusters to point defects in graphene: adsorption, morphology, and electronic structure. J Phys Chem C 116(11): 6543–6555
36. Liu M, Zhang R, Chen W (2014) Graphene-supported nanoelectrocatalysts for fuel cells: synthesis, properties, and applications. Chem Rev 114(10):5117–5160
37. Huang H, Chen H, Sun D, Wang X (2012) Graphene nanoplate-Pt composite as a high performance electrocatalyst for direct methanol fuel cells. J Power Sources 204:46–52
38. Qian W, Hao R, Zhou J, Eastman M, Manhat BA, Sun Q, Andrea MG, Jiao J (2013) Exfoliated graphene-supported Pt and Pt-based alloys as electrocatalysts for direct methanol fuel cells. Carbon 52:595–604
39. Peng K-J, Lai J-Y, Liu Y-L (2016) Nanohybrids of graphene oxide chemically-bonded with Nafion: preparation and application for proton exchange membrane fuel cells. J Membr Sci 514:86–94
40. Zhou X, Tang S, Yin Y, Sun S, Qiao J (2016) Hierarchical porous N-doped graphene foams with superior oxygen reduction reactivity for polymer electrolyte membrane fuel cells. Appl Energy 175:459–467

41. Kirubaharan CJ, Santhakumar K, Gnanankumar G, Senthilkumar N, Jang J (2015) Nitrogen doped graphene sheets as metal free anode catalysts for the high performance microbial fuel cells. Int J Hydrogen Energy 40(38):13061–13070
42. Rezakazemi M, Zhang Z (2018) Desulfurization materials A2—Dincer, Ibrahim comprehensive energy systems. Elsevier, Oxford, pp 944–979
43. Georgakilas V, Perman JA, Tucek J, Zboril R (2015) Broad family of carbon nanoallotropes: classification, chemistry, and applications of fullerenes, carbon dots, nanotubes, graphene, nanodiamonds, and combined superstructures. Chem Rev 115(11):4744–4822
44. Akbari E, Buntat Z (2017) Benefits of using carbon nanotubes in fuel cells: a review. Int J Energ Res 41(1):92–102
45. Ghasemi M, Ismail M, Kamarudin SK, Saeedfar K, Wan Daud WR, Hassan SHA, Heng LY, Alam J, Oh SE (2013) Carbon nanotube as an alternative cathode support and catalyst for microbial fuel cells. Appl Energy 102:1050–1056
46. Mehdinia A, Ziaei E, Jabbari A (2014) Multi-walled carbon nanotube/SnO_2 nanocomposite: a novel anode material for microbial fuel cells. Electrochim Acta 130:512–518
47. Tohidian M, Ghaffarian SR (2017) Polyelectrolyte nanocomposite membranes with imidazole-functionalized multi-walled carbon nanotubes for use in fuel cell applications. J Macromol Sci B 56(10):725–738
48. Zhang R, He S, Lu Y, Chen W (2015) Fe Co, N-functionalized carbon nanotubes in situ grown on 3D porous N-doped carbon foams as a noble metal-free catalyst for oxygen reduction. J Mater Chem A 3(7):3559–3567
49. Mishra P, Jain R (2016) Electrochemical deposition of MWCNT-MnO_2/PPy nano-composite application for microbial fuel cells. Int J Hydrogen Energy 41(47):22394–22405
50. Mirzaei F, Parnian MJ, Rowshanzamir S (2017) Durability investigation and performance study of hydrothermal synthesized platinum-multi walled carbon nanotube nanocomposite catalyst for proton exchange membrane fuel cell. Energy 138:696–705
51. Liew KB, Wan Daud WR, Ghasemi M, Loh KS, Ismail M, Lim SS, Leong JX (2015) Manganese oxide/functionalised carbon nanotubes nanocomposite as catalyst for oxygen reduction reaction in microbial fuel cell. Int J Hydrogen Energy 40(35):11625–11632
52. Mourya V, Inamdar NN, Tiwari A (2010) Carboxymethyl chitosan and its applications. Adv Mater Lett 1(1):11–33
53. Vakili M, Rafatullah M, Salamatinia B, Abdullah AZ, Ibrahim MH, Tan KB, Gholami Z, Amouzgar P (2014) Application of chitosan and its derivatives as adsorbents for dye removal from water and wastewater: a review. Carbohyd Polym 113:115–130
54. Srinivasa P, Tharanathan R (2007) Chitin/chitosan—safe, ecofriendly packaging materials with multiple potential uses. Food Rev Int 23(1):53–72
55. Lim S-H, Hudson SM (2003) Review of Chitosan and its derivatives as antimicrobial agents and their uses as textile chemicals. J Macromol Sci C Polym Rev 43(2):223–269
56. Mourya V, Inamdar NN (2009) Trimethyl chitosan and its applications in drug delivery. J Mater Sci Mater Med 20(5):1057
57. Ilium L (1998) Chitosan and its use as a pharmaceutical excipient. Pharm Res 15(9): 1326–1331
58. Ma J, Sahai Y (2013) Chitosan biopolymer for fuel cell applications. Carbohydr Polym 92 (2):955–975
59. Bai H, Zhang H, He Y, Liu J, Zhang B, Wang J (2014) Enhanced proton conduction of chitosan membrane enabled by halloysite nanotubes bearing sulfonate polyelectrolyte brushes. J Membr Sci 454:220–232
60. Ma J, Sahai Y, Buchheit RG (2012) Evaluation of multivalent phosphate cross-linked chitosan biopolymer membrane for direct borohydride fuel cells. J Power Sources 202:18–27
61. Hasani-Sadrabadi MM, Dashtimoghadam E, Mokarram N, Majedi FS, Jacob KI (2012) Triple-layer proton exchange membranes based on chitosan biopolymer with reduced methanol crossover for high-performance direct methanol fuel cells application. Polymer 53 (13):2643–2651

62. Zhou T, He X, Song F, Xie K (2016) Chitosan modified by polymeric reactive dyes containing quanternary ammonium groups as a novel anion exchange membrane for alkaline fuel cells. Int J Electrochem Sci 11(1):590–608
63. Li P-C, Liao G-M, Rajeshkumar S, Shih C-M, Yang C-C, Wang D-M, Lue SJ (2016) Fabrication and characterization of chitosan nanoparticle-incorporated quaternized poly (vinyl alcohol) composite membranes as solid electrolytes for direct methanol alkaline fuel cells. Electrochim Acta 187:616–628
64. Santamaria M, Pecoraro CM, Di Quarto F, Bocchetta P (2015) Chitosan–phosphotungstic acid complex as membranes for low temperature H_2–O_2 fuel cell. J Power Sources 276: 189–194
65. Noroozifar M, Motlagh MK, Kakhki M-S, Roghayeh K-M (2014) Enhanced electrocatalytic properties of Pt–chitosan nanocomposite for direct methanol fuel cell by $LaFeO_3$ and carbon nanotube. J Power Sources 248:130–139
66. Linlin M, Mishra AK, Kim NH, Lee JH (2012) Poly (2,5-benzimidazole)–silica nanocomposite membranes for high temperature proton exchange membrane fuel cell. J Membr Sci 411:91–98
67. Chang Y-N, Lai J-Y, Liu Y-L (2012) Polybenzimidazole (PBI)-functionalized silica nanoparticles modified PBI nanocomposite membranes for proton exchange membranes fuel cells. J Membr Sci 403:1–7
68. Hooshyari K, Javanbhakt M, Shabanikia A, Enhessari M (2015) Fabrication $BaZrO_3$/PBI-based nanocomposite as a new proton conducting membrane for high temperature proton exchange membrane fuel cells. J Power Sources 276:62–72
69. Devrim Y, Devrim H, Eroglu I (2016) Polybenzimidazole/SiO_2 hybrid membranes for high temperature proton exchange membrane fuel cells. Int J Hydrogen Energy 41(23):10044–10052
70. Wu J-F, Lo C-F, Li H-Y, Chang C-M, Liao K-S, Hu C-C, Liu Y-L, Lue S-J (2014) Thermally stable polybenzimidazole/carbon nano-tube composites for alkaline direct methanol fuel cell applications. J Power Sources 246:39–48
71. Özdemir Y, Özkan N, Devrim Y (2017) Fabrication and characterization of cross-linked polybenzimidazole based membranes for high temperature PEM fuel cells. Electrochim Acta 245:1–13
72. Hogarth WH, Da Costa JD, Lu GM (2005) Solid acid membranes for high temperature proton exchange membrane fuel cells. J Power Sources 142(1):223–237
73. Tripathi BP, Shahi VK (2007) SPEEK–zirconium hydrogen phosphate composite membranes with low methanol permeability prepared by electro-migration and in situ precipitation. J Colloid Interface Sci 316(2):612–621
74. Lim SS, Wan Daud WR, Jahim J, Ghasemi M, Chong PS, Ismail M (2012) Sulfonated poly (ether ether ketone)/poly (ether sulfone) composite membranes as an alternative proton exchange membrane in microbial fuel cells. Int J Hydrogen Energy 37(15):11409–11424
75. Jiang Z, Zhao X, Manthiram A (2013) Sulfonated poly (ether ether ketone) membranes with sulfonated graphene oxide fillers for direct methanol fuel cells. Int J Hydrogen Energy 38 (14):5875–5884
76. Heo Y, Im H, Kim J (2013) The effect of sulfonated graphene oxide on sulfonated poly (ether ether ketone) membrane for direct methanol fuel cells. J Membr Sci 425:11–22
77. Wang J, Bai H, Zhang H, Zhao L, Yifan Li C (2015) Anhydrous proton exchange membrane of sulfonated poly (ether ether ketone) enabled by polydopamine-modified silica nanoparticles. Electrochim Acta 152:443–455
78. Gang M, He G, Li Z, Cao K, Li Z, Yin Y, Wu H, Jiang Z (2016) Graphitic carbon nitride nanosheets/sulfonated poly (ether ether ketone) nanocomposite membrane for direct methanol fuel cell application. J Membr Sci 507:1–11
79. Mossayebi Z, Saririchi T, Rowshanzamir S, Parnian MJ (2016) Investigation and optimization of physicochemical properties of sulfated zirconia/sulfonated poly (ether ether ketone) nanocomposite membranes for medium temperature proton exchange membrane fuel cells. Int J Hydrogen Energy 41(28):12293–12306

80. Rahnavard A, Rowshanzamir S, Parnian MJ, Amirkhanlou GR (2015) The effect of sulfonated poly (ether ether ketone) as the electrode ionomer for self-humidifying nanocomposite proton exchange membrane fuel cells. Energy 82:746–757
81. Salarizadeh P, Javanbhakht M, Abdollahi M, Naji L (2013) Preparation, characterization and properties of proton exchange nanocomposite membranes based on poly (vinyl alcohol) and poly (sulfonic acid)-grafted silica nanoparticles. Int J Hydrogen Energy 38(13):5473–5479
82. Liew C-W, Ramesh S, Arof A (2014) A novel approach on ionic liquid-based poly (vinyl alcohol) proton conductive polymer electrolytes for fuel cell applications. Int J Hydrogen Energy 39(6):2917–2928
83. Beydaghi H, Javanbakht M, Kowsari E (2014) Synthesis and characterization of poly (vinyl alcohol)/sulfonated graphene oxide nanocomposite membranes for use in proton exchange membrane fuel cells (PEMFCs). Ind Eng Chem Res 53(43):16621–16632
84. Mollá S, Compañ V (2015) Nanocomposite SPEEK-based membranes for direct methanol fuel cells at intermediate temperatures. J Membr Sci 492:123–136
85. Yang J-M, Wang N-C, Chiu H-C (2014) Preparation and characterization of poly (vinyl alcohol)/sodium alginate blended membrane for alkaline solid polymer electrolytes membrane. J Membr Sci 457:139–148
86. Huang C-Y, Lin J-S, Pan W-H, Shih C-M, Liu Y-L, Lue S-J (2016) Alkaline direct ethanol fuel cell performance using alkali-impregnated polyvinyl alcohol/functionalized carbon nano-tube solid electrolytes. J Power Sources 303:267–277

Rubber Clay Nanocomposites

Mariajose Cova Sanchez, Alejandro Bacigalupe, Mariano Escobar and Marcela Mansilla

List of Abbreviations

AFM	Atomic force microscopy
APTES	(3—aminopropyl) triethoxysilane
CB	Carbon black
CEC	Cation exchange capacity
CIIR	Chlorobutyl rubber
CL	Concentrated Natural Rubber Latex
Dim-Br	o-xylylenebis (triphenylphosphoniumbromide)
DMA	Dynamic mechanical analysis
DSC	Differential scanning calorimetry
DTG	Derivative thermogravimetric analysis
EDX	Electron dispersive X-ray spectroscopy
EG	Expanded graphite
EPDM	Ethylene propylene diene rubber
FL	Fresh Natural Rubber Latex
FTIR	Fourier transform infrared
Hal	Hallosyte
$HDTMA^+$	Hexadecyl trimethylammonium
IIR	Isobutylene isoprene rubber
M_H	Maximum torque
M_t	Montmorillonite
NBR	Acrylonitrile butadiene rubber

M. C. Sanchez · A. Bacigalupe
Instituto de Investigación e Ingeniería Ambiental, Universidad Nacional de San Martín (UNSAM 3iA), Francia 34, 1650 San Martín, Argentina

M. C. Sanchez · A. Bacigalupe · M. Escobar (✉) · M. Mansilla
Instituto Nacional de Tecnología Industrial (INTI), Centro de Caucho, Av. General Paz 5445, B1650WAB San Martín, Argentina
e-mail: mescobar@inti.gob.ar

M. Escobar · M. Mansilla
Consejo Nacional de Investigaciones Científicas y Técnicas (CONICET), Godoy Cruz 2290, C1425FQB CABA, Argentina

Inamuddin et al. (eds.), *Sustainable Polymer Composites and Nanocomposites*,
https://doi.org/10.1007/978-3-030-05399-4_21

NK	Nanokaolin
NR	Natural rubber
OC	Organoclay
$ODTMA^+$	Octadecyl trimethylammonium
OMt	Organomodified montmorillonite
PAS	Positron annhilation lifetime spectroscopy
PA6	Polyamide 6
phr	per hundred of rubber
PLA	Olylactide acid
SANS	Small angle neutrón scattering
SAXS	Small Angle X-Ray Scattering
SBR	Styrene Butadiene Rubber
SEM	Scanning electron microscopy
SI	Silica
tan δ	Loss tangent
t_{c90}	Cure time
TEM	Transmission electron microscopy
T_g	Glass transition temperature
TGA	Thermogravimetric analysis
t_{s2}	Scorch time
WAXS	Wide Angle X-Ray Scattering
W_c	Water content
XRD	X-ray diffraction

1 Introduction

The use of nanofillers allows the development of nanocomposites with improved properties and novel applications [1–7]. The technological goal is possible due to the new compounding method that allows a particle dispersion in the nanometer scale increasing the specific surface area [8].

This chapter focuses on explaining the role that nanoclays play as exceptional reinforcing particles in rubber matrices. The complexity of the analysis comes from the fact that it is a multicomponent system, which contains, for example, curing agents, co-agents, processing aids, reinforcements and fillers, affecting the final structure of a rubber/layered silicate nanocomposite. This was likely the reason for its late development compared to thermoplastic or thermoset clay nanocomposites.

The use of nanoclay in rubber allows to improve the mechanical properties (such as tensile strength and elastic modulus) [1, 8], barrier properties [9], thermal resistance [6] and, in some applications, antimicrobial properties [10]. Examples and cases of most used rubber matrices in the industrial field are presented. The methods of clays incorporation into rubber matrices as well as the final properties reached are described.

2 Reinforcement Particles

A wide variety of ingredients are used in the rubber industry for various purposes, being the most important: reinforcement, fillers and process aids. The principal reinforcement effects are increase of strength, abrasion resistance, hardness and modulus. The extents to which these changes occur strongly depend on:

(i) particle filler size
(ii) rubber-filler interaction
(iii) filler-filler interaction
(iv) filler shape and structure
(v) filler concentration
(vi) filler dispersion in the matrix.

CB and precipitate SI, with a mean diameter of 10–30 and 30–100 nm respectively, are the conventional fillers in the rubber industry, with a content that in some compounds could reach up to 40 phr of CB. Considering compounds containing carbon black (CB) as reinforcement, it is well known the so-called *bound rubber,* which refers to a microstructure seen as a gel of reinforcement particles held together in a three-dimensional lattice by polymer molecules: the amount and morphology of the bound rubber depend on the listed variables [11]. Moreover, bound rubber type also shows a dependence on the processing conditions of the compound. Bound rubber content is measured by extracting the unbound rubber with solvent, and the result is influenced by the solvent nature and the extraction temperature.

Other nanofillers have received attention for reinforcing of rubbers like organoclay (OC), nano silica, carbon nanotubes (CNT) and nano calcium carbonate due to the high aspect ratio and surface area as compared to conventional fillers. These characteristics are important to develop high-performance materials with low nanofiller content.

2.1 *Carbon Black*

CB is by far the most popular filler for rubber modification. CB is elemental carbon in the form of spherical particles that coalesced into aggregates and agglomerates, which is obtained by the combustion or thermal decomposition of hydrocarbons. The degree of aggregation is denoted by the term “structure”: a low-structure may have an average of 30 particles while a high structure may have around 200 particles per aggregate. The aggregates have a tendency to cluster together to form agglomerates due to van der Waals forces. During the mixing process, polymer free radicals are formed due to the shear forces generated and such free radicals react with CB to form bound rubber as CB acts as radical acceptors [12].

About 90% of the worldwide production of CB is used in the tire industry, in which acts as reinforcing fillers to improve tear strength, modulus, and wear characteristics. Commercially, CB fillers are available with different levels of structure, particle size, chemical reactivity and pH that lead to different levels of reinforcement.

2.2 *Silica*

SI (silicon dioxide) consists of silicon and oxygen arranged in a tetrahedral structure of a three-dimensional lattice. The use of silica as rubber reinforcement started around 1950 as an alternative to CB, shoeing sole the first applications due to the possibility of obtaining a non-black product. The incorporation of SI in heavy-service tires started in the 70's due to the improvement of cutting and chipping resistance of the compound and, also rubber adhesion to textile and metal.

The presence of hydroxyl groups in the silanol (Si-OH) on the SI results in strong filler-filler interactions and adsorption of polar materials by hydrogen bonds. Since these intermolecular bonds are very strong, the SI particles aggregate tightly, reducing the dispersion of silica in the rubber compound.

Around 1970 started the introduction of silane coupling agents such as the bis(3-(triethoxysilyl)-propyl)-tetrasulfide (TESPT) and other bifunctional silanes in order to prevent adsorption of curatives on the silica surface [13]. In these cases, the silane coupling agent reacts with the silanol on the silica surface to form a siloxane bond. The silane molecule bonded to the silica surface reacts with the rubber chain to form a crosslink between the silica and the polymer.

Silica is available as fumed SI and precipitated SI, which is obtained by coagulation of SI particles from an aqueous medium under the influence of high salt concentration or other coagulants [12, 14].

In tire treads, silica yields a lower rolling resistance at equal wear resistance and wet grip than CB [15]. Rattanasom et al. [16] analyzed the reinforcement of Natural Rubber with SI/CB hybrid filler in a total amount of 50 phr. The results reveal that the compounds with 20 and 30 phr of silica in hybrid filler exhibit the better overall mechanical properties.

2.3 *Clay*

Clays have been known for many decades as a non-black filler type for rubber compounds [4]. These fillers are inorganic in nature and therefore incompatible with organic polymer matrices. Thus, the reinforcing effect of these fillers was lower than that of CB.

Clays have a low price and many attractive structural features such as hydroxyl groups, Lewis acidity, exchangeable interlayer cations, differences of Si–O tetrahedral sheet and Al-OH octahedral sheets [17]. Also, layered silicate nano-fillers have already been developed: the silicate layer surfaces have been effectively modified to render them organophilic so that the organically modified nano-fillers can significantly enhance the critical performance properties of polymer–clay composites. This is possible when the silicates are dispersed in nanometer- instead of micrometre-scale.

3 Layered Silicates

Galimberti et al. [18] define layered silicates as inexpensive natural mineral fillers which promote high reinforcements on the rubber matrices thanks to their high aspect ratio. They present a very thin plate-shape with low thickness (<10 nm) and width (<2 mm) [19], which permits to incorporate a relatively low amount of the filler to obtain superior mechanical properties, thermal stability, flame retardancy and gas barrier properties.

There are many works that show significant improvements in elastic modulus, tensile strength and elongation at break [8], thermal resistance [20], reduction in water/gas permeability [21] and flammability [20] for clays in their natural an modified form. The nanocomposite structure depends on the clay mineral–polymer compatibility and on processing conditions [22].

Siririttikrai et al. [19] show that clays in their natural form are easily exfoliated in aqueous suspension, which can be convenient for certain processing conditions. Depending on the compatibility with the rubber matrix, using clays in natural or pristine form can reduce the steps and cost involved.

3.1 Structure and Physical Characteristics

Figure 1 presents a typical structure of lamellar clays. Moreover, clays can be defined depending on the organization of the silicon and oxygen atoms, as well as the arrangement of the atoms within the laminar structure. The main structure consists of building blocks of tetrahedral sheets in which silicon is surrounded by four oxygen atoms and octahedral sheets in which a metal such as aluminium is between eight oxygen atoms [23, 24].

Silicate layers present oxygen atoms and hydroxide groups, giving it a hydrophilic nature. However, most rubber compounds possess a hydrophobic structure. Galimberti et al. [25] explain that, in order to achieve a compatibilization with the rubber matrix, there must be a chemical modification in the clay surface. This

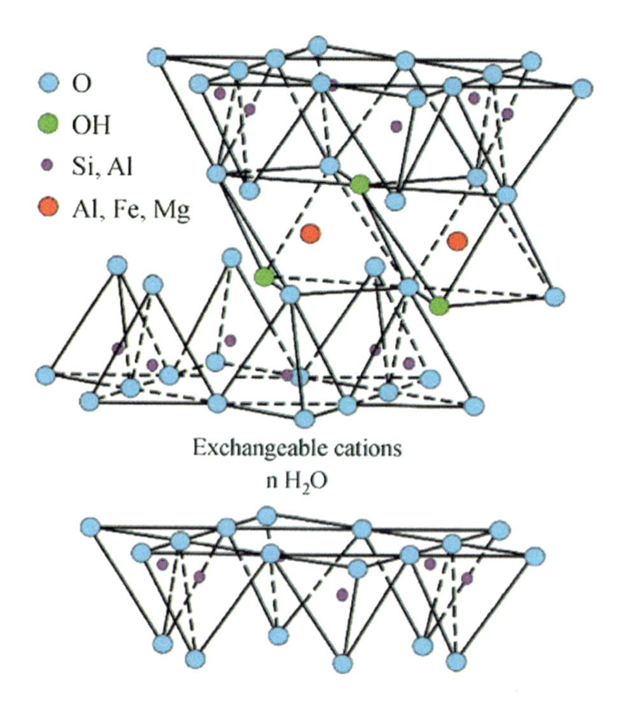

Fig. 1 General layered silicate structure [24]. Reprinted from Kievani and Edrak [24]

technique consists in an exchange reaction of alkali and alkaline earth cations, which are located inside the clay galleries with different cations, modifying the clay surface polarity and increasing the interlayer distance [25]. This process of chemical modification will be described later in the chapter. Pavlidou et al. [23] define the first structure as 1:1, in which a tetrahedral sheet is fused with an octahedral sheet, whereby the oxygen atoms are shared.

The 2:1 layered silicates consists in two-dimensional layers where a central octahedral sheet of alumina is fused to two external silica tetrahedra by the tip, so the oxygen ions of the octahedral sheet also belong to the tetrahedral sheets. The layer thickness is around 1 nm and the lateral dimensions may vary from 300Å to several microns, therefore, the aspect ratio of this kind of layers (ratio length/thickness) is high, having values greater than 1000 [23]. The 2:1 structure is called pyrophyllite. Furthermore, since inside the structure, there are no substitution atoms and it has only an external surface area and no internal one, the layers do not expand in water.

When the silicon in the tetrahedral sheet is substituted with aluminium, the structure is called mica. With this type of substitution, the mineral is characterized by a negative surface charge, which is balanced by cations, such as potassium. Unlike the structure of pyrophyllite, mica does not swell in water, because due to the large size of potassium, the structure collapses and the sheets are held together only by electrostatic forces [23].

Montmorillonite (Mt) presents a structure similar to that of pyrophyllite, in which the trivalent Al-cation in the octahedral layer is partially substituted by

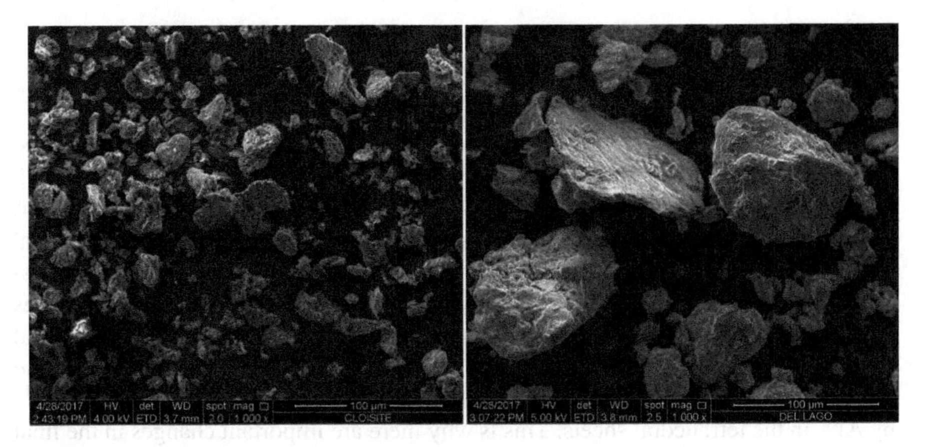

Fig. 2 SEM images of Na Mt (Left) and Raw Mt (Right)

magnesium cation. This group is called "smectites". In addition, the overall negative charge is balanced by sodium and calcium ions, which are present in a hydrated form inside the interlayer. Since the ions are not part of the tetrahedral structure, the sheets are held together by weak interacting forces, which allow the easy entry of water molecules into the clay structure [23].

As previously mentioned, depending on the type of clay different structures can be presented, which affect the size and texture of the clay. Figure 2 compares the structural differences between two types of Mts for 1000X magnification, one Na–Mt (Left) and a Raw Mt (Right). Na–Mt clay has a smaller particle size compared to the raw Mt since it undergoes a purification process to eliminate any impurities. One of the most interchangeable cations is Na^+, which is responsible for its swelling in presence of water. Moreover, the presence of Na^+ ions lets absorb a great number of water molecules resulting in the formation of a gel. If Ca^{+2} are the cations present in the interlayer, a lower amount of water molecule can be absorbed, so the clay is not able to swell. Depending on the cation presents in the interlayer, it will be determined the final application of the clay, as well as its compatibility with the rubber matrix.

3.2 *Classification of Clays*

The clays are classified according to the amount of silica present and the combination of tetrahedral and octahedral sites in the final structure. It can be separated into 5 main families:

(a) Kaolinite
(b) Serpentines

(c) Smectites
(d) Illites
(e) Chlorites.

Mt, hectorite and saponite belong to the smectites family and are the most commonly used clay minerals for the preparation of polymer nanocomposites [22]. Hectorite is a colloidal clay widely used in the synthetic form. Dispersed in water, Hectorite forms a colloidal suspension used in many industrial and technological applications, such as rheology modifiers in several products, like paints and cosmetics. It has a very particular anisotropy due to the nanodisc shape of the particles and to the inhomogeneity of the sheet charges [26]. The Mt presents a dioctahedral structure with predominantly octahedral substitution. On the other hand, saponite presents a trioctahedral structure with the mainly isomorphous substitution of Si^{4+} by Al^{3+} in the tetrahedral sheets. This is why there are important changes in the final structure configuration; the Mt has hexagonal lamellae, while the saponite has a ribbon shape. Throughout the chapter, the discussion will focus on the use of Mt as a reinforcing, compatibilizing or property modifying agent since most of the published works use this type of clay.

- *Kaolinite*

Kaolinite is characterized by an isomorphic variation with temperature: up to 700 C, the Al^{+3} presents a tetrahedral structure (what is known as amorphous kaolin), while above this temperature has an octahedral one.

- *Montmorillonite*

It is part of the group of smectites. Their structure consists of two silica tetrahedral sheets sandwiching an edge-shared octahedral sheet of Al or Mg hydroxide. Normally the layer thickness is around 1 nm, and the lateral dimensions may vary from 30 nm to several microns or larger. Ambre et al. [27] explain how the stacking of the clay layers leads to a van der Waals gap between the platelets called the interlayer or gallery. Also, the author described that the isomorphic substitutions within the layers generate negative charges that are counterbalanced by alkali and alkaline earth cations inside the clay galleries, and these cations are replaced by organic ones such as alkyl ammonium ions via an exchange reaction.

As it will be discussed, interlayer distance is one of the most important parameters since it defines the possibility for the polymer chains to enter inside the interlayer to effectively achieve an intercalation or exfoliation of the clay.

4 Chemical Modification of Clays

Natural clays can be well dispersed in hydrophilic polymers, such as poly (ethylene oxide) and poly(vinyl alcohol). In order to increase the compatibility with hydrophobic polymers, an ion exchange is performed with cationic-organic

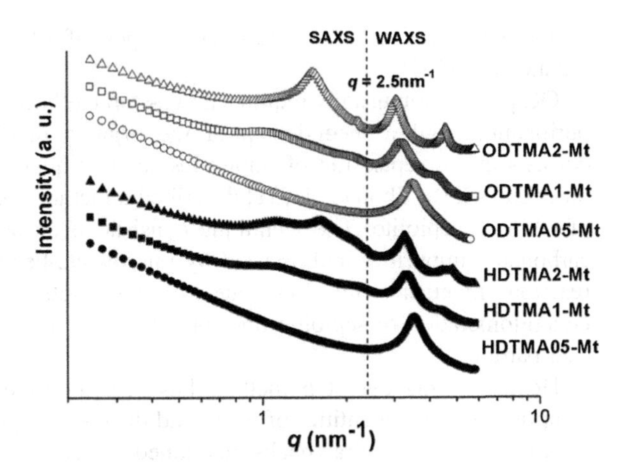

Fig. 3 Variation of the interlayer distance q of Mt and OMt according to the modifier concentration, determined by SAXS and WAXS. Reprinted from Bianchi et al. [28]

surfactants. Pavlidou et al. [23] referred the alkylammonium ions as the most used, but other "onium" salts can be used, such as sulfonium and phosphonium. The objective of the modification is to decrease the surface energy of the silicate and improve wetting with the polymer matrix. The long organic chains of the surfactants are tethered to the surface of the negatively charged silicate layers, promoting an increase in the interlayer gallery and therefore favouring the entrance of polymer chain between interlayer space in order to improve the mechanical properties of the nanocomposite.

During the modification process, the excess of negative charge should be considered since it is related to the ability of clay to exchange ion, this characteristic is defined as the cation-exchange capacity (CEC), expressed in meq/100 g (milliequivalents/100grams). Pavlidou et al. [23] explain how this property is highly dependent on the nature of the isomorphous substitutions in the tetrahedral and octahedral layers.

Figure 3 shows an analysis of organo modified montmorillonite (OMt) by SAXS and WAXS. In their work, Bianchi et al. characterized a Mt two with different organic modifiers: octadecyl trimethylammonium ($ODTMA^+$) and hexadecyl trimethylammonium ($HDTMA^+$) in different proportions: 0.5, 1 and 2 CEC of Mt. They obtained important changes in the final interlayer distance of the clay, which determine the final interaction with a rubber or polymeric matrix [28].

4.1 Synthesis of Organoclays

Due to the hydrophobic behaviour of most elastomeric polymers, raw natural clays needs to be modified in order to use them as reinforcing materials and increase the compatibility between the matrix and the nanoparticles [29]. The most common technique is by using long tallow organic ammonium cations through ion exchange of the interlayer cations. These new types of filler are commonly known as organoclays (OC).

OC preparation usually requires a two-step method, which includes a first step of purification, and a second step of the organic modification itself. Purification consists in the separation of impurities such as quartz, calcite, dolomite, talc and other clays [30]. Zhou et al. [31] described a method for purification and defibering of Chinese sepiolite. The technique consists of an acid treatment to decompose carbonated impurities and assisted by microwave-heating in order to improve the reaction kinetics with low energy consumption. Defibering aims to the de-agglomeration of sepiolite fiber bundles with the objective of obtaining nanoscale particles.

However, this two-step method has several disadvantages such as the consumption of large quantities of water and the energy spent on repetitive techniques. In order to avoid the drawbacks mentioned, Zhuang et al. [30] proposed a one-step method combining the purification and organic modification of sepiolite into a simple and more environmentally friendly procedure. The authors stated that organic surfactants can not only be used as an organic modifier but also as flocculants, therefore organo-sepiolite was easier to be separated from the dispersion.

Daitx et al. [29] described a method for the modification of Mt and halloysite (Hal) with (3- aminopropyl) triethoxysilane (APTES). Both OC were prepared by dispersion of the clay in an ethanol solution. In the case of Mt, the pH was adjusted to 3 with acetic acid for the protonation of the amino group and to promote the cationic exchange with Na^+ located in the interlayer space. APTES was added to the clay/ethanol suspension and stirred for 2 h and later filtered and washed with 96% ethanol. Lastly, the filtered cake was dried under vacuum at 70 °C for 8 h. In this work, the modification of the nanoparticles with aminosilane increased the interaction between clays minerals and the polymer due to the amino groups that can freely interact with the matrix.

5 Characterization of Clay Nanoparticles

The characterization of the clays is quite important to determine certain properties that are fundamental to be considered before being included into a rubber matrix. For example, mineralogical and thermal properties are important to determine the composition of the clay, particularly to know the nature of the clay (hydrophilic or hydrophobic nature). Some of the most used characterization methods include X-Ray diffraction (XRD), Infrared spectroscopy (IR), Inductively coupled plasma atomic emission spectroscopy (ICP-OES), thermal analysis (DTA, TGA), Specific surface area (SSA), CEC determination, microscopic techniques (TEM and SEM), gravimetric and grain size measurements [32–34]. In this chapter, we will show some techniques that can be useful to characterize any type of clay.

5.1 X-Ray Diffraction (XRD)

XRD is a very useful technique for the study of layered type nanoparticles. Furthermore, XRD is often used for the determination of the d_{001} plane corresponding to the basal space between the layered silicates which can be calculated by the Bragg's law:

$$\lambda = 2d \sin \theta$$

were λ is the wavelength of the incident wave, θ is the scattering angle and d is the distance between the atoms of a crystalline system.

Gamoudi and Srasra [33] studied the characterization of natural clays suitable for pharmaceutical and cosmetic applications by XRD analysis. Authors determined the presence of smectite in the composition by studying the characteristic peaks of the d_{001} reflection at 12.57 Å. Moreover, they also detected the presence of kaolinite by the d_{001} = 7.12 Å and d_{002} = 3.56 Å.

Successful organic modification of clay particles produces a shift of the d_{001} plane to lower angles values. This can be explained by taking into account the increase in the basal space due to the volume of long tallow organic molecules compared to the exchangeable cations. Ezquerro et al. [35] studied the modification of Mt with organic divalent phosphonium cations. Na–Mt showed a basal reflection at 7.78° and an interlayer spacing of 1.14 nm. The organic modification with Dim-Br (o-xylylenebis (triphenylphosphoniumbromide)) shifted the basal reflection to lower angles due to the intercalation of the surface between the layers of the OC. Moreover, an increase in the concentration of Dim-Br produced a shift to lower 2θ angles.

Alves et al. [36] studied the organic modification of Mt with ionic and non-ionic surfactants. They attributed the correct intercalation to a shift of the d_{001} to lower 2θ angles. Moreover, the basal space increase when surfactant amount increases and decreased after washed.

5.2 Microscopy

Scanning electron microscopy (SEM) and transmission electron microscopy (TEM) can provide information about the morphology of the clay platelets. Also, energy-dispersive X-Ray spectroscopy (EDX) can be used to study the composition after the modification with organic molecules.

Ezquerro et al. [35] studied the changes in the morphology of the Mt clay platelets. They claimed that pristine Mt shows soft surfaces and platelets tend to agglomerate. While surface modification of Na^+ for a phosphorous compound generates a rougher surface and breaks the agglomeration of the lamellas. Furthermore, EDX analysis was also conducted to study the yield of the organic modification of the interlayer cation. In their studies, the Na^+ content decreases and the phosphorous content increases with the surfactant concentration.

Figure 4 presents SEM images of Mt and OMt by Bianchi et al. [28], there can be seen the morphology modifications of the Mt surface structure induced by organic modification with alkylammonium salts ($ODTMA^+$ and $HDTMA^+$), with alters the surface structure, the curvature of the plates, and the aggregate formation.

Zhuang et al. [30] studied the purification and organic modification of sepiolite by TEM analysis, which was used to detect the presence of impurities, before and after the purification, such as calcite, quartz and talc. Also, the morphology of sepiolite fibers was studied by TEM. Sepiolite fibers form crystal bundles and rods due to hydrogen bonding and van der Waals' forces between the fibers.

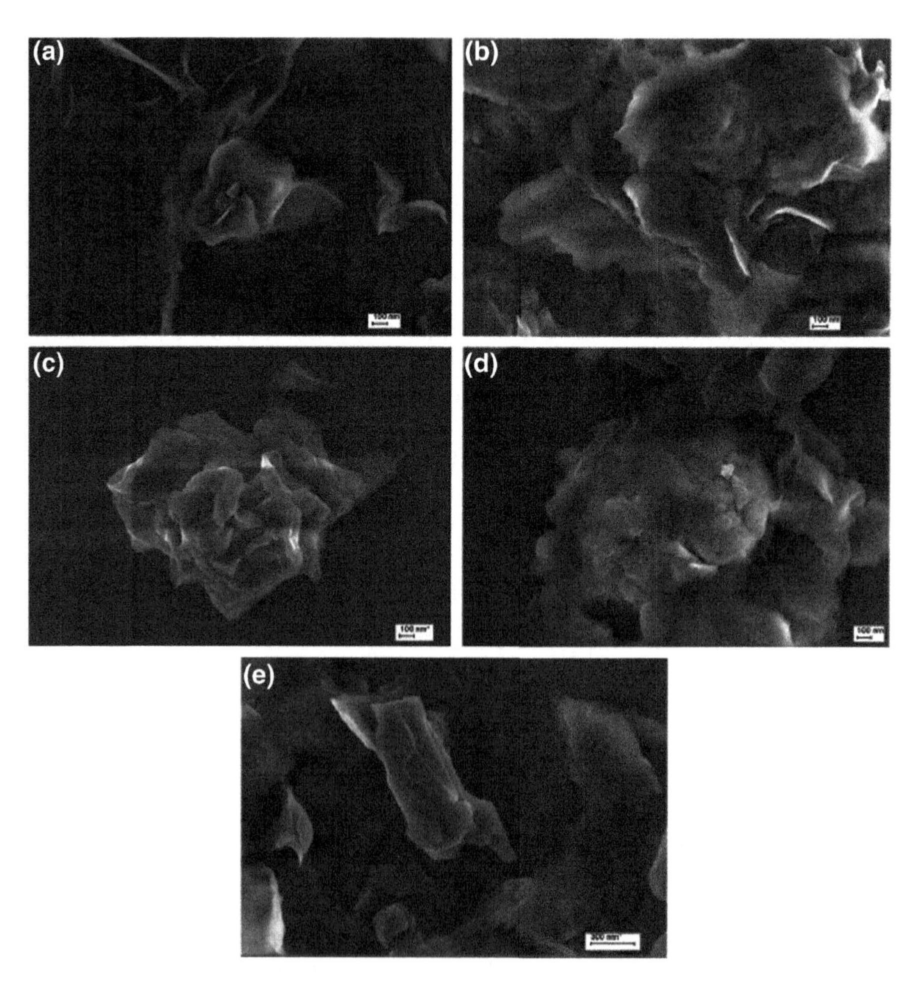

Fig. 4 SEM micrographs of Mt (A) and OMt, using as modifiers HDTMA (B and C) and ODTMA (D and F). Reprinted from Bianchi et al. [28]

5.3 Fourier Transform Infrared Spectroscopy

Fourier transform infrared spectroscopy (FTIR) can be used to characterize aluminium silicates materials. The most important bands present in clays are OH stretching (3800–3200 cm^{-1}), Si-O symmetrical and asymmetrical stretching (1200–900 cm^{-1}) and Al-O stretching (800 cm^{-1}). Moreover, FTIR could provide information about the substitution of cations from the interlayer space by organic molecules.

Hojiyev et al. [37] made an FTIR analysis of purified and cationic surfactant modified. They identified the bands corresponding to stretching vibrations of OH groups, Si-O groups, Al–Al-OH groups, and Al–Fe-OH groups. Moreover, Na–Mt spectra show the existence of free and interlayer water at $\sim$ 3400 cm^{-1} and adsorbed water at $\sim$ 1630 cm^{-1}. In addition, the spectra of cationic modified Mt show new peaks corresponding to asymmetric stretching vibration, symmetric stretching vibration and bending stretching vibration of methylene groups (2924, 2849 and 1469 cm^{-1} respectively). Also, the broad peak at 3404 cm^{-1} corresponding to free and interlayer water was removed upon modification due to the hydrophobic nature of the cationic surfactant.

5.4 Thermal Properties

Thermal analysis is also a useful tool to characterize OC due to the difference between the thermal stability of the inorganic cation and the organic surfactants. Within the thermal characterization techniques, the most commonly used ones are thermogravimetric analysis (TGA), derivative thermogravimetric analysis (DTG) and differential scanning calorimetry (DSC).

Sookyung et al. [38] quantified the modification degree of Na–Mt with octadecylamine by TGA/DTG as the ratio between the mass fraction of organic weight loss owing to alkylammonium ions, and the molecular mass of the organic cation exchanged on the clay platelets surfaces. In the TGA/DTG thermograms, three different regions can be distinguished, as is shown in Fig. 5. Region I appear below 150 °C related to the loss of surface-adsorbed water and gas. Moreover, Na–Mt presents the higher weight loss in this region caused by the hydrophilic nature of sodium cation. Region II occurs in the temperature range of 150–500 °C and is attributed to the decomposition of the octadecyl amine. Therefore, region II was used to the estimation of the modifying content in the OC. Finally, region III emerge at temperatures higher than 500 °C as a result of the loss of structural water released by the dehydroxylation of the clays layers [38]. Also, the authors made DTG characterization for further study of OC in region II. They stated that samples with low contents of modifying agent present only one peak, while at higher CEC concentrations a second peak appears. This difference is caused by the location of the modifying agent in the OC.

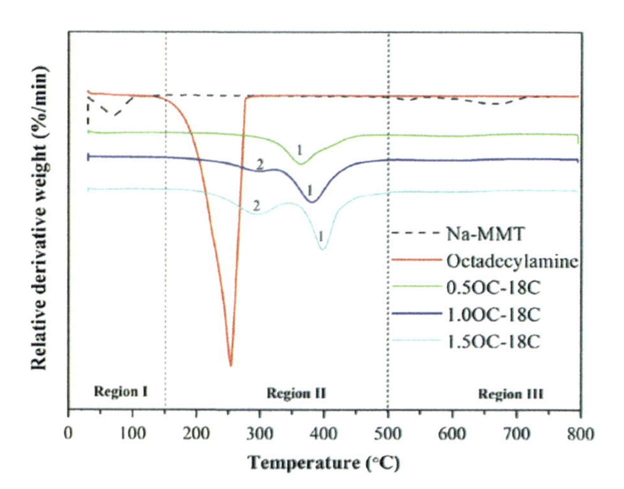

Fig. 5 DTG thermograms of Na–Mt, octadecylamine and OMt with various concentrations of modifying agent: 0.5, 1.0 and 1.5 CEC times of Na-MMT. Reprinted from Sookyung et al. [38]

Soares et al. [39] analyzed the thermal behaviour of modified anionic and cationic clays with imidazolium ionic liquid by TGA. The decomposition mass below 200 °C was attributed to the absorbed water and volatile molecules. Around 320 °C, a second weight loss region appears corresponding to the decomposition of the imidazolium ionic liquid. Also, the content of the modifying agent was calculated as the difference of loss of mass between the raw clay and the organoclay.

6 Elastomeric Clay Composites

During the last 25 years, the development of nanocomposites was intensively explored by researchers, achieving remarkable improvements in mechanical, thermal and physicochemical properties compared with those of conventional composites materials. In this section, a review of some recent works with rubber as matrix reinforced with some type of clay are presented.

6.1 Natural Rubber

Natural Rubber (NR) is obtained from the sap of different trees, almost entirely from the plant *Hevea brasiliensis*, therefore is regarded as a sustainable polymer. In the latex state, NR contains approximately 30% dry rubber content. The latex is normally concentrated, mostly by centrifugation, increasing the dry rubber content approximately up to 60% before it is distributed. NR is a hydrocarbon diene monomer whose repeating unit is cis-isoprene. The outstanding strength of natural rubber has maintained its position as the preferred material in many engineering applications. It has a long fatigue life, good creep and stress relaxation resistance and low cost [40].

Clay/NR composites have been successfully prepared by several procedures: melt mixing, latex compounding and in solution.

George et al. [6] prepared NR reinforced with different organo modified Cloisite clays (10A, 15A, 25A, 30B and 93B) in a two-roll mixing mill and studied the effect of concentration (from 1 up to 15 phr), type of clay and vulcanizing systems on the mechanical properties. They concluded that the conventionally vulcanized samples exhibited higher tensile strength and elongation at break compared to efficient and peroxide vulcanized samples. The maximum tensile strength was reached by the composite containing 5 phr of nanoclay. Among the different samples, NR containing the cloisite 25A exhibited the better properties.

Carli et al. [41] prepared composites of NR with 2, 4 and 8 phr of Cloisite 15A as a substitute for conventional silica (SiO_2). Based on mechanical properties, Authors concluded 6that 50 phr of silica can be replaced by 4 phr of organoclay without affecting the final properties even after ageing.

Siririttikrai et al. [19] prepared compounds from fresh NR latex (FL) and from concentrated NR latex (CL) with unmodified montmorillonite, due to that mixing in the latex state before being coagulated could be advantageous and this process can be carried out at the rubber plantation or at collection points nearby. Different methods yielded different results because during centrifugation process not only water is removed, also some water-soluble materials (like proteins), therefore the chemical composition of natural latex and concentrated latex are not exactly the same, an explanation of how this could affect the material properties is proposed in the scheme shown in Fig. 6. They concluded that MMT can be used to reinforce NR effectively by using fresh latex and MMT water dispersion as the starting point.

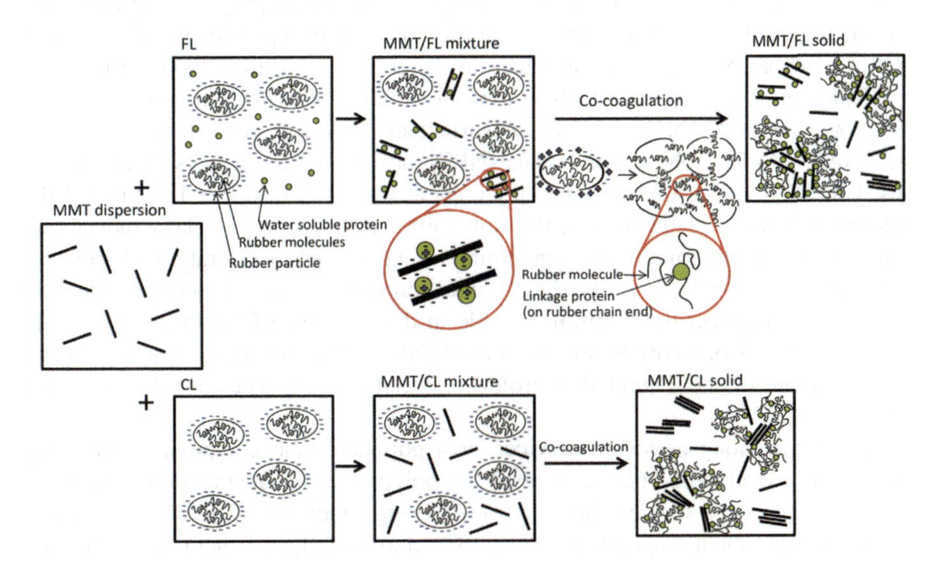

Fig. 6 Schematic illustration of Mt and rubber latex mixing and the co-coagulation processes for different types of NR latex. Reprinted from Siririttikrai et al. [19]

Sookyung et al. [38] made composites of NR with Na–Mt nanoclay modified with different concentrations of octadecylamine at 0.5, 1.0 and 1.5 times the CEC of Na–Mt, the obtained that at higher concentration value a larger *d*-spacing and a higher degree of clay dispersion in the matrix were observed. With a concentration of 1.5 times of the CEC, a faster curing reaction and a higher crosslink density were obtained and the mechanical and dynamic properties and thermal stability were enhanced.

6.2 Styrene Butadiene Rubber

Styrene-butadiene rubber (SBR) is an elastomeric copolymer consisting of styrene and 1,3-butadiene. Most SBR is produced by emulsion polymerization as well as by solution polymerization. SBR is characterized by a good abrasion resistance, ageing stability and low-temperature properties. The tire industry consumes approximately 70% of total SBR production. SBR formulations usually require a high amount of reinforcing filler, being CB the most used [40].

Praveen et al. [42] prepared SBR nanocomposites reinforced with a hybrid system of CB (N330) and OMt using octadecylamine as a modifier. The study included the variation of clay content (5, 10, 15 and 20 phr of OMt) in gum rubber and in rubber filled with 20 phr of CB, preparing in a two-roll mill. The presence of intercalated, aggregated and partially exfoliated structures were revealed through XDR and TEM. Samples containing 10 phr of OMt resulted in an increase of 153% in tensile strength, 157% in elongation at break and 144% of modulus 100%. The obtained results open up a new prospect in developing CB–OMt-hybrid nanocomposites, facilitating the possibility of partial replacement of CB with OMt in rubber products without affecting the critical performance properties.

Sadek et al. [43] studied SBR composites reinforced with sodium bentonite as well as modified with two types of surfactants: dodecyl benzene sulfonic acid (DBSA) and nonyl phenol ethoxylate (NPE), also a mixture of the surfactants was used: DBSA/NPE (50/50%). The clay content was 2, 4, 6, 8 and 10 phr, and the compounds were prepared in a two-roll mill. The presence of clays induces an increase in the minimum and maximum torque values, acceleration of the vulcanization process, and improved mechanical properties with organoclay content up to 6 phr. This effect was more noticeable in the presence of the treated clay with DBSA/NPE. Also, incorporation of 6 phr of DBSA/NPE-clay resulted in significant improvement of the degradation profile of the nanocomposites at 90 ± 1 °C for 4 days.

During the long period of service, most polymers (and composites) gradually lose their useful properties as a result of polymer chain degradation. The main harmful agents are oxygen, heat, and high energy radiation. The effects of these agents depend mainly upon the chemical structure of polymer chain. Youssef et al. [44] analyzed the effect of gamma radiation on the final properties of NR/Na–Mt and SBR/Na–Mt nanocomposites prepared by coagulating the mixture of rubber latex and clay aqueous suspension. The clay concentration was varied from 2 to 10 phr and the irradiation of the samples was carried out using a cobalt-60 gamma cell source with doses of 10, 25, 50, 75, and 100 kGy at a dose rate 5 kGy h^{-1}. XRD results indicated well dispersion of rubber latex into clay stacked layers and the platelets have a preferential orientation forming exfoliated NR/Na^{+}-MMT nanocomposites while SBR/Na^{+}-MMT form intercalated nanocomposites. Author explains that SBR have bulky benzene groups, which partially restrict the chains to intercalate into the gallery gap of the clays. Second, NR contains a higher number of unsaturated double bonds than SBR, thus the polarity of NR is higher than SBR, as a result, NR could form exfoliated whereas SBR are intercalated. Overall irradiation dose range together with clay loading, an improve of the mechanical properties of rubber/Na–Mt nanocomposites was obtained. However, SBR nanocomposite, in particular, attains its higher value at 50 kGy, then decreases; also, the thermal stability at 50 kGy is higher than 100 kGy.

6.3 Nitrile-Butadiene Rubber

Nitrile-butadiene rubber (NBR) is a polar elastomer, with a carboxylic group and a double bond in the rubber backbone. The polar nature of the NBR allows a higher resistance to non-polar substances such as hydrocarbons and oils. Depending on the acrylonitrile content, important improvements in abrasion resistance are obtained, therefore it has a great performance in the seal industry. Other advantages of NBR are its low cost and good processability. However, some of the negative aspects include a low ageing resistance because of the unsaturated backbone present in the chemical structure, thus is sensitive to environmental factors [45, 46].

Costa et al. [47] explain the effect of Mt in polar rubbers as NBR. The presence of the Mt offers a crosslinking effect defined as "ion cluster". It was demonstrated that there is a chemical interaction between the NBR chains and the clay particles, especially in cases where there are metal oxides in the rubber vulcanization formula like ZnO. This reaction can be explained taking into account the acid-base reaction, where acidic ACOOH group of the NBR reacts with the ZnO producing an ionic salt and releasing water.

De Sousa et al. [48] formulated nanocomposites of XNBR with different types of clays (Na- Mt, OMt and a natural bentonite). Using XRD and mechanical properties, they concluded that the organo-modifier agent and the molar mass of rubber had a significant influence on the dispersion of the clay after the mixing process on a two-roll mill.

Finally, Yu et al. [49] presented a very interesting work combining OMt and NBR by emulsion process. TEM images of the nanocomposites exhibited uniform dispersions of the clay sheets inside the NBR matrix. Also, the XRD patterns confirmed the intercalation of the NBR polymeric chains inside the interlayer of the OMt. The final nanocomposites showed an improvement of mechanical properties and solvent resistance comparing to pure NBR. Moreover, thermogravimetry indicates an increase in the maximum decomposition rate of the nanocomposites in comparison with that of pure NBR were observed, indicating an enhancement of the thermal stability. There was also a faster curing rate and a higher glass transition temperature, storage modulus and loss modulus.

6.4 Ethylene-Propylene-Diene Rubbers

Ethylene propylene diene terpolymer (EPDM) is an unsaturated polyolefin rubber with wide applications. Due to its good mechanical properties, very low unsaturation and associated resistance to ageing and ozone deterioration, it has become extensively used in making automotive tire sidewalls, cover strips, wires, etc. In a non-polar rubber like EPDM, the presence of stearic acid in the formulation favours the dispersion of clays due to the chemical reactions leading to ester formation involving carbonyl groups of acid and hydroxyl group of clay [50]. The early works on preparation and characterization of EPDM nanocomposites included that of Usuki et al. [51] and Zheng et al. [52].

Liu et al. [17] analyzed the reinforcement of nano kaolin (NK) powder into four types of rubber: SBR, NR, BR and EPDM. The reinforcing effects were compared with those from precipitated silica (PS). The compounds were made in an open roll mill. The results showed that NK can greatly improve the vulcanizing process by shortening the time to optimum cure (t_{c90}) and prolonging the setting-up time (t_{10}), which improves production efficiency and operational security. The rubber composites filled with nano kaolin enhances the mechanical properties and thermal stability. The tensile strength of the rubber/NK composites is close to those of rubber/PS composites. The tear strength and modulus presented a lower performance compared with those containing precipitated silica.

Chang et al. [53] showed that the oxygen barrier properties of EPDM/OMt clay nanocomposites are better than that of pristine EPDM. The oxygen permeability of 10 phr layered clay filled EPDM nanocomposite was reduced to 60% as compared with the pristine EPDM compound. The gas permeability was ascribed due to the uniform dispersion of the impermeable clay layer with the planar orientation in the EPDM matrix.

Zhang et al. [54] synthesized an intercalation agent containing double bonds for modifying the interface of EPDM/clay nanocomposites, using a Na–Mt. In order to compare with the OMt containing double bonds (VMMT), the OMt modified with cetyl trimethyl ammonium bromide (CTAB) without double bonds was also prepared; in both cases, the content used were 0, 3, 5 and 10 phr. Authors analyzed the relaxation dynamics via broadband dielectric relaxation spectroscopy. In polymer/clay nanocomposites, the environment of the polymer chains inside the silicate gallery will greatly affect the molecular relaxation and mobility. According to the literature, two distinct types of dynamic behaviour may exist in such systems. One is the slower relaxation dynamics associated with glass-transition temperature (T_g) and is attributed to the large interlayer distance and the strong polymer-filler interactions. In contrast, when the polymer-filler interactions are weak, the polymer segments will exhibit the faster relaxation mode, thereby leading to the depression of T_g value. According to the XRD analysis, the VMMT containing double bonds was exfoliated, and the crosslinking reaction between VMMT and EPDM chains promotes the exfoliation of the organic layered compounds. The dielectric relaxation spectra show that the segmental relaxation of EPDM shifted to higher temperature with increasing VMMT content, and a new relaxation, which was attributed to the interfacial relaxation being confined by the MMT platelets, was detected at a much higher temperature and lower frequency, whereas that did not appear when adding OMt.

6.5 *Role of Nanofiller as Compatibilizer in Rubber Blends*

Natural and synthetic rubbers are not used very often in an isolated form. An elastomer is mixed with another one, especially in the tire industry, for three main reasons: to improve properties of the material, to improve processing or to reduce costs. Most blends of rubbers are immiscible, which leads to a non-uniform dispersion of the components in the rubber matrix. For example, in Fig. 7 can be seen images taken by TEM of unfilled NR/SBR blend with a different ratio: on the left a compound with 25 phr of NR and 75 phr of SBR (25NR/75SBR), and on the right the complementary sample (75NR/25SBR), being SBR the dark phase and NR the clear one [55].

Nanoclay also shows preferential migration to specific rubbers when used in blends, due to organoclay is more compatible with polar rubbers. By the introduction of functional groups, the polarity can be changed so as to have better filler dispersion and hence compatibilizing action.

The different solubility of two rubber is related to different polarity and level of unsaturation in the rubber components, leading to an unequal dispersion but also to

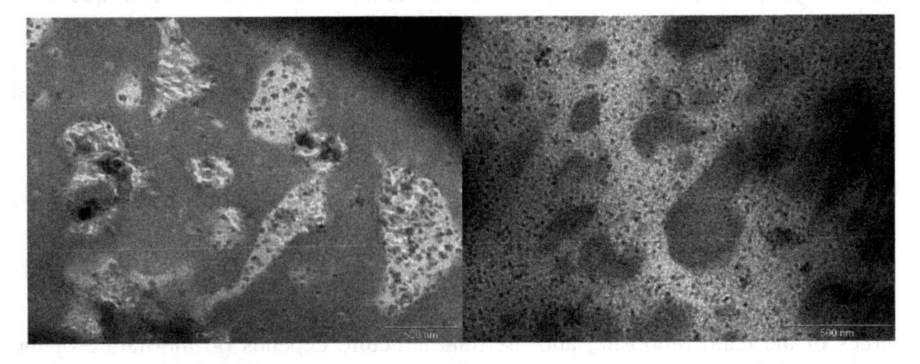

Fig. 7 TEM micrographs of vulcanized NR/SBR blends (left) 75NR/25SBR and (right) NR/SBR, being SBR the dark phase and NR the clear one

a diffusion of the components during the mixing or during the vulcanization process [56, 57], which could generate an uneven crosslinking. The presence of reinforcement particles in some cases can complicate this situation even more.

The properties of rubber blend/clay nanocomposites are strongly affected by the location and distribution of the clay in addition to the size, shape and dispersion of each rubber phase. Depending on the location of the clay in the matrix or dispersed phase, a separated or encapsulated morphology could be formed. All this can cause a narrow interphase and poor physical or chemical interactions between the two phases, influencing the mechanical properties of rubber blend.

OC particles added at a low loading level could play as interfacial compatibilizers for immiscible rubber blends reducing the domains sizes. This is due to nano level dispersion, generating a reduction of the interfacial tension between the two rubber phases [58].

Compatibilizers act through a chemical reaction (reactive compatibilization) or by its capability for interaction with blend components or by its interaction with chemicals present at the interface of blend components. Most of the interactions are intermolecular forms of attraction such as van der Waals and hydrogen bonding, based on the polarity of materials (nonreactive compatibilization). The surfactants present in organoclay also contribute to compatibilization. It is also possible that two immiscible polymer chains can intercalate the same clay platelet and play the role of a block or graft copolymer. Generally, graft or block copolymers have similar segments to blend components and hence act as compatibilizers in blends [12].

Rajasekar et al. [59] used clays as a compatibilizing agent for NR and NBR blends. They incorporated an OMt by solution mixing to promote a uniform dispersion inside the NR matrix. After that, they incorporate the first NR/OMt composite to a sulfur formulated NBR matrix, observing important changes in morphology, curing characteristics, mechanical properties, swelling and compression set. They established how the improvements in properties come from the distribution of the clay inside the rubber matrix. By their morphological studies, it was possible to determined intercalation in the whole NR/NBR composite. Also, faster scorch time and cure time, as well as an increase in maximum torque were obtained compared to the neat NBR matrix. Increases in storage modulus and lesser damping characteristics for the compounds with OMt were measured. The final reinforcing efficiency was determined comparing enhancements in mechanical properties and higher swelling resistance in oil and solvents.

7 Preparation of Nanocomposites

The incorporation process of clays into a rubber matrix influences the final properties of the nanocomposite, and the final structure depends on the nature of the rubber and the type of clay employed. Maiti et al. [4] explain the mechanism by which the nanofiller aid the rubber formulations to improve their properties [1].

The optimum improvement of the nanocomposite properties depends upon the incorporation technique, the dispersion of the clay and the concentration of the reinforcement within the rubber matrix [60].

Several techniques were used to prepare rubber/silicate nanocomposites. The most known are [1]:

- Melt mixing
- Solution blending
- Latex blending/latex compounding
- Sol-gel processing
- Emulsion polymerization.

7.1 Melt Mixing

The melt mixing technique is one of the most utilized to prepare elastomer/OC nanocomposites. The method involves internal mixers and open two-roll mills [4], which consists of blending the layered silicate with the polymer matrix in the molten state. The full exfoliated rubber/clay nanocomposites are difficult to obtain by this method because of the high viscosity of the rubber in the molten state but if the layer surfaces of the clay are sufficiently compatible with the constituent polymer of rubber, the polymer chains can crawl into the interlayer space and form an intercalated or an exfoliated nanocomposite (see Fig. 8b) [23].

The intercalation/exfoliation processes are governed by the modification of elastomers and also the chemistry involved in compounding and curing state to form elastomer-clay nanocomposites with improved properties [4]. Maiti et al. also explained how the temperature and shear rate are important variables to promote an intercalation/exfoliation phenomena. Ray et al. [61] discussed how the melt intercalation is highly specific to the polymer, leading to new rubber formulations that were previously inaccessible. Normally, polymers containing polar groups (capable of associative interactions, such as Lewis-acid/base interactions or hydrogen bonding) favour intercalation phenomena because the surfactant molecule in the OC has the same non-polar character as the polymer [62].

Several authors have utilized this technique to prepare nanocomposites. For example, George et al. [6] investigated the mechanical properties of natural rubber/OC systems by melt mixing using a two-roll mill. Authors have studied the effect of clay concentration, type of clay and vulcanizing systems on the mechanical properties. Their results show how the clay interlayer distance is increased in all samples after incorporation into the rubber matrix, as great improvements in tensile strength, elongation at break and modulus.

Most of the published works present the reinforcement of clay nanocomposites based on acrylonitrile butadiene rubber (NBR), ethylene propylene diene (EPDM) and NR [4]. Wang et al. [63] studied NBR/expanded graphite (EG)/CB micro and nanocomposites systems to compare with NBR/CB systems. Their results show an

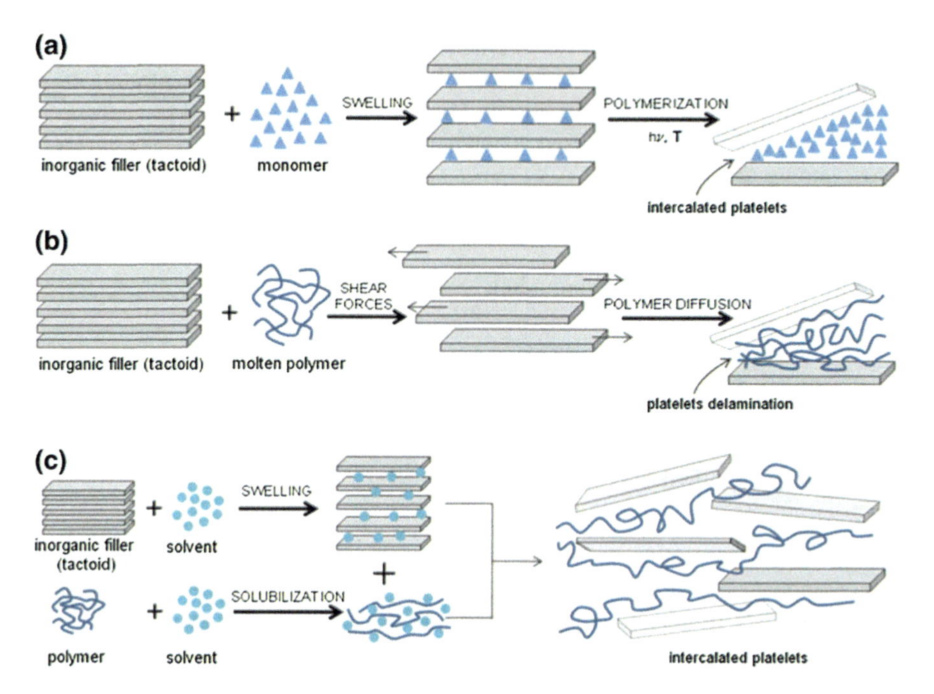

Fig. 8 Schematic representation of **a** in situ polymerization, **b** melt processing, and **c** solution casting [89]. Reprinted from Unalan et al. [90]

effect on the properties depending on the EG and CB loading content. By melt mixing, there is a better dispersion of EG, showing remarkable enhancement on tensile and dynamic mechanical properties. Przybyek et al. [63] prepared NBR reinforced with an antibacterial Mt, presenting important improvements in mechanical and thermal properties.

Melt mixing is also the preferred mode of incorporation of oxides like nano titanium dioxide, nano zinc oxide and others [4].

7.2 *Solution Blending*

Solution blending includes the solubilization of rubber in a proper solvent and then, the dispersion of the filler into the solution [4]. Later the solvent is evaporated after reaching a good dispersion of the nanofiller, usually, under vacuum or precipitation [4], the process is shown in Fig. 8c) and also in Fig. 9. The principal feature of this technique is that the layered silicates can be exfoliated into single layers using a solvent in which the polymer is soluble, overcoming the weak forces that stack the layers together [23]. The solvent allows the elastomer chains to be uncoiled and disentangled and easily enter between the layers of the clay or interact with the

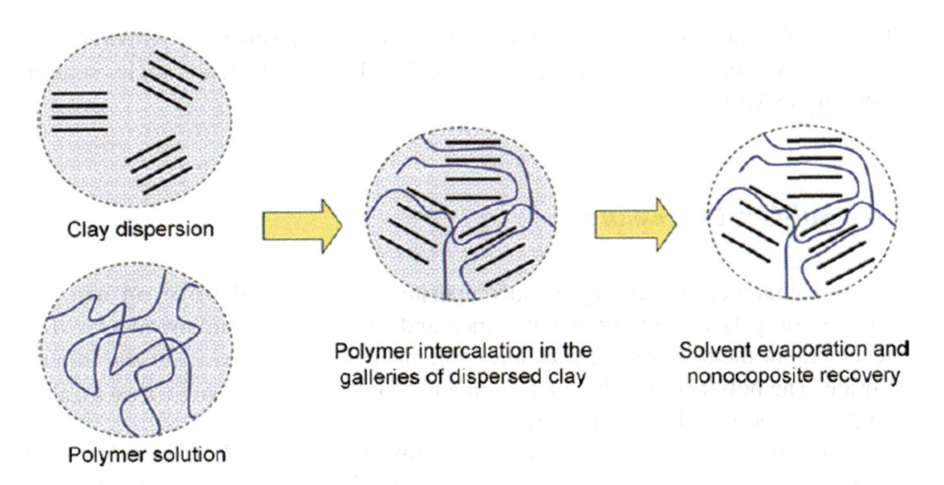

Fig. 9 Schematic representation of solution blending process. Reprinted from Pavlidou et al. [23]

particulate nanofillers [4]. In this method, the ultrasonic dispersion of clay is indispensable to increase the interfacial adhesion between clay and elastomer matrix in order to reach a high-performance rubber–clay nanocomposite. The main drawback of this method is the disposal of large quantities of solvent [23].

Usha Devi et al. [20] swelled an OMt in toluene to promote an effective method to fabricate an SBR nanocomposite, observing a complete rate of dispersion in the elastomer.

7.3 Latex Blending/Latex Compounding

The latex compounding method was developed to solve the problem of incompatibility between the rubber and fillers, improving the dispersion of fillers in the rubber matrix [3]. Latex is defined as a stable aqueous dispersion of fine rubber particles, with a particle size around 300–500 nm [1]. Most of the elastomers are available in latex form, so pristine clay can be added directly or in its aqueous dispersion (slurry). Moreover, an additional advantage is that the layered silicates can absorb water from the suspension, which aid to expand the interlayer space [4].

After the combination of latex and layered silicates, a co-coagulation process is carried out to produce a masterbatch product. This method allows the fabrication of latex film products, such as gloves, condoms, and coatings if the latex is mixed with the layered silicates and curatives in suspension [1].

If there is a good compatibility between rubber and clay, the formation of structures known as "house of cards" or "skeleton" can occur because of the presence of rubber particles that penetrate the clay interlayer gallery due to a polarity mismatch [64]. Rezende et al. [8] used the latex compounding method to obtain an NR/clay composite obtaining improvements in mechanical properties and characterizing the nanocomposite structure by TEM and small angle neutron scattering (SANS).

7.4 Sol-Gel Processing

In this technique, rubber/clay nanocomposite is synthesized using an aqueous solution (or gel) containing the polymer and the silicate building blocks. The precursors for the reaction are clay silica sol, magnesium hydroxide sol and lithium fluoride. The polymer aids the nucleation and growth of the inorganic host crystals and gets inside the clay layers [23].

Some authors have developed sol-gel systems. Brantseva et al. [65] used clays to improve the adhesive capacity of rubber compounds showing how both structural and pressure-sensitive adhesives (PSAs) were improved. With the sol-gel method, they ensured exfoliated or at least intercalated conditions. George et al. [6] prepared SBR/NBR/OMMT composites and notice great improvements in thermal stability and swelling properties, attributed to the barrier characteristic the clays impart to the rubber matrix.

Some of the problems this method presents are [23]:

- The synthesis of clay minerals generally requires high temperatures, which can decompose the rubber chains.
- There is an aggregation tendency of the growing silicate layers.

7.5 Emulsion Polymerization

In this method, the monomers of the polymer are dispersed in water together with an emulsifying agent and the clays, as can be seen in Fig. 8a). It is beneficial to achieve a good interaction between the rubber matrix and the reinforcement, since there is a joint polymerization between the rubber chains and the clays, leaving a nanocomposite where some of the clay sheets are embedded in the rubber particles, while some polymer chains are adsorbed on the surface of the clay particles [66], due to this, a better dispersion of the reinforcement can be obtained [67].

Polymerization is promoted with temperature, and the clay swelling process occurs directly in the polymerization medium (water) and in some cases, by the low molecular weight of liquid monomers [67]. One of the drawbacks of this method is that requires a certain time period to form the nanocomposite and depends extensively on the polarity of the monomers, the type and surface of the filler and the initiation temperature of the polymerization reaction [4].

Distler et al. [68] explained a simple procedure to perform an emulsion polymerization. The base in this type of process consists of mixing: deionized water, surfactant (for example sodium dodecyl sulfate), the reinforcement particles and the monomer. Within the process, the agitation is fundamental so that phase separation does not occur. As mentioned above, the polymerization initiator (for example sodium persulfate) is also fundamental, since the speed and the form of initiation of the process depend on it.

The polymerization system consists of a reactor with a stirrer, a reflux condenser, two lines for content the monomer emulsion and the initiator solution, and a temperature control unit which normally is rinsed with nitrogen. The reaction initiates with heat (approximately 70°C) and takes a few hours to complete. The final product is then filtered and then characterized by solid content and particle size [69].

Among the aforementioned methods, in situ polymerization and melt intercalation are considered as commercially attractive approaches for preparing polymer/clay nanocomposites [23]. In addition, the absence of a solvent makes direct melt intercalation an environmentally and an economically favourable method for industries from a waste perspective [62]. Moreover, it is compatible with the current industrial processes, such as extrusion and injection moulding. However, the dispersion of the clay in the polymer prepared by the melt mixing is not as good as that obtained by the latex compounding [69]. The main drawback is the poor dispersion within the matrix and the lack of affinity between the layered silicate and the organic polymers.

The latex compounding process is recommended when using hydrophilic clays because it occurs in an aqueous medium. While the process of melt blending is recommended when using hydrophobic clays (OMt) since with chemical modification they become more compatible with the elastomer.

8 Rubber/Clay Nanocomposites Properties

As it was mentioned above, there are numerous variables that define the final performance of a composite material, particularly when it is about a complex combination such as a vulcanized nanocomposite of rubber and clays.

The type of reinforcing the material, the shape of the particle as well as its structure, the type of rubber matrix and the preparation method play a fundamental role in the final properties of the compound. Depending on the application, it is necessary to comply with certain mechanical standards; therefore, it is necessary to consider those tests that conveniently help to determine the properties that describe how the behaviour of the material will be during its life in use.

8.1 Vulcanization Variables

The presence of nanoclay has a strong influence on the vulcanization process. The cure characterization of nanocomposites gives information about the interaction between filler and the matrix, and also the extent of filler-filler inter aggregations. George et al. [6] investigated NR reinforced with organo modified Cloisite clay and found that the t_{c90} value decreases when the concentration of clay is higher. This indicates that organophilic clays accelerate the vulcanization process in these samples due to the formation of a transition metal complex in which sulfur and amine groups of the intercalate layers may involve [65].

Siririttikrai et al. [19] studied the cure characteristics of NR containing different amounts of Mt prepared from fresh or concentrated NR latex (FL and CL respectively). Figure 10 shows the cure curves of the compounds for different Mt content and Table 1 summarizes the characteristic cure parameters. It can be seen that, for the Mt/FL series, the maximum torque (M_H) increases with increasing Mt content. No effects were observed on scorch time (t_{s2}). The Mt/CL series displays a different behaviour: the sample made from CL with 4 phr of Mt displays the maximum value for t_{s2} and also for M_H, then these values decrease when the amount of Mt is higher. When the amount of Mt is the same, the Mt/FL compound displays slightly shorter cure times (t_{c90}) than does Mt/CL compound.

Several authors proposed that high energy ionizing radiation can be considered as a cost-effective and additive–free technique [70, 71]. Moreover, it can be carried out at room temperature, so it decreases the toxic volatile releases. Shoushtari Zadeh Naseri and Jalali-Arani [72] studied SBR/EPDM blends with and without OC vulcanized with gamma radiation. The effects of the radiation dose on the interaction between phases, crosslink density, gel content, and microstructure of the prepared OC containing samples compared to those of the unfilled blends were investigated. Authors used gel content as a criterion to estimate the crosslink density and found that the OC induces an enhancement of gel content respect to the system without reinforcement (Fig. 11), which would be related with the improvement of radical-radical interaction and formation of physical crosslinks [71]. Sookyung et al. [38] analyzed compounds of NR reinforced with 5 phr of Na–Mt modified with octadecylamine in concentration values from 0.5 up to 1.5 times

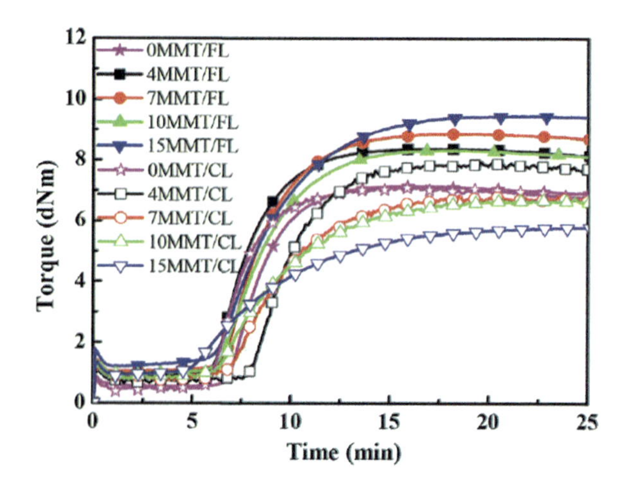

Fig. 10 Cure curves of Mt/ fresh latex (FL) and Mt/ concentrated latex (CL) compounds with the indicated amount of Mt. Reprinted from Siririttikrai et al. [19]

Table 1 Cure parameters of different Mt/latex compounds

Compound	M_L (dN m)	M_H (dN m)	t_{S2} (min)	t_{C90} (min)
0Mt/FL	0.54	7.15	7.60	10.88
4Mt/FL	0.91	8.37	6.83	10.65
7Mt/FL	1.04	8.86	7.33	11.67
10Mt/FL	1.00	8.32	7.32	11.93
15Mt/FL	1.20	9.42	7.15	13.21
0Mt/CL	0.44	7.08	6.65	10.15
4Mt/CL	0.71	7.94	8.79	13.10
7Mt/CL	0.78	6.72	8.18	13.86
10Mt/CL	0.83	6.76	7.75	14.82
15Mt/CL	0.94	5.79	7.43	15.09

The number in the compound indicates Mt amount in phr. Reprinted from Siririttikrai et al. [19]

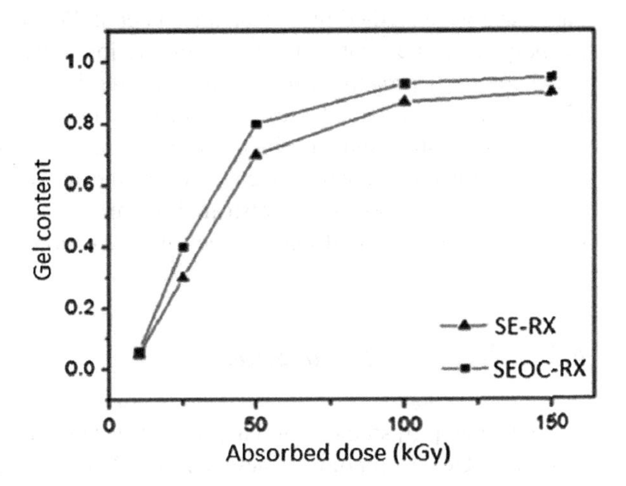

Fig. 11 Gel content of unfilled SBR/EPDM (SE) and SBR/EPDM with OC (SEOC) samples as a function of absorbed dose (RX). Reprinted from Shoushtari Zadeh Naseri and Jalali-Arani [72]

the CEC value of Na–Mt. From the rheometric curves, they obtained that t_{s2} and t_{c90} decrease when the amount of the modifying agent increases. This behaviour is attributed to the effect of the ammonium groups of the octadecylamine that causes an increase of the zinc complex content, and therefore acceleration in the vulcanization reaction. Also, a higher value of M_H was obtained with a higher content of the modifying agent, which indicates a higher interaction between the clay and the rubber chains: a higher loading of the modifying agent gives a higher interlayer spacing of OC, allowing the penetration of rubber chains or their molecular segments into the silicate. This allows to explain the improvements in mechanical and thermal properties, the tensile strength and modulus at 100 and 300%; elongation and hardness of the composites were improved with the concentration of the modifying agent.

8.2 Rheological Properties

Microstructural changes of the nanocomposites can be followed by dynamic frequency sweep tests. Maroufkhani et al. [58] performed measurements within the linear viscoelastic region ($\Upsilon = 1$) in blends of polylactide acid (PLA) and NBR reinforced with 4 phr of Cloisite 10A (OC), prepared via melt compounding. They investigated the influence of the acrylonitrile (ACN) content of NBR on the dispersion and localization of the OC; three different ACN levels content in NBR were used: low (19%), medium (33%) and high (51%). It was found that the presence of nanoclays increases the G' values of the samples in the low-frequency region. The plateau in this region is caused by physical interactions between the nanolayers and the rubber chains. In addition, the presence of quaternary ammonium surfactants of OC promotes a good compatibility with PLA by hydrogen bonding. Finally, this phenomenon was not altered by the ACN content in the polymer blend.

Dynamic mechanical analysis (DMA) of XNBR/NR blends reinforced with nanoclays were studied by Satyanarayana et al. [73] in order to determine the T_g of the polymers. They stated that the incorporation of clays into a rubber matrix affects the rheological properties by shifting the $tan\ \delta$ peak to lower temperature values [74]. As it was explained before, this effect is influenced by the filler type and load, the type of matrix and the filler-matrix interaction. Furthermore, the authors also reported that the presence of clay lowers the height of the $tan\ \delta$ peak due to a reduction of the amount of rubber mobile during the dynamic transition process as consequence of the reinforcing effect of clay nanoparticles.

8.3 Mechanical Properties

Jahromi et al. [75] studied a ternary hybrid system consisting of polyamide 6 (PA6), NBR and OC. Compounds included an NBR and NBR activated with glycidyl methacrylate (GMA) groups. The Young modulus and tensile strength of the hybrid systems were improved by the increase of nanoclay content due to the effective reinforcing role of the inorganic filler [76]. However, the reinforcing effect was more effective for the nanocomposites including a reactive elastomer, which stems obviously from the better dispersion of nanoparticles along with nanoplatelet confinement in the thermoplastic phase domains.

A distinctive property to analyze the elastomer-filler interaction and filler-filler interaction is the non-linear viscoelastic response of filled systems, otherwise known as Payne effect [77]. Payne suggested that the formation of a three-dimensional structure was due to the filler incorporated in rubbers, which leads to higher modulus values at lower strains. Regarding the use of nanofiller, the surface area, surface modification and activity are some of the crucial factors that affect the non-linear viscoelastic response. A competition between filler-filler and polymer-filler interactions takes place in filled elastomers. If filler-filler interaction predominates, the Payne effect is more pronounced. Zachariah et al. [78] evaluated the dynamic shear storage modulus and loss modulus of NR and Chlorobutyl Rubber (CIIR) containing OC. The authors studied these systems because of their application in the manufacture of automotive inner tubes and inner liners. They found that the nanoclay loading in both matrices increases the modulus values at lower strain due to the formation of filler-filler and filler-polymer networks. These networks could break at higher strains, which results in the reduction of the modulus.

8.4 Barrier Properties

Many researchers reported on the permeability behaviour of nanoclay loaded elastomeric composites proposing that this property increases due to the creation of a tortuous path in the microstructure for the transport of gas molecules [79].

Mohan et al. [80] investigated the water uptake of NR/SBR blends reinforced with OC. The water content, W_c, in the sample was measured as % weight increase in the sample. W_c was measured until the composite specimen attained equilibrium water uptake content. The following equation was used to calculate the W_c in the specimen:

$$W_c = \frac{(W_t - W_o)}{W_o} \times 100$$

where W_t is the weight of specimen at time t and W_o is the initial weight of the sample before placing in water. Figure 12 presents the water mass uptake as a function of time (in days) of all samples, the sample that containing 3 wt% nanoclay has the lower water content value. Also, static and dynamic mechanical analysis (DMA) properties were less affected in the nanoclay filled rubber than in the base rubber. Chao et al. [1] studied gas-barrier properties of IIR/clay nanocomposites obtained via the latex method. Moreover, they compared samples prepared using latex and co-coagulating methods. It is important to remark that TEM images revealed both partially exfoliated and intercalated structures for the isobutylene-isoprene rubber (IIR)/Mt nanocomposite prepared using the latex method and, on the other hand, a purely intercalated structure in the sample prepared using the co-coagulating method. It was found that the oxygen transmission of the IIR/clay nanocomposites decreases progressively as the silicate content increase. The oxygen permeability of the compound with 10 phf of clay was reduced to 60% as compared with the neat IIR. Layered silicate is composed of many monolayers that are 1 nm thick and 200–300 nm long; this aggregated platelet structure of the layered silicates in the IIR matrix could provide an excellent barrier to the diffusion of oxygen because of the increased tortuosity of the diffusing path [1].

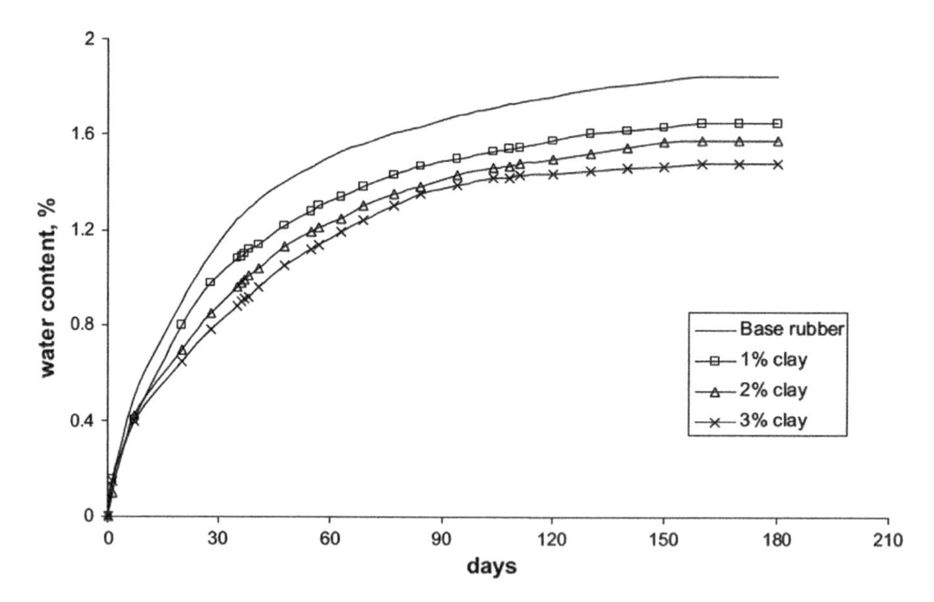

Fig. 12 Water mass uptake of rubber and rubber-nanoclay series in NR/SBR blends. Reprinted from Mohan et al. [80]

Wang et al. [81] stated through positron annihilation lifetime spectroscopy (PAS) and DSC that gas permeability in rectorite/SBR nanocomposites is mainly influenced by the free volume fraction and tortuous diffusional path effects attributed to the clay platelet-like morphology.

9 Applications

The incorporation of clays into rubber matrices comes from the need to improve the mechanical properties, such as tensile strength, elongation at break or elongation module [6, 8, 38]. There is also a need to improve vulcanization times [82, 83], as well as to extend the range of decomposition temperature [84]. There are also other properties that can be improved using rubber/clay compounds.

An important area of application of rubber/clay nanocomposites is gas and moisture barriers. The layers of clays that filled a polymer matrix serve as an impermeable medium for the gas and moisture. Although there is information about barrier properties of clay filled thermoset and thermoplastics are available in the literature, those related to barrier properties of rubber filled with clay nanoparticles are relatively few. Recently, several works were reported on the synthesis, characterization, mechanical and thermal properties.

The improvement of fatigue strength of rubber composites is also considered within automotive applications. Woo et al. [85] explain how the interest of fatigue life evaluation for rubber components such as in engine parts was increasing according to extend the warranty period of the automotive components. In their study, they developed a rubber material environmentally friendly with superior physical properties and fatigue life using a rubber-clay nanocomposite.

Other applications include structural and pressure-sensitive adhesives (PSAs). Brantseva et al. presented a paper about the uses of clays in the adhesive industry as rheological modifiers where an enhancement of barrier properties, thermal resistance and mechanical properties were found [86–89]. For example, clays have been used to increase shear strength and heat resistance of waterborne acrylic PSAs as an environmentally friendly replacement of solvent-borne acrylic PSAs [87]. Furthermore, there are reports of the use of Na–Mt and OMt to improve rheological and mechanical properties of polyisobutylene (PIB) based adhesive [87]. Results show better compatibility between PIB/OMt systems rather than PIB/Na–Mt, which was evidenced as an increase in viscosity with a 10 wt% of OMt content. On the other hand, Na–Mt did not show significant improvement of the rheological properties.

Finally, within the most innovative applications, the use of modified clays with bactericidal agents to formulate rubber mixtures with antimicrobial properties was analyzed. Przybyłek et al. developed an elastomeric nanocomposite that exhibits antibacterial and antifungal activity for the polish company Spoiwo (Spoldzielnia Pracy Chemiczno–Wytworczej 'Spoiwo', Radom, Poland) [20]. They used a rubber blend matrix and modified bentonite clay (Nanobent® ZR2), and noticed that by adding 1–3% of bentonite nanoparticles there was an enhancement in the elongation and tensile stress at break; they also observed improvements in thermal properties. Their work resulted significantly in the medicine, biomedical engineering and in the food industry.

10 Final Remarks

In this chapter, a review of recent works including rubber/clay nanocomposites were made. This review covers the structure and physical characteristics of layered silicates used as fillers in rubber matrices as well as the chemical and physical characteristic of the type of rubber polymers, and rubber blends, which are more likely used for nanocomposites development. Moreover, it describes several techniques for organic modification of clays nanoparticles to enhance hydrophobicity and improve the compatibility by intercalation or exfoliation of the polymer chains into the clay interlaminar space. Finally, the chapter reviews the rubber/clay nanocomposites characteristics (i.e. vulcanization and rheological, mechanical and barrier properties), and the nowadays applications, like the tire, adhesive and biomedical industries.

Acknowledgements Authors wish to thank the financial support from the National Agency of Scientific and Technological Promotion (ANPCyT PICT-2015-0027) of the Minister of Science and Technology and Productive Innovation (MinCyT) of Argentina.

References

1. Chao CC, Lin GG, Tsai HC, Lee YL, Chang PH, Cheng WT, Hsiue GH (2015) Isobutylene-isoprene rubber/layered silicate nanocomposites prepared using latex method: direct casting versus melt mixing after coagulation. J Reinf Plast Compos 34(21):1791–1803
2. Conzatti L, Stagnaro P, Colucci G, Bongiovanni R, Priola A, Lostritto A, Galimberti M (2012) The clay mineral modifier as the key to steer the properties of rubber nanocomposites. Appl Clay Sci 61:14–21
3. Gui Y, Zheng J, Ye X, Han D, Xi M, Zhang L (2016) Preparation and performance of silica/SBR masterbatches with high silica loading by latex compounding method. Compos Part B Eng 85:130–139
4. Maiti M, Bhattacharya M, Bhowmick AK (2008) Elastomer Nanocomposites. Rubber Chem Technol 81(3):384–469
5. Osman AF, Abdul Hamid AR, Rakibuddin, M, Khung Weng G, Ananthakrishnan R, Ghani, SA, Mustafa Z (2017) Hybrid silicate nanofillers: impact on morphology and performance of EVA copolymer upon in vitro physiological fluid exposure. J Appl Polym Sci 134(12)
6. George SC, Rajan R, Aprem AS, Thomas S, Kim SS (2016) The fabrication and properties of natural rubber-clay nanocomposites. Polym Test 51:165–173
7. Pal K, Pal SK, Das CK, Kim JK (2010) Influence of fillers on NR/SBR/XNBR blends. Morphology and wear. Tribol Int 43(8):1542–1550
8. Rezende CA, Bragança FC, Doi TR, Lee LT, Galembeck F, Boué F (2010) Natural rubber-clay nanocomposites: mechanical and structural properties. Polym (Guildf) 51 (16):3644–3652
9. Zhang Y, Liu Q, Zhang S, Zhang Y, Cheng H (2015) Gas barrier properties and mechanism of kaolin/styrene-butadiene rubber nanocomposites. Appl Clay Sci 111:37–43
10. Przybyłek M, Bakar M, Mendrycka M, Kosikowska U, Malm A, Worzakowska M, Szymborski T, Kędra-Królik K (2017) Rubber elastomeric nanocomposites with antimicrobial properties. Mater Sci Eng, C 76:269–277
11. Choi SS (2002) Difference in bound rubber formation of silica and carbon black with styrene-butadiene rubber. Polym Adv Technol 13(6):466–474
12. Alex R (2010) Nanofillers in rubber-rubber blends. In: Thomas S, Stephen R (eds) Rubber nanocomposites: preparation, properties and applications. Wiley, Chichester, UK, pp 209–234
13. Hashim AS, Azahari B, Ikeda Y, Kohjiya S (1998) The effect of bis (3-triethoxysilylpropyl) tetrasulfide on silica reinforcement of styrene—butadiene rubber. Rubber Chem Technol 71 (2):289–299
14. Gatos KG, Karger-Kocsis J (2010) Rubber/clay nanocomposites: preparation, properties and applications. In: Thomas S, Stephen R (eds) Rubber nanocomposites: preparation, properties and applications. Wiley, Chichester, UK, pp 169–190
15. ten Brinke JW, Debnath SC, Reuvekamp LAEM, Noordermeer JWM (2003) Mechanistic aspects of the role of coupling agents in silica-rubber composites. Compos Sci Technol 63 (8):1165–1174
16. Rattanasom N, Saowapark T, Deeprasertkul C (2007) Reinforcement of natural rubber with silica/carbon black hybrid filler. Polym Test 26(3):369–377
17. Liu Q, Zhang Y, Xu H (2008) Properties of vulcanized rubber nanocomposites filled with nanokaolin and precipitated silica. Appl Clay Sci 42(1–2):232–237
18. Galimberti M, Agnelli S, Cipolletti, V (2016) Hybrid filler systems in rubber nanocomposites. Elsevier Ltd

19. Siririttikrai N, Thanawan S, Suchiva K, Amornsakchai T (2017) Comparative study of natural rubber/clay nanocomposites prepared from fresh or concentrated latex. Polym Test 63:244–250
20. Usha Devi, KS, Ponnamma, D, Causin, V, Maria HJ, Thomas S (2015) Enhanced morphology and mechanical characteristics of clay/styrene butadiene rubber nanocomposites. Appl Clay Sci 114, 568–576
21. Nawani P, Burger C, Rong L, Hsiao BS, Tsou AH (2015) Structure and permeability relationships in polymer nanocomposites containing carbon black and organoclay. Polym (United Kingdom) 64:19–28
22. Botana A, Mollo M, Eisenberg P, Torres Sanchez RM (2010) Effect of modified montmorillonite on biodegradable PHB nanocomposites. Appl Clay Sci 47(3–4):263–270
23. Pavlidou S, Papaspyrides CD (2008) A review on polymer-layered silicate nanocomposites. Prog Polym Sci 33(12):1119–1198
24. Kievani MB, Edrak M (2015) Synthesis, characterization and assessment thermal properties of clay based nanopigments. Front Chem Sci Eng 9:40–45
25. Galimberti M, Senatore S, Lostritto A, Giannini L, Conzatti L, Costa G, Guerra G (2009) Reinforcement of diene elastomers by organically modified layered silicates. E-Polymers 57:1–16
26. Marques FADM, Angelini R, Ruocco G, Ruzicka B (2017) Isotopic effect on the gel and glass formation of a charged colloidal clay: laponite. J Phys Chem B 121(17):4576–4582
27. Ambre A, Jagtap R, Dewangan B (2009) ABS nanocomposites containing modified clay. J Reinf Plast Compos 28(3):343–352
28. Bianchi AE, Fernández M, Pantanetti M, Viña R, Torriani I, Sánchez RMT, Punte G (2013) ODTMA + and HDTMA + organo-montmorillonites characterization: new insight by WAXS, SAXS and surface charge. Appl Clay Sci 83–84:280–285
29. Daitx TS, Carli LN, Crespo JS, Mauler RS (2015) Effects of the organic modification of different clay minerals and their application in biodegradable polymer nanocomposites of PHBV. Appl Clay Sci 115:157–164
30. Zhuang G, Gao J, Chen H, Zhang, Z (2018) A new one-step method for physical purification and organic modification of sepiolite. Appl. Clay Sci 153(November 2017), 1–8
31. Zhou F, Yan C, Zhang Y, Tan J, Wang H, Zhou S, Pu S (2016) Purification and defibering of a Chinese sepiolite. Appl Clay Sci 125, 119–126
32. Milošević M, Logar M, Kaluderović L, Jelić I (2017) Characterization of clays from Slatina (Ub, Serbia) for potential uses in the ceramic industry. Energy Proc 125:650–655
33. Gamoudi S, Srasra E (2017) Characterization of Tunisian clay suitable for pharmaceutical and cosmetic applications. Appl Clay Sci 146(May):162–166
34. Peyne J, Gharzouni A, Sobrados I, Rossignol S (2018) Identifying the differences between clays used in the brick industry by various methods: iron extraction and NMR spectroscopy. Appl Clay Sci (October 2017):0–1
35. Ezquerro CS, Ric GI, Miñana CC, Bermejo JS (2015) Characterization of montmorillonites modified with organic divalent phosphonium cations. Appl Clay Sci 111:1–9
36. Alves JL de T. V. e. Rosa P, Morales, AR (2017) Evaluation of organic modification of montmorillonite with ionic and nonionic surfactants. Appl Clay Sci 150(June):23–33
37. Hojiyev R, Ulcay Y, Çelik MS (2017) Development of a clay-polymer compatibility approach for nanocomposite applications. Appl Clay Sci 146(April):548–556
38. Sookyung U, Nakason C, Venneman N, Thaijaroend W (2016) Influence concentration of modifying agent on properties of natural rubber/organoclay nanocomposites. Polym Test 54:223–232
39. Soares BG, Ferreira SC, Livi S (2017) Modification of anionic and cationic clays by zwitterionic imidazolium ionic liquid and their effect on the epoxy-based nanocomposites. Appl Clay Sci 135:347–354
40. Verdejo R, Lopez-Manchado MA, Valentini L, Kenny JM (2010) Carbon nanotube reinforced rubber composites. In: Thomas S, Stephen R (eds) Rubber nanocomposites: preparation, properties and applications. Wiley, Chichester, UK, pp 147–162

41. Carli LN, Roncato CR, Zanchet A, Mauler RS, Giovanela M, Brandalise RN, Crespo JS (2011) Characterization of natural rubber nanocomposites filled with organoclay as a substitute for silica obtained by the conventional two-roll mill method. Appl Clay Sci 52(1–2):56–61
42. Praveen S, Chattopadhyay PK, Albert P, Dalvi VG, Chakraborty BC, Chattopadhyay S (2009) Synergistic effect of carbon black and nanoclay fillers in styrene butadiene rubber matrix: development of dual structure. Compos Part A Appl Sci Manuf 40(3):309–316
43. Sadek EM, El-Nashar DE, Ahmed SM (2015) Effect of organoclay reinforcement on the curing characteristics and technological properties of styrene-butadiene rubber. Polym Compos 36(7):1293–1302
44. Youssef HA, Abdel-Monem YK, Diab WW (2017) Effect of gamma irradiation on the properties of natural rubber latex and styrene-butadiene rubber latex nanocomposites. Polym Compos 38(2):E189–E198
45. Liu J, Li X, Xu L, Zhang P (2016) Investigation of aging behavior and mechanism of nitrile-butadiene rubber (NBR) in the accelerated thermal aging environment. Polym Test 54 (2016):59–66
46. Xue X, Yin Q, Jia H, Zhang X, Wen Y, Ji Q, Xu Z (2017) Enhancing mechanical and thermal properties of styrene-butadiene rubber/carboxylated acrylonitrile butadiene rubber blend by the usage of graphene oxide with diverse oxidation degrees. Appl Surf Sci 423:584–591
47. Costa FR, Pradhan S, Wagenknecht U, Bhowmick AK, Heinrich G (2010) XNBR/LDH nanocomposites: effect of vulcanization and organic modifier on nanofiller dispersion and strain-induced crystallization. J Polym Sci, Part B: Polym Phys 48(22):2302–2311
48. de Sousa F, Mantovani G, Scuracchio C (2011) Mechanical properties and morphology of NBR with different clays. Polym Testing 30:819–825
49. Yu Y, Gu Z, Song G, Li P, Li H, Liu W (2011) Structure and properties of organo-montmorillonite/nitrile butadiene rubber nanocomposites prepared from latex dispersions. Appl Clay Sci 52(4):381–385
50. Ma Y, Li Q-F, Zhang L-Q, Wu Y-P (2006) Role of stearic acid in preparing EPDM/clay nanocomposites by melt compounding. Polym J 39(1):48–54
51. Usuki A, Tukigase A, Kato M (2002) Preparation and properties of EPDM-clay hybrids. J Appl Polym Sci 43:2185–2189
52. Zheng H, Zhang Y, Peng Z, Zhang Y (2004) Influence of clay modification on the structure and mechanical properties of EPDM/montmorillonite nanocomposites. Polym Test 23 (2):217–223
53. Chang YW, Yang Y, Ryu S, Nah C (2002) Preparation and properties of EPDM/organomontmorillonite hybrid nanocomposites. Polym Int 51(4):319–324
54. Zhang F, Zhao Q, Liu T, Lei Y, Chen C (2018) Preparation and relaxation dynamics of ethylene–propylene–diene rubber/clay nanocomposites with crosslinking interfacial design. J Appl Polym Sci 135(1):1–8
55. Mansilla MA, Valentín JL, López-Manchado MA, González-Jiménez A, Marzocca AJ (2016) Effect of entanglements in the microstructure of cured NR/SBR blends prepared by solution and mixing in a two-roll mill. Eur Polym J 81:365–375
56. Hess WM, Herd CR, Vegvari PC (1993) Characterization of immiscible elastomer blends. 330–372
57. Groves S (1998) Crosslink density distributions in NR/BR blends: effect of cure temperature and time. Rubber Chem Technol 44:958–965
58. Maroufkhani M, Katbab AA, Zhang J (2018) Manipulation of the properties of PLA nanocomposites by controlling the distribution of nanoclay via varying the acrylonitrile content in NBR rubber. Polym Test 65:313–321
59. Rajasekar R, Pal K, Heinrich G, Das A, Das CK (2009) Development of nitrile butadiene rubber-nanoclay composites with epoxidized natural rubber as compatibilizer. Mater Des 30 (9):3839–3845

60. Kanny K, Mohan TP (2017) Rubber nanocomposites with nanoclay as the filler. In: Thomas S, Maria HJ (eds) Progress in rubber nanocomposites. Woodhead Publishing, Duxford, United Kingdom, pp 153–177
61. Sinha Ray S, Okamoto M (2003) Polymer/layered silicate nanocomposites: a review from preparation to processing. Prog Polym Sci 28(11), 1539–1641
62. Theng BKG (2012) Polymer-clay nanocomposites, 2nd ed., vol. 4. Elsevier B.V
63. Wang LL, Zhang LQ, Tian M (2012) Mechanical and tribological properties of acrylonitrile-butadiene rubber filled with graphite and carbon black. Mater Des 39:450–457
64. Varghese S, Karger-Kocsis J (2003) Natural rubber-based nanocomposites by latex compounding with layered silicates. Polym (Guildf) 44(17):4921–4927
65. Brantseva TV, Antonov SV, Gorbunova IY (2018) Adhesion properties of the nanocomposites filled with aluminosilicates and factors affecting them: a review. Int J Adhes Adhes 82:263–281
66. Fawaz J, Mittal V (2015) Synthesis of polymer nanocomposites: review of various techniques. In: Mittal V (ed) Synthesis techniques for polymer nanocomposites, 1st edn. Wiley-VCH, Weinheim, pp 1–30
67. Ponnamma D Maria HJ, Chandra AK, Thomas S (2013) Rubber nanocomposites: latest trends and concepts. In: Visakh PM, Thomas S, Chandra A (ed) Advances in elastomers II. Advanced structured materials, vol. 12, April, Springer, Berlin, Heidelberg, pp 69–107
68. Distler D, Neto WS (2017) Machado F emulsion polymerization. In: Reference module in materials science and materials engineering, June, Elsevier, pp 35–56
69. Tan J, Wang X, Luo Y, Jia D (2012) Rubber clay nanocomposites by combined latex compounding. pp 825–831
70. Chaudhari CV, Dubey KA, Bhardwaj YK, Sabharwal S (2012) Radiation processed styrene-butadiene rubber/ethylene-propylene diene rubber/multiple-walled carbon nanotubes nanocomposites: effect of MWNT addition on solvent permeability behavior. J Macromol Sci Part B Phys 51(5):839–859
71. Dubey KA, Bhardwaj YK, Chaudhari CV, Bhattacharya S, Gupta SK, Sabharwal S (2006) Radiation effects on SBR–EPDM blends: a correlation with blend morphology. J Polym Sci, Part B: Polym Phys 44(12):1676–1689
72. Shoushtari Zadeh Naseri A, Jalali-Arani A (2015) A comparison between the effects of gamma radiation and sulfur cure system on the microstructure and crosslink network of (styrene butadiene rubber/ethylene propylene diene monomer) blends in presence of nanoclay. Radiat Phys Chem 115, 68–74
73. Satyanarayana MS, Bhowmick AK, Dinesh Kumar K (2016) Preferentially fixing nanoclays in the phases of incompatible carboxylated nitrile rubber (XNBR)-natural rubber (NR) blend using thermodynamic approach and its effect on physico mechanical properties. Polym (United Kingdom) 99:21–43
74. Bandyopadhyay A, Thakur V, Pradhan S, Bhowmick AK (2010) Nanoclay distribution and its influence on the mechanical properties of rubber blends. J Appl Polym Sci 115:1237–1246
75. Ebrahimi Jahromi A, Ebrahimi Jahromi HR, Hemmati F, Saeb MR, Goodarzi V, Formela K (2016) Morphology and mechanical properties of polyamide/clay nanocomposites toughened with NBR/NBR-g-GMA: a comparative study. Compos Part B Eng 90, 478–484
76. Wang C, Su JX, Li J, Yang H, Zhang Q, Du RN, Fu Q (2006) Phase morphology and toughening mechanism of polyamide 6/EPDM-g-MA blends obtained via dynamic packing injection molding. Polym (Guildf) 47(9):3197–3206
77. Yang R, Song Y, Zheng Q (2017) Payne effect of silica-filled styrene-butadiene rubber. Polym (United Kingdom) 116:304–313
78. Zachariah AK, Chandra AK, Mohammed PK, Parameswaranpillai J, Thomas S (2016) Experiments and modeling of non-linear viscoelastic responses in natural rubber and chlorobutyl rubber nanocomposites. Appl Clay Sci 123:1–10
79. Zachariah AK Transport properties of polymeric membranes: gas permeability and theoretical modeling of elastomers and its nanocomposites. Chapter 21, p 441

80. Mohan TP, Kuriakose J, Kanny K (2012) Water uptake and mechanical properties of natural rubber-styrene butadine rubber (nr-sr)—nanoclay composites. J Ind Eng Chem 18(3):979–985
81. Wang ZF, Wang B, Qi N, Zhang HF, Zhang LQ (2005) Polymer, 46(3):719–724
82. Qureshi MN, Qammar H (2010) Mill processing and properties of rubber-clay nanocomposites. Mater Sci Eng, C 30(4):590–596
83. Mathew G, Rhee JM, Lee YS, Park DH, Nah C (2008) Cure kinetics of ethylene acrylate rubber/clay nanocomposites. J Ind Eng Chem 14(1):60–65
84. Zhang W, Ma Y, Xu Y, Wang C, Chu F (2013) Lignocellulosic ethanol residue-based lignin-phenol-formaldehyde resin adhesive. Int J Adhes Adhes 40(2013):11–18
85. Woo CS, Kim WD, Do Kwon J (2008) A study on the material properties and fatigue life prediction of natural rubber component 483–484(1–2) C, 376–381
86. Brantseva TV, Antonov SV, Gorbunova, IY Adhesion properties of the nanocomposites filled with aluminosilicates and factors affecting them: a review. Int J Adhes Adhes 82(December 2017):263–281, 2018
87. Ahsan T (2007) Composition of bulk filler and epoxy-clay nanocomposite; U.S. Patent 7163973
88. Long-acting waterborne nanometer attapulgite clay/epoxy anticorrosive coating material and preparing method thereof, Chinese patent CN 102676028A, 2012
89. Gazeley KF, Wake WC (1990) Natural rubber adhesives Handbook of adhesives 3rd ed. Skeist I (ed) Chapman & Hall, NY, pp 167–84
90. Unalan IU, Cerri Gi, Marcuzzo E, Cozzolino CA, Farris S (2014) Nanocomposite films and coatings using inorganic nanobuilding blocks (NBB): current applications and future opportunities in the food packaging sector. RSC Adv 4:29393

Organic/Silica Nanocomposite Membranes Applicable to Green Chemistry

Mashallah Rezakazemi, Amir Dashti, Nasibeh Hajilary and Saeed Shirazian

1 Introduction

Organic-inorganic composites have been thoroughly used for various applications; these are referred to as nanocomposites if nano-size inorganic building blocks are used in the composite structure [1–5]. Organic-inorganic nanocomposites are commonly made of organic polymers incorporated with inorganic nanoscale building blocks such as nano silica [6–11]. They possess the beneficial features of both the inorganic substance (e.g., stiffness, thermal stability) and the organic polymer (e.g., ductility, flexibility, processability, and dielectric) [12, 13]. Furthermore, they generally include unique characteristics of fillers resulting in composites with enhanced characteristics. A primary property of nanocomposites is the tiny dimension of the nanofillers contributes to a large rise in the interfacial area in comparison to conventional composites. Also, at reduced filler content, the interfacial area growth generates a substantial increase in the volume of interfacial polymer with properties distinctive from the polymer [14].

Among the several inorganic/organic nanocomposites, polymer/Si composites are the most prevail quoted [15]. These nanocomposites have attracted considerable

M. Rezakazemi (✉)
Faculty of Chemical and Materials Engineering, Shahrood University of Technology, Shahrood, Iran
e-mail: mashalah.rezakazemi@gmail.com

A. Dashti
Department of Chemical Engineering, Kashan University, Kashan, Iran

N. Hajilary
Department of Chemical Engineering, Golestan University, Gorgan, Iran

S. Shirazian
Department of Chemical Sciences, Bernal Institute, University of Limerick, Limerick, Ireland

Inamuddin et al. (eds.), *Sustainable Polymer Composites and Nanocomposites*,
https://doi.org/10.1007/978-3-030-05399-4_22

interests recently and have been used in many different applications. Organic-inorganic nanocomposites can be fabricated by numerous synthesis techniques, showing the different strategies offered to present each phase. The organic part can be presented as (i) a precursor (oligomer or monomer), (ii) a preformed linear polymer (in emulsion, molten or solution states), or (iii) a polymer network, either chemical (e.g., elastomers, thermosets,), physical (semicrystalline linear polymer) or cross-linked [16]. The inorganic portion can be presented as (i) preformed nanoparticles or (ii) a precursor (e.g. TEOS) for further treatment. Organic or inorganic polymerization typically is required if at least one of the starting moieties is a precursor.

The addition of inorganic additives to the polymeric membranes has received the attention of numerous scientists in the recent years. Many organic-inorganic nanocomposite membranes have been studied in the literature. A number of the nanoparticles utilized in the membrane matrix contain Si [17], ZrO_2 [18], heteropolyacids [19], TiO_2 [20, 21] and carbon nanotubes [22].

Nano-si has been broadly investigated due to their reduced cost, poor electrical conductivity and enhanced water uptake characteristics compared with other nanomaterials. Phase conflict involving the organic matrix and Si was resolved using the adjustment of the Si surfaces applying various agents.

Polymer Electrolyte Membrane Fuel Cells (PEMFCs) signify an encouraging tool for the technology of power in the 21st century due to their escalated efficiency in contrast to coal combustion engines and eco-environment [23]. Among different parts within the fuel cell systems, proton exchange membranes (PEMs) are influential parts of PEMs functionality. Currently, the PEMFCs are based on perfluoro sulfonic acid membranes like Nafion®, and Flemion, which have particular disadvantages like huge fuel crossover and high cost. Therefore, scientists have concentrated on creating PEMs with escalated proton conductivity, longevity, thermal stability, maximum power density, minimal fuel crossover, and reduced cost.

Organic-inorganic nanocomposite PEMs consist of nanosized inorganic fillers in the polymer by the molecular level of hybridization. This structure has revealed the opportunity to integrate into an individual solid, the appealing attributes of a thermally and mechanically stable inorganic backbone and the particular chemical reactivity, dielectric, processability, ductility and flexibility of the organic polymer. The preparation, properties characterization, and polymer/Si nanocomposites applications are becoming a rapidly increasing research subject. Many publications and review articles [24, 25] have presented which are partially dedicated to the polymer/Si nanocomposites. Silicate nanoparticles were employed in various polymeric membranes applicable to PEMFCs. The aim of this account is to introduce the fabrication of silicates and nanocomposite membranes. The silicates impact on the various properties of nanocomposites such as thermal and mechanical properties, proton conductivity, water uptake and cell performance are investigated.

1.1 Challenges in Synthesizing Organic/Si Nanocomposite Membranes

PEMFCs have proved a great environmental friendly technology to satisfy the prevailing energy prerequisites of the recent years. Nafion is the extensively approved to date and has great electrochemical characteristics under 80 °C and in extreme humidity. However, a reduction in the proton conductivity of Nafion in lower humidity and over 80 °C, as well as high membrane price, has motivated the improvement of novel techniques and membranes for fuel cell applications. The incorporation of inorganic nanofillers, particularly Si-based nanofillers, into the polymeric membrane was employed, to resolve these restrictions. This is a result of availability, reduced cost, higher thermal stability and high hydrophilicity of the inorganic silicates. The addition of inorganic nanofillers into the polymeric membranes has received the attention of several research studies over the last few decades, plus lowering cost and enhanced water retention characteristics [26]. Inorganic nanofillers enhance the functioning temperature of the hybrid membranes because of their hydrophilicity [27, 28]. The polymer-nanocomposite membranes preparation for PEMFC can be achieved by using two separate approaches, namely, ex situ or in situ approach. In the ex situ method, silicate nanofiller or its precursor is dispersed in the matrix of polymer, while in situ technique includes the dispersion of nano-silicates or their precursor in the monomer followed by a polymerization step [12].

The goodness of dispersion involving the inorganic silicates and polymer matrix has important effects on enhancing the main characteristics and performance of the prepared composite membranes [4, 29, 30]. Accordingly, various methods and solvents are employed to enhance the dispersion levels of filler and the physical and chemical attributes of the nanocomposite membranes. General topics crucial to any PEMs involve: (1) elevated proton conductivity, (2) low water transportation through electro-osmosis and diffusion phenomena, (3) low fuel and oxidant permeability, (4) reduced electronic conductivity, (5) good mechanical stability in both hydrated and dry states, (6) hydrolytic and oxidative stableness, (7) low price and (8) capacity for fabrication into membrane electrode assemblies (MEAs). Because of the increased sorption of water in the membrane, mechanical properties, and water crossover emerge as critical problems. Developing proton-conducting devices with minimum or without water can be a significant challenge for novel membranes. Novel membranes with decreased methanol permeability and water transfer (by electro-osmotic drag or diffusion) in addition to appropriate conductivity and stabilities are necessary for Direct-Methanol Fuel Cells (DMFCs). Fuels diffusion from the anode to cathode sides reduces the voltage efficiency and performance of fuel cell. The intricacy and membrane resistivity were improved, with minimal energy density, by employing diluted methanol as the fuel. Conductivity and water absorption of the membrane rely upon the ionic group concentration, mostly sulphonic acids. Though, an elevated density of ionic groups motivates the increased membrane swelling that diminishes membrane durability and mechanical performance. Therefore, the essential properties for PEMs (IEC, proton conductivity and water uptake) must be improved under the operating circumstances. The organic-inorganic nanocomposite PEMs development appears to be promising in overcoming the aforementioned operational problems.

1.2 Possible Methods to Overcome the Challenges

As exhibited in Fig. 1, showing the common preparation techniques of polymer/Si nanocomposites, generally in the blending and in situ polymerization approaches, Si nanoparticles are directly introduced into the polymer [12], while Si precursors are employed in the sol-gel processes, in which the common precursors are Tetraethyl orthosilicate (TEOS(, Tetramethyl orthosilicate (TMOS) and silicon alkoxides [31]. Sometimes, alkoxysilane-containing polymers [32, 33] as Si precursors are also employed in the sol-gel method. Furthermore, in particular instances, Si in the nanocomposites is formed from precursors such as perhydropolysilazane (PHPS) [34], soluble glass [35], Na_2SiO_3 [36], silicic acid [37], etc. [38–40].

Research studies accomplished over the recent years on the subject can be split into four classes: (i) introducing inorganic proton conductors in PEMs; (ii) covalently bonded inorganic segments with a polymer; (iii) nanocomposites by sol-gel approach; and (iv) acid-based PEM nanocomposites. The fabrication of hybrid-nanocomposite membranes for PEMFC can be achieved by using two separate techniques, namely ex situ or in situ approach.

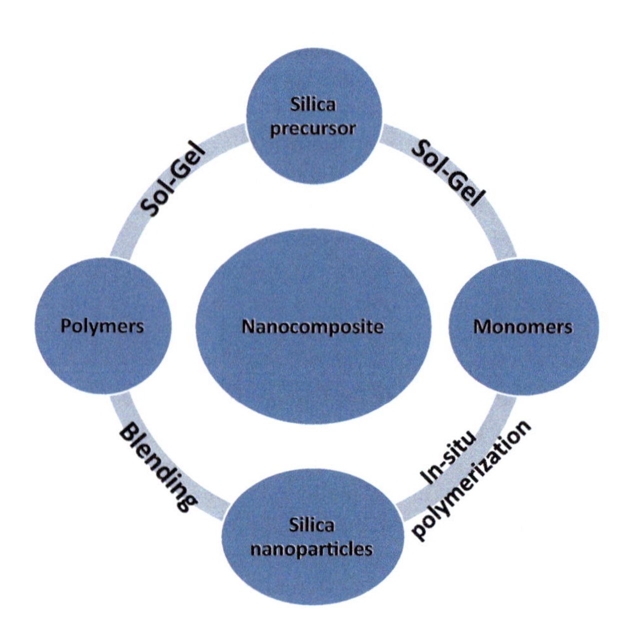

Fig. 1 The common methods to prepare polymer/Si nanocomposites

1.3 Ex Situ Technique

Melt blending and solution mixing, are generally accepted ways for the nanocomposites preparation through the ex situ approach. Melt blending is recognized as a green approach where the nanoparticles are mixed with the molten polymer. The drawbacks of the method are the raw polymers decomposition, poor filler dispersion, and the surface modifiers degradation. Therefore, the solution mixing method is the commonly accepted method, specifically for the fabrication of nanocomposites on the laboratory-scale [41, 42].

The beneficial characteristics of the solution mixing approach are the homogeneous dispersion of nanofillers in the polymer matrix. This approach consists of the nanoparticles blending (dispersed in a specific solvent) with the polymer solution, accompanied by casting and drying [42–44]. Several strategies and solvents have been chosen to improve nanoparticles dispersion in the polymer matrix and to enhance the nanocomposite membranes properties.

1.4 In Situ Technique

In situ techniques include the nanoparticles dispersion or their precursors in a low viscosity monomer solution. Superior dispersion, in addition to excellent physical properties, is observed in in situ technique. However, the intricacy of the method means that researches in this field are not enough.

Poly(styrene-co-methacrylate)-Si covalent membrane was made by blending an aqueous solution of TEOS/nitric acid with azobisisobutyronitrile (AIBN), 2-hydroxyethyl methacrylate (HEMA) and styrene (STY). At a temperature of 40 °C, the mixture was cast in a mould to start the free radical polymerization of HEMA and STY combined with the sol-gel reaction of TEOS. The produced membrane was processed at various temperatures to 150 °C. Various membranes were sulfonated via immersion in a 0.3 M solution of sulfur-chloridic acid in ethylene dichloride for 6 h [45]. In an identical experiment, the researchers used phosphotungstic acid (PWA) rather than nitric acid, and the temperature of free radical polymerization was adjusted to 55 °C [46]. In other procedure, sol 1 (including PWA or 1 N HCl and 3-methacryloxypropyl trimethoxysilane (MPS) in ethanol) and sol 2 (made up of AIBN and HEMA) were blended, cast in a mould at 60 °C, dried and post-treated at 150 °C for 24 h [26]. The post-treatment of the membranes changed the membrane colour to brown as a result of the cross-linked structure in the membrane [19, 47].

A combination of nethylimidazolium trifluoro methanesulfonate ([EIm][TfO]) (40 wt%), acrylonitrile (40 wt%), STY (20 wt%), divinylbenzene (6 wt% to the formulation, based on the monomers weight), and Si and benzoin ethyl ether as a photo initiator was mixed and ultrasonicated to acquire a homogeneous solution. At room temperature, the solution was cast onto a glass mould and using irradiation by UV light, photo cross-linked for 30 min [48]. Given that Nafion® is a commercial polymer, nanocomposite membranes made from it are commonly prepared by an ex situ preparation approach. Many other polymers with high performance (such as PPSU, PEEK, and PBI) require inert atmosphere and monotonous preparation process. The introduction of extremely hydrophilic nano silicate into the monomer makes the polymerization complex, leading to decreased molecular weight (M_w) and interfering crosslinking in the polymer structure. Limited selection of solvent media for the formation of nanocomposite by in situ synthesis adds further intricacy to the system. Therefore, the previous researches for the fabrication of Si-based nanocomposite using the in situ approach are scant. New synthetic procedures and proposed new types of block copolymers can be too advantageous to the aforementioned issues.

2 Si Preparation

Two types of techniques have been proposed for nano-Si preparation: the sol-gel approach and the microemulsion technique [49]. Stöber et al. [50] suggested an easy method for preparation of mono-dispersed spherical Si via hydrolysis of TEOS solution in ethanol in the acidic environment. Symmetrical amorphous Si nanoparticles were conveniently acquired by employing different reactants concentrations. Subsequently, the Stöber technique was developed by researchers and seems like the most convenient and efficient approach [51].

Osseoasare et al. [52] synthesized monodisperse nano-Si using limited TEOS hydrolysis in an inverse microemulsion. Moreover, this micro-emulsion technique is commonly employed to prepare nano-Si. These particles are now produced commercially, generally in the form of powder or colloid.

NanoSi powder is chiefly industrially synthesized using the precipitation technique and fuming approach. Fumed Si is an amorphous, fine, tasteless, white and odourless powder. This powder is prepared using a vapour process at an elevated temperature where $SiCl_4$ is hydrolyzed in an O_2–H_2 combustion process according to the following reaction [53]:

$$SiCl_4 + 2H_2 + O_2 \rightarrow SiO_2 + 4HCl$$

The nano-Si has a 3D structure. Siloxane and silanol groups are formed on the surface of Si, causing the nanoSi to be hydrophilic.

The Si surfaces are peculiarly ceased by three silanol species: isolated or free silanols, vicinal silanols or hydrogen-bonded and geminal silanols (Fig. 2) [51]. The silanols occupy the near molecules, in proper order, creating H_2 bonds and causing the aggregation as demonstrated in Fig. 3. The aforementioned bonds maintain single fumed Si jointly and the aggregations stay unchanged even in good blending situation if the vigorous interaction between filler and matrix does not exist [54]. Nanoparticles dispersion in the matrix of the polymer has a noticeable effect on the nanocomposites properties.

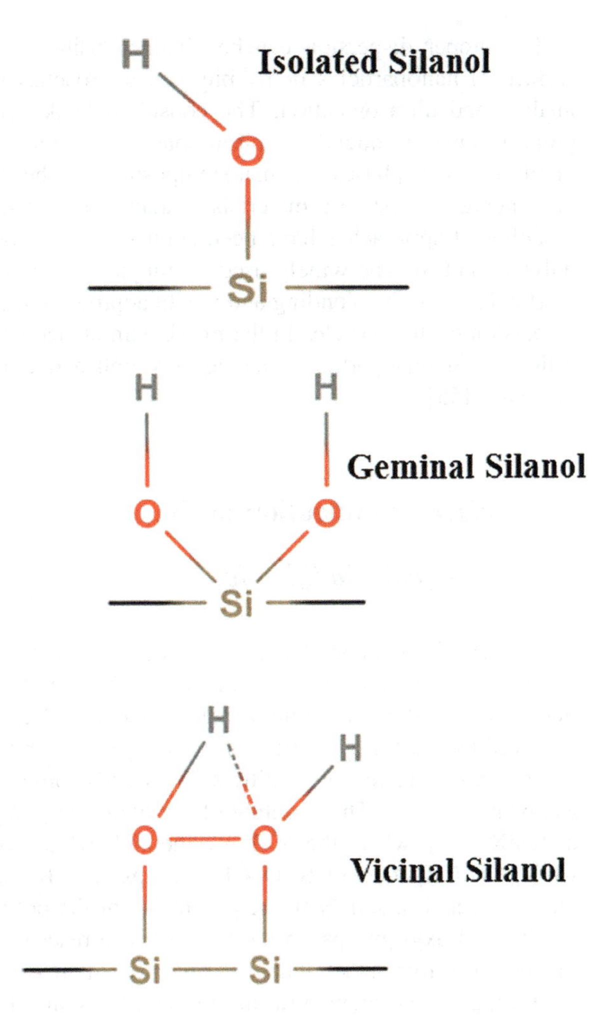

Fig. 2 Three surface silanols types

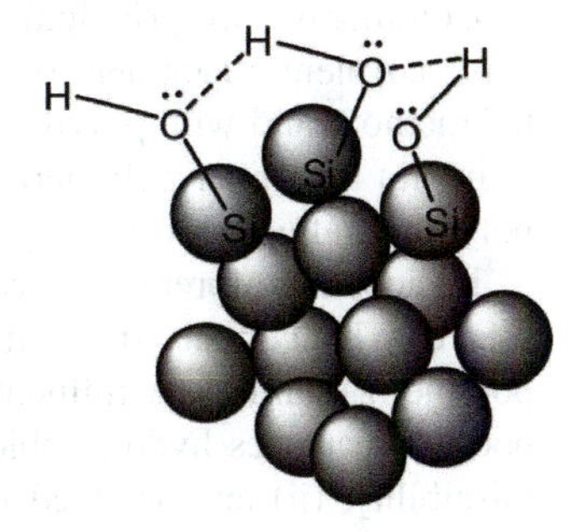

Fig. 3 Aggregate creation among adjacent molecules of fumed Si via H_2 bonding between the silanols

The proper dispersion can be obtained using a chemical modification of the surface of nanoparticles or by physical approaches like high-energy ball-milling method and ultrasonication. The considerable deviations in the Si and polymer properties may frequently result in phase separation. Hence, the surface interaction involving two phases of nanocomposites is the highly determining element influencing the resulting materials characteristics [14].

Different approaches have been applied to improve the congruity involving the polymer and Si. The widely used technique is to treat the Si nanoparticles surface (particularly for the blending and in situ approach), that in addition can enhance the dispersion of the particles in the matrix simultaneously. Generally, surface modification of Si nanoparticles can be accomplished either by physical or chemical processes [55].

3 Surface Modification of Si

3.1 *Chemical Modification*

A number of researches have been focused on the surface modification of the Si nanoparticles by chemical modification because it can eventuate to a vigorous interaction involving modifier agent and nano-Si. These chemical techniques can be achieved by grafting polymers or by using modifier agents. Silane coupling agents are the prevailing sort of modifiers. Their ends commonly are organofunctional and can be hydrolyzed. The structure of the silane coupling agents can be demonstrated as $RSiX_3$ [56], where the X shows the ends which can be hydrolyzed that can be Cl–R, CH_3CH_2O–R, or R–O–CH_3 groups. The R can possess different functionalities selected to satisfy the requirements of the polymer.

The hydroxyl groups on the SiO_2 surface react with functional group X, while the polymer may react with the alkyl chain in order to form, hydrophobic Si. Besides, polymer chains grafting to nano-Si is an influential approach to raise the hydrophobicity of the Si and to produce adjustable interfacial interactions.

Commonly, two principal methods of chemical joining the polymer to a surface exist: covalent attachment of end-functionalized polymers to the surface ("grafting to" method) and with polymer chains monomer growth from immobilized initiators in in situ monomer polymerization ("grafting from" method). In some respect, the polymer grafted nano-Si may be considered as polymer-nanoSi hybrid composites.

Besides the aforementioned chemical methods, polymers grafting to nanoparticles can be conceived by irradiation. The nanoparticles modification via graft polymerization is very influential in the construction of nanocomposites due to (i) a boost in particles hydrophobicity which is favourable to the nanoparticle/polymer miscibility, (ii) an enhanced interfacial interaction obtained by the molecular surrounding of the matrix polymer and the grafting polymer on the nanoparticles, and (iii) a modifiable structure-properties relationship obtained by varying the grafting monomers and the grafting conditions because different grafting polymers might provide various interfacial characteristics.

Si functionalized particles (Fig. 4) have been produced by uniting sulfonated aromatic bis-hydroxy compounds onto fumed Si surface. Firstly, by bromophenyltrimethoxysilane reaction with a fumed Si, a bromophenyl group was formed on the Si surface. Then, by nucleophilic substitution reactions, compounds of sulfonated bis-hydroxy aromatic were connected to the Si surface chemically [57].

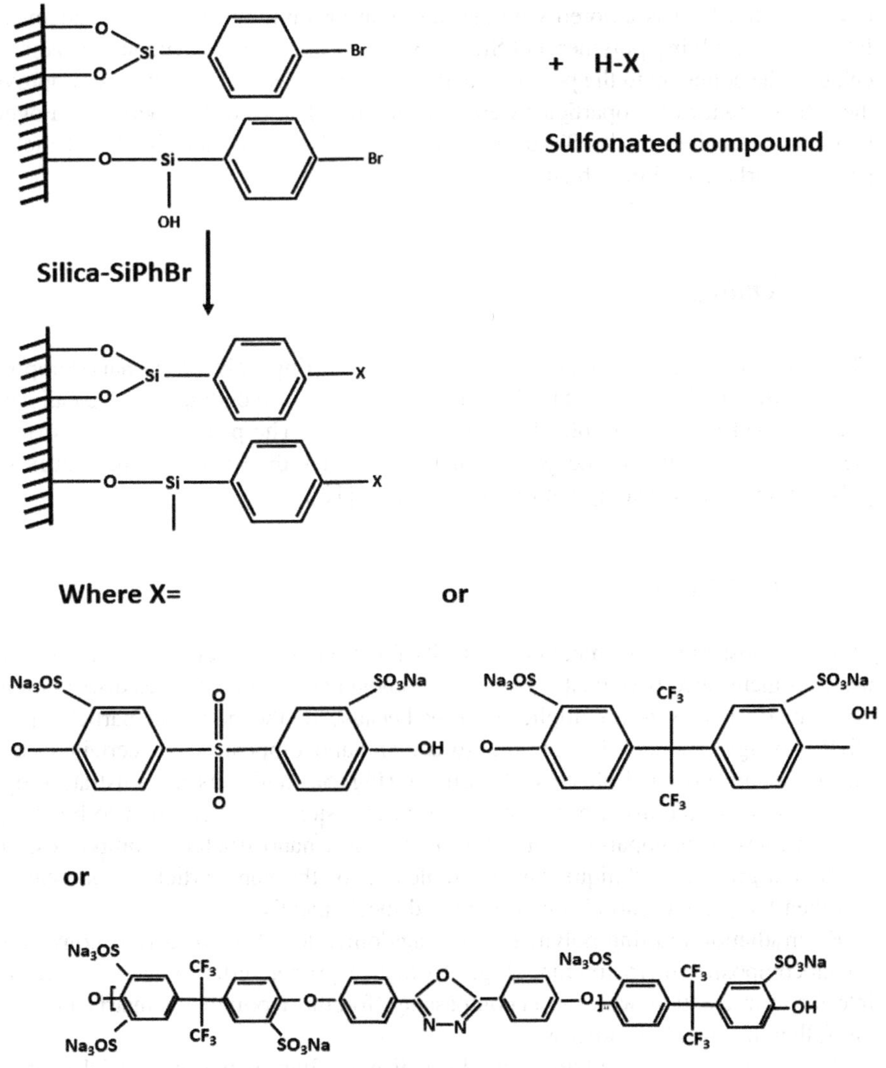

Fig. 4 Functionalized Si surface

3.2 Modification by Physical Interaction

Tailoring the surface physically is commonly carried out utilizing adsorption of macromolecules or surfactants onto the Si particles surface. Modification based on surfactant t is based on the preferential sorption of a surfactants polar group onto the Si surface via electrostatic forces.

A surfactant decreases the interaction among the nano-Sis inside agglomerates via decreasing the physical attraction and may be combined with a polymer without any difficulty. Si was tailored with cetrimonium bromide to enhance the chemical interaction involving polymer and SiO_2 [58]; SiO_2 were treated with stearic acid to enhance the adhesion to the polymer matrix and additives [59, 60] and also improve their dispersion; Si nanoparticles were treated with oleic acid that, with a hydrogen bond, was attached to the Si surface. Polymers adsorption can also boost the Si particles surface hydrophobicity.

3.3 Blending

The simplest and conventional technique for fabricating polymer-Si nanocomposites is a direct blending of the Si and a polymer. The blending can typically be accomplished by solution blending or melt blending. The principal complexity of the mixing procedure is the perfect Si nanoparticles dispersion in the matrix of polymer since they usually tend to agglomerate [12].

3.3.1 Melt Blending

This is a most prevalent method due to its functionality, effectualness, and cleanness. As mentioned previously, providing homogeneous nanoparticles dispersion in a polymeric matrix is a complicated task because of the powerful particles proclivity to agglomerate. Hence, the so-called nanocomposites, in certain cases, include many loosened clusters of particles (Fig. 5a) and thus demonstrate properties inferior to the usual nanoparticle/polymer systems [61]. In order to break up agglomerates of nanoparticle and to form uniform nanostructural composites, an irradiation grafting technique can be employed for the nanoparticles modification and then the grafted particles can be mixed mechanically.

By irradiation grafting polymerization, agglomerates of nanoparticles convert to a nanocomposite microstructure (Fig. 5b) that in proper order creates a powerful interfacial interaction with the encompassing, foremost polymeric matrix through the following blending process.

As various grafting polymers produce distinct filler/matrix interfacial characteristics, properties and microstructures of the final nanocomposites can be modified. It was discovered that the nanoparticles toughening and fortifying impacts on

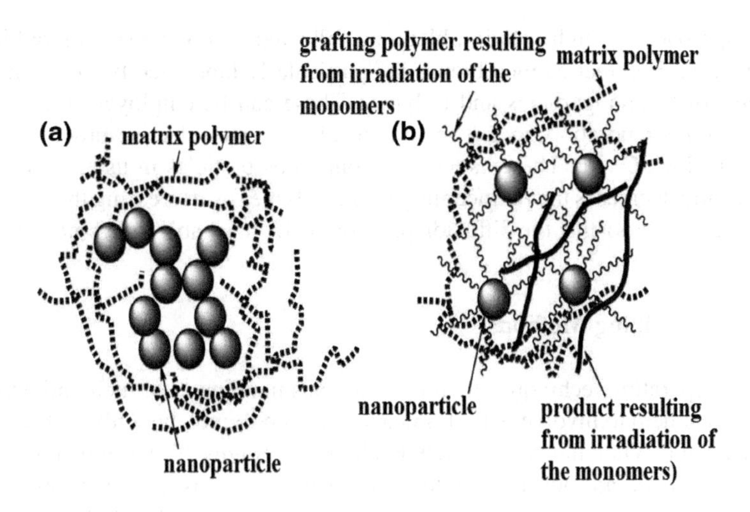

Fig. 5 **a** Schematic of an agglomerated nanoparticles dispersed in a matrix of polymer and **b** the probable structure of grafted nanoparticles dispersed in a polymer matrix [61]. Reprinted with permission from Elsevier

the matrix and could play a major role at low loadings of filler (typically lower than 3 vol.%) as compared with usual composites with larger particles. The method has several benefits, like low cost, easy to perform, easily controllable and generally practical [61]. Twice percolation of stress volumes in the vicinity of the particles and the agglomerates, that was attributed to the visibly attached shear yielded networks all over the nanocomposite, described the toughening and fortifying impacts of the processed nanoparticles [62].

An industrial-scale injection moulding machine [63] and twin-screw extruder were used instead of pilot-scale compression moulding and a screw extruder, PE [64] instead of PP, and precipitated nanoSi [65] instead of fumed nanoSi. All researchers suggested that the approach was still excellent. As grafting pre-treatment and drawing methods were integrated with melt blending to synthesize the nanocomposites, the nanoparticles separation was evoked, β-crystals in the PP matrix were generated, and the obtained PP-based nanocomposites were very stronger than the bare polymer matrix [66].

3.3.2 Solution Blending

Solution blending is a powder treatment technique in liquid-state that provides an excellent molecular size of blending and extensively is applied in material processing and preparation. Many of the restrictions of melt blending may be surmounted if the nanoparticles and polymer are dispersed or dissolved in solution but the solvent loss and its recovery must be considered [25, 55].

An approach through solution blending followed by compression moulding can also be used. Polymer-Si membranes are particularly fabricated by solution casting mixtures of Si nanoparticles and polymer. These can be employed in liquid separation like pervaporation, in gas separation like reverse-selective process, and as a PEM for PEMFC, etc. The existence of nonporous particles in usually composites significantly decreases the permeability of the polymer by decreasing the free volume for transfer and boosting the diffusion path tortuosity reachable to gas molecules [12].

3.3.3 Cryomilling Methods

Classic preparation techniques treat the materials in solution or melt and depend on surface treatment to involve particles in a matrix of polymer, but only a little progress was achieved. When the nanoparticle loading in a polymer is too much, the melt or solution methods are not applicable, as the solution is viscous or it never occurs. A solution to these issues is to treat the polymer in the solid state that hinders the thermal and solvent issues met in conventional methods but improving design and operation straightforwardness. Cryogenic ball milling (cryomilling) is a solid-state approach that efficiently enhances mixing characteristics. Poly(ethylene terephthalate) (PET)/SiO_2 nanocomposites can be synthesized using cryomilling [67]. A 3-step model (Fig. 6) to show the creation mechanism of PET-SiO_2 nanocomposite was proposed [67, 68]. The initial step was the huge decrease in nanoparticle dimension and the conversion of large PET blocks to flakes; in the meantime, the SiO_2 conglomerations were collapsed and scattered in PET flakes creating the foremost nanocomposite particle. The next step was described by the slow dispersion of nano SiO_2 to PET flakes and the creation of the second nanocomposite particle as a result of a conglomeration of the improved PET/SiO_2 initial nanocomposite particles. The next step (3) was described by the stable dimension of the second nanocomposite particles following by uniform dispersion of nano SiO_2 in the polymer. This was demonstrated that, through cryomilling around 10 h, SiO_2 were properly segregated into individual nanoparticles (approximately 30 nm) which are dispersed in polymer suitably. The originally obtained PET-SiO_2 particles formed in the shape of flake ($\sim$400 nm). These primary particles of the composite agglomerated to create second nanocomposite particles with a mean size around 7.6 μm. Well-dispersion of nano-SiO_2 in the polymer was so much higher than traditional approaches, that was attributed to solid-state treatment, the elevated mechanical energy of ball milling, and cryogenic temperature.

3.3.4 Thermal Spraying

Another superior approach to the restrictions of polymer/Si nanocomposites treatments is thermal spraying like the employment of solvent. In thermal spraying, a composite is under the heating, acceleration, and propulsion. In the propulsion procedure, a jet with a high heating rate propels a material via a restricting nozzle

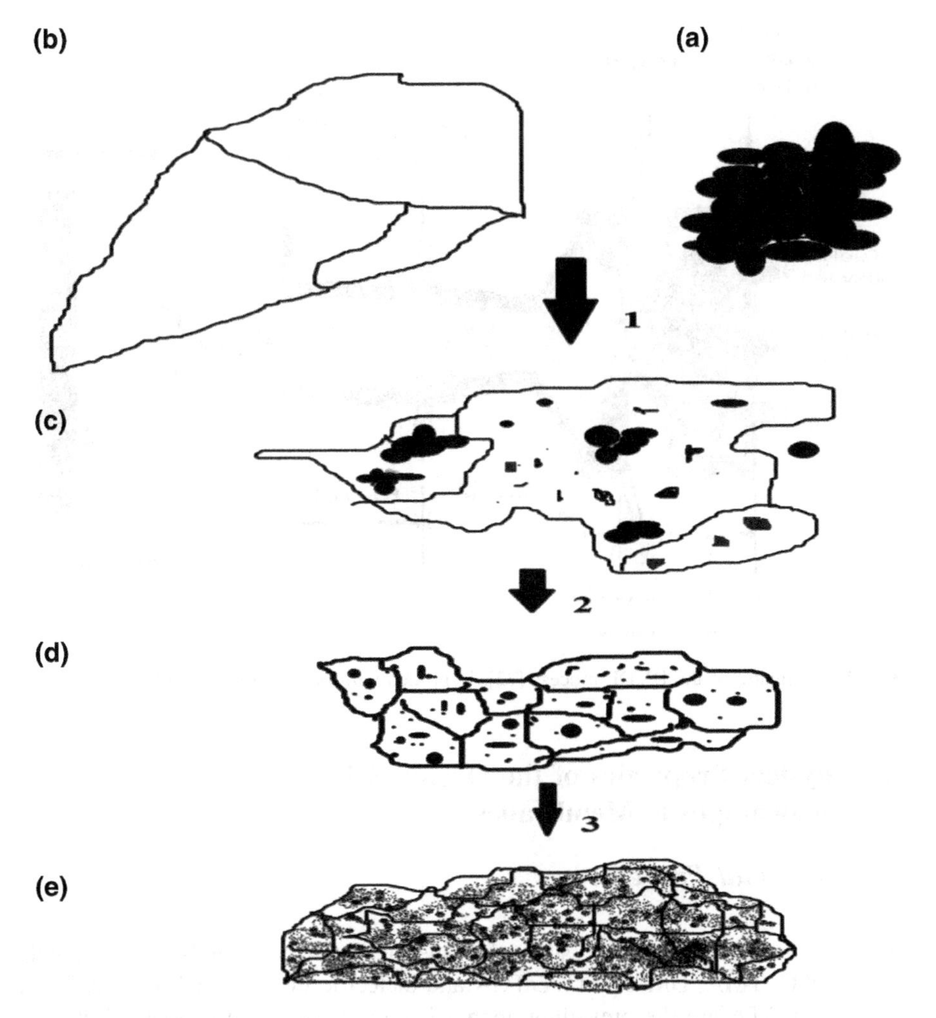

Fig. 6 Schematic of nanocomposites formation mechanism through cryo-milling, and **a** shows SiO_2 particles, **b** shows PET particles, **c** shows the initial milling step, **d** shows the secondary milling step, and **e** shows the final milling step [67]

towards a surface. The softened or single molten droplets crash, sprawl, lose heat and become solids to create uninterrupted and uniform covers. High-velocity oxy-fuel (HVOF) (Fig. 7) supplies heat given via fuels reaction with air. Petrovicova et al. [69–71] fabricated nylon 11 coatings loaded with nanoSi by the HVOF treatment. The composite powder contributed the fillers dispersion in the coating and the concurrent powder feeding into the HVOF spray jet.

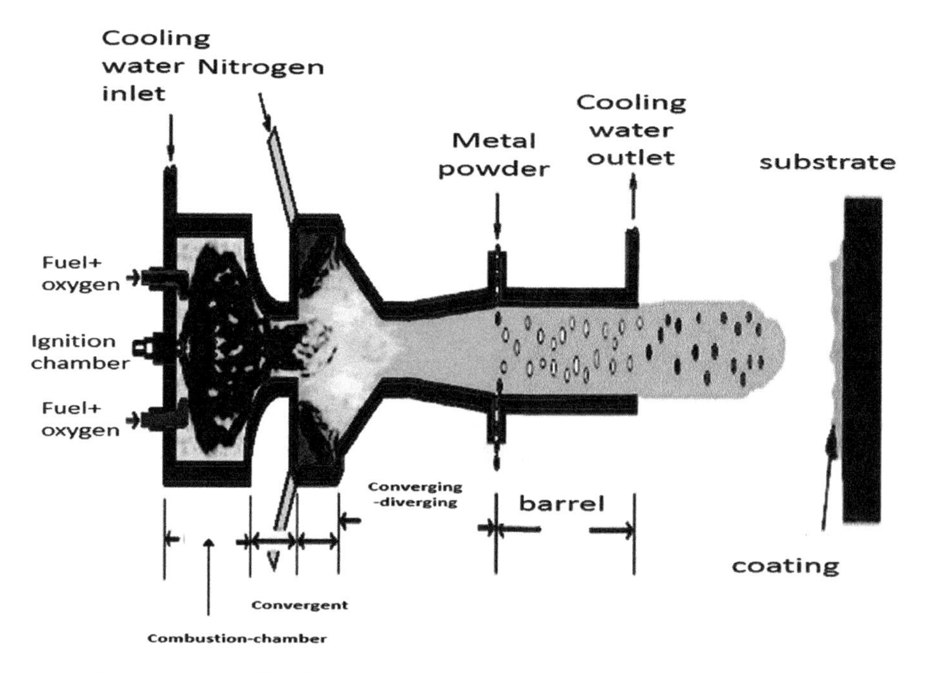

Fig. 7 Schematic of the HVOF process [72]. Reprinted with permission from Elsevier

4 Physical Properties of the Organic/Si Nanocomposite Membranes

4.1 Thermal Properties

Thermal properties can be investigated by thermal analysis methods, which include DTA, DSC, TMA, DMA/DMTA, TGA and dielectric thermal analysis, etc. DSC, TGA, and DTA are the prevailing approaches for recognizing of thermal characteristics of organic/inorganic composites. TGA exhibits the thermal stability, the degradation initiation, and the Si % present in the polymer. DSC employs to ascertain the behaviour of thermal transformation in organic/inorganic composites expeditiously. Moreover, the Coefficient of Thermal Expansion (CTE), which is used to analyze the dimensional stability of composites, is investigated by TMA. Additionally, thermal mechanical characteristics analyzed by DMA/DMTA are substantial in understanding the nanocomposites viscoelasticity. The storage and loss modulus investigate the stored energy, demonstrating the elastic share, and the energy spread as heat, demonstrating the viscous share. Moreover, the dielectric analysis is beneficial in understanding the nanocomposites viscoelasticity.

Generally, the addition of inorganic nanoparticles to the organic matrix can improve thermal stability by performing as a better insulator and mass transfer hurdle to the volatile materials formed through decomposition [73]. Moreover, this is very impressive in reducing the CTE of the polymer. The thermal decomposition temperatures (T_d) obtained using TGA, the CTEs obtained using TMA, and the T_g obtained using DSC of PI/SiO_2 composites are shown in Table 1 [74]. Table 1 shows that the hybrid show more thermal stabilities and reduced CTEs as compared to the counterparts. Also, the T_d and T_g of a hybrid enhance with the addition of more Si.

The composite with coupling agent demonstrated increased T_g's. Because firstly, the coupling agent reinforced the interaction between the organic polymer matrix and the inorganic mineral particles, which caused an elevated confining strength of SiO_2 on the PI molecules; second, the coupling agent reduced SiO_2 particles size and by that means highly extended the interfacial region for Si. Besides, the decreased dimension of the SiO_2, somewhat, causing a rise in the cross-linking degree. Subsequently, increased T_g and reduced CTE for the PI/SiO_2 composites with a coupling agent as compared with other composites are obtained. Sometimes, the nanocomposites thermal stability is diminished by increasing the Si content. TGA characterization of the nanocomposite of Si-PMMA demonstrated that increasing the nanoSi particles addition almost decreased the thermal stability at reduced temperatures [75].

4.2 Mechanical Properties

Given that one of the foremost justifications for incorporating inorganic materials to polymers is to enhance their mechanical properties [76, 77]. The basic prerequisite of this kind of nanocomposites is the ability to maximize the trade-off involving the stiffness and the toughness [78]. Hence, it is often essential to investigate the mechanical performance from various standpoints.

Several standards, including impact strength, tensile strength, hardness, flexural strength and fracture toughness have been employed to analyze the nanocomposites. Table 2 shows the mechanical properties of PP nanocomposites filled with SiO_2 particles grafted with different polymers at a constant SiO_2 content [61]. Although the grafting polymers monomers must have various miscibilities with PP, PEA in contrast with other grafting polymers demonstrated an intensifying impact on the tensile strength of the nanocomposites.

Table 1 Coupling agents affect the PI/SiO_2 thermal properties [74]

Run	SiO_2^a (wt%)	GOTMS/TEOS	T_d^b (°C)	CTE ($\times 10^{-5}$ K^{-1})	T_g^c (°C)
1	0	0	561	5.41	289
2	10	0	581	4.86	294
3	20	0	588	3.45	301
4	30	0	600		310
5	10	1/10	572	2.53	298
6	20	1/10	576		309
7	30	1/10	592		316

[a]Calculated Si contents in hybrid films. [b]T_d calculated via TGA in N_2, on-set. [c]T_g calculated by DSC

Table 2 Mechanical properties of PP based nanocomposites, filled with various polymer-grafted SiO_2^a, SiO_2 content = 3.31 vol.% [61]

Properties	Nanocomposites						
Grafting polymers	PS	PBA	PVA	PEA	PMMA	PMA	Neat PP
Tensile strength (MPa)	34.1	33.3	33.0	26.8	35.2	33.9	32.0
Young's modulus (GPa)	0.92	0.86	0.81	0.88	0.89	0.85	0.75
Elongation break (%)	9.3	12.6	11.0	4.6	12.0	11.9	11.7
Area under tensile stress strain curve (MPa)	2.4	3.3	2.3	0.8	3.2	2.9	2.2
Unnotched Charpy impact strength (KJ/m^2)	19.8	19.4	22.9	14.6	20.5	4.7	8.0

[a]Irradiation dose = 10 Mrad; the monomer/SiO_2 weight ratio = 20/100; the systems employed acetone as a solvent throughout the irradiation, except for methyl acrylic acid/SiO_2 system with ethanol as solvent

It was found that entanglement and inter-diffusion in contrast to the miscibility between the grafting polymer segments with the matrix, has the most contribution to the interfacial interaction. It can be inferred that high M_w PP molecules must be entangled efficaciously with the SiO_2 agglomerates, in order to create an elevated tensile strength growth.

The tensile stress-strain figure of PP and its mixed matrix nanocomposite are demonstrated in Fig. 8, showing that a toughening and reinforcing impact of the fillers on the PP were significant.

4.3 *Proton Conductivity*

PEMFC functionality is a result of protons transfer from the anode to the cathode via the PEM. Thus, the membranes proton conductivity is an essential factor for the efficient utilization of PEMFC. The proton conductivities of numerous membranes under various conditions are shown in Table 3.

4.4 *Water Uptake*

Water uptake in the PEM represents an important effect in the proton conductivity in PEMFC. The formula employed for the measurement of water uptake per cent of PEM at various temperatures is:

$$\text{Water uptake } (\%) = \frac{W_t - W_d}{W_d}$$

where W_t is the weight of the membrane following swelling at a specific temperature for 24 h and W_d is the dry weight of the membrane. The water uptakes of different polymer membranes are presented in Table 4.

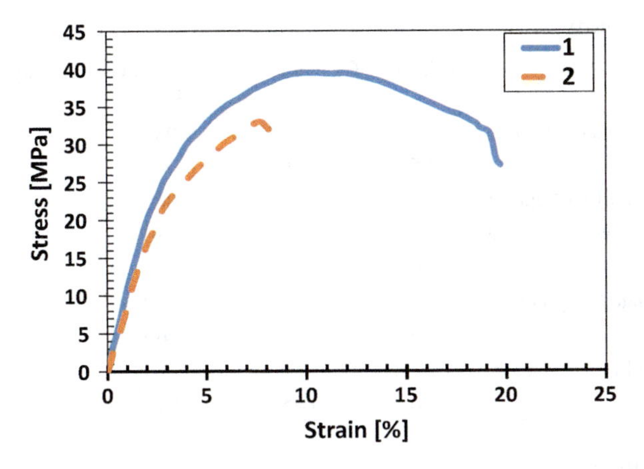

Fig. 8 Tensile stress-strain curves of (1) the neat PP matrix resin; and (2) the one filled with SiO_2-*g*-PS (SiO_2 content = 1.96 vol.%) [61]. Reprinted with permission from Elsevier

Table 3 Proton conductivities of different membranes

Membrane	Temp. (°C)	RH (%)	Proton conductivity (S/cm)	References
Nafion®	80	100	0.080	[79]
Nafion®–Si			0.089	
Nafion®–phosphonated Si			0.152	
Nafion®	80	50	0.008	
Nafion®–Si			0.014	
Nafion®–phosphonated Si			0.049	
Nafion®-117	80	100	0.120	[80]
Nafion®-Si (3 nm)			0.197	
Nafion®-Si (90 nm)			0.190	
Nafion®-Si(1 μm)			0.075	
Nafion®	60	98	0.140	[81]
Nafion®–sulfonated Si			0.130	
Nafion®	60	80	0.070	
Nafion®–sulfonated Si			0.080	[82]
Nafion® 212	80	80	0.080	
PFSA nanofiber			0.100	
PFSA/SPOSS/PAA			0.200	
Recast Nafion®	80	90	0.067	[18]
Nafion®–Si			0.068	
Nafion®–Si–ZrO2			0.100	
Nafion®	90	98	0.200	[83]
Nafion®–sulfonated MMT			0.160	

(continued)

Table 3 (continued)

Membrane	Temp. (°C)	RH (%)	Proton conductivity (S/cm)	References
Nafion®–protonated MMT			0.085	
Nafion®	95	98	0.064	[84]
Nafion®–unmodified Laponite			0.065	
Nafion®–sulfonated Laponite			0.080	
SPSU	90	100	0.170	[85]
SPSU–sulfonated Laponite			0.220	
SPAES	30	95	0.060	[86]
SPAES–SPOSS			0.150	
SPEEK	120	50	0.018	[87]
SPEEK–Si			0.019	
SPEEK–Si–HPMC			0.020	
PBI	Dry	140	0.001	[88]
PBI–Si			0.004	
PBI	120	50	0.010	[89]
PBI–SiO_2–Im			0.050	
PBI–SBA–15			0.070	
SiPANP	100	80	0.005	[90]
SiPANP–PWA			0.041	

4.5 Cell Performance Investigation

The ultimate utilization of the PEM in PEMFC is established employing an investigation of the PEMFC performance. In this kind of investigation, the membranes power density is determined at a specific voltage, temperature, and RH. Table 4 shows the current densities tested for various PEMs at various operational conditions.

5 Summary and Future Direction

The recent advances in the fabrication, properties, characterization, and application of polymer/Si nanocomposite are investigated. Primarily, three approaches for the polymer/Si hybrids preparation are employed: the sol-gel, blending and in situ polymerization.

All of these techniques are studied thoroughly. Apart from the properties of the single components in a nanocomposite, the interfacial interaction and nanoparticles dispersion in the matrix have significant impacts in improving or restricting the performance of the nanocomposites.

Several directions were discovered but no global trends of the polymeric nanocomposites behaviour may be inferred generally. The polymer/Si nanocomposites

Table 4 Water uptake and cell performance of different membranes

Membrane	Water uptake (%)	Temp. (°C)	Cell voltage (V)	Current density (mA/cm^2)	References
Nafion®-117	23.0	80	0.2	600	[80]
Nafion®–Si (3 nm)	28.0			1130	
Nafion®–Si (90 nm)	25.0			1030	
Nafion®–Si (1 μm)	21.0			500	
Nafion®	–	140	0.4	133	[91]
Nafion®–Si	–			225	
Nafion®–mesoporous Si	48	120	0.6	540	[92]
Recast Nafion®	13.9	60	0.4	540	[81]
Nafion®–sulfonated Si	24.7			1040	
Nafion®–Si–TBS	–	80	0.6	1600	[93]
Nafion®-115	32	110	0.4	95	[94]
Nafion®–Si	34			320	
Nafion®–Si–PWA	38			540	
Recast Nafion®	50.0	80	0.6	600	[84]
Nafion®–Laponite	87.0			–	
Nafion®–sulfonated Laponite	70.0			720	
Recast Nafion®	13.5	60	60	450	[83]
Nafion®–H + MMT	13.1	–			
Nafion®–sulfonated MMT	20.1			800	
SPAES	15	–	–	–	[86]
SPAES–SPOSS	21	–	–	–	
SPEEK	30.2	120	0.1	150	[87]
SPEEK–SiO_2	38.9			250	
SPEEK–SiO_2–HPMC	48.5			700	
SPEEK (70°C)	100	60	0.6	80	[95]
SPEEK–Laponite clay	30			370	
PBI	16.6	–	–	–	[88]
PBI–Si	21.6	–	–	–	
PBI–sulfonated Si	23.8	–	–	–	
BIS	8.0	–	–	–	[96]
BIS–MCM	14.0	–	–	–	
BIS–sulfonated MCM	12.5	–	–	–	
SDF	100–190	93	0.3	1200	[97]
SDF–Si	187–385			1700	

properties, however, are usually better than the neat polymer matrix and microcomposites. Particularly, they generally show enhanced thermal stability and mechanical properties whatever of the preparation method. By taking advantage of this huge interfacial volume and area, distinctive combinations of polymer nanocomposites properties can be produced. Although several studies have been performed on polymer-Si nanocomposites, the extra investigation is necessary to comprehend the complicated structure-property relationships. Modifying the filler/matrix interfacial interaction contributes to an improved recognition of the relationships.

References

1. Rezakazemi M, Shahidi K, Mohammadi T (2012) Hydrogen separation and purification using crosslinkable PDMS/zeolite A nanoparticles mixed matrix membranes. Int J Hydrogen Energy 37:14576–14589
2. Rezakazemi M, Sadrzadeh M, Mohammadi T, Matsuura T (2017) Methods for the preparation of organic–inorganic nanocomposite polymer electrolyte membranes for fuel cells. In: Inamuddin D, Mohammad A, Asiri AM (eds) Organic-inorganic composite polymer electrolyte membranes: preparation, properties, and fuel cell applications. Springer International Publishing, Cham, pp 311–325
3. Baheri B, Shahverdi M, Rezakazemi M, Motaee E, Mohammadi T (2015) Performance of PVA/NaA mixed matrix membrane for removal of water from ethylene glycol solutions by pervaporation. Chem Eng Commun 202:316–321
4. Shahverdi M, Baheri B, Rezakazemi M, Motaee E, Mohammadi T (2013) Pervaporation study of ethylene glycol dehydration through synthesized (PVA–4A)/polypropylene mixed matrix composite membranes. Polym Eng Sci 53:1487–1493
5. Rezakazemi M, Ebadi Amooghin A, Montazer-Rahmati MM, Ismail AF, Matsuura T (2014) State-of-the-art membrane based CO_2 separation using mixed matrix membranes (MMMs): an overview on current status and future directions. Prog Polym Sci 39 817–861
6. Rostamizadeh M, Rezakazemi M, Shahidi K, Mohammadi T (2013) Gas permeation through H_2-selective mixed matrix membranes: Experimental and neural network modeling. Int J Hydrogen Energy 38:1128–1135
7. Rezakazemi M, Mohammadi T (2013) Gas sorption in H_2-selective mixed matrix membranes: Experimental and neural network modeling. Int J Hydrogen Energy 38:14035–14041
8. Rezakazemi M, Dashti A, Asghari M, Shirazian S (2017) H_2-selective mixed matrix membranes modeling using ANFIS, PSO-ANFIS, GA-ANFIS. Int J Hydrogen Energy 42:15211–15225
9. Rezakazemi M, Shahidi K, Mohammadi T (2012) Sorption properties of hydrogen-selective PDMS/zeolite 4A mixed matrix membrane. Int J Hydrogen Energy 37:17275–17284
10. Rezakazemi M, Vatani A, Mohammadi T (2015) Synergistic interactions between POSS and fumed silica and their effect on the properties of crosslinked PDMS nanocomposite membranes. RSC Adv 5:82460–82470
11. Rezakazemi M, Vatani A, Mohammadi T (2016) Synthesis and gas transport properties of crosslinked poly(dimethylsiloxane) nanocomposite membranes using octatrimethylsiloxy POSS nanoparticles. J Nat Gas Sci Eng 30:10–18
12. Zou H, Wu S, Shen J (2008) Polymer/silica nanocomposites: preparation, characterization, properties, and applications. Chem Rev 108:3893–3957
13. Rezakazemi M, Maghami M, Mohammadi T (2018) High loaded synthetic hazardous wastewater treatment using lab-scale submerged ceramic membrane bioreactor. Periodica Polytech, Chem Eng 62:299–304

14. Schadler LS, Kumar SK, Benicewicz BC, Lewis SL, Harton SE (2007) Designed interfaces in polymer nanocomposites: a fundamental viewpoint. MRS Bull 32:335–340
15. Rezakazemi M, Shahverdi M, Shirazian S, Mohammadi T, Pak A (2011) CFD simulation of water removal from water/ethylene glycol mixtures by pervaporation. Chem Eng J 168:60–67
16. Rezakazemi M, Sadrzadeh M, Matsuura T (2018) Thermally stable polymers for advanced high-performance gas separation membranes. Prog Energy Combust Sci 66:1–41
17. Mura F, Silva R, Pozio A (2007) Study on the conductivity of recast Nafion®/montmorillonite and Nafion®/TiO 2 composite membranes. Electrochim Acta 52:5824–5828
18. Park KT, Jung UH, Choi DW, Chun K, Lee HM, Kim SH (2008) ZrO 2–SiO 2/Nafion® composite membrane for polymer electrolyte membrane fuel cells operation at high temperature and low humidity. J Power Sources 177:247–253
19. Aparicio M, Mosa J, Etienne M, Durán A (2005) Proton-conducting methacrylate–silica sol–gel membranes containing tungstophosphoric acid. J Power Sources 145:231–236
20. Di Vona M, Sgreccia E, Donnadio A, Casciola M, Chailan J, Auer G, Knauth P (2011) Composite polymer electrolytes of sulfonated poly-ether-ether-ketone (SPEEK) with organically functionalized TiO 2. J Membr Sci 369:536–544
21. Zhengbang W, Tang H, Mu P (2011) Self-assembly of durable Nafion/TiO 2 nanowire electrolyte membranes for elevated-temperature PEM fuel cells. J Membr Sci 369:250–257
22. Rezakazemi M, Zhang Z (2018) 2.29 Desulfurization Materials A2—Dincer, Ibrahim. In: Comprehensive energy systems. Elsevier, Oxford, pp 944–979
23. Hashemi F, Rowshanzamir S, Rezakazemi M (2012) CFD simulation of PEM fuel cell performance: Effect of straight and serpentine flow fields. Math Comput Model 55:1540–1557
24. Cong H, Radosz M, Towler BF, Shen Y (2007) Polymer–inorganic nanocomposite membranes for gas separation. Sep Purif Technol 55:281–291
25. Ajayan PM, Schadler LS, Braun PV (2006) Nanocomposite science and technology. Wiley
26. Xing D, He G, Hou Z, Ming P, Song S (2011) Preparation and characterization of a modified montmorillonite/sulfonated polyphenylether sulfone/PTFE composite membrane. Int J Hydrogen Energy 36:2177–2183
27. Cho Y-H, Kim S-K, Kim T-H, Cho Y-H, Lim JW, Jung N, Yoon W-S, Lee J-C, Sung Y-E (2011) Preparation of MEA with the polybenzimidazole membrane for high temperature PEM fuel cell. Electrochem Solid-State Lett 14:B38–B40
28. Tago T, Kuwashiro N, Nishide H (2007) Preparation of acid-functionalized poly (phenylene oxide) s and poly (phenylene sulfone) and their proton conductivity. Bulletin of the Chem Soc Jpn 80:1429–1434
29. Sodeifian G, Raji M, Asghari M, Rezakazemi M, Dashti A (2018) Polyurethane-SAPO-34 mixed matrix membrane for CO2/CH4 and CO2/N2 separation. Chin J Chem Eng
30. Rezakazemi M, Razavi S, Mohammadi T, Nazari AG (2011) Simulation and determination of optimum conditions of pervaporative dehydration of isopropanol process using synthesized PVA–APTEOS/TEOS nanocomposite membranes by means of expert systems. J Membr Sci 379:224–232
31. Gómez-Romero P, Sanchez C, Functional hybrid materials. Wiley (2006)
32. Zhang S, Xu T, Wu C (2006) Synthesis and characterizations of novel, positively charged hybrid membranes from poly (2, 6-dimethyl-1, 4-phenylene oxide). J Membr Sci 269:142–151
33. Wu C, Xu T, Yang W (2005) Synthesis and characterizations of novel, positively charged poly (methyl acrylate)–SiO 2 nanocomposites. Eur Polymer J 41:1901–1908
34. Saito R, Kobayashi S-I, Hayashi H, Shimo T (2007) Surface hardness and transparency of poly(methyl methacrylate)-silica coat film derived from perhydropolysilazane. J Appl Polym Sci 104:3388–3395
35. Shen L, Du Q, Wang H, Zhong W, Yang Y (2004) In situ polymerization and characterization of polyamide-6/silica nanocomposites derived from water glass. Polym Int 53:1153–1160
36. Ding X, Jiang Y, Yu K, Hari B, Tao N, Zhao J, Wang Z (2004) Silicon dioxide as coating on polystyrene nanoparticles in situ emulsion polymerization. Mater Lett 58:1722–1725
37. Laugel N, Hemmerlé J, Porcel C, Voegel J-C, Schaaf P, Ball V (2007) Nanocomposite silica/polyamine films prepared by a reactive layer-by-layer deposition. Langmuir 23:3706–3711

38. Grund S, Kempe P, Baumann G, Seifert A, Spange S (2007) Nanocomposites prepared by twin polymerization of a single-source monomer. Angew Chem Int Ed 46:628–632
39. Suffner J, Schechner G, Sieger H, Hahn H (2007) In-situ coating of silica nanoparticles with acrylate-based polymers. Chem Vap Deposition 13:459–464
40. Senkevich JJ, Desu SB (1999) Near-room-temperature thermal chemical vapor deposition of poly(chloro-p-xylylene)/SiO_2 nanocomposites. Chem Mater 11:1814–1821
41. Mishra AK, Chattopadhyay S, Rajamohanan P, Nando GB (2011) Effect of tethering on the structure-property relationship of TPU-dual modified Laponite clay nanocomposites prepared by ex-situ and in-situ techniques. Polymer 52:1071–1083
42. Seo W, Sung Y, Han S, Kim Y, Ryu O, Lee H, Kim WN (2006) Synthesis and properties of polyurethane/clay nanocomposite by clay modified with polymeric methane diisocyanate. J Appl Polym Sci 101:2879–2883
43. Mishra AK, Rajamohanan P, Nando GB, Chattopadhyay S (2011) Structure–property of thermoplastic polyurethane–clay nanocomposite based on covalent and dual-modified Laponite. Adv Sci Lett 4:65–73
44. Mishra A, Nando G, Chattopadhyay S (2008) Exploring preferential association of laponite and cloisite with soft and hard segments in TPU-clay nanocomposite prepared by solution mixing technique. J Polym Sci, Part B: Polym Phys 46:2341–2354
45. Aparicio M, Durán A (2004) Hybrid organic/inorganic sol-gel materials for proton conducting membranes. J Sol-Gel Sci Technol 31:103–107
46. Aparicio M, Castro Y, Duran A (2005) Synthesis and characterisation of proton conducting styrene-co-methacrylate–silica sol–gel membranes containing tungstophosphoric acid. Solid State Ionics 176:333–340
47. Tillet G, Boutevin B, Ameduri B (2011) Chemical reactions of polymer crosslinking and post-crosslinking at room and medium temperature. Prog Polym Sci 36:191–217
48. Lin B, Cheng S, Qiu L, Yan F, Shang S, Lu J (2010) Protic ionic liquid-based hybrid proton-conducting membranes for anhydrous proton exchange membrane application. Chem Mater 22:1807–1813
49. Darbandi M, Thomann R, Nann T (2007) Hollow silica nanospheres: in situ, semi-in situ, and two-step synthesis. Chem Mater 19:1700–1703
50. Stöber W, Fink A, Bohn E (1968) Controlled growth of monodisperse silica spheres in the micron size range. J Colloid Interface Sci 26:62–69
51. Bronstein LM, HCD, Kim G (Ed) (2004) Dekker encyclopedia of nanoscience and nanotechnology. Taylor & Francis, New York, pp 1–10
52. Osseo-Asare K, Arriagada F (1990) Preparation of SiO_2 nanoparticles in a non-ionic reverse micellar system. Colloids Surf 50:321–339
53. Vassiliou AA, Papageorgiou GZ, Achilias DS, Bikiaris DN (2007) Non-Isothermal Crystallisation Kinetics of In Situ Prepared Poly (ε-caprolactone)/Surface-Treated SiO_2 Nanocomposites. Macromol Chem Phys 208:364–376
54. Jana SC, Jain S (2001) Dispersion of nanofillers in high performance polymers using reactive solvents as processing aids. Polymer 42:6897–6905
55. Nalwa HS (2003) Handbook of organic-inorganic hybrid materials and nanocomposites. In: Zhang MQR, MZ, Friedrich K (Ed) American Scientific Publishers, California, pp 113–150
56. Blum FD (2004) Encyclopedia of polymer science and technology, concise. In: Kroschwitz JI (Ed) Wiley, pp 38–50
57. Gomes D, Buder I, Nunes SP (2006) Sulfonated silica-based electrolyte nanocomposite membranes. J Polym Sci, Part B: Polym Phys 44:2278–2298
58. Wu T-M, Chu M-S (2005) Preparation and characterization of thermoplastic vulcanizate/silica nanocomposites. J Appl Polym Sci 98:2058–2063
59. Ahn SH, Kim SH, Lee SG (2004) Surface-modified silica nanoparticle–reinforced poly (ethylene 2, 6-naphthalate). J Appl Polym Sci 94:812–818
60. Lai YH, Kuo MC, Huang JC, Chen M (2007) On the PEEK composites reinforced by surface-modified nano-silica. Mater Sci Eng, A 458:158–169

61. Rong MZ, Zhang MQ, Zheng YX, Zeng HM, Walter R, Friedrich K (2001) Structure–property relationships of irradiation grafted nano-inorganic particle filled polypropylene composites. Poly 42:167–183
62. Rong MZ, Zhang MQ, Zheng YX, Zeng HM, Friedrich K (2001) Improvement of tensile properties of nano-SiO_2/PP composites in relation to percolation mechanism. Polymer 42:3301–3304
63. Wu CL, Zhang MQ, Rong MZ, Friedrich K (2002) Tensile performance improvement of low nanoparticles filled-polypropylene composites. Compos. Sci. Technol. 62:1327–1340
64. Zhang MQ, Rong MZ, Zhang HB, Friedrich K (2003) Mechanical properties of low nano-silica filled high density polyethylene composites. Polym Eng Sci 43:490–500
65. Wu CL, Zhang MQ, Rong MZ, Friedrich K (2005) Silica nanoparticles filled polypropylene: effects of particle surface treatment, matrix ductility and particle species on mechanical performance of the composites. Compos. Sci. Technol. 65:635–645
66. Ruan WH, Huang XB, Wang XH, Rong MZ, Zhang MQ (2006) Effect of drawing induced dispersion of nano-silica on performance improvement of poly(propylene)-based nanocomposites. Macromol Rapid Commun 27:581–585
67. Zhu Y, Li Z, Zhang D, Tanimoto T (2006) PET/SiO_2 nanocomposites prepared by cryomilling. J Polym Sci, Part B: Polym Phys 44:1161–1167
68. Zhu Y-G, Li Z-Q, Zhang D, Tanimoto T (2006) Thermal behaviors of poly(ethylene terephthalate)/SiO_2 nanocomposites prepared by cryomilling. J Polym Sci, Part B: Polym Phys 44:1351–1356
69. Petrovicova E, Knight R, Schadler L, Twardowski T (2000) Nylon 11/silica nanocomposite coatings applied by the HVOF process. II. Mechanical and barrier properties. J. Appl. Polym. Sci. 78:2272–2289
70. Petrovicova E, Knight R, Schadler L, Twardowski T (2000) Nylon 11/silica nanocomposite coatings applied by the HVOF process. I. Microstructure and morphology. J. Appl. Polym. Sci. 77:1684–1699
71. Schadler LS, Laut KO, Smith RW, Petrovicova E (1997) Microstructure and mechanical properties of thermally sprayed silica/nylon nanocomposites. J Therm Spray Technol 6:475–485
72. Jafari H, Emami S, Mahmoudi Y (2017) Numerical investigation of dual-stage high velocity oxy-fuel (HVOF) thermal spray process: a study on nozzle geometrical parameters. Appl Therm Eng 111:745–758
73. Sinha Ray S, Okamoto M (2003) Polymer/layered silicate nanocomposites: a review from preparation to processing. Prog. Polym. Sci. 28:1539–1641
74. Shang X-Y, Zhu Z-K, Yin J, Ma X-D (2002) Compatibility of soluble polyimide/silica hybrids induced by a coupling agent. Chem Mater 14:71–77
75. Kashiwagi T, Morgan AB, Antonucci JM, VanLandingham MR, Harris RH, Awad WH, Shields JR (2003) Thermal and flammability properties of a silica–poly(methylmethacrylate) nanocomposite. J Appl Polym Sci 89:2072–2078
76. Crosby AJ, Lee JY (2007) Polymer Nanocomposites: the “Nano” effect on mechanical properties. Polym Rev 47:217–229
77. Mammeri F, Bourhis EL, Rozes L, Sanchez C (2005) Mechanical properties of hybrid organic-inorganic materials. J Mater Chem 15:3787–3811
78. Lach R, Kim G-M, Michler GH, Grellmann W, Albrecht K (2006) Indentation fracture mechanics for toughness assessment of PMMA/SiO_2 nanocomposites. Macromol Mater Eng 291:263–271
79. Joseph J, Tseng C-Y, Hwang B-J (2011) Phosphonic acid-grafted mesostructured silica/Nafion hybrid membranes for fuel cell applications. J Power Sources 196:7363–7371
80. Kumar GG, Kim A, Nahm KS, Elizabeth R (2009) Nafion membranes modified with silica sulfuric acid for the elevated temperature and lower humidity operation of PEMFC. IJHE 34:9788–9794
81. Choi Y, Kim Y, Kim HK, Lee JS (2010) Direct synthesis of sulfonated mesoporous silica as inorganic fillers of proton-conducting organic–inorganic composite membranes. J Membr Sci 357:199–205

82. Choi J, Wycisk R, Zhang W, Pintauro PN, Lee KM, Mather PT (2010) High Conductivity Perfluorosulfonic Acid Nanofiber Composite Fuel-Cell Membranes. Chemsuschem 3:1245–1248
83. Kim Y, Choi Y, Kim HK, Lee JS (2010) New sulfonic acid moiety grafted on montmorillonite as filler of organic–inorganic composite membrane for non-humidified proton-exchange membrane fuel cells. J Power Sources 195:4653–4659
84. Bébin P, Caravanier M, Galiano H (2006) Nafion®/clay-SO 3 H membrane for proton exchange membrane fuel cell application. J Membr Sci 278:35–42
85. Buquet CL, Fatyeyeva K, Poncin-Epaillard F, Schaetzel P, Dargent E, Langevin D, Nguyen QT, Marais S (2010) New hybrid membranes for fuel cells: plasma treated laponite based sulfonated polysulfone. J Membr Sci 351:1–10
86. Choi J, Lee KM, Wycisk R, Pintauro PN, Mather PT (2010) Sulfonated polysulfone/POSS nanofiber composite membranes for PEM fuel cells. JElS 157:B914–B919
87. Zhang Y, Wang S, Xiao M, Bian S, Meng Y (2009) The silica-doped sulfonated poly (fluorenyl ether ketone) s membrane using hydroxypropyl methyl cellulose as dispersant for high temperature proton exchange membrane fuel cells. IJHE 34:4379–4386
88. Liu Y-L (2009) Preparation and properties of nanocomposite membranes of polybenzimidazole/sulfonated silica nanoparticles for proton exchange membranes. J Membr Sci 332:121–128
89. Quartarone E, Magistris A, Mustarelli P, Grandi S, Carollo A, Zukowska G, Garbarczyk J, Nowinski J, Gerbaldi C, Bodoardo S (2009) Pyridine-based PBI composite membranes for PEMFCs. Fuel Cells 9:349–355
90. Cui X, Zhong S, Wang H (2007) Organic–inorganic hybrid proton exchange membranes based on silicon-containing polyacrylate nanoparticles with phosphotungstic acid. J Power Sources 173:28–35
91. Adjemian K, Lee S, Srinivasan S, Benziger J, Bocarsly A (2002) Silicon oxide nafion composite membranes for proton-exchange membrane fuel cell operation at 80–140 C. JElS 149:A256–A261
92. Pereira F, Vallé K, Belleville P, Morin A, Lambert S, Sanchez C (2008) Advanced mesostructured hybrid silica−nafion membranes for high-performance PEM fuel cell. Chem Mater 20:1710–1718
93. Mulmi S, Park CH, Kim HK, Lee CH, Park HB, Lee YM (2009) Surfactant-assisted polymer electrolyte nanocomposite membranes for fuel cells. J Membr Sci 344:288–296
94. Shao Z-G, Joghee P, Hsing I-M (2004) Preparation and characterization of hybrid Nafion–silica membrane doped with phosphotungstic acid for high temperature operation of proton exchange membrane fuel cells. J Membr Sci 229:43–51
95. Chang J-H, Park JH, Park G-G, Kim C-S, Park OO (2003) Proton-conducting composite membranes derived from sulfonated hydrocarbon and inorganic materials. J Power Sources 124:18–25
96. Wilhelm M, Jeske M, Marschall R, Cavalcanti WL, Tölle P, Köhler C, Koch D, Frauenheim T, Grathwohl G, Caro J (2008) New proton conducting hybrid membranes for HT-PEMFC systems based on polysiloxanes and SO 3 H-functionalized mesoporous Si-MCM-41 particles. J Membr Sci 316:164–175
97. Kim YM, Choi SH, Lee HC, Hong MZ, Kim K, Lee H-I (2004) Organic–inorganic composite membranes as addition of SiO_2 for high temperature-operation in polymer electrolyte membrane fuel cells (PEMFCs). Electrochim Acta 49:4787–4796

Extraction of Cellulose Nanofibers and Their Eco-friendly Polymer Composites

M. Hazwan Hussin, Djalal Trache, Caryn Tan Hui Chuin, M. R. Nurul Fazita, M. K. Mohamad Haafiz and Md. Sohrab Hossain

List of Abbreviations

%	Percentage
[BMIM]Cl	1-butyl-3-methylimidazolium chloride
[BMIM]HSO_4	1-butyl-3-methylimidazolium hydrogen sulfate
[EMIM][OAc]	1-ethyl-3-methyllimidazolium acetate
[SBMIM]HSO_4	1-(4-sulfobutyl)-3-methylimidazolium hydrogen sulfate
AFM	Atomic force microscopy
AGU	Anhydroglucose unit
ANC	Amorphous nanocellulose
BC	Bacterial cellulose
cm	Centimeter
CNC	Cellulose nanocrystal
CNF	Cellulose nanofibrils
CNM	Cellulose nanomaterials
CNY	Cellulose nanoyarn
CrI	Crystallinity index
D	Apparent crystallite size
DSC	Differential scanning calorimetry
FTIR	Fourier Transform
HEBM	High-energy bead milling
kg/day	Kilogram per day

M. Hazwan Hussin (✉) · C. T. H. Chuin
Materials Technology Research Group (MaTReC), School of Chemical Sciences, Universiti Sains Malaysia, 11800 Minden, Penang, Malaysia
e-mail: mhh@usm.my; mhh.usm@gmail.com

D. Trache
UER Procédés Energétiques, Ecole Militaire Polytechnique, BP 17, Bordj El-Bahri, Algiers, Algeria

M. R. Nurul Fazita · M. K. Mohamad Haafiz · Md. S. Hossain
School of Industrial Technology, Universiti Sains Malaysia, 11800 Minden, Penang, Malaysia

Inamuddin et al. (eds.), *Sustainable Polymer Composites and Nanocomposites*, https://doi.org/10.1007/978-3-030-05399-4_23

nm	Nanometer
PLA	Polylactide
SEM	Scanning electron microscope
TAPPI	Technical Association of the Pulp and Paper Industry
TBAA	Tetrabutylammonium acetate
TEM	Transmission electron microscope
TEMPO	2,2,6,6-tetramethyl-1-piperidinyloxy
TGA	Thermogravimetric analysis
T_{max}	Melting point temperature
XRD	X-ray Diffraction
λ	X-ray wavelength

1 Introduction

Polymer-based materials are an important and promising area of research exhibiting strong developments [1, 2]. They play a prominent role in the modern civilization and find application in different industries related to electrical and electronic equipment, chemicals, automotive, spacecraft, energy storage in batteries and supercapacitors and medical to cite a few [3–5]. The polymeric materials have substituted the employment of metal, ceramics, and glass in various fields owing to their availability, low weight, low cost, chemical inertness, strength and ease of processing [6, 7]. However, for particular applications, some thermal, physical and mechanical properties of polymer materials appeared to be insufficient. As a way to avoid these limitations, the utilization of polymer as a matrix with along the incorporation of fibers and fillers for the creation of composites became widespread. These composite materials demonstrate interesting thermal, physicochemical, barrier and swelling properties, and mechanical features with respect to the conventional materials [8, 9]. On the other hand, rapidly depletion global petroleum resources, along with awareness of global environmental and health issues as well as the end-of-life disposal challenges, have proved the way to switch toward renewable and sustainable materials. In this regards, bio-based materials such as lignocelluloses and their derivatives can form the basis for various eco-efficient and sustainable products and can prevent the widespread dependence on fossil fuels [10–12].

Lignocelluloses are mainly composed of cellulose nano-fibrils, which can be isolated by chemical, mechanical and biological methods in order to get cellulosic nanofibers [13, 14]. Furthermore, cellulosic nanofibers can also be synthesized by the bacterial procedure. These nanofibers are an emerging class of nanomaterials with various desirable properties. These features are mainly depending on the original source and the extraction procedure. The possibility of obtaining cellulose nanofibers with different properties remains a quite interesting topic, which can bring valorization of residual or unexplored biomass.

There are mainly two kinds of cellulose nanofibers: cellulose nanofibrils (CNF) and cellulose nanocrystal (CNC). The three-dimensional hierarchical structures that compose cellulose nanofibers at different scales, the combination of the physicochemical properties of cellulose, together with the principal advantages of nanomaterials (a high specific surface area, aspect ratio) opens new opportunities in several fields, ranging from electronics to medical applications. Cellulose nanofibers can be employed to reinforce polymers, papers, and membranes. These nanofibers have unique properties including high elastic modulus, dimensional stability, low thermal expansion coefficient, outstanding reinforcing potential and transparency [15, 16]. Figure 1 shows the comparison of specific strength and elastic modulus of various materials [11]. Furthermore, Owing to their –OH side

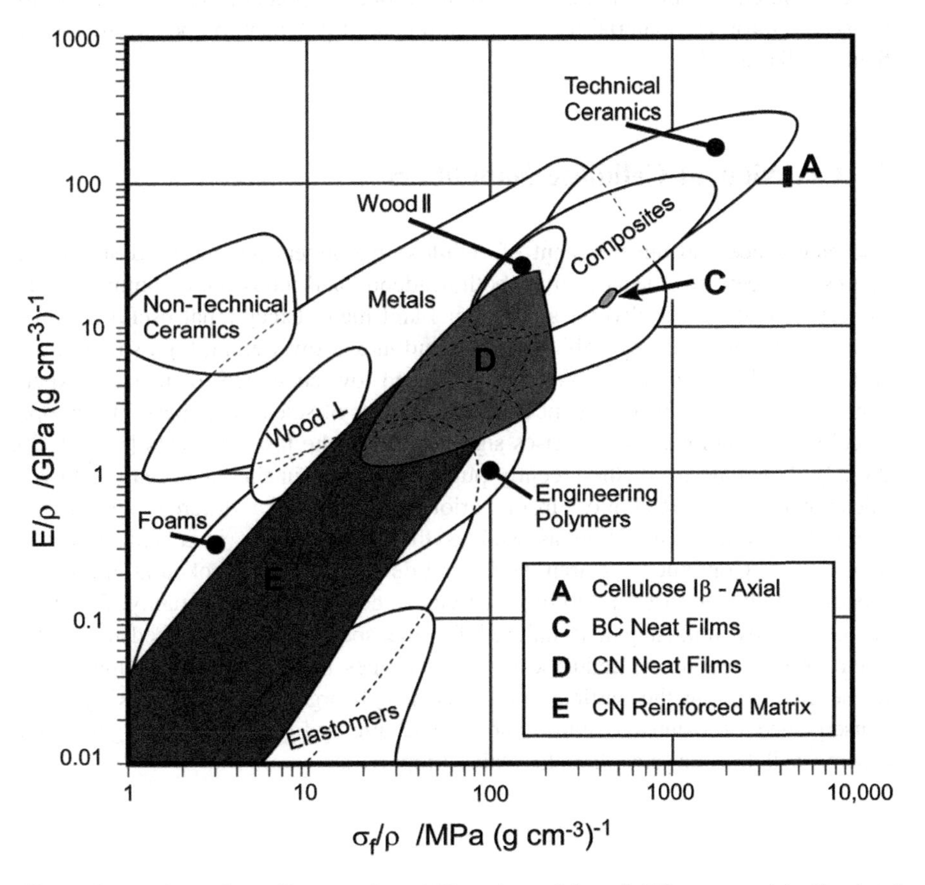

Fig. 1 Comparison of specific strength and Young's modulus of different materials. Reprinted with permission from [11], Copyright © The Royal Society of Chemistry

groups on the surface, cellulose nanofibers can be readily modified into different forms through surface functionalization allowing the tailoring of particle surface chemistry to ease self-assembly, controlled dispersion within a wide range of matrix polymers, and control of both the particle-particle and particle-matrix bond strength [17–20]. Currently, CNF and CNC have been extensively added to polymer composites as reinforcing elements. CNC have exhibited to be preferable as load-bearing component owing to their ability to enhance toughness along with strength and stiffness through the interface interaction between CNC-matrix, whereas CNF has revealed to impart greater reinforcement owing to their highly flexible fibrils and the ease of the interconnection to obtain rigid web-like fibrils networks [4, 5, 21, 22]. Thus, the extraction process of cellulose nanofibers from various sources has a particular interest in their employment as reinforcing agents.

In this review, we describe an overview on the recent research developments on principal cellulose sources followed by the principal methodologies employed for its extraction. The isolation procedures and characterization of cellulose nanofibers are considered and provided as well. In addition, the potential use of these nanofibers as reinforcing material for the development of polymer composites in various fields is discussed.

2 Overview of Cellulose Nanofibers

The emergence and development of cellulose nanofibers has attracted significant interest in recent two decades from both academic and industrial communities due to its potential for the diverse applications and many exceptional useful features, including abundance, renewability, eco-friendliness, low weight, high strength and stiffness, high surface area-to-volume ratio and low coefficient of thermal expansion. This interest is well evident from the number of scientific papers on the topic of cellulose nanofibers, which rises significantly in the last five years [13, 17, 20, 23–28]. The majority of the recently published works in this field deals with cellulose nanofibers preparation from various sources using different approaches, characterization, modification as well as their employment in a wide range of applications. Continued research works and development are looked at optimizing processes to lower costs and to improve yields, consistency, and quality. Processes are commonly multistep and tailored for the specific cellulose feedstock. One common trait among all cellulose nanofibers types is the parallel stacking of cellulose chains along the particle length. This high organization of stacking is the consequence of extensive intra- and inter-chain hydrogen bonding, generating ordered cellulose an exceptionally stable biopolymer.

2.1 Cellulose: Structure and Chemistry

Cellulose, the most abundant carbohydrate polymer in the earth, represents about fifty percent of natural biomass having a yearly fabrication estimated to be over 7.5×10^{10} tons [9, 13]. This ubiquitous renewable natural polymer is regarded as an inexhaustible source of raw materials for the increasing demand for biocompatible and environmentally friendly materials. It is present in various kinds of living species as well as wood, annual plants, tunicates, algae, fungi and some bacteria. Its empirical formula was determined since 1838 by the French chemist Anselm Payen by isolating a white powder from plant tissue [29, 30]. In 1839 he invented the term cellulose for the first time. The structure of cellulose was established late, in 1920, by Hermann Staudinger [31]. Several books [4, 17, 26, 27, 32, 33] and review papers [6, 9, 11, 13, 29, 34–36] have already summarized the state of current knowledge on this fascinating and innovative biopolymer. Hence, only some important details will be provided in the present chapter to avoid duplication. The structural levels of organization of cellulose from its source to the basic molecule are graphically depicted in Fig. 2.

Generally, cellulose is commonly produced with a desired size by the top-down enzymatic, mechanical and/or chemical treatments of cellulosic precursors, in which cotton, wood, annual plants or other agricultural residues can be utilized [9]. In contrast, cellulose can be fabricated by a bottom-up approach, where cellulose is

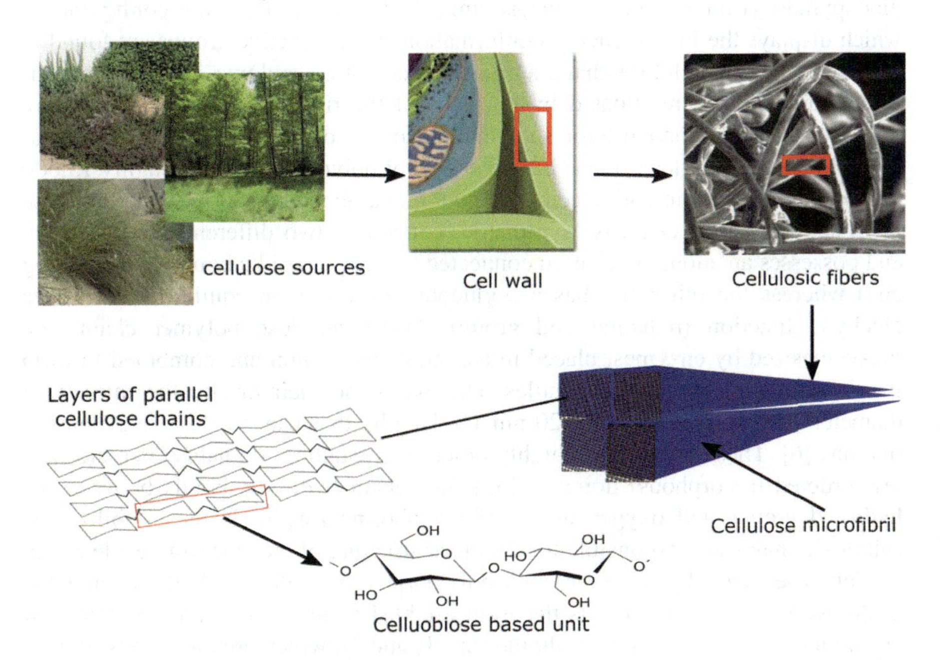

Fig. 2 Structural levels of organization of cellulose from the source to the molecule. Reprinted with permission from [9], Copyright © Elsevier Limited

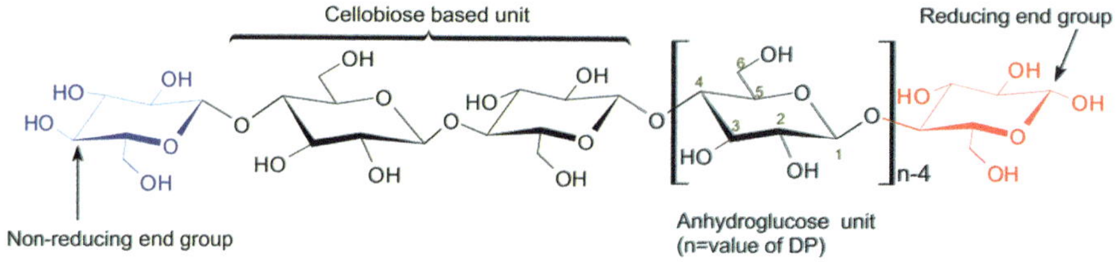

Fig. 3 Molecular structure of cellulose showing the numbering of carbon atoms, the reducing end with a hemiacetal, and the non-reducing end with a free hydroxyl at C4. Reprinted with permission from [9], Copyright © Elsevier Limited

biosynthesized from glucose using the direct action of specific bacterial strains belonging to genus such as *Alcaligenes*, *Achromobacter,* and *Gluconacetobacter* among which only *Gluconacetobacter xylinus* (*K. xylinus*) has been revealed to fabricate cellulose at commercial production levels [37]. Bacterial cellulose is typically produced in a pure form without needing intensive processing to remove unwanted impurities or contaminants [28]. It is worth noting that the quantities and the characteristics of cellulose depend closely on the isolation process, the origin and the natural source [6]. Cellulose can occur in pure form in lignocellulosic sources but it is usually accompanied by lignins, hemicelluloses, and comparably small quantities of extractives and trace elements.

Cellulose (Fig. 3) is defined as a linear polymer of repeating β (1,4)-bound D-glucopyranosyl units (anhydroglucose unit, AGU) in the $^4C_1$-chain configuration, which displays the lowest energy conformation. Three reactive groups are found in each AGU within cellulose chain, a primary group at C6 and two secondary groups at C2 and C3 that are situated in the plane of the ring [38]. The monomers are related together by condensation such that glycosidic oxygen bridges link the sugar rings. In nature, cellulose chains have a degree of polymerization of roughly 15,000 glucopyranose units in native cellulose cotton and about 10,000 in wood cellulose [9]. Each polymeric chain is asymmetric, containing two different end-units; one end possesses an anomeric C atom connected by the glycosidic bonds (nonreducing end) whereas the other end has a D-glucopyranose unit in equilibrium with the aldehyde function (reducing end group). These cellulose polymer chains are biosynthesized by enzymes, placed in a continuous fashion and combined to form microfibrils, long threadlike bundles. Depending on their origin, the microfibril diameters range from about 2–20 nm for lengths that can reach several tens of microns [6]. These microfibrils highly ordered (crystalline) domains alternate with less ordered (amorphous) domains [39]. Interchain hydrogen bonds between the hydroxyl groups and oxygen atoms of neighboring ring molecules stabilize the cellulosic chains are responsible for the linear structure of the macromolecule chain.

Cellulose generally has four main polymorphs vis. cellulose I, II, III, and IV. Cellulose I, native cellulose, is the form established in nature, and it is the most crystalline type existing in two allomorphs, I_α and I_β, which are analogous to each other whereas their packing pattern in the lattice is different because of the different extent of hydrogen bonding developed between the chains. Cellulose II, or regenerated cellulose, has an antiparallel arrangement of chains and it comes out after regeneration by solubilization and subsequent recrystallization or mercerization with aqueous sodium hydroxide. Cellulose III_I and III_{II} can be produced by ammonia treatment from either cellulose I or cellulose II, respectively. Cellulose IV_1 and cellulose IV_2 can be obtained from the corresponding form of cellulose III_1 and III_2 by heating in glycerol [38, 40].

The rigidity of chains and the existence of both polar and non-polar groups make the molecule insoluble in most common solvents. Cellulose does not melt below degradation temperature as well. This solvent resistance and thermal stability make cellulose an attractive polymer, but at the same time prevent direct application of industrial process developed for commodity polymers [41]. Furthermore, the crystallinity of cellulose makes it recalcitrant to acid and base-catalyzed hydrolysis too, thereby making the chemical processing of cellulose difficult. Because of these situations, an appropriate combination of various chemical, mechanical and enzymatic treatments of cellulose for further processing have been investigated to produce cellulose nanofibers from the early 2000s until now.

2.2 Cellulose Nanofibers

2.2.1 Types of Cellulose Nanofibers

Cellulose nanomaterials (CNM), as biopolymers, have recently gained substantial interest and have widely reported in the literature owing to their unique structural features and impressive physicochemical properties. CNM is regarded as a type of nano-objects where the term nano-object is reserved to material with one, two or three external dimensions in the nanoscale according to ISO publications [42, 43]. CNM is a term commonly used to define nanoscale of cellulosic materials, which is considered to be in the nanoscale range if the fibril particle diameters or width is between 1 and 100 nm. CNM combine crucial cellulose features—such as hydrophilicity, biodegradability, renewability, broad chemical modification capacity and the formation of versatile semicrystalline fiber morphologies as well as the specific properties of nanoscale materials, which are due to the very large specific area. Different types of CNM can be divided into various subcategories based on their preparation method, dimension, shape, function, which in turn primarily depend on the cellulosic source and processing conditions. It is worthy to note that ambiguities still exist regarding the terminology and nomenclature applied to cellulose nanomaterials [4–6, 11, 25, 30, 34, 44, 45]. More recently, the Technical Association of the Pulp and Paper Industry (TAPPI) has established a Nanotechnology Division devoted to the standardization of CNM nomenclature. A draft version of nanomaterials standard (TAPPI WI 3021: Standard Terms and Their Definition for Cellulose Nanomaterials) has been made [46], but comments on this standard are still under review. The existing literature encouraged that a

Table 1 Types of cellulose nanofibers and their particle sizes

Terminology and nomenclature of cellulose nanofibers	Width (nm)	Length (nm)	Aspect ratio (length/width)	Reference
Cellulose nanocrystals, nanocrystalline cellulose, cellulose nanocrystals	4–70	100–6000	10–100	[13]
Cellulose nanofibril, nanofibrillar cellulose, nanofbrillated cellulose	20–100	>10,000	>1000	[41]
Bacterial cellulose, microbial cellulose	10–50	>1000	100–150	[24]
Amorphous cellulose	50–120[a]	50–120[a]	~1	[4]
Cellulose nanoyarn	100–1000	>10,000	>100	[4]

[a]Diameter of spherical or elliptical nanoparticles

number of terminologies have been and are currently used to refer to cellulose nanomaterials, which unfortunately reflects some misunderstanding and anomalies. Various terms have been employed to designate cellulose nanomaterial elements including nanocellulose, cellulose nanofibers, nanoscale cellulose, cellulose microfibrils, nanocellulosic fibrils, cellulose nanofibrils, cellulose nanoparticles, and nano-sized cellulose fibrils. Generally, CNM can be split up into nanostructured materials and nanofibers. The nanostructured materials are categorized into cellulose microcrystals (or microcrystalline cellulose) and cellulose microfibrils (TAPPI WI 3021), whereas, the cellulose nanofibers (Table 1) are mainly sub-grouped into two types. The first type concerned cellulose nanofibrils (CNF) with various nomenclature, including nanofibrilated cellulose, cellulosic fibrillar fines, nanofibrillar cellulose, nanoscale-fibrillated cellulose, nanofibrils, nanofibers, fibril aggregates and sometimes microfibrillated cellulose or microfibrils, while the second type involved cellulose nanocrystals (CNC) with various terminologies, comprising nanocrystalline cellulose, nanorods, cellulose whiskers, cellulose nanowhiskers, rodlike cellulose crystals, nanowires and cellulose crystallites [13]. Other families of cellulose nanofibers materials that could be considered are bacterial cellulose discussed above, amorphous nancellulose and cellulose nanoyarn [4, 47]. The nomenclature that will be utilized in the present chapter is in accordance with the TAPPI standard recommendations.

2.2.2 Feedstock

Cellulose nanofibers have exciting physicochemical and biological properties such as good stability, high strength, low degradation, and nontoxicity. Various cellulose nanofibers could be prepared, depending on the source, origin, maturity, processing procedures and reaction conditions. Broadly, cellulose nanofibers derived from lignocellulosic materials is obtained by the top-down chemical and/or mechanical treatment. In contrast, these nanofibers can be produced by biosynthesis by some bacteria, giving rise to bacterial cellulose, which is obtained directly as a fibrous network, contains no lignin, pectin, hemicelluloses, or other biogenic products; it is very highly crystalline and possesses a high degree of polymerization (DP). Several researchers have beautifully compiled detailed studies on various sources for production/isolation of cellulose [4, 6, 13, 23, 24, 38, 40, 48–50]. Thus, only a concise overview of cellulose nanofibers sources will be displayed here.

Basically, any source of cellulose could be employed for cellulose fibers preparation. Woody and non-woody plants are considered as excellent raw materials for the production of several materials that have been demonstrated by the number of patents, peer reviewed articles and books, besides the number of products already marketed [11, 14, 16, 39, 44, 45, 51–55]. It is evident that pulps obtained from softwood and hardwood are nowadays the most important sources for cellulose nanofibers. However, agricultural residues and annual plants have recently received more attention due to costs and environmental concerns. These lignocellulose sources can be broadly classified upon the origin of the plants: (1) bast or stem, (2) seed or fruit, (3) leaf, (4) grass, and (5) straw fibers [13]. They can be defined as cellular hierarchical bio-composites naturally created in which lignin/hemicellulose/extractives and traces elements play a role of the matrix, whereas cellulose microfibrils serve as reinforcement. An effective removal process of non-cellulosic components gives rise to pure cellulose. Usually, the non-woody plants comprise less lignin than wood. Therefore, delignification and bleaching methods are less chemical and energy consuming [9]. The isolation of cellulose nanofibers forms different sources is relevant since it dictates the overall properties. Cellulose nanofibers produced from various lignocellulosic sources of miscellaneous provenance employing different extraction approaches and conditions commonly vary in their degree of polymerization, geometric dimensions, morphology, surface charge, surface area, porosity, crystallinity, thermal stability, mechanical properties, etc. [4, 5, 17]. Since there are many reviews and books dealing with cellulose nanofibers from several sources, Table 2 aims to summarize the most commonly utilized feedstock for their production.

Table 2 Lignocellulosic sources for the production of cellulose nanofibers

Source group	Sources
Hardwood	Eucalyptus, Elm, Birch
Softwood	Pine, Spurce, Cedar
Annual plants/ Agricultural residues	Oil palm, Hemp, Jute, Sisal, Alfa, Kenaf, Begasse, Corn, Sunflower, Bamboo Canola, Wheat, Rice, pineapple leaf and coir, Peanut shells, Potato peel, Garlic straw residues
Animal	Tunicates
Bacteria	*Alcaligenes*, *Achromobacter* and *Gluconacetobacter*
Algae	Green, gray, red, yellow-green, brown algae

Although lignocellulosic materials are regarded as the most important sources of cellulose, other living organisms including animals (tunicates), bacteria (mainly *K. xylinus*) and some types of algae (green, gray, red, yellow-green, etc.) can be used to prepare cellulose nanofibers as well [13].

2.2.3 Preparation of Cellulose Nanofibres

Production of cellulose nanofibers (Fig. 4) has gained increasing attention during the last decades, as very recently reviewed by Trache et al. [13] and Nechyporchuk et al. [14] for CNC and CNF, respectively. The production of cellulose nanofibers is mainly a top-down process, where lignocellulosic materials are involved and are broken down into nanocellulosic materials. Broadly, these lignocellulosic sources are first submitted to different pretreatments. Here we are not going to deeply touch pulping processes for breaking lignocellulosic materials by chemical, biological, mechanical or combined methods since several books and review articles have recently treated the subject and our main concern is for extracting cellulose nanofibers [4, 5, 13, 27, 50, 56, 57]. In nature, lignocellulosic are bio-composites containing nanoscale domains of cellulose, lignin, hemicellulose, extractives, and contaminants. From a technological point of view, lignin amount is an important parameter that should be considered to correctly optimize the pretreatment procedure required to isolate a pure cellulose pulp. Indeed, lignin is regarded as the hardest chemical compound to be eliminated from lignocellulosic sources [9]. For this purpose, the dewaxing, delignification and bleaching processes have often been utilized as a pretreatment to simplify the production of CNC and CNF. On the other

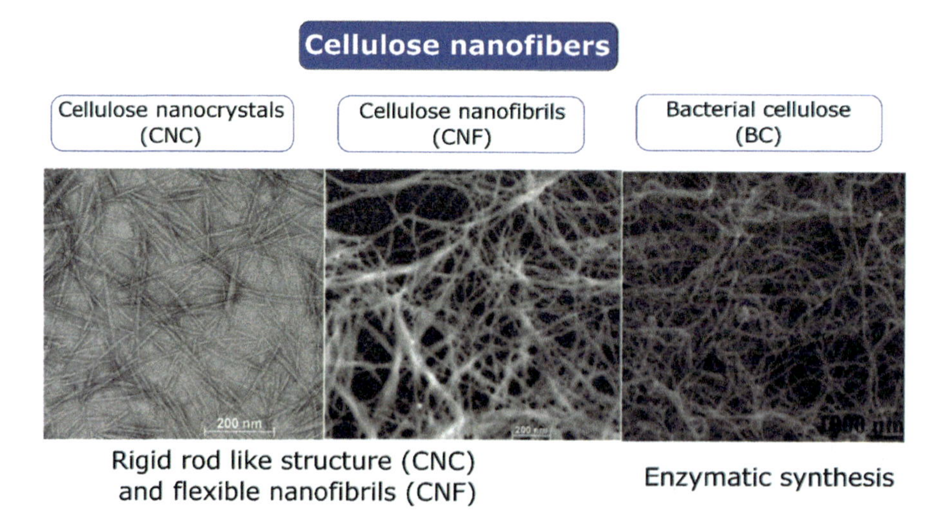

Fig. 4 Transmission electron micrographs of different cellulose nanofibers. Reprinted with permission from [98], Copyright © Elsevier Limited; [34] Copyright © WILEY-VCH

hand, bacterial cellulose production is a bottom-up process, where bacteria generate glucose, and cellulose is obtained by connecting glucose units [58–61]. This approach does not necessitate chemical or mechanical actions to remove hemicellulose and lignin, as is the case for lignocellulosic since BC is considered as a highly hydrated and pure cellulose membrane. More recently, many research works have focused on optimizing processes to lower energy consumption and other costs and to improve quality, consistency, and yields. Table 3 displays some recent examples of different cellulose nanofibers that can be obtained through the application of various methodologies for nanofibers extraction.

Table 3 Examples of cellulose nanofibers extracted by using various methodologies

Source	Form	Isolation procedure	Reference
Plum seed shells	CNC	Cellulose extraction, H_2SO_4 30% at 40 °C hydrolysis, ultrasonication, freeze drying	[99]
Waste cotton cloth	CNC	Cellulose extraction, treatment with sulfuric acid/hydrochloric acid/water mixture (3:1:11) at 55 °C assisted by ultrasonic wave, centrifugation, dialysis, freeze drying	[100]
Bleached sugarcane bagasse pulp	CNC	ultrasonic assisted TEMPO[a]/sodium bromide mediated oxidation, centrifugation, freeze drying	[101]
Bleached kraft eucalyptus dry lap pulp	CNC	Anhydrous organic acid hydrolysis at 90–120 °C, dilution, filtration, washing, centrifugation, dialysis	[65]
Commercial microcrystalline cellulose	CNC	Citric/hydrochloric acid hydrolysis	[102]
Bacterial cellulose	CNC	Washing, homogenization, drying, grinding, treatment with H_2SO_4/HCl mixture at 45 °C, dilution, centrifugation, dialysis, ultrasonication	[59]
Bleached kraft pulp	CNC	Pre-soaking in water, grinding, centrifugation, treatment with commercial enzymes or termite cellulose and incubated at intervals from 6–72 h at 35 °C, washing, lyophilization	[68]
Cotton cellulose fibers	CNC	Swelling in 1-butyl-3-methylimidazolium chloride and 1-(4-sulfobutyl)-3-methylimidazolium hydrogen sulfate followed by quenching by adding cold water, washing, centrifugation, freeze drying	[72]
Commercial microcrystalline cellulose	CNC	Water hydrolysis at 120 °C and pressure of 20.3 MPa, filtration with a Pyrex® Buchner funnel with glass fritted disc, dialysis, ultrasonication	[76]
Oil palm empty fruit bunch microcrystalline cellulose	CNC	Sono-assisted TEMPO-oxidation, followed by sonication, washing, centrifugation, drying	[103]

(continued)

Table 3 (continued)

Source	Form	Isolation procedure	Reference
Needle-leaf	CNF	Kraft process, refining, Extrusion (twin screw)	[104]
Softwood pulp	CNF	Sulfite process, Blending, refining, enzymatic/ carboxylation (TEMPO), grinding and/or homogenization	[105]
Hardwood pulp	CNF	Bleaching, solvent-based system (deep eutectic solvent), Homogenization (microfluidizer)	[106]
Ushar seed fiber	CNF	Cellulose isolation, TEMPO/sodium bromide/ NaClO mediated oxidation, centrifugation, homogenization, sonication	[107]
Fluff pulp	CNF	Alkaline pretreatment, high pressure homogenization, dialysis, filtration	[85]
Microalgae strains	CNF	Cellulose isolation, TEMPO/sodium bromide/ NaClO mediated oxidation, centrifugation,	[108]
Corn husks, oat hulls	CNF	Cellulose isolation, TEMPO/sodium bromide/ NaClO mediated oxidation, washing, filtration	[109]
Fluff cellulose pulp	CNF	Drying, grinding	[110]
Raft wood	CNF	Cellulose isolation, micro-fibrillation	[3]
Wood species	CNF	Cellulose isolation, defibrillation with wet disk-milling	[111]

[a]TEMPO: 2,2,6,6-tetramethylpiperidine-1-oxyl

Preparation of Cellulose Nanocrystals

Cellulose nanocrystals consist of elongated, cylindrical and rod-like particles with widths and length of 4–70 nm and between 100 nm and several micrometers, respectively [24, 34, 62]. CNCs can be isolated from several feedstocks that initially require following various pretreatment processes for the complete/partial elimination of the non-cellulosic materials (hemicellulose, lignin, waxes, fats, proteins, etc.). The naturally occurring cellulose consists of highly ordered crystalline regions and amorphous domains in different proportions. A proper combination of chemical, mechanical, oxidation and/or enzymatic treatment can be applied to remove the disordered regions to recover the CNC product [4, 13, 17, 27, 33, 49]. The paracrystalline domains of cellulose act as structural defects and are responsible for the transverse cleavage of the cellulose fibers into short nanocrystals under hydrolysis process. The transformation involves the disruption of the disorder parts surrounding the microfibrils, in addition to those embedded between them leaving the crystalline parts intact. Several procedures have been developed and continue to appear in order to prepare CNC in an economic/sustainable way with desired properties [13, 21]. Until recently, the most important methods reported in the literature for the extraction of CNC are chemical acid hydrolysis (solid/liquid/ gaseous/organic/inorganic acids) [17, 53, 63–67], enzymatic hydrolysis [68–70], mechanical refining [27, 68–71], ionic liquid treatment [72–75], subcritical water hydrolysis [76, 77], oxidation method [25, 78–81] and combined processes [82–84]. The advantages and drawbacks of each procedure have been recently reviewed by Trache et al. [13]. After each process, some post-treatments of solvent elimination, washing, neutralization, centrifugation, sonication, filtration, purification, fractionation, stabilization and surface modifications are commonly required to get the final CNC product [13, 34]. It is worth noting that the acid hydrolysis treatment remains the most commonly used technique for the separation of CNC. These nanofibers may display diverse geometries, depending on their biological source; for instance, algal cellulose membrane shows a rectangular structural arrangement, whereas both bacterial and tunicate cellulose fibers present a twisted-ribbon geometry [34].

Recently, considerable research programs on the fabrication of CNC have been initiated at the industrial scale. Four commercial entities producing cellulose nanocrystals at capacities beyond pilot plant scale: CelluForce (Canada, 1000 kg/day), American process (USA, 500 kg/day), Melodea/Holmen (Sweden, 100 kg/day), and Alberta Innovates (Canada, 10 kg/day). Further research facilities are currently fabricating CNC as well [13].

Preparation of Cellulose Nanofibrils

Cellulose nanofibrils (CNF) are flexible and have entangled network structure with a diameter of approximately 1–100 nm consisting of alternating crystalline and amorphous domains [85]. This kind of cellulosic nanofibers was introduced by Turbak et al. [86] and Herrick et al. [87] who prepared cellulose in nanoscale range by passing softwood pulp aqueous suspension several times through a high-pressure mechanical homogenizer and giving rise to CNF due to high shearing forces. Depending on the preparation conditions, cellulose fibers can be disintegrated to flexible CNF with lateral dimensions starting from ca. 5 nm, demonstrating elementary fibrils, to tens of nanometers, which correspond to single microfibrils and their bundles. Production of CNF (Fig. 5) has received much attention over the past few decades, as recently reviewed by Nechyporchuk et al. [14]. CNF can be extracted by the disintegration of cellulosic fibers along their long axis. The processes include either conventional mechanical methods (e.g. homogenization, grinding) or a combination of enzymatic/chemical pretreatments with mechanical processing [5, 11, 14, 23, 57, 88].

Broadly, cellulose is present in lignocellulosic materials in combination with hemicellulose, lignin, and extractives. The latter are generally eliminated by various processes before the production of CNF. A simple mechanical disintegration includes homogenization, grinding, refining, extrusion, blending, ultrasonication, cryocrushing, steam explosion, ball milling and aqueous counter collision [5, 11, 14, 23, 57, 88, 89]. Among these processes homogenization (using homogenizers and microfluidizers), grinding and refining are the most common techniques employed in mechanical isolation of CNF. These techniques are considered more efficient for CNF extraction thanks to high shear delamination and are appropriate for scaling up. Thus, they are utilized nowadays for industrial production of CNF.

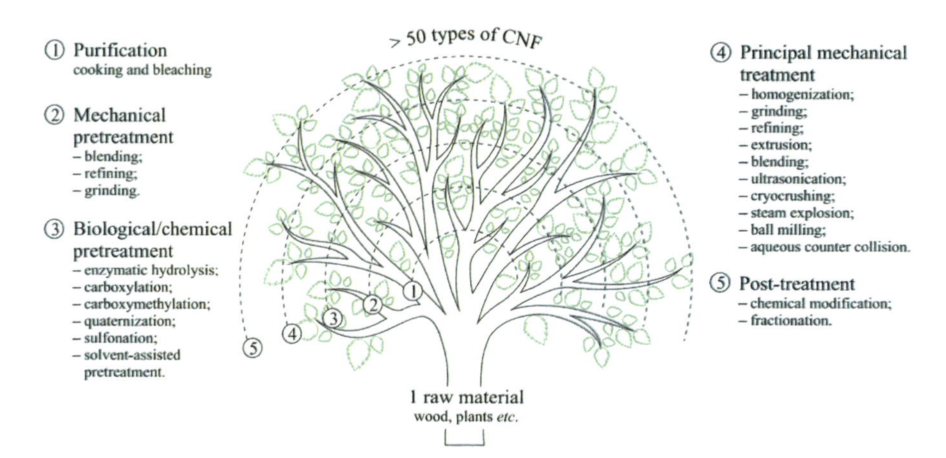

Fig. 5 Schematic diagram of CNF production "tree". Reprinted from [14], Copyright © Elsevier Limited

However, the main drawback that should be overcome for successful commercialization of CNF is the high-energy consumption needed for mechanical disintegration involved in processing pure cellulosic fibers [90]. Therefore, chemical and/or enzymatic pretreatments have helped to decrease the energy consumption. NFC was extracted by an environmentally friendly procedure by combining enzymatic hydrolysis and mechanical shearing of lignocellulosic material pulps. Effective chemical pretreatments were also attempted prior the mechanical disintegration, by TEMPO or periodate-chlorine oxidation of cellulose, sulfonation, carboxymethylation, quaternization or solvent-assisted pretreatment [14]. After each mechanical shearing, some post-treatments such as surface chemical modification and fractionation are needed to isolate the final CNF product.

Currently, around ten companies are producing CNF at commercial/pre-commercial scale, including Forest Products Laboratory (cooperating with the University of Maine), Paperlogic, Innventia (Sweden), American Process (all USA), Borregaard (Norway), Oji Paper (both Japan), Nippon Paper, CTP/FCBA (France) and others. Moreover, several organizations have developed pilot scale production of various CNF [14].

Preparation of Other Families of Cellulose Nanofibers Materials

Bacterial cellulose (BC), known as microbial cellulose, is typically obtained from bacteria, as a separate molecule and does not necessitate processing to eliminate contaminants. In the biosynthesis of BC, the glucose chains are supplied inside the bacteria body and expelled out through minor pores existing on the cell wall. A 20–100 nm ribbon-shaped nanofibers with micrometer lengths, entangled to form stable network structures [58, 91–93].

Another category, amorphous nanocellulose (ANC), can be formed through acid hydrolysis of regenerated cellulose with subsequent ultrasound disintegration [4, 94, 95]. The obtained particles have spherical to elliptical shapes with diameters ranging from 80 to 120 nm, depending on the cellulose source, isolation procedure, and extraction conditions. Because of its amorphous structure, ANC displays specific characteristics, such as enhanced sorption, high accessibility, increased functional group content. However, this kind of nanofibers presents poor mechanical features and is unsuitable for utilization as reinforcing nanofillers. Hence, the main use of ANC as carries is for bioactive substances and thickening agent in different aqueous systems.

Cellulose nanoyarn (CNY) remains the less investigated cellulosic nanofibers. It is commonly produced by electrospinning of a solution containing cellulose or cellulose derivative [4, 49, 96]. Different parameters, such as the electric field strength, tip-to-collector distance, solution feed rate are broadly employed to control the morphological characteristics of the electrospun nanofibers [97]. The obtained electrospun nanofibers present diameters ranging from 100 to 1000 nm.

Nevertheless, these three categories of cellulose nanofibers are out of this chapter's scope.

3 Cellulose Modification

3.1 *Acid Hydrolysis*

Acid hydrolysis is the traditional method and widely used to manufacture microcrystalline cellulose into nano-cellulose as this method can remove the amorphous region [68]. Ultrasonication is applied after harsh acid treatment to produce cellulose nano whiskers. The 'harsh' acid treatment usually referred to sulfuric acid which is highly acidic compared to hydrochloric acid even though it also works and the size of nano whisker reduced with the increasing of acidity. This is because during hydrochloric acid hydrolysis, there is a weaker charge density from the entire process and it leads to poorer dispersion in an organic solvent [112]. Hydrocholoric acid resulted in the formation of poor nano-cellulose aqueous dispersion with the present of hydroxyl and carboxyl group [68]. While in sulfuric acid treatment, the anionic sulfonate ester group maintain resulted in the production of high crystallinity of nano whiskers that are easy to disperse in aqueous solvents [68, 112]. Also, during the process, the disordered or paracrystalline regions in cellulose are hydrolysed while the crystalline regions have higher resistance towards acid attack and remain intact [29]. Table 4 shows different types of acid used to form cellulose derivatives from various sources.

Table 4 Different types of acid used to form cellulose derivatives from various sources

Sources	Process	References
Norway spruce	H_2SO_4 hydrolysis	[168]
Sisal fibres	60% H_2SO_4 hydrolysis,	[170]
Pea hull	64% H_2SO_4 hydrolysis	[114]
Filter paper	HCl hydrolysis	[29]
Filter paper	64–65% H_2SO_4 hydrolysis	[123]
Coconut husks	64% H_2SO_4 hydrolysis	[139]
Coconut husks	64% H_2SO_4 hydrolysis	[170]
Sugarcane bagasse	60% H_2SO_4 hydrolysis	[124]
Ethanol bio-residue	H_2SO_4 hydrolysis	[155]
Bamboo fibres	64% H_2SO_4 hydrolysis	[171]
Sisal fibres	64% H_2SO_4 hydrolysis	[171]
Curauá fibres	64% H_2SO_4 hydrolysis	[171]
Eucalyptus fibres	64% H_2SO_4 hydrolysis	[171]
Cotton pulp fibres	64% H_2SO_4 hydrolysis	[172]
Rice straw	64–64% H_2SO_4 hydrolysis	[122]
Rice husks	H_2SO_4 hydrolysis, 64% H_2SO_4 hydrolysis	[126, 138]
Mengkuang leaves	60% H_2SO_4 hydrolysis	[127]
Dried *Eucalyptus* pulp	H_2SO_4 hydrolysis	[146]
Phormium tenax fibres	64% H_2SO_4 hydrolysis	[173]
Raw cotton linter	60% H_2SO_4 hydrolysis	[174]
Soy hulls	64% H_2SO_4 hydrolysis	[141]
Pineapple leaf	64% H_2SO_4 hydrolysis	[175]
Corncob	H_2SO_4 hydrolysis	[140]
Filter Paper	85% H_3PO_4 hydrolysis	[176]
Coconut husks	60% H_2SO_4 hydrolysis	[129]
Oil palm empty fruit bunch	64% H_2SO_4 hydrolysis	[120]
Spruce bark	60% H_2SO_4 hydrolysis	[177]
Hardwood pulp	$H_3PW_{12}O_{40}$ hydrolysis (phosphotungstic acid)	[66]
Sugarcane bagasse	64% H_2SO_4 hydrolysis	[119]
Hardwood pulp	64% H_2SO_4 hydrolysis	[66]
Banana plant	H_2SO_4 hydrolysis	[178]
Unriped coconut husks	30% H_2SO_4 hydrolysis	[130]
Posidonia oceanica	H_2SO_4 hydrolysis	[179]
Water hyacinth	40% H_2SO_4 hydrolysis	[180]
Oil palm trunk	64% H_2SO_4 hydrolysis	[131]
Recycled newspaper	65% H_2SO_4 hydrolysis	[182]
Arecanut husks	CH_3COOH hydrolysis	[133]
Eucalyptus wood	60% H_2SO_4 hydrolysis	[142]
Groundnut shells	65% H_2SO_4 hydrolysis	[183]
Mulberry pulp	50% H_2SO_4 hydrolysis	[100]

3.2 Enzyme Hydrolysis

Nano-cellulose can be also produced by enzyme hydrolysis process and the biomasses of cellulose materials used for these enzymatic processes up to date were bacterial cellulose, cotton fibers, and crystalline cellulose powder [69]. According to Meyabadi et al. [113] and Anderson et al. [68], the evaluation of new and established methods to produce new products can be obtained through enzyme application. Enzymatic hydrolysis that produces nano-cellulose was an alternative, environmentally sustainable, and cheap method that reduced water consumption. Furthermore, the digestion of enzyme omitted the use of harsh chemicals such as sulphuric acid which must be washed and properly disposed after traditional production of nano-cellulose. The usage of energy is minimized due to less mechanical processing for fibrillation and lower temperature for heating. The amorphous region of cellulose fibres was selectively degraded by the enzyme but not the crystalline region resulted in the presence of unmodified hydroxyl group surface which is similar to hydrochloric acid digestion in the production of nano-cellulose [68]. In enzyme, there is a class which called cellulases and the sub-groups are cellobiohydrolase and endoglucose. The function of cellulases is acting as a catalyst for the hydrolysis of cellulose while for cellobiohydrolase and endogluconase are attacked in the crystalline region of cellulose and catalyzed the hydrolysis of the amorphous region in cellulose. Thus, the chain length of cellulose will decrease during the enzymatic hydrolysis in the cellulose [113]. *Aspergillus oryzae* [69], fungus T. *reesei* [125], Cellulose L from Novozyme [113] and *Aspergillus niger* [68] enzymes were used because due to low level of β-glucosidase enzyme and this condition prevent from complete hydrolysis to glucose. *Trichoderma Viride G* grows in diverse environments, safe to use as an enzyme to hydrolze nano-cellulose due to their non-toxic nature and anti-metabolic repression ability [70].

3.3 Ionic Liquid

Another alternative nano-cellulose extraction is using ionic liquids treatment. The stabilities of ionic liquid in chemical and thermal condition, relatively low melting point, non-volatile and negligible vapour pressure proved that ionic liquid treatment has the ability to dissolve cellulose as well as allow envisaging safe and low energy consumption [72, 73, 75]. Ionic liquid has the potential at the swelling and portrays as reactants and catalysts which proved during hydrolyzing polysaccharides into nanofibers in the cellulose [74]. Apart from that, the ionic liquid is an eco-friendly solvent which can be easily recovered and maintained its activity upon reuse after the regeneration of cellulose by a simple method such as evaporation, salting out and reverse osmosis [73, 74]. According to Lazko et al. [72] considering their cost and energy, the recyclability and reuse of ionic liquid undeniable to be essential for the conception of eco-friendly and economically save to apply for extraction nano-cellulose. The lower amount of sulfonic groups and can be cast in transparent layered films of slightly lower hydrophilicity shown ionic liquid treatment for nano-cellulose compared to the traditional method which is by concentrated sulphuric acid hydrolysis [74]. Ionic liquid that are normally used are 1-butyl-3-methylimidazolium hydrogen sulfate ([BMIM]HSO_4) [74] 73], 1-butyl-3-methylimidazolium chloride ([BMIM]Cl) [72, 75], 1-(4-sulfobutyl)-3-methylimidazolium hydrogen sulfate ([SBMIM]HSO_4) [72], 1-ethyl-3-methyllimidazolium acetate ([EMIM][OAc]) [115] Tetrabutylammonium acetate (TBAA) [116]. [BMMIM]Cl treatment on cellulose gives limited impact on the length of cellulose fiber as it does not support any hydrolytic phenomena. So, this shows that [BMMIM]Cl reduces destructuring and unfolding of cellulose fibers to nano-fibrils and it maintains the crystallinity and morphology of nano-cellulose [75]. Apart from that, Tan et al. [73] revealed that [BMMIM]Cl recovery rate can be up to 99.5% by evaporating of anti-solvent and this condition showed that ionic liquid can again be desirable solvent as well as a catalyst with excellent properties. On the other hand, ([BMIM]HSO_4) treatment gives higher yields and high crystallinity of nano-cellulose from microcrystalline cellulose [73, 74].

3.4 Mechanical Treatment

To date, mechanical treatment such as microfluidization, high pressure homogenization, ultrasonication, pulp beating, cryocrushing and ball milling have been applied for the preparation of nano-cellulose. Those methods are low cost, simple and environment-friendly can be used as a part of combination of acid hydrolytic, enzymatic, and oxidative treatments [71, 117]. Ultrasonication is the application of sound energy towards chemical and physical systems. The acoustic cavitation which is the formation, growth and collapse of bubbles in a liquid form during the chemical effects of ultrasonication and the polysaccharide linkages are break down. Thus, cellulose fibers are being treated and improved on the accessibility and reactivity of cellulose [117]. High-energy bead milling (HEBM) gives better impact compared to ball milling and ultrasonication as HEBM can micronize the particles. Beads are forced to rotate around the mill when centrifugal force is generated in the mill. The centrifugal force will also goes in the opposite direction and gives advantage to the transition balls to roll over the opposite walls of the mill when the reverse rotation of disc is applied. Thus, this condition gives impact on the micronize the material in between the bead. In addition, HEBM is also cost-saving method to produce nano-cellulose effectively with great thermal stability in a bigger yield [71].

3.5 Subcritical Water

The conventional method such as acid hydrolysis method to produce nano-cellulose consume a lot of time to wash it. Thus, there is another new method which used only pressured hot water also known as subcritical water. This new method is new, greener towards the environment, and less expensive as the consumption of chemical and electrical energy is theoretically evaluated to produce nano-cellulose [77]. Hydrolyzation of amorphous and semi-crystalline region of cellulose can be enhanced by subcritical water method as it is absolutely water which act as the hydrolysing agent at higher pressurized reactor. At higher pressure in the reactor, the yield of nano-cellulose will increased linearly [76]. The opening of restrictor valve of the reactor decreased the pressure and water injection with precision pump increased the pressure inside it. Those two functions controlled the pressure inside the reactor effectively [77]. The internal pressure gives huge impact on the feasibility of the steps ad will as the hydrolyzation of cellulose which normally used in the removal of hemicellulose [76].

3.6 2,2,6,6-Tetramethyl-1-Piperidinyloxy

The 2,2,6,6-tetramethyl-1-piperidinyloxy (TEMPO) radical is a suitable solvent which is selectively oxidized the primary alcohol groups in aqueous solution [81]. 2,2,6,6-tetramethyl-1-piperidinyloxy (TEMPO) gives positive results in the fabrication of nanofibrils cellulose by stretching the stands of nanosized cellulose that consists of the amorphous and crystalline region [80].

3.7 Combined Method

Combined method is a mixed method to obtain better cellulose derivatives. Table 5 shows different types of combined method used to form cellulose derivatives from various sources.

4 Characterization of Cellulose Nanofibers

4.1 Fourier Transform Infrared

The changes in structural and chemical composition after the effect of chemical treatment is studied using infrared spectroscopy [118, 119].

Table 5 Different types of combined method used to form cellulose derivatives from various sources

Source	Combined method	References
Nordic paper	HCl hydrolysis and 2,2,6,6-tetramethylpiperidine-1-oxyl radical (TEMPO)	[184]
Oil palm empty fruit bunch	4-Acetamido-TEMPO (2,2,6,6-tetramethylpiperidin-1-oxyl) and ultrasonication	[103]
Commercial MCC	H_2SO_4 hydrolysis and Subcritical water treatment	[185]
Jute	Microwave-assisted alkaline peroxide and ultrasonication	[82]
Cotton linter	HCl hydrolysis and Subcritical water treatment	[186]
Cotton linter	H_2SO_4 hydrolysis and Enzymatic (*Cerrena sp.* fungus) treatment	[186]
Cotton linter	H_2SO_4 hydrolysis and Enzymatic (cellulase from Fungal Bioproducts, Spain) treatment	[187]
Sugarcane Bagasse	Enzymatic (commercial enzyme extract Cellic® CTec2) treatment, purification, and H_2SO_4 hydrolysis	[188]
Ciona intestinalis	Enzymatic treatment, TEMPO-mediated oxidation and H_2SO_4 hydrolysis	[189]
Cotton fabric	H_2SO_4 hydrolysis and ultrasonication	[190]
Miscanthus Giganteus	Acid (HCl and H_2SO_4) hydrolysis and TEMPO oxidation	[191]
Cotton linter	$H_3PW_{12}O_{40}$ hydrolysis (phosphotungstic acid) and ultrasonication	[192]
Wood flour	Ethanol and peroxide solvothermal and ultrasonication	[193]
Filter paper	H_2SO_4 hydrolysis, Ultrasonication, and Microwave-assisted	[84]
Bamboo pulp sheet	$FeCl_3$-catalyzed hydrolysis and ultrasonication	[194]
Oil palm empty fruit bunch	H_2SO_4 hydrolysis and Ultrasonication	[195]
Hardwood paper pulp	Acid (HCl and HCOOH) hydrolysis	[196]
Commercial MCC	H_2SO_4 hydrolysis and Ultrasonication	[83]
Old corrugated container (OCC)	H_3PO_4 hydrolysis, Enzyamatic treatment, and ultrasonication	[197]
Commercial MCC	HCl hydrolysis under hydrothermal conditions	[198]

Mid-region band between 4000 and 400 cm^{-1} in the transmittance mode has been widely used. Microcrystalline cellulose and nanowhisker cellulose have the almost identical chemical composition in Fourier Transform Infrared (FTIR) spectra. Absorption of water by cellulose appeared in the peak at 1635–1645 cm^{-1} while the intermolecular hydrogen attraction at C_6 group happened in between 1428 and 1433 cm^{-1} [120, 121]. There was a broad absorption band between 3000 and 3600 cm^{-1} which proven the presence of hydroxyl groups (–OH stretching intermolecular hydrogen bonds) and absorption at 2900 cm^{-1} was due to stretching of

–CH group from variation of biomass which were sugarcane bagasse, coconut husks, rice husks, rice straw, banana peel, cotton fabrics, mengkuang leaves, hardwood pulp, arecanut husk fibre, oil palm trunk, oil palm empty fruit bunch, pea hull, and filter paper [66, 114, 119–133].

As an example, in sugarcane bagasse and oil palm empty fruit bunch, there was a slight difference in the peak at 1105 cm^{-1} and the band between 1155 and 1167 cm^{-1} which were corresponded to C–O–C glycosidic ether band and C–C band gradually disappeared from microcrystalline cellulose to nanowhisker cellulose. The disappearance was due to the reduction in molecular weight during hydrolysis treatment [120, 124]. Same goes to nanocrystalline cellulose which was extracted from banana peel which showed that the obvious peak at 1030 cm^{-1} corresponded to C–O–C pyranose ring skeletal vibration stretching. This resulted in the presence of xylans which associated with hemicellulose [121]. The polysaccharide of glycoside bond or crystallinity band of cellulose presence at peak 896 or 897 cm^{-1} [66, 130, 133].

4.2 X-ray Diffraction Analysis

X-ray diffraction (XRD) is a fascinating instrument which used to examine the changes of crystallographic structure of the materials after hydrolysis treatment [128]. On top of that, XRD can provide the in-depth information of the crystalline solids based on their atomic-scale structure of materials [51].

The crystallinity index is calculated using the formulation reported by Segal et al. [134]:

$$\mathrm{CrI}\,(\%) = \frac{(I002 - I_{am})}{I002} \times 100$$

where CrI is percentage crystallinity index, I_{002} is the maximum intensity of the peak at 22° which is a crystalline part, and I_{am} is the intensity of diffraction of the amorphous part at 18°.

According to Chandrahasa et al. [129], acid hydrolysis treatment produced a higher degree of crystallinity index and the average size of hydrolyzed fibers will be almost 25 nm.

Scherrer equation reported by Revol et al. [135] is used to calculate the crystallite size in the following form:

$$\mathrm{D} = \frac{\mathrm{k}\lambda}{(\beta \cos \theta)}$$

where D is the apparent crystallite size, β is a full-width half maximum of lattice plane reflection in radian, k is Scherrer constant (0.94), λ is X-ray wavelength (0.15418 nm), and θ is corresponding Bragg angle (reflection angle).

Table 6 XRD analysis for crystallinity index (CrI) and crystallite size (L) of nano-cellulose from different biomass

Samples	Reflection at 2θ (°)	CrI (%)	L (nm)	References
Coconut husks	22–24	76	N/A	[129]
Sisal fibres	18, 26	75	N/A	[169]
Pineapple leaves	20.4, 22.7	74	N/A	[137]
Coconut husks	15.6, 22.7, 34.6	66	N/A	[139]
Bioethanol-residue	14.2, 16, 22.1	77	N/A	[155]
Cotton fibres	14.7, 16.4, 22.6	86	6.3	[125]
Bamboo fibres	15, 17, 21, 22.6	87	5.7	[171]
Eucalyptus fibres	15, 17, 21, 22.6	89	6.1	[171]
Sisal fibres	15, 17, 21, 22.6	78	5.9	[171]
Curauá fibres	15, 17, 21, 22.6	87	5.0	[171]
Rice straw	14.7, 16.4, 22.6	86	7.4	[122]
Rice husks	16, 22, 35	59	N/A	[126]
Mengkuang leaves	22.6	70	N/A	[127]
Cotton fabrics	14.9, 16.5, 22.6	84	8.7	[128]
Oil palm empty fruit bunch	22.6, 15.0	88	N/A	[120]
Raw cotton linter	17.4, 19.2, 26.5	90	N/A	[174]
Soy hulls	17, 21,F 23, 34	74	2.7	[141]
Corncob	17, 21, 23, 34	84	N/A	[140]
Sugarcane bagasse	16.5, 22.5, 34.6	73	3.5	[119]
Hardwood pulp	16.4, 22.7, 34.5	85	5.8	[66]
Unriped coconut husks	17, 19, 26, 41	82	5.0	[130]
Oil palm trunk	19, 22	70	N/A	[131]
Arecanut husk	15, 22	73	4.3	[133]
Banana peel	16, 22	64	N/A	[121]

Numerous studies reported that the crystallinity index and the sizes of nano-whisker or nanocrystalline cellulose showed the highest value compared to microcrystalline cellulose when tested under XRD with different lignocellulosic biomass (Table 6). The increase in crystallinity structure in nano-cellulose appeared after acid treatment because there was hydrolytic cleavage of glycosidic bonds which lead to rearrangement of molecules in cellulose [51, 137]. Prolonged acid treatment with the increasing of time period severe enough to destroy not only the amorphous structure but also some part of the crystalline structure in cellulose. This resulted in a slight decrease in crystallinity index [139–141].

4.3 Scanning Electron Microscope

The surface morphology is evaluated by scanning electron microscope (SEM). According to Luykx et al. [143] and Goldstein et al. [144], SEM contains a shadow-relief effect of secondary and backscattered electron contrast, produce a good image and greater view of the three-dimensional structure of the sample. The hydrolysis treatment has affected the surface morphology in terms of surface smoothness and the size of cellulose fibers [130, 131]. In Table 7, the majority of the samples have shortening in fiber and rough surface as this indicates that the cementing materials around the material of the fibre are partially removed. The cementing materials are hemicellulose, lignin, wax, and pectin [126, 136]. According to Nascimento et al. [130], bleaching process increased the surface area of the cellulose fiber.

4.4 Transmission Electron Microscope

Transmission electron microscope (TEM) is a unique and powerful nanoscale imaging with higher resolution to analyze the structure and size of nano-samples [131, 145]. The different nano-cellulose samples which stated in Table 8 shows in either rod liked the shape, needle shape or ribbon shape. Those shape formed is because of hydrolysis treatment breaks down the linkage that linked along the

Table 7 SEM analysis for surface morphology of nano-cellulose from different biomass

Samples	Surface morphology	References
Sisal fibres	Nano-ordered chain	[169]
Filter paper	Sphere-shaped structure, shortening of fibres	[123]
Pineapple leaf	Rough surface, defibrillation of fibres	[136]
Sugarcane bagasse	Refinement of fibrillar structure	[124]
Cotton pulp fibres	Rod-like shape, shortening fibres	[172]
Rice husk	Rough surface, shortening fibres, rolled shape	[126, 138]
Rice straw	Shortening fibres	[122]
Oil palm empty fruit bunch	Aggregation of fibres structure, rod-like shape, smooth surface	[120]
Raw cotton linter	Curled, soft-flat shape, rough surface with some pits	[174]
Soy hulls	Micro-sized fibres, irregular shape and size	[141]
Corncob	Micro-sized fibres, irregular shape and size	[140]
Hardwood pulp	Porous network structure	[66]
Unriped coconut husk	Loosely fibres, rough surface	[130]
Oil palm trunk	Less irregular and impurities	[131]
Arecanut husk fibres	Defibrillation of fibres, rough surface	[133]
Banana peel	Reduced and shortening fibres	[121]

Table 8 TEM analysis for surface morphology, length and width of nano-cellulose from different biomass

Samples	Surface morphology	Length, L (nm)	Width, D (nm)	References
Norway spruce	Bundle rod-shaped whisker	200–400	<10	[168]
Pineapple leaves	Non-woven ribbon	200–300	5-25	[137]
Coconut husks	Rod-liked shape	179 ± 59	5.5 ± 1.4	[139]
Sugarcane bagasse	Rod-shaped cellulose	170	35	[124]
Bamboo fibres	Rod-liked shape with individual nanowhiskers	100 ± 28	8 ± 3	[171]
Eucalyptus	Rod-shaped in bundle form	100 ± 33	7 ± 1	[171]
Sisal fibres	Rod-shaped with individual form	119 ± 45	6 ± 1	[171]
Curauá	Rod-like shaped in bundle form	129 ± 32	5 ± 1	[171]
Rice husks	Needle-like structure	–	10–15	[126]
Rice straw	Rod-like crystal structure	50–700	10–65	[122]
Rice husks	Needle like structure	100–400	6–14	[138]
Mengkuang leaves	Rod-liked shape	200	5–25	[127]
Soy hulls	Needle-liked nanowhiskers	122.7 ± 39.4	4.43 ± 1.2	[141]
Oil palm empty fruit bunch	Rod-liked structure	>100	<10	[120]
Sugarcane bagasse	Agglomerated rod-liked nanocrystals	250–480	20–60	[119]
Hardwood pulp	Rod-liked shape	100	15–40	[66]
Unriped coconut husks	Long, disordered needles like nanowhisker	178 ± 88	8 ± 3	[130]
Oil palm trunk	Individual, needle-liked nanocrystals in fibre bundles	397.03	7.67	[131]
Arecanut husks	Long fibrils	120–150	1–10	[133]
Banana peel	Needle-liked and wire nanofibrils	263.9 ± 7.2	20 ± 5.2	[121]

microfibrils by amorphous domains. Thus, the amorphous region which portrays as structural defeats and responsible for the transverse cleavage of microfibrils into nano-cellulose [139].

4.5 *Atomic Force Microscopy*

Similar to TEM, atomic force microscopy (AFM) is used to characterize the surface morphology, topography and measure dimensional image of different nano-samples

Table 9 AFM analysis for surface morphology, length and width of nano-cellulose from different biomass

Samples	Surface morphology	Length, L (nm)	Width, D (nm)	References
Sisal fibres	Nano-ordered chain	–	30.9 ± 12.5	[169]
Pineapple leaves	Few lateral association occur between adjacent nanofibres	–	5–15	[136]
Sugarcane bagasse	Reduced in fibres	–	70–90	[124]
Bioethanol residue	Rod-liked shape	–	10–20	[155]
Cotton fibres	Narrow and well separated distribution of nanofibrils	287.24 ± 79.75	29.69 ± 5.07	[125]
Bamboo fibres	Rod-liked nanowhiskers	–	4.5 ± 0.9	[171]
Eucalyptus	Rod-liked nanowhiskers	–	4.5 ± 1.0	[171]
Sisal fibres	Rod-liked nanowhiskers	–	3.3 ± 1.0	[171]
Curauá	Rod-liked nanowhiskers	–	4.7 ± 1.0	[171]
Rice straws	Rod-liked structure	30.7	5.95	[122]
Rice husks	Needle-liked structure	100–400	6–14	[138]
Soy hulls	Aggregation of nanocrystals	–	2.77 ± 0.67	[141]
Corncob	Aggregation of needle-liked nanocrystal	287.3 ± 75.5	4.9 ± 1.34	[140]
Sugarcane bagasse	Rod-likes crystal fibres	250–480	20–60	[119]

which shown in Table 8 [119, 124, 141] AFM provides more precise and accurate characterization of the thickness of the individual crystallites as well as detailed structure compared to TEM [141]. Through transverse height profiles, AFM allows the discernment of individual whiskers of agglomerated structures [140, 141]. Table 9 displays about the AFM analysis for surface morphology, length and width of nano-cellulose from different biomass.

4.6 Thermal Behaviour

According to Azubuike and Okhamafe [147], thermogravimetric analysis (TGA) and differential scanning calorimetry (DSC) used to analyze the thermal and degradation properties of cellulose fibers. Information regarding on the transition phase, weight loss and decomposition pattern are provided by TGA while information on maximum temperature for sample degradation retrieved from DSC. Measurement of heat capacity needed to increase the temperature and heat energy consumed by the sample is provided by DSC. Thus, both instruments able to provide a clarification on the thermal stability and compatibility of the sample.

There are two stages in every TG curves from different biomass which indicates the weight loss of the sample. The first stage is the water removal within the cellulose in the region between 60 and 100 °C. The second stage is the degradation of hemicellulose and lignin in cellulose followed by the char residue formation in the range of 200–400 °C. Melting point temperature (T_{max}) examined by using DSC and the endothermic peak appeared which is due to volatilization of cellulose [39]. The results of TGA and DSC are constructed in Table 10. Evolution of the volatile compounds occurred when cellulose degrades while the degradation of lignin corresponds to thermal degradation properties [148].

Table 10 Comparison of thermal properties using TGA and DSC from different biomass

Samples	TG analysis						DSC analysis T_{max} (°C)	References
	Onset (°C)	Degradation temperature (°C)			Residual weight loss at 400 °C (%)	DTG peak temperature T_{max} (°C)		
		T_{10}	T_{15}	T_{20}				
Sisal fibres	190	200	280	300	30	296	347	[169]
Pea hulls	100	180	–	235	17	299	–	
Banana plant	228	300	319	328	18	–	–	[114, 118]
Filter paper	290	330	340	349	9	–	330	[123]
Coconut husks	200	240	281	300	20	160	–	[139]
Sugarcane bagasse	220	250	260	280	17	345	253	[124]
Bio-residue	215	265	275	280	35	283	–	[155]
Rice husks	215	220	230	235	38	240	–	[126]
Mengkuang leaves	261	281	310	320	18	370	–	[127]
Cotton fabric	223	256	248	256	26	281	–	[128]
Phormium tenax fibres	238	280	300	320	10	90	–	[173]
Oil palm empty fruit bunch	270	350	375	390	71	–	–	[120]
Raw cotton linter	201	213	230	240	42	219	–	[174]
Soy hulls	170	190	210	251	23	294	–	[141]
Coconut husks	211	240	300	310	24	319	–	[129]
Sugarcane bagasse	236	240	260	270	31	250	–	[119]
Hardwood pulp	260	330	340	348	25	–	–	[66]

(continued)

Table 10 (continued)

Samples	TG analysis						DSC analysis T_{max} (°C)	References
	Onset (°C)	Degradation temperature (°C)			Residual weight loss at 400 °C (%)	DTG peak temperature T_{max} (°C)		
		T_{10}	T_{15}	T_{20}				
Unriped coconut husks	230	260	270	288	35	310	–	[130]
Oil palm empty fruit bunch	190	230	260	280	32	310	–	[132]
Arecanut husks	250	309	320	330	20	358	–	[133]
Banana peel	261	–	–	–	–	–	–	[121]

5 The Recent Development of Cellulose Nanofibre as Filler in Polymer Composite

The sustainable development of the renewable polymeric materials has received great attention in recent years, mainly due to the increasing demand for the environmentally friendly materials. Unfortunately, the thermo-mechanical properties of the natural polymer or commercially available biopolymers are often inferior when compare with the traditional petroleum-based polymer [149]. For instance, polylactide (PLA) is, a commercially available biopolymer, is brittle and process at low distortion temperature [150]. Many research has been conducted to improve the mechanical and overcome the existing shortcoming of biopolymeric materials applications. Studies reported that the preparation of polymeric composite materials by combing bio-based polymers with cellulosic materials could be one of the effective solutions to match or exceed the mechanical and deterioration performance of polymeric materials with commonly used petroleum-based engineering polymers [149–151].

CNF is considered as the most promising engineering materials, which has the potentiality to be used as filler in polymeric composite because of its abidance availability, low cost, low density, renewability and biodegradability [152]. Cellulosic fibers isolated from various lignocellulosic sources such as jute, kenaf, hemp, sisal wood, flax were applied utilized as a filler in polymeric composites due to its distinct attractive properties when compared to the conventional fibers. Utilization of the CNF in polymeric composite has gained increasing attention for the last decades because of the application of CNF as filler in composite enhance mechanical, thermal and biodegradation properties [153, 154]. Although, CNF is viewed as promising engineering materials in various industrial fields, but the distinct weakness of CNF impregnated polymeric composites such as high moisture absorption, poor wettability lack of good interfacial adhesion, low melting point and the poor compatibility between CNF and polymer matrix have limited CNF applications as a filler in polymeric composite [154–156] These drawbacks of CNF filled polymeric composite has exhilarated researcher to focus on the modification of CNF surface to enhance the composites physicochemical and thermal properties. Generally, CNF is hydrophilic in nature, so it urges to enhance surface roughness of CNF by surface modification for the development of composites with enhanced properties. Studies reported that the exemplary stiffness and strength of CNF could be gained with a strong interface from CNF to polymeric matrix by the CNF medication [156].

The CNF has a tendency to agglomerate with a matrix to form forms larger particles [154]. Poor dispersion of the CNF as a filler into the polymer matrix affects the composite's mechanical properties [157]. Further, the –OH group present on the surface of CNF form hydrogen bonding with the adjacent –OH groups, results in agglomeration of the CNF. The presence of abundance –OH groups on the surface of CNF attracts to scientist and engineers interest to modify the surface of CNF by targeting the –OH via different chemical modifications, such as *Acetylation*, Silylation, carboxylation, esterification and polymer grafting. The purpose of chemical modification of CNF is to [158]:

i. obtain better CNF dispersion to polymeric matrix by introducing stable electrostatic charges on the surface of CNFs; and
ii. improve the CNF compatibility in conjunction with hydrophobic polymer matrices in composites.

The major challenge of surface modification of CNF *via* chemical functionalization to maintain the integrity of nanocellulose crystal and avoid any polymorphic conversion for preserving the original morphology CNF. Thus, the selection of the chemical functionalization methods is crucial. A brief description of some impotent chemical functionalization processes is presented below.

Acetylation is chemical functionalization process, which involves the replacement of –OH group on the surface of CNF with an acetyl group (CH_3–CO–) to yield a specific ester. Wherein, acetic anhydride is most commonly utilized as acetylene agent. Generally, acetylation is a type of chemical modification, which changes the hydrophilic surface of CNF to more hydrophobic and thereby increase the CNF compatibility to be used as filler non-polar polymer matrices [158]. Ashori et al. [151] modified CNF derived from wood and noon-wood plants using acetic anhydride in presence of pyridine to change the surface properties, minimize the hydrophilic nature of CNF and enhance compatibility with non-polar polymer matrices. The study revealed that the acetylation treatment of CNF changes the surface characteristics, improve the thermal stability and enhance the crystallinity of cellulose chains [151]. However, the performance of CNF's surface modification *via* acetylation depends on susceptibility and accessibility of –OH groups in the crystalline and amorphous regions within the cellulose polymer chain [158].

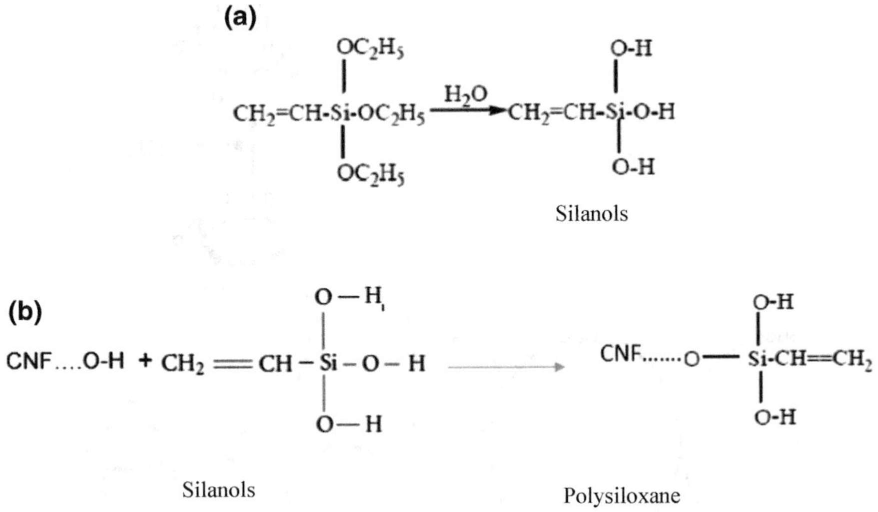

(a)

$$CH_2{=}CH\text{-}Si(OC_2H_5)_3 \xrightarrow{H_2O} CH_2{=}CH\text{-}Si(O\text{-}H)_3$$

Silanols

(b)

$$CNF....O\text{-}H + CH_2{=}CH\text{-}Si(O\text{-}H)_3 \longrightarrow CNF.......O\text{-}Si(O\text{-}H)_2\text{-}CH{=}CH_2$$

Silanols Polysiloxane

Fig. 6 **a** Silane hydrolysis, **b** Silylation of cellulosic nanofiber [160]

Silanation is an effective process for the surface modification CNF using a saline-coupling agent, generally, non-polar organic solvent. The silanation treatment improves the compatibility of CNF by eliminating surface hydroxyl groups with the non-polar organic solvent [159]. Nonetheless, the silane agent (Hydrolyzable alkoxy group) undergoes hydrolysis reaction in the presence of moisture to form silanols. Subsequently, the silanols react with the surface hydroxyl groups of CNF and forms to a Polysiloxane. The details chemical reaction of silylation process is presented in Fig. 6 [160]. Therefore, the hydrocarbon chains forms by the covalent bond of silane protect the fiber swelling by creating a crosslinked network between the matrix and the fiber [159, 161].

Carboxylation is another effective approach for the cellulosic fiber modification. In carboxylation process, TEMPO is most commonly utilized as an oxidizing agent replace surface hydroxyl groups by imparting carboxyl acid groups on the surface of CNF [162, 163]. This oxidation reaction is often performed in alkaline media in the presence of NaBr and NaOCl [163]. Figure 7 represents the TEMPO-mediated oxidation reaction for imparting carboxyl acid groups in place of the surface hydroxyl groups of CNF [164]. In carboxylation process using the TEMPO-mediated oxidizing agents involves a topologically confined reaction sequence with CNF, consequently, it forms to the 2-fold screw axis of the CNF chain. Wherein, one part of the CNF chain react with the accessible hydroxymethyl groups and the other part lies within the crystalline particle (Fig. 7). The study reported that the TEMPO-mediated oxidation maintained the CNF's initial surface morphology and morphological integrity and imparted negative charges at the CNF surface and thus induced electrostatic stabilization [164].

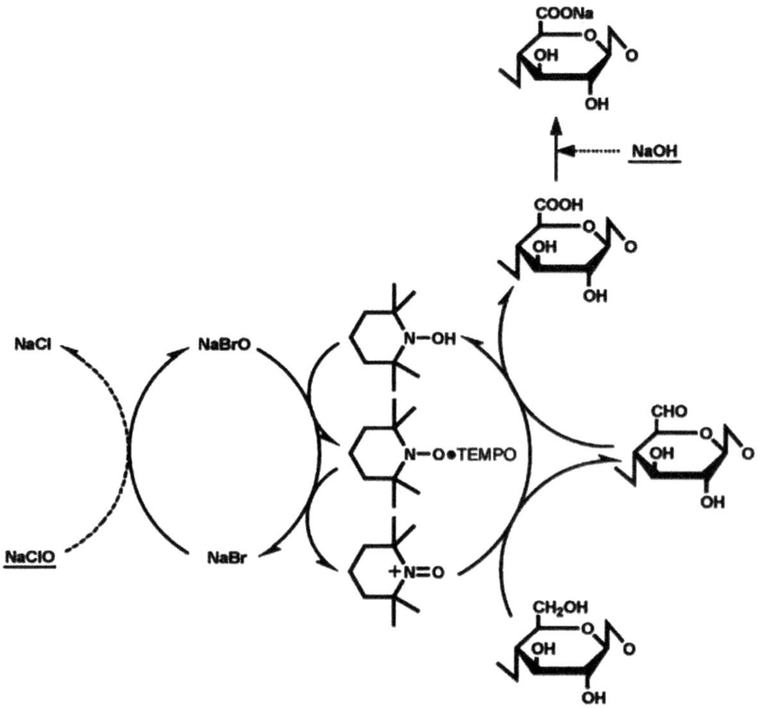

Fig. 7 TEMPO mediated oxidation for surface modification of CNF. Adapted from Missoum et al. [164]

Polymer grafting into the CNF surface is one of the research areas of development in order to fully realize the advantage of a long aliphatic chain of CNF modification for being utilized as a filler in the polymeric composite. In polymer grafting, the CNF's surface modification can be attained by covalently attaching small polymeric molecules. The main purpose of the polymer grafting is to enhance the apolar characteristics of the CNF to have better compatibility with hydrophobic polymer matrices [165]. However, the polymer chain can be covalently bounded on the surface of CNF either "grafting ono" or "grafting from" approaches [166]. In the "grafting on" approach, CNF is mixing with a coupling agent or an existing polymer to assign the polymer on the surface of the CNF so that the polymer and the CNF grafting one to another one. The main advantage of the "grafting on" approach is that the molecular weight of the polymer can be determined before grafting and therefore, the grafting materials properties can be controlled. However, it is difficult to obtain high grafting from this approach because of the steric interference and high viscosity of the reaction medium due to the presence of macromolecular chains [167]. On the other hand, the "grafting from" approach, consists the cellulosic nanoparticles activation with an initiator agent and a monomer to induce polymerization of the monomer from the CNF surface.

This "grafting from" approach has been reported to be a very effective method to generate high grafting densities on the CNF surface because of steric hindrance limitation and lower viscosity. However, this process limits to a low degree of polymerization because it is unable to determine the molecular weight of the grafted polymer [164].

6 Concluding Remarks

In this chapter, a comprehensive state-of-art of several aspects of cellulose nanofibers and their importance was described. Different approaches including preparation and modification techniques were discussed in the preparation of cellulose nanofibers. Several properties of this renewable material such as chemical composition, molecular weight, crystallinity, morphology and thermal properties were also discussed. Owing to its interesting characteristics, various uses continue to be developed by scientists all around the world. With regards to cellulose nanofiber composites applications, we only report and discuss several selected examples since the number of polymers used and their applications in various fields increase continuously. However, more research should be performed, which focuses on efficient cellulose nanofiber isolation procedures, treatments, and drying. Furthermore, to date, the engineering properties and cellulose nanofiber-based composites performances are still being developed. In order to satisfy the criteria of employing cellulose nanofiber for widespread use with higher efficiency, more effort and developments are required to expand the use of cellulose nanofiber for science and technology.

Acknowledgements Authors wish to thank their parental institutes (Universiti Sains Malaysia through USM Research University Incentive, RUI Grant 1001/PKIMIA/8011077 and Short Term Grant 304/PKIMIA/6315100) for providing the necessary facility to accomplish this work.

References

1. Sadeghi A et al (2018) Predictive construction of phase diagram of ternary solutions containing polymer/solvent/nonsolvent using modified Flory-Huggins model. J Mol Liq 263:282–287
2. Rezakazemi M, Sadrzadeh M, Matsuura T (2018) Thermally stable polymers for advanced high-performance gas separation membranes. Prog Energy Combust Sci 66:1–41
3. Kumode MMN et al (2017) Microfibrillated nanocellulose from balsa tree as potential reinforcement in the preparation of 'green'composites with castor seed cake. J Clean Prod 149:1157–1163
4. Kargarzadeh H et al (2017) Handbook of nanocellulose and cellulose nanocomposites. Wiley Online Library
5. Jawaid M, Boufi S, Abdul KH et al (2017) Cellulose-reinforced nanofiber composites. Elsevier

6. Trache D (2017) Microcrystalline cellulose and related polymer somposites: synthesis, characterization and properties. In: Handbook of composites from renewable materials, Thakur VK, Kumari Thakur M, Kessler MR et al (eds). Scrivener Publishing LLC, pp 61–92
7. Kargarzadeh H et al (2017) Recent developments on nanocellulose reinforced polymer nanocomposite: A review polymer
8. Mariano M, El Kissi N, Dufresne A (2014) Cellulose nanocrystals and related nanocomposites: review of some properties and challenges. J Polym Sci, Part B: Polym Phys 52(12): 791–806
9. Trache D et al (2016) Microcrystalline cellulose: isolation, characterization and bio-composites application—A review. Int J Biol Macromol 93(Pt A):789–804
10. Satyanarayana KG, Arizaga GG, Wypych F (2009) Biodegradable composites based on lignocellulosic fibers—An overview. Prog Polym Sci 34(9):982–1021
11. Moon RJ et al (2011) Cellulose nanomaterials review: Structure, properties and nanocomposites. Chem Soc Rev 40(7):3941–3994
12. Haafiz MM et al (2015) Bionanocomposite based on cellulose nanowhisker from oil palm biomass-filled poly (lactic acid). Polym Test 48:133–139
13. Trache D et al (2017) Recent progress in cellulose nanocrystals: Sources and production. Nanoscale 9(5):1763–1786
14. Nechyporchuk O, Belgacem MN, Bras J (2016) Production of cellulose nanofibrils: a review of recent advances. Ind Crops Prod 93:2–25
15. Oun AA, Rhim J-W (2016) Isolation of cellulose nanocrystals from grain straws and their use for the preparation of carboxymethyl cellulose-based nanocomposite films. Carbohyd Polym 150:187–200
16. Chirayil CJ, Mathew L, Thomas S (2014) Review of recent research in nano cellulose preparation from different lignocellulosic fibers. Rev Adv Mater Sci 37:20–28
17. Thakur VK (2015) Nanocellulose polymer nanocomposites: fundamentals and applications. Wiley
18. Qin X et al (2015) Tuning glass transition in polymer nanocomposites with functionalized cellulose nanocrystals through nanoconfinement. Nano Lett 15(10):6738–6744
19. Boujemaoui A et al (2015) Preparation and characterization of functionalized cellulose nanocrystals. Carbohyd Polym 115:457–464
20. Kim J-H et al (2015) Review of nanocellulose for sustainable future materials. Int J Prec Eng Manufact-Green Technol 2(2):197–213
21. Ng H-M et al (2015) Extraction of cellulose nanocrystals from plant sources for application as reinforcing agent in polymers. Compos B Eng 75:176–200
22. Xu X et al (2014) Comparison between cellulose nanocrystal and cellulose nanofibril reinforced poly (ethylene oxide) nanofibers and their novel shish-kebab-like crystalline structures. Macromolecules 47(10):3409–3416
23. Singla R et al (2016) Nanocellulose and nanocomposites. In: Nanoscale materials in targeted drug delivery, theragnosis and tissue regeneration, Springer, pp 103–125
24. Moon RJ, Schueneman GT, Simonsen J (2016) Overview of cellulose nanomaterials, their capabilities and applications. JOM 68(9):2383–2394
25. Vazquez A et al (2015) Extraction and production of cellulose nanofibers. In: Handbook of polymer nanocomposites. Processing, performance and application. Springer, pp 81–118
26. Thakur VK (2015) Lignocellulosic polymer composites: processing, characterization, and properties. Wiley
27. Pandey J et al (2015) Handbook of polymer nanocomposites. Processing, performance and application. Springer
28. Lin N, Dufresne A (2014) Nanocellulose in biomedicine: current status and future prospect. Eur Polymer J 59:302–325
29. Habibi Y, Lucia LA, Rojas OJ (2010) Cellulose nanocrystals: chemistry, self-assembly, and applications. Chem Rev 110(6):3479–3500
30. Siqueira G, Bras J, Dufresne A (2010) Cellulosic bionanocomposites: a review of preparation, properties and applications. Polym 2(4):728–765

31. Borges J et al (2015) Cellulose-based liquid crystalline composite systems. In: Thakur VK (ed) Nanocellulose polymer nanocomposites: fundamentals and applications. Wiley-Scrivener, pp 215–235
32. Wertz J-L, Mercier JP, Bédué O (2010) Cellulose science and technology. CRC Press, Switzerland
33. Postek MT et al (2013) Production and applications of cellulose. Tappi Press, Peachtree Corners
34. Klemm D et al (2011) Nanocelluloses: a new family of nature-based materials. Angew Chem Int Ed 50(24):5438–5466
35. Klemm D et al (2005) Cellulose: fascinating biopolymer and sustainable raw material. Angew Chem Int Ed 44(22):3358–3393
36. Eyley S, Thielemans W (2014) Surface modification of cellulose nanocrystals. Nanoscale 6(14):7764–7779
37. Wei H et al (2014) Environmental science and engineering applications of nanocellulose-based nanocomposites. Environ Sci: Nano 1(4):302–316
38. Heinze T (2016) Cellulose: structure and properties. In: Cellulose chemistry and properties: Fibers, nanocelluloses and advanced materials. Springer, pp 1–52
39. Trache D et al (2014) Physico-chemical properties and thermal stability of microcrystalline cellulose isolated from Alfa fibres. Carbohyd Polym 104:223–230
40. Gupta V et al (2016) Cellulose: a review as natural, modified and activated carbon adsorbent. Biores Technol 216:1066–1076
41. Oksman K et al (2014) Handbook of green materials: Processing technologies, properties and applications (in 4 volumes), vol 5. World Scientific
42. ISO/TS80004–1 (2010) International organization for standardization. ISO technical specification ISO/TS80004-1, Nanotechnologies—Vocabulary—Part 1: Core terms
43. ISO/TS27687 (2008) International organization for standardization. ISO technical specification ISO/TS 27687, Nanotechnologies—Terminology and definitions for nano-objects-Nanoparticle, nanofiber and nanoplate
44. Brinchi L et al (2013) Production of nanocrystalline cellulose from lignocellulosic biomass: technology and applications. Carbohyd Polym 94(1):154–169
45. Charreau H, Foresti ML, Vázquez A et al (2013) Nanocellulose patents trends: a comprehensive review on patents on cellulose nanocrystals, microfibrillated and bacterial cellulose. Recent patents on nanotechnology, 7(1), pp 56–80
46. TAPPI (2017) Standard terms and their definition for cellulose nanomaterial. WI 3021, http://www.tappi.org/content/hide/draft3.pdf. Accessed 01 Dec 2017
47. Gama M, Gatenholm P, Klemm D et al (2012) Bacterial nanocellulose: a sophisticated multifunctional material. CRC Press
48. Thakur VK, Voicu SI (2016) Recent advances in cellulose and chitosan based membranes for water purification: a concise review. Carbohyd Polym 146:148–165
49. Dufresne A (2013) Nanocellulose: from nature to high performance tailored materials. Walter de Gruyter
50. Agbor VB et al (2011) Biomass pretreatment: fundamentals toward application. Biotechnol Adv 29(6):675–685
51. Trache D et al (2016) Physicochemical properties of microcrystalline nitrocellulose from Alfa grass fibres and its thermal stability. J Therm Anal Calorim 124(3):1485–1496
52. Ummartyotin S, Manuspiya H (2015) A critical review on cellulose: from fundamental to an approach on sensor technology. Renew Sustain Energy Rev 41:402–412
53. Jonoobi M et al (2015) Different preparation methods and properties of nanostructured cellulose from various natural resources and residues: a review. Cellul 22(2):935–969
54. Dufresne A, Belgacem MN (2013) Cellulose-reinforced composites: from micro-to nanoscale. Polímeros 23(3):277–286
55. Dufresne A (2013) Nanocellulose: a new ageless bionanomaterial. Mater Today 16(6): 220–227

56. Abdul Khalil H et al (2014) Production and modification of nanofibrillated cellulose using various mechanical processes: a review. Carbohyd Polym 99:649–665
57. Lavoine N et al (2012) Microfibrillated cellulose—Its barrier properties and applications in cellulosic materials: a review. Carbohyd Polym 90(2):735–764
58. Gama M, Dourado F, Bielecki S et al (2016) Bacterial nanocellulose: from biotechnology to bio-economy. Elsevier
59. Vasconcelos NF et al (2017) Bacterial cellulose nanocrystals produced under different hydrolysis conditions: properties and morphological features. Carbohyd Polym 155:425–431
60. Campano C et al (2016) Enhancement of the fermentation process and properties of bacterial cellulose: a review. Cellul 23(1):57–91
61. Anwar B, Bundjali B, Arcana IM (2015) Isolation of cellulose nanocrystals from bacterial cellulose produced from pineapple peel waste juice as culture medium. Procedia Chem 16:279–284
62. George J, Sabapathi S (2015) Cellulose nanocrystals: synthesis, functional properties, and applications. Nanotechnol Sci Appl 8:45
63. Kontturi E et al (2016) Degradation and crystallization of cellulose in hydrogen chloride vapor for high-yield isolation of cellulose nanocrystals. Angew Chem Int Ed 55(46):14455–14458
64. Du H et al (2016) Preparation and characterization of thermally stable cellulose nanocrystals via a sustainable approach of FeCl3-catalyzed formic acid hydrolysis. Cellulose 1–19
65. Chen L et al (2016) Highly thermal-stable and functional cellulose nanocrystals and nanofibrils produced using fully recyclable organic acids. Green Chem
66. Liu Y et al (2014) A novel approach for the preparation of nanocrystalline cellulose by using phosphotungstic acid. Carbohyd Polym 110:415–422
67. Tang L-R et al (2011) Manufacture of cellulose nanocrystals by cation exchange resin-catalyzed hydrolysis of cellulose. Biores Technol 102(23):10973–10977
68. Anderson SR et al (2014) Enzymatic preparation of nanocrystalline and microcrystalline cellulose. TAPPI J, vol 13, pp 35–41
69. Xu Y et al (2013) Feasibility of nanocrystalline cellulose production by endoglucanase treatment of natural bast fibers. Ind Crops Prod 51:381–384
70. Chen X et al (2012) Controlled enzymolysis preparation of nanocrystalline cellulose from pretreated cotton fibers. BioRes 7(3):4237–4248
71. Amin KNM et al (2015) Production of cellulose nanocrystals via a scalable mechanical method. RSC Adv 5(70):57133–57140
72. Lazko J et al (2016) Acid-free extraction of cellulose type I nanocrystals using Brønsted acid-type ionic liquids. Nanocomposites 2(2):65–75
73. Tan XY, Hamid SBA, Lai CW (2015) Preparation of high crystallinity cellulose nanocrystals (CNCs) by ionic liquid solvolysis. Biomass Bioenerg 81:584–591
74. Mao J et al (2015) Cellulose nanocrystals production in near theoretical yields by 1-butyl-3-methylimidazolium hydrogen sulfate ([Bmim] HSO_4)–mediated hydrolysis. Carbohyd Polym 117:443–451
75. Lazko J et al (2014) Well defined thermostable cellulose nanocrystals via two-step ionic liquid swelling-hydrolysis extraction. Cellulose 21(6):4195–4207
76. Novo LP et al (2016) A study of the production of cellulose nanocrystals through subcritical water hydrolysis. Indus Crops Prod
77. Novo LP et al (2015) Subcritical water: a method for green production of cellulose nanocrystals. ACS Sustain Chem Eng 3(11):2839–2846
78. Miller J (2015) Cellulose nanomaterials production-state of the industry.[cited 2017 05-12-2017]; Available from http://www.tappinano.org/media/1114/cellulose-nanomaterials-production-state-of-the-industry-dec-2015.pdf
79. Sun B et al (2015) Sodium periodate oxidation of cellulose nanocrystal and its application as a paper wet strength additive. Cellulose 22(2):1135–1146

80. Visanko M et al (2014) Amphiphilic cellulose nanocrystals from acid-free oxidative treatment: Physicochemical characteristics and use as an oil—water stabilizer. Biomacromol 15(7):2769–2775
81. Cao X et al (2012) Cellulose nanowhiskers extracted from TEMPO-oxidized jute fibers. Carbohyd Polym 90(2):1075–1080
82. Chowdhury ZZ, Hamid SBA (2016) Preparation and characterization of nanocrystalline cellulose using ultrasonication combined with a microwave-assisted pretreatment process. BioRes 11(2):3397–3415
83. Tang Y et al (2014) Preparation and characterization of nanocrystalline cellulose via low-intensity ultrasonic-assisted sulfuric acid hydrolysis. Cellulose 21(1):335–346
84. Lu Z et al (2013) Preparation, characterization and optimization of nanocellulose whiskers by simultaneously ultrasonic wave and microwave assisted. Biores Technol 146:82–88
85. Lee H et al (2018) Improved thermal stability of cellulose nanofibrils using low-concentration alkaline pretreatment. Carbohyd Polym 181:506–513
86. Turbak AF, Snyder FW, Sandberg KR et al (1983) Microfibrillated cellulose, a new cellulose product: Properties, uses, and commercial potential. J Appl Polym Sci: Appl Polym Symp (U S). ITT Rayonier Inc., Shelton, WA
87. Herrick FW et al (1983) Microfibrillated cellulose: Morphology and accessibility. J Appl Polym Sci: Appl Polym Symp (U S). ITT Rayonier Inc., Shelton, WA
88. Osong SH, Norgren S, Engstrand P (2016) Processing of wood-based microfibrillated cellulose and nanofibrillated cellulose, and applications relating to papermaking: a review. Cellul 23(1):93–123
89. Lee H, Sundaram J, Mani S et al (2017) Production of cellulose nanofibrils and their application to food: a review, in nanotechnology. Springer, pp 1–33
90. Rol F et al (2017) Pilot-Scale twin screw extrusion and chemical pretreatment as an energy-efficient method for the production of nanofibrillated cellulose at high solid content. ACS Sustain Chem Eng 5(8):6524–6531
91. Yan H et al (2017) Synthesis of bacterial cellulose and bacterial cellulose nanocrystals for their applications in the stabilization of olive oil pickering emulsion. Food Hydrocolloids 72:127–135
92. Sacui IA et al (2014) Comparison of the properties of cellulose nanocrystals and cellulose nanofibrils isolated from bacteria, tunicate, and wood processed using acid, enzymatic, mechanical, and oxidative methods. ACS Appl Mater Interfaces 6(9):6127–6138
93. Keshk SM (2014) Bacterial cellulose production and its industrial applications. J Bioprocess Biotechniques
94. Ioelovich M (2013) Nanoparticles of amorphous cellulose and their properties. Am J Nanosci Nanotechnol 1(1):41–45
95. Ioelovich M (2014) Peculiarities of cellulose nanoparticles. Tappi J 13(5):45–51
96. Quan S-L, Kang S-G, Chin I-J (2010) Characterization of cellulose fibers electrospun using ionic liquid. Cellulose 17(2):223–230
97. Rebouillat S, Pla F (2013) State of the art manufacturing and engineering of nanocellulose: a review of available data and industrial applications. J Biomater Nanobiotechnol 4(02):165
98. Nascimento SA, Rezende CA (2018) Combined approaches to obtain cellulose nanocrystals, nanofibrils and fermentable sugars from elephant grass. Carbohyd Polym 180:38–45
99. Frone AN et al (2017) Isolation of cellulose nanocrystals from plum seed shells, structural and morphological characterization. Mater Lett 194:160–163
100. Wang Z et al (2017) Reuse of waste cotton cloth for the extraction of cellulose nanocrystals. Carbohyd Polym 157:945–952
101. Zhang K et al (2016) Extraction and comparison of carboxylated cellulose nanocrystals from bleached sugarcane bagasse pulp using two different oxidation methods. Carbohyd Polym 138:237–243
102. Yu H-Y et al (2016) New approach for single-step extraction of carboxylated cellulose nanocrystals for their use as adsorbents and flocculants. ACS Sustain Chem. Eng 4(5): 2632–2643

103. Rohaizu R, Wanrosli W (2017) Sono-assisted TEMPO oxidation of oil palm lignocellulosic biomass for isolation of nanocrystalline cellulose. Ultrason Sonochem 34:631–639
104. Ho TTT et al (2015) Nanofibrillation of pulp fibers by twin-screw extrusion. Cellul 22(1): 421–433
105. Nechyporchuk O, Pignon F, Belgacem MN (2015) Morphological properties of nanofibrillated cellulose produced using wet grinding as an ultimate fibrillation process. J Mater Sci 50(2):531–541
106. Sirviö JA, Visanko M, Liimatainen H (2015) Deep eutectic solvent system based on choline chloride-urea as a pre-treatment for nanofibrillation of wood cellulose. Green Chem 17(6): 3401–3406
107. Oun AA, Rhim J-W (2016) Characterization of nanocelluloses isolated from Ushar (Calotropis procera) seed fiber: effect of isolation method. Mater Lett 168:146–150
108. Lee H-R et al (2018) A new method to produce cellulose nanofibrils from microalgae and the measurement of their mechanical strength. Carbohyd Polym 180:276–285
109. Valdebenito F et al (2017) On the nanofibrillation of corn husks and oat hulls fibres. Ind Crops Prod 95:528–534
110. Lee H, Mani S (2017) Mechanical pretreatment of cellulose pulp to produce cellulose nanofibrils using a dry grinding method. Ind Crops Prod 104:179–187
111. Park C-W et al (2017) Preparation and characterization of cellulose nanofibrils with varying chemical compositions. BioRes 12(3):5031–5044
112. Eichhorn SJ (2011) Cellulose nanowhiskers: promising materials for advanced applications. Soft Matter 7(2):303–315
113. Meyabadi TF et al (2014) Spherical cellulose nanoparticles preparation from waste cotton using a green method. Powder Technology, vol 261, pp 232–240
114. Chen Y et al (2009) Bionanocomposites based on pea starch and cellulose nanowhiskers hydrolyzed from pea hull fibre: effect of hydrolysis time. Carbohyd Polym 76(4):607–615
115. Abushammala H, Krossing I, Laborie MP (2015) Ionic liquid-mediated technology to produce cellulose nanocrystals directly from wood. Carbohyd Polym 134:609–616
116. Miao J et al (2016) One-pot preparation of hydrophobic cellulose nanocrystals in an ionic liquid. Cellulose 23(2):1209–1219
117. Li W, Yue J, Liu S (2012) Preparation of nanocrystalline cellulose via ultrasound and its reinforcement capability for poly (vinyl alcohol) composites. Ultrason Sonochem 19(3): 479–485
118. Elanthikkal S et al (2010) Cellulose microfibres produced from banana plant wastes: isolation and characterization. Carbohyd Polym 80(3):852–859
119. Kumar A et al (2014) Characterization of cellulose nanocrystals produced by acid-hydrolysis from sugarcane bagasse as agro-waste. J Mater Phy Chem 2(1):1–8
120. Haafiz MM et al (2014) Isolation and characterization of cellulose nanowhiskers from oil palm biomass microcrystalline cellulose. Carbohyd Polym 103:119–125
121. Khawas P, Deka SC (2016) Isolation and characterization of cellulose nanofibers from culinary banana peel using high-intensity ultrasonication combined with chemical treatment. Carbohyd Polym 137:608–616
122. Lu P, Hsieh YL (2012) Preparation and characterization of cellulose nanocrystals from rice straw. Carbohyd Polym 87(1):564–573
123. Lu P, Hsieh YL (2010) Preparation and properties of cellulose nanocrystals: rods, spheres, and network. Carbohyd Polym 82(2):329–336
124. Mandal A, Chakrabarty D (2011) Isolation of nanocellulose from waste sugarcane bagasse (SCB) and its characterization. Carbohyd Polym 86(3):1291–1299
125. Satyamurthy P et al (2011) Preparation and characterization of cellulose nanowhiskers from cotton fibres by controlled microbial hydrolysis. Carbohyd Polym 83(1):122–129
126. Johar N, Ahmad I, Dufresne A (2012) Extraction, preparation and characterization of cellulose fibres and nanocrystals from rice husk. Ind Crops Prod 37(1):93–99
127. Sheltami RM et al (2012) Extraction of cellulose nanocrystals from mengkuang leaves (*Pandanus tectorius*). Carbohyd Polym 88(2):772–779

128. Xiong R et al (2012) Comparing microcrystalline with spherical nanocrystalline cellulose from waste cotton fabrics. Cellulose 19(4):1189–1198
129. Chandrahasa R, Rajamane NP, Jeyalakshmi et al (2014) Development of cellulose nanofibres from coconut husks. Int J Emerg Technol Adv Eng 4(4):2250–2259
130. Nascimento DM et al (2014) A novel green approach for the preparation of cellulose nanowhiskers from white coir. Carbohyd Polym 110:456–463
131. Lamaming J et al (2015) Cellulose nanocrystals isolated from oil palm trunk. Carbohyd Polym 127:202–208
132. Indarti E, Marwan, Wanrosli WD et al (2015) Thermal stability of oil palm empty fruit bunch (OPEFB) nanocrystalline cellulose: effects of post-treatment of oven drying and solvent exchange techniques. J Phys: Conf Ser 622(1):12–25
133. Chandra J, George N, Narayanankutty SK (2016) Isolation and characterization of cellulose nanofibrils from arecanut husk fibre. Carbohyd Polym 142:158–166
134. Segal LGJMA et al (1959) An empirical method for estimating the degree of crystallinity of native cellulose using the X-ray diffractometer. Text Res J 29(10):786–794
135. Revol JF, Dietrich A, Goring DAI (1987) Effect of mercerization on the crystallite size and crystallinity index in cellulose from different sources. Can J Chem 65(8):1724–1725
136. Cherian BM et al (2011) Cellulose nanocomposites with nanofibres isolated from pineapple leaf fibers for medical applications. Carbohyd Polym 86(4):1790–1798
137. Cherian BM et al (2010) Isolation of nanocellulose from pineapple leaf fibres by steam explosion. Carbohyd Polym 81(3):720–725
138. Rosa SM et al (2012) Chlorine-free extraction of cellulose from rice husk and whisker isolation. Carbohyd Polym 87(2):1131–1138
139. Rosa MF et al (2010) Cellulose nanowhiskers from coconut husk fibers: Effect of preparation conditions on their thermal and morphological behavior. Carbohyd Polym 81(1):83–92
140. Silvério HA et al (2013) Extraction and characterization of cellulose nanocrystals from corncob for application as reinforcing agent in nanocomposites. Ind Crops Prod 44:427–436
141. Neto WPF et al (2013) Extraction and characterization of cellulose nanocrystals from agro-industrial residue—soy hulls. Ind Crops Prod 42:480–488
142. Neto WPF et al (2016) Comprehensive morphological and structural investigation of cellulose I and II nanocrystals prepared by sulphuric acid hydrolysis. RSC Adv 6(79): 76017–76027
143. Luykx DM et al (2008) A review of analytical methods for the identification and characterization of nano delivery systems in food. J Agric Food Chem 56(18):8231–8247
144. Goldstein J et al (2012) Scanning electron microscopy and X-ray microanalysis: a text for biologists, materials scientists, and geologists. Springer Science and Business Media, (2)
145. Wang ZL (2000) Transmission electron microscopy of shape-controlled nanocrystals and their assemblies. J Phys Chem B 104(6):1153–1175
146. Wang QQ et al (2012) Approaching zero cellulose loss in cellulose nanocrystal (CNC) production: recovery and characterization of cellulosic solid residues (CSR) and CNC. Cellulose 19(6):2033–2047
147. Azubuike CP, Okhamafe AO (2012) Physicochemical, spectroscopic and thermal properties of microcrystalline cellulose derived from corn cobs. Inter J Recycl Org Waste Agric 1:1–7
148. Adel AM et al (2011) Characterization of microcrystalline cellulose prepared from lignocellulosic materials. Part II: Physicochemical Properties. Carbohyd Polym 83:676
149. Lee KY et al (2014) On the use of nanocellulose as reinforcement in polymer matrix composites. Compos Sci Technol 105:15–27
150. Yang S, Bai S, Wang Q (2018) Sustainable packaging biocomposites from polylactic acid and wheat straw: enhanced physical performance by solid state shear milling process. Compos Sci Technol 158:34–42
151. Ashori A et al (2014) Solvent-free acetylation of cellulose nanofibers for improving compatibility and dispersion. Carbohyd Polym 102:369–375

152. Abdul Khalil HPS et al (2016) A review on nanocellulosic fibres as new material for sustainable packaging: process and applications. Renew Sustain Energy Rev 64:823–836
153. Frone AN et al (2013) Morphology and thermal properties of PLA–cellulose nanofibers composites. Carbohyd Polym 91(1):377–384
154. Oksman K et al (2016) Review of the recent developments in cellulose nanocomposite processing. Compos Part A: Appl Sci Manuf 83:2–18
155. Oksman K et al (2011) Cellulose nanowhiskers separated from a bio-residue from wood bioethanol production. Biomass Bioenergy 35(1):146–152
156. Lu Y et al (2017) Synthesis of new polyether titanate coupling agents with different polyethyleneglycol segment lengths and their compatibilization in calcium sulfate whisker/poly(vinyl chloride) composites. RSC Adv 7(50):31628–31640
157. Poveda RL, Gupta N (2016) Mechanical properties of CNF/polymer composites carbon nanofiber reinforced polymer composites. Cham: Springer, pp 27–42
158. Kobe R et al (2016) Stretchable composite hydrogels incorporating modified cellulose nanofiber with dispersibility and polymerizability: Mechanical property control and nanofiber orientation. Polym 97:480–486
159. Ng HM et al (2015) Extraction of cellulose nanocrystals from plant sources for application as reinforcing agent in polymers. Composites Part B: Eng 75:176–200
160. Kalia S et al (2011) Cellulose-Based Bio- and Nanocomposites: a review. Int J Polym Sci
161. Kalia S et al (2014) Nanofibrillated cellulose: Surface modification and potential applications. Colloid Polym Sci 292(1):5–31
162. Ishii D, Saito T, Isogai A (2011) viscoelastic evaluation of average length of cellulose nanofibers prepared by TEMPO-mediated oxidation. Biomacromolecules 12(3):548–550
163. Qing Y et al (2013) A comparative study of cellulose nanofibrils disintegrated via multiple processing approaches. Carbohyd Polym 97(1):226–234
164. Missoum K, Belgacem M, Bras J (2013) Nanofibrillated cellulose surface modification: a review. Mater 6(5):1745
165. Ahmadi M et al (2017) Topochemistry of cellulose nanofibers resulting from molecular and polymer grafting. Cellulose 24(5):2139–2152
166. Roy D et al (2009) Cellulose modification by polymer grafting: a review. Chem Soc Rev 38(7):2046–2064
167. Safdari F et al (2017) Enhanced properties of poly(ethylene oxide)/cellulose nanofiber biocomposites. Cellulose 24(2):755–767
168. Bondeson D, Mathew A, Oksman K (2006) Optimization of the isolation of nanocrystals from microcrystalline cellulose by acid hydrolysis. Cellulose 13(2):171–180
169. Morán JI et al (2008) Extraction of cellulose and preparation of nanocellulose from sisal fibers. Cellulose 15(1):149–159
170. Fahma F et al (2011) Effect of pre-acid-hydrolysis treatment on morphology and properties of cellulose nanowhiskers from coconut husk. Cellulose 18:443–450
171. Brito BS et al (2012) Preparation, morphology and structure of cellulose nanocrystals from bamboo fibers. Cellulose 19(5):1527–1536
172. Fan JS, Li YH (2012) Maximizing the yield of nanocrystalline cellulose from cotton pulp fiber. Carbohyd Polym 88(4):1184–1188
173. Fortunati E et al (2013) Extraction of cellulose nanocrystals from Phormium tenax fibres. J Polym Environ 21(2):319–328
174. Morais JPS et al (2013) Extraction and characterization of nanocellulose structures from raw cotton linter. Carbohyd Polym 91(1):229–235
175. Santos RMD et al (2013) Cellulose nanocrystals from pineapple leaf, a new approach for the reuse of this agro-waste. Ind Crops Prod 50:707–714
176. Espinosa CS et al (2013) Isolation of thermally stable cellulose nanocrystals by phosphoric acid hydrolysis. Biomacromolecules 14(4):1223–1230
177. Le Normand M, Moriana R, Ek M (2014) Isolation and characterization of cellulose nanocrystals from spruce bark in a biorefinery perspective. Carbohyd Polym 111:979–987

178. Mueller S, Weder C, Foster EJ (2014) Isolation of cellulose nanocrystals from pseudostems of banana plants. RSC Adv 4(2):907–915
179. Bettaieb F et al (2015) Mechanical and thermal properties of Posidonia oceanica cellulose nanocrystal reinforced polymer. Carbohyd Polym 123:99–104
180. Devi RR (2015) Fabrication of cellulose nanocrystals from agricultural compost. Compost Sci Utilization 23(2):104–116
181. Mohamed MA et al (2015) Physicochemical properties of "green" nanocrystalline cellulose isolated from recycled newspaper. RSC Adv 5(38):29842–29849
182. Dungani R et al (2016) Preparation and fundamental characterization of cellulose nanocrystal from oil palm fronds biomass. J Poly Environ 1:1–9
183. Salajková M, Berglund LA, Zhou Q (2012) Hydrophobic cellulose nanocrystals modified with quaternary ammonium salts. J Mater Chem 22(37):19798–19805
184. Pan M, Zhou X, Chen M (2013) Cellulose nanowhiskers isolation and properties from acid hydrolysis combined with high pressure homogenization. BioRes 8(1):933–943
185. Savadekar NR et al (2015) Preparation of cotton linter nanowhiskers by high-pressure homogenization process and its application in thermoplastic starch. Appl Nanosci 5(3): 281–290
186. Beltramino F et al (2015) Increasing yield of nanocrystalline cellulose preparation process by a cellulase pretreatment. Biores technol 192:574–581
187. Beltramino F et al (2016) Optimization of sulfuric acid hydrolysis conditions for preparation of nanocrystalline cellulose from enzymatically pretreated fibers. Cellulose 23(3):1777–1789
188. Camargo LA et al (2016) Feasibility of manufacturing cellulose nanocrystals from the solid residues of second-generation ethanol production from sugarcane bagasse. BioEnergy Res 9(3):894–906
189. Zhao Y et al (2015) Tunicate cellulose nanocrystals: preparation, neat films and nanocomposite films with glucomannans. Carbohydr Polym 117:286–296
190. Csiszar E et al (2016) The effect of low frequency ultrasound on the production and properties of nanocrystalline cellulose suspensions and films. Ultrason Sonochem 31:473–480
191. Cudjoe E et al (2017) *Miscanthus Giganteus*: a commercially viable sustainable source of cellulose nanocrystals. Carbohyd Polym 155:230–241
192. Hamid SBA et al (2016) Synergic effect of tungstophosphoric acid and sonication for rapid synthesis of crystalline nanocellulose. Carbohyd Polym 138:349–355
193. Li Y et al (2016) Facile extraction of cellulose nanocrystals from wood using ethanol and peroxide solvothermal pretreatment followed by ultrasonic nanofibrillation. Green Chem 18(4):1010–1018
194. Lu Q et al (2014) Preparation and characterization of cellulose nanocrystals via ultrasonication-assisted $FeCl_3$-catalyzed hydrolysis. Cellulose 21(5):3497–3506
195. Lim YH et al (2016) NanoCrystalline cellulose isolated from oil palm empty fruit bunch and its potential in cadmium metal removal. In: MATEC web of conferences, vol 59. EDP Sciences
196. Sun B et al (2016) Single-step extraction of functionalized cellulose nanocrystal and polyvinyl chloride from industrial wallpaper wastes. Ind Crops Prod 89:66–77
197. Tang Y et al (2015) Extraction of cellulose nano-crystals from old corrugated container fiber using phosphoric acid and enzymatic hydrolysis followed by sonication. Carbohyd Polym 125:360–366
198. Yu H et al (2013) Facile extraction of thermally stable cellulose nanocrystals with a high yield of 93% through hydrochloric acid hydrolysis under hydrothermal conditions. J Mater Chem A 1(12):3938–3944

Recyclable and Eco-friendly Single Polymer Composite

Mohd Azmuddin Abdullah, Muhammad Afzaal, Safdar Ali Mirza, Sakinatu Almustapha and Hanaa Ali Hussein

1 Introduction

Composites based on metal, ceramic and polymer matrices have ubiquitous applications in aerospace, automotive, electrical appliances, microelectronics, infrastructure and construction, medical and chemical industries [1–3]. Interest has shifted towards synthetic and biopolymers and polymer composites as replacements to conventional composite materials as they are far more economical and easier to process [3, 4]. To meet the requirements for, The polymeric material properties can be modified and tuned for specific applications using fillers and fibers [1, 2] for more well-defined physicochemical properties, enhanced mechanical strength and stiffness, with low specific gravity but high thermal and chemical resistance [5–7]. The two main constituents of a polymeric composite are the polymer matrix which is the continuous phase, and the reinforcing material which is the discontinuous or dispersed phase. The properties are governed by the properties of the matrix and the reinforcing material such as the aspect ratio, chemical nature, purity, distribution, orientation and geometry; and the amount and the interfacial adhesion between the matrix and the reinforcement material [4].

M. A. Abdullah (✉) · H. A. Hussein
Institute of Marine Biotechnology, Universiti Malaysia Terengganu, 21030, Kuala Nerus, Terengganu, Malaysia
e-mail: azmuddin@umt.edu.my; joule1602@gmail.com

M. Afzaal
Department of Sustainable Development Study Center (Environmental Sciences), GC University, Lahore, Punjab, Pakistan

S. A. Mirza
Department of Botany, GC University, Lahore, Punjab, Pakistan

S. Almustapha
Department of Basic and Applied Sciences, Hassan Usman Katsina Polytechnic, P.M.B 2052 Katsina State, Nigeria

Inamuddin et al. (eds.), *Sustainable Polymer Composites and Nanocomposites*,
https://doi.org/10.1007/978-3-030-05399-4_24

The global issues of sustainability and climate change, industrial ecology, eco-efficiency, and green chemistry are the way forward to develop the new route and next generation of materials, products, and processes [8–10]. The recovery and recycling of petroleum-based plastics remain insufficient when the global plastics production has risen from 245 million tonnes in 2008 [11], to 299 million tonnes in 2013 [12]. Yet the knowledge of human and environmental hazards and risks from the chemicals associated with the diversity of plastic products is limited when most chemicals in plastic polymers production have their origin from the crude oil. Several are known to be hazardous which may be released during the production, use, and disposal of the plastic product [12], with millions of tonnes, end up in the landfills and oceans. Approximately 10–20 million tonnes of plastic end up in the oceans each year, and an estimated 5.25 trillion plastic particles weighing a total of 268,940 tons are currently floating in the world's oceans with the debris damaging the marine ecosystems [12]. Natural fibers have therefore attracted greater attention as these could be obtained from agro-wastes which are abundant, low-cost, low density, biodegradable, non-abrasive but tunable for high specific properties [3, 10, 11]. The combination of biofibers such as oil palm, kenaf, hemp, or flax, to produce composite materials with non-renewable and renewable polymer matrices are competitive with synthetic composites but may need to address the biofiber—matrix interface and novel processing routes [10].

The natural fiber—reinforced polypropylene (PP) composites or natural fiber—polyester composites have attracted attraction as this could remove some portions of the plastics from the environment [13], but these are not sufficiently eco-friendly because of the non-biodegradable nature of the polymer matrix and also the production of the synthetic polymers (like the PP or polyethylene, PE) are fossil fuel-based, and non-renewable [10, 14], which has often been singled out as the major cause of climate change and furthermore subjected to fluctuating prices and declining reserves. The natural fibers, although can be incinerated, may however not be compatible with the hydrophobic polymer matrix as they may form aggregates during processing, have poor resistance to moisture and are generally not mechanically recyclable and reused after the end-of-life [3, 14, 15]. Not all polymers, especially the thermosets, are equally easy to recycle and those composites having glass or carbon fibers can only be recycled into new fiber reinforced grades [2]. Thermoplastic composites can be re-melted and cooled to solidify for an infinite number of cycles but each melting, cooling and re-melting process necessitates high energy requirements, causing the material to eventually degrade. Composites recycling is achievable by a single polymer composite (SPC) with specific economic and ecological advantages [1, 16, 17]. SPC consists of mono-component systems or single polymer, or a minimum of different, compatible polymers [2]. This is possible by making use of the noticeable difference in melting temperature between the matrix and the reinforcement material [15]. The high-density polyethylene (HDPE—crystallized conventionally) matrix and the HDPE reinforcement (containing aligned and extended molecular chains) is an example of the SPC based on the mono-component system [15]. Also, using the natural fibers with polymers based on renewable resources, or the green bio-composites solve many environmental issues to be solved, by embedding the biofibers with biopolymers such as cellulosic plastics, polylactides (PLA), starch plastics, polyhydroxyalkanoates (bacterial polyesters), and soy-based plastics [10]. This review article highlights the fundamental aspects of the polymer and the fibers, the production methods, different components of the SPC and the fabrication methods, and the issues and challenges in the manufacturing and the applications. The term "eco-friendly SPCs" is extended to include those chemical-based, recyclable SPCs and the natural-fibre based SPCs.

2 Recyclable Single Polymer Composite

2.1 Basic Polymer Chemistry

Polymers are macromolecules joined with each other by covalent bonding along the backbone chain and lead to properties, such as the ability to form fibers and elastomers, which is not achievable with the small molecules [18]. Polymeric materials and new polymers are being discovered with unique sets of properties including from a single atom, such as sulfur, or from different building blocks arranged in a specific sequence as in proteins or nucleic acids [19]. The long chains are flexible, joined together by primary covalent bonding forces, and each chain can have side groups, branches, and copolymeric chains or blocks, through crosslinks between chains, or by van der Waals and hydrogen bonds. The linear polymers are partially crystalline or semicrystalline, composed of disordered non-crystalline (amorphous) regions and ordered crystalline regions, with the combination of folded and extended chains. Linear polymers are much easier to crystallize than the branched or cross-linked polymers [20]. Copolymerization, in which two or more homopolymers (one type of repeating unit throughout its structure) are chemically combined, disrupts the regularity of polymer chains, thus forming the non-crystalline structure. The arrangements of the side groups (X) can be in the form of atactic, isotactic, and syndiotactic (Fig. 1) [21]. The isotactic polymer is in which the pendant groups are all on the same side of the polymer backbone, as in isotactic PP. In the syndiotactic arrangement, the pendant groups are regularly spaced like the methyl groups in PP, or the HDPE (a linear polymer produced by the polymerization of ethylene in the presence of Ziegler-Natta or Phillips catalysts). The atactic arrangement is in which the side groups are randomly distributed and not packed well on each side of the chain and is normally amorphous as in the atactic PP [21]. The polymer crystallizes easily if the side groups are small like the PE (X = H) and the isotactic and syndiotactic arrangements are crystalline at ordinary temperatures, or even when the side groups are large because the chains are linear and are either in alternating positions or on one side of the main chain [20]. In the case of polyvinyl chloride (PVC) (X = Cl) and polystyrene (PS) ($X = C_6H_5$), the side groups are large and are randomly distributed along the chains (atactic) and therefore form a non-crystalline structure [20].

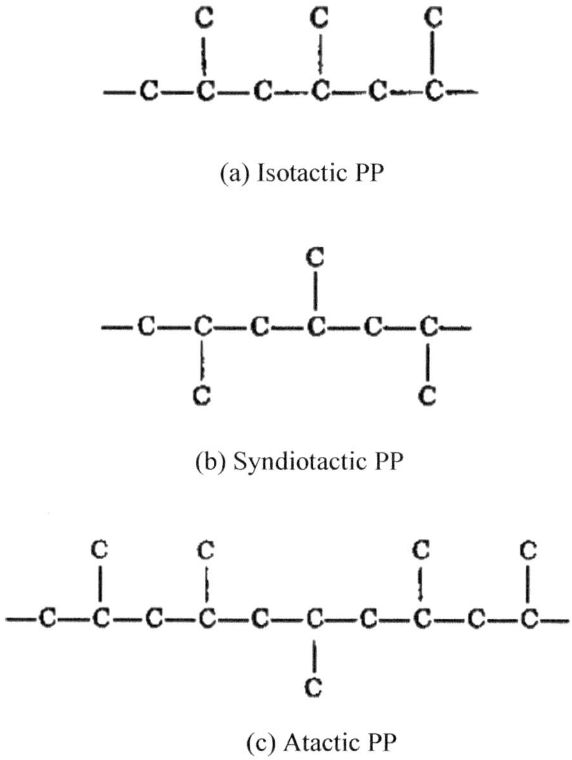

Fig. 1 Skeletal formulas of **a** isotactic, **b** syndiotactic, and **c** atactic (random) PP. (Adapted from [21])

(a) Isotactic PP

(b) Syndiotactic PP

(c) Atactic PP

The differences between all polymers including plastics, fibers, and elastomers or rubbers, are determined primarily by the intermolecular forces between the molecules and the intramolecular forces, and by the functional groups present [21]. The polymer properties are therefore controlled among others by the covalent and non-covalent intra-chain forces. Atoms in individual polymer molecules are joined to each other by strong covalent bonds with the bond energies in the order of 80–90 kcal/mol (320–370 kJ/mol) for the carbon-carbon bonds [21]. Hydrogen bonding between a positively charged hydrogen with an electronegative element nitrogen [OH···N], oxygen [OH···O, NH···O], or fluorine [OH···F] are the strongest non-covalent forces, followed by the dipole-dipole interaction, the magnitude of which depends on the electro-negativity difference of the two atoms involved in the polar bond, and the Van der Waals forces, the small attractive forces and the weakest interchain interactions between all atoms, whether with or without dipole moment, but remain important as they hold non-polar liquids and solids together [20, 21]. The thermoplastic material such as PE, PP, nylon, acetal resin, polycarbonate and polyethylene terephthalate (PET), either linear or branched structure, can be heated up, dissolved in a suitable solvent, and re-processed or remoulded when it softens or melts [22, 23]. Thermoplastics have relatively low tensile moduli, low densities and transparent which are ideal for consumer and medical products [22].

The cross-linked polymers such as phenolic resins, polyesters, and epoxy resins are thermosets, widely used in composite materials, and reinforced with stiff fibres such as fibreglass and aramids. Crosslinking stabilizes the thermoset matrix resulting in the physical properties similar to steel, but with lower densities and suffer less from fatigue and therefore ideal for lightweight structures and for safety-critical parts under regular stress [22]. Elastomers have low moduli and exhibit reversible extension when strained, valuable properties for vibration absorption and damping. Elastomers can be thermoplastic elastomers, vulcanized thermoplastic or crosslinked (as in the case of rubber tyres), and rubbers [melt-processable, natural and synthetic such as nitrile rubber, polychloroprene, polybutadiene, styrene-butadiene and fluorinated rubbers (Viton)] [22, 23].

2.2 Structural Modification

The basic building blocks of polymers are repeating monomer units, normally in covalently linked chains. By having high numbers of functional/hydrophilic groups, high surface/volume ratio, and mechanical stresses, but low molecular weight/ crystallization and low or no cross-links, hydrolysis of polymers can be increased. Polymers can be formed by the addition polymerization or free radical addition, the condensation polymerization or step reaction polymerization, and the ring-opening/ cyclic polymerization. Addition polymerization is rapid, with the initiated species continue to propagate until termination. It yields linear polymers and achieved by rearranging the bonds within each monomer which should have at least one double bond, for each monomer to share at least two covalent electrons with other monomers. A free radical such as benzoyl peroxide ($H_5C_6COO–OOCC_6H_5$), or the cations, anions, and coordination (stereospecific) catalysts are initiators that break the double bond which can be activated by heat, ultraviolet light, and other chemicals. During propagation, the free radicals or initiators react with the monomers, and then with another monomer and continue on until termination stage where two free radicals are combined through transfer, or by disproportionate processes. Some monomers can use two or more of the initiation processes while others may use only one process [20, 21]. Addition polymers including vinyl, aldehyde, and acetylene polymers are prepared via addition polymerization reaction such as polyethers, poly(vinyl ether)s, polystyrene derivatives, polyolefins, poly-methacrylate, polymethacryloylamine, and polyacethylene and [24–29]. To stiffen and retain their shape, and also make them insoluble in solvents, crosslinking agents which contain two or more double bonds per molecule can be used to produce cross-linked linear chain polymers [20].

The condensation polymerization involves two different monomer units, each containing two functional groups joining together to form a larger molecule by the elimination of a small molecule such as water [20]. While an addition polymer has the same atoms as the monomer in its repeat unit and the atoms in the polymer backbone are usually carbon atoms, the condensation polymers contain fewer atoms within the polymer repeat unit than the reactants because of the formation of byproducts during the polymerization process, and the polymer backbone usually contains atoms of more than one element [21]. Condensation polymerization has received considerable attention for the preparation of polymeric materials as these are used in a vast array of applications including the synthesis of chiral polymers [24, 30]. Nature uses chirality as one of the key structural factors to perform functions such as molecular recognition and catalytic activities. Optically active polymers often play important roles as key fundamental materials for well-defined polymers with specific secondary and/or tertiary structures aimed at the application in optoelectronic devices and chiral recognition materials including asymmetric catalyst, chiral stationary phase in HPLC, enantioselective permeation membrane, chiral adsorbent for separation of racemates and in liquid crystals [24]. The optically active polymers based on condensation polymers are prepared by a polycondensation reaction of chiral monomers, which is more versatile, inexpensive though generally poor in control than addition polymerization. Cross-linked gels possessing chiral cavities are prepared and their chiral recognition ability established. The synthesis of the gels is based on the molecular imprinting technique. Two distinctive methods have been developed, that is polymerizing a monomer having a removable chiral template moiety with a cross-linking agent and removing the template groups from the products; and polymerizing the monomer with the crosslinking agent in the presence of a non-polymerizable template molecule before removing the template [24, 31]. One major drawback of condensation polymerization is the tendency for the reaction to cease before the chains grow to a sufficient length due to the decreased mobility of the chains and reactant chemical species as polymerization progresses, resulting in short chains. It is possible in the case of nylon, that the chains are polymerized to a sufficiently large extent before this occurs resulting in the nylon physical properties preserved [20].

Polylactic glycolic acid polymers can be obtained either by condensation from lactic acid, glycolic acid, and light condensates or by ring-opening polymerization (ROP) of the related cyclic dimers, namely, lactide and glycolide. The ROP route allows for a much higher control of the polymerization and remains by far the most widely used method for the synthesis of well-defined materials [32]. This may involve different strategies following the nature of their ancillary ligands such as the Coordination—Insertion Anionic Polymerization, Nucleophilic or Cationic Polymerization, with different catalysts or enzymes and stereoregulation of Lactide ROP [32]. For the cyclic polymerization, a variety of precisely controlled, branched and single-cyclic (ring) and multi-cyclic topologies can be realized through a bimolecular process, a unimolecular process with asymmetric telechelic, a unimolecular process with symmetric telechelic, a ring-expansion polymerization process and an electrostatic self-assembly and covalent fixation (ESA-CF) process [33]. Conventional synthetic protocol for ring polymers has been an end-to-end linking reaction using a linear polymer precursor having reactive groups, telechelic, and an equimolar amount of a bifunctional coupling reagent [34–36]. The combination of the classical ring-expansion polymerization process, in which the initiator fragment is included within the ring polymer structure, with the end-to-end polymer cyclization process could provide a new effective means for large size ring polymers. The new ring-expansion polymerization processes make use of a metathesis catalyst having cyclic ligand, an N-heterocyclic carbene initiator, a cyclic stannate and subsequent intramolecular cross-linking. For the effective synthesis of not only cyclic but also functionalized cyclic and multicyclic polymers, the ESA-CF process has been introduced, where linear or star telechelic polymers having cyclic ammonium salt groups accompanying small or polymeric plurifunctional carboxylate counter-anions are employed as key precursors [33].

Table 1 summarizes the four most important commodity thermoplastic polymers —PP, PE, polystyrene (PS) and polyvinyl chloride (PVC); two thermosetting polymers—polyurethanes, and unsaturated polyesters; and two elastomers—rubber and acrylonitrile-butadiene-styrene (ABS); their route of polymerization and characteristics [20, 21]. The reaction temperature, pressure, and time in the presence of a catalyst(s) control the degree of polymerization, while the chemical composition and the arrangement of chains affect the physical properties such that the polymers can be tailored and tuned to meet the end user. Amorphous polymers undergo substantial changes in properties as a function of temperature. The temperature at which the transition from the glassy region where the polymer is relatively stiff to the rubbery region where it is very compliant is termed the glass transition temperature, T_g. It is the temperature at which there is a discontinuity in slope in the profile of volume change versus temperature. The melting point, T_m, is the temperature range where the total or whole polymer chain mobility occurs and the T_m is called a first-order transition temperature while the T_g is sometimes called a second-order transition temperature. The former is usually 33–100% greater than the latter and the symmetrical polymers like HDPE exhibit the greatest difference between T_m and T_g [21]. Increasing the size of the side groups in the linear polymers such as PE and cross-linking of the main chains will decrease the T_m due to the imperfect molecular packing from the steric hindrance of the side chains, resulting in a decreased mobility and reduced crystallization rate. For nylon polymer, the NH groups make hydrogen bonds with the C=O groups of another chain and each bond worth only about 15–20 kJ/mole, as compared to 300–400 kJ/mole for a covalent bond. Nevertheless, a lot of interchain hydrogen bonds add up to make the nylon polymer becoming rather stiff with a T_g of 57 °C [20]. The high melting point of nylon-66 (T_m = 265 °C) is, therefore, a result of the combined Van der Waals, dipole-dipole, and hydrogen bonding forces between the polyamide chains. The corresponding polyester, which cannot make hydrogen bonds, has a T_g of −40 °C [20, 21]. The dipole-dipole interactions typically only about 5 kJ/mole, which is weaker than the hydrogen bonds, can again add up so that the PVC has a T_g of 81 °C, whereas that of polyethylene is −125 °C. [20].

Table 1 Different types of polymers, structure and characteristics (Adapted from [20, 21, 152]

Types of polymer	Structure	Reaction and types
1 Thermoplastics Polyethylene (PE)		• The world's most commonly used polymer for plastic goods and products • Molecular weight of a linear PE polymer can range from 200,000 to 500,000 • Available commercially in five major grades: (I) high density (HDPE), (2) low density (LDPE), (3) linear low density (LLDPE). (4) Very low density (VLDPE), and (5) ultra high molecular weight (UHMWPE) • HDPE is polymerized at low temperature (60–80 °C), and low pressure ($\approx$ 40 kg/cm^2) using metal catalysts. A highly crystalline, linear polymer with a density ranging from 0.94 to 0.965 g/cm^3 is obtained—pharmaceutical bottles, nonwoven fabrics, and caps • LDPE is derived from a high temperature (150–300 °C) and pressures (1000–3000 kg/cm^2) using free radical initiators. A highly branched polymer with lower crystallinity and densities ranging from 0.915 to 0.935 g/cm^3 is obtained—flexible container applications, nonwoven-disposable and laminated (or coextruded with paper) foil, and polymers for packaging • LLDPE (density 0.91–0.94 g/cm^3) and VLDPE (density: 0.88–0.89 g/cm^3), are linear polymers, polymerized under low pressures and temperatures using metal catalysts with co-monomers such as 1-butene, 1-hexene, or 1-octene to obtain the desired physical properties and density ranges. LLDPE—pouches and bags due to its excellent puncture resistance; VLDPE—extruded tubes • UHMWPE (MW > 2 $\times$ 10^6 g/mol)—orthopedic implant fabrications
Polypropylene (PP)		• Used widely in plastic industry for its remarkable level of resistance against chemical and mechanical weathering, tear and abrasion and low thermal expansion coefficient • polymerized by a Ziegler-Natta stereo specific catalyst which controls the isotactic position of the methyl group. Thermal (*Tg*: −12 °C, *Tm*: 125–167 °C and density: 0.85–0.98 g/cm^3) • Physical properties similar to PE; average molecular weight of from 2.2 to 7.0 $\times$ 10^5 g/mol and with wide molecular weight distribution (poly-dispersity) from 2.6 to 12 • Additives include antioxidants, light stabilizer, nucleating agents, lubricants, mold release agents; antiblock, and slip agents to improve the physical properties and processability • Exceptional high flex life and excellent environment stress cracking resistance; gas and water vapor permeability in between those of LDPE and HDPE

(continued)

Table 1 (continued)

Types of polymer	Structure	Reaction and types
Polystyrene (PS)	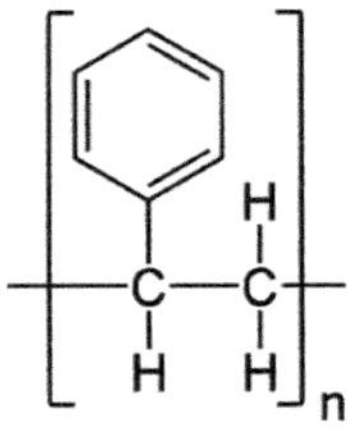	• Polymerized by free radical polymerization and is usually atactic • Available in three grades: (1) unmodified general purpose PS (GPPS, T: 100 °C), (2) high impact PS (HIPS), (3) PS foam • GPPS has good transparency, lack of color, ease of fabrication, thermal stability, low specific gravity (1.04–1.12 g/cm^3), and relatively high modulus—tissue culture flasks, roller bottles, vacuum canisters, and filterware • HIPS contain a rubbery modifier which forms chemical bonding with the growing PS chains; enhanced ductility, impact strength and resistance to environmental stress-cracking; mainly processed by injection molding at 180–250 °C; additives such as stabilizers, lubricants, and mold releasing agents are formulated
Polyvinylchloride (PVC)	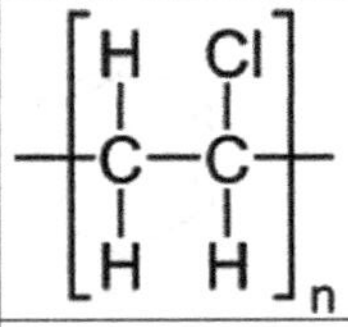	• An amorphous rigid polymer due to the large Cl side group; T_g of 75 to 105 °C; high melt viscosity, hence difficult to process • To prevent thermal degradation (HCl could be released), thermal stabilizers such as metallic soaps or salts are incorporated; lubricants formulated to prevent adhesion to metal surfaces and facilitate the melt flow during processing; plasticizers used at 10–100 parts per 100 parts of PVC resins to increase flexibility
2 Thermoset Polyurethanes (PU)	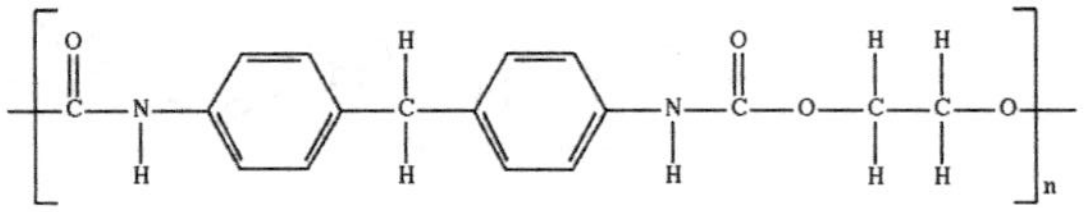	• Contain the urethane linkage • PU rubbers produced by reacting a prepared pre polymer chain with an aromatic di-isocyanate to make very long chains possessing active iso-cyanate groups for cross-linking; can be tailored for many applications by changing the chemical constituents • PU rubber is strong and has good resistance to oil and chemicals
Polyesters	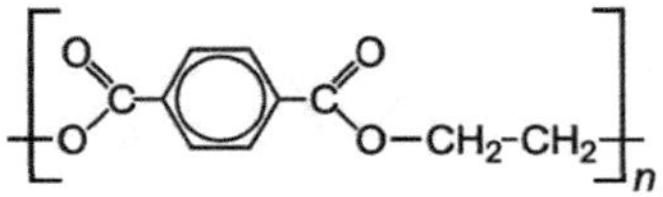	• Polyesters such as polyethyleneterephthalate are highly crystalline; high *Tm*: 265 °C; hydrophobic and resistant to hydrolysis in dilute acids • Can be converted by conventional techniques into molded articles

(continued)

Table 1 (continued)

Types of polymer	Structure	Reaction and types
3. Elastomer Rubbers	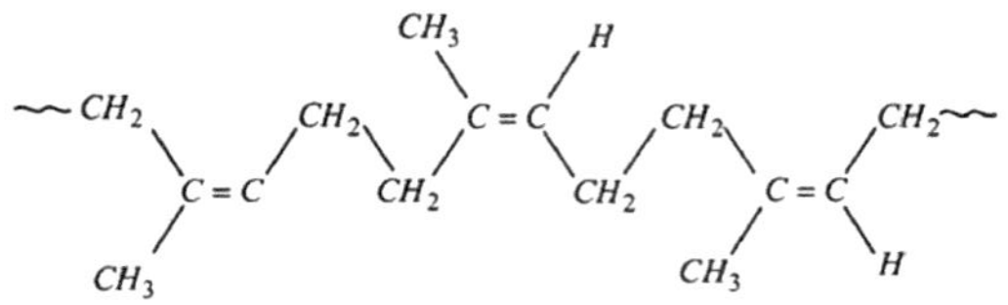	• Silicone, natural, and synthetic rubbers have been used for the fabrication of implants • Natural rubber mostly from the latex of the *Hevea brasiliensis* tree and the chemical formula is similar to *cis*-1.4 polyisoprene; compatible with blood in its pure form that synthetic rubbers being developed to substitute the natural rubber • The Ziegler-Natta types of stereo-specific polymerization techniques made the variety possible. The synthetic rubbers rarely used to make implants; physical properties vary widely due to the wide variations in the preparation recipes
Acrylonitrile-butadiene-styrene (ABS)	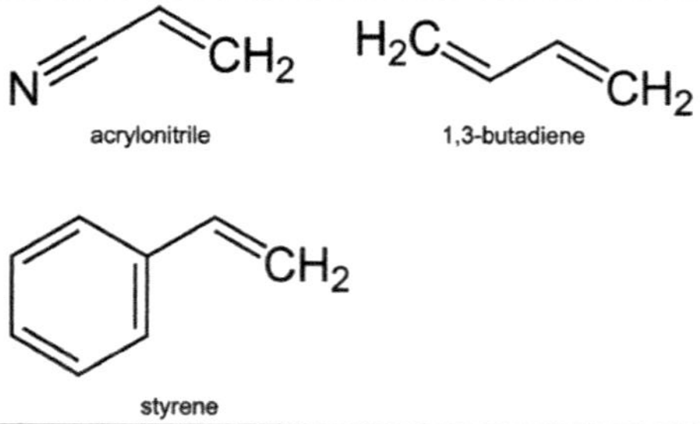 acrylonitrile 1,3-butadiene styrene	• Produced by three monomers: acrylonitrile, butadiene, and styrene • The desired physical and chemical properties with a wide range of functional characteristics can be controlled by changing the ratio of the monomers • Resistant to the common inorganic solutions, have good surface properties, and dimensional stability

NB All chemical structures are adapted from Wikipedia

2.3 Production of Polymeric Fibers

Synthesis of SPC involves two phases—the production of polymeric fibers, and the fabrication of the fibers with the matrix. Table 2 summarizes the different techniques used for the production of different polymeric fibres and the fabrication of recyclable SPCs. In the production phase, the fibers are produced from the polymeric raw material by the spinning technique which is based upon the extrusion principle where the liquefied/molten form of the material is forced through a die to acquire desired cross sections. The different types of spinning techniques are the dry, wet, gel, melt or electrospinning method. The wet spinning technique is the oldest production method, where the polymer is dissolved in a non-volatile solvent and the fibers are extruded through a spinneret placed in a liquid filled coagulation container which contains a non-solvent for the polymer but can dissolve out the solvent used for the polymer dissolution. Once the polymer solvent and the non-solvent coagulate together, the polymer fibers are precipitated out. The polymer solvent can be recovered later which can be costly. The major drawback is that direct extrusion through a liquid phase results in larger drag force/friction which slows down the process and increasing the extrusion speed weakens the fibers by causing micro-level cavities throughout the filament. Unlike the wet method, dry spinning utilizes evaporation to extrude the fibers. The polymer is dissolved in a solvent for extrusion after which the fibers are exposed to hot air to make the solvent evaporate and to solidify the fibers. No chemical reaction or dilution involved and the evaporating solvent can be recovered via condensation without any interference to the extruded fibers [37–40].

The gel spinning/dry-wet spinning method is used when the polymeric fibers with high strength or special properties are required such as the ultra-high-molecular-weight polyethylene (UHMWPE) fibers [41]. The polymer chains, due to the close contact with each other, develop intra- and interchain linkages in a liquid crystal form, resulting in a high tensile strength fiber. After extrusion, the polymeric fibers obtained in highly oriented form with liquid crystals aligned along the length/ long axis of the fibers, are immediately exposed to air and then subjected to a liquid bath for further cooling, hence the name the "dry-wet spinning". It can be further

Table 2 Different techniques used for the production of different polymeric fibres and the fabrication of recyclable SPCs [37, 39]

Fabrication method	Polymeric materials
Wet spinning	Rayon, spandex, aramid and acrylic
Dry spinning	Spandex, vinyon, modacrylic, triacetate, acrylic and acetate
Gel spinning	Aramid and polyethylene; UHMWPE
Melt-spinning technique	Nylon, polyester, sulfar, PP, PET, PLA and PA6
Electrospinning	PE, PP, PET, PMMA and PLA
Hot compaction	PP, nylon 6, 6, PE, PMMA and crystalline polymer fibers
Film stacking method	Homocomposites of PP, PE, UHMWPE, iPP

modified into "dry-jet wet spinning" method where the polymer solution is extruded and then exposed to pressurized hot air. After heat treatment, a coagulation bath is done which is followed by further heat and drawing treatments, all under nitrogenous environment to prevent the oxidization of the extruded fibers [41–43].

Melt spinning is the most commonly used commercial process for the production of synthetic polymeric fibers and have been reported for PP, polyamides 6, and PET fibers [2], PLA [17] and isotactic PP (iPP) fibers [15, 44]. It involves direct melting of the polymeric granules and the flow of the melted polymer to the spin head, controlled through a metering pump. The polymer melt is extruded through small orifices in the spinneret and drawn into thin fibers by a uniaxial drawing process. The spinneret is submerged in the liquid coagulation bath and the emerging filaments are coagulated in a precipitating bath or a series of baths of increasing precipitant concentration or exposed to air as a cooling down step. To prevent irregularity in diameter or cross-section of the fibers, the molten polymer is passed through a filter which filters out un-melted granules. Fibers of different cross-sections can be tuned as per their applications such as round, pentagonal, trilobal, octagonal or hexagonal [39, 40, 45, 46]. Electrospinning method uses electrical charges where the extrusion material is in the form of a charged polymer solution or a polymer melt. The charged solution/melt is ejected through a nozzle onto an oppositely charged substrate (a charged metal target). In the case of polymer solution, the fibers can be spun into nonwoven structures which are porous where solvent is evaporated via hot air, and in the case of polymer melt, there is no drying step to ensure purity and mainly non-porous, with the diameter of sub-micron level or nanoscale and high surface areas [39, 47]. PE, PP, PET, poly (methyl methacrylate) (PMMA) and poly(lactic acid) (PLA) are among the polymer materials that have been successfully electrospun into fibers of different diameters [15, 48–55].

2.4 Fabrication of the Polymeric Fibres with the Matrix

The main challenge in producing an SPC is in combining the fiber and the matrix into one composite, as both may have the same chemical structure and hence the melting temperature [15]. The molecular orientation may change during spinning and drawing and from the heating and cooling processes which makes it difficult to retain the properties of the oriented polymer molecules in the final composite for enhanced mechanical properties [56]. High mechanical properties of the SPCs also require the development of high strength polymeric fibers which are recyclable, economical, have low density, and good interfacial bonding [57]. The fabrication phase involves melt or powder impregnation, hot compaction, overheating, film stacking or co-extrusion method [15, 39]. Impregnation method is a traditional method, initially developed for the fabrication of PE and PP homo composites [58, 59]. There are two types, depending on the physical state of the impregnation material, the powder impregnation and melt (solution) impregnation method. In melt impregnation method, the fibers are impregnated with molten polymer either cross-head extruder which provides molten polymer in a die for the fibers to pass or a molten resin bath where the fibers are passed through to increase the permeability of the polymer contained in the tow. In the dry-powder impregnation method, the fibers such as the glass fibers are exposed to a dry a bed of loosely packed thermoplastic powder which is processed by heating to sinter the powder particles onto the fibers. Electrostatic attraction causes the powder particles to stick onto the fiber surface resulting in macroscopic impregnation where the fiber clusters are coated with the powder particles rather than the individual fibers [15, 39]. Impregnation is slow and not economical and more suitable for the polymers of low molecular weight. The impregnation material can be highly viscous and must fully wet the fibers, which may cause thermal degradation and loss of mechanical properties of the reinforcement material/fibers. The forces applied to the die pressure of the crosshead extruder and the partial dissolving or melting during impregnation may cause damage to the fibers' integrity [56, 60].

Hot compaction subjects the polymer fibers to a temperature range high enough to melt down the outer surface of the fiber bundle. The molten surface, upon cooling, becomes the binding phase serving as the matrix of the composite. The semi-crystalline polymers' broad melting temperature range is manipulated via prefabrication under the effect of constraints. The oriented polymer fibers are compacted to an oriented polymer sheet under suitable conditions of temperature and pressure. The major concern of this process is that the temperature should be almost 5 °C below the T_m and any overheating may result in relaxation and a loss of the fiber orientation [15]. The hot compaction method has been used to prepare the SPCs from PP [61, 62], PE [63], poly(ethylene naphthalate) (PEN) [64], and PMMA [65]. In the overheating method, the polymer fibers are embedded in a molten polymer matrix of the same grade, overheated considerably above their T_m whilst being constrained to prevent shrinkage and the loss of mechanical properties. The constraining of the fiber shifts the melting temperature to higher values to provide larger temperature window for the SPCs processing of the SPCs [2, 15].

In the film stacking technique, the reinforcement fibers are sandwiched between the films of the matrix and the composite material is produced by hot pressing. The matrix films ensure a wide temperature processing window for hot pressing, freedom to select the material with no expensive pre-production [15]. It has been applied for PP [66], PE [67], PEN [68], UHMWPE [69], PLA [17], and iPP fibers [44]. In co-extrusion technique, a multilayer web can be produced without the need for initial production of individual webs and a separate combining step. The melted polymers are fed together carefully to produce a layered melt, which is then processed in conventional ways to produce a plastic film or sheet [70]. Two different types of polymer tapes such as the random PP copolymer/PP homopolymer, each at a different melting temperature, are cold drawn to increase the mechanical properties and the reinforcement tapes are consolidated. Once oriented, the polymer tapes can be constrained by the molding pressure during consolidation to increase the melting temperature of the oriented core material, and further extend the processing window of about 20–40 °C, a high volume fraction of reinforcement (90%)

and an excellent bonding between the tapes due to the co-extrusion process [15]. All-PP tapes (PP homopolymer—core; random PP copolymer—skin) fabricated by coextrusion into tapes of different melting temperatures are superior to glass mat-reinforced thermoplastics or natural fiber mat-reinforced in terms of good mechanical properties with low density [60].

In any composite, the stress absorbed by the matrix (the weaker phase due to the weaker inter-chain interaction) is transferred to the reinforcement phase (the stronger phase) through an interphase [71]. For this reason, polymeric fibers with high performance are very important in SPC development as these high-performance polymer fibers can be designed to give SPC high mechanical features. To ensure efficient transfer of load from the matrix to the fibers, overlapping of the polymer chains is required and for this, polymers of high molecular weight with extended chains are preferred [2]. Other than strength and stiffness, the low processing cost, ease of production and recycling to meet the environmental requirements, low densities and the naturally strong interfacial bonding between the molecules of the same polymer with less requirement for surface treatments, are among major factors to be considered in the SPC development [3]. Although produced from a single type of polymer or polymers related to a single family of polymers, an SPC is still a two-phase system. Taking advantage of the high stiffness and strength, the reinforcement phase is the anisotropic semicrystalline forms of the polymers, and the matrix is either amorphous or semicrystalline but isotropic form of the same polymer [71]. Carbon and glass fibers are the common reinforcement material used in the composite materials to strengthen and enhance the efficiency of the polymer matrix. However, carbon fibre especially is wasteful to produce and both are difficult to recycle, mechanically and thermally, and for ultimate disposal, making them not environmentally-friendly [15, 60, 72]. Furthermore, the density of the polymer fibers such as PE, PP, All-PP tapes and UHMWPE is less than that of the glass and carbon fibers, and their high mechanical properties induced during drawing are major advantages in applications where a high strength-to-weight ratio is required [15]. Since the development of the first SPC based on PE [1], different matrix-fiber combinations based on PE, PET, PMMA, PLA, PEN, nylon and polyester have been reported (Table 3) [15].

The major advantage of SPCs over heterogeneous composites is the chemical compatibility for improved interfacial bonding and recyclability. PE can be classified into different categories which do not depend entirely on its density and branching, and those main forms of PE divided based on density and branching which include HDPE, High molecular weight HDPE (HMW HDPE), UHMW-HDPE, linear low-density polyethylene (LLDPE), and very low-density polyethylene (VLDPE). The most used PE grades are HDPE, low-density polyethylene (LDPE) and medium-density polyethylene (MDPE) [73]. During fabrication, the main hurdle to overcome is the small difference in T_m between the fiber and the matrix. The T_m for HDPE matrix and fibers are 132 and 139 °C, respectively [74], and for UHMWPE are in the range of 5–9 °C [75]. Under normal processing conditions of the SPCs, the PE fibers annealed at the temperature close to its T_m will have a much-reduced modulus towards that of the bulk HDPE [74].

Table 3 Matrix-fiber combinations for SPCs development (Adapted from [15])

Matrix	Fiber
HDPE	HDPE
HDPE	UHMWPE
UHMWPE	UHMWPE
LDPE	UHMWPE
PP	PP
PPE random copolymer	PP
PET	PET
PMMA	PMMA
PLA	PLA
Poly(ethylene naphthalate)	Poly(ethylene naphthalate)
Nylon 6,6	Nylon 6,6
Vectran M	Vectran HS

HDPE High density polyethylene, *UHMWPE* Ultra-high molecular weight polyethylene, *LDPE* Low density poly-ethylene, *PP* Polypropylene, *PLA* Polylactide, *PET* Polyethylene terephthalate, *PMMA* Polymethyl methacrylate and *PPE* Propylene ethylene

To enlarge the process window, the incorporation of polymers with the same chemical composition but different chemical structures have been explored such as the HDPE matrix reinforced by UHMWPE fibers [76, 77], and LDPE matrix reinforced by HDPE fibers [74] or UHMWPE fibers [77]. These composites have the manufacturability enhanced but with reduced interfacial adhesion than the original HDPE homocomposite [74, 78], a result of different molecular weight and chain configurations particularly the length of branched chains of HDPE, LDPE, and UHMWPE which affects the compatibility and miscibility [79]. The level of short chain branching (SCB) in LDPE has a strong influence on its miscibility with linear HDPE [79]. The difference between the T_m of high-performance PE fibre and LDPE matrix and the high chemical compatibility of the two components permit the use of the fibre in a composite fabrication [78]. The chemically-treated PE fibres surface substituting for the glass fibres in the PE matrix has markedly increased the adhesion but the thermal processing conditions of the composite material and the surface treatment cause a reduction in the mechanical property of the PE fibre [78]. With greater concern over climate change and sustainable development goals, the shift is now geared towards the development of SPCs from renewable and bioresources.

3 Eco-friendly Single Polymer Composites

Continuous and rapid exhaustion of natural and non-renewable resources has made it imperative for the utilization of sustainable and eco-friendly material. Regulatory and environmental authorities are making it compulsory for the manufacturers to review their processes and the impact of their products on the environment.

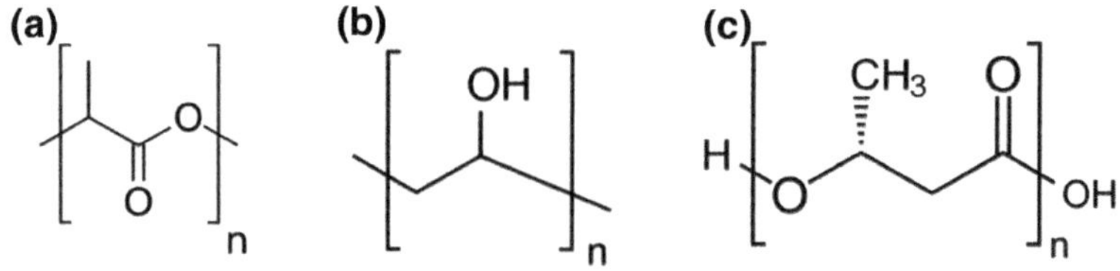

Fig. 2 Structure of **a** PLA, **b** PVA, **c** PHB. (Adapted from wikipedia)

In SPCs, the use of biopolymer matrix re-enforced with the same polymer in the reinforcement phase is the way forward for the sustainable approach as both the matrix and the fibers are from renewable resources. Despite this, the biopolymer is not the polymer of choice for SPC production due to its instability against mechanical recycling and thermal degradation during recycling process [60, 80]. Biodegradable polymers can be categorized into natural polymers (e.g. cellulose, alginate, starch), synthetic polymers [e.g. PLA, polycaprolactone, polyvinyl alcohol (PVA)], and microbial polymers [e.g. polyhydroxybutyrate (PHB) and polyhydroxybutyrate valerate (PHBV)] (Fig. 2) [81]. The two commonly reported biodegradable SPCs—PLA and PVA, are highlighted.

3.1 PLA-Based

3.1.1 PLA Synthesis

PLA is a thermoplastic, high-strength, high-modulus polymer, belonging to the family of aliphatic polyesters, and is commonly made from α-hydroxy acids which include polyglycolic acid or polymandelic acid. The basic building block for PLA is lactic acid (2-hydroxy propionic acid) which is the simplest hydroxy acid with an asymmetric carbon atom and exists in two optically active configurations. The $_{\rm L}(+)$-isomer which is produced in humans and other mammals, and both the $_{\rm D}(-)$- and $_{\rm L}(+)$-enantiomers which are produced in the bacterial systems [82]. The basic building block for PLA is lactic acid (2-hydroxy propionic acid) which is the simplest hydroxy acid with an asymmetric carbon atom and exists in two optically active configurations—the $_{\rm L}(+)$-isomer which is produced in humans and other mammals, and both the $_{\rm D}(-)$- and $_{\rm L}(+)$-enantiomers which are produced in the bacterial systems [82]. The commercial batch fermentation may take 3–6 days at 5–10% sugar concentrations with the production rate of 2 g L^{-1} h^{-1}. The standard fermentation conditions are relatively low to neutral pH, temperatures around 40 °C, and low oxygen concentrations and the sugars such as glucose, maltose, and dextrose can be obtained from corn or potato starch, or sucrose from cane or beet sugar, or lactose from the cheese whey. The supplements may include corn steep liquor, yeast extract, cotton seed flour, or soy flour, to provide proteins, B-vitamins, amino acids and nucleotides [83–86]. The strains that produce the $_{\rm L}(+)$-isomer are *Lactobacilli amylophilus, L. bavaricus, L. casei, L. maltaromicus*, and *L. salivarius* and the strains that produce the D(−)-isomer or the mixtures of both are *L. delbrueckii, L. jensenii*, or *L. acidophilus,* with 90% yield (1.8 mol Lactic acid mol^{-1} glucose) [82, 85, 87]. Calcium hydroxide or calcium carbonate is used to neutralize the fermentation acid and give soluble calcium lactate broth which is filtered, evaporated, recrystallized, and acidified with sulphuric acid to yield the crude lactic acid. Higher purity is obtained by distillation of the acid as the methyl or ethyl ester, followed by hydrolysis back to the acid [83, 86, 84, 85] (Fig. 3).

The synthesis of lactic acid into high-molecular-weight PLA can follow two different routes of polymerization (Fig. 4). The lactic acid undergoes condensation polymerization to yield a low-molecular-weight, brittle, which can be made useful for applications if the molecular weight is increased through the use of esterification-promoting adjuvants such as bis(trichloromethyl) carbonate, dicyclohexylcarbodiimide, and carbonyl diimidazole; and chain-extending agents such as isocyanates, acid chlorides, anhydrides, epoxides, thiirane, and oxazoline [82, 87, 88]. Another route is via the ring-opening polymerization (ROP) of the lactide to yield high molecular weight PLA (M_w = 100,000). Lactide, the cyclic dimer of lactic acid is obtained by the depolymerization of low-molecular-weight PLA under reduced pressure to give a mixture of three stereoisomers (L, L)-lactide, (D, D)-lactide and *meso*(D, L)-lactide, the composition of which depends on the lactic acid isomer feedstock, temperature and catalyst [87]. Without racemization reactions, polymerization of (L, L)-lactide (LLA) and its enantiomer (D, D)-lactide (D-LA) give

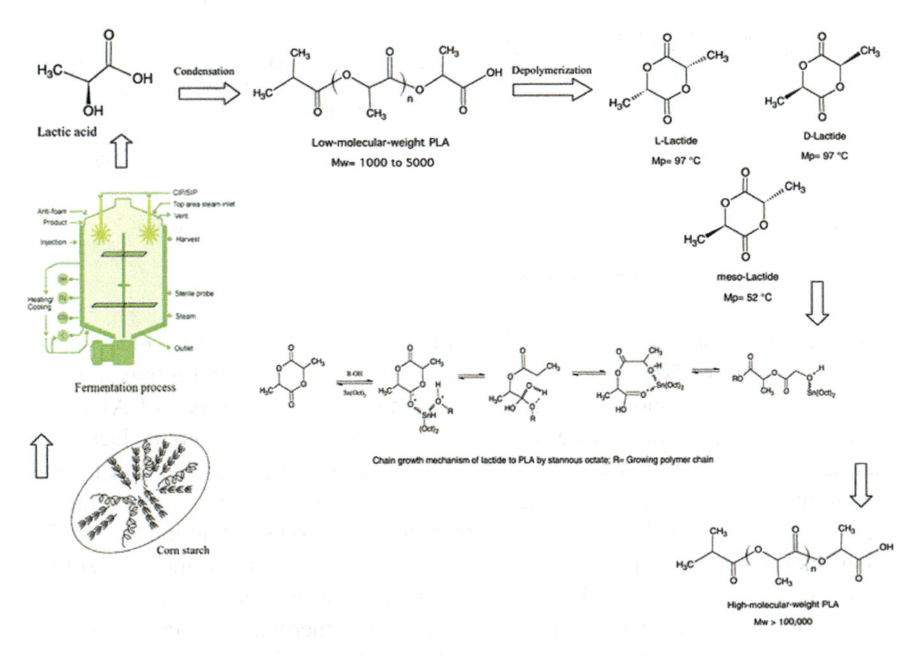

Fig. 3 Production steps for PLA. (Adapted from [153])

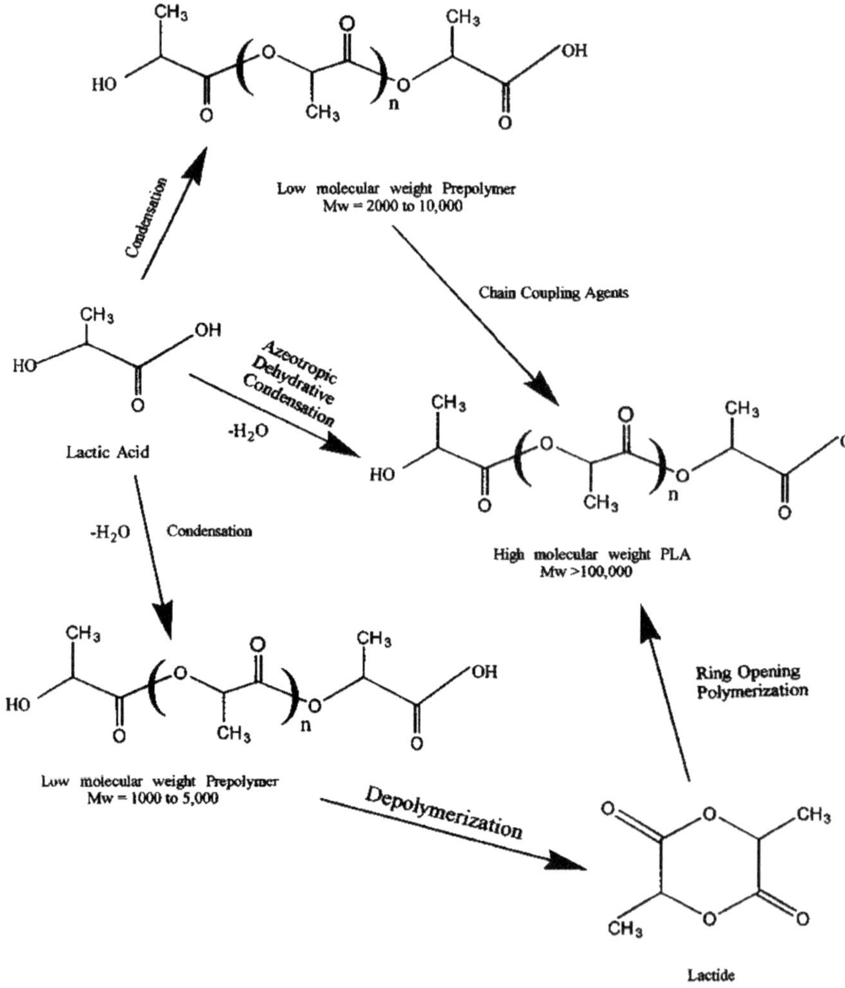

Fig. 4 Synthesis methods for high M_w PLA. (Adapted from [82])

isotactic semicrystalline polymers. The polymerization of *meso* (D, L)-lactide, or a racemic mixture of 50% of D-LA and 50% of L-LA gives an amorphous polymer, while the polymerization of optically pure monomers (L-LA or D-LA) gives a semi-crystalline polymeric material [89]. The synthesis of PLA from the lactide ring can take place via cationic ROP or anionic ROP and the D-lactide and L-lactide enantiomers can form a 1:1 racemic stereo complex (D, L-lactide), which melts at 126–127 °C [87, 90]. The cationic initiators to polymerize lactide in cationic ROP are trifluoromethane-sulfonic acid (triflic acid) and methyl trifluoromethane-sulfonic acid (methyl triflate) [91–93]. The polymerization proceeds via triflate ester end-groups at 100 °C to yield an optically active polymer without racemization. The propagation begins with the cleavage of the positively charged lactide ring at the alkyl-oxygen bond by an S_N2 attack of the triflate anion. In another S_N2 type of attack, the triflate end-group reacts with the second molecule of the lactide to yield a positively charged lactide that is opened, and then the triflate anion again opens the charged lactide, and the polymerization proceeds [94].

The anionic lactide polymerizations proceed through the nucleophilic reaction of the anion with the carbonyl and the subsequent cleavage of the acyl-oxygen to produce an alkoxide end-group that continues to propagate [87]. For a large-scale commercial application, the use of bulk melt polymerizations with lower levels of non-toxic catalysts metal carboxylates oxides, and alkoxides are preferred. The high-molecular-weight PLA is easily polymerized in the presence of tin(II) and zinc to produce pure polymers due to their covalent metal-oxygen bonds and free *p* or *d* orbitals [95, 96] and the best results are achieved with tin oxide and octoate at 120–150 °C with 90% conversions and less than 1% racemization [97]. The tin(II) bis-2-ethyl hexanoic acid (tin or stannous octoate) is the catalyst of choice in the PLA synthesis mainly due to the FDA approval and high catalytic activity with low toxicity, and its solubility in many lactones with the ability to give high molecular-weight polymers with low racemization [82, 92, 98]. The commercial route is where the lactic acid and the catalyst are azeotropically dehydrated in a refluxing, high-boiling, aprotic solvent under reduced pressures to obtain much higher molecular weight PLA ($M_w > 300{,}000$) [99–102]. The azeotropic condensation polymerization utilizes reduced pressure distillation of lactic acid for 2–3 h at 130 °C to remove the majority of the condensation water and produces high-molecular-weight polymer without the use of esterification-promoting adjuvants or chain extending agents. The catalyst and diphenyl ether are added, a tube packed with 3-A° molecular sieves attached to the reaction vessel and the refluxing solvent returned to the vessel via the molecular sieves for an additional 30–40 h at 130 °C. The polymer can finally be isolated as is or dissolved and precipitated for further purification [99–104]. Though producing high molecular weight polymer, the technique may result in high catalyst impurities which can cause problems such as unwanted degradation, uncontrolled hydrolysis rates, toxicity and differing slow-release properties. The catalyst level should, therefore, be reduced to 10 ppm or less, deactivated or filtered out after the reaction [82, 87, 105].

Newly developed PLA synthesis techniques based on polycondensation and ROP on the basis of industrial technique modifications and advanced laboratory research including the various solvents, heating programs, reaction and catalyst systems have been reported. The four synthesis methods are direct polycondensation (DP), azeotropic polycondensation (AP), solid state polymerization (SSP) and ROP (Fig. 5). In the case of PLA, polycondensation of lactic acid connects carboxyl and hydroxyl groups to produce water by-product simultaneously. Due to the difficulty in removing the byproducts from the highly viscous reaction mixture, the polymer produced through DP is usually <50,000 $g{\cdot}mol^{-1}$ and low quality [106]. However, with the newly developed using more efficient catalyst, solvent system, temperature, pressure, and duration have improved the polymer processing and properties, PLA with M_w of 90,000 $g{\cdot}mol^{-1}$ is achievable even without any catalyst, initiator or solvent. However, this may require a long heating time (>100 h at

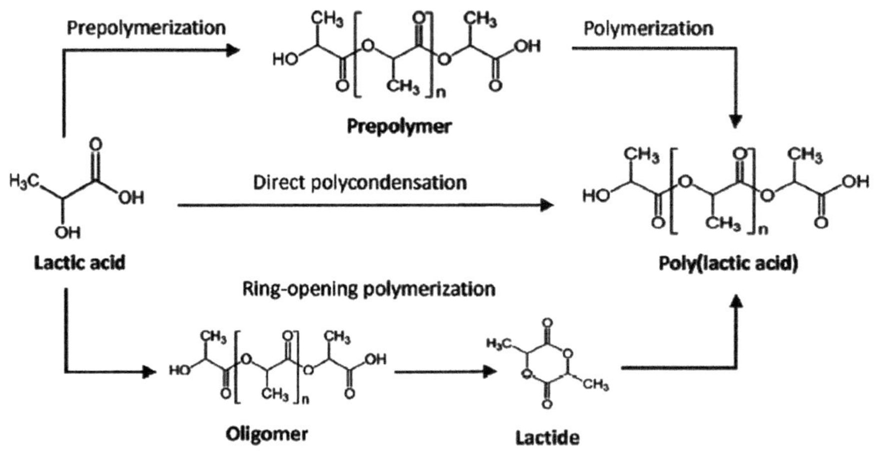

Fig. 5 Routes of PLA synthesis from lactic acid. (Adapted from [106])

200°C) resulting in high energy consumption [107]. Improved heating with microwave-assisted synthesis is more efficient with PLA (M_w 16,000 g•mol^{-1}) produced within 30 min using the enhanced catalytic effect of binary $SnCl_2$ or *p*-TsOH catalyst [108].

For AP where the water is removed efficiently by the azeotropic solvents, the appropriate solvent is critical to the performance and polymer properties. The equilibrium between the monomer and the polymer is manipulated in the organic solvent to produce relatively high M_w polymer in one step at the temperature lower than the polymer T_m, to avoid any impurities caused by the depolymerisation and racemization [109]. Besides azeotropic solvents, a Soxhlet extractor with molecular sieve (3 Å) inside simultaneously removes trace water from the refluxed solvent to produce 33,000 g•mol^{-1} M_w polymer [110]. The SSP consists of the melt state to produce oligomer at high temperature (150–200 °C) and the solid state to further increase the M_w between the T_g and the T_m. In the second step before heating, the prepolymer of relatively low M_w is pulverized into semi-crystalline powder, chip, pellet or fiber of diameter less than 150 μm, and thoroughly dried for highly efficient and homogeneous heat transfer and distribution resulting in high molecular weight and improved polymer properties and purity [111]. The heating program starts at 130 °C and rises to 160 °C stepwise as the polymer T_m increases, resulting in PLA with M_w of 202,000 g mol^{-1} without any catalyst and solvent used [111]. A starting mixture of L-PLA and D-PLA at 1:1 ratio in solid-state polycondensation could improve the polymer T_m from 160–170 °C to over 200 °C, suggesting reinforced thermal stability [112].

PLA leading producers, such as Cargill Dow (USA) and Shimadzu (Japan) applies ROP as the propagation process of the intermediate lactide [113–115] where the terminal end of the polymer classifies the mechanism into anionic ROP, cationic ROP and radical ROP [116]. Controlling the purity of lactide and synthesis conditions, without chain coupling agent or azeotropic system, could lead to a polymer with specific and desirable properties such as refractive index and a wider range of molecular weight, making ROP of lactide the route to the synthesis of high molecular weight PLA [117]. A twin-screwed extruder conducts the reactive extrusion, improving the conversion yield to 99% and shortening the duration time to only 7 min, significantly improving the efficiency [118]. More hydroxyl groups as co-initiators could lead to higher M_w (>400,000 g mol^{-1}) and faster polymerization (<5 h), without affecting the polymer thermal properties, with M_w of 160,000 g mol^{-1} at 200 °C for 1 h without solvent as co-initiator [119]. Solvent-free polymerization could actually achieve higher molecular weight PLA with higher efficiency [120]. PLA with high M_w (100,000 g mol^{-1}) is attained at 140 °C for 10 h, but a longer period of preparation may be required at extremely low pressure (0.001 kPa) and dry condition [121]. The metal-free organocatalyst has also attracted increasing interest in polymerization where guanidine and amidine organocatalyst have proven highly effective towards ROP of cyclic esters such as lactide [122, 123]. The organocatalyst 1,5,7-Triazabicyclo[4.4.0]dec-5-ene (TBD) shortens the ROP reaction time of with high conversion yield of 95–99% [122]. Potent organocatalysts could contribute towards reaction efficiency improvement, ROP under atmospheric pressure at room temperature and preventing residual organometallic catalyst contamination, but the resultant polymer may be of lower M_w (10,000–50,000 g mol^{-1}) and some may possess acute toxicity and high cost [123]. There is a need to develop a safe operating procedure and environmental assessment, proper catalyst recovery method and the economic flexibility for industrial application [106].

3.1.2 PLA SPCs

PLAs are biodegradable and compostable and degraded by simple hydrolysis of the ester bond without the need for enzymes but the rate of degradation is dependent on the size and shape of the article, the isomer ratio, and the temperature of hydrolysis. The current market and applications of PLA are 70% in the industrial packaging sector or the biocompatible/bioabsorbable medical device market [87, 124]. Polycondensation may produce PLA of low M_w using basic equipment and process, while ROP produces a wider range of M_w by controlling the purity of lactide and its polymerization, but the selection of the specific method should be based on its intended final application. Low M_w PLA, for example, may be suitable for drug release materials [106], so that it could be tuned for controlled-release and high M_w PLA for packaging and textile products [106]. High-molecular-weight PLA is colourless, glossy, stiff with properties similar to PS. The T_g of PLA is about 55 °C and the T_m is 175 °C, and the processing temperatures should be in the excess of 185–190 °C [125]. PLA degrades at temperatures above 200 °C, and can be caused by hydrolysis, lactide reformation, oxidative main chain scission, and inter- or intramolecular transesterification reactions, depending on time, temperature, low-M_w impurities, and the catalyst concentration. Catalysts and oligomers though decrease the degradation temperature may actually increase the degradation rate of

PLA, cause viscosity and rheological changes, fuming during processing, and poor mechanical properties [82, 126].

Several approaches such as copolymerization (block and stereoblock copolymers), microstructure and architecture control, and stereocomplexation, have been developed for designing new PLA-based polymers with a broad range of properties and improved processability [125]. Multiblock copolymers with alternating "soft" and "hard" segments, synthesized over a broad range of chemical compositions, show properties ranging from hard plastics to elastomers. Stereoblock copolymers with alternating amorphous and semicrystalline PLA blocks combine the advantages of PLA homopolymers (crystallinity) and random copolymers (processability). Independent control of polymer architecture and microstructure allows for the synthesis of star polymers with various arm morphologies. The stereocomplex formation between L-PLA and D-PLA, combining in situ polymerization with stereocomplexation takes advantage of the chirality of the lactide monomer, retention of configuration during polymerization, living nature of the lactide ROP in the presence of active hydrogen groups such as OH and NH_2, and control of the level of transesterification reactions [125]. PLA homopolymers have a very narrow processing window and the most widely used method for improving PLA processability is based on T_m depression by random incorporation of small amount of D-lactide to the L-lactide to obtain PDLLA, but this could lead to a significant decrease in crystallinity and crystallization rates [125]. The mixture (50/50) of preformed chains of P(L, L)LA and P(D, D)LA gives a stereocomplex with physicochemical and structural properties different from the corresponding PLA with the T_m of 230 °C, almost 50–60 °C higher than the PLA homopolymers [127, 128]. However, processing at these temperatures may lead to thermal degradation and the loss of M_w.

Compared to the bulk material, drawn fibers can exhibit a shift of the T_m and an increased enthalpy of melting. If the same grade of polymer is used for the matrix and the reinforcement, the shift in the T_m of the drawn fiber is not always sufficient enough for the production of SPCs. The concept of "overheating" is one of the methods in the manufacturing of SPCs. Both post-drawing temperature and the ultimate draw ratio have a significant influence on the degree of overheating. The enthalpy of a polymer is determined by the interaction forces between the molecular chains, while the entropy is determined by the conformation possibilities of a molecular chain. Controlled changing of either enthalpy or entropy, or both, would alter the crystalline T_m. Another way is by changing the conformation of the molecular chains and constraining the chains upon heating would shift the T_m towards higher temperatures. For the production of single fiber model composites, both ends of the polymer fibers are fixed on glass slides in order to prevent relaxation during heating. Pellets from the same polymer grade are isothermally hot pressed and the resulting thin films are placed on the same glass slide as the fiber. These stacked samples are heated in a hot-stage, melting only the matrix material but not the constrained fiber. The samples are then either air-cooled or isothermally crystallized in a hot-stage [2]. The overheating behavior of constrained fibers has resulted in the T_m shifts of about 10 °C [129–132]. Overheating of different polymers is due to the decreased conformational entropy of constrained amorphous phase upon melting which may depend on the crystallinity level, the crystal size, and the kinetics of crystal melting and on the scan rate. A shift higher than 20 °C of the T_m has been shown with highly extended iPP (draw ratios > 14) and the 10 °C overheating observed in the ultra-drawn PE upon constraining mainly due to the change in chain mobility for PE in the hexagonal phase. iPP, and UHMWPE are apolar polymers where the interchain interactions are relatively weak which leads to a high degree of drawability [2]. The polar polymers (e.g. PET, PA) on the other hand have relative strong interchain interactions and are therefore less drawable. These are exhibited in PET and PA6 which show the draw ratios of only 4 with the temperature shifts of about 10 °C for the constrained fibers as compared to the unconstrained fibers [2]. To achieve the wide processing window, the drawing temperature should be optimized to avoid relaxation processes in the amorphous phase while at the same time induce orientation and improvement of the crystal size and perfection. The draw ratio should be above 7 to have chain unfolding and perfectly oriented especially for drawable polymers, and the chain mobility should be relative low for effective constraining. For high performance fibers and effective constraining, highly extended chains are therefore of considerable importance [2].

PLA SPCs have been prepared using PLLA fibers as reinforcement and poly (D,L-lactic acid) (PDLLA) as matrix and the PLLA SPCs prepared by partially fusing together softened PLLA fibers in the pressurized cylindrical mold [133]. PLA SPCs prepared using the same method makes use of PLA fiber for both the matrix and reinforcement such that the melt processing window is very narrow [134, 135]. The preparation of a poly(lactic acid) SPC consisting of amorphous sheets as matrix and highly crystalline fibres, yarns and fabrics have been made on the basis of PLA's slowly crystallizing characteristics. The crystallinity of the PLA sheets and fibers used are about 5 and 40%, respectively, and the T_g and T_m of the PLAs used for the sheets and the fabric are approximately 60 and 167 °C, respectively. The amorphous PLA sheets and the crystalline PLA fibers/fabrics are laminated and compression-molded to form an SPC at a processing temperature substantially lower than the PLA's T_m. The processing temperature plays a profound role in affecting the fiber–matrix bonding properties where an increase in the processing temperature results in drastic improvement in the interfacial bonding at around 135 °C, indicating the lower boundary of the process window. The compression-molded SPC exhibits enhanced mechanical properties with the tearing strength of the fabric-reinforced SPC almost an order higher than the non-reinforced PLA. The SPC with 25 wt% yarns achieves a significant improvement with Young's modulus of 3.7 GPa and the tensile strength of 58.6 MPa [17]. The amorphous PLA could be used as a matrix material but there are two competing processes for amorphous PLA, fusing and crystallizing, exist during heating. During PLA SPCs preparation, the fusion should be promoted while the crystallization in the matrix should be restricted. The amorphous PLA must therefore be rapidly heated to a suitable temperature between T_g and T_m so that it does not have time to crystallize. However, the PLA SPCs with large thickness may require a long time for thermal energy transfer from the SPC surface to its center [136] and therefore may not be cost-effective. Highly oriented PLA tapes as the reinforcement and isotropic PLA film as matrix have also been explored to prepare the PLA SPCs. The highly oriented PLA tapes are pre-tensioned during hot-pressing to restrict the relaxation of the molecular chain so that the T_m of the PLA tape shifts to a higher temperature [137]. This widens the melt processing window of the PLA SPCs but it is difficult to prepare PLA SPCs by the normal hot-pressing device as this requires an additional constraining device [136].

Heat-bondable PLA core-shell fibre is a patented technology whereby the overlapping fibre-shells can be fused at a temperature significantly lower than the T_m of fibre-cores [138]. Mats of core-shell fibres of semi-crystalline poly(L-lactic acid) (core) and amorphous poly(D,L-lactic acid) (shell) produced through coaxial electrospinning have been used to prepare fibre reinforced SPC yarns and films [139]. The internal molecular arrangement (fine structure) within the fibres and the potential to enhance the crystallinity of the core and the heat-bond neighbouring fibres within the thermal operating window, between where the shell components fused and the T_m of the core components are established. Annealing/heat-pressing the core-shell fibres has been found to fuse the shells while enhancing the crystallinity of the cores. Heat-pressing plied fibre mats into a film has resulted in enhanced crystallinity (53%). significantly larger than the PLLA yarn (27%) although the mass component of the PLLA in the film is only 44% of that of the pure PLLA yarn. Thermal treatments therefore increase the crystallinity and the mechanical strength of the composite yarns. These core-shell fibres allow for continuous fibre reinforcement of biodegradable materials and offer a simple route to disperse nano-fibres homogeneously in a transparent matrix resin (as compared to the solvent casting impregnation methods) [139, 140]. Polylactide (PLA) SPCs have also been prepared using PLA nonwovens made of core-sheath PLA fibers as raw materials by hot-pressing. The core and sheath materials are poly(L-lactide) (PLLA) and PLA with D-lactide of about 10 mol% (PLA90), respectively. The melt processing window of PLA SPCs reaches more than 40 °C. The effects of hot-pressing temperature on the crystallinity, the crystal size distribution, the lamellar thickness and the mechanical property of PLA SPCs are more significant than the hot-pressing pressure treatment. The high mechanical properties of PLA SPCs are attributable to the strong interfacial adhesion between matrix and reinforcement. The tensile strength (σ_b), elongation at break (ε_b), the work of rupture (W) and impact strength (αcU) of the SPCs prepared at 130–160 °C are 47–65 MPa, 15–32%, 9–22 J, and 14.9–67.2 kJ m^{-2}, respectively, while those of pure PLAs prepared at 180 °C, are only 19 MPa, 1.8%, 0.15 J, and 2.8 kJ m^{-2}, respectively. The increase in hot-pressing temperature also reduces the σ_b, W and αcU [136].

3.2 PVA-Based

Poly(vinyl alcohol) (PVOH) is the only known water-soluble, carbon-carbon backbone polymer that is biodegradable under both aerobic and anaerobic conditions, applicable in a wide range of applications such as films, fibres, adhesives, textile sizing, emulsifiers, paper coating [141–143]. The hydroxyl groups in its main backbone provide for the strong intra- and intermolecular hydrogen bonds, conferring high tensile strength, excellent adhesive properties, abrasion resistance, chemical resistance and gas barrier properties [144]. PVOH can be prepared by using natural gas as raw materials, and produced on an industrial scale by hydrolysis (methanolysis) of poly(vinyl acetate) through a one-pot reaction, to obtain PVOH of various grades depending upon the degree of hydrolysis [145]. Commercial production of PVOH fibres is carried out by wet spinning or dry spinning, where PVOH chips are dissolved in hot water and the solution is extruded through a spinneret. The extrudates are then coagulated to form continuous filaments and heat treated to gain adequate mechanical properties. The water resistance of the fibres can be improved by a heat treatment followed by acetalization and the thermal stability can be enhanced by plasticizers such as glycerol, ethylene glycol, amine alcohols and polyvalent hydroxyl compounds [146]. PVOH possesses excellent mechanical properties because of the high crystallinity with high tensile strength and a greater modulus of elasticity than the regular concrete, where the PVOH fibres develop a chemical bond with the cement during hydration and curing. Hence, PVOH fibres are effective in controlling the shrinkage and the fatigue cracking of the concretes [147], with wide-ranging industrial applications including as reinforcement in rubber hosing and geogrid, and in paper and non-woven applications [141].

The PVOH-based SPCs have been prepared through a melt compounding process. The PVOH chips are utilized as the matrix evaluated against three types of commercial PVOH Kuralon® fibres (WN2, WN4, WN8) as the reinforcement. The fibres are provided in the form of staples and differ in their T_m and tensile properties. PVOH fibres are dried at 60 °C for 6 h before being compounded with PVOH chips in a Thermo Haake internal mixer operating at 180 °C for 6 min. These provide good dispersion of the fibres whilst avoiding thermal degradation of the matrix. Square sheets of the composite samples with a mean thickness of about 1 mm are prepared by compression moulding the resulting materials at 180 °C for 5 min in a Carver laboratory press under a pressure of about 1 MPa. Samples are sealed in vacuum plastic bags under vacuum to prevent moisture absorption [141]. Based on the cryo-fractured surface morphology, the reinforcement structure and shape is maintained only when high fibres T_m are used. The introduction of PVOH fibres increases proportionally the stiffness, the yield properties and the Vicat softening temperature of the neat PVOH matrix with marked improvements of the viscoelastic properties of the composites. The storage modulus and the T_g of the PVOH-SPC increases with respect to the neat matrix but with progressive reduction of the elongation at break of the filled samples and strong reduction of the creep compliance values at all the tested temperatures [141].

Table 4 Ranking of different polymers based on Green design and Life-cycle assessment [154]

Material	Green design rank	LCA rank
PLA (NatureWorks)	1	6
PHA (Utilizing Stover)	2	4
PHA (General)	2	8
PLA (General)	4	9
High Density Polyethylene	5	2
Polyethylene Terephthalate	6	10
Low Density Polyethylene	7	3
Bio-polyethylene Terephthalate	8	12
Polypropylene	9	1
General Purpose Polystyrene	10	5
Polyvinyl chloride	11	7
Polycarbonate	12	11

4 Conclusion and Future Outlook

SPC preparation, morphology, and mechanical behavior based on semicrystalline polymers, amorphous—amorphous or amorphous—semicrystalline systems have been greatly developed in the past 40 years [15, 148]. The 'toolbox' to create SPCs include resolving the issues of matrix and reinforcement (the M_w and nucleation), tacticity (polymorphism, melting, crystallization, chain branching, copolymers, overheating), and interfacial bonding and adhesion [149]. SPCs have exhibited high levels of mechanical properties at lower densities and efficient thermal recyclability. Depending on the reinforcing component dimensions, SPC can be produced as micro- or nanocomposites. The same materials used in the SPCs matrix and the reinforcements results in enhanced interfacial adhesion and confers the composite fully recyclable by reprocessing. However, this has also led to the major issue in SPCs synthesis and fabrication which is the narrow temperature range of processing that even the slightest overheating of the fibrous material could irreversibly degrade its reinforcing properties. This can be resolved by employing the techniques of copolymerization and polymorphism of matrix and the reinforcement fibers can be improved by using nano-fillers of high aspect ratios. The SPCs are typically prepared by melt-processing techniques through the hot compaction, wherein the polymer fibers are consolidated by applying heat and pressure. A partial melting of the outer surface of the fibers allows for the matrix formation whilst the inner part remains unmelted, and highly oriented to act as reinforcement [150]. The concept of overheating of constrained fibers could resolve the problem associated with hot compaction. The key for cost-effective preparation of SPCs with optimum impregnation of the reinforcements by the matrix material of the same chemical composition can be the significant decrease of the viscosity of the matrix. This is possible when the thermoplastic matrix is obtained in situ, through polymerization of low-viscosity monomers or oligomers in the presence of the reinforcements such as the ROP [19], where the ring-shaped monomer molecules are opened and transformed into high M_w polymers without release and accumulation of by-products [150]. Complex structures of SPCs also can be developed for more versatile applications in different fields such as for fire retardant or injection moldable grades [149, 151]. Table 4 shows that PLA attains the top spot in terms of green design ranking suggesting the bright future lies in the development of bio-based SPCs.

References

1. Capiati NJ, Porter RS (1975) The concept of one polymer composites modelled with high density polyethylene. J Mater Sci 10(10):1671–1677
2. Barkoula NM, Peijs T, Schimanski T, Loos J (2005) Processing of single polymer composites using the concept of constrained fibers. Polym Compo 26(1):114–120
3. Saheb DN, Jog JP (1999) Natural fiber polymer composites: a review. Adv Polym Technol 18(4):351–363
4. Herrera-Franco PJ, Valadez-Gonzalez A (2004) Mechanical properties of continuous natural fibre-reinforced polymer composites. Compos A Appl Sci Manuf 35(3):339–345
5. Thakur VK, Thakur MK, Raghavan P, Kessler MR (2014) Progress in green polymer composites from lignin for multifunctional applications: a review. ACS Sustain Chem Eng 2 (5):1072–1092
6. Thakur VK, Thakur MK, Gupta RK (2014) Raw natural fiber—based polymer composites. Int J Polym Anal Charact 19(3):256–271
7. Klapiszewski Ł, Tomaszewska J, Skórczewska K, Jesionowski T (2017) Preparation and characterization of eco-friendly Mg (OH) 2/Lignin hybrid material and its use as a functional filler for Poly (Vinyl Chloride). Polym 9(7):258
8. Abdullah MA, Nazir MS, Ajab H et al (2017) Advances in eco-friendly pre-treatment methods and utilization of agro-based lignocelluloses. In: Thangadurai D, Sangeetha J (eds) Industrial biotechnology: sustainable production and bioresource utilization. Apple Academic Press, USA, pp 371–420
9. Nazir MS, Abdullah MA, Raza MR (2017) Polypropylene composite with oil palm fibers: method development, properties and applications. Polypropylene-Based Biocomposites and Bionanocomposites 287
10. Mohanty AK, Misra M, Drzal LT et al (2002) Sustainable bio-composites from renewable resources: Opportunities and challenges in the green materials world. J Polym Environ 10 (1–2):19–26, Environ 10(112):19–20
11. Lithner D, Larsson Å, Dave G (2011) Environmental and health hazard ranking and assessment of plastic polymers based on chemical composition. Sci Total Environ 409 (18):3309–3324
12. Gourmelon G (2015). Global plastic production rises, recycling lags. New Worldwatch Institute analysis explores trends in global plastic consumption and recycling. Recuperado de http://wwwworldwatch.org
13. Abdullah MA, Nazir MS, Raza MR, Wahjoedi BA, Yussof AW (2016) Autoclave and ultra-sonication treatments of oil palm empty fruit bunch fibers for cellulose extraction and its polypropylene composite properties. J Clean Prod 126:686–697
14. Singh AA, Afrin S, Karim Z (2017) Green composites: versatile material for future. In Green biocomposites. Springer, Cham, pp 29–44
15. Matabola KP, De Vries AR, Moolman FS, Luyt AS (2009) Single polymer composites: a review. J Mater Sci 44(23):6213
16. Peijs T (2003) Composites for recyclability. Mater Today 6(4):30–35

17. Li R, Yao D (2008) Preparation of single poly (lactic acid) composites. J Appl Polym Sci 107(5):2909–2916
18. Huo M, Yuan J, Tao L, Wei Y (2014) Redox-responsive polymers for drug delivery: from molecular design to applications. Polym Chem 5(5):1519–1528
19. Sanjay MR, Madhu P, Jawaid M, Senthamaraikannan P, Senthil S, Pradeep S (2018) Characterization and properties of natural fiber polymer composites: a comprehensive review. J Clean Prod 172:566–581
20. Lee HB, Khang G, Lee JH (2013) Polymeric biomaterials. In: Wong JY, Bronzino JD, Peterson DR (eds) Biomaterials: principles and practices. CRC Press, Boca Raton. Florida, USA
21. Carraher Jr CE (2003) Seymour/Carraher's polymer chemistry. CRC Press
22. Shade Y (2016) Polymer engineering. White Word Publications, NY, USA
23. Harper CA, Petrie EM (2003) Plastics materials and processes: a concise encyclopedia. Wiley
24. Mallakpour S, Zadehnazari A (2011) Advances in synthetic optically active condensation polymers—a review. Express Polym Lett 5(2):142–181
25. Cao J, Yang NF, Wang PD, Yang LW (2008) Optically active polyethers from chiral terminal epoxides with bulky group. Polym Int 57(3):530–537
26. Chiellini E, Senatori L, Solaro R (1988) A new chiral poly (alkyl vinyl ether): synthesis and chiroptical properties. Polym Bull 20(3):215–220
27. Marvel CS, Overberger CG (1944) An optically active styrene derivative and its polymer1. J Am Chem Soc 66(3):475–477
28. Bailey WJ, Yates ET (1960) Polymers. III. Synthesis of optically active stereoregular polyolefins1-3. J Org Chem 25(10):1800–1804
29. Pino P, Ciardelli F, Lorenzi GP, Natta G (1962) Optically active vinyl polymers. VI. Chromatographic resolution of linear polymers of (R)(S)-4-methyl-1-hexene. J Am Chem Soc 84(8):1487–1488
30. Rogers ME, Long TE (eds) (2003). Synthetic methods in step-growth polymers. Wiley
31. Nakano T (2001) Optically active synthetic polymers as chiral stationary phases in HPLC. J Chromatogr A 906(1–2):205–225
32. Dechy-Cabaret O, Martin-Vaca B, Bourissou D (2004) Controlled ring-opening polymerization of lactide and glycolide. Chem Rev 104(12):6147–6176
33. Yamamoto T, Tezuka Y (2011) Topological polymer chemistry: a cyclic approach toward novel polymer properties and functions. Polymer Chem 2(9):1930–1941
34. Adachi K, Tezuka Y (2009) Topological polymer chemistry in pursuit of elusive polymer ring constructions. J Synth Org Chem Jpn 67(11):1136–1143
35. Endo K (2008) Synthesis and properties of cyclic polymers. New Frontiers in Polymer Synthesis. Springer, Berlin, Heidelberg, pp 121–183
36. Yamamoto T, Tezuka Y (2012) Multicyclic polymers. In Synthesis of polymers: new structures and methods. Wiley-VCH, Weinheim
37. McKetta Jr JJ (1976) Encyclopedia of chemical processing and design: volume 1-abrasives to acrylonitrile. CRC press
38. Brody H (1994).Synthetic fibre materials. Longman
39. Kricheldorf HR, Nuyken O, Graham S (2005) Handbook of polymer synthesis. Marcel Dekker, New York
40. Zhang X (2014) Fundamentals of fiber science. DEStech Publications, Inc
41. http://www.tikp.co.uk/knowledge/technology/fibre-and-filament-production/dry-jet-wet-spinning/. Accessed 19 May 2018
42. Kristiansen M, Tervoort T, Smith P (2003) Synergistic gelation of solutions of isotactic polypropylene and bis-(3, 4-dimethyl benzylidene) sorbitol and its use in gel-processing. Polymer 44(19):5885–5891
43. Lemstra PJ, Kirschbaum R (1985) Speciality products based on commodity polymers. Polymer 26(9):1372–1384

44. Loos J, Schimanski T, Hofman J, Peijs T, Lemstra PJ (2001) Morphological investigations of polypropylene single-fibre reinforced polypropylene model composites. Polymer 42 (8):3827–3834
45. Takayanagi M, Imada K, Kajiyama T et al (1967) Mechanical properties and fine structure of drawn polymers. J Polym Sci: Polym Symp 15(1):263–281. (Wiley Subscription Services, Inc., A Wiley Company)
46. Peterlin A (1971) Molecular model of drawing polyethylene and polypropylene. J mater sci 6(6):490–508
47. Gupta B, Revagade N, Hilborn J (2007) Poly (lactic acid) fiber: an overview. Prog Polym Sci 32(4):455–482
48. Huang ZM, Zhang YZ, Kotaki M, Ramakrishna S (2003) A review on polymer nanofibers by electrospinning and their applications in nanocomposites. Compos sci technol 63 (15):2223–2253
49. Ma J, Zhang Q, Mayo A, Ni Z, Yi H, Chen Y, Li D (2015) Thermal conductivity of electrospun polyethylene nanofibers. Nanoscale 7(40):16899–16908
50. Berber E, Horzum N, Hazer B, Demir MM (2016) Solution electrospinning of polypropylene-based fibers and their application in catalysis. Fibers Polym 17(5):760–768
51. Biazar E, Ahmadian M, Heidari S, Gazmeh A, Mohammadi SF, Lashay A, Hashemi H (2017) Electro-spun polyethylene terephthalate (PET) mat as a keratoprosthesis skirt and its cellular study. Fibers Polym 18(8):1545–1553
52. Sereshti H, Amini F, Najarzadekan H (2015) Electrospun polyethylene terephthalate (PET) nanofibers as a new adsorbent for micro-solid phase extraction of chromium (VI) in environmental water samples. RSC Adv 5(108):89195–89203
53. Matabola KP, De Vries AR, Luyt AS, Kumar R (2011) Studies on single polymer composites of poly (methyl methacrylate) reinforced with electrospun nanofibers with a focus on their dynamic mechanical properties
54. Casasola R, Thomas NL, Trybala A, Georgiadou S (2014) Electrospun poly lactic acid (PLA) fibres: effect of different solvent systems on fibre morphology and diameter. Polymer 55(18):4728–4737
55. Casasola R, Thomas NL, Georgiadou S (2016) Electrospinning of poly (lactic acid): theoretical approach for the solvent selection to produce defect-free nanofibers. J Polym Sci, Part B: Polym Phys 54(15):1483–1498
56. Alcock B, Cabrera NO, Barkoula NM, Loos J, Peijs T (2006) The mechanical properties of unidirectional all-polypropylene composites. Compos A Appl Sci Manuf 37(5):716–726
57. Houshyar S, Shanks RA, Hodzic A (2005) The effect of fiber concentration on mechanical and thermal properties of fiber-reinforced polypropylene composites. J Appl Polym Sci 96 (6):2260–2272
58. Houshyar S, Shanks RA (2006) Mechanical and thermal properties of flexible poly (propylene) composites. Macromol Mater Eng 291(1):59–67
59. Lacroix FV, Lu HQ, Schulte K (1999) Wet powder impregnation for polyethylene composites: preparation and mechanical properties. Compos A Appl Sci Manuf 30(3):369–373
60. Cabrera N, Alcock B, Loos J, Peijs T (2004) Processing of all-polypropylene composites for ultimate recyclability. Proc Inst Mech Eng, Part L: J Mat: Des Appl 218(2):145–155
61. Hine PJ, Olley RH, Ward IM (2008) The use of interleaved films for optimising the production and properties of hot compacted, self reinforced polymer composites. Compos Sci Technol 68(6):1413–1421
62. Jordan ND, Bassett DC, Olley RH, Hine PJ, Ward IM (2003) The hot compaction behaviour of woven oriented polypropylene fibres and tapes. II. Morphology of cloths before and after compaction. Polymer 44(4):1133–1143
63. Shavit-Hadar L, Khalfin RL, Cohen Y, Rein DM (2005) Harnessing the melting peculiarities of ultra-high molecular weight polyethylene fibers for the processing of compacted fiber composites. Macromol Mater Eng 290(7):653–656

64. Hine PJ, Astruc A, Ward IM (2004) Hot compaction of polyethylene naphthalate. J Appl Polym Sci 93(2):796–802
65. Wright-Charlesworth DD, Lautenschlager EP, Gilbert JL (2005) Hot compaction of poly (methyl methacrylate) composites based on fiber shrinkage results. J Mater Sci—Mater Med 16(10):967–975
66. Wang J, Chen J, Dai P, Wang S, Chen D (2015) Properties of polypropylene single-polymer composites produced by the undercooling melt film stacking method. Compos Sci Technol 107:82–88
67. Lacroix FV, Loos J, Schulte K (1999) Morphological investigations of polyethylene fibre reinforced polyethylene. Polymer 40(4):843–847
68. Wang J, Chen J, Dai P (2014) Polyethylene naphthalate single-polymer-composites produced by the undercooling melt film stacking method. Compos Sci Technol 91:50–54
69. Porras A, Tellez J, Casas-Rodriguez JP (2012) Delamination toughness of ultra high molecular weight polyethylene (UHMWPE) composites. In EPJ Web of Conferences, vol 26, p 02016. EDP Sciences
70. Goswami TK, Mangaraj S (2011) Advances in polymeric materials for modified atmosphere packaging (MAP). In Multifunctional and nanoreinforced polymers for food packaging, pp 163–242
71. Karger-Kocsis J, Siengchin S (2014) Single-polymer composites: concepts, realization and outlook. KMUTNB Int J Appl Sci Technol 7(1):1–9
72. https://www.theguardian.com/sustainable-business/2017/mar/22/carbon-fibre-wonder-material-dirty-secret
73. Khanam PN, AlMaadeed MAA (2015) Processing and characterization of polyethylene-based composites. Adv Manuf: Polym Compos Sci 1(2):63–79
74. Mead WT, Porter RS (1978) The preparation and tensile properties of polyethylene composites. J Appl Polym Sci 22(11):3249–3265
75. Mosleh M, Suh NP, Arinez J (1998) Manufacture and properties of a polyethylene homocomposite. Compos A Appl Sci Manuf 29(5–6):611–617
76. Lacroix FV, Werwer M, Schulte K (1998) Solution impregnation of polyethylene fibre/ polyethylene matrix composites. Compos A Appl Sci Manuf 29(4):371–376
77. Teishev A, Incardona S, Migliaresi C, Marom G (1993) Polyethylene fibers-polyethylene matrix composites: Preparation and physical properties. J Appl Polym Sci 50(3):503–512
78. Devaux E, Caze C (1999) Composites of ultra-high-molecular-weight polyethylene fibres in a low-density polyethylene matrix: II. Fibre/matrix adhesion. Compos Sci Technol 59 (6):879–882
79. Hameed T, Hussein IA (2004) Effect of short chain branching of LDPE on its miscibility with linear HDPE. Macromol Mater Eng 289(2):198–203
80. Houshyar S, Shanks RA (2003) Morphology, thermal and mechanical properties of Poly (propylene) fibre-matrix composites. Macromol Mater Eng 288(8):599–606
81. Rhim JW, Park HM, Ha CS (2013) Bio-nanocomposites for food packaging applications. Prog Polym Sci 38(10–11):1629–1652
82. Garlotta D (2001) A literature review of poly (lactic acid). J Polym Environ 9(2):63–84
83. Benninga H (1990) A history of lactic acid making: a chapter in the history of biotechnology, vol 11. Springer Science & Business Media
84. Datta R, Tsai SP, Bonsignore P, Moon SH, Frank JR (1995) Technological and economic potential of poly (lactic acid) and lactic acid derivatives. FEMS Microbiol Rev 16(2–3):221–231
85. Kharas GB, Sanchez-Riera F, Severson DK (1994) In: Mobley DP (ed) Plastics from microbes. Hanser-Gardner, Munich, pp 93–137
86. Van Ness JH (1981) Kirk-Othmer encyclopedia of chemical technology. 3rd ed, vol 13. Wiley, New York, pp 80–103
87. Hartmann MH (1998) High molecular weight polylactic acid polymers. In: Biopolymers from renewable resources, pp 367–411. Springer, Berlin, Heidelberg
88. Buchholz B (1994) U.S. Patent No. 5,302,694. U.S. Patent and Trademark Office, Washington, DC

89. Tsuji H, Ikada Y (1999) Physical properties of polylactides. Curr Trends Polym Sci 4:27
90. Lunt J (1998) Large-scale production, properties and commercial applications of polylactic acid polymers. Polym Degrad Stab 59(1–3):145–152
91. Kricheldorf HR, Kreiser I (1987) Polylactones, 11. Cationic copolymerization of glycolide with l, l-dilactide. Die Makromolekulare Chemie. Macromol Chem Phys 188(8):1861–1873
92. Kricheldorf HR, Sumbel M (1989) Polylactones—18. Polymerization of l, l-lactide with Sn (II) and Sn (IV) halogenides. Eur Polymer J 25(6):585–591
93. Dittrich VW, Schulz RC (1971) Kinetik und Mechanismus der ringöffnenden Polymerisation von L (−)-Lactid. Die Angewandte Makromolekulare Chemie: Applied Macromolecular Chemistry and Physics 15(1):109–126
94. Kricheldorf HR, Dunsing R (1986) Polylactones, 8. Mechanism of the cationic polymerization of L, L-dilactide. Die Makromolekulare Chemie. Macromol Chem Phys 187(7):1611–1625
95. Kricheldorf HR, Kreiser-Saunders I (1990) Polylactones,19. Anionic polymerization of L-lactide in solution. Die Makromolekulare Chemie: Macromol Chem Phys 191(5):1057–1066
96. Dahlmann J, Rafler G, Fechner K, Mehlis B (1990) Synthesis and properties of biodegradable aliphatic polyesters. Polym Int 23(3):235–240
97. Kricheldorf HR, Serra A (1985) Polylactones. Polym Bull 14(6):497–502
98. Kohn FE, Van Den Berg JWA, Van De Ridder G, Feijen J (1984) The ring-opening polymerization of D, L-lactide in the melt initiated with tetraphenyltin. J Appl Polym Sci 29 (12):4265–4277
99. Enomoto K, Ajioka M, Yamaguchi A (1994) U.S. Patent No. 5,310,865. U.S. Patent and Trademark Office, Washington, DC
100. Kashima T, Kameoka T, Higuchi C, Ajioka M, Yamaguchi A (1995) U.S. Patent No. 5,428,126. U.S. Patent and Trademark Office, Washington, DC
101. Ichikawa F, Kobayashi M, Ohta M, Yoshida Y, Obuchi S, Itoh H (1995) U.S. Patent No. 5,440,008. U.S. Patent and Trademark Office, Washington, DC
102. Ohta M, Obuchi S, Yoshida Y (1995) U.S. Patent No. 5,444,143. U.S. Patent and Trademark Office, Washington, DC
103. Ajioka M, Enomoto K, Suzuki K, Yamaguchi A (1995) The basic properties of poly (lactic acid) produced by the direct condensation polymerization of lactic acid. J Environ Polym Degradat 3:225–234
104. Ajioka M, Enomoto K, Suzuki K, Yamaguchi A (1995) Basic properties of polylactic acid produced by the direct condensation polymerization of lactic acid. Bull Chem Soc Jpn 68 (8):2125–2131
105. Suizu H, Takagi M, Ajioka M, Yamaguchi A (1996) U.S. Patent No. 5,496,923. U.S. Patent and Trademark Office, Washington, DC
106. Hu Y, Daoud WA, Cheuk KKL, Lin CSK (2016) Newly developed techniques on polycondensation, ring-opening polymerization and polymer modification: focus on poly (lactic acid). Materials 9(3):133
107. Achmad F, Yamane K, Quan S, Kokugan T (2009) Synthesis of polylactic acid by direct polycondensation under vacuum without catalysts, solvents and initiators. Chem Eng J 151 (1–3):342–350
108. Nagahata R, Sano D, Suzuki H, Takeuchi K (2007) Microwave-assisted single-step synthesis of poly (lactic acid) by direct polycondensation of lactic acid. Macromol Rapid Commun 28(4):437–442
109. Gupta AP, Kumar V (2007) New emerging trends in synthetic biodegradable polymers–Polylactide: a critique. Eur Polymer J 43(10):4053–4074
110. Kim KW, Woo SI (2002) Synthesis of high-molecular-weight poly (L-lactic acid) by direct polycondensation. macromolecular chemistry and physics 203(15):2245–2250
111. Fukushima K, Kimura Y (2008) An efficient solid-state polycondensation method for synthesizing stereocomplexed poly (lactic acid) s with high molecular weight. J Polym Sci, Part A: Polym Chem 46(11):3714–3722

112. Fukushima K, Furuhashi Y, Sogo K, Miura S, Kimura Y (2005) Stereoblock poly (lactic acid): synthesis via solid-state polycondensation of a stereocomplexed mixture of poly (L-lactic acid) and poly (D-lactic acid). Macromol Biosci 5(1):21–29
113. www.jimluntllc.com/pdfs/PolylacticAcidPolymersFromCorn. Accessed 26 May 2018
114. https://www.natureworksllc.com. Accessed 26 May 2018
115. Oota M, Ito M (1998) U.S. Patent No. 5,821,327. U.S. Patent and Trademark Office, Washington, DC
116. Nuyken O, Pask SD (2013) Ring-opening polymerization—an introductory review. Polymers 5(2):361–403
117. Stridsberg KM, Ryner M, Albertsson AC (2002) Controlled ring-opening polymerization: polymers with designed macromolecular architecture. In: Degradable aliphatic polyesters, pp. 41–65. Springer, Berlin
118. Jacobsen S, Fritz HG, Degée P, Dubois P, Jérôme R (2000) New developments on the ring opening polymerisation of polylactide. Ind Crops Prod 11(2–3):265–275
119. Korhonen H, Helminen A, Seppälä JV (2001) Synthesis of polylactides in the presence of co-initiators with different numbers of hydroxyl groups. Polymer 42(18):7541–7549
120. Zhong Z, Dijkstra PJ, Feijen J (2002) [(salen) Al]-mediated, controlled and stereoselective ring-opening polymerization of lactide in solution and without solvent: synthesis of highly isotactic polylactide stereocopolymers from racemic d, l-lactide. Angew Chem Int Ed 41 (23):4510–4513
121. Kaihara S, Matsumura S, Mikos AG, Fisher JP (2007) Synthesis of poly (L-lactide) and polyglycolide by ring-opening polymerization. Nat Protoc 2(11):2767
122. Lohmeijer BG, Pratt RC, Leibfarth F, Logan JW, Long DA, Dove AP et al (2006). Guanidine and amidine organocatalysts for ring-opening polymerization of cyclic esters. Macromolecules 39(25):8574–8583
123. Kamber NE, Jeong W, Waymouth RM, Pratt RC, Lohmeijer BG, Hedrick JL (2007) Organocatalytic ring-opening polymerization. Chem Rev 107(12):5813–5840
124. Rusu D, Boyer SE, Lacrampe MF, Krawczak P (2001). Bioplastics and vegetal fiber reinforced bioplastics for automotive applications. In: Pilla S (ed) Handbook of bioplastics and biocomposites engineering applications. Scrivener Publishing, Massachusetts 2011:397–449
125. Spinu M, Jackson C, Keating MY, Gardner KH (1996) Material design in poly (lactic acid) systems: block copolymers, star homo-and copolymers, and stereocomplexes. J Macromol Sci Part A Pure Appl Chem 33(10):1497–1530
126. Farah S, Anderson DG, Langer R (2016) Physical and mechanical properties of PLA, and their functions in widespread applications—A comprehensive review. Adv Drug Deliv Rev 107:367–392
127. Ikada Y, Jamshidi K, Tsuji H, Hyon SH (1987) Stereocomplex formation between enantiomeric poly (lactides). Macromolecules 20(4):904–906
128. Tsuji H (2000) Stereocomplex from enantiomeric polylactides. Res Adv Macromol 1:25–28
129. Kirschbaum R, van Dingenen JLJ (1989) Integration of fundamental polymer science and technology, vol 3. In: Lemstra PJ, Kleintjes LA (eds) Elsevier, London, p 178
130. Bastiaansen CMW, Lemstra PJ (1989) Macromolecular chemistry. Macromol Symp 28: p. 73)
131. Samuels RJ (1979) High strength elastic polypropylene. J Polym Sci, Part B: Polym Phys 17 (4):535–568
132. Tanaka H, Takagi N, Okajima S (1974) Melting behavior of highly stretched isotactic polypropylene film. J Polym Sci, Part A: Polym Chem 12(12):2721–2728
133. Tormala P, Rokkanen P, Laiho J, Tamminmaki M, Vainionpaa S (1988) U.S. Patent No. 4,743,257. U.S. Patent and Trademark Office, Washington, DC
134. Jia W, Gong RH, Hogg PJ (2014) Poly (lactic acid) fibre reinforced biodegradable composites. Compos B Eng 62:104–112
135. Wu N, Liang Y, Zhang K, Xu W, Chen L (2013) Preparation and bending properties of three dimensional braided single poly (lactic acid) composite. Compos B Eng 52:106–113

136. Liu Q, Zhao M, Zhou Y, Yang Q, Shen Y, Gong RH, Deng B (2018) Polylactide single-polymer composites with a wide melt-processing window based on core-sheath PLA fibers. Mat Des 139:36–44
137. Mai F, Tu W, Bilotti E, Peijs T (2015) Preparation and properties of self-reinforced poly (lactic acid) composites based on oriented tapes. Compos A Appl Sci Manuf 76:145–153
138. Barrows TH (1999) Bioabsorbable fibers and reinforced composites produced there from PCT. US 6,511,748 B1. WO99/34750
139. Kriel H, Sanderson RD, Smit E (2013) Single polymer composite yarns and films prepared from heat bondable poly (lactic acid) core-shell fibres with submicron fibre diameters. Fibres & Textiles in Eastern Europe
140. Chen LS, Huang ZM, Dong GH, He CL, Liu L, Hu YY, Li Y (2009) Development of a transparent PMMA composite reinforced with nanofibers. Polym Compos 30(3):239–247
141. Dorigato A, Pegoretti A (2012) Biodegradable single-polymer composites from polyvinyl alcohol. Colloid Polym Sci 290(4):359–370
142. Matsumura S, Toshima K (1996) Biodegradation of poly (vinyl alcohol) and vinyl alcohol block as biodegradable segment
143. Matsumura S, Tomizawa N, Toki A, Nishikawa K, Toshima K (1999) Novel poly (vinyl alcohol)-degrading enzyme and the degradation mechanism. Macromolecules 32(23):7753–7761
144. Chen N, Li L, Wang Q (2007) New technology for thermal processing of poly (vinyl alcohol). Plast, Rubber Compos 36(7–8):283–290
145. Chiellini E, Corti A, D'Antone S, Solaro R (2003) Biodegradation of poly (vinyl alcohol) based materials. Progr Polym Sci 28(6):963–1014
146. Petrushenko EF, Voskanyan PS, Pakharenko V (1988) Rheological Properties of PVA Based Compositions. Plast Massy 11:23–24
147. Shao Y, Shah SP (1997) Mechanical properties of PVA fiber reinforced cement composites fabricated by extrusion processing. ACI Mater J 94(6):555–564
148. Fakirov S (2013) Nano-and microfibrillar single-polymer composites: a review. Macromol Mater Eng 298(1):9–32
149. Karger-Kocsis J, Bárány T (2014) Single-polymer composites (SPCs): Status and future trends. Compos Sci Technol 92:77–94
150. Dencheva N, Denchev Z, Pouzada AS, Sampaio AS, Rocha AM (2013) Structure–properties relationship in single polymer composites based on polyamide 6 prepared by in-mold anionic polymerization. J Mat Sci 48(20):7260–7273
151. Bocz K, Toldy A, Kmetty Á, Bárány T, Igricz T, Marosi G (2012) Development of flame retarded self-reinforced composites from automotive shredder plastic waste. Polym Degrad Stab 97(3):221–227
152. Lukkassen D, Meidell A (2003) Advanced materials and structures and their fabrication processes. Narrik University College, Hin
153. Jamshidian M, Tehrany EA, Imran M, Jacquot M, Desobry S (2010) Poly-lactic acid: production, applications, nanocomposites, and release studies. Compr Rev Food Sci Food Safety 9(5):552–571
154. https://www.epa.gov/greenchemistry/presidential-green-chemistry-challenge-winners. Accessed 26th May 2018